T0394044

Managing Water Resources and Hydrological Systems

Environmental Management Handbook, Second Edition

Edited by
Brian D. Fath and Sven E. Jørgensen

Volume 1
Managing Global Resources and Universal Processes

Volume 2
Managing Biological and Ecological Systems

Volume 3
Managing Soils and Terrestrial Systems

Volume 4
Managing Water Resources and Hydrological Systems

Volume 5
Managing Air Quality and Energy Systems

Volume 6
Managing Human and Social Systems

Managing Water Resources and Hydrological Systems

Second Edition

Edited by
Brian D. Fath and Sven E. Jørgensen

Assistant to Editor
Megan Cole

CRC Press
Taylor & Francis Group
Boca Raton London New York

CRC Press is an imprint of the
Taylor & Francis Group, an **informa** business

Cover photo: Forsand, Norway, N. Fath

Second edition published 2021
by CRC Press
6000 Broken Sound Parkway NW, Suite 300, Boca Raton, FL 33487-2742

and by CRC Press
2 Park Square, Milton Park, Abingdon, Oxon, OX14 4RN

© 2021 Taylor & Francis Group, LLC

First edition published by CRC Press 2013

CRC Press is an imprint of Taylor & Francis Group, LLC

ISBN: 978-1-138-34266-8 (hbk)
ISBN: 978-1-003-04504-5 (ebk)

Typeset in Minion
by codeMantra

Contents

Preface xi

Editors xiii

Contributors xv

SECTION I APC: Anthropogenic Chemicals and Activities

1 Aquatic Communities: Pesticide Impacts 3
   David P. Kreutzweiser and Paul K. Sibley

2 Coastal Water: Pollution 17
   Piotr Szefer

3 Groundwater: Mining Pollution 37
   Jeff Skousen and George Vance

4 Groundwater: Nitrogen Fertilizer Contamination 45
   Lloyd B. Owens and Douglas L. Karlen

5 Groundwater: Pesticide Contamination 59
   Roy F. Spalding

6 Lakes and Reservoirs: Pollution 65
   Subhankar Karmakar and O.M. Musthafa

7 Mines: Acidic Drainage Water 81
   Wendy B. Gagliano and Jerry M. Bigham

8 Rivers and Lakes: Acidification 87
   Agniezka Gałuszka and Zdzistaw M. Migaszewski

9 Rivers: Pollution 105
   Bogdan Skwarzec

10 Sea: Pollution 111
   Bogdan Skwarzec

SECTION II   COV: Comparative Overviews of Important Topics for Environmental Management

11   Rain Water: Harvesting 123
K.F. Andrew Lo

12   Water Harvesting 129
Gary W. Frasier

13   Groundwater: Saltwater Intrusion...133
Alexander H.-D. Cheng

14   Irrigation Systems: Water Conservation 151
Pai Wu, Javier Barragan, and Vince Bralts

15   Irrigation: Erosion...155
David L. Bjorneberg

16   Irrigation: River Flow Impact 165
Robert W. Hill and Ivan A. Walter

17   Irrigation: Saline Water 171
B.A. Stewart

18   Irrigation: Sewage Effluent Use 175
B.A. Stewart

19   Irrigation: Soil Salinity 179
James D. Rhoades

20   Managing Water Resources and Hydrological Systems................185
Honglin Zhong, Zhuoran Liang, and Cheng Li

21   Runoff Water 195
Zaneta Polkowska

22   Salt Marsh Resilience and Vulnerability to Sea-Level Rise
and Other Environmental Impacts..215
Daria Nikitina

23   The Evolution of Water Resources Management 225
Francine van den Brandeler and Joyeeta Gupta

24   Wastewater and Water Utilities..241
Rudolf Marloth

25   Wastewater: Municipal...253
Sven Erik Jørgensen

26   Water Quality and Quantity: Globalization................................265
Kristi Denise Caravella and Jocilyn Danise Martinez

27   Water: Cost..277
Atif Kubursi and Matthew Agarwala

28   Wetlands: Methane Emission 293
     Anna Ekberg and Tørben Røjle Christensen

SECTION III   CSS: Case Studies of Environmental Management

29   Alexandria Lake Maryut: Integrated Environmental Management 301
     Lindsay Beevers

30   Aral Sea Disaster 317
     Guy Fipps

31   Chesapeake Bay 323
     Sean M. Smith

32   Giant Reed (*Arundo donax*): Streams and Water Resources 327
     Gretchen C. Coffman

33   Inland Seas and Lakes: Central Asia Case Study 337
     Andrey G. Kostianoy

34   Oil Pollution: The Baltic Sea 349
     Andrey G. Kostianoy

35   Status of Groundwater Arsenic Contamination in the GMB Plain 369
     Abhijit Das, Antara Das, Meenakshi Mukherjee, Bhaskar Das, Subhas Chandra
     Mukherjee, Shyamapada Pati, Rathindra Nath Dutta, Quazi Quamruzzaman, Khitish
     Chandra Saha, Mohammad Mahmudur Rahman, Dipankar Chakraborti, and Tarit
     Roychowdhury

36   Yellow River 383
     Zixi Zhu, Ynuzhang Wang, and Yifei Zhu

SECTION IV   DIA: Diagnostic Tools: Monitoring, Ecological Modeling, Ecological Indicators, and Ecological Services

37   Groundwater: Modeling 389
     Jesus Carrera

38   Groundwater: Numerical Method Modeling 399
     Jesus Carrera

39   Nitrogen (Nitrate Leaching) Index 407
     Jorge A. Delgado

40   Nitrogen (Nutrient) Trading Tool 415
     Jorge A. Delgado

41   The Accounting Framework of Energy–Water Nexus in
     Socioeconomic Systems 423
     Saige Wang and Bin Chen

42   Water Quality: Modeling 427
     Richard Lowrance

SECTION V ELE: Focuses on the Use of Legislation or Policy to Address Environmental Problems

43 Drainage: Hydrological Impacts Downstream.. 433
Mark Robinson and D.W. Rycroft

44 Drainage: Soil Salinity Management... 439
Glenn J. Hoffman

45 Lakes: Restoration.. 445
Anna Rabajczyk

46 Wastewater Use in Agriculture: Policy Issues..461
Dennis Wichelns

47 Water: Total Maximum Daily Load..481
Robin Kundis Craig

48 Watershed Management: Remote Sensing and GIS.......................................491
A.V. Shanwal and S.P. Singh

49 Wetlands: Conservation Policy.. 497
Clayton Rubec

SECTION VI ENT: Environmental Management Using Environmental Technologies

50 Irrigation Systems: Subsurface Drip Design 505
Carl R. Camp, Jr. and Freddie L. Lamm

51 Recent Approaches to Robust Water Resources Management under
Hydroclimatic Uncertainty... 511
J. Pablo Ortiz-Partida, Mahesh L. Makey, and Alejandra Virgen-Urcelay

52 Rivers: Restoration 519
Anna Rabajczyk

53 Waste: Stabilization Ponds...535
Sven Erik Jørgensen

54 Wastewater Treatment Wetlands: Use in Arctic Regions 5-Year Update 543
Colin N. Yates, Brent Wootton, and Stephen D. Murphy

55 Wastewater Treatment: Biological...561
Shaikh Ziauddin Ahammad, David W. Graham, and Jan Dolfing

56 Wastewater Treatment: Conventional Methods 577
Sven Erik Jørgensen

57 Water and Wastewater: Filters 583
Sandeep Joshi

58 Wetlands: Constructed Subsurface... 603
    Jan Vymazal

59 Wetlands: Sedimentation and Ecological Engineering.................................617
    Timothy C. Granata and J.F. Martin

60 Wetlands: Treatment System Use 621
    Kyle R. Mankin

SECTION VII   NEC: Natural Elements and Chemicals Found in Nature

61 Cyanobacteria: Eutrophic Freshwater Systems 631
    Anja Gassner and Martin V. Frey

62 Estuaries 635
    Claude Amiard-Triquet

63 Everglades...651
    Kenneth L. Campbell, Rafael Munoz-Carpena, and Gregory Kiker

64 Water Quality: Range and Pasture Land... 655
    Thomas L. Thurow

65 Water: Drinking 661
    Marek Biziuk and Matgorzata Michalska

66 Water: Surface 679
    Victor de Vlaming

67 Wetlands... 685
    Ralph W. Tiner

SECTION VIII   PRO: Basic Environmental Processes

68 Eutrophication... 693
    Sven Erik Jørgensen

69 Wastewater Use in Agriculture...701
    Manzoor Qadir, Pay Drechsel, and Liqa Raschid-Sally

70 Wetlands: Biodiversity 709
    Jean-Claude Lefeuvre and Virginie Bouchard

71 Wetlands: Carbon Sequestration 715
    Virginie Bouchard and Matthew Cochran

Index...721

Preface

Given the current state of the world as compiled in the massive Millennium Ecosystem Assessment Report, humans have changed ecosystems more rapidly and extensively during the past 50 years than in any other time in human history. These are unprecedented changes that need certain action. As a result, it is imperative that we have a good scientific understanding of how these systems function and good strategies on how to manage them.

In a very practical way, this multivolume *Environmental Management Handbook* provides a comprehensive reference to demonstrate the key processes and provisions for enhancing environmental management. The experience, evidence, methods, and models relevant for studying environmental management are presented here in six stand-alone thematic volumes, as follows:

VOLUME 1 – Managing Global Resources and Universal Processes
VOLUME 2 – Managing Biological and Ecological Systems
VOLUME 3 – Managing Soils and Terrestrial Systems
VOLUME 4 – Managing Water Resources and Hydrological Systems
VOLUME 5 – Managing Air Quality and Energy Systems
VOLUME 6 – Managing Human and Social Systems

In this manner, the handbook introduces in the first volume the general concepts and processes used in environmental management. The next four volumes deal with each of the four spheres of nature (biosphere, geosphere, hydrosphere, and atmosphere). The last volume ties the material together in its application to human and social systems. These are very important chapters for a wide spectrum of students and professionals to understand and implement environmental management. In particular, features include the following:

- The first handbook that demonstrates the key processes and provisions for enhancing environmental management.
- Addresses new and cutting-edge topics on ecosystem services, resilience, sustainability, food–energy–water nexus, socio-ecological systems, etc.
- Provides an excellent basic knowledge on environmental systems, explains how these systems function, and gives strategies on how to manage them.
- Written by an outstanding group of environmental experts.

Since the handbook covers such a wide range of materials from basic processes, to tools, technologies, case studies, and legislative actions, each handbook entry is further classified into the following categories:

APC: Anthropogenic chemicals: The chapters cover human-manufactured chemicals and their activities
COV: Indicates that the chapters give comparative overviews of important topics for environmental management

CSS: The chapters give a case study of a particular environmental management example

DIA: Means that the chapters are about diagnostic tools: monitoring, ecological modeling, ecological indicators, and ecological services

ELE: Focuses on the use of legislation or policy to address environmental problems

ENT: Addresses environmental management using environmental technologies

NEC: Natural elements and chemicals: The chapters cover basic elements and chemicals found in nature

PRO: The chapters cover basic environmental processes.

Volume 4, *Managing Water Resources and Hydrological Systems*, has extensive coverage in over 80 entries of water supply, water treatment, wetlands, lakes, and other natural water systems. New entries cover the evolution of water management, with application of optimization tools, and the innovative move toward integrating the energy and water nexus. Case studies include the Aral Sea, Chesapeake Bay, Baltic Sea, and Yellow River to name a few. Policy implications regarding wetland conservation, use of remote sensing and GIS, and agricultural water use are included.

Brian D. Fath
Brno, Czech Republic
December 2019

Editors

Brian D. Fath is Professor in the Department of Biological Sciences at Towson University (Maryland, USA) and Senior Research Scholar at the International Institute for Applied Systems Analysis (Laxenburg, Austria). He has published over 180 research papers, reports, and book chapters on environmental systems modeling, specifically in the areas of network analysis, urban metabolism, and sustainability. He has co-authored the books *A New Ecology: Systems Perspective* (2020), *Foundations for Sustainability: A Coherent Framework of Life–Environment Relations* (2019), and *Flourishing within Limits to Growth: Following Nature's Way* (2015). He is also Editor-in-Chief for the journal *Ecological Modelling* and Co-Editor-in-Chief for *Current Research in Environmental Sustainability*. Dr. Fath was the 2016 recipient of the Prigogine Medal for outstanding work in systems ecology and twice a Fulbright Distinguished Chair (Parthenope University, Naples, Italy in 2012 and Masaryk University, Czech Republic in 2019). In addition, he has served as Secretary General of the International Society for Ecological Modelling, Co-Chair of the Ecosystem Dynamics Focus Research Group in the Community Surface Modeling Dynamics System, and member and past Chair of Baltimore County Commission on Environmental Quality.

Sven E. Jørgensen (1934–2016) was Professor of environmental chemistry at Copenhagen University. He received a doctorate of engineering in environmental technology and a doctorate of science in ecological modeling. He was an honorable doctor of science at Coimbra University (Portugal) and at Dar es Salaam (Tanzania). He was Editor-in-Chief of *Ecological Modelling* from the journal inception in 1975 until 2009. He was Editor-in-Chief for the *Encyclopedia of Environmental Management* (2013) and *Encyclopedia of Ecology* (2008). In 2004, Dr. Jørgensen was awarded the Stockholm Water Prize and the Prigogine Medal. He was awarded the Einstein Professorship by the Chinese Academy of Sciences in 2005. In 2007, he received the Pascal Medal and was elected a member of the European Academy of Sciences. He has published over 350 papers, and has edited or written over 70 books. Dr. Jørgensen gave popular and well-received lectures and courses in ecological modeling, ecosystem theory, and ecological engineering worldwide.

Contributors

Matthew Agarwala
Bennett Institute for Public Policy
University of Cambridge
Cambridge, United Kingdom

Shaikh Ziauddin Ahammad
School of Civil Engineering and Geosciences
Newcastle University
Newcastle, United Kingdom

Claude Amiard-Triquet
French National Center for Scientific Research
(CNRS)
University of Nantes
Nantes, France

Javier Barragan
Department of Agroforestry Engineering
University of Lleida
Lleida, Spain

Lindsay Beevers
Lecturer in Water Management
School of the Built Environment
Heriot Watt University
Edinburgh, United Kingdom

Jerry M. Bigham
School of Environment and Natural Resources
Ohio State University
Columbus, Ohio

Marek Biziuk
Chemical Faculty
Department of Analytical Chemistry
Gdansk University of Technology
Gdansk, Poland

David L. Bjorneberg
Northwest Irrigation and Soil Research
 Laboratory
Agricultural Research Service (USDA-ARS)
U.S. Department of Agriculture
Kimberly, Idaho

Virginie Bouchard
School of Natural Resources
Ohio State University
Columbus, Ohio

Vince Bralts
Agricultural and Biological Engineering
Purdue University
West Lafayette, Indiana

Carl R. Camp, Jr.
Agricultural Research Service (USDA-ARS)
U.S. Department of Agriculture
Florence, South Carolina (Retired)

Kenneth L. Campbell
Agricultural and Biological Engineering
 Department
University of Florida
Gainesville, Florida

Kristi Denise Caravella
Florida Atlantic University
Boca Raton, Florida

Jesus Carrera
Technical University of Catalonia (UPC)
Barcelona, Spain

Dipankar Chakraborti
School of Environmental Studies
Jadavpur University
Calcutta, India

Bin Chen
 State Key Joint Laboratory of Environmental
   Simulation and Pollution Control
School of Environment
Beijing Normal University
Beijing, China

Alexander H.-D. Cheng
Department of Civil Engineering
University of Mississippi
Oxford, Mississippi

Tørben Røjle Christensen
Climate Impacts Group
Department of Ecology
Lund University
Lund, Sweden

Matthew Cochran
School of Natural Resources
Ohio State University
Columbus, Ohio

Gretchen C. Coffman
Department of Environmental Science
University of San Francisco
San Francisco, California

Robin Kundis Craig
Attorneys' Title Professor of Law and Associate
   Dean for Environmental Programs
Florida State University College of Law
Tallahassee, Florida

Abhijit Das
Vijoygarh Jyotish Ray College
University of Calcutta
Kolkata, India

Antara Das
School of Environmental Studies
Jadavpur University
Kolkata, India

Bhaskar Das
School of Environmental Studies
Jadavpur University
Kolkata, India

Jorge A. Delgado
Soil Management and Sugar Beet Research
   Unit
Agricultural Research Service (USDA-ARS)
U.S. Department of Agriculture
Fort Collins, Colorado

Jan Dolfing
School of Civil Engineering and Geosciences
Newcastle University
Newcastle, United Kingdom

Pay Drechsel
International Water Management Institute
   (IWMI)
Colombo, Sri Lanka

Rathindra Nath Dutta
Department of Dermatology
Institute of Post Graduate Medical Education and
   Research
SSKM Hospital
Kolkata, India

Anna Ekberg
Department of Ecology
Lund University
Lund, Sweden

Guy Fipps
Agricultural Engineering Department
Texas A&M University
College Station, Texas

Gary W. Frasier
U.S. Department of Agriculture (USDA)
Fort Collins, Colorado

Martin V. Frey
Department of Soil Science
University of Stellenbosch
Matieland, South Africa

Wendy B. Gagliano
Clark State Community College
Springfield, Ohio

Agniezka Gałuszka
Division of Geochemistry and the
   Environment
Institute of Chemistry
Jan Kochanowski University
Kielce, Poland

Anja Gassner
Institute of Science and Technology
University of Malaysia-Sabah
Kota Kinabalu, Malaysia

David W. Graham
School of Civil Engineering and Geosciences
Newcastle University
Newcastle, United Kingdom

Timothy C. Granata
Department of Civil and Environmental
   Engineering and Geodetic Science
Ohio State University
Columbus, Ohio

Joyeeta Gupta
Department of Human Geography
Planning and International Development
Amsterdam Institute of Social Science Research
University of Amsterdam
Amsterdam, the Netherlands

and

IHE Institute for Water Education in Delft
Delft, the Netherlands

Robert W. Hill
Biological and Irrigation Engineering
   Department
Utah State University
Logan, Utah

Glenn J. Hoffman
Biological Systems Engineering
University of Nebraska–Lincoln
Lincoln, Nebraska

Sven Erik Jørgensen
Section of Environmental Chemistry
Copenhagen University
Copenhagen, Denmark

Sandeep Joshi
Shrishti Eco-Research Unit (SERI)
Pune, India

Douglas L. Karlen
U.S. Department of Agriculture (USDA)
Ames, Iowa

Subhankar Karmakar
Center for Environmental Science and
   Engineering (CESE)
Indian Institute of Technology Bombay
Mumbai, India

Gregory Kiker
University of Florida
Gainesville, Florida

Andrey G. Kostianoy
P.P. Shirshov Institute of Oceanology
Russian Academy of Sciences
Moscow, Russia

David P. Kreutzweiser
Canadian Forest Service
Natural Resources Canada
Sault Sainte Marie, Ontario, Canada

Atif Kubursi
Department of Economics
McMaster University
Hamilton, Ontario, Canada

Freddie L. Lamm
Northwest Research-Extension Center
Kansas State University
Colby, Kansas

Jean-Claude Lefeuvre
Laboratory of the Evolution of Natural and
   Modified Systems
University of Rennes
Rennes, France

Cheng Li
Guangdong Key Laboratory of Agricultural
    Environment Pollution Integrated Control
Guangdong Institute of Eco-Environmental
    Science and Technology
Guangzhou, China

Zhuoran Liang
Hangzhou Meteorological Bureau
Hangzhou, China

K.F. Andrew Lo
Department of Natural Resources
Chinese Culture University
Taipei, Taiwan

Richard Lowrance
Agricultural Research Service (USDA-ARS)
U.S. Department of Agriculture
Tifton, Georgia

Mahesh L. Makey
University of California
Oakland, California

Kyle R. Mankin
Department of Biological and Agricultural
    Engineering
Kansas State University
Manhattan, Kansas

Rudolf Marloth
San Diego State University
San Diego, California

J.F. Martin
Department of Food, Agricultural, and Biological
    Engineering
Ohio State University
Columbus, Ohio

Jocilyn Danise Martinez
University of South Florida
Tampa, Florida

Matgorzata Michalska
Institute of Maritime and Tropical Medicine
Gdynia, Poland

Zdzistaw M. Migaszewski
Division of Geochemistry and the Environment
Institute of Chemistry
Jan Kochanowski University
Kielce, Poland

Meenakshi Mukherjee
School of Environmental Studies
Jadavpur University
Kolkata, India

Subhas Chandra Mukherjee
Department of Neurology
Medical College
Kolkata, India

Rafael Munoz-Carpena
University of Florida
Gainesville, Florida

Stephen D. Murphy
Faculty of Environment
University of Waterloo
Waterloo, Ontario, Canada

O.M. Musthafa
Center for Pollution Control and Environmental
    Engineering
Pondicherry University
Pondicherry, India

Daria Nikitina
Department of Earth and Space Sciences
West Chester University of Pennsylvania
West Chester, Pennsylvania

J. Pablo Ortiz-Partida
Union of Concerned Scientists
Cambridge, Massachusetts

Lloyd B. Owens
U.S. Department of Agriculture (USDA)
Coshocton, Ohio

Shyamapada Pati
Department of Obstetrics and Gynaecology
Calcutta National Medical College
Kolkata, India

Zaneta Polkowska
Gdansk University of Technology
Gdansk, Poland

Manzoor Qadir
Institute for Water, Environment and Health
 (UNU-INWEH)
Hamilton, Ontario, Canada

Quazi Quamruzzaman
Dhaka Community Hospital
Dhaka, Bangladesh

Anna Rabajczyk
Independent Department of Environment
 Protection and Modeling
Jan Kochanowski University of Humanities and
 Sciences
Kielce, Poland

Mohammad Mahmudur Rahman
School of Environmental Studies
Jadavpur University
Kolkata, India

Liqa Raschid-Sally
International Water Management Institute
 (IWMI)
Colombo, Sri Lanka

James D. Rhoades
Agricultural Salinity Consulting
Riverside, California

Mark Robinson
Center for Ecology and Hydrology
Wallingford, United Kingdom

Tarit Roychowdhury
School of Environmental Studies
Jadavpur University
Calcutta, India

Clayton Rubec
Center for Environmental Stewardship and
 Conservation
Ottawa, Ontario, Canada

D.W. Rycroft
Department of Civil and Environmental
 Engineering
Southampton University
Southampton, United Kingdom

Khitish Chandra Saha
School of Environmental Studies
Jadavpur University
Kolkata, India

A.V. Shanwal
Department of Soil Science
Chaudhary Charan Singh Haryana Agricultural
 University
Hisar, India

Paul K. Sibley
School of Environmental Sciences
University of Guelph
Guelph, Ontario, Canada

S.P. Singh
National Bureau of Soil Survey and Land Use
 Planning
Indian Agricultural Research Institute
New Delhi, India

Jeff Skousen
Division of Plant and Soil Sciences
West Virginia University
Morgantown, West Virginia

Bogdan Skwarzec
Faculty of Chemistry
University of Gdansk
Gdansk, Poland

Sean M. Smith
Ecosystem Restoration Center
Maryland Department of Natural Resources
Annapolis, Maryland

Roy F. Spalding
Water Science Laboratory
University of Nebraska–Lincoln
Lincoln, Nebraska

B.A. Stewart
Dryland Agriculture Institute
West Texas A&M University
Canyon, Texas

Piotr Szefer
Department of Food Sciences
Medical University of Gdansk
Gdansk, Poland

Thomas L. Thurow
Department of Renewable Resources
University of Wyoming
Laramie, Wyoming

Ralph W. Tiner
National Wetlands Inventory Program
U.S. Fish and Wildlife Service
Hadley, Massachusetts

Francine van den Brandeler
Department of Human Geography
Planning and International Development
University of Amsterdam
Amsterdam, the Netherlands

George Vance
Department of Renewable Resources
University of Wyoming
Laramie, Wyoming

Alejandra Virgen-Urcelay
University of British Columbia
Vancouver, Canada

Victor de Vlaming
Aquatic Toxicology Laboratory
University of California—Davis
Davis, California

Jan Vymazal
Faculty of Environmental Sciences
Department of Landscape Ecology
Czech University of Life Sciences
Prague, Czech Republic

Ivan A. Walter
Ivan's Engineering, Inc.
Denver, Colorado

Saige Wang
State Key Joint Laboratory of Environmental
   Simulation and Pollution Control
School of Environment
Beijing Normal University
Beijing, China

Ynuzhang Wang
Academy of Yellow River Conservancy Science
Zhengzhou, China

Dennis Wichelns
International Water Management Unit
Colombo, Sri Lanka

Brent Wootton
Center for Alternative Wastewater Treatment
Fleming College
Lindsay, Ontario, Canada

Pai Wu
College of Tropical Agriculture and Human
   Resources
University of Hawaii
Honolulu, Hawaii

Colin N. Yates
Centre for Research and Innovation
Fanshawe College
London, Ontario, Canada

Honglin Zhong
Department of Geographical Sciences
University of Maryland
College Park, Maryland

Yifei Zhu
Gemune LLC
Fremont, California

Zixi Zhu
Henan Institute of Meteorology
Zhengzhou, China

I

APC: Anthropogenic Chemicals and Activities

1

Aquatic Communities: Pesticide Impacts

Introduction 3
Measuring Impacts on Aquatic Communities 4
Assessing Risk of Pesticide Impacts on Aquatic Communities.................5
Some Examples of Pesticide Impacts on Aquatic Communities................7
Reducing Risk of Pesticide Impacts on Aquatic Communities.................9
Recent Advances and Outstanding Issues...9
Conclusions..11
References...12

David P.
Kreutzweiser and
Paul K. Sibley

Introduction

A biotic community can be defined as an assemblage of plant or animal species utilizing common resources and cohabiting a specific area. Examples could include a fish community of a stream, an insect community of a forest pond, or a phytoplankton community of a lake. Interactions among species provide ecological linkages that connect food webs and energy pathways, and these interconnections provide a degree of stability, or balance, to the community. Community balance can be described as a state of dynamic equilibrium in which species and their population dynamics within a community remain relatively stable, subject to changes through natural adjustment processes. Toxic effects of pesticides can disrupt these processes and linkages and thereby cause community balance upsets. For example, this can occur when a pesticide has a direct impact on a certain species in a community and reduces its abundance while other unaffected species increase in abundance in response to the reduced competition for food resources or increased habitat availability. Some of the best examples of pesticide impacts on biological communities are found in freshwater studies. Freshwater aquatic communities are usually contained within distinct boundaries or systems, and this generates a high degree of connectivity among species, thereby increasing their susceptibility to pesticide-induced disturbances at the community level.

We examine traditional and developing methods for measuring pesticide impacts on freshwater communities, with emphasis on recent improvements in risk assessment approaches and analyses, and provide some examples for illustration. We then describe some advances in impact mitigation strategies and discuss some ongoing issues pertaining to understanding, assessing, and preventing pesticide impacts including probabilistic risk assessment (PRA), population and ecological modeling, and pesticide interactions with multiple stressors. The integration of improved risk assessment and mitigation approaches and technologies together with information generated from the numerous impact studies available will provide a sound scientific basis for decisions around the use and regulation of pesticides in and near water bodies.

Measuring Impacts on Aquatic Communities

Changes in aquatic communities can be measured directly in water bodies by a number of quantitative and qualitative sampling methods. Descriptions of those methods can be found in any up-to-date text or handbook (e.g., Hauer and Lamberti[1]). Measurements can be in terms of community structure (species composition) or community function (a measurable ecosystem process attributable to a biotic community that causes a change in condition) and can include both direct and indirect effects.[2,3] Community structure is a measure of biodiversity in its most general sense, that is, the number of species or other taxonomic units and their relative abundances. Some community functions are referred to as environmental or ecosystem services. Examples include organic matter breakdown and nutrient cycling that is largely mediated by microbial communities, or water uptake, filtration, and flood control mediated by shoreline plant communities.[4] Both community structure (biodiversity) and function (ecosystem services) are being increasingly valued by society and global economies,[5,6] and therefore sustaining healthy aquatic communities will be an important driver of pesticide impact mitigation efforts.

Detecting impacts of pesticides typically involves repeated sampling and a comparison of community attributes among contaminated and uncontaminated test units over time, or across a gradient of pesticide concentrations. The test units can range from petri dishes to natural ecosystems, with a trade-off between experimental control in small test units and environmental realism in field-level testing and whole ecosystems.[7] In an effort to incorporate both experimental control and environmental realism in pesticide impact testing, the use of microcosms or model ecosystems for measuring impacts on aquatic communities has increased over the past couple of decades.[8,9] Model ecosystems for community-level pesticide testing can be quite simple at lower-trophic levels such as with microbial communities (e.g., Widenfalk et al.[10]) but will necessarily be more complex for testing higher-order biological communities (e.g., Wojtaszek[11]). Regardless of the test units, an important consideration for measuring pesticide impacts will be an assessment of the duration of impact or rate of recovery. A rapid return to pre-pesticide or reference (nopesticide) community condition will reduce the long-term ecological consequences of the pesticide disturbance.[12]

Traditional measures of community-level impacts have focused on structure and have usually been expressed in terms of single-variable indices such as species richness, diversity, or abundance. These indices are useful descriptors of community structure but suffer from the fact that they reduce complex community data to a single summary metric and may miss subtle or ecologically important changes in species composition across sites or times. Over the last couple of decades, ecotoxicologists have increasingly turned to multivariate statistical techniques for analyzing community response data.[13] A variety of multivariate statistical techniques and software are available and are usually considered superior for the analysis of community data because they retain and incorporate the spatial and temporal multidimensional nature of biological communities.[14] This includes various ordination techniques that can provide graphical representation of spatiotemporal patterns in community structure in which points that lie close together in the ordination plot represent communities of similar composition (richness, abundance), while communities with dissimilar species composition are plotted further apart.

Figure 1 illustrates the use of an ordination plot generated by nonmetric multidimensional scaling for detecting differences among aquatic insect communities in four control and eight insecticide-treated streams. These data have been adjusted for illustrative purposes but are based on real invertebrate community responses to an insecticide in outdoor stream channels.[15] At both concentrations of the insecticide, the community structure of stream insects clearly shifted away from the natural community composition in control streams as depicted by the separation of treated streams (T1 and T2) from controls (C) in the ordination bi-plot. The plot also illustrates that the variability among treated streams (relative distance between points) was greater than that among control streams, that the low-concentration streams (T1) and high-concentration streams (T2) tended to separate along axis 1, and that the T2 streams were further removed from controls than the T1 streams, indicating a differential response by the insect communities to the two test concentrations. Canonical correspondence analysis

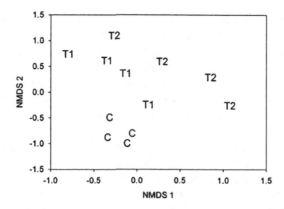

FIGURE 1 Ordination by nonmetric multidimensional scaling of aquatic insect communities in stream channels. Each point represents the community structure of control channels (C) and channels treated with a neem-based insecticide at a low (T1) and high (T2) concentration.
Source: Adapted from Kreutzweiser et al.[15]

and redundancy analysis have also been commonly used to assess aquatic community responses to pesticide contamination.[16,17] A useful refinement of an ordination technique for detecting and interpreting pesticide impacts on aquatic communities is principal response curves (PRCs).[18] PRC is derived from redundancy analysis, and time-dependent responses in the treatments are expressed as deviations from the control or reference system allowing for clear visualization of pesticide effects.

Assessing Risk of Pesticide Impacts on Aquatic Communities

The likelihood or risk of harmful effects on aquatic communities from exposure to pesticides will depend on the exposure concentration, bioavailability, exposure duration, rate of uptake, inherent species sensitivities, community composition, and other community attributes. All of these must be measured, estimated, modeled, or predicted to derive an assessment of risk to aquatic communities for any given pesticide. Formalized risk assessment frameworks and guidelines for pesticides have been developed in the United States,[19] the European Union,[20] Canada,[21] and elsewhere and can be consulted for detailed information on the various components of a risk assessment. In brief, pesticide risk assessments typically include the following phases: 1) defining the problem by determining the pesticide use patterns and developing conceptual models and hypotheses around how it is expected to behave, the anticipated exposure regimes, the kinds of organisms that are likely to be at risk, the community or entity that is to be protected, and the level of protection that will be acceptable; 2) developing the measurement endpoints for assessing risk of harm by establishing which response measurements are relevant and applicable, and how the measurements will be made; 3) outlining the risk assessment process by specifying the kinds of data to be used and how they will be derived including simulation modeling, empirical laboratory, microcosm or field testing, their appropriate spatial and temporal scales, and their statistical analyses; 4) applying the risk assessment by running models or collecting data, completing analyses, summarizing outputs, and providing risk estimates; 5) conducting risk communication and management by answering questions posed in the problem formulation, suggesting risk mitigation strategies if necessary, and communicating those to appropriate users; and 6) conducting follow-up monitoring to evaluate the success of mitigation strategies and to implement adaptive management to address deficiencies if or when necessary.[22,23]

Traditionally, pesticide risk assessments have relied on standardized, single-species toxicity tests to predict effects on communities, the underlying assumption being that protecting the most sensitive species will protect whole communities. In this case, the selection and relevance of test species are critically important to a successful and meaningful risk assessment.[24] However, the accuracy and relevance of estimating the potential risk to aquatic communities can be greatly improved by consideration of specific species or community attributes. In particular, attribute information can improve the ecological relevance and predictive capabilities of conceptual models and the generation of hypotheses in the risk assessment process. Insofar as these attributes affect exposure, sensitivity, or both, they can increase or decrease risk beyond what could be determined from toxicity estimates or species sensitivity distributions alone.

Behavioral attributes can elevate the risk of pesticide effects on species by increasing the likelihood of intercepting the stressor. For example, young-of-the-year bluefish (*Pomatomus saltatrix*) typically feed in estuaries during their early life stages where agricultural runoff can elevate concentrations of pesticides in food items. This feeding behavior can result in bioaccumulation and in adverse effects such as reduced migration, overwinter survival, and recruitment success in fish communities.[25] Incorporating this kind of information into conceptual models and risk hypotheses will generate more realistic risk assessments. In addition, behavioral attributes themselves can be relevant measurement endpoints if the pesticide mode of action indicates risk of sublethal behavioral effects at expected concentrations. For example, some pesticides have been shown to impair the ability to capture prey in fish[26] and the ability to avoid predators in zooplankton.[27] These types of adverse effects can disrupt trophic linkages and reduce survival or reproduction, thus impacting community balance.

Inclusion of life history information into conceptual models and risk hypotheses can also refine and improve the risk assessment process. Life history strategies can influence a species susceptibility to a stressor through effects on a population's resilience or ability to recover from disturbance.[28] Different species exposed to the same pesticide and experiencing similar levels of effect in terms of population declines do not necessarily recover at the same rates when recovery is dependent on reproduction or dispersal. Populations of organisms with short regeneration times (e.g., several generations per year) and/or high dispersal capacity have higher likelihood of recovery from pesticide-induced population declines than those with longer regeneration periods and limited dispersal capacity. These differential life history strategies and their influences on community response and recovery from pesticide effects have been demonstrated empirically (e.g., van den Brink et al.[29] and Kreutzweiser et al.[30]) and through population modeling.[31] These community balance upsets could not have been predicted from screening-level toxicity data or from species sensitivity data; thus, inclusion of life history information in conceptual models can improve risk hypotheses and direct the assessment to focus on species at higher risk owing to specific life history strategies.

Life history attributes can also influence the risk of pesticide effects through differential life-stage sensitivity or susceptibility. Early life stages are often (but not exclusively) more sensitive to pesticides than later stages. An organism's life stage can also influence its susceptibility to a pesticide by increasing or decreasing the likelihood of intercepting the stressor. If a contaminant is present in the environment at effective concentrations during a period in which the particular life stage of a species is present, then the risk to that species is increased. For some amphibians, aquatic (larval) stages could be at higher risk of direct and indirect effects of pesticides than their terrestrial (adult) life stages when their larval stage coincides with pesticide contamination of water bodies.[32] Thus, while a species sensitivity and geographical distribution may indicate potential risk, the life-stage information coupled with pesticide use pattern, timing, or fate information may indicate little likelihood of exposure to the pesticide and the risk assessment can be adjusted accordingly.

Functional attributes may also be important for refining or improving pesticide risk assessments. Protection goals for populations and communities often include the safeguarding of critical biological processes or ecosystem function. Measuring ecosystem function integrates responses of component populations and can be a relevant measurement endpoint when species loss affects ecosystem function such as energy transfer and organic matter cycling.[33] However, most ecosystems are complex and it may not be clear which functional attributes are critical for sustaining ecological processes or the extent to which they can sustain changes in structural properties (e.g., population levels, diversity) without adversely affecting ecosystem function. Neither is it clear if functional endpoints are more or less sensitive than structural endpoints for detecting ecosystem disturbance. Some studies investigating the relationship between species diversity and ecosystem function have indicated that ecosystems can tolerate some species loss because of functional redundancy.[34] Functional redundancy is thought to occur when several species perform similar functions in ecosystems such that some may be eliminated with little or no effect on ecosystem processes. Others have suggested that redundant species are required to ensure ecosystem resilience to disturbance as a form of biological insurance, especially at large spatial scales.[35]

Given these discrepancies, measurement endpoints based on functional attributes are not typically used in pesticide risk assessments because it is generally accepted that protection of community structure will protect ecosystem function. However, when specific functional attributes can be identified and are known or suspected to be at risk from a pesticide, they can be included in the data requirements for a risk assessment. An example would be the risk of adverse effects on leaf litter decomposition (a critical ecosystem function in forest soils and water bodies) posed by a systemic insecticide for control of wood-boring insects in trees.[36] In that case, the protection goal was maintaining leaf litter decomposition, the community at risk was decomposer invertebrates feeding on leaves from insecticide-treated trees, and the selection of test species was directed to a specific functional group because of the unique route of exposure to decomposer organisms identified in the risk hypotheses.

Some Examples of Pesticide Impacts on Aquatic Communities

A few examples will serve to illustrate how pesticides can cause disruptions to aquatic communities. DeNoyelles et al.[37] reviewed studies into pesticide impacts on aquatic communities and reported that herbicides like atrazine, hexazinone, and copper sulfate were directly toxic to most species of phytoplankton (waterborne algae). After herbicide applications, reductions in phytoplankton caused secondary reductions in herbivorous zooplankton, resulting from a depleted food source for the zooplankton. They further showed that direct adverse effects on phytoplankton can also cause disruptions to the bacterial-based energy pathways by reducing carbon flow from phytoplankton to bacteria, and ultimately to grazing protozoans and zooplankton. Boyle et al.[38] found that applications of the insecticide diflubenzuron to small ponds reduced populations of several aquatic invertebrate species. This in turn resulted in indirect effects on algae (increased productivity because of release from grazing pressure by the invertebrates) and on juvenile fish populations (reduced production because of limited invertebrate prey availability). George et al.[39] used a novel approach to predict effects of pesticide mixtures on zooplankton communities and then tested the predictions in outdoor microcosms. Responses among zooplankton populations within the community differed, depending on the pesticide mixture, and those differences appeared to reflect the relative susceptibilities among specific taxa within groups. Cladocerans declined but were less sensitive than copepods to a chlorpyrifos-dominated mixture, while rotifers actually increased after application in response to release from competition or predation pressures.

Kreutzweiser et al.[40] applied a neem-based insecticide to forest pond enclosures and measured effects on zooplankton community structure, respiration, and food web stability. Significant concentration-dependent reductions in numbers of adult copepods were observed, but immature copepods and cladocerans were unaffected (Figure 2). There was no evidence of recovery of adult copepods within the sampling season. During the period of maximal impact (about 4 to 9 weeks after the applications), total plankton community respiration was significantly reduced, and this contributed to significant concentration-dependent increases in dissolved oxygen and decreases in specific conductance. The reductions in adult copepods resulted in negative effects on zooplankton food web stability through elimination of a trophic link and reduced interactions and connectance.

Van Wijngaarden et al.[41] evaluated the responses of aquatic communities in indoor microcosms to a suite of pesticides used for bulb crop protection. At pesticide concentrations equivalent to 5%

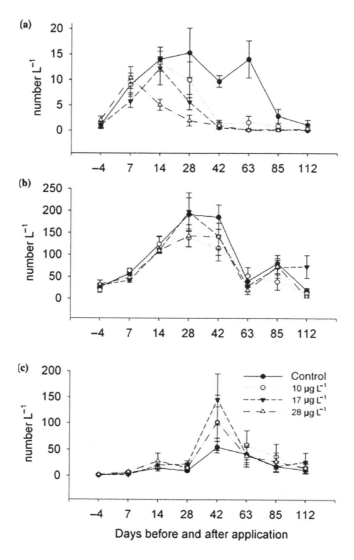

FIGURE 2　Mean abundance (±1 SE, n = 5) of (a) adult copepods, (b) immature copepods, and (c) cladocerans in natural pond microcosms (controls) and microcosms treated at three different rates of a neem-based insecticide
Source: Taken from Kreutzweiser et al.[40]

spray drift deposition, zooplankton taxa within communities showed significant changes relative to non-treated controls, reflecting taxon-specific sensitivities. Some copepods and rotifers in particular showed significant declines for at least 13 weeks, while many other rotifers and cladocerans were unaffected or increased weeks, while many other rotifers and cladocerans were unaffected or increased. Several macroinvertebrate taxa were negatively affected, and this contributed to significant declines in leaf litter decomposition among treated microcosms. The herbicide asulam was among the suite of pesticides, and it induced significant reduction of the macrophyte *Elodea nuttallii*. This in turn caused significant changes in water chemistry (decreases in dissolved oxygen and pH, increases in alkalinity and specific conductance) and increases in phytoplankton biomass from decreased competition for nutrients. Increased phytoplankton and reduced zooplankton predators combined to support higher abundance of less sensitive zooplankton taxa. The authors point out that most of these effects were not measurable at more realistic rates of spray drift deposition.

Relyea and Hoverman[42] investigated impacts of the insecticide malathion on aquatic communities in microcosms designed to mimic a simple aquatic food web that can be found in ponds and wetlands. The insecticide generally reduced zooplankton abundance, and these reductions stimulated increases in phytoplankton, decreases in periphyton (attached algae), and decreases in growth of frog tadpoles. While invertebrate predator survival was not affected, amphibian prey survival increased with insecticide concentration, apparently the result of insecticide-induced impairment of predation success by the invertebrates. Overall, the study demonstrated that realistic concentrations of an insecticide can interact with natural predators to induce large changes in aquatic community balance.

Reducing Risk of Pesticide Impacts on Aquatic Communities

For pesticides applied to crops and forests, exposure to aquatic communities can be minimized by the implementation of vegetated spray buffers or setbacks to intercept off-target spray drift and runoff.[43] Pesticide runoff can be further reduced by using formulations that are less prone to wash-off, leaching, and mobilization. Recent advances in spray drift reduction and improved spray guidance systems can also significantly reduce the off-target movement of pesticides to water bodies.[44] Examples include new technologies in map-based automated boom systems for row crops[45] and Geographical Information System (GIS)-based landscape analysis for predicting off-target pesticide movement.[46]

The risk of adverse effects on aquatic communities may also be decreased by intentional selection and use of pesticides that are inherently safer to the environment. This would include so-called reduced-risk pesticides that are bioactive compounds usually with unique modes of action and derived from microbial, plant, or other natural sources. These are generally thought to be less persistent and toxic to non-target organisms than conventional synthetic pesticides.[47] Examples include the bacteria-derived insecticide *Bt (Bacillus thuringiensis),* the plant-derived insecticide neem, and the microbe-derived herbicide phosphinothricin. However, Thompson and Kreutzweiser[48] caution that it cannot be assumed that this group of pesticides is inherently safer or more environmentally acceptable than synthetic counterparts and that full environmental risk evaluations must be conducted to ensure their environmental safety.

These types of technologies combined with the use of non-pesticide approaches to pest management form the basis of integrated pest management (IPM) strategies. IPM strategies are those in which the judicious use of pesticides is only one of several concurrent methods to control or manage losses from pest damage. This can include the use of natural enemies and parasites, biological control agents, insect growth regulators, confusion pheromones, sterile male releases, synchronizing with weather patterns known to diminish pest populations, and cultivation methods and crop varieties to improve conditions for natural enemies or degrade conditions for pest survival.[49] Increasing the use of IPM approaches can reduce reliance on pesticides and thus reduce the risk of pesticide impacts overall.

Recent Advances and Outstanding Issues

Pesticide risk assessments and risk reductions have recently been advanced in terms of ecological realism and effectiveness through some developing methods and techniques. Traditional risk assessments have estimated hazards from pesticides by comparing the expected environmental concentration (often predicted from worst-case scenarios) to the toxic threshold for the most sensitive test species. When the expected concentration is higher than the toxicity threshold, the pesticide is considered to have potential for environmental effects. These so-called hazard or risk quotient approaches are still widely used in pesticide risk assessment and regulation, but more recently, PRA and probabilistic hazard assessment (PHA) approaches are being adopted. In these approaches, pesticide exposure levels and the likelihood of toxic effects are estimated from probability distributions based on all reliable data available.[50] In PRA, exposure and effects distributions are developed from modeling or measurements in laboratory, microcosm, or field studies and used to improve the accuracy and relevance of the estimated likelihood

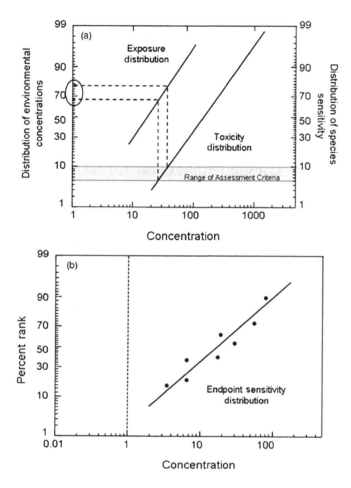

FIGURE 3 Schematic illustrating the principle of PRA (a) and PHA (b). PRA is based on a comparison of expo-sure and effects distributions using a predetermined criterion typically in the range of 5%-10% (shaded area and dashed lines in panel A) to determine the probability of exceeding the criterion (ellipse on y-axis); PHA is based on a comparison of an endpoint-derived sensitivity distribution within a test species to a threshold value such as a hazard quotient (dashed line in panel B).

of environmental effects compared to the traditional worst-case (hazard/risk quotient) approach (e.g., Solomon[51]). In PHA, a distribution approach is also used, except that the probability of hazard is estimated from distributions built on the relative sensitivity of interspecies endpoints rather than species sensitivity itself.[52] Figure 3 illustrates the principles of PRA (Figure 3a) and PHA (Figure 3b). Regardless of the approach, one important aspect of PRA that is ongoing is the development and use of uncertainty analysis to quantify variability and uncertainty in exposure and effects estimates. Characterizing and quantifying uncertainty will provide more meaningful risk assessments and improved decision making for minimizing potential risk of pesticide impacts in or near water.[53]

Efforts at incorporating population or ecological modeling into pesticide risk assessments have also improved their accuracy and relevance for predicting, and therefore mitigating, risk of harm to aquatic communities.[54] The use of ecological models to incorporate a suite of factors including lethal and sub-lethal effects and their influences on the risks to organisms, populations, or communities can provide useful insights into receptor/pesticide interactions and can thereby improve risk assessments and direct mitigation measures. Population models that account for differential demographics and population growth rates within communities have been shown to provide more accurate assessments of potential pesticide impacts on populations and communities than what conventional lethal concentration estimates can provide.[55] Ecological and population modeling combined with pesticide exposure modeling and case-based reasoning (drawing on past experience or information from similar chemical exposures) can provide further refinements and improve risk assessment for aquatic communities.[56] Another recent advancement in ecological modeling to predict pesticide effects is the use of trait-based information such as organism morphology, life history, physiology, and feeding ecology in risk assessments.[57] This approach includes some of the functional attributes and concepts described above in the section on "Assessing Risk of Pesticide Impacts on Aquatic Communities" and has the advantage of formally expressing communities as combinations of functional traits rather than as groups of species, thereby yielding a more meaningful description of community structure and function. Taken together, these modeling approaches that incorporate probability distributions, toxicological sensitivities, population dynamics, ecological information, and functional trait attributes can be integrated into improved risk assessments that will inform mitigation and prevention strategies for pesticide use.[58]

Two additional issues that present challenges to pesticide risk assessment and mitigation are pesticide mixtures and the combined or cumulative effects of multiple stressors on pesticide impacts. Pesticides frequently occur as mixtures in aquatic systems, particularly in agricultural regions, and methods to assess and/or predict pesticide mixture toxicity under laboratory conditions have been relatively well developed. However, there are still large uncertainties associated with the prediction of pesticide mixture toxicity, and additional studies are needed to evaluate the performance of mixture models when evaluating community-level endpoints and toxicity thresholds over long-term exposures.[59] Secondly, whereas most pesticide assessment data are derived from tests or experiments under controlled or semicontrolled environmental conditions, pesticides in natural environments may interact with a number of other natural or human-caused stressors that can substantially alter the likelihood and magnitude of pesticide impacts.[60] Other stressors could include overarching effects of climate change that can influence water temperature and quality; land use activities that result in chemical, sediment, and nutrient pollution of waterways; and biotic interactions with invasive species in aquatic communities. A number of studies have examined the combined effects of a pesticide with other stressors, but they have usually been single stressor effects tested at the single-species level. Examples of studies that examined combined effects include pesticide interactions with water temperature,[61] pH,[62] dissolved organic matter,[63] UV radiation,[64] predators,[65] competitors,[66] food availability,[67] elevated sediments,[68] and other chemical stressors.[69] However, potential multiple stressors and their interactions with pesticides can be myriad and testing or extrapolating to community-level impacts is onerous at best. Sorting out and mitigating pesticide impacts from among these multiple stressors continues to be a challenge, and the suggestion by Laskowski et al.[70] to include studies of toxicant interactions with a range of environmental conditions in risk assessments seems warranted.

Conclusions

Because of the high degree of connectivity among species in an aquatic community, pesticides pose a risk of harm to the community stability or balance. The community structure can be altered by direct effects, indirect effects, or both, and this can cause disruptions to the interactions and linkages among species and to their ecological function. This risk of harm will depend on exposure concentration, bioavailability, exposure duration, rate of uptake, species sensitivities, community composition, and other community attributes. Recent advances in pesticide risk assessment for aquatic communities have improved the ecological relevance and predictive capabilities for determining, and thus mitigating, potential harmful impacts. Pesticide impacts on aquatic communities can be minimized by the use of improved application technologies to reduce application rates and to decrease off-target movement to water bodies. Potential impacts can be further minimized through the selection and use of pesticides that are demonstrated to be inherently safer to the environment and through the application of IPM strategies. Given the preponderance of pesticide impact studies in freshwater aquatic ecosystems, the improved risk assessment frameworks and regulatory requirements for pesticide evaluations, and the recent advances in mitigation technologies, many decisions around the use of pesticides can be made on a sound scientific basis rather than on misinformed perceptions or politically driven agendas. Integrated, science-based pest management strategies including the prudent use of appropriate pesticides will contribute to ensuring the sustainability of aquatic communities in areas subjected to pest management programs.

References

1. Hauer, F.R.; Lamberti, G.A. *Methods in Stream Ecology,* 2nd Ed.; Elsevier: Amsterdam, 2006.
2. Fleeger, J.W.; Carman, K.R.; Nisbet, R.M. Indirect effects of contaminants in aquatic ecosystems. Sci. Tot. Environ. **2003**, *317,* 207–233.
3. Rohr, J.R.; Crumrine, P.W. Effects of an herbicide and an insecticide on pond community structure and processes. Ecol. Appl. **2005**, *15,* 1135–1147.
4. Daily, G.C. Introduction: What are ecosystem services? In *Nature's Services: Societal Dependence on Natural Ecosystems*; Daily, G.D., Ed.; Island Press: Washington, DC, 1997; 1–10.
5. Anderson, J.; Gomez, W.C.; McCarney, G.; Adamowicz, W.; Chalifour, N.; Weber, M.; Elgie, S.; Howlett, M. *Ecosystem Service Valuation, Market-based Instruments and Sustainable Forest Management: A Primer*; Sustainable Forest Management Network: Edmonton, Alberta, 2010.
6. Bayon, R.; Jenkins, M. 2010. The business of biodiversity. Nature **2010**, *466,* 184–185.
7. Sibley, P.K.; Chappel, M.J.; George, T.K.; Solomon, K.R.; Liber, K. Integrating effects of stressors across levels of biological organization: Examples using organophosphorus insecticide mixtures in field-level exposures. J. Aquat. Eco-syst. Stress Recovery **2000**, 7, 117–130.
8. Campbell, P.J.; Arnold, D.J.S.; Brock, T.C.M.; Grandy, N.J.; Heger, W.; Heimbach, F.; Maund, S.J.; Streloke, M. *Guidance Document on Higher-Tier Aquatic Risk Assessment for Pesticides;* SETAC Europe: Brussels, Belgium, 1999.
9. Kennedy, J.H.; LaPoint, T.W.; Balci, P.; Stanley, J.K.; Johnson, Z.B. Model aquatic ecosystems in ecotoxicological research: Considerations of design, implementation, and analysis. In *Handbook of Ecotoxicology,* 2nd Ed.; Hoffman, D.J.; Rattner, B.A.; Burton, G.A., Jr.; Cairns, J., Jr., Eds.; Lewis Publishers: Boca Raton, Florida, 2003; 45–74.
10. Widenfalk, A.; Svensson, J.M.; Goedkoop, W. Effects of the pesticides captan, deltamethrin, isoproturon, and pirimi-carb on the microbial community of a freshwater sediment. Environ. Toxicol. Chem. **2004**, *23,* 1920–1927.
11. Wojtaszek, B.F.; Buscarini, T.M.; Chartrand, D.T.; Stephenson, G.R.; Thompson, D.G. Effect of Release® herbicide on morality, avoidance response, and growth of amphibian larvae in two forest wetlands. Environ. Toxicol. Chem. **2005**, *24,* 2533–2544.
12. Barnthouse, L.W. Quantifying population recovery rates for ecological risk assessment. Environ. Toxicol. Chem. **2004**, 23, 500–508.
13. Maund, S.; Chapman, P.; Kedwards, T.; Tattersfield, L.; Matthiessen, P.; Warwick, R.; Smith, E. Application of multivariate statistics to ecotoxicological field studies. Environ. Toxicol. Chem. **1999**, *18,* 111–112.
14. Clarke, K. R.; Warwick, R.M. Change in Marine Communities: An Approach to Statistical Analysis and Interpretation, 2nd Ed.; PRIMER-E: Plymouth, U.K., 2001.
15. Kreutzweiser, D.P.; Capell, S.S.; Scarr, T.A. Community-level responses by stream insects to neem products containing azadirachtin. Environ. Toxicol. Chem. **2000**, *19,* 855–861.
16. Frieberg, N.; Lindstrom, M.; Kronvang, B.; Larsen, S.E. Macroinvertebrate/sediment relationships along a pesticide gradient in Danish streams. Hydrobiologia **2003**, *494,* 103–110.

17. Berenzen, N.; Kimke, T.; Schulz, H.K.; Schulz, R. Macroinvertebrate community structure in agricultural streams: Impact of run-off-related pesticide contamination. Ecotoxicol. Environ. Saf. **2005**, *60*, 37–46.
18. van den Brink, P.J.; ter Braak, C.J.F. Principal response curves: Analysis of time-dependent multi-variate responses of biological community to stress. Environ. Toxicol. Chem. **1999**, *18*, 138–148.
19. USEPA. *Guidelines for Ecological Risk Assessment*; United States Environmental Protection Agency, Risk Assessment Forum: Washington, DC, 1998.
20. EUFRAM. *Introducing Probabilistic Methods into the Ecological Risk Assessment of Pesticides, Version 6*; European Framework for Risk Assessment of Pesticides (EUFRAM): York, U.K., 2005.
21. Delorme, P.; Francois, D.; Hart, C.; Hodge, V.; Kaminski, G.; Kriz, C.; Mulye, H.; Sebastien, R.; Takacs, P.; Wandel-maier, F. *Final Report for the PMRA Workshop: Assessment Endpoints for Environmental Protection*; Environmental Assessment Division, Pest Management Regulatory Agency, Health Canada: Ottawa, Ontario, 2005.
22. Suter, G.W.; Barnthouse, L.W.; Bartell, S.M.; Mill, T.; Mackay, D.; Patterson, S. *Ecological Risk Assessment*; Lewis Publishers: Boca Raton, Florida, 1993.
23. Reinert, K.H.; Bartell, S.M.; Biddinger, G.R., Eds. *Ecological Risk Assessment Decision-Support System: A Conceptual Design;* SETAC Press: Pensacola, Florida, 1998.
24. Maltby, L.; Blake, N.; Brock, T.C.M.; van den Brink, P.J. Insecticide species sensitivity distributions: Importance of test species selection and relevance to aquatic ecosystems. Environ. Toxicol. Chem. **2005**, *24*, 379–388.
25. Candelmo, A.C.; Deshpande, A.; Dockum, B.; Weis, P.; Weis, J.S. The effect of contaminated prey on feeding, activity, and growth of young-of-the-year bluefish, *Pomato-mus saltatrix*, in the laboratory. Estuaries Coasts **2010**, *33*, 1025–1038.
26. Baldwin, D.H.; Spromberg, J.A.; Collier, T.K.; Scholz, N.L. A fish of many scales: Extrapolating sublethal pesticide exposures to the productivity of wild salmon populations. Ecol. Appl. **2009**, *19*, 2004–2015.
27. Pestana, J.L.T.; Loureiro, S.; Baird, D.J.; Soares, A.M.V.M. Pesticide exposure and inducible anti-predator responses in the zooplankton grazer, *Daphnia magna* Straus. Chemo-sphere **2010**, *78*, 241–248.
28. Stark, J.D.; Banks, J.E.; Vargas, R.I. How risky is risk assessment: The role that life history strategies play in susceptibility of species to stress. Proc. Natl. Acad. Sci. U. S. A. **2004**, *101*, 732–736.
29. van den Brink, P.J.; Hattink, J.; Bransen, F.; van Donk, E.; Brock, T.C.M. Impact of the fungicide carbendazim in freshwater microcosms. II. Zooplankton, primary producers and final conclusions. Aquat. Toxicol. **2000**, *48*, 251–264.
30. Kreutzweiser, D.P.; Back, R.C.; Sutton, T.M.; Pangle, K.L.; Thompson, D.G. Aquatic mesocosm assessments of a neem (azadirachtin) insecticide at environmentally realistic con-centrations—2: Zooplankton community responses and recovery. Ecotoxicol. Environ. Saf. **2004**, *59*, 194–204.
31. Wang, M.; Grimm, V. Population models in pesticide risk assessment: Lessons for assessing population-level effects, recovery, and alternative exposure scenarios from modeling a small mammal. Environ. Toxicol. Chem. **2010**, *29*, 1292–1300.
32. Brodman, R.; Newman, W.D.; Laurie, K.; Osterfeld, S.; Lenzo, N. Interaction of an aquatic herbi-cide and predatory salamander density on wetland communities. J. Herpetol. **2010**, *44*, 69–82.
33. Rosenfeld, J.S. Functional redundancy in ecology and conservation. Oikos **2002**, *98*, 156–162.
34. Lawton, J.H.; Brown, V.K. Redundancy in ecosystems. In *Biodiversity and Ecosystem Function*; Schulze, E.D., Mooney, H.A., Eds.; Springer: New York, 1993; 255–268.
35. Naeem, S.; Li, S. Biodiversity enhances ecosystem stability. Nature **1997**, *390*, 507–509.
36. Kreutzweiser, D.P.; Good, K.P.; Chartrand, D.T.; Scarr, T.A.; Thompson, D.G. Are leaves that fall from imidacloprid-treated maple trees to control Asian longhorned beetles toxic to non-target decomposer organisms? Journal of Environmental Quality **2008**, *37*, 639–646.

37. deNoyelles, F., Jr.; Dewey, S.L.; Huggins, D.G.; Kettle, W.D. Aquatic mesocosms in ecological effects testing: Detecting direct and indirect effects of pesticides. In *Aquatic Mesocosm Studies in Ecological Risk Assessment*; Graney, R.L., Kennedy, J.H., Rodgers, J.H., Jr., Eds.; Lewis Publishers: Boca Raton, 1994; 577–603.

38. Boyle, T.P.; Fairchild, J.F.; Robinson-Wilson, E.F.; Haver-land, P.S.; Lebo, J.A. Ecological restructuring in experimental aquatic mesocosms due to the application of diflubenzuron. Environ. Toxicol. Chem. **1996**, *15*, 1806–1814.

39. George, T.K.; Liber, K.; Solomon, K.R.; Sibley, P.K. Assessment of the probabilistic ecological risk assessment-toxic equivalent combination approach for evaluating pesticide mixture toxicity to zooplankton in outdoor microcosms. Archives of Environmental Contamination and Toxicology **2003**, *45*, 453–461.

40. Kreutzweiser, D.P.; Sutton, T.M.; Back, R.C.; Pangle, K.L.; Thompson, D.G. Some ecological implications of a neem (azadirachtin) insecticide disturbance to zooplankton communities in forest pond enclosures. Aquat. Toxicol. **2004**, *67*, 239–254.

41. van Wijngaarden, R.P.A.; Cuppen, J.G.M.; Arts, G.H.P.; Crum, S.J.H.; van den Hoorn, M.W.; van den Brink, P.J.; Brock, T.C.M. Aquatic risk assessment of a realistic exposure to pesticides used in bulb crops: a microcosm study. Environ. Toxicol. Chem. **2004**, *23*, 1479–1498.

42. Relyea, R.A.; Hoverman, J.T. Interactive effects of predators and a pesticide on aquatic communities. Oikos **2008**, *117*, 1647–1658.

43. Zhang, X.; Liu, X.; Zhang, M.; Dahlgren, R.A.; Eitzel, M. A review of vegetated buffers and a metaanalysis of their mitigation efficacy in reducing nonpoint source pollution. J. Environ. Qual. **2010**, *39*, 76–84.

44. van de Zande, J.C.; Porskamp, H.A.; Michielsen, J.M.; Holterman, H.J.; Juijsmans, J.M. Classification of spray applications for driftability to protect surface water. Aspects Appl. Biol. **2000**, *57*, 57–64.

45. Luck, J.D.; Zandonadi, R.S.; Luck, B.D.; Shearer, S.A. Reducing pesticide over-application with map-based automatic boom section control on agricultural sprayers. Trans. Am. Soc. Agric. Biol. Eng. **2010**, *53*, 685–690.

46. Pfleeger, T.G.; Olszyk, D.; Burdick, C.A.; King, G.; Kern, J.; Fletcher, J. Using a geographical information system to identify areas with potential of off-target pesticide exposure. Environ. Toxicol. Chem. **2006**, *25*, 2250–2259.

47. PMRA. *Regulatory Directive DIR2002–02: The PMRA Initiative for Reduced Risk Pesticides*; Pest Management Regulatory Agency, Health Canada Information Services: Ottawa, Ontario; 2002, availabe at http://www.hc-sc.gc.ca/pmra-arla/english/pdf/dir/dir2002–02-e.pdf. (accessed September 2010).

48. Thompson, D.G.; Kreutzweiser, D.P. A review of the environmental fate and effects of natural "reduced-risk" pesticides in Canada. In *Crop Protection Products for Organic Agriculture: Environmental, Health, and Efficacy Assessment, ACS Symposium Series 947*; Felsot, A.S., Racke, K. D., Eds.; American Chemical Society: Washington, DC, 2007; 245–274.

49. van Emden, H. Integrated pest management. In *Encyclopedia of Pest Management*; Pimentel, D., Ed.; Marcel Dekker Inc.: New York, 2002; 413–415.

50. Solomon, K.R.; Takacs, P. Probabilistic ecological risk assessment using species sensitivity distributions. In *Species Sensitivity Distributions in Ecotoxicology*; Posthuma, L., Suter, G.W., Traas, T.P., Eds.; Lewis Publishers: Boca Raton, Florida, 2002; 285–314.

51. Solomon, K.R.; Baker, D.B.; Richards, R.P.; Dixon, K.R.; Klaine, S.J.; La Point, T.W.; Kendall, R.J.; Weisskopf, C.P.; Giddings, J.M.; Giesy, J.P.; Hall, L.W.; Williams, W.M. Ecological risk assessment of atrazine in North American surface waters. Environ. Toxicol. Chem. **1996**, *15*, 31–76.

52. Hanson, M.L.; Solomon, K.R. New technique for estimating thresholds of toxicity in ecological risk assessment. Environ. Sci. Technol. **2002**, *36*, 3257–3264.

53. Warren-Hicks, W.J.; Hart, A., Eds. *Application of Uncertainty Analysis to Ecological Risks of Pesticides*; CRC Press: Boca Raton, Florida, 2010.

54. Thorbek, P.; Forbes, V.E.; Heimbach, F.; Hommen, U.; Thulke, H.; van den Brink, P.; Wogram, J.; Grimm, V. *Ecological Models for Regulatory Risk Assessments of Pesticides*; SETAC Press: Pensacola, Florida, 2010.

55. Stark, J.D.; Banks, J.E. Population-level effects of pesticides and other toxicants on arthropods. Annu. Rev. Entomol. **2003**, *48*, 505–519.

56. van den Brink, P.J.; Roelsma, J.; van Nes, E.H.; Scheffer, M.; Brock, T.C.M. PERPEST model: A case-based reasoning approach to predict ecological risks of pesticides. Environ. Toxicol. Chem. **2002**, *21*, 2500–2506.

57. Baird, D.J.; Baker, C.J.O.; Brua, R.B.; Hajibabaei, M.; McNicol, K.; Pascoe, T.J.; de Zwart, D. Towards a knowledge infrastructure for traits-based ecological risk assessment. Integr. Environ. Assess. Manage. 2010, online DOI 10.1002/ieam.129 (accessed September 2010).

58. van den Brink, P.J. Ecological risk assessment: From bookkeeping to chemical stress ecology. Environ. Sci. Technol. **2008**, 42, 8999–9004.

59. Beldon, J.B.; Gilliom, R.J.; Lydy, M.J. How well can we predict the toxicity of pesticide mixtures to aquatic life? Integr. Environ. Assess. Manage. **2007**, 3, 364–372.

60. Heugens, E.H.W.; Hendricks, A.J.; Dekker, T.; van Straalen, N.M.; Admiraal, W. A review of the effects of multiple stressors on aquatic organisms and analysis of uncertainty factors for use in risk assessment. Crit. Rev. Toxicol. **2001**, *31*, 247–285.

61. Lydy, M.J.; Lohner, T.W.; Fisher, S.W. Influence of pH, temperature and sediment type on the toxicity, accumulation and degradation of parathion in aquatic systems. Aquat. Toxicol. **1990**, 17, 27–44.

62. Howe, G.E.; Marking, L.L.; Bills, T.D.; Rach, J.J.; Mayer, F.L. Jr. Effects of water temperature and pH on toxicity of terbufos, trichlorfon, 4-nitrophenol and 2,4-dinitrophenol to the amphipod *Gammarus pseudolimnaeus* and rainbow trout (*Oncorhynchus mykiss*). Environ. Toxicol. Chem. **1994**, *13*, 51–66.

63. Yang, W.; Spurlock, F.; Liu, W.; Gan, J. Effects of dissolved organic matter on permethrin bioavailability to *Daphnia* species. J. Agric. Food Chem. **2006**, *54*, 3967–3972.

64. Puglis, H.J.; Boone, M.D. Effects of technical-grade active ingredient vs. commercial formulation of seven pesticides in the presence or absence of UV radiation on survival of green frog tadpoles. Arch. Environ. Contam. Toxicol. **2010**, online DOI 10.1007/s00244–010–9528-z (accessed September 2010).

65. Sandland, G.J.; Carmosini, N. Combined effects of a herbicide (atrazine) and predation on the life history of a pond snail, *Physa gyrina*. Environ. Toxicol. Chem. **2006**, *25*, 2216–2220.

66. Davidson, C.; Knapp, R.A. Multiple stressors and amphibian declines: Dual impacts of pesticides and fish on yellowlegged frogs. Ecol. Appl. **2007**, *17*, 587–597.

67. Barry, M.J.; Logan, D.C.; Ahokas, J.T.; Holdway, D.A. Effect of algal food concentration on toxicity of tow agricultural pesticides to *Daphnia carinata*. Ecotoxicol. Environ. Saf. **1995**, *32*, 273–279.

68. Wu, Q.; Riise, G.; Pflugmacher, S.; Greulich, K.; Steinberg, C.E.W. Combined effects of the fungicide propiconazole and agricultural runoff sediments on the aquatic bryophyte *Vesicularia dubyana*. Environ. Toxicol. Chem. **2005**, *24*, 2285–2290.

69. Boone, M.D.; Bridges, C.M.; Fairchild, J.F.; Little, E.E. Multiple sublethal chemicals negatively affect tadpoles of the green frog, *Rana clamitans*. Environ. Toxicol. Chem. **2005**, *24*, 1267–1280.

70. Laskowski, R.; Bednarska, A.J.; Kramarz, P.E.; Loureiro, S.; Schell, V.; Kudlek, J.; Holmstrup, M. Interactions between toxic chemicals and natural environmental factors— A meta-analysis and case studies. Sci. Tot. Environ. **2010**, *408*, 3763–3774.

2

Coastal Water: Pollution

Introduction 17
Heavy Metals and Metalloids 18
Radionuclides 20
Organic Compounds 22
Biological Pollution 24
　　　Eutrophication and Algal Bloom • Invasive Species
Fertilizers and Pesticides 26
Sewage Effluents 26
Oils 26
Marine Debris and Plastics 27
Noise Pollution 27
Light Pollution 28
Conclusion 28
References 29

Piotr Szefer

Introduction

The anthropogenic activity of man in coastal regions and even in areas located far inland is responsible for generating a huge amount of pollutants that are transported to marine ecosystems directly or by means of coastal watersheds, rivers, and precipitation from air.

Therefore, water pollution is a key global problem that has threatened marine organisms, including edible ones, and marine life in general.

There are two types of water pollutants, i.e., point source and nonpoint source.

The point source type is attributable to harmful contaminants released directly to the aquatic environment while nonpoint source delivers pollutants indirectly to the site of their approach.

The former one is a single, well-localized source, e.g., directly discharging sewage or industrial waste to the sea, whereas in the latter, the source of pollution is not well defined. Examples of such nonpoint source are agricultural runoff and windblown debris. Nonpoint sources are considered to be much more difficult to control and regulate as compared to point source pollutants.

The following are the classifications of other sources of pollution in coastal waters:

- Discharge of sewage and industrial waste
- Exploration and exploitation of the seabed
- Accidental pollution by oil and other pollutants from the land via air and other routes

Among these sources of pollutants dominate those connected with the discharge of municipal sewage and industrial wastes into coastal or estuarine regions, especially in the case of their inadequate treatment to remove persistent and harmful compounds. However, natural (and not anthropogenic) phenomena (e.g., volcanoes, storms, algae blooms, earthquakes, and geysers) could also be responsible for polluting aquatic systems. Their influence causes crucial changes in the ecological status of aquatic ecosystems.

The following are factors that determine the severity of pollutants:[1]

- Chemical structure
- Concentration
- Persistence

Independent on their sources in the water, pollutants may be classified as those for which the environment has some or little/no absorptive capacity. They are named "stock pollutants" (e.g., persistent synthetic chemicals, non-biodegradable plastics, and heavy metals).[2]

Most marine pollutants have land origin. They are often transported via rivers from agricultural sources and also via atmospheric trajectory. A lot of pollutants may be taken up by various compartments (biotic and abiotic) of aquatic environments; some of them could be biomagnified along the successive members of the food chain. A good example of having such ability to biomagnify is mercury. Such biomagnification could have negative effect on the quality of the water and hence on the health of the plants, animals, and humans whose lives depend on the quality of aquatic environments.

It should be emphasized that coastal areas are generally damaged from pollution, resulting in considerable impact on commercial coastal and marine fisheries. The pollution problem is very complex because of its interactions, interconnectedness, and uncertainty.[3–5] Pollutants, independent of their origin (e.g., air, water, land), enter the ocean, whether earlier or later.[3] Spatial distribution patterns of contamination concentrations exhibit a trend of their increase during transition from the south to the northern part of all oceans, i.e., to areas neighboring with both industrial centers and concentration of main pollution sources.[6]

The following are considered major pollutants:[6–11] fertilizers, pesticides, and agrochemicals; domestic and municipal wastes and sewage sludge; oil and ship pollution; trace elements; radionuclides; organic compounds; plastics; sediments; eutrophication and algal bloom; biological pollution; noise pollution; and light.

Heavy Metals and Metalloids

In contrast to organic pollutants, e.g., Polichlorinated Biphenyls (PCBs), heavy metals occur as natural elements of particular abiotic and biotic components of continental and aquatic ecosystems. They are present at a natural background level in rocks, soils, sediments, water, and biota. Human industrial and agricultural activities result in the elevation of this natural level to sometimes significantly higher values.

Typical metal concentrations are generally observed in open waters of marine ecosystems, although these remote regions can be affected by elevated levels of trace elements of anthropogenic or volcanic origin. For instance, the atmosphere affects the oceans and continental matter facilitating metal fluxes between these two compartments. Therefore, the atmosphere is a very important component as it makes it possible to transport metals that are natural in origin into distances far from their sources, e.g., from areas closest to forest fires as well as windblown dust, vegetation, and sea aerosols.[12,13] These sources are responsible for contributing metals to the lower troposphere, and therefore, their transport is associated to local and regional wind patterns, in contrast to specific sources such as volcano eruption, which can be responsible for injecting particulate metals not only into the troposphere but also into the stratosphere. In the latter case, particulate metals can be transported long distances under the appropriate circumstances.[13] Another example of long-distance transport of metals from their sources is dust carried from the Sahara Desert resulting in deposition of Fe, Mn, Al, and trace elements across the Mediterranean Sea, Atlantic Ocean, and Caribbean Sea.[13,14] Therefore, wind-driven dust transports particulate metals far offshore, in contrast to riverine flux carrying greater pole of mineral components from continental material to the coastal waters. These metals are promptly deposited to bottom sediments or taken up by biota, especially by phytoplankton in the surface waters, and transferred next to the food chain, recycles or settled to the bottom.[13]

The mass of metal of anthropogenic origin emitted to ecosystems is now equal to or greater than the mass introduced to the natural cycle on a global scale.[13] Some metals, e.g., Pb, Hg, and Cd, owing to their great toxicity, pose a high health risk; therefore, great attention has been paid to estimate their inputs to marine ecosystems, particularly to coastal waters. For instance, the largest masses of Pb are emitted to the atmosphere during processing of the metal (smelters) or from combustion-related sources like motor vehicles. Lead from motor vehicle exhaust was identified not only in the atmosphere but also in remote surface waters as well as in remote terrestrial areas. A ban on Pb usage in vehicle fuel has resulted in the effective reduction of metal inputs to the ecosystem since the 1970s. The decline in atmospheric Pb detected over a time scale of 10 years (from 1979 to 1989) because of the reduction of leaded gasoline in western European countries should be reflected in decreasing Pb levels in surficial water and biota. In fact, the temporal negative trend for cod in the Baltic Sea seems to support this argumentation.[15]

According to Mason et al.,[16] preindustrial fluxes and reservoirs of Hg pose one-third of its fluxes in the civilization era. It is suggested that modern emission of Hg to the atmosphere increased considerably, even 4 to 5 times, due to the human activities. The extremely elevated levels of Hg are frequently associated with Hg mining.

As has been reported, ca. 300 metric tons of dissolved Cd annually enter the oceans from rivers while ca. 400 to 700 metric tons of dissolved and particulate Cd are annually deposited to the oceans from the atmosphere.[17,18] It is estimated that human activities have contributed to increased Cd inputs to the ocean by 60% in the 1980s. It is also found that the higher proportion of land deposition of Cd is associated to the rapid removal of this metal from the atmosphere near inputs of air pollution. A substantial pole of Cd transported by rivers is deposited in estuaries and continental margins of oceans. It is found that increased concentrations of Cd occur locally, reflecting its mosaic contamination, especially near mining and industrial point sources—not managed.

A significant fraction of Zn entering the oceans is derived from atmospheric deposition.[18] Soils and sediments are main natural reservoirs of Zn. Zinc, like Cd, is not distributed evenly across the Earth's surface, since its increased concentrations occur locally in the vicinity of increased inputs, i.e., specific points of source inputs.

The concentrations of many trace elements, e.g., Pb, Cd, Hg, Cu, Zn, Se, and As in coastal and estuarine waters, especially in highly industrialized areas, are generally significantly greater than those in open oceanic waters.[13,18–20] The waters of harbors and marinas around the world contain variable concentrations of tributyltin (TBT),[13] but its extremely high levels may be characteristic for marinas in southwest England.[21]

Human industrial and agricultural activities affect inputs of several metals to reservoir/reservoirs and hence increase their concentrations, even sometimes very dramatically, above natural background levels. There are numerous examples of worldwide events leading to serious contamination of coastal waters by heavy metals and metalloids.[15] Therefore, relationships between man and ecosystem health have been explored, especially in relation to perturbed ecosystems. This includes the pollution status of coastal regions harmed by some catastrophes, large-scale pollution, environmental accidents and episodes, etc. High risk groups consume extremely high quantities of trace metals present in specific assortment of seafood or offal and it concerns seaside populations. Marine fish and shellfish may be the dominant dietary sources of Hg for local populations.[22–25] A notable example of aquatic pollution by a toxic metal is the Minamata incident, commencing in 1953 and resulting in fish, shellfish, and bird mortalities in waters of the partially landlocked Minamata Bay.[15] Dogs, pigs, and especially cats were also victims of this incident. By the end of 1974, 107 of 798 officially verified patients had died. According to Tomiyasu et al.,[26] the sediments from the Minamata Bay contained levels of Hg that highly exceeded its background level. Among other incidents resulting in the release of Hg compounds to the environment, the most significant ones happened in the 1960s and early 1970s in Sweden, Canada and the United States, northern Iraq, Guatemala, Pakistan, and Ghana.[27–31] MeHg in aquatic ecosystems, especially those that bioaccumulated in fish, is a major public health problem all over the world.[32] Its levels in the hair of fishermen represent the critical group for dietary exposure. For instance, the concentrations of Hg (total and MeHg) in the hair of fishermen from Kuwait were 2 times higher than the "normal" level according to the World Health Organization.[33] Biomass burning in tropical forests also seems to have contributed significantly to the Hg input to the atmosphere. Approximately 31% of the Hg concentrations were associated with the vegetation fire component.[34] It is postulated (based on long-range air mass trajectory analyses) that Hg occurs in the Amazon basin over two main routes: to the South Atlantic and to the Tropical Pacific, over the Andes.[34]

Global emission flux estimates exhibited that biomass burning could be major contributor of heavy metals and black carbon to the atmosphere.[15] It is estimated that savannah and tropical forest biomass burning could emit huge amounts of Cu, Zn, and black carbon to the atmosphere, corresponding to 2%, 3%, and 12%, respectively, of the global level of these elements.[35]

The toxic effects of TBT were first indicated towards the end of the 1970s in Arcachon Bay, France, as the "TBT problem."[36,37] The release of TBT (from antifouling paints) to the area resulted in shell abnormalities and reduced growth and settlement in oysters, *Crassostrea gigas,* cultured in the vicinity of marinas. In much polluted water, oyster production was severely affected by the absence of reproduction, resulting in a strong decline in the marketable value of the remaining stock.[37] Imposex, i.e., the development of male sexual characteristics in female marine mesogastropoda and neogastropoda caused by TBT pollution, is a widespread phenomenon concerning several coastal species and, more recently, offshore species as well.[23,38,39] Subsequent regulations in 1990 that prohibit the use of TBT-based antifoulants on vessels less than 25 m in length have been highly effective in reducing TBT levels in coastal waters. However, larger vessels have continued the release of TBT, and major harbors remained pollution hot spots.[40] The Organotin Antifouling Paint Control Act restricted in the United States the use of TBT paints to vessels greater than 25 m in length.[41] The voluntary stoppage of TBT production in January 2001 by major U.S. and European manufacturers resulted in the decline of its presence in marine biota, but TBT paint is still being used in most Asian countries. The International Maritime Organization (IMO) imposed an international ban for the use of organotin compounds in antifouling treatments on ships longer than 25 m. The target is to prohibit their application starting 2003 and to require the removal of TBT from ships' hulls by January 1, 2008.[41,42]

The extensive flooding, especially occurred in river area of former or operating metalliferous mining can be responsible for wide-spreading of heavy metals and metalloids far distance from pollution source. An example of such environmental events is the flooding of the Severn catchment (United Kingdom) in January 1998.[43]

Radionuclides

Physicochemical aspects and applications of radioactivity in the environment were extensively presented in a book by Valcovic.[8] There are numerous papers reporting on problems resulting from radionuclide pollution and their sources in different ecosystems.[15,44] One of the first low- level emissions of radioactivity took place in the Hanford reactors (Columbia River, Washington, United States), which released radionuclides (mainly ^{60}Co, ^{51}Cr, and ^{65}Zn) to its environs from 1940 to 1971.[28] The nuclear reactors in Cumbria (northwest England) have also been responsible for discharging quantities of radioisotopes, i.e., ^{144}Ce, ^{137}Cs, ^{95}Nb, ^{106}Ru, and ^{95}Zr, to the marine environment. Although these emissions have been diminished recently, discharges from nuclear power stations such as Sellafield (formerly named Windscale) could still be identified, even in distances far away from their source.[45,46] Significant quantities of artificial radionuclides (^{137}Cs, ^{134}Cs, ^{90}Sr, ^{99}Tc) have been transported to the North Atlantic and Arctic from Sellafield, together with measurable amounts of Pu and Am.[46–48] The nuclear reprocessing plant at La Hague in France emitted ^{137}Cs and $^{239+240}$Pu to the environment, although this plant mainly supplies ^{129}I and ^{125}Sb.[28,48,49] Besides the expected emission of radionuclides from nuclear and reprocessing facilities, significant quantities of radioisotopes contaminate aquatic and terrestrial environments from either nuclear weapons testing or nuclear reactor accidents.[15] For instance, the thermonuclear detonation that took place in 1954 at Bikini Atoll resulted in the contamination of a large area of the Marshall Islands. A number of atmospheric tests (520 in total) were mostly carried out in the Northern Hemisphere, including eight underwater tests, with a total yield of 542 Mt. Moreover, there have been a total of 1352 underground tests with a total yield of 90 Mt.[50]

A number of nuclear incidents were concernedly noted, including those affecting the crew of the Japanese fishing vessel "Fukuru Maru".[28] Plutonium released from the Kyshtym accident in the Urals has been much probably detected in deep basins of the Arctic Ocean.[51] In 1968, an aircraft from the U.S. Strategic Air Command crashed near the Thule Airbase in NW Greenland, releasing to the marine environment ca. 1 TBq $^{239+240}$Pu.[52] As a consequence, marine sediments as well as benthic organisms, i.e., bivalves, shrimps, and sea stars, have been contaminated by Pu, although their levels rapidly decreased.[53] A number of American and Russian nuclear submarines have been lost in the world's oceans. For instance, the Soviet Komsomolets submarine sank at a depth of 1700 m at Bear Island in the eastern part of the Norwegian Sea. The estimated radioactivity in the wreck was 2.8 PBq ^{90}Sr and 3 PBq ^{137}Cs.[54] Some nuclear powered satellites can incidentally be sources of radioactivity. They can burn up in the upper atmosphere, resulting in the contamination of the ocean. For instance, such an accident happened in 1964 when a SNAP-9A nuclear power generator containing 0.6 PBq ^{238}Pu aboard a U.S. satellite re-entered the atmosphere in the Southern Hemisphere. The estimated ^{238}Pu/$^{239+240}$Pu ratio in this region was higher than that in the ocean water from the Northern Hemisphere.[55,56]

Sea dumping was carried out since the late 1940s to mid-1960s mainly by the United States in the Atlantic Ocean and Pacific Ocean as well as by the United Kingdom in the Northeast Atlantic Ocean.[56] In 1967, an international operation was initiated by the former European Nuclear Energy Agency that contributed to the deposition of ca. 0.3 PBq solid waste at a depth of 5 km in the eastern Atlantic Ocean. Other international operations were continued until 1982 when ca. 0.7 PBq α activity, 42 PBq β activity, and 15 PBq tritium activity have been dumped in the North Atlantic.[57] It has been assessed that the radiological impact of the NEA (former European Nuclear Energy Agency) dumping activities resulted in some releases of Pu from the dumped waste.[15] This source would be responsible for only a part of the total body burden radioactivity in local benthic organisms, e.g., sea cucumbers; the remainder has been attributed to fallout.[58] According to Consortium for Risk Evaluation with Stakeholder Participation (CRESP) evaluation, the individual dose of a critical group consuming seafood such as molluscs from the Antarctic Ocean was estimated to be 0.1 μSv yr^{-1}, in effect labeling ^{239}Pu and ^{241}Am as critical radionuclides. The indefinite collective dose to the world's population coming from sea dumping was estimated at 40,000 manSv with predominance of ^{14}C and ^{239}Pu.[56,58]

U.S. weapons production facilities account for a large fraction of radiocaesium discharges during the 1950s.[15] A striking incident occurred at Chernobyl in the former USSR where an explosion of a reactor core of the nuclear plant took place in April 1986. The Baltic countries and a large part of central and western Europe have been contaminated principally by ^{131}I, ^{134}Cs, and ^{137}Cs.[28,59] It is found that a significant part of the activity fell over the European marginal seas from which the Baltic Sea was the most affected by contamination.[56,60] It has been mainly responsible for additional inflow of the radioactive contaminants to the Northeast Atlantic Ocean.[56] Due to the Chernobyl accident, significant levels of ^{137}Cs were also found in the Black Sea. The outflow from this Sea has been the major source of additional ^{137}Cs in the Mediterranean Sea.[56] In the summer of 1987, the Chernobyl-derived ^{137}Cs was also detected in surficial waters of the Greenland Sea, Norwegian Sea, and Barents Sea as well as in the west coast of Norway and the Faroe Islands. According to Aarkrog,[56] the total Chernobyl ^{137}Cs input to the world's oceans was relatively significantly smaller than that estimated for nuclear weapons fallout because of the tropospheric nature of this accident that has contaminated the surrounding European continental areas.[15]

After the 2011 Tōhoku earthquake and tsunami, the radiation effects from the Fukushima Daiichi nuclear disaster resulted in the release of radioactive isotopes from the crippled Fukushima Daiichi

Nuclear Power Plant. The total amount of ^{131}I and ^{137}Cs released into the atmosphere has been estimated to exceed 10% of the emissions from the Chernobyl disaster. Large amounts of radioactive isotopes have also been released into the Pacific Ocean.[61]

Organic Compounds

The high lipophilicity of many persistent organic pollutants (POPs) enhances their bioconcentration/biomagnification, resulting in potential health hazards on predators at higher trophic levels, including humans. These xenobiotics occur widely in coastal waters and oceans from the Arctic to the Antarctic and from intertidal to abyssal. It should be emphasized that most of these compounds exist at a very low concentration level, and hence, their threat to marine biota is still not well recognized. However, it is well known that exposure to extremely low levels of halogenated hydrocarbons, e.g., PCBs, Dichlorodiphenyltrichloroethane (DDT), and TBT, may disrupt the normal metabolism of sex hormones in fish, birds, and marine mammals. Moreover, sublethal effects of these organic chemicals over long-term exposure may result in serious damage to marine populations since some of these POPs may impair reproduction functions of organisms while others may show carcinogenic, mutagenic, or teratogenic activity.[6] Some of the effects of these compounds have been reported by Goldberg.[62] For instance, very low levels of TBT (as endocrine disruptor) cause a significant disruption in sex hormone metabolism, resulting in the malformation of oviducts and suppression of oogenesis in female whelks, e.g., *Nucella lapillus*.[63] As a consequence, sex imbalance leads to species decline if not species extinction in some field populations.[64] Butyltins may be responsible for mass mortality events of bottlenose dolphins in Florida through suppression of the immune system.[65] Trace environmental levels of other compounds like chlorinated hydrocarbons, organophosphates, and diethylstilbestrol may be responsible for significant endocrine disruption and reproductive failure in different groups of animals, i.e., marine invertebrates, fish, birds, reptiles, and mammals.[6] For instance, high levels of DDT, PCBs, and organochlorines in the Baltic Sea significantly reduced the hatching rates of the fish-eating whitetailed eagle *(Haliaeetus albicilla)* in the 1960s and the 1970s.[66] Another example of the toxic impact of POPs is organochlorine contamination in different cetacean species dependent upon their diet, sex, age, and behavior. Many of these compounds, as endocrine disruptors, reduce reproduction and/or suppress immune function. DDT and PCBs are known as compounds affecting steroid reproductive hormones and can increase mammalian vulnerability to bacterial and viral diseases. Jepson et al.[67] reported a statistically significant relationship between elevated PCB level and infectious disease mortality of harbor porpoises *(Phocoena phocoena)*.

The assessment and monitoring of existing and emerging chemicals in the European marine and coastal environment have been overviewed based on numerous, most recent worldwide references.[5] From this report, the extensive range of chemicals that are capable of disrupting the endocrine systems of animals can be categorized into the following: environmental estrogens (e.g., bisphenol A, methoxychlor, octylphenol, and nonylphenol), environmental anti-estrogens (e.g., dioxin, endosulfan, and tamoxifen), environmental anti-androgens [e.g., dichlorodiphenyldichloroethylene (DDE), procymidone, and vinclozolin], chemicals that reduce steroid hormone levels (e.g., fenarimol and ketoconazole), chemicals that affect reproduction primarily through effects on the central nervous system (e.g., dithiocarbamate pesticides, and methanol), and chemicals with multiple mechanisms of endocrine action (e.g., phthalates and TBT). There is a high level of international concern regarding developmental and reproductive impacts on marine organisms from exposure to endocrine- disrupting chemicals. This is the case for "new" substances such as alkylphenols; there is also renewed interest for some "old" organochlorines such as DDT and its metabolites. Brominated flame retardants (BFRs), particularly the brominated diphenyl ethers (BDEs) and hexabromocyclododecane (HBCD), have been detected in the European marine environment. It has been reported that the input of BDEs into the Baltic Sea through atmospheric deposition now exceeds that of PCBs by almost a factor of 40. BDEs are found in fish from various geographic regions. This resulted from the long-range atmospheric transport and deposition of these substances.[5] HBCD was detected in liver and blubber samples from harbor seals and harbor porpoises from the Wadden Sea and the North Sea. It is found that environmental concentrations of these BFRs in Japan and South China increased significantly during the last decades. PBDE levels in marine mammals and sediments from Japan, after showing peak concentrations in the 1990s, appear to have leveled off in recent years. Furthermore, in recent years, HBCD concentrations in marine mammals from Japanese waters appear to exceed those of PBDEs, presumably reflecting the increasing use of HBCDs over PBDEs. Pentabromotoluene (PBT) and Decarbomodiphenyl (DBDPE), for example, have been found in Arctic samples remote from sources of contamination. It is an indication of their potential for long-range atmospheric transport, showing a tendency for accumulation in top predators. Polymeric BFRs may be a source of emerging brominated organic compounds to the environment. Medium- and short-chain chlorinated paraffins (SC- CPs) are ubiquitous in the environment and tend to behave in a similar way to POPs. They have been found in water as well as in fish and marine mammals.[5]

Perfluorinated compounds (PFCs), namely, perfluorooctane sulfonate (PFOS), have been detected in marine mammals.[5] They are globally distributed anthropogenic contaminants. PFCs, such as PFOS, have been industrially manufactured for more than 50 years and their production and use have increased considerably since the early 1980s. The main producer of PFOS voluntarily ceased its production in 2002. Furthermore, the large-scale use of PFOS has been restricted. PFOS has been used in many industrial applications such as fire-fighting foams and consumer applications such as surface coatings for carpets, furniture, and paper. PFCs are released into the environment during the production and use of products containing these compounds. About 350 polyfluorinated compounds of different chemical structures are known.[5] The most widely known are PFOS ($C_8F_{17}SO_3$) and perfluorooctanoic acid (PFOA; $C_8F_{15}O_2$), which are chemically stable and thus may be persistent (substance dependent). PFCs do not accumulate in lipid but instead accumulate in the liver, gallbladder, and blood, where they bind to proteins. PFCs have been detected worldwide, including the Arctic Ocean and Antarctic Ocean, in almost all matrices of the environment. High concentrations of PFCs have been found in marine mammals.[5] A screening project in Greenland and the Faroe Islands indicated high biomagnification of PFCs, with elevated concentrations in polar bear liver. A time trend study (1983–2003) showed increasing concentrations for all PFCs for ringed seals from East Greenland. In the United Kingdom, a study on stranded and by-catch harbor porpoise liver (1992 and 2003) found PFOS at up to 2420 pg kg^{-1} wet weight. There is a decreasing trend going from south to north.[5]

Antifouling paint booster biocides were recently introduced as alternatives to organotin compounds in antifouling products.[5] These replacement products are generally based on copper metal oxides and organic biocides. Commonly used biocides in today's antifouling paints are as follows: Irgarol 1051, diuron, Sea-Nine 211, dichlofluanid, chlorothalonil, zinc pyrithione, TCMS (2,3,3,6-tetra- chloro-4-methylsulfonyl) pyridine, TCMTB [2-(thiocyanomethylthio) benzothiazole], and zineb. It has been reported that the presence of these biocides in coastal environments around the world is a result of their increased use (notably in Australia, the Caribbean, Europe, Japan, Singapore, and the United States). For example, Irgarol 1051, the Irgarol 1051 degradation product GS26575, diuron, and three diuron degradation products [1-(3-chlorophenyl)-3,1-di- methylurea (CPDU), 1-(3,4-dichlorophenyl)-3-methylu- rea (DCPMU), and 1-(3,4-dichlorophenyl)urea (DCPU)] were all detected in marine surface waters and some sediments in the United Kingdom. Risk assessments indicate that the predicted levels of chlorothalonil, Sea-Nine 211, and dichlofluanid, in contrast to Irgarol 1051, in marinas represent a risk to marine invertebrates. Finally, non-eroding silicone-based coatings can effectively reduce fouling of ship hulls and are an alternative to biocidal and heavy- metal-based antifouling paints. Although polydimethylsiloxanes (PDMSs) are unable to bioaccumulate in marine organisms and their soluble fractions have low toxicity to marine biota, undissolved silicone oil films or droplets can cause physical–mechanical effects such as trapping and suffocation of organisms.[5]

Human and veterinary pharmaceuticals are designed to have a specific mode of action, affecting the activity of, e.g., an enzyme, ion channel, receptor, or transporter protein.[5] Clotrimazole, dextro-propoxyphene, erythromycin, ibuprofen, propranolol, tamoxifen, and trimethoprim were detected in U.K. coastal waters and in U.K. estuaries. Concentrations of some pharmaceutical compounds are effectively reduced during their passage through a tertiary wastewater treatment works, while others are sufficiently persistent to end up in estuaries and coastal waters.[5] Compared with mammalian and freshwater organisms, there is a lack of experimental data on the impacts of pharmaceuticals in marine and estuarine species. However, there is experimental evidence that selected pharmaceuticals have the potential to cause sublethal effects in a variety of organisms. It has been concluded that antibiotic substances in marine ecosystems can pose a potential threat to bacterial diversity, nutrient recycling, and removal of other chemical pollutants. Although data on the occurrence of pharmaceuticals and antibiotics in the marine environment are becoming more available, the true extent of the potential risks posed by this group of contaminants cannot, at present, be assessed, mainly due the lack of effect data.[5]

Several studies showed that among personal care products (PCPs), synthetic musks (nitromusks, polycyclic musks, and macrocyclic musks) are widespread in marine and freshwater environments and bioaccumulate in fish and invertebrates.[5] There were identified products such as benzotriazole organic UV filters, namely, UV-320 [2-(3,5- di-i-butyl-2-hydroxyphenyl)benzotriazole], UV-326 [2-(3- i-butyl-2-hydroxy-5-ethylphenyl)-5-chlorobenzotriazole], UV-327 [2,4-di-t-butyl-6-(5-chloro-2H-benzotriazol-2-yl) phenol], and UV-328 [2-(2H-benzotriazol-2yl)-4,6-di-t- pentylphenol]. Their relatively high concentrations were found in marine organisms collected from waters of western Japan. There are indications that marine mammals and seabirds accumulate UV-326, UV-328, and UV-327. Benzotriazole UV filters were also detected in surface sediments from this area. The results suggest a significant bioaccumulation of UV filters through the marine food webs and a strong adsorption to sediments. Although a full risk assessment of some of these has been performed (e.g., musks), for most PCPs, there is little data on their occurrence and their effects in the marine environment.[5]

Biological Pollution

Eutrophication and Algal Bloom

Nutrient loadings in coastal waters cause direct responses such as changes in chlorophyll, primary production, macro- and microalgal biomass, sedimentation of organic matter, altered nutrient ratios, and harmful algal blooms. The indirect responses of nutrient loadings are responsible for changes in benthos biomass, benthos community structure, benthic macrophytes, habitat quality, water transparency, sediment organic matter, sediment biogeochemistry, dissolved oxygen, mortality of aquatic organisms, food web structure, etc. Moreover, increase in phytoplankton biomass and attributing decrease in transparency and light intensity limit growth of submerged vascular plants.[6,68] Generally speaking, eutrophication leads to major changes in qualitative and quantitative species composition, structure, and function of marine communities over large areas. As for phytoplankton communities, such changes are connected with an increase in biomass and productivity.[69] For instance, a general shift from diatoms to dinoflagellates, as well as dominance of small-size nanoplankton (microflagellates, coccoids), has been reported. Similar trends were observed in the case of zooplankton communities, indicating replacement of herbivorous copepods by small- size zooplankton.[70,71] Some examples of consequences of eutrophication have been reported based on worldwide references.[15] The harmful deoxygenation of water giving rise to fish kills was producing nutrient-derived large mats of macroalgae in the Peel-Harvey Estuary, Western Australia.[72] Similar events took place in the northern Adriatic Sea where diatom blooming in summer resulted in the production of mucilage, affecting tourism in northeastern Italy and reducing fish catch.[28,73,74] Insufficient water exchange and increasing production of organic matter during this century caused depletion of O_2 in all deep waters of the Baltic Proper.[15] It resulted in devastating consequences for marine biota, leading to the replacement of O_2 by H_2S in these bottom waters.[75] Although eutrophication generally leads to an increase in fish productivity, it can also cause negative environmental changes in fish populations. Fish such as cod and plaice are threatened by O_2 depletion in Baltic deep basins, causing decreasing fish catch in Koge Bay in the Sound.[75]

The blooms of blue-green algae as well as *Nodularia* produce a toxic peptide hepatoxin under particular conditions, which can pass through the food web, affecting top consumers, e.g., man. The toxin is responsible for the degeneration of liver cells, promoting tumors and causing death from hepatic hemorrhage.[75] Paralytic shell poisoning (PSP) and/or ciguatera has/have been identified predominantly in the subtropical and tropical zones such as Australia[76–80] and especially in other Indo-Pacific regions, e.g., India, Thailand, Indonesia, Philippines, and Papua New Guinea.[81,82] Principal toxic dinoflagellate species, i.e., *Pyrodinium bahamense* var. *compressa,* killed many fish and shellfish from these regions.[15] The consumption of seafood in the Indo-Pacific area posed considerable public health problems.[28] The significant PSP incidences also took place in temperate zones. For instance, in May 1968, a poisoning episode affected 78 persons inhabiting Britain after consumption of soft tissue of the blue mussel *Mytilus edulis.*[83] Another dinoflagellate-poisoning event again happened in northeast England in the summer of 1990, possibly attributed to a specific combination of elevated nutrient inputs from rivers and exceptionally warm weather conditions, which could be favorable for algae growing.[28]

It has been reported that anthropogenically derived atmospheric N deposition to the North Atlantic Ocean was strictly responsible for harmful algal bloom expansion.[84] This event concerned especially the Eastern Gulf of Mexico, U.S. Atlantic coastal waters, the North Sea, and the Baltic Sea.[84–95] Expanding blooms of the noxious dinoflagellate *Alexandrium tamarense* have been observed along the Northeast U.S. Atlantic coastline.[84,92] There are numerous examples of specific harmful algal bloom expansions in coastal and off-shore waters in case of significant atmospheric deposition of N, e.g., in the North Sea, Adriatic Sea, Western Mediterranean Sea, and Baltic Sea.[84,96] Great attention has been paid to toxic hypoxia- inducing dinoflagellate blooms in the North Sea and the Western Baltic.[84] In the summer of 1991, a very extensive bloom of *Nodularia spumigena* in the open Baltic Sea and along the southern and southeastern Swedish coasts was observed. Dogs' mortalities caused by toxic *Nodularia* blooms have been observed in Denmark, Gotland, and the Swedish coastal waters.[15] In other Baltic areas, horses, cows, sheep, pigs, cats, birds, and fish also suffered from this event. *Nodularia* blooms have caused human health problems such as stomach complaints, headaches, eczema, and eye inflammation.[75] In the Skagerrak and Kattegat, harmful algal bloom expansion of toxic algae species such as *Prorocentrum, Dinophysis, Dichtyocha, Prymnesium,* and *Chrysochromulina* has taken place.[88] The recent blooms mostly killed pelagic organisms and the phyto- and zoobenthic organisms.

Invasive Species

The impacts of introduction and invasion of species throughout the world have recently been identified. There are an increasing number of reports that document this phenomenon taking place in coastal, estuarine, and marine waters.[6] For instance, the Chinese mitten crab *(Eriocheir sinensis),* as invasive species, now inhabits coastal regions in northwestern Europe, and it has caused damage to flood defense walls by burrowing, affecting local community structure.

Worldwide fish species introduction is connected with various consequences.[97] It has been pointed out that many aquaculture species are recently genetically modified. Such modified populations are frequently released and mixed with the natural populations and are breeding with them. It causes biological pollution from a molecular level to community and ecosystem levels. An example of such events is the flooding in Central Europe that caused the release of hybrid and modified fish like sturgeon *(Acipenser* spp.) from aquaculture installations.[98] The local populations of fish are generally not resistant to the pathogenic organisms carried by the introduced species and vice versa. Therefore, deliberate genetic selection and breeding for a long time may have numerous consequences in the aquaculture unit itself as well as the loss of the natural stock for numerous species in a global scale.[6,98,99]

Fertilizers and Pesticides

Agricultural activity as an important pollution source has contributed to significant enrichment of nutrients (mainly ammonium ion and nitrates) in coastal marine waters. It is found[100] that wastes, manures, and sludges provide soils with significantly more hazardous substances as compared to fertilizers for achieving the equivalent plant nutrient content. The worldwide use of fertilizers, including organic fertilizers like manure, is huge. In the case of intensively monocultivated areas, a relatively small number of pesticides have been widely used in spite of their variety.[6]

The large mass of pesticide residues is accumulated in the environment since they are not rapidly degradable. The total global DDT production from the 1940s to 2004 was estimated as ca. 4.5 Mt.[101] Duursma and Marchand[102] estimated the world production of DDT to be ca. 2.8 Mt, of which 25% is assumed to be released to the ocean. According to Shahidul Islam and Tanaka,[6] the total emission of DDT through agricultural applications amounts to 1030 kt between 1947 and 2000. Organochlorine pesticides (OCPs) originating mostly in temperate and warmer areas of the world can be transported to coastal waters and even via atmospheric long-range transport and ocean currents to the Arctic. Owing to their bioaccumulative abilities (as lipophilic compounds) and biomagnification along the sequential trophic levels of the food chain, pesticides are classified as one of the most destructive agents for marine organisms. As a consequence, their very high levels can be observed among top predators, including man. Their toxic effects to marine organisms are often complex because they may be associated with the combination of exposure to pesticides and other POPs with environmental stresses such as eutrophication and pathogens.[6]

Sewage Effluents

Sewage effluents contain industrial, municipal, and domestic wastes; animal remains; etc. The huge amounts of these effluents generated in big cities are transported by drainage systems into rivers or other aquatic systems, e.g., coastal waters. It is estimated that the annual production of sewage amounts to ca. 1.8×10^8 m^3 for a population of 800,000. This load is equivalent to an annual release of 3.6×10^3 tons of organic matter.[6] Sewages pose significant effects on coastal marine ecosystems because they contain POPs (heavy metals/trace elements, organic pollutants) as well as viral, bacterial, and protozoan pathogens and organic substances subjected to bacterial decay. In case of such bacterial activity, the content of oxygen in water is reduced, resulting in the destruction of proteins and other nitrogenous compounds. Releasing hydrogen sulfide and ammonia exhibits toxic activity to marine biota, even at low levels. As for pathogens, domestic sewage released to coastal waters contains such harmful pathogens as *Salmonella* spp., *Escherichia coli*, *Streptococcus* sp., *Staphylococcus aureus*, *Pseudomonas aeruginosa*, the fungi *Candida*, and viruses such as enterovirus, hepatitis, poliomyelitis, influenza, and herpes.[6] Different bacteria and viruses can be transferred to some representatives of marine fauna, e.g., marine mammals.

Oils

The recently observed increase in tanker operations and oil use as well as marine tanker catastrophes has been responsible for the presence of excessively large amounts of oil spillage in coastal and marine ecosystems. It is estimated that ca. 2.7 million tons of oil pollution enter the ocean each year. The tanker accidents between 1967 and 2007 released ca. 4.5 million tons of oil to seawater. Notable examples of ecological catastrophes are the huge spill from a drilling platform in Gulf of Mexico (Mexico) in 1979 and the Deepwater Horizon drilling rig explosion in the Gulf of Mexico (United States) (April 20 to July 15, 2010), resulting in massive amounts of oil in the gulf. Another similar example took place during the Persian Gulf War in 1991, where ca. 2 million tons of oil was spilled, resulting to the death of many species of marine biota.[7,103,104] Therefore, oil pollution poses serious adverse effects on aquatic environment and marine organisms represented different trophic levels from primary producers to the top predators.[6]

Although aerial and flying birds (e.g., gulls, gannets) are not seriously exposed to oil toxicity, birds that spend most of their time in contact with oil on the water surface (e.g., ducks, auks, divers, penguins) are at greater risk of oil toxicity. According to Smith,[105] the annual release of hydrocarbon can range from 0.6 to 1 million tons. Coastal refineries can be an important source of oil pollution since millions of gallons of crude oil and its fractions are processed and stored there. During their operation, pollutants are continuously released by way of leakages, spills, etc.

Marine Debris and Plastics

Marine debris, especially plastics, is one of the most pervasive pollution problems. Nets, food wrappers, bottles, resin pellets, etc., have serious impacts on humans and marine biota. Medical and personal hygiene debris can enter coastal water through direct sewage outflows, posing a serious threat to human health and safety. Contact with water contaminated with these pollutants and pathogens (e.g., *E. coli*) can result in infectious hepatitis, diarrhea, bacillary dysentery, skin rashes, typhoid, and cholera.[106]

There are numerous reviews devoted to an important topic such as pollution by marine debris.[106–110] Entanglement in marine debris such as nets, fishing line, ropes, etc., can hamper an organism's mobility, prevent it from eating, inflict wounds, and cause suffocation or drowning. It was estimated that 136 marine species have been involved in entanglement incidents, including some species of seabirds, marine mammals, and sea turtles.[111] The decline in the population of the northern sea lion (*Eumetopias jubatus*), endangered Hawaiian monk seal (*Monachus schauinslandi*), and northern fur seal has been explained by entanglement of young specimens in lost or discarded nets and packing bands.[112] Abandoned fishing gear, e.g., fishing net, can contribute to catching and killing marine animals. This process called ghost fishing or ghost net can kill a huge number of commercial species.[108] An example of another serious pollution problem is ingestion of debris by marine animals. Plastic pellets and plastic shopping bags can be swallowed and lodged in animals' throats and digestive tracts, causing some animals to stop eating and slowly starve to death.[106] According to the U.S. Marine Mammal Commission,[111] ingestion incidents concerned 111 species of seabirds, 26 species of marine mammals, and 6 species of turtles. For instance, plastic cups were found in the gut of some species of fish from British coastal waters; the ingested cups were eventually responsible for their deaths.[112] Even Antarctic and sub-Antarctic seabirds, e.g., Wilson's storm-petrel (*Oceanites oceanicus*) and white-faced storm-petrel (*Pelagodroma marina),* are at risk for this ingestion hazard.[112–115] It is reported that the proportion of plastic debris among litter increases with distance from source because it is transported more easily as compared to a denser material like glass or metal and because it lasts longer than other low-density materials (paper). Floating plastic articles (material less dense than water, e.g., polyamide, polyterephthalate, polyvinyl chloride) pose a global problem because they can contaminate even the most remote islands.[107,116] Drift plastics can increase the range of some marine organisms or introduce unwanted and aggressive alien taxa species into an environment. It could be risky to littoral, intertidal zones, and the shoreline.[112,117] There is also potential danger to marine ecosystems from the accumulation of plastic debris (material more dense than water) on the seafloor. Such bottom accumulation of plastic can inhibit the gas exchange between overlying waters and the pore water. This process can result in hypoxia or anoxia in the benthic fauna, altering the makeup of life on the sea bottom.[6] Another threat is connected with potential entanglement and ingestion hazards for pelagic and benthic animals.[62,112,118] Plastic can adsorb and concentrate some pollutants in coastal waters, including PCBs, DDE, nonylphenyl, and phenanthrene. It has been reported that these sorbed POPs could subsequently be released if the plastics are ingested.[109,110] For instance, PCBs in tissues of great shearwaters (*Puffinus gravis)* were derived from ingested plastic debris.[119]

Noise Pollution

In recent years, the marine biota has been affected by noise pollution. Natural sources of underwater noise may be physical and biological in character. Physical sources include wind, waves, rainfall, thunder and lighting, earthquake-generated seismic energy, and the movement of ice. Biological sources include marine mammal vocalizations and sounds produced by fish and invertebrates.[120,121]

Anthropogenic sound sources can be grouped into six categories, namely, shipping, seismic survey-ing, sonars, explosions, industrial activity, and miscellaneous.[122] Vessel traffic significantly contributes to underwater noise, mainly at low frequencies. Commercial shipping vessels generate noise mainly in areas confined to ports, harbors, and shipping lanes.[122] In contrast to wide geographic distribution of shipping industry, the oil and gas industry activities have taken place along continental margins in specific worldwide areas. Such resources exploration activities have been typically observed in shallow waters less than 200 m in depth. Other activities, in spite of their geographically widespread range, are also confined to near-shore coastal regions, namely, pile driving, dredging, operation of land- and ocean-based wind power turbines, power plant operations, and typical harbor and shipyard activities.[120] Offshore wind turbines may have significantly contributed noise to the underwater ecosystem bearing in mind that the relatively recent growth in offshore wind development has increased. It has been sug-gested that marine mammals may be indirectly affected by noise from offshore wind turbines, e.g., prey fish avoiding the sound source as well as the masking of marine mammals' mating and communication calls. On the other hand, a number of mass stranding of marine mammals, especially whales, found on worldwide beaches may be associated with the use of concurrent military sonar.[120] Another example of noise pollution affecting marine animals is continued exposure to anthropogenic noise pressure in vital sea turtle habitats, resulting in potential impact on its behavior and ecology. Brown shrimp exposed to higher pressure levels of noise in experimental area exhibited increased aggression, higher mortality rates, and significant reduction in their food uptake, growth, and reproduction. Sound exhibits measur-able damage to sensory cells in the ears of fish.[123]

Light Pollution

A remarkable recent interest concerns the introduction of light to the coastal zone and nearshore envi-ronment. It is estimated that at least 3351 cities in the coastal zones all over the world are illuminated. It is expected that artificial light will be continuously intensified not only by population growth but also by dramatically increasing the number of locations of high-intensity artificial light. According to the United Nations World Tourism Organization (UN- WTO), there were ca. 900 million international tourist arriv-als all over the world.[9] Tourist visits to beaches cause light pollution along the coastline since tracking the movement of population over time by research using satellite imagery showed that wherever human population density increases, the use of artificial light at night also increases. Living organisms are mostly sensitive to changes in the quality and intensity of natural light in the ecosystem. For instance, for algae and seaweeds, photosynthetic activity is highly dependent on available light, i.e., different cycles in natural light intensity and quality.[9] Light pollution takes place when biota is exposed to artificial light, especially in coastal areas, resulting in damaging effects on marine species in seas. The behavior, reproduction, and survival of marine invertebrates, amphibians, fish, and birds have been influenced by artificial lights. Light pollution disrupts the migration patterns of nocturnal birds and can result in hatchling sea turtles to head inland, away from the sea, which could be eaten by predators or run over by cars.[124] Ecological effects of light pollution concern disruption of predator–prey relationship. For instance, artificial light disturbs natural vertical migrations of zooplankton in the water column in accordance with the day-night cycle when natural light helps to reduce their predation by fish and other animals.[125]

Conclusion

The anthropogenic activity of man in coastal regions and even in offshore areas is responsible for emis-sion of a huge amount of pollutants that are transported to marine ecosystems directly or by means of coastal watersheds, rivers, and precipitation from air. A lot of pollutants may be taken up by various compartments, i.e., biotic and abiotic, of aquatic environments and some of them could be biomagnified along the successive members of the food chain. Therefore, water pollution could have a negative effect on the quality of the water and hence on the health of the plants, animals, and humans whose lives depend on the quality of aquatic environments. Coastal areas are generally damaged from pollution, resulting in considerable impact on commercial coastal and marine fisheries.

There are numerous examples of worldwide events leading to serious contamination of coastal waters by persistent pollutants. Therefore, these areas have been extensively explored, especially in relation to perturbed ecosystems by heavy metals, radionuclides, POPs, oils, etc.

Elevated levels of nutrients in coastal waters resulted in eutrophication and proliferation of toxic algal blooms. The recently observed increase in tanker operations and oil use as well as marine tanker catastrophes has been responsible for the presence of excessively large amounts of oil spillage in coastal and marine ecosystems. Marine debris, especially plastics, is one of the most pervasive pollution problems. Marine pollutants are generally present in increased concentrations in the enclosed seas and coastal areas than in the open seawaters. Spatial distribution patterns of contamination concentrations exhibit a trend of their increase during transition from the south to the northern part of all oceans, i.e., in areas neighboring with industrial centers and concentration of main pollution sources.

References

1. *Pollutant,* 2011, available at http://en.wikipedia.org/wiki/Pollution (accessed 2011).
2. *Pollutant,* 2011, available at http://en.wikipedia.org/wiki/Pollutant#a-Stock_pollutants (accessed 2011).
3. Williams, C. Combating marine pollution from land-based activities: Australian initiatives. Ocean Coastal Manage. **1996**, *33*, 87–112.
4. Falandysz, J.; Trzosinska, A.; Szefer, P.; Warzocha, J.; Draganik, B. The Baltic Sea, especially southern and eastern regions. In *Seas at the Millennium: An Environmental Evaluation, Vol. I: Europe, The Americas and West Africa;* Sheppard, C.R.C., Ed.; Pergamon, Elsevier: Amsterdam, 2000; 99–120.
5. Albaiges, J.; Bebianno, M.J.; Camphuysen, K.; Cronin, M.; de Leeuw, J.; Gabrielsen, G.; Hutchinson, T.; Hyl- land, K.; Janssen, C.; Jansson, B.; Jenssen, B.M.; Roose, P.; Schulz-Bull, D.; Szefer, P. *Monitoring Chemical Pollution in Europe's Seas-Programmes, Practices and Priorities for Research;* Position Paper 16, Marine Board—European Science Foundation: Ostend, 2011.
6. Shahidul Islam, Md.; Tanaka, M. Impacts of pollution on coastal and marine ecosystems including coastal and marine fisheries and approach for management: A review and synthesis. Mar. Pollut. Bull. **2004**, *48*, 624–649.
7. Laws, E.A. *Aquatic Pollution;* John Wiley and Sons: New York, 2000.
8. Valcovic, V. Radioactivity in the Environment—Physicochemical Aspects and Applications; Elsevier: Amsterdam, 2000.
9. Depledge, M.H.; Godard-Codding, C.A.J.; Bowen, R.E. Light pollution in the sea. Mar. Pollut. Bull. **2010**, *60*, 1383–1385.
10. Sheppard, C.R.C., Ed. Seas in the Millennium: An Environmental Evaluation, Vol. I: Europe, The Americas and West Africa; Pergamon, Elsevier: Amsterdam, 2000.
11. Tanabe, S., Ed. *Mussel Watch—Marine Pollution Monitoring in Asian Waters;* Center for Marine Environmental Studies, Ehime University: Japan, 2000.
12. Nriagu, J.O. Global inventory of natural and anthropogenic emissions of trace metals to the atmosphere. Nature **1979**, *279*, 409–411.
13. Luoma, S.N.; Rainbow, P.S. Metal Contamination in Aquatic Environments: Science and Lateral Management; Cambridge University Press: Cambridge, 2008.
14. Prospero, J.M. African dust in America. Geotimes. **2001**, 24–27, available at http://www.rsmas.miami.edu/assets/pdfs/mac/fac/Prospero/Publications/Prospero_Geotimes_African%20Dust_2001.pdf (accessed 2011).

15. Szefer, P. Metals, Metalloids and Radionuclides in the Baltic Sea Ecosystem; Elsevier Science B.V.: Amsterdam, 2002.

16. Mason, R.P.; Fitzegerald, W.F.; Morel, F.M.M. The biogeochemical cycling of elemental mercury: Anthropogenic influences. Geochim. Cosmochim. Acta **1994**, *58*, 3191–3198.

17. Jickells, T. Atmospheric inputs of metals and nutrients to the oceans: Their magnitude and effects. Mar. Chem. **1995**, *48*, 199–214.

18. Neff, J.M. Bioaccumulation in Marine Organisms: Effects of Contaminants from Oil Well Produced Water; Elsevier: Amsterdam, 2002.

19. Bryan, G.W.; Gibbs, P.E. Impact of low concentrations of tributyltin (TBT) on marine organisms: A review. In *Metal Ecotoxicology: Concepts and Applications;* Newman, M.C.; McIntosh, A.W., Eds.; Lewis Publishers: Ann Arbor, MI, 1991; 323–361.

20. Kabata-Pendias, A.; Mukherjee, A.B. *Trace Elements from Soil to Human;* Springer: Berlin, 2007.

21. Bryan, G.W.; Langston, W.J. Bioavailability, accumulation and effects of heavy metals in sediments with special reference to United Kingdom estuaries: A review. Environ. Pollut. **1992**, *76*, 89–131.

22. US EPA (US Environmental Protection Agency). *Health Effects Assessment of Mercury;* Environmental Criteria and Assessment Office: Cincinnati, Ohio, 1984.

23. dos Santos, M.M.; Vieira, N.; Santos, A.M. Imposex in the dogwhelk *Nucella lapillus* (L.) along the Portuguese coast. Mar. Pollut. Bull. **2000**, *40*, 643–646.

24. Gray, J.E.; Theodorakos, P.M.; Bailey, E.A., Turner, R.R. Distribution, speciation, and transport of mercury in stream- sediment, stream-water, and fish collected near abandoned mercury mines in southwestern Alaska, USA. Sci. Total Environ. **2000**, *260*, 21–33.

25. Maurice-Bourgoin, L.; Quiroga, I.; Chincheros, J.; Courau, P. Mercury distribution in waters and fishes of the upper Madeira rivers and mercury exposure in riparian Amazonian populations. Sci. Total Environ. **2000**, *260*, 73–86.

26. Tomiyasu, T.; Nagano, A.; Yonehara, N.; Sakamoto, H.; Rifardi; Oki, K; Akagi, H. Mercury contamination in the Yat- sushiro Sea, south-western Japan: Spatial variations of mercury in sediment. Sci. Total Environ. **2000**, *257*, 121–132.

27. Förstner, U.; Wittmann, G.T.W. *Metal Pollution in the Aquatic Environment,* 2nd Ed.; Springer-Verlag: Berlin, 1983.

28. Phillips, D.J.H.; Rainbow, P.S. *Biomonitoring of Trace Aquatic Contaminants;* Elsevier Science Publishers Ltd.: London, 1993.

29. Akagi, H.; Malm, O.; Kinjo, Y.; Harada, M.; Branches, F.J.P.; Pfeiffer, W.C.; Kato, H. Methylmercury pollution in the Amazon, Brazil. Sci. Total Environ. **1995**, *175*, 85–95.

30. Harada, M. Characteristics of industrial poisoning and environmental contamination in developing countries. Environ. Sci. **1996**, *4*, 157–169.

31. Harada, M.; Nakachi, S.; Cheu, T.; Hamada, H.; Ohno, Y.; Tsuda, T.; Yanagida, K.; Kizaki, T.; Ohno, H. Monitoring of mercury pollution in Tanzania: Relation between head hair and health. Sci. Total Environ. **1999**, *227*, 249–256.

32. Wheatley, B.; Wheatley, M.A. Methylmercury and the health of indigenous peoples: A risk management challenge for physical and social sciences and for public health policy. Sci. Total Environ. **2000**, *259*, 23–29.

33. Al-Majed, N.B.; Preston, M.R. Factors influencing the total mercury and methyl mercury in the hair of the fishermen of Kuwait. Environ. Pollut. **2000**, *109*, 239–250.

34. Artaxo, P.; Calixto de Campos, R.; Fernandes, E.T.; Martins, J.V.; Xiao, Z.; Lindquist, O.; Fernandez-Jimenez, M.T.; Maenhaut, W. Large scale mercury and trace element measurements in the Amazon basin. Atmos. Environ. **2000**, *34*, 4085–4096.

35. Yamasoe, M.A.; Artaxo, P.; Miguel, A.H.; Allen, A.G. Chemical composition of aerosol particles from direct emissions of vegetation fires in the Amazon Basin: Water- soluble species and trace elements. Atmos. Environ. **2000**, *34*, 1641–1653.

36. Alzieu, C. TBT detrimental effects on oyster culture in France—Evolution since antifouling paint regulation. In Proceedings of Oceans 86 Conference Record. Organotin Symposium; 4 Institute of Electrical and Electronics Engineers: New York, 1986; 1130–1134.

37. Alzieu, C. Environmental impact of TBT: The French experience. Sci. Total Environ. **2000**, *258*, 99–102.

38. Shim, W.J.; Kahng, S.H.; Hong, S.H.; Kim, N.S.; Kim, S.K.; Shim, J.H. Imposex in the rock shell, *Thais clavigera*, as evidence of organotin contamination in the marine environment of Korea. Mar. Environ. Res. **2000**, *49*, 435–451.

39. Hung, T.-C.; Hsu, W.-K.; Mang, P.-J.; Chuang, A. Organo- tins and imposex in the rock shell, *Thais clavigera*, from oyster mariculture areas in Taiwan. Environ. Pollut. **2001**, *112*, 145–152.

40. Evans, S.M.; Nicholson, G.J. The use of imposex to assess tributyltin contamination in coastal waters and open seas. Sci. Total Environ. **2000**, *258*, 73–80.

41. Batt, J.M. *The world of organotin chemicals: Applications, substitutes, and the environment*; 2006, available at http://www.ortepa.org/WorldofOrganotinChemicals.pdf (accessed 2011).

42. Rumengan, I.F.; Ohji, M.; Arai, T.; Harino, H.; Arfin, Z.; Miyazaki, N. Contamination status of butyltin compounds in Indonesian coastal waters. Coastal Mar. Sci. **2008**, *32*, 116–126.

43. Zhao, Y.; Marriott, S.; Rogers, J.; Iwugo, K. A preliminary study of heavy metal distribution on the floodplain of the River Severn, U.K. by a single flood event. Sci. Total Environ. **1999**, *243/244*, 219–231.

44. Skwarzec, B. *Radiochemia Srodowiska i Ochrona Radiolo- giczna* (in Polish), Environmental Radiochemistry and Radiological Protection; Wydawnictwo DJ sc.: Gdansk, 2002.

45. ISSG. *The Irish Sea: An Environmental Review*; Report of the Irish Sea Study Group; Liverpool University Press: Liverpool, 1990.

46. Kershaw, P.J.; McCubbin, D.; Leonard, K.S. Continuing contamination of North Atlantic and Arctic waters by Sellafield radionuclides. Sci. Total Environ. **1999**, *237/238*, 119–132.

47. Aarkrog, A.; Dahlgaard, H.; Hansen, H.; Holm, E.; Hall- stadius, L.; Rioseco, J.; Christensen, G. Radioactive tracer studies in the surface waters of the northern North Atlantic including the Greenland, Norwegian and Barents Seas. Rit. Fiskideildar **1985**, *9*, 37–42.

48. Kershaw, P.J.; Baxter, A.J. The transfer of reprocessing wastes from north-west Europe to the Arctic. Deep Sea Res. **1995**, *42*, 1413–1448.

49. Förstner, U. Inorganic pollutants, particularly heavy metals in estuaries In *Chemistry and Biogeochemistry of Estuaries*; Olausson, E., Cato, I., Eds.; John Wiley & Sons: New York, 1980; 307–348.

50. UNSCEAR. *Sources and effects of ionizing radiation*; Report to the General Assembly by the United Nations Scientific Committee on the Effects of Atomic Radiation; United Nations: New York, 1993.

51. Beasley, T.M.; Cooper, L.W.; Grebmeier, J.M.; Orlandini, K.; Kelley, J.M. Fuel reprocessing Pu in the Arctic Ocean Basin: evidence from mass spectrometry measurements. In Proc. Conf. on Environmental Radioactivity in the Arctic, Oslo, August 1995.

52. Aarkrog, A.; Dahlgaard, H.; Nilsson, K.; Holm, E. Studies of plutonium and americum at Thule, Greenland. Health Phys. **1984**, *46*, 29–44.

53. Smith, J.N.; Ellis, K.M.; Aarkrog, A.; Dahlgaard, H.; Holm, E. Sediment mixing and burial of the [239,240]Pu pulse from the 1968 Thule, Greenland nuclear weapons accident. J. Environ. Radioact. **1994**, *25*, 135–159.

54. Joint Russian–Norwegian Expert Group. Radioactive Contamination of Dumping Sites for Nuclear Wastes in the Kara Sea. Results from the 1993 Expedition; Norwegian Radiation Protection Authority: Østerås, 1994.

55. National Academy of Sciences. *Radioactivity in the Marine Environment*; National Academy of Sciences: Washington, DC, 1971.

56. Aarkrog, A. A retrospect of anthropogenic radioactivity in the global marine environment. Radiat. Prot. Dosim. **1998**, *75*, 23–31.

57. Commission of the European Communities. The Radiological Exposure of the Population of the European Community from Radioactivity in North European Marine Waters— Project MARINA, Report EUR 12483; EU, 1989.
58. NEA. Co-ordinated Research and Environmental Surveillance Programme Related to Sea Disposal of Radioactive Waste, CRESP Final Report, 1981–1995; OECD Paris, 1996.
59. INSAG. *Post Accident Review Meeting on the Chernobyl Accident;* Summary Report; International Atomic Energy Agency: Vienna, 1986.
60. WHO. Health hazards from radiocaesium following the Chernobyl nuclear accident. Report on a WHO Working Group. J. Environ. Radioact. **1989,** *10,* 257–259.
61. Radiation effects from Fukushima Daiichi nuclear disaster, available at http://en.wikipedia.org/wiki/Radiation_effects_from_Fukushima_Daiichi_nuclear_disaster (accessed 2011).
62. Goldberg, E.D. Emerging problems in the coastal zone for the 21th century. Mar. Pollut. Bull. **1995,** *31,* 152–158.
63. Gibbs, P.E. Oviduct malformation as a sterilising effect of tributyltin-induced imposex in *Ocenebra erinacea* (Gastropoda: Muricidae). J. Molluscan Stud. **1996,** *62,* 403–413.
64. Cadee, G.C.; Boon, J.P.; Fischer, C.V.; Mensink, B.P.; Tjabbes, C.C. Why the whelk *Buccinum undatum* has become extinct in the Dutch Wadden Sea. Neth. J. Sea Res. **1995,** *34,* 337–339.
65. Jones, P. TBT implicated in mass dolphin deaths. Mar. Pollut. Bull. **1997,** *34,* 146.
66. HELCOM. Batlic Sea Environment Proceedings No. 64B. Third Periodic Assessment on the State of the Marine Environment of the Baltic Sea, 1989–93 Background Document. Helsinki Commission, Baltic Marine Environment Protection Commission, 1996; 252.
67. Jepson, P.D.; Bennet, P.M.; Allchin, C.R.; Law, R.J.; Kui- ken, T.; Baker, J.R.; Rogan, E.; Kirkwood, J.K. Investigating potential associations between chronic exposure to polychlorinated biphenyls and infectious disease mortality in harbour porpoises from England and Wales. Sci. Total Environ. **1999,** *243–244,* 339–348.
68. Cloern, J.E. Our evolving conceptual model of the coastal eutrophication problem. Mar. Ecol. Prog. Ser. **2001,** *210,* 223–253.
69. Riegman, R. Nutrient-related selection mechanisms in marine plankton communities and the impact of eutrophication on the plankton food web. Water Sci. Technol. **1995,** *32,* 63–75.
70. Kimor, B. Impact of eutrophication on phytoplankton composition. In *Marine Coastal Eutrophication;* Vollenweider, R.A., Marchetti, R., Vicviani, R., Eds.; Elsevier: Amsterdam, 1992; 871–878.
71. Zaitsev, Y.P. Recent changes in the trophic structure of the Black Sea. Oceanography **1992,** *1,* 180–189.
72. Birch, P.B.; Forbes, G.G.; Schofield, N.J. Monitoring effects of catchment management practices on phosphorus loads into the eutrophic Peel-Harvey Estuary, Western Australia. Water Sci. Technol. **1986,** *18,* 53–61.
73. Justic, D. Long-term eutrophication of the Northern Adriatic Sea. Mar. Pollut. Bull. **1987,** *18,* 281–284.
74. Degobbis, D. Increased eutrophication of the Northern Adriatic Sea. Second act. Mar. Pollut. Bull. **1989,** *20,* 452–457.
75. Forsberg, C. *Eutrophication of the Baltic Sea;* The Baltic Sea Environment: Uppsala, Sweden, 1993; 32 pp.
76. Gillespie, N. Ecological and epidemiological aspects of ciguatera fish poisoning. In Proc. of the Red Tide Workshop, Cronulla, June 18–20, 1984; Australian Department of Science: Canberra.
77. Hallegraeff, G.M.; Sumner, C. Toxic plankton blooms affect shellfish farms. Aust. Fish. **1986,** *45,* 15–18.
78. Holmes, P.R.; Lam, C.W.Y. Red tides in Hong Kong waters—Response to a growing problem. Asian Mar. Biol. **1985,** *2,* 1–10.

79. Phillips, D.J.H. Monitoring and control of coastal water quality. In *Pollution in the Urban Environment, POLMET 85;* Chan, M.W.H., Hoare, R.W.M., Holmes, P.R., Law, R.J.S., Reed, S.B., Eds.; Elsevier Applied Science Publishers: London, 1985; 559–565.

80. Morton, B.S. Pollution of the coastal waters of Hong Kong. Mar. Pollut. Bull. **1989**, *20,* 310–318.

81. Maclean, J.L. Indo-Pacific red tides, 1985–1988. Mar. Pollut. Bull. **1989**, *20, 304–310.*

82. Maclean, J.L.; White, A.W. Toxic dinoflagellate blooms in Asia: A growing concern. In *Toxic Dinoflagellates;* Anderson, D.M., White, A.W., Baden, D.G., Eds.; Elsevier: New York, 1985; 517–520.

83. Ayres, P.A. Mussel poisoning in Britain with special reference to paralytic shellfish poisoning. Environ. Health **1975,** *July,* 261–265.

84. Paerl, H.W., Whitall, D.R. Anthropogenically-derived atmospheric nitrogen deposition, marine eutrophication and harmful algal bloom expansion: Is there a link? Ambio **1999**, *28,* 307–311.

85. Paerl, H.W. Enhancement of marine primary productivity by nitrogen enriched rain. Nature **1985,** *315,* 747–749.

86. Paerl, H.W. Coastal eutrophication in relation to atmospheric nitrogen deposition: Current perspectives. Ophelia **1995**, *41,* 237–259.

87. Anderson, D.M. Toxic algal blooms and red tides. A global perspective. In *Red Tides: Biology, Environmental Science and Toxicology;* Okaichi, T., Anderson, D.M., Nemoto, T., Eds.; Elsevier Science Publishing Co., Inc.: New York, 1989; 11–21.

88. Aksnes, D.L.; Aure, J.; Furnes, G.K.; Skjoldal, H.R.; Sae- tre, R. Analysis of the *Chrysochromulina polylepis* bloom in the Skagerrak. Environmental conditions and possible causes. Bergen Scientific Centre Publication 1989, No. BSC 89/1.

89. Tester, P.A.; Stumpf, R.P.; Vukovich, F.M.; Fowler, P.K.; Turner, J.T. An expatriate red tide bloom: Transport, distribution, and persistence. Limnol. Oceanogr. **1991**, *36,* 1053–1061.

90. Buskey, E.J.; Stockwell, D.A. Effects of a persistent "brown tide" on zooplankton populations in the Laguna Madre of South Texas. In *Toxic Phytoplankton Blooms in the Sea,* Proc. 5th Intern. Conf. on Toxic Marine Phytoplankton; Elsevier, 1993; 659–665.

91. Hallegraeff, G.M. A review of harmful algal blooms and their apparent global increase. Phycologia **1993**, *32,* 79–99.

92. Anderson, D.M.; Kulis, D.M.; Doucette, G.J.; Gallagher, J.C.; Balech, E. Biogeography and toxic dinoflagellates in the genus *Alexandrium* from the northeastern United States and Canada. Mar. Biol. **1994**, *120,* 467–478.

93. ECOHAB. *The Ecology and Oceanography of Harmful Algal Blooms;* A National Research Agenda, US N.S.F./ N.O.O.A. Publication; Woods Hole Oceanographic Inst.: Mass., USA, 1995.

94. Howarth, R.W.; Billen, G.; Swaney, D.; Townsend, A.; Ja- worski, N.; Lajtha, K.; Downing, J.A.; Elmgren, R.; Caraco, N.; Jordan, T.; Berendse, F.; Freney, J.; Kudeyarov, V.; Murdoch, P.; Zhu, Z.-L. Regional nitrogen budgets and riverine N and P fluxes for the drainages to the North Atlantic Ocean: Natural and human influences. Biogeochemistry **1996**, *35,* 75–139.

95. Prospero, J.M.; Barret, K.; Church, T.; Detener, F.; Duce, R.A.; Galloway, J.N.; Levy, H.; Moody, J.; Quinn, P. Atmospheric deposition of nutrients to the North Atlantic basin. Biogeochemistry **1996**, *35,* 27–73.

96. Paerl, H.W. Coastal eutrophication and harmful algal blooms: Importance of atmospheric deposition and groundwater as "new" nitrogen and other nutrient sources. Limnol. Oceanogr. **1997**, *42,* 1154–1165.

97. Mills, E.L.; Holeck, K.T. Biological pollutants in the Great Lakes. Clearwaters **2001**, *31* (1), available at http://www. nywea.org/clearwaters/pre02fall/311010.html (accessed 2011).

98. Elliott, M. Biological pollutants and biological pollution— An increasing cause for concern. Mar. Pollut. Bull. **2003**, *46,* 275–280.

99. FAO. Precautionary Approach to Fisheries. Part 1. Guidelines on the Precautionary Approach to Capture Fisheries and Species Introductions; Food and Agriculture Organization of the United Nations: Rome, 1995.

100. Joly, C. Plant nutrient management and the environment. In *Prevention of Water Pollution by Agriculture and Related Activities*, Proceedings of the FAO Expert Consultation, Santiago, Chile, October 20–23, 1992; Water Report 1,FAO: Rome.

101. Li, Y.F.; Macdonald, R.W. Sources and pathways of selected organochlorine pesticides to the Arctic and the effect of pathway divergence on HCH trends in biota: A review. Sci. Total Environ. **2005**, *342*, 87–106.

102. Duursma, E.K.; Marchand, M. Aspects of organic marine pollution. Oceanogr. Mar. Biol. Ann. Rev. **1974**, *12*, 315–431.

103. *Oil spill*; 2011, available at http://en.wikipedia.org/wiki/ Oil_spill (accessed 2011).

104. *Oil spills and disasters*; 2011, available at http://www.info- please.com/ipa/A0001451.html (accessed 2011).

105. Smith, N. The problem of oil pollution of the sea. Adv. Mar. Biol. **1970**, *8*, 215–306.

106. Sheavly, S.B.; Register, K.M. Marine debris and plastics: Environmental concerns, sources, impacts and solutions. J.Polym. Environ. **2007**, *15*, 301–305.

107. Ryan, P.G.; Moore, C.J.; van Franeker, J.A.; Moloney, C.L. Monitoring the abundance of plastic debris in the marine environment. Philos. Trans. R. Soc. **2009**, *364*, 1999–2012.

108. Moore, C.J. Synthetic polymers in the marine environment: A rapidly increasing. Environ. Res. **2008**, *108*, 131–139.

109. Barnes, K.A.; Galgani, F.; Thompson, R.C.; Barlaz, M. Accumulation and fragmentation of plastic debris in global environments. Philos. Trans. R. Soc. **2009**, *364*, 1985–1998.

110. Teuten, E.L.; Saquing, J.M.; Knappe, D.R.U.; Barlaz, M.A.; Jonsson, S.; Björn, A.; Rowland, S.J.; Thompson, R.C.; Galloway, T.S.; Yamashita, R.; Ochi, D.; Watanuki, Y.; Moore, C.; Viet, P.H.; Tana, T.S.; Prudente, M.; Boonyatumanond, R.; Zakaria, M.P.; Akkhavong, K.; Ogata, Y.; Hirai, H.; Iwasa, S.; Mizukawa, K.; Hagino, Y.; Imamura, A.; Saha, M.; Takada, H. Transport and release of chemicals from plastics to the environment and to wildlife. Philos. Trans. R. Soc. B **2009**, *364*, 2027–2045.

111. US Marine Mammal Commission. *Marine Mammal Commission Annual Report to Congress*; Effects of Pollution on Marine Mammals: Bethesda, MD, 1996.

112. Derraik, J.G.B. The pollution of the marine environment by plastic debris: A review. Mar. Pollut. Bull. **2002**, *44*, 842–852.

113. Slip, D.J.; Green, K.; Woehler, E.J. Ingestion of anthropogenic articles by seabirds at Macquarie Island. Mar. Orni- thol. **1990**, *18*, 74–77.

114. Van Frakener, J.A.; Bell, P.J. Plastic ingestion by petrels breeding in Antarctica. Mar. Pollut. Bull. **1988**, *19*, 672–674.

115. Bourne, W.R.P.; Imber, M.J. Plastic pellets collected by a prion on Gough Island, Central South Atlantic Ocean. Mar. Pollut. Bull. **2001**, *13*, 20–21.

116. Mato, Y.; Isobe, T.; Takada, H.; Kanehiro, H.; Ohtake, C.; Kaminuma, T. Plastic resin pellets as a transport medium for toxic chemicals in the marine environment. Environ. Sci. Technol. **2001**, *35*, 318–324.

117. Gregory, M.R. Plastics and South Pacific Island shores: Environmental implications. Ocean Coastal Manage. **1999**, *42*, 603–615.

118. Hess, N.A.; Ribic, C.A.; Vining, I. Benthic marine debris, with an emphasis in fishery-related items, surrounding Kodiak Island, Alaska 1994–1996. Mar. Pollut. Bull. **1999**, *38*, 885–890.

119. Ryan, P.G.; Connell, A.D.; Gardener, B.D. Plastic ingestion and PCBs in seabirds: Is there a relationship? Mar. Pollut. Bull. **1988**, *19*, 174–176.

120. Firestone, J.; Jarvis, C. Response and responsibility: Regulating noise pollution in the marine environment. J. Int. Wildl. Law Policy **2007**, *10*, 109–152.

121. Hatch, L.T.; Wright, A.J. A brief review of anthropogenic sound in the oceans. Int. J. Comp. Psychol. **2007**, *20*, 121–133.

122. National Research Council (NRC). *Ocean Noise and Marine Mammals;* National Academy Press: Washington, DC, 2003.

123. Samuel, Y.; Morreale, S.J.; Clark, C.W.; Greene, C.H.; Richmond, M.E. Underwater, low-frequency noise in coastal sea turtle habitat. J. Acoust. Soc. Am. **2005**, *117*, 1465–1472.

124. Gallaway, T.; Olsen, R.N.; Mitchell, D.M. The economics of global light pollution. Ecol. Econ. **2010**, *69*, 658–665.

125. Gliwicz, Z.M. A lunar cycle in zooplankton. Ecology **1986**, *67*, 883–897.

3

Groundwater: Mining Pollution

Introduction 37
Groundwater Resources 38
Groundwater Contaminants...39
Groundwater Analysis 41
Strategies for Remediating Contaminated Groundwaters.......................42
References..43

Jeff Skousen and
George Vance

Introduction

Mining activities can impact the quantity, quality, and usability of groundwater supplies. Underground mining for coal by longwall or room and pillar mining methods often interrupts and depletes groundwater, and can also alter its quality. Surface mining can enhance the introduction of surface water with dissolved solids into groundwater systems through fractures or other conduits. The type and nature of the mining activity, the disturbed geologic strata, and alteration of surface and subsurface materials will determine how groundwater supplies will be impacted. As waters contact and interact with disturbed geologic materials, constituents such as salts, metals, trace elements, and organic compounds become mobilized [1,2]. The dissolved substances can leach into deep aquifers and cause groundwater quality impacts [3]. In addition to concerns due to naturally occurring contaminants from disturbance activities, mining operations may also contribute to groundwater pollution from leaking underground storage tanks, improper disposal of lubricants and solvents, and contaminant spills. Blasting and hydraulic fracking activities can provide additional connection to surface water inputs, and underground injection of wastes can also occur during these operations [4].

In the United States, the Clean Water Act (CWA) and its subsequent amendments establish the authority for all water pollution control actions at the federal level [5] and regulate discharges into surface streams, wetlands, and oceans. Mining operations must acquire National Pollutant Discharge Elimination System (NPDES) permits for discharges to surface waters. Groundwater quality in the United States is regulated by the Safe Drinking Water Act (SDWA), which was originally enacted in 1974 and amended in 1996. The SDWA was passed to protect drinking water supplies by requiring discharges into groundwaters to meet the use standard or the ambient condition, whichever is of higher quality [6]. This is done by legislating maximum contaminant levels (MCLs) above which waters are considered unsafe for human consumption. The Office of Water within the Environmental Protection Agency provides guidance, specifies scientific methods and data collection requirements, and performs oversight for entities that supply drinking water including groundwater. Examples of some water contaminants with specified MCLs associated with mining activities are listed in Table 1 [7].

Because mining activities can result in poor-quality groundwaters, enforcement of regulations is needed to minimize and/or eliminate potential problems. The Surface Mining Control and Reclamation

TABLE 1 Selected Contaminants in Drinking Waters That May Be Influenced by Mining Activities [7]

Contaminant	MCL (mg/L)	MCLG
Inorganics		
Arsenic	0.010	0
Cadmium	0.005	0.005
Chromium	0.1	0.1
Copper	LV	1.3
Cyanide	0.2	0.2
Fluoride	4.0	4.0
Lead	LV	0
Mercury	0.002	0.002
Nitrate (NO_3-N)	10	10
Selenium	0.05	0.05
Sulfate	500	500
Radionuclides		
Radium	5 pCi/L	0
Uranium	30 ug/L	0
Organics		
Benzene	0.005	0
Carbon tetrachloride	0.005	0
Pentachlorophenol	0.001	0
Toluene	1	1
Xylenes	10	10
Microbiological		
Total coliforms	LV	0
Viruses	LV	0

MCL, Maximum contaminant levels permissible for a contaminant in water that is delivered to any user of a public water system; MCLG, Maximum contaminant level goals of a drinking water contaminant that is protective of adverse human health effects and which allows for an adequate margin of safety; LV, Lowest value that can be achieved using the best available technology.

Act (SMCRA) of 1977 identifies policies and practices for mining and reclamation to minimize water quality impacts [8]. SMCRA requires that specific actions be taken to protect the quantity and quality of both on- and off-site groundwaters. All mines are required to meet either state or federal groundwater guidelines, which are generally related to priority pollutant standards described in the CWA and SDWA.

Groundwater Resources

Groundwater resources are the world's third largest source of water behind oceans (97%) and glaciers (2%), and represent 0.6% of the earth's water content [9]. Approximately 53% of the US population uses groundwater as a drinking water source, but this percentage increases to almost 97% for rural households. In areas of low rainfall, weathering and translocation of dissolved constituents are relatively slow compared to high rainfall areas. For example, only 12% of precipitation will recharge underground water supplies in a dry coal mining area like Gillette, Wyoming, while almost 47% of precipitation was available for recharge in coal mining areas of Tennessee [10]. Transport of contaminants from surface

TABLE 2   Important Hydrogeological Characteristics of a Site That Determine Groundwater Quantity and Quality

Geological

Type of water-bearing unit or aquifer (rock type, overburden).

Thickness and areal extent of water-bearing units and aquifers.

Type of porosity (primary, such as intergranular pore space, or secondary, such as bedrock discontinuities, e.g., fracture or solution cavities).

Presence or absence of impermeable units or confining layers.

Depths to water tables; thickness of vadose zone.

Permeability and connectivity to other voids or conduits.

Hydraulic

Hydraulic properties of water-bearing unit or aquifer (hydraulic conductivity, transmissivity, storability, permeability, dispersivity).

Pressure conditions (confined, unconfined, leaky confined).

Groundwater flow directions (hydraulic gradients, both horizontal and vertical), volumes (specific discharge), rate (average linear velocity).

Recharge and discharge areas.

Groundwater or surface water interactions; areas of groundwater discharge to surface water or vice versa.

Seasonal variations of groundwater conditions.

Groundwater Use

Existing or potential underground sources of drinking water.

Existing or near-site use of groundwater.

and subsurface environments to groundwaters is generally accelerated as the amount of percolating water increases.

Infiltrating water moves through the vadose zone (unsaturated region) into groundwater zones (saturated region). The upper boundary of the groundwater system (e.g., water table) fluctuates depending on the amount of water received or removed from the groundwater zone. Groundwater movement is a function of hydraulic gradients and hydraulic conductivities, which represent the combined forces with which water moves as a function of gravitational, osmotic, and pressure forces and the permeability of geologic strata. Groundwater moves faster in coarse-textured materials and where hydraulic gradients are high. Aquifers are groundwater systems that have sufficient porosity and permeability to supply enough water for specific purposes. For an aquifer to be useful, it must be able to store, transmit, and yield sufficient amounts of good-quality water. Important hydrogeological characteristics of a site that determine groundwater quantity and quality are listed in Table 2.

Groundwater Contaminants

Several types of substances affect groundwater quality [1,11]. Water contaminants include inorganic, organic, and biological materials. Some have a direct impact on water quality, while others indirectly cause physical, chemical, or biological changes that make the water unsuitable for its designated use. Substances that degrade groundwaters include nutrients, salts, heavy metals, trace elements, and organic chemicals, as well as contaminants such as radionuclides, carcinogens, pathogens, and petroleum wastes (Table 3, [12]). Several types of organic chemicals entering groundwaters are less dense than water and tend to move to and along the surface of the water table. Changes can also occur in groundwaters due to temperature fluctuations and odors. Some groundwaters near coal seams contain natural organic substances (such as dissolved methane gas) and synthetic organic chemicals. Methane gas can be extracted from coal beds where underground and surface mining operations are projected,

TABLE 3   Different Classes of Groundwater Pollutants and Their Causes [12]

Water Pollutant Class	Contributions
Inorganic chemicals	Toxic metals and acidic substances from mining operations and various industrial wastes
Organic chemicals	Petroleum products, pesticides, and materials from organic wastes industrial operations
Infectious agents	Bacteria and viruses
Radioactive substances	Waste materials from mining and processing of radioactive substances or from improper disposal of radioactive isotopes

and this extraction can alter methane gas concentrations in groundwaters [10]. Organic contamination may also result from leaking gas tanks, oil spills, or runoff from equipment-serving areas. In these cases, the source of the contamination must be identified and removed. Gasoline, diesel, or oil-soaked areas should be immediately excavated and disposed of by approved methods.

The chemistry of groundwaters and potential levels of naturally occurring contaminants are related to (1) groundwater hydrologic conditions, (2) mineralogy of the mined and locally impacted geological materials, (3) mining operations (e.g., extent of disturbed materials and its exposure to atmospheric conditions), and (4) time. Movement of metal contaminants in groundwaters varies depending on the chemical of concern. Solubility considerations include metals such as cobalt, copper, nickel, and zinc being more mobile than silver and lead, and gold and tin being even less mobile [1]. As conditions such as pH, redox, and ionic strength change over time, dissolved constituents in groundwaters may decrease due to adsorption, precipitation, and chemical speciation reactions and transformations.

Acid mine drainage (AMD) is the most prevalent groundwater quality concern at inactive and abandoned surface and underground mine sites. If geologic strata containing reduced S minerals (e.g., pyrite (FeS_2)) are exposed to weathering conditions, such as when pyritic overburden materials are brought to the surface during mining activities and then reburied, high concentrations of sulfuric acid (H_2SO_4) can develop and form acid waters with pH levels below 2 [2]. Neutralization of some or all of the acidity produced during the oxidation of reduced S compounds can occur when carbonate minerals in proximity to the acid-producing materials dissolve [3]. Neutralization can also occur when silicate minerals dissolve, but sometimes high levels of potentially toxic metals such as Al, Cu, Cd, Fe, Mn, Ni, Pb, and Zn may be released. For example, mining of coal in the Toms Run area of northwestern Pennsylvania resulted in groundwater contamination by AMD containing high concentrations of Fe and sulfate (SO_4) that leached into the underlying aquifer through joints, fractures, and abandoned oil and gas wells.

The Gwennap Mining District in the United Kingdom contained numerous mines that operated over several centuries to extract various mineral resources. One of these mines, the Wheal Jane metal mine in Cornwall, extracted ores that included cassiterite (Sn-containing mineral), chalcopyrite (Cu), pyrite (Fe), wolframite (tungsten, W), arsenopyrite (arsenic, As), in addition to smaller deposits of Ag, galena (Pb), and other minerals. After closure in the early 1990s, extensive voids remaining in the Wheal Jane mine that contained oxidized and weathered minerals were flooded. Initial groundwater quality was poor with a pH of 2.9 and a total metal concentration of 5000 mg/L, which contained elevated levels of Fe, Zn, Cu, and Cd. Water quality worsened with depth, and at 180 m, the groundwater had a pH of 2.5 and a metal concentration of 7000 mg/L. Treatment of discharge waters originating from the mine involves an expensive process that will continue long term to preserve environmental quality in surface and groundwaters in the region. A similar situation occurred when a Zn mine in southwestern France was closed. In this case after flooding, discharge mine waters had a solution pH near neutral, but the water still contained high concentrations of Zn, Cd, Mn, Fe, and SO_4.

Within the Coeur D' Alene District of Idaho at the Bunker Hill Superfund site, groundwater samples were found to contain high concentrations of Zn, Pb, and Cd [13]. The contamination originated from the leaching of old mine tailings deposited on a sand and gravel aquifer. When settling ponds were constructed to catch the runoff from the tailings, water from the ponds infiltrated into the aquifer and caused an increase in metal concentration in the local groundwater system [14].

Gold mining operations have used cyanide as a leaching agent to solubilize Au from ores, which often contain arsenopyrite (As, Fe, and S) and pyrite [1]. Unfortunately, cyanide, in addition to being toxic on its own, is a powerful nonselective solvent that solubilizes numerous substances that can be environmental contaminants. These ore waste materials are often stored in tailing ponds and, depending on the local geology and climate, the cyanide present in the tailings can exist as free cyanide (CN^-, HCN); inorganic compounds containing cyanide (NaCN, $HgCN_2$); metal-cyanide complexes with Cu, Fe, Ni, and Zn; and/or the compound CNS. Because cyanide species are mobile and persistent under certain conditions, a large potential exists for trace element and cyanide migration into groundwaters. For example, a tailings dam failure resulted in cyanide contamination of groundwater at a gold mining operation in British Columbia, Canada [1].

Arsenic and uranium (U) contamination has resulted from extensive mining and smelting of ores containing various metals (Ag, Au, Co, Ni, Pb, and Zn) and/or nonmetals (As, P, and U). Arsenic-contaminated groundwaters have been a source of surface recharge and drinking water supplies. At one site, a nearby river had As levels 7 and 13 times greater than the recommended national and local drinking water standards, respectively [1]. Arsenic is known as a carcinogen and has been the contributing cause of death to humans in several parts of the world that rely on As-contaminated drinking water [11]. Waters from dewatering a U mine in New Mexico had elevated levels of U and radium (Ra) activities as well as high concentrations of dissolved Mo and Se, which were detected in stream water 140 km downstream from the mine.

Groundwater Analysis

Both the remediation and prevention of groundwater contamination by nutrients, salts, heavy metals, trace elements, organic chemicals (natural and synthetic), pathogens, and other contaminants require the evaluation of the composition and concentration of these constituents either *in situ* or in groundwater samples [2,10]. Monitoring may require the analysis of physical properties, inorganic and organic chemical compositions, and/or microorganisms according to well-established protocols for sampling, storage, and analysis [15]. For example, if groundwater will be used for human or animal consumption, the most appropriate tests would be nitrate-nitrogen (NO_3-N), trace metals, pathogens, and organic chemicals. Several common constituents measured in groundwaters are listed in Table 4. However, other tests can be conducted on waters including tests for hardness, electrical conductivity (EC), chlorine, radioactivity, water toxicity, and odors [16].

Recommendations based on interpretation of the groundwater test results should be related to the ultimate use of the water [2]. The interpretation and recommendation processes may be as simple as determining that a drinking water well exceeds the established MCLs for NO_3-N and recommending the

TABLE 4 Groundwater Quality Parameters and Constituents Measured in Some Testing Programs [16]

Physical	Metals and Trace Elements	Nonmetallic Constituents	Organic Chemicals	Microbiological Parameters
Conductivity	Al, Ag, As, Ca, Cd, Cr,	pH, acidity, alkalinity, dissolved oxygen,	Methane	Fecal coliforms
Salinity	Cu, Fe, Mg, Mn, Na,	carbon dioxide, bicarbonate, B, Cl, CN, F,	Oil and grease	Bacteria
Sodicity	Ni, Pb, Se, Sr, Zn	I, ammonium, nitrite, nitrate, P, Si, sulfate	Organic acids	Viruses
Dissolved solids			Volatile acids	
Temperature			Organic C	
Odors			Pesticides	
			Phenols	
			Surfactants	

well should not be used as a drinking water source or that a purification system be installed. However, interpretations of most groundwater analyses can be quite complicated and require additional information for proper interpretation. If a contaminant exceeds an acceptable concentration, all potential sources contributing to the pollution and pathways by which the contaminant moves must be identified. In many cases, multiple groundwater contaminants are present at different concentrations. Because the interpretation of water analyses is a complex process, recommendations should be based on a complete evaluation of the water's physical, chemical, and biological properties. Integrating water analyses into predictive models that can assess the effects of mining activities on water quality is needed in the long term to determine the most effective means to preserve and restore water quality.

Strategies for Remediating Contaminated Groundwaters

Mine sites that have been contaminated generally contain mixtures of inorganic and/or organic constituents, so it is important to understand these multi-component systems in order to develop remediation strategies. Therefore, a proper remediation program must consider identification, assessment, and correction of the problem [17,18]. Identification of a potential problem site requires that the past history of the area and activities that took place are known, or when a water analysis indicates a site has been contaminated. Assessment addresses questions such as (1) what is the problem, (2) where and to what extent is the problem, and (3) who and what is affected by the problem. Afterward, a remediation action plan must be developed that will address the specific problems identified. A remediation action program may require that substrata materials (e.g., backfill) and groundwater be treated.

If remedial action is considered necessary, then three general options are available: (1) containment, (2) *in situ* treatment, or (3) pump-and-treat method (Figure 1). The method(s) used for the containment of contaminants are beneficial for restricting contaminant transport and dispersal. Of the remediation techniques, *in situ* treatment measures are the most appealing because they generally do less surface damage, require a minimal amount of facilities, reduce the potential for human exposure to contaminants, and when effective, reduce or remove the contaminant so that the groundwater can be utilized again [18]. *In situ* remediation can be achieved by physical, chemical, and/or biological techniques. Biological *in situ* techniques used for groundwater bioremediation can either rely on the indigenous (native) microorganisms to degrade organic contaminants or on amending the groundwater environment with specialized microorganisms (bioaugmentation). The pump-and-treat method, however, is one of the more commonly used processes for remediating contaminated groundwaters [17]. With the pump-and-treat methods, the contaminated waters are pumped to the surface where one of the

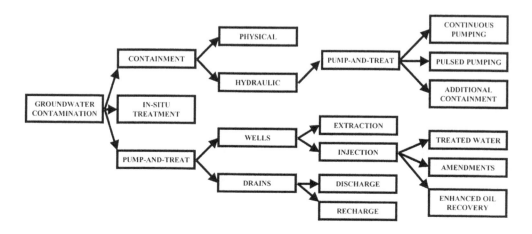

FIGURE 1   Remedial options to consider if cleanup of contaminated groundwater is deemed necessary.

many treatment processes can be utilized. A major consideration in the pump-and-treat technology is the placement of wells, which is dependent on the contaminant and site characteristics (see Table 2). Extraction wells are used to pump the contaminated water to the surface where it can be treated and re-injected or discharged. Injection wells can be used to re-inject the treated water, water containing nutrients and other substances that increase the chances for chemical alteration or microbial degradation of the contaminants, or materials for enhanced oil recovery.

Treatment techniques can be grouped into three categories, namely, physical, chemical, and biological methods [2,18].

Physical methods include several techniques. Adsorption methods physically adsorb or trap contaminants on various types of resins. Separation treatments include physically separating contaminants by forcing water through semipermeable membranes (e.g., reverse osmosis). Flotation, or density separation, is commonly used to separate low-density organic chemicals from groundwaters. Air and steam stripping can remove volatile organic chemicals. Isolation utilizes barriers placed above, below, or around sites to restrict movement of the contaminant. Containment systems should have a permeability of 10^{-7} cm/s or less.

Chemical methods are also numerous. Chemical treatment involves addition of chemical agent(s) in an injection system to neutralize, immobilize, and/or chemically modify contaminants. Extraction (leaching) of contaminants uses one of the several different aqueous extracting agents such as an acid, base, detergent, or organic solvent miscible in water. Oxidation and reduction of groundwater contaminants are commonly done using air, oxygen, ozone, chlorine, hypochlorite, and hydrogen peroxide. Ionic and nonionic exchange resins can adsorb contaminants, thus reducing their leaching potential.

Biological methods for contaminant remediation are less extensive than physical and chemical techniques. Land treatment is an effective method for treating groundwaters by applying the contaminated waters to lands using surface, overland flow, or subsurface irrigation. Activated sludge and aerated surface impoundments are used to precipitate or degrade contaminants present in water and include both aerobic and anaerobic processes. Biodegradation is one of the several biological-mediated processes that transform contaminants, and it utilizes vegetation and microorganisms.

References

1. Ripley, E.A., R.E. Redmann, and A.A. Crowder. 1996. *Environmental Effects of Mining.* St. Lucie Press, Delray Beach, FL. 356 pp.
2. Pierzynski, G.M., J.T. Sims, and G.F. Vance. 2005. *Soils and Environmental Quality.* 3rd Edition. CRC Press, Inc., Boca Raton, FL. 584 pp.
3. Prokop, G., P. Younger, and K. Roehl. 2003. Groundwater management in mining areas. *Proceedings of the 2nd IMAGE TRAIN Advanced Study Course.* Available at http://www.umweltbundesamt.at/fileadmin/site/publikationen/CP035.pdf.
4. U.S. Environmental Protection Agency. 2016. EPA's study of hydraulic fracturing for oil and gas and its potential impact on drinking water resources. Available at https://www.epa.gov/hfstudy.
5. U.S. Congress. 1977. Clean Water Act. Public Law 95-217. Available at https://www.epa.gov/laws-regulations/summary-clean-water-act.
6. U.S. Congress. 1996. Safe Drinking Water Act and Amendments. Public Law 104-182. Available at https://www.epa.gov/laws-regulations/summary-safe-drinking-water-act.
7. U.S. Environmental Protection Agency. 2002. Potential impacts of hydraulic fracturing of coalbed methane wells on underground sources of drinking water. Available at http://permanent.access.gpo.gov/lps21800/www.epa.gov/safewater/uic/cbmstudy.html.
8. U.S. Congress. 1977. Surface Mining Control and Reclamation Act. Public Law 95-87. Available at https://www.osmre.gov/lrg/docs/SMCRA.pdf.
9. National Groundwater Association (NGWA). 2019. Information on earth's water. Available at https://www.ngwa.org/what-is-groundwater/About-groundwater/information-on-earths-water.

10. National Academy of Sciences. 1990. Surface Coal Mining Effects on Ground Water Recharge. Committee on Ground Water Recharge in Surface-Mined Areas, Water Science and Technology Board, National Research Council. 170 pp. Available at http://www.nap.edu/catalog.php?record_id=1527.
11. Manahan, S.E. 2017. *Environmental Chemistry.* 10th Edition. Lewis Publishers, Chelsea, MI.
12. Nielsen, D.M. 2005. *Practical Handbook of Groundwater Monitoring.* 2nd Edition. Lewis Publishers, Boca Raton, FL.
13. Todd, D.K., D.E.O. McNulty. 1975. *Polluted Groundwater.* Water Information Center, Inc., Huntington, NY.
14. National Research Council. 2005. *Superfund and Mining Megasites: Lessons learned from the Coeur D' Alene River Basin.* The National Academies Press, Washington, DC. https://doi.org/10.17226/11359.
15. Rice, E.W., R.B. Baird, and A.D. Eaton (Eds). 2017. *Standard Methods for the Examination of Water and Wastewater.* 23rd Edition. American Public Health Association, Washington DC.
16. U.S. Environmental Protection Agency. 2009. Drinking Water Contaminants. Available at https://www.epa.gov/sdwa/drinking-water-regulations-and-contaminants
17. Interstate Technology & Regulatory Council. 2005. Overview of Groundwater Remediation Technologies for MTBE and TBA. Available at https://www.itrcweb.org/GuidanceDocuments/MTBE-1.pdf.
18. Hyman, M.H. 1999. Groundwater and Soil Remediation, pp. 684–712. In: Meyers, R.A., (ed.), *Encyclopedia of Environmental Pollution and Cleanup.* Volume 1. John Wiley & Sons, Inc., New York, NY.

4

Groundwater: Nitrogen Fertilizer Contamination

Introduction 45
Why N in Groundwater Is a Problem 46
    Human Health Impacts • Environmental Impacts
Agricultural Practices Contributing to Groundwater NO_3 47
    Row Crops • Grasslands/Turf • Containerized Horticultural Crops
Practices That Can Mitigate NO_3 Leaching from Agriculture 50
    Testing, Timing, Rates of Application, and Nitrification Inhibitors •
    Winter Cover Crops, Diversified Crop Rotations, and Reduced Tillage •
    Use of Alternate Grassland/Turf Management
Conclusions 52
References 53

Lloyd B. Owens and
Douglas L. Karlen

Introduction

Groundwater is widely used for domestic and public water supplies, particularly in rural areas.[1] When it resurfaces, groundwater also becomes surface water that ultimately flows through streams and lakes to the oceans. Generally, groundwater is quite pure unless it contains natural contaminants such as high iron, sulfur, or possibly an intrusion of saltwater if the aquifer is located close to an ocean or estuary. Such contaminants usually result from the geological formations through which the groundwater source is flowing or residing in. Another groundwater contaminant, nitrate-nitrogen (NO_3-N), can originate naturally but is most often associated with areas where human intervention and the management practices being used have significantly increased the amount of NO_3-N that is available for leaching as water moves through soil and the underlying geology into groundwater aquifers.

Why is NO_3-N in groundwater a problem? When water with high nitrate (NO_3) concentrations is consumed by humans, it can cause several adverse health problems. One of the most common is known as "blue baby syndrome" or methemoglobinemia, an illness that arises when an infant's blood is unable to carry enough oxygen to body cells and tissue. Consumption of high-NO_3 water has also been associated with increased levels of nitrosamine and some types of cancer. High levels of NO_3-N in streams, lakes, and oceans can stimulate excess growth of plants and bacteria, which upon death and decay subsequently deplete much of the oxygen in water. This causes fish kills and "dead zones," such as the hypoxic area in the Gulf of Mexico.

Elevated levels of NO_3-N in groundwater have been reported in many parts of the world.[2,3] Although this can occur naturally, leaching of N fertilizer is often a major factor responsible for high NO_3-N concentrations in groundwater. This occurs because nitrogen (N) is an essential plant nutrient, and to achieve optimum crop yields, the amount of N that becomes available through natural cycling must be supplemented with additional N from either inorganic (i.e., fertilizer) or organic (i.e., manure or legume) sources. When the total amount of N provided through natural and supplemental sources exceeds that removed by harvested portions of the crop, excess N can accumulate in the soil profile and become available for leaching into subsurface drainage lines or directly into groundwater aquifers. The challenge of providing adequate plant-available N without creating an excess that is available for leaching has been studied for decades in attempts to determine the economically optimum N level (EON). Such studies have been conducted not only for agronomic crops (e.g., maize, wheat, cotton, and rice) but also for turf grasses and containerized horticultural crops, which usually receive frequent and high N fertilizer rates. Despite these efforts to identify EON rates, there are some situations when there will still be a positive yield response to N fertilizer at rates exceeding the amount of N that the plants will utilize. When this occurs, excess NO_3-N accumulates in the soil profile and becomes available for leaching, thus resulting in NO_3-N concentrations that exceed the maximum contaminant levels (MCLs) for drinking water (i.e., 10 mg L^{-1}). Many approaches can be used to reduce the NO_3 leaching potential. This includes reducing overall N fertilizer inputs; applying N in split applications, coinciding with the time when plants can use the N most efficiently; using slow-release N fertilizers; and growing cover crops to capture part of the residual N accumulating in the soil profile before it can be leached to the groundwater. This entry focuses on groundwater and how it is impacted by NO_3-N leaching. Several factors contributing to leaching are addressed and supported by accompanying references, but space does not permit this to be a comprehensive literature review. Our primary emphasis is given to alternate management practices for row crops, grasslands/turf, and containerized horticultural crops that can be used to reduce potential N loss to groundwater resources.

Why N in Groundwater Is a Problem

Human Health Impacts

Groundwater is a major source of water for human consumption. When contaminated, the N is usually in the form of NO_3, which can pose major human health concerns at high levels, especially for infants. The link between high NO_3 in polluted water and serious blood changes in infants was first reported in 1945. From 1947 to 1950, 139 cases of methemoglobinemia were reported, including 14 deaths in Minnesota alone. In response to this documented threat to human health, a MCL standard has been set stating that NO_3 in excess of 45 mg L^{-1} (10 mg L^{-1} NO_3-N) is considered hazardous to human health.[4]

Even though the number of reported cases of methemoglobinemia has greatly decreased, recent studies indicate possible adverse impacts on human health at NO_3 levels below MCL.[5–7] Occurrences of bladder and ovarian cancer[8] and non-Hodgkin's lymphoma[9] have been linked to people with long-term exposure to public water supplies with NO_3-N concentrations of 2 to 4 mg L^{-1}.

Environmental Impacts

As part of the hydrologic cycle, a portion of what is classified as groundwater, especially shallow groundwater, resurfaces to feed streams, rivers, reservoirs, and eventually, estuaries and oceans. Nutrients and pollutants in the groundwater are thus transported in surface waters as well. A major challenge associated with groundwater contamination is that its flow is generally very slow. It can take years or decades for water to move through an aquifer from recharge areas to discharge areas and for land management changes/chemical applications to be reflected in "downstream" groundwater quality.[1,10,11] Green et al.[12] concluded that current fertilizer management practices in the United States will likely affect regional groundwater quality for decades to centuries.

Over the past 40 years, there has been an eightfold increase in the use of synthetic N fertilizers.[13] This has led to increased NO_3-N contamination in both groundwater and surface water bodies.[2,14,15] High NO_3-N levels can cause excess plant and bacterial growth in aquatic systems. Subsequently, the decay of this organic matter can deplete much of the oxygen in the water, causing fish kills and dead zones to appear. Phosphorus receives much of the attention with regard to eutrophication in fresh waters because it often is the limiting nutrient, but as water systems become more brackish, N often becomes the limiting factor.[16] Well-known examples of this situation include hypoxia in the Gulf of Mexico,[17] Danish coastal waters,[18] the western Indian continental shelf,[19] and the Changjiang Estuary in the East China Sea.[20] Hypoxia occurs when the concentration of dissolved oxygen is less than 2 mg L^{-1}. Nitrogen contributions to the Gulf of Mexico, and other large bodies of water, come from multiple sources, including surface runoff, subsurface drainage, and resurfacing of groundwater. There are several sources of N, e.g., fertilizer, animal manures, septic tanks, atmospheric deposition, land application of treated wastewater and biosolids, and mineralization of organic matter that can contribute to NO_3-N contamination. For example, in a 960 km^2 basin in Florida, it was estimated that fertilizer applied to cropland, lawns, and pine stands contributed 51% of the annual N load to groundwater in the basin.[21]

Agricultural Practices Contributing to Groundwater NO_3

Row Crops

High levels of NO_3-N in subsurface drainage from row crops, especially corn (*Zea mays* L.), are well documented. Nitrate-nitrogen concentrations in tile drainage from silt loam soils in Iowa that were fertilized for either continuous corn or corn grown in rotation with soybean [*Glycine max* (L.) Merr.] were reported to exceed 10 mg L^{-1} more than two decades ago.[22-24] Similar findings have been reported for clay loam soil in Minnesota,[25-27] silty clay loam in Illinois,[28] silt loam in Indiana,[29] silt loam and silty clay loam in Ohio,[30] and clay over silty clay loam and fine sand over clay in Ontario, Canada.[31] Analyses of subsurface water collected with monolith lysimeters[32,33] and ceramic porous-cup samplers[34,35] are in agreement with these findings.

Nitrate-nitrogen concentrations in tile drainage have been studied frequently because of the widespread use of artificial drainage throughout the U.S. Corn and Soybean Belt and the relative ease of collecting water samples. The majority of NO_3-N moves in the subsurface water during the late autumn, winter, and early spring recharge period[28,29,32] in this region. There are several factors influencing the amount of N exported via tile drainage, including timing, rate, source, and area receiving N fertilization.[28] Variable weather patterns, especially rainfall amounts, have a major influence on the amount of drainage that occurs. Increasing subsurface drain spacing can decrease NO_3-N losses,[25,29] although the NO_3-N concentration in the drainage water may change very little.[29] Furthermore, even though the increased spacing may reduce NO_3-N losses through the drainage water, it probably increases NO_3-N losses in seepage below the drains. Model simulation studies show that reducing N fertilization rates will have much greater impact for reducing NO_3-N losses than changing tile drain spacing or depth.[25]

There are many management factors contributing to the leaching loss of NO_3-N. Nitrogen is an essential plant nutrient that exhibits easily recognizable visual symptoms (e.g., yellow or light green plants, slow growth and development, decreased yield) when plant-available supplies are low. Also, until recently, N fertilizer was relatively inexpensive, and therefore, to reduce the risk of encountering a deficiency, it was often applied at rates in excess of crop need to ensure that inadequate N would not limit crop yield. Another factor is the difficulty in synchronizing N application with crop need because of the narrow window of time that producers may have if weather conditions are not optimum. As a result, it has been shown conclusively that there is a direct relationship between NO_3-N loss by leaching and N application rates that exceed crop needs.[34,36,37] Soil NO_3-N in excess of current plant needs can result from overapplication of N fertilizer or manure or from not accurately accounting for residual N from previous years, mineralization of organic N, or decomposition of legume crops. The latter is especially true following a dry year because if water limits crop growth, the plant will not utilize as much N as when a "normal" amount of water is available.[27] When this occurs, there is an increased amount of residual N to begin the next cropping season. Nitrogen management is therefore very difficult, and even when EON practices are used, it is not unusual to find NO_3-N concentrations in subsurface water that exceed the 10 mg L^{-1} MCL standard for drinking water.[34–36] This leads to the conclusion that optimum corn production will likely result in elevated NO_3-N concentrations in groundwater.[36] Stated in another manner, to achieve NO_3-N concentrations in groundwater that are less than the MCL, N fertilizer rates will need to be below the level associated with normal crop production recommendations.[37]

Nitrogen fertilizer management is also difficult within irrigated production systems, as evidenced by high NO_3-N concentrations found in subsurface water beneath sprinkler-irrigated crops in Spain[38] and Nebraska.[39] High NO_3-N concentrations also occurred in North Dakota[40] with sprinkler-irrigated corn and intermittent soybean and potato (*Solanum tuberosum* L.). The different crops did not directly affect NO_3-N concentrations. The most important factor leading to increased NO_3-N moving through the root zone and into groundwater was the amount of fall residual NO_3-N in the soil profile. Flood-irrigated wheat in Arizona[41] also produced high NO_3-N concentrations in groundwater, and even with best management practices (BMPs) for flood irrigation, NO_3-N concentrations in groundwater in excess of the MCL can be expected.

Animal manures are generally land applied for both disposal and their fertilizer nutrient value. This is especially true for organic production systems. High rates of manure application can result in high groundwater NO_3-N concentrations,[42,43] but even at low rates, manure management can be very difficult. For example, when liquid swine manure was applied at rates to meet the N requirements for either continuous corn or the corn phase of a corn/soybean rotation in Iowa, NO_3-N concentrations in tile flow from manure and urea ammonium NO_3 treatments consistently exceeded the MCL for drinking water. Furthermore, NO_3-N concentrations from the manure treatment exceeded those from the fertilizer treatments.[42] In another Iowa study, swine manure applications to both corn and soybean in a corn/soybean rotation were compared with application to only the corn phase of the rotation.[44] Average flow-weighted NO_3-N concentrations and leaching losses to subsurface drainage water were more than 50% greater when manure was applied to both corn and soybean. This response was credited to greater total N application when manure was applied every year compared with every other year. The studies also identified some of the challenges associated with using manure as a nutrient source. These include variability in manure composition and the importance of knowing the composition at the time of application.[43] Because of these factors, knowing the quantity of manure-applied N is much more difficult than for mineral fertilizer. Therefore, manure is often applied at a target rate to meet crop P requirements, with additional N subsequently being applied using another source.

Although leaching of NO_3-N dominates N loss to surface and groundwater, movement of dissolved organic N (DON) has also been recognized for more than 100 years.[45] Numerous studies have shown that DON leaching from forest ecosystems can be substantial, but DON leaching from agricultural soil has received little attention. Significant amounts of DON leaching were reported for agricultural soils in England,[46] and for cropped soils in Germany, DON leaching accounted for 6–21% of the total N flux.[47] In Ohio, DON accounted for 32–37% of the total N leaching from corn/soybean rotations,[48] with the primary source of DON in agricultural soils being crop residues and soil organic matter.[45] Just as for mineral N dynamics,[46] several factors, including leaching, mineralization, immobilization, and plant uptake, also affect DON leaching.

Grasslands/Turf

Because of the animal component, NO_3-N leaching in grazed grassland is very complex,[49,50] and even on highly fertilized pastures, much of the N loss has been attributed to excreta.[51,52] Considerable NO_3-N leaching can occur from the feces and urine in a management-intensive grazing system[53] and can result in a greater impact on water quality than moderate N fertilization.[54] A New Zealand study showed that when subsurface drainage occurred immediately following intensive grazing, drainage water had higher total N concentrations as a result of total organic N and NH_4-N from direct drainage or preferential flow of cattle urine.[55] However, those elevated concentrations were short lived and contributed a relatively small amount to the total N loss. Studies in England,[56,57] the Netherlands,[58] and the eastern United States[59,60] have shown that NO_3-N concentrations in subsurface water are often greater than the MCL when >100 kg N ha^{-1} was applied annually to grazed grasslands. In grazed bahiagrass (*Paspalum notatum* Fluegge) pastures in Florida that were fertilized with 76.5 kg N ha^{-1} (the recommended rate), NO_3-N leaching was not harmful to water quality.[61] Other processes, including accumulation of fertilizer N during drought or release of N from decaying plant material following tillage or chemically killing the sod in preparation for reseeding, may affect N leaching from pastures more than the loss from small areas affected by urine.[51] Once again, management is the key as demonstrated by a field experiment in the Netherlands showing that cattle slurry can replace some or most of the mineral N fertilizer for cut grassland on wet sandy soils without increasing NO_3-N leaching if the slurry is applied during the growing season and at rates that do not exceed crop uptake.[62] This was not the case on dry sandy soil, perhaps because denitrification in the wet soil minimized the amount of N available for leaching. Nevertheless, long-term leaching risk may increase as soil organic N increases unless mineral N applications are decreased accordingly.[61] In some nongrazed systems, NO_3-N leaching from highly fertilized systems has been reported to be very low, e.g., 29 kg N ha^{-1} lost from ryegrass (*Latium perenne* L.) receiving 420 kg N ha^{-1}.[63]

Fertilized turf, whether for home lawns or golf courses, is another N source with potential environmental issues. Annual applications up to 244 kg N ha^{-1} to turfgrass on sandy loam soils in Rhode Island did not appear to pose a threat to drinking water aquifers,[64] but overwatering did increase N loadings to bays and estuaries in coastal areas. A literature review by the Horsley Witten Group[65] concluded that a 20% loading rate to groundwater was an adequate estimate for modeling N leaching in the Cape Cod, Massachusetts, area. The excess N movement was most prevalent with late-summer N applications. Several studies have shown that late-fall applications of N fertilizer to turf result in higher N leaching rates during the late fall and winter.[66–69] Some late-season (October through December) N fertilization recommendations are for the purpose of maintaining/improving turf color, but increased N leaching from such applications indicated they should be managed to achieve acceptable water quality and not to maximize turf color.[67,70] Slow-release formulations of N fertilizer have also shown great reductions in the rate of N leaching.[66,69] The differences were even greater during a high-precipitation year than during a low-or normal-precipitation year,[69] especially with high autumn precipitation.[68] Irrigation management is critical, as evidenced by studies showing that increasing irrigation from 70% to 140% replacement of daily pan evaporation increased N leaching by almost 400%[71] and that flushing porous golf greens with high rates of irrigation can leach elevated rates of N.[72] Grass species is an important factor influencing N leaching, with less occurring beneath grasses with greater aboveground biomass and deeper root systems, e.g., bentgrass species (*Agrostis* spp.). High-quality grass could also reduce fertilizer N leaching under sand-based putting greens.[73]

Turfgrass is most vulnerable for NO_3-N leaching during establishment. Even though this period represents a small period of the average turf's life,[65] the soil disturbance, limited root biomass and N uptake, and tendency of turf managers to overfertilize during this period cause the NO_3-N leaching losses to exceed the MCL.[74,75] A mature turf may have actual N requirements below recommended levels. Research in Michigan on a 10-year-old Kentucky bluegrass (*Poa pratensis* L.) compared annual applications of 245 kg N ha^{-1} (49 kg N ha^{-1} per application) and 98 kg N ha^{-1} (24.5 kg N ha^{-1} application).[76] With resultant NO_3-N leachate concentrations often greater than 20 mg L^{-1} and below 5 mg L^{-1}, respectively, the conclusion was that applying high rates of water-soluble N to mature turfgrass should be avoided to minimize the potential for NO_3-N leaching. Returning clippings to the turfgrass ecosystem reduced the N fertilizer requirement by 25–60%, with the reduction increasing as the time after establishment increased from 10 to more than 50 years.[77]

Grass sod has the capacity to use large amounts of N, with 85–90% of fertilizer N being retained in the turf-soil ecosystem.[78] Roots and thatch can represent a large N pool that becomes available for mineralization and subsequent leaching if disturbed.[79] Reseeding and sod establishment within 2 mo of "turfdeath" can stabilize this N pool.[78] High rates of NO_3-N leaching can occur at very high N fertilizer rates, e.g., 450 kg N ha^{-1} per year. Although most NO$_3$-N leaching occurred during autumn and winter, it was the accumulation of all N fertilizer applications and not just the autumn application[68] that determined the actual loss. Excess soil NO$_3$ in the fall is the driving force that causes leaching, regardless of N source or time of application. Therefore, high rates of N application to turf should be avoided in the fall because it can result in high NO$_3$-N leaching rates. A significant portion of the N leached can be in the organic form.[71] A survey of several golf courses across the United States indicated that NO$_3$-N concentrations above the MCL occurred in only 4% of the samples,[80] with most of these being attributed to prior agricultural land use. Pollution of groundwater by NO3-N leaching from N-fertilized turf should be minimal with good management, which includes consideration of soil texture, N source, rate and timing, and irrigation/rainfall.[81]

Containerized Horticultural Crops

Although the acreage for containerized horticultural crops is small compared with row crops or grasslands, the production intensity is great, and "hot spots" of potential NO$_3$-N leaching can develop. Assuming 80,000 pots ha^{-1} for a typical foliage plant nursery and using a soluble granular fertilizer, over 650 kg N ha^{-1} could be lost annually through leaching.[82] During a 10-week greenhouse study of potted flowers, average NO$_3$-N in the leachate ranged from 250 mg NL^{-1} to 450 mg NL^{-1}.[83] Irrigation and fertilization are critical management components of this intensive industry. Trickle irrigation has been shown to move less water and leach less N than overhead irrigation,[84] but precipitation could nullify this difference.[85] Increasing the irrigation rate, e.g., from 1 to 2 cm day^{-1}, increased the amount of water lost and N leached[86] even though N concentrations were decreased. The use of controlled-release fertilizers (CRFs) is one practice that can significantly reduce leaching losses.[82,87] Even with the use of CRF, NO$_3$-N concentrations can be high in leachate[88]; if the use of CRF is combined with large irrigation volumes, NO$_3$-N can move into the soil profile beneath containers.[89] Water management is crucial, as illustrated by one experiment that compared a low leaching fraction (0.0 to 0.2 of the irrigation water) with a high leaching fraction (0.4 to 0.6). The lower leaching fraction reduced irrigation volume, effluent volume, and NO$_3$-N in the effluent by 44%, 63%, and 66%, respectively.[90] However, these gains in efficiency resulted in a 10% loss in total plant growth (shoots and roots). Thus, establishing an acceptable balance among the level of plant growth, water and nutrient use efficiencies, and the potential environmental impacts becomes a management decision. Although beyond the scope of this entry, vegetable crops also have a high N demand but low apparent N recovery, as illustrated by sweet peppers,[91] which can leave large amounts of N in the soil and residues at harvest.

Practices That Can Mitigate NO$_3$ Leaching from Agriculture

Nitrate contamination of groundwater is not caused by any single factor, because non-agricultural and agricultural practices contribute to the problem. Fertilizer management decisions (i.e., rates, formulations, timing, etc.) are an agricultural contributing factor, but so are tillage, crop selection, soil organic matter levels, and drainage.[92] Temperature and precipitation patterns also combine with these factors. Likewise, various strategies and approaches are needed to reduce NO$_3$-N loss to groundwater. This includes using appropriate N fertilizer rates, proper timing, soil testing and plant monitoring, nitrification inhibitors, cover crops, diversified crop rotations, and reduced tillage[92] as well as various combinations of these and other management practices.

Testing, Timing, Rates of Application, and Nitrification Inhibitors

Preplant N tests or pre-side-dress N tests[93] can assess soil N levels from cover crops and help provide adequate N credits for legume or fertilizer carryover from prior crops. For example, in Iowa, use of the late-spring NO$_3$ test reduced fertilizer N application[94] and resulted in up to a 30% decrease in NO$_3$-N in discharge waters compared with using traditional fall application. Testing plant tissue for N concentration can determine in-season crop N status and be used to guide supplemental N fertilizer applications. Calibrated chlorophyll meter readings have been correlated to plant N status[95–97] and used to increase N fertilizer efficiency and reduce leaching.[96] Splitting N applications also has the potential to reduce the total application rate and to apply the fertilizer N when it is needed by the crop. However, one disadvantage of applying N fertilizer after initial crop growth is that weather conditions may create a very small window of opportunity for side-dress N applications.

Nitrification inhibitors used with ammoniacal N sources can slow the rate of oxidation to NO_3-N and thus decrease the amount of NO_3-N available for leaching,[98] especially with fall N applications. In a Minnesota study, NO_3-N losses in subsurface drainage from a corn/soybean rotation were reduced 14% by spring N application compared with late-fall anhydrous ammonia application and 10% by late-fall N application using the nitrification inhibitor nitrapyrin.[99] The use of nitrapyrin with spring-applied N showed no further reduction of N losses. However, even with these improved practices, it may be necessary to reduce the N fertilization rate to below the EON level to achieve NO_3-N concentrations in groundwater that are below the MCL[37] for drinking water.

Winter Cover Crops, Diversified Crop Rotations, and Reduced Tillage

Winter cover crops have been shown to be an effective strategy for reducing NO_3-N leaching.[100–103] A variety of crops including annual grasses, cereals, and legumes have been used as cover crops with varying degrees of success—often depending upon the specific soil and climatic pattern of cropping sequence being used. Winter rye (*Secale cereale* L.) is the most common cover crop used to reduce NO_3-N leaching following corn or corn-soybean rotations in the United States.[100,101,104–106] Winter cover crops have also significantly reduced NO_3-N leaching for broccoli (*Brassica oleracea* L.) crops[107] and potato (*S. tuberosum* L.)–based rotations[108] in the United States. For potato, spring wheat (*Triticum aestivum* L.), sugar beet (*Beta vulgaris* L.), and oat (*Avena sativa* L.) rotation in the Netherlands, adding cover crops helped to decrease NO_3- N concentrations in leachate to near or below the European Union standard (11.3 mg L^{-1}).[109] Care needs to be exercised with long-term cover crops because if they are disturbed, some of the accumulated N may become mineralized and actually increase NO_3-N leaching.[110] Even in short-term situations, the efficacy of cover crops to reduce NO_3-N leaching is relatively low when considering the entire crop succession, and N saved by the cover crop generally does not increase N utilization by the next crop.[111] Deep-rooted cover crops may help capture N leached to deeper soil layers,[112] and rapid establishment of a deeprooting system is one factor influencing the efficacy of cover crops for reducing NO_3-N leaching.[113] Although sometimes cover crops are considered to be a BMP for reducing NO_3-N leaching,[114] this is not always true. For example, on the Delmarva Peninsula in the mid-Atlantic United States, a rye winter cover crop following corn did not reduce NO_3-N leaching, presumably because the existing corn crop did not allow the rye to be seeded early enough in the autumn. Similarly, owing to the restricted time for cover crop growth in a Wisconsin study, winter rye did not utilize significant amounts of fertilizer N from the previous crop residues or soil.[115] However, in temperate regions with mild winters that favor long growing seasons and winter N mineralization, cover crops have been demonstrated as a valuable tool to reduce NO_3-N leaching.[116]

Diversifying cropping systems to include perennial crops is another strategy that will probably reduce NO_3-N leaching.[117] Compared with annual crops, perennials have an extended period of growth and therefore greater N utilization. However, having an adequate market for satisfactory economic return on such crops can be problematic. Fortunately, the emerging biofuel industry provides a potential market that may make such cropping systems more profitable.

Reduced tillage is a BMP for reducing soil erosion, but it is not usually an option for N management. Tillage promotes N mineralization from soil organic matter, which decreases soil quality and makes more N available for oxidation and subsequent leaching. Although water movement through the soil profile tends to be greater with no-till than conventional or limited tillage, NO_3-N concentrations are usually lower with no-till. Therefore, the amount of NO_3-N leaching can be either similar or less with no-till.[118] Overall, differences are small, and it is often concluded that tillage has less impact on NO_3-N leaching than factors such as crop rotation.[92,117,119]

Use of Alternate Grassland/Turf Management

Demonstrated management options to reduce NO3-N leaching from grasslands include using grass–legume mixtures instead of highly fertilized grass[52,120]; ensuring total N input does not exceed 100 kg ha^{-1}[59,60]; coordinating fertilizer application rate and timing with other N sources (e.g., manure applications); avoiding excessive N application rates[121]; using irrigation, especially during dry periods, to encourage N uptake[52]; and integrating forage cutting and grazing for optimum management throughout the year. Concentrations of NO_3-N in groundwater associated with high-fertility, high-stocking-density grazing systems can also be reduced by continuous grazing or haying if external N inputs are reduced or eliminated.[122]

In areas where NO_3-N contamination from turf is a concern, late-summer N fertilizer applications should be reduced, and the amount of irrigation water should be limited.[64] Other management practices for reducing N leaching from turf include using slow-release fertilizers,[66,69] avoiding excess irrigation, and using grass species with greater aboveground biomass and deeper root systems.[73] Because the vulnerability for leaching during turf establishment, irrigation and fertilizer rates should be limited until the turf root system is well developed. Throughout the midwestern United States, excess soil NO_3-N in the fall is the driving force responsible for NO_3-N leaching. Therefore, using any or all of the management practices mentioned above should help avoid this buildup and thus help reduce NO_3-N leaching.

Conclusions

High NO_3-N concentrations in drinking water can pose human health problems, and when found in surface water bodies, they can create many environmental problems. Several factors have contributed to high groundwater NO_3-N concentrations, including the increased use of relatively inexpensive N fertilizer to minimize the yield risk associated with encountering nutrient deficiencies. The unintended consequence is that when N is applied at rates exceeding levels to which crops respond, the potential for excess N to accumulate in the soil and leach is increased. Thus, EON rates may often produce NO_3-N levels in groundwater that exceed MCLs. This has been measured with grain crops, grasslands, and turf, which receive high rates of fertilizer N, and nursery container crops.

There are several management practices that can help reduce excess N application. These include splitting applications of N so that it is most available when plants need it; using preplant, pre-side-dress, and plant tissue N tests to determine appropriate rates of application; applying slow-release fertilizer sources so that the N is available as plants need it instead of all at once; improving timeliness of irrigation to help plants take up N during drought periods and to avoid excess irrigation, which will reduce leaching; avoiding late-season fertilizer applications, especially common for turf; and using cover crops to capture excess N that remains in the soil profile after growth of the crop so that NO_3-N leaching is reduced.

Research challenges to improve N management and reduce NO_3-N leaching are multidisciplinary. Although *precision agriculture* is a term generally applied to using soil tests, yield results, and Geographic Information system (GIS) mapping to match fertilizer rates with yield responses in a field, it also has the potential to reduce NO_3-N leaching by encouraging the use of variable N rates instead of a constant rate for an entire field. Traditionally, the sole focus for agriculture has been on crop yield, but to simultaneously address groundwater quality, effects of climate change, and other emerging factors, interactions with other crops must also be included. Collectively, these practices can improve overall N use efficiency and have both economic and environmental benefits.

References

1. Debrewer, L.M.; Ator, S.W.; Denver, J.M. Temporal trends in nitrate and selected pesticides in mid-Atlantic ground water. J. Environ. Qual. **2008**, *37* (5), 296–308.

2. Rupert, M.G. Decadal-scale changes of nitrate in ground water of the United States, 1988–2004. J. Environ. Qual. **2008**, *37* (5), 240–248.

3. Spalding, R.F.; Exner, M.E. Occurrence of nitrate in groundwater—A review. J. Environ. Qual. **1993**, *22* (3), 392–402.

4. U.S. Public Health Service. *Public Health Drinking Water Standards*, U.S. Public Health Pub I. 956; U.S. Government Printing Office: Washington, DC, 1996.

5. Brender, J.D.; Olive, J.M.; Felkner, J.M.; Suarez, L.; Hendricks, K.; Marckwardt, W. Intake of nitrates and nitrites and birth defects in offspring. Epidemiology **2004**, *15* (4), 184.

6. Brender, J.D.; Olive, J.M.; Felkner, J.M.; Suarez, L.; Marckwardt, W.; Hendricks, K.A. Dietary nitrites and nitrates, nitrosatable drugs, and neural tube defects. Epidemiology **2004**, *15* (3), 330–336.

7. DeRoos, A.J.; Ward, M.H.; Lynch, C.F.; Cantor, K.P. Nitrate in public water systems and the risk of colon and rectum cancers. Epidemiology **2003**, *14* (6), 640–649.

8. Weyer, P.J.; Cerhan, J.R.; Kross, B.C.; Hallberg, G.R.; Kantamneni, J.; Breuer, G.; Jones, M.P.; Zheng, W.; Lynch, F. Municipal drinking water nitrate level and cancer risk in older women: The Iowa Women's Health Study. Epidemiology **2001**, *12* (3), 327–338.

9. Ward, M.H.; Mark, S.D.; Cantor, K.P.; Weisenburger, D.D. Drinking water nitrate and the risk of non-Hodgkin's lymphoma. Epidemiology **1996**, *7* (5), 465–471.

10. Bohlke, J.K. Groundwater recharge and agricultural contamination. Hydrogeol. J. **2002**, *10* (1), 153–179.

11. Puckett, L.J.; Hughes, W.B. Transport and fate of nitrate and pesticides: Hydrogeology and riparian zone processes. J. Environ. Qual. **2005**, *34* (6), 2278–2292.

12. Green, C.T.; Puckett, L.J.; Bohlke, J.K.; Bekins, B.A.; Phillips, S.P.; Kauffman, L.J.; Denver, J.M.; Johnson, H.M. Limited occurrence of denitrification in four shallow aquifers in agricultural areas of the United States. J. Environ. Qual. **2008**, 37 (3), 994–1009.

13. Mulvaney, R.L.; Khan, S.A.; Ellsworth, T.R. Synthetic nitrogen fertilizers deplete soil nitrogen: A global dilemma for sustainable cereal production. J. Environ. Qual. **2009**, *38* (6), 2295–2314.

14. Goolsby, D.A.; Battaglin, W.A. Long-term changes in concentrations and flux of nitrogen in the Mississippi River Basin, USA. Hydrol. Processes **2001**, *15* (7), 1209–1226.

15. Zhu, J.G.; Liu, G.; Han, Y.; Zhang, Y.L.; Xing, G.X. Nitrate distribution and denitrification in the saturated zone of paddy field under rice/wheat rotation. Chemosphere **2003**, *50* (6), 725–732.

16. Correll, D.L. The role of phosphorus in the eutrophication of receiving waters: A review. J. Environ. Qual. **1998**, *27* (2), 261–266.

17. Burkart, M.R.; James, D.E. Agricultural-nitrogen contributions to hypoxia in the Gulf of Mexico. J. Environ. Qual. **1999**, *28* (3), 850–859.

18. Conley, D.J.; Carstensen, J.; Aertebjerg, G.; Christensen, P.B.; Dalsgaard, T.; Hansen, J.L.S.; Josepson, A.B. Longterm changes and impacts of hypoxia in Danish coastal waters. Ecol. Appl. *17* (5), S165–S184.

19. Naqvi, S.W.A.; Jayakumar, D.A.; Narvekar, P.V.; Naik, H.; Sarma, V.V.S.S.; D'Souza, W.D.; Joseph, S.; George, M.D. Increased marine production of N_2O due to intensifying anoxia on the Indian continental shelf. Nature *408* (6810), 346–349.

20. Wei, H.; He, Y.; Li, Q.; Liu, Z.; Wang, H. Summer hypoxia adjacent to the Changjiang Estuary. J. Mar. Syst. *67* (3–4), 252–303.

21. Katz, B.G.; Sepulveda, A.A.; Verdi, R.J. Estimating nitrogen loading to ground water and assessing vulnerability to nitrate contamination in a large karstic springs basin, Florida. J. Am. Water Res. Assoc. **2009**, *45* (3), 607–627.

22. Baker, J.L.; Campbell, H.P.; Johnson, H.P.; Hanway, J.J. Nitrate, phosphorus, and sulfate in subsurface drainage water. J. Environ. Qual. **1975**, *4* (3), 406–412.
23. Baker, J.L.; Johnson, R.P. Nitrate-nitrogen in tile drainage as affected by fertilization. J. Environ. Qual. **1981**, *10* (4), 519–522.
24. Kanwar, R.S.; Cruse, R.M.; Ghaffarzadeh, M.; Bakhsh, A.; Karlen, D.L; Bailey, T.B. Corn-soybean and alternative cropping systems effects on NO_3-N leaching losses in subsurface drainage water. Appl. Eng. Agric. **2005**, *21* (2), 181–188.
25. Davis, D.M.; Gowda, P.R.; Mulla, D.J.; Randall, G.W. Modeling nitrate nitrogen leaching in response to nitrogen fertilizer rate and tile drain depth or spacing for southern Minnesota, USA. J. Environ. Qual. **2000**, *29* (5), 1568–1581.
26. Randall, G.W.; Huggins, D.R.; Russelle, M.P.; Fuchs, D.J.; Nelson, W.W.; Anderson, J.L. Nitrate losses through subsurface tile drainage in Conservation Reserve Program, alfalfa, and row crop systems. J. Environ. Qual. **1997**, *26* (5), 1240–1247.
27. Randall, G.W.; Iragavarapu, T.K. Impact of long-term tillage systems for continuous corn on nitrate leaching to tile drainage. J. Environ. Qual. **1995**, *24* (2), 360–366.
28. Gentry, L.E.; David, M.B.; Smith, K.M.; Kovacic, D.A. Nitrogen cycling and tile drainage nitrate loss in corn/soybean watershed. Agric. Ecosyst. Environ. **1998**, *68* (1–2), 85–97.
29. Kladivko, E.J.; Grochulska, J.; Turco, R.F.; Van Scoyoc, G.E.; Eigel, J.D. Pesticide and nitrate transport into subsurface tile drains of different spacings. J. Environ. Qual. **1999**, *28* (3), 997–1004.
30. Logan, T.J.; Schwab, G.O. Nutrient and sediment characteristics of tile effluent in Ohio. J. Soil Water Conserv. **1976**, *31* (1), 24–27.
31. Miller, M.H. Contribution of nitrogen and phosphorus to subsurface drainage water from intensively cropped mineral and organic soils in Ontario. J. Environ. Qual. **1979**, *8* (1), 42–48.
32. Cookson, W.R.; Rowatth, J.S.; Cameron, K.E. The effect of autumn applied^{15}N-labelled fertilizer on nitrate leaching in a cultivated soil during winter. Nutr. Cycling Agroecosyst. **2000**, *56* (2), 99–107.
33. Owens, L.B.; Malone, R.W.; Shipitalo, M.J.; Edwards, W.M.; Bonta, J.V. Lysimeter study on nitrate leaching from a corn-soybean rotation. J. Environ. Qual. **2000**, *29* (2), 467–474.
34. Andraski, T.W.; Bundy, L.G.; Brye, K.R. Crop management and corn nitrogen rate effects on nitrate leaching. J. Environ. Qual. **2000**, *29* (4), 1095–1103.
35. Steinheimer, T.R.; Scoggin, K.D.; Kramer, L.A. Agricultural chemical movement through a field-size watershed in Iowa: Subsurface hydrology and distribution of nitrate in groundwater. Environ. Sci. Technol. **1998**, *32* (8), 1039–1047.
36. Jemison, J.M., Jr.; Fox, R.H. Nitrate leaching from nitrogen-fertilized and manured corn measured with zero-tension pan lysimeters. J. Environ. Qual. **1994**, *23* (2), 337–343.
37. Lawlor, P.A.; Helmers, M.J.; Baker, J.L.; Melvin, S.W.; Lemke, D.W. Nitrogen application rate effect on nitrateitrogen concentration and loss in subsurface drainage for a corn-soybean rotation. Trans. ASABE **2008**, *51* (1), 83–94.
38. Diez, J.A.; Caballero, R.; Roman, R.; Tarquis, A.; Cartegena, M.E.; Vallejo, A. Integrated fertilizer and irrigation management to reduce nitrate leaching in central Spain. J. Environ. Qual. **2000**, *29* (5), 1539–1547.
39. Klocke, N.L.; Watts, D.G.; Schneekloth, J.P.; Davison, D.R.; Todd, R.W.; Parkhurst, A.M. Nitrate leaching in irrigated corn and soybean in a semi-arid climate. Trans. ASAE **1999**, *42* (2), 1621–1630.
40. Derby, N.E.; Casey, F.X.M.; Knighton, R.E. Long-term observations of vadose zone and groundwater nitrate concentrations under irrigated agriculture. Vadose Zone J. **2009**, *8* (2), 290–300.
41. Ottman, M.J.; Tickes, B.R.; Husman, S.H. Nitrogen-15 and bromide tracers of nitrogen fertilizer movement in irrigated wheat production. J. Environ. Qual. **2000**, 29 (5), 1500–1508.
42. Bakhsh, A.; Kanwar, R.S.; Karlen, D.L. Effects of liquid swine manure applications on NO_3-N leaching losses to subsurface drainage water from loamy soils in Iowa. Agric. Ecosyst. Environ. **2005**, *109* (1–2), 118–128.

43. Karlen, D.L.; Cambardella, C.A.; Kanwar, R.S. Challenges of managing liquid swine manure. Appl. Eng. Agric. **2004**, *20* (5), 693–699.

44. Bakhsh, A.; Kanwar, R.S.; Baker, J.L.; Sawyer, J.; Malarino, A. Annual swine manure applications to soybean under corn-soybean rotation. Trans. Am. Soc. Agric. Biol. Eng. **2009**, *52* (3), 751–757.

45. Van Kessel, C.; Clough, T.; van Groenigen, J.W. Dissolved organic nitrogen: An overlooked pathway of nitrogen loss from agricultural systems? J. Environ. Qual. **2009**, *38* (2), 393–401.

46. Murphy, D.V.; Macdonald, A.J.; Stockdale, E.A.; Goulding, K.W.T.; Fortune, S.; Gaunt, J.L.; Poulton, P.R.; Wakefield, J.A.; Webster, C.P.; Wilmer, W.S. Soluble organic nitrogen in agricultural soils. Biol. Fertil. Soils **2000**, *30* (5–6), 373–387.

47. Siemens, J.; Kaupenjohann, M. Contribution of dissolved organic nitrogen to N leaching from four German agricultural soils. J. Plant Nutr. Soil Sci. **2002**, *165* (6), 675–681.

48. Shuster, W.D.; Shipitalo, M.J.; Subler, S.; Aref, S.; McCoy, E.L. Earthworm additions affect leachate production and nitrogen losses in typical Midwestern agroecosystems. J. Environ. Qual. **2003**, 32 (2), 2132–2139.

49. Ball, P.R.; Ryden, J.C. Nitrogen relationships in intensively managed temperate grasslands. Plant Soil **1984**, *76* (1–3), 23–33.

50. Ryden, J.E.; Ball, P.R.; Garwood, E.A. Nitrate leaching from grassland. Nature **1984**, *311* (5981), 50–53.

51. Cuttle, S.P.; Scurlock, R.V.; Davies, B.M.S. A 6-year comparison of nitrate leaching from grass/clover and N fertilized grass pastures grazed by sheep. J. Agric. Sci., Cambridge **1998**, *131* (1), 39–50.

52. Whitehead, D.E. Leaching of nitrogen from soils. In *Grassland Nitrogen;* CAB International: Wallingford, 1995; 129–151.

53. Stout, W.L.; Fales, S.A.; Muller, L.D.; Schnabel, R.R.; Priddy, W.E.; Elwinger, G.F. Nitrate leaching from cattle urine and feces in northeast USA. Soil Sci. Soc. Am. J. **1997**, *61* (6), 1787–1794.

54. Decau, M.L.; Simon, J.C.; Jacquet, A. Nitrate leaching under grassland as affected by mineral nitrogen fertilization and cattle urine. J. Environ. Qual. **2004**, *33* (2), 637–644.

55. Houlbrooke, D.; Hanly, J.; Horne, D.; Hedley, M. Nitrogen losses in artificial drainage and surface runoff from pasture following grazing by dairy cattle. In *SuperSoil 2004*, Proceedings of the 3rd Australian New Zealand Soil Conference, Sydney, Australia, Dec 5–9, 2004; The Regional Institute Ltd., http://www.regional.org.au/au/asssi/, 8 pp.

56. Haigh, R.A.; White, R.E. Nitrate leaching from a small, underdrained, grassland, clay catchment. Soil Use Manage. **1986**, *2* (2), 65–70.

57. Roberts, G. Nitrogen inputs and outputs in a small agricultural catchment in the eastern part of the United Kingdom. Soil Use Manage. **1987**, *3* (4), 148–154.

58. Steenvoorden, J.H.A.M.; Fonck, H.; Oosterom, H.P. Losses of nitrogen from intensive grassland systems by leaching and surface runoff. In *Nitrogen Fluxes in Intensive Grassland Systems;* van der Meer, H.G., Ryden, J.E., Ennik, G.E., Eds.; Martinus Nijhoff Publ.: Dordrecht, The Netherlands, 1986; 85–97.

59. Owens, L.B.; Van Keuren, R.W.; Edwards, W.M. Nitrogen loss from a high-fertility, rotational pasture program. J. Environ. Qual. **1983**, *12* (3), 346–350.

60. Owens, L.B.; Edwards, W.M.; Van Keuren, R.W. Nitrate levels in shallow groundwater under pastures receiving ammonium nitrate or slow-release nitrogen fertilizer. J. Environ. Qual. **1992**, *21* (4), 607–613.

61. Sigua, G.C.; Hubbard, R.K.; Coleman, S.W.; Williams, M. Nitrogen in soils, plants, surface water and shallow groundwater in a bahiagrass pasture of Southern Florida, USA. Nutr. Cycling Agroecosyst. **2010**, *86* (2), 175–187.

62. Schroder, J.J.; Assinck, F.B.T.; Uenk, D.; Velthof, G.L. Nitrate leaching from cut grassland as affected by the substitution of slurry with nitrogen mineral fertilizer on two soil types. Grass Forage Sci. **2010**, *65* (1), 49–57.

63. Garwood, E.A.; Ryden, J.E. Nitrate loss through leaching and surface runoff from grassland: Effects of water supply, soil type and management. In *Nitrogen Fluxes in Intensive Grassland Systems;* van der Meer, H.G., Ryden, J.C., Ennik, G.E., Eds.; Martinus Nijhoff Publ.: Dordrecht, Netherlands, **1986**; 99–113.

64. Morton, T.G.; Gold, A.J.; Sullivan, W.M. Influence of overwatering and fertilization on nitrogen losses from home lawns. J. Environ. Qual. **1988**, *17* (1), 124–130.

65. Horsley Witten Group. *Evaluation of Turfgrass Nitrogen Fertilizer Leaching Rates in Soils on Cape Cod, Massachusetts*; MA Department of Environmental Protection: Hyannis, MA, 2009; 33 pp.

66. Guillard, K.; Kopp, K.L. Nitrogen fertilizer form and associated nitrate leaching from cool-season lawn turf. J. Environ. Qual. **2004**, *33* (5), 1822–1827.

67. Mangiafico, S.S.; Guillard, K. Fall fertilization timing effects on nitrate leaching and turfgrass color and growth. J. Environ. Qual. **2006**, *35* (1), 163–171.

68. Roy, J.W.; Parkin, G.W.; Wagner-Riddle, E. Timing of nitrate leaching from turfgrass after multiple fertilizer applications. Water Qual. Res. J. Canada **2000**, *35* (4), 735–752.

69. Petrovic, A.M. Nitrogen source and timing impact on nitrate leaching from turf. In *I International Conference on Turfgrass Management and Science for Sports Fields*, Athens, Greece, June 2–7, 2003. ISHS Acta Hortic. *661*, 427–432.

70. Mangiafico, S.S.; Guillard, K. Nitrate leaching from Kentucky bluegrass soil columns predicted with anion exchange membranes. Soil Sci. Soc. Am. J. **2007**, *71* (1), 219–224.

71. Barton, L.; Wan, G.G.Y.; Colmer, T.D. Turfgrass (*Cynodon dactylon* L.) sod production on sandy soils: II. Effects of irrigation and fertilizer regimes on N leaching. Plant Soil **2006**, *284* (1–2), 147–164.

72. Shuman, L.M. Normal and flush irrigation effects on nitrogen leaching from simulated golf greens in the greenhouse. Commun. Soil Sci. Plant Anal. **2006**, *37* (3–4), 605–619.

73. Pare, K.; Chantigny, M.H.; Carey, K.; Johnston, W.J.; Dionne, J. Nitrogen uptake and leaching under annual bluegrass ecotypes and bentgrass species: A lysimeter experiment. Crop Sci. **2006**, *46* (2), 847–853.

74. Geron, E.A.; Danneberger, T.K.; Traina, S.J.; Logan, T.J.; Street, J.R. The effects of establishment methods and fertilization practices on nitrate leaching from turfgrass. J. Environ. Qual. **1993**, *22* (1), 119–125.

75. Easton, A.M.; Petrovic, A.M. Fertilizer source effect on ground and surface water quality in drainage from turfgrass. J. Environ. Qual. **2004**, *33* (2), 645–655.

76. Frank, K.W.; O'Reilly, K.M.; Crum, J.R.; Calhoun, R.N. The fate of nitrogen applied to a mature Kentucky bluegrass turf. Crop Sci. **2006**, *46* (1), 209–215.

77. Qian, Y.L.; Bandaranayake, W.; Parton, W.J.; Mecham, B.; Harivandi, M.A.; Mosier, A.R. Long-term effects of clipping and nitrogen management in turfgrass on soil organic carbon and nitrogen dynamics: The CENTURY model simulation. J. Environ. Qual. **2003**, *32* (5), 1694–1700.

78. Bushoven, J.T.; Jiang, Z.; Ford, H.J.; Sawyer, E.D.; Hull, R.J.; Amador, J.A. Stabilization of soil nitrate by reseeding with ryegrass following sudden turf death. J. Environ. Qual. **2000**, *29* (5), 1657–1661.

79. Jiang, Z.; Bushoven, J.T.; Ford, H.J.; Sawyer, E.D.; Amador, J.A.; Hull, R.J. Mobility of soil nitrogen and microbial responses following the sudden death of established turf. J. Environ. Qual. **2000**, *29* (5), 1625–1631.

80. Cohen, S.; Svrjcek, A.; Durborow, T.; Barnes, N.L. Water quality impacts by golf courses. J. Environ. Qual. **1999**, *28* (3), 798–809.

81. Petrovic, A.M. The fate of nitrogenous fertilizers applied to turfgrass. J. Environ. Qual. **1990**, *19* (1), 1–14.

82. Broschat, T.K. Nitrate, phosphorus, and potassium leaching from container-grown plants fertilized by several methods. HortScience **1995**, *30* (1), 74–77.

83. McAvoy, RJ. Nitrate nitrogen movement through the soil profile beneath a containerized greenhouse crop irrigated with two leaching fractions and two wetting agent levels. J. Am. Soc. Hort. Sci. **1994**, *119* (3), 446–451.

84. Rathier, T.M.; Frink, C.R. Nitrate in runoff water from container grown juniper and Alberta spruce under different irrigation and N fertilization regimes. J. Environ. Hort. **1989**, *7* (1), 32–35.

85. Colangelo, D.J.; Brand, M.H. Nitrate leaching beneath a containerized nursery crop receiving trickle or overhead irrigation. J. Environ. Qual. **2001**, *30* (5), 1564–1574.

86. Million, J.; Yeager, T.; Albano, J. Consequences of excessive overhead irrigation on runoff during container production of sweet viburnum. J. Environ. Hort. **2007**, *25* (3), 117–125.

87. Million, J.; Yeager, T.; Albano, J. Effects of container spacing practice and fertilizer placement on runoff from overhead-irrigated sweet viburnum. J. Environ. Hort. **2007**, *25* (2), 61–72.

88. Newman, J.; Albano, J.; Merhaut, D.; Blythe, E. Nutrient release from controlled-release fertilizers in neutral-pH substrate in an outdoor environment: I. Leachate electrical conductivity, pH, and nitrogen, phosphorous, and potassium concentrations. HortScience **2006**, *41* (7), 1674–1682.

89. Colangelo, D.J.; Brand, M.K. Effect of split fertilizer application and irrigation volume on nitrate-nitrogen concentration in container growing area soil. J. Environ. Hort. **1997**, *15* (4), 205–210.

90. Tyler, H.H.; Warren, S.L.; Bilderback, T.E. Reduced leaching fractions improve irrigation use efficiency and nutrient efficacy. J. Environ. Hort. **1996**, *14* (4), 199–204.

91. Tei, F.; Benincasa, P.; Guiducci, M. Nitrogen fertilisation of lettuce, processing tomato and sweet pepper: Yield, nitrogen uptake and the risk of nitrate leaching. Welles- bourne, Warwick, United Kingdom. December 1999. Acta Hort. **1999**, *506,* 61–67.

92. Dinnes, D.L.; Karlen, D.L.; Jaynes, D.B.; Kaspar, T.C.; Hatfield, J.L.; Colvin, T.S.; Cambardella, C.A. Nitrogen management strategies to reduce nitrate leaching in tile-drained Midwestern soils. Agron. J. **2002**, *94* (1), 153–171.

93. Magdorff, F.R.; Ross, D.; Amadon, J. A soil test for nitrogen availability to corn. Soil Sci. Soc. Am. J. **1984**, *48* (6), 1301–1304.

94. Jaynes, D.B.; Dinnes, D.L.; Meek, D.W.; Karlen, D.L.; Cambardella, C.A.; Colvin, T.S. Using the Late Spring Nitrate Test to reduce nitrate loss within a watershed. J. Environ. Qual. **2004**, *33* (2), 669–677.

95. Schepers, J.S.; Francis, D.D.; Vigil, M.; Below, F.E. Comparison of corn leaf nitrogen concentration and chlorophyll meter readings. Commun. Soil Sci. Plant Anal. **1992**, *23* (17–20), 2173–2187.

96. Varvel, G.E.; Schepers, J.S.; Francis, D.D. Ability for in-season correction of nitrogen deficiency in corn using chlorophyll meters. Soil Sci. Soc. Am. J. **1997**, *61* (4), 1233–1239.

97. Hawkins, J.A.; Sawyer, J.E.; Barker, D.W.; Lundvall, J.P. Using relative chlorophyll meter values to determine nitrogen application rates for corn. Agron. J. **2007**, *99* (4), 1034–1040.

98. Owens, L.B. Nitrate leaching losses from monolith lysimeters as influenced by nitrapyrin. J. Environ. Qual. **1987**, *16* (1), 34–38.

99. Randall, G.W.; Vetsch, J.A. Nitrate losses in subsurface drainage from a corn-soybean rotation as affected by fall and spring application of nitrogen and nitrapyrin. J. Environ. Qual. **2005**, *34* (2), 590–597.

100. McCracken, D.V.; Smith, M.S.; Grove, J.H.; MacKown, E.T.; Blevins, R.L. Nitrate leaching as influenced by cover cropping and nitrogen source. Soil Sci. Soc. Am. J. **1994**, *58* (5), 1476–1483.

101. Rasse, D.P.; Ritchie, J.T.; Peterson, W.R.; Wei, J.; Smucker, A.J. Rye cover crop and nitrogen fertilization effects on nitrate leaching in inbred maize fields. J. Environ. Qual. **2000**, *29* (1), 298–304.

102. Zhou, X.; MacKenzie, A.F.; Madramootoo, C.A.; Kaluli, J.W.; Smith, D.L. Management practices to conserve soil nitrate in maize production systems. J. Environ. Qual. **1997**, *26* (5), 1369–1374.

103. Ball-Coelho, B.R.; Roy, R.C. Overseeding rye into corn reduces NO_3 leaching and increases yield. Can. J. Soil Sci. **1997**, *77,* 443–451.

104. Kaspar, T.C.; Jaynes, D.B.; Parkin, T.B.; Moorman, T.B. Rye cover crop and gamagrass strip effects on NO_3 concentration and load in tile drainage. J. Environ. Qual. **2007**, *36* (5), 1503–1511.

105. Li, L.; Malone, R.W.; Ma, L.; Kaspar, T.C.; Jaynes, D.B.; Saseendran, S.A.; Thorp, K.R.; Yu, Q.; Ahuja, L.R. Winter cover crop effects on nitrate leaching in subsurface drainage as simulated by RZWQM-DSSAT. Trans. ASABE **2008**, *51* (5), 1575–1583.

106. Strock, J.S.; Porter, P.M.; Russelle, M.P. Cover cropping to reduce nitrate loss through subsurface drainage in the northern U.S. corn belt. J. Environ. Qual. **2004**, *33* (3), 1010–1016.

107. Wyland, L.J.; Jackson, L.E.; Chaney, W.E.; Klonsky, K.; Koike, S.T.; Kimple, B. Winter cover crops in a vegetable cropping system on nitrate leaching, soil water, crop yield, pests and management costs. Agric. Ecosyst. Environ. **1996**, *59* (1–2), 1–17.

108. Weinert, T.L.; Pan, W.L.; Moneymaker, M.R.; Santo, G.S.; Stevens, R.G. Nitrogen recycling by nonleguminous winter cover crops to reduce leaching in potato rotations. Agron. J. **2002**, *94* (2), 365–372.

109. Vos, J.; van der Putten, P.E.L. Nutrient cycling in a cropping system with potato, spring wheat, sugar beet, oats and nitrogen catch crops. II. Effect of catch crops on nitrate leaching in autumn and winter. Nutr. Cycling Agroecosyst. **2004**, *70* (1), 23–31.

110. Hansen, E.M.; Djurhuus, J.; Kristensen, K. Nitrate leaching as affected by introduction or discontinuation of cover crop use. J. Environ. Qual. **2000**, *29* (4), 1110–1116.

111. Herrera, J.M.; Liedgens, M. Leaching and utilization of nitrogen during a spring wheat catch crop succession. J. Environ. Qual. **2009**, *38* (4), 1410–1419.

112. Kristensen, H.L.; Thorup-Kristensen, K. Root growth and nitrate uptake of three different catch crops in deep soil layers. Soil Sci. Soc. Am. J. **2004**, *68* (2), 529–537.

113. Herrera, J.M.; Feil, B.; Stamp, P.; Liedgens, M. Root growth and nitrate-nitrogen leaching of catch crops following spring wheat. J. Environ. Qual. **2010**, *39* (3), 845–854.

114. Ritter, W.F.; Scarborough, R.W.; Christie, A.E.M. Winter cover crops as a best management practice for reducing nitrogen leaching. J. Contam. Hydrol. **1998**, *34* (1), 1–15.

115. Bundy, L.G.; Andraski, T.W. Recovery of fertilizer nitrogen in crop residues and cover crops on an irrigated sandy soil. Soil Sci. Soc. Am. J. **2005**, *69* (3), 640–648.

116. Hooker, K.V.; Coxon, C.E.; Hackett, R.; Kirwan, L.E.; O'Keeffe, E.; Richards, K.G. Evaluation of cover crop and reduced cultivation for reducing nitrate leaching in Ireland. J. Environ. Qual. **2008**, *37* (1), 138–145.

117. Randall, G.W.; Mulla, D.J. Nitrate nitrogen in surface waters as influenced by climatic conditions and agricultural practices. J. Environ. Qual. **2001**, *30* (2), 337–344.

118. Kanwar, R.S.; Baker, J.L.; Baker, D.G. Tillage and split N-fertilization effects on subsurface drainage water quality and crop yields. Trans. ASAE **1988**, *31* (2), 453–461.

119. Zhu, Y.; Fox, R.H.; Toth, J.D. Tillage effects on nitrate leaching measured by pan and wick lysimeters. Soil Sci. Soc. Am. J. **2003**, *67* (5), 1517–1523.

120. Owens, L.B.; Edwards, W.M.; Van Keuren, R.W. Groundwater nitrate levels under fertilized grass and grass-legume pastures. J. Environ. Qual. **1994**, *23* (4), 752–758.

121. Jarvis, S.C. Progress in studies of nitrate leaching from grassland soils. Soil Use Manage. **2000**, *16* (s1), 152–156.

122. Owens, L.B.; Bonta, J.V. Reduction of nitrate leaching with haying or grazing and omission of nitrogen fertilizer. J. Environ. Qual. **2004**, *33* (4), 1230–1237.

5

Groundwater: Pesticide Contamination

Introduction 59
Pesticide Use 59
Associated Pesticide Behavior in Soils and Water 60
    Insecticides • Fungicides and Fumigants • Herbicides • Groundwater
    Contamination
Management of Point Sources of Groundwater Contamination 62
Management of Non-Point Sources of Groundwater Contamination 62
    Irrigation Management
Future Research 63
References 63

Roy F. Spalding

Introduction

Trace concentrations of most of the commonly used pesticides have been confirmed in groundwaters of the United States. Since groundwater is the source of 53% of the potable water, the more toxic pesticides and their transformation products are a concern from the standpoint of human health. Others are a risk to the environment in areas where contaminated groundwater enters surface water. Through toxicological testing, the USEPA has established Maximum Contaminant Levels (MCLs) or lifetime Health Advisory Levels (HALs) for several pesticides (Table 1).

The EPA also has a separate list of unregulated compounds, including newly registered pesticides and their transformation products, such as acetochlor and alachlor ESA, that are presently being evaluated or being considered for toxicological evaluation. Based on the results of the EPAs National Pesticide Assessment,[2] 10.4% of 94,600 community systems contained detectable concentrations of at least one pesticide. Evaluation of these results led to an estimated 0.6% of rural domestic wells containing one or more pesticides above the MCL.

Pesticide Use

In the United States about 80% of pesticide usage is in agriculture. The remainder is used by industry, homeowners, and gardeners. About 500 million pounds of herbicide, 180 million pounds of insecticide, and 70 million pounds of fungicide were applied for agricultural purposes in 1993.[3] Several maps of the United States delineate usage patterns of several pesticides.[4] The majority of the triazine and amide herbicides are applied to fields in the north central corn belt states of Michigan, Wisconsin, Minnesota, Nebraska, Iowa, Illinois, Indiana, and Ohio. Commonly used organophosphorus insecticides are more heavily applied to fields in California and along the southeastern seaboard than in the northern corn belt. Carbamate and thiocarbamate pesticides are heavily used in potato growing areas of northern

TABLE 1   U.S. Maximum Contaminant Levels for Drinking Water

Organic Chemical Name	MCL (mg/L)	Organic Chemical Name	MCL (mg/L)	Organic Chemical Name	MCL (mg/L)
2,4,5-TP (Silvex)	0.05	Chlordane	0.002	Heptachlor	0.0004
2,4-D	0.07	Dalapon	0.2	Heptachlor Epoxide	0.0002
Alachlor	0.002	Dinoseb	0.007	Lindane	0.0002
Aldicarb	0.007	Diquat	0.02	Methoxychlor	0.04
Aldicarb sulfone	0.007	Endothall	0.1	Oxamyl (Vydate)	0.2
Aldicarb sulfoxide	0.004	Endrin	0.002	Picloram	0.5
Atrazine	0.003	Ethylene dibromide	0.00005	Simazine	0.004
Carbofuran	0.04	Glyphosate	0.7	Toxaphene	0.003
Carbon tetrachloride	0.005				

Source: U.S. Environmental Protection Agency.[1]

Maine, Idaho, the Delmarva Peninsula, and vegetable fields of California and the southeastern coastal states. Fungicide use is concentrated in high humidity and irrigated areas of the coastal states and to some extent along the Great Lakes and Mississippi River Valley. The fumigants carbon tetrachloride and ethylene dibromide (EDB) were used heavily in the past at grain storage elevators throughout the Midwest and elsewhere in the United States.

Associated Pesticide Behavior in Soils and Water

Although pesticide use is a dominant factor in groundwater contamination, leaching variability among pesticides exhibiting similar behaviors is striking and explains why several heavily used pesticides seldom if ever are detected in groundwater. In general, pesticides within a class have similar chemical characteristics upon which soil leaching predictions can be made based on persistence, solubility, and mobility. Pesticide class relationships with soils and water transport described in the following text are detailed in Weber.[5] Individual frequencies of groundwater pesticide detection, in parenthesis next to commonly used products, are calculated from the Pesticide Groundwater Data Base (PGWDB)[4] and the National Water Quality Assessment (NAWQA) database.[6] High frequencies of detection identify those pesticides with a disposition to leach.

Insecticides

Chlorinated hydrocarbons are one of the oldest chemical classes of insecticides. Some of the best-known compounds include aldrin, dieldrin, DDE, DDT, endrin, and toxaphene. Although banned since the 1960s, their extremely persistent nature precludes their detection in very trace quantities in groundwater of the upper Midwest. On the other hand, heavily used organophosphates like malathion, methylparathion, disulfoton, and others have been extensively surveyed during several groundwater monitoring studies and have not been detected. The organophosphate insecticides, parathion (not reported (NR), <1), % occurrence from PGWDB, % occurrence from NAQWA data. terbufos (<1, <1), fonofos (<1, <1), and chlorpyrifos (<1, <1), which are heavily used on corn and sorghum, were also seldom detected. Diazinon (1.1, 1.3), the common garden insecticide, is occasionally detected in groundwater. Generally, the organic phosphates are rapid degraders and are strongly retained on soils.

For the most part, carbamates and thiocarbamates are very sparingly soluble and exhibit low to moderate soil retention; however, a small number have high solubility and low soil retention. Most carbamates are characterized as having short longevities. Generally, pesticides in this group having half-lives of 30 days or more have the potential to leach. The thiocarbamates butylate (<1, <1) and

EPTC (2.6, <1) are extensively used in agriculture and have relatively short half-lives. Aldicarb (<1, <1) and carbofuran (14.7, <1) are at the high end for solubility and longevity in their class. Their metabolites have been frequently detected beneath high use crops, such as potatoes in the potato growing regions of the United States.

The pyrethroid insecticides have low solubilities, short half-lives, and high soil retentions that make them unlikely to leach. Yet, permethrin (<1, <1) is occasionally detected in very trace quantities in groundwater.

Fungicides and Fumigants

Fungicides are non-volatile organometallic compounds with low aqueous solubility that inhibit growth of actinomycetes and many fungi. The best-known fungicides zineb (not detected (ND), NR) and captan are zinc-based, and maneb (ND, NR) is manganese-based. Some, like bordeaux, are copper sulfate-based. Although their detection frequency is very low, fungicides have not been analyzed in many surveys.

Fumigants are very volatile halogenated compounds that generally are knifed below the soil surface. These compounds have high aqueous solubility and very low soil retention. The fumigants EDB and 1,3-dichloropropene have been frequently detected in the subsurface and in groundwater in high-use regions, such as California.[7] Ethylene dibromide and carbon tetrachloride were also used in grain storage facilities during the 1950s and 1960s. Spills, leaks, and improper handling resulted in 400 reported groundwater contamination sites in Kansas and Nebraska.

Herbicides

There are at least eight major chemical classes of herbicides. These include: quaternary N, basic, acidic, carboxylic acid, hydroxy and aminosulfonyl, amide and anilide, dinitroaniline, and phenylurea herbicides.[5] Several herbicide classes have similar behaviors with respect to soil and water.

Both quaternary N and dinitroaniline herbicides are very highly retained by soils and are not expected to be detected in groundwater. However, paraquat, pendimethalin, and trifluralin have been reported several times in groundwater. Their presence indicates that transport is dependent on factors not directly related to compound longevity, solubility, and mobility. Vertical transport by preferential flow through macropores is a commonly accepted mechanism used to explain these detections. In some instances, compounds have been described as preferentially transported attached to colloidal material.

Carboxylic, hydroxy, and aminosulfonyl acids, and thiocarbamate herbicides have very low to low soil retention and very short to moderate longevity. Thus, the more heavily used and persistent pesticides in these groups are the ones most generally detected in groundwater. They include the acids, dicamba (2.0, NR), picloram (2.5, <1), bromacil (1.8, 1.0), and dinoseb (1.4, <1).

Phenylurea herbicides have low to high soil retentions and short to moderate longevity. Linuron (16.7, <1) and diuron (<1, 1.9) are the most frequently detected in groundwater and both have moderately long half-lives ranging from 60 to 90 days.

Amide and anilide herbicides have low soil retention and short to moderate longevity. Several amide herbicides and their transformation products have been detected in groundwater. The commonly used amides in the Midwestern corn belt, namely alachlor (1.7, 2.7), metolachlor (<1, 12), propachlor (1.2, <1), and acetochlor (NR, <1), are the most frequent offenders because they are relatively persistent.

As suggested by the name, basic herbicides behave as bases. The group contains several subclasses including aniline, formamidine, imidazole, pyrimidine, thiadiazole, triazines, and triazole. Basic herbicides have low to high soil retention and very short to moderate longevity. Again, it is generally the most persistent and heavily used pesticides that are more frequently found in groundwater. The most frequently detected compounds in the group are the triazines, namely atrazine (5.6, 30), metribuzin (4.2, 1.9), cyanazine (2.0, 1.4), simazine (2.0, 14.8), and prometon (2.1, 11.6).

Groundwater Contamination

It stands to reason that there are generally good associations between pesticide use and their detection in groundwater. Since groundwater flows very slowly at rates normally ranging from 0.1 ft/day to 3 ft/day, pesticide sources are generally very near the monitored well. Thus, high frequencies of triazine and acetamide detections are reported in the states of the northern corn belt. More fungicides and fumigants were detected in warm humid states of California and Florida where vegetable and fruit crops dominate the landscape. In an analysis of the 20 NAWQAs for pesticides, frequencies of pesticide detection in groundwater were significantly related to the estimated amount of agricultural use within a 1km radius of the sampled site.[6] They also emphasized that pesticides were detected beneath both agricultural (60.4%) and urban areas (48.5%). Discontinued used pesticides have been detected numerous times in shallow aquifers.

In general, families of pesticides have similar chemical characteristics from which predictions have been made as to the product's potential for contamination of groundwater; however, differences in the leaching behavior of pesticides exhibiting similar chemistry can be appreciable and is the reason several heavily used pesticides are seldom, if ever, detected in groundwater.

Management of Point Sources of Groundwater Contamination

Important steps are being taken to reduce water quality pollution by pesticides occurring from spills and back siphoning events (point sources). Since it is easier to resolve point than non-point sources, laws have been enacted to eliminate contamination of surface water bodies, which may be in hydraulic contact with groundwater, from used pesticide containers and rinseate from chemical wash downs. Check valves are mandatory when pesticides are mixed and/or diluted and prevent backflow to groundwater. Soils at and adjacent to agrichemical supply facilities have been surveyed in several states and found to be highly contaminated with pesticide residues. The herbicides, atrazine, alachlor, metolachlor, cyanazine, and metribuzin are the worst offenders from the standpoint of pesticide mass in the soils at sites in Wisconsin and Illinois.[4] Many of these sites and those in other states are now involved in soil cleanups, which are designed to protect underlying groundwater from further pollution.

Management of Non-Point Sources of Groundwater Contamination

Normal farm chemical applications of pesticides are generally considered potential non-point sources of groundwater contamination because they are dispersed over large areas ranging from fields to watersheds. Management strategies are in place to reduce leaching of field applied chemicals.[8] These strategies vary from regulatory restrictions to outright bans on application in areas deemed more vulnerable to leaching. Integrated pest management, fostered by the office of pesticide management at the USEPA, is designed to reduce chemical applications. The practice of banding applications has reduced amounts applied. Both target more efficient pesticides and genetically engineered plants sensitive only to specific herbicidal action have been and are being developed. These new pesticides and pesticide-plant combinations require less chemical than in the past, and the altered plants allow for pest control with more environmentally sensitive chemicals. The USEPA has announced a plan to reduce the mass of applied chemicals from commonly used triazines and amides that are frequently detected in groundwater.

Irrigation Management

Irrigation practices can influence pesticide leaching. Atrazine was vertically transported deeper and faster when using flood rather than sprinkler irrigation.[19] Sprinkler systems allow for much more uniform and efficient water management practices than furrow irrigation, and recent studies have shown that they reduce chemical leaching.[10] In the Nebraska's Platte Valley[11] and in the Walnut Creek watershed in Iowa,[12] peak herbicide concentrations were strongly related to rapid flushing beneath drainage areas where surface water ponds during heavy rainfall events on the cropped fields. Application of excess irrigation water also was reported to increase herbicide leaching.[9,11]

Future Research

More research is necessary to evaluate the health risks of transformation products from heavily used pesticides that are frequently detected in groundwater. Research needs to focus on precision application of pesticides to specific field problem areas as a potential mechanism to reduce chemical application.

There is a need to evaluate the environmental cost/benefit of safer product replacements used in conjunction with genetically altered crops. As new products are registered to replace more persistent and mobile pesticides, long-term fate studies, including the monitoring of the transformation product impact, on groundwater quality are necessary.

References

1. U.S. Environmental Protection Agency. *Drinking Water Standards and Health Advisories,* EPA 822-B-00-001; USEPA Office of Water: Washington, DC, Summer, 2000.
2. U.S. Environmental Protection Agency. *National Survey of Pesticides in Drinking Water Wells: Phase I Report,* EPA 570/9-90-015; USEPA Office of Pesticide and Toxic Substance: Washington, DC, November, 1990.
3. Aspelin, A.L. *Pesticide Industry Sales and Usage, 1992 and 1993 Estimates;* Economic Analysis Branch Report 733-K-92-001; USEPA Office of Pesticide Programs, Biological and Economic Analysis Division, 1994; 1–37.
4. Barbash, J.E.; Resek, E.A. *Pesticides in Ground Water. Distribution, Trends, and Governing Factors*; Ann Arbor Press, Inc.: Chelsea, MI, 1996; 1–588.
5. Weber, J. Properties and behavior of pesticides in soils. In *Mechanisms of Pesticide Movement into Ground Water,* Honeycutt, R.C., Schabacker, D.J., Eds.; Lewis Publishers: London, 1994; 15–41.
6. Kolpin, D.; Barbash, J.E.; Gilliom, R.J. Pesticides in ground water of the United States, 1992–1996. Ground Water **2000**, *38* (6), 858–863.
7. Troiano, J.; Weaver, D.; Marade, J.; Spurlock, F.; Pepple, M.; Nordmark, C.; Bartkowiak, D. Summary of well water sampling in California to detect pesticide residues resulting from nonpoint-source applications. J. Environ. Qual. **2001**, *30* (2), 448–459.
8. Guyot, C. Strategies to minimize the pollution of water by pesticides. In *Pesticides in Ground and Surface Water,* Börner, H., Ed.; Springer: Berlin, 1994; 87–148.
9. Troiano, J.; Garretson, C.; Krauter, C.; Brownwell, J.; Huston, J. Influence of amount and method of irrigation water application on leaching of atrazine. J. Environ. Qual. **1993**, *22*, 290–298.
10. Spalding, R.F.; Watts, D.G.; Schepers, J.S.; Burbach, M.E.; Exner, M.E.; Poreda, R.J.; Martin, G.E. Controlling nitrate leaching in irrigated agriculture. J. Environ. Qual. **2001**, *30* (4), 1184–1194.
11. Spalding, R.F.; Watts, D.G.; Snow, D.D.; Cassada, D.A.; Exner, M.E.; Schepers, J.S. Herbicide loading to shallow ground water beneath Nebraska's management systems evaluation area. J. Environ. Qual. **2003**, *32*, 84–91.
12. Moorman, T.B.; Jaynes, D.B.; Cambardella, C.A.; Hatfield, J.L.; Pfeiffer, R.L.; Morrow, A.J. Water quality in walnut creek watershed: herbicides in soils, subsurface drainage and groundwater. J. Environ. Qual. **1999**, *28* (1), 35–45.

6

Lakes and Reservoirs: Pollution

Introduction 65
Classification of Lakes and Reservoirs..66
    Freshwater/Saline Lakes • Trophic Status
Problems Associated with Lakes and Reservoirs 68
    Eutrophication • Toxic Materials • Sedimentation • Acidification •
    Fish Depletion • Stratification
Sources of Lakes and Reservoir Pollution 72
Water Quality Monitoring..72
    Types of Monitoring
Protective and Restorative Measures 74
    Eutrophication Model Framework
Conventions for the Protection of Lakes and Reservoirs............................77
    Lakes and Wetlands—Ramsar Convention, Iran, February 2, 1971 • UNECE
    Water Convention, Helsinki, March 17, 1992 • Protocol on Water and Health,
    London, June 17, 1999
Conclusion 77
References..78

Subhankar
Karmakar and
O.M. Musthafa

Introduction

Surface water is one of the most important natural resources in the world. It has been explicitly established that water of good quality is a fundamental element to sustainable socioeconomic development. It is the habitat for a large number of species and is a crucial component for metabolic activities of plants and animals. Aquatic ecosystems are endangered on a worldwide scale by a multitude of pollutants as well as damaging land-use or water-management practices. Some problems have been present for a long time but have only recently reached a critical level, while the rest are recently emerging. Oxygen balance in the aquatic systems is severely affected by organic pollution, which often results in severe pathogenic contamination. Enrichment of aquatic systems with nutrients from various origins, predominantly domestic sewage, agricultural runoff, and agro-industrial effluents, results in enhanced eutrophication, of which lakes and reservoirs are affected the most.

Lakes and reservoirs are the major resources of fresh surface water. They are larger and deeper than ponds and are not part of the ocean. Lakes and reservoirs are major resources as these hold about 90% of the world's fresh surface water and are the key freshwater resources for agriculture, fisheries, domestic, industrial, recreational, landscape entertainment, and energy production. Natural lakes are bodies of water, created by volcanic, tectonic, or glacial activity, whereas reservoirs are artificial impoundments. A lake is a relatively large lentic freshwater or saltwater body, which is localized in a basin surrounded by land. Natural lakes are generally found in mountainous areas, rift zones, and areas with ongoing glaciations. In some parts of the world, there are many lakes formed due to chaotic drainage patterns left over from the last ice age. They are the habitats of a variety of flora and fauna, making them a source of fish and a destination for migratory birds to reproduce or rest. A reservoir, which is known as an artificial lake, is constructed for the benefit of man's water needs, sometimes for one particular purpose, but more recently for multiple purposes. Reservoirs are different from lakes in many ways. They have usually larger drainage basins than lakes, and many are located in watersheds with extensive agricultural activities.

Direct contamination of surface waters with metals in discharges from mining, smelting, and industrial manufacturing is a long-lasting phenomenon. The emission of airborne pollutants has now reached such proportions that long-range atmospheric transport causes contamination, not only in the vicinity of industrial regions but also in more remote areas. Similarly, precipitation of acid rain occurs, when moisture in the atmosphere combines with gases such as sulfur dioxide, which are produced when fossil fuels are burnt. This may cause significant acidification of surface waters, especially lakes and reservoirs. Contamination of water by synthetic organic micropollutants and emerging contaminants results either from direct discharge into the surface waters through runoff or after transport through the atmosphere.

This entry briefly presents classification of lakes and reservoirs based on the flow of water in and out of the system and their utility, respectively, followed by the problems associated with lakes and reservoirs causing deterioration of water quality. A discussion on monitoring of water quality and various protective and restorative measures is also made through a literature survey, which may be useful to future research aspirants and water resource professionals in identifying economically and environmentally sustainable lake and reservoir management strategy.

Classification of Lakes and Reservoirs

All lakes are temporary over geologic time scales, as they will slowly fill in with sediments or spill out of the basin.[1] Water enters into lakes from a variety of sources such as seepage through groundwater storage, runoff from watershed, direct precipitation into the lake, and other surface waters bodies (like streams or rivers). Water may drain out from lakes through deep percolation to join groundwater table or through surface water flow and evaporation. Natural lakes can be classified into four major types based on how water enters and exits the lake. Water may enter into the lake through one source or multiple sources. The water quality of a lake and its biodiversity are significantly influenced by the type of lake. Depending upon the way of entrance and exit of water, lakes can be classified into four categories as shown in Figure 1.

Seepage lakes do not have a distinct inlet or an outlet, and occasionally overflow. The major sources of water are direct precipitation, surface runoff from the immediate drainage area, and seepage through groundwater storage, as seepage lakes are landlocked water bodies. Since seepage lakes are sensitive to groundwater levels and local rainfall patterns, water levels may fluctuate seasonally. These lakes may have a less diverse fishery as the direct source of water is not a flowing water body or stream. Seepage lakes also have a smaller drainage area, which may help to account for lower nutrient levels. *Spring lakes* have no distinct inlet, but do have an outlet. The major source of water for spring lakes is groundwater flowing into the bottom of the lake from inside and outside the immediate surface drainage area. *Groundwater drained lakes* have no inlet, but similar to spring lakes, these may have an uninterruptedly flowing outlet. Drained lakes are not groundwater fed and their principal sources of water are precipitation and direct drainage from the surrounding land. The water levels in drained lakes fluctuate frequently depending on the supply of water. Under severe conditions, the outlets from drained lakes may become intermittent. *Drainage lakes* have both an inlet and an outlet where the main water source is stream drainage. These lakes support fish populations that are not necessarily identical to the streams connected to them. Drainage lakes mostly have higher nutrient levels than natural seepage or spring lakes.[2] Depending on the utility, reservoirs can be classified into four classes as shown in Figure 2.

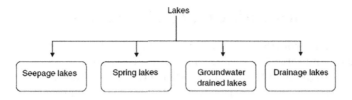

FIGURE 1 Classification of lakes.

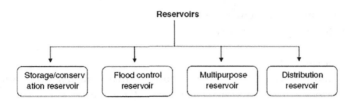

FIGURE 2 Classification of reservoirs.

Storage/conservation reservoirs retain excess water supplies during peak flows and release water gradually during low flows as and when needed. A *flood control reservoir* stores a portion of the flood flows so that it can minimize the flood peaks in the areas to be protected downstream. A *multipurpose reservoir* is meant for serving multiple purposes such as water supply, flood and soil erosion control, hydroelectric power generation, recreation, irrigation, etc. A *distribution reservoir* is connected with a network of primary water supply and is used to supply water to the end users according to fluctuations in demand over a short time period and serves as local storage in the case of emergency. Such reservoirs, thus, support the water treatment plants to work at a uniform rate and can store water when there is less demand and, thus, supply water during high-demand periods. Water quality in both lakes and reservoirs is influenced by many factors such as water body type, ecosystem characteristics, land use and land cover, and human activities.[3]

Freshwater/Saline Lakes

Most lakes hold freshwater, but some, particularly those where water cannot discharge via a river, can be salty. As a matter of fact, some lakes such as the Great Salt Lakes are saltier than the oceans. Lakes whose salinity content is more than 3 g/L are considered as saline lakes. They are prevalent and present on all regions, including Antarctica(e.g., Caspian Sea, Dead Sea, etc.). Though the inland saline water constitutes around 45% of the total inland water, a few deep lakes (mainly the Caspian Sea) occupy a significant volume of these saline waters. Salinity has a great influence on the freezing point of water, amount of dissolved oxygen, etc.[4]

Trophic Status

Lakes can be classified based on their trophic state as "eutrophic," "mesotrophic," and "oligotrophic." The word "trophic" means nutrition or growth. A eutrophic lake is characterized by the presence of a high concentration of plant nutrients and associated excess plant growth. On the other hand, an oligotrophic lake is characterized by low nutrient concentrations and low plant growth. Mesotrophic lakes fall in between these two. The major factors that regulate the trophic status of a lake are as follows: 1) rate of nutrient supply; 2) climatic condition (sunlight, temperature, precipitation, lake basin turnover time, etc.), and 3) morphometry/shape of lake basin (mean and maximum depth, volume and surface area, watershed-to-lake surface area ratio, etc.).[5]

Problems Associated with Lakes and Reservoirs

Although more than three-fourths of the earth is occupied with water, only less than 0.3% (including surface water and groundwater) is available for human consumption.[6] The total area of the lakes on earth amounts to approximately 2.5×10^6 km^2 or 1.8% of the continental area, containing 1.2×10^5 km^3 of water. The 253 largest lakes (larger than 500 km^2) of the planet contain an estimated 78% of the world's unfrozen fresh surface water and thus represent an essential global life support system.[7] Though the amount of water in lakes and reservoirs is very small, because of its rapid renewal, these habitats are the primary sustainable supplier of freshwater for most regions. Since freshwater on the earth is scarce, sincere efforts are required for the conservation of existing water resources to ensure the availability of sufficiently good quality water. The surface water resources are more susceptible to pollution as compared to the groundwater resources due to the ease of access of pollutants and contaminants in the former.[8,9]

Most of the lakes and reservoirs around the world face environmental stress, and the appropriate functioning of various vital ecosystems is in danger. Due to the explosion in human population, supplementary demand-related damage is forced on lakes and reservoirs. Water levels become lower as a result of higher consumption by households and industries; a growing number of human population results in shrinking and altering their water resources; inappropriate use of the land, particularly hills and mountains, results in increased sedimentation on their basins. Ultimately, pollution from agricultural lands and from domestic and industrial sources may produce eutrophication, resulting in undesirable effects such as the presence of toxic algae, reduction of oxygen, and generation of foul odor. The proliferation of contaminants within lake and reservoir systems can deteriorate water quality significantly. In many regions, lake ecosystems have already degraded and restoration to desirable water quality needs enormous effort and high cost. Following the present status, none can predict for how long these resources can serve as renewable sources of pure surface water for domestic, industrial, and agricultural uses, or as sources of protein-rich food.[10] Lakes and reservoirs have a more vulnerable and complex ecosystem than rivers as they do not have a self purifying ability and, hence, readily accumulate pollutants. Because of their importance, their beauty, their religious and cultural significance, and their relative susceptibility to degradation, lakes and reservoirs require more concerted attention than is paid generally to river systems.[11] The next few paragraphs elaborate the major pollution problems of lakes and reservoirs.

Eutrophication

It can be defined as the process of enrichment of waters with plant nutrients that causes raw water quality loads, such as high primary production, low oxygen concentrations, and increased concentrations of hydrogen sulfide, carbon dioxide, dissolved iron, and manganese in the hypolimnion.[12] It is considered as one of the serious negative effects faced by the lentic water bodies such as lakes, ponds, and reservoirs and is one of the major water quality problems worldwide[13] and the most serious challenge for water management professionals in densely populated areas.[14] Water from eutrophic reservoirs may have meager taste, odor, and color, and some have high concentrations of naturally occurring organic compounds that may form trihalomethanes, which are carcinogenic and mutagenic and other by-products of disinfection.[15,16]

Seasonal pattern is prominently influenced by the availability of solar insolation and nutrients. Usually, the shallower the lake, the less the check of internal nutrient recycling by thermal stratification and light availability by critical depth as compared to mixing depth. During summer wind events, the lakes that are having an Osgood Index (OI)[17] less than 7 will show a strong tendency for mixing, as a result of which, nutrients from sediments/hypolimnion undergo recycling and enter into the photic zone. Such events entrain nutrients and cause summer blooms.[18] Thus, in highly dynamic lakes, the pattern of phytoplankton abundance/species composition differs with wind events.

Toxic Materials

It includes mainly heavy metals, toxin-producing microphytes, and pesticides from agricultural land. Toxic substances may enter water bodies directly as land runoff from urban streets and mining areas or as agricultural runoff including forestry drainage, discharge of inadequately treated sewage, and industrial effluents, and through deposition of airborne pollutants. Toxic substances can also be created in drinking water when chemicals from treatment plants interact with organic molecules in the raw water to form carcinogenic compounds such as trihalomethanes. A range of heavy metals such as mercury, arsenic, lead, and cadmium; chlorinated substances such as dichlorodiphenyltrichloroethane (DDT), dichlorodiphenyldichloroethylene (DDE), and polychlorinated biphenyls (PCBs); organic substances such as polyaromatic hydrocarbons (PAHs); Dieldrin; etc., create toxic conditions.

A large diversity of organic pollutants poses a serious threat to aquatic ecosystems. PCBs and 3,4-benzpyrene were confirmed as jeopardizing Lake Constance.[19] However, pesticide is the major group of contaminants in the top layers of sediments. The recognition of DDT as being harmful to human and animals served as the eye opener to serious consequences of various pesticides. PCBs are also considered as persistent organic pollutants and, thus, have huge potential hazards. The main sources of pesticides in lakes and reservoirs are 1) agriculture and forestry; 2) actions against aquatic weeds (e.g., water fern *Salvinia molesta, Eichhornia crassipes*); 3) actions against parasites and waterborne diseases (e.g., malaria and schistosomiasis); and 4) regulation of fish populations with rotenone (an active ingredient of derris, which is used as a fish poison for centuries). Atrazine, which is used for the protection of corn from weeds, is demonstrated as detrimental to human, and hence, it has recently been proscribed in many countries. Likewise, lindane, a persistent organochlorine insecticide, is toxic to fish in concentrations as low as 1 ppm. Pesticides are applied for curbing undesirable weeds such as water fern *Salvinia* sp. and *Azolla* sp., water hyacinth *E. crassipes*, water lettuce *Pistia stratiotes*, etc. Application of the pesticides should only be considered in cases where mechanical and biological approaches fail.[19]

Sedimentation

Sedimentation of lakes is a common phenomenon; in general, it takes place very slowly. Any process in a lake watershed that disturbs the soil can significantly hasten this process. Most of these activities are anthropogenic and comprised farming on fragile soils and on steep slopes, surface mining, and construction activities. Suspended sediment and sedimentation have many degrading effects on lakes and shoreline ecosystems. Peripheral wetlands can be completely covered by silt, eradicating their value as nutrient sinks, wave absorbers, nursery areas for fishers, and habitat. As silt settles in a lake, spawning areas are covered and lake volume is reduced, and this causes degradation in fishery production. As a result of reduced storage capacity, both the volume and extent of flooding can increase. In the case of hydroelectric amenities, generating capacity may reduce significantly. The increase in shoal areas as a result of sedimentation can enhance increased macrophyte growth and can interfere with recreational activities such as boating and fishing. Soil loss and suspended silt also contribute to eutrophication and lake contamination since the silt generally includes attached nutrients, herbicides, pesticides, and other chemicals. This increases water treatment costs and maintenance problems of water treatment plants.

Acidification

Atmospheric pollution by sulfur dioxide and nitrous oxides is the major cause of acidification in aquatic systems, through precipitation (dry or wet deposition) and, to a larger extent, through leaching from affected land. Other sources of acid deposits include industrial effluents and mining wastes.

Fish Depletion

Fish provides a substantial portion of animal protein consumed by humans. In tropical developing countries, 60% of the people depend on fish and 40% or more of total protein intake comes from fish.[20] The majority of the world's landed fish catch (87%) comes from marine areas. New developments in watersheds, such as dams and reservoirs to control the annual distribution of water, frequently cause large losses in floodplain fertility and species diversity. Floodplains are needed by many species of fish for reproduction and for refuge. The natural seasonal flooding and drying are signals used by some river-dwelling animals to begin to reproduce or migrate. Proper timing of water level drawdowns in upstream reservoirs is therefore important, especially to enhance spawning of certain species.

Stratification

Stratification is a significant feature influencing water quality in reasonably stagnant, deep waters, such as lakes and reservoirs, which occurs mainly because of the difference in temperature, leading to a variation in density. Occasionally, it can be due to the difference in solute concentrations. Water quality in various layers of stratified water body is subjected to different influences. Solar insolation will be more in the upper layer while the lower layer is physically detached from the atmosphere and may be in touch with decaying sediments that exert an oxygen demand. Because of these varying influences, the lower layer will usually have a reduced oxygen concentration relative to the upper layer. The anoxic condition thus produced will enhance the diffusion of constituents from sediments and form various compounds such as ammonia, nitrate, phosphate, sulfide, silicate, iron, and manganese.

During summer and spring, the surface layer of the water body in temperate regions becomes warmer and hence less dense. A resistance to vertical mixing is formed, because of the existence of warm water over cold water. The top warmer surface layer is known as the epilimnion and the colder water stuck underneath is the hypolimnion. There is a shallow zone called metalimnion or the thermocline, in between epilimnion and hypolimnion, where the temperature changes from warmer epilimnion to colder hypolimnion. Normally, wind-and surface-current-induced mixing is limited to epilimnion, while hypolimnion remains stagnant. The density difference between two layers (viz., epilimnion and hypolimnion) is diminished gradually when the weather becomes cooler. This will enhance the wind-induced vertical mixing between these two layers, which will result in the phenomenon known as "overturn," which can occur quite rapidly. The frequency of overturn and mixing governs predominantly on climate (temperature, solar insolation, and wind) and the characteristics of the lake and its surroundings (depth and exposure to wind). Lakes may be classified according to the frequency of overturn as follows:

- Monomictic: once a year—temperate lakes that do not freeze
- Dimictic: twice a year—temperate lakes that do freeze.
- Polymictic: several times a year—shallow, temperate, or tropical lakes
- Amictic: no mixing—arctic or high-altitude lakes with permanent ice cover, and underground lakes
- Oligomictic: Poor mixing—deep tropical lakes
- Meromictic: incomplete mixing—mainly oligomictic lakes but sometimes deep monomictic and dimictic lakes Because of the action of crosswind and the flow of water, thermal stratification is not seen in lakes, which have depths less than 10 m. In shallow tropical lakes, complete mixing occurs several times a year, whereas in very deep lakes, stratification may continue all over the year, even in tropical and equatorial regions. This stable stratification results in "meromixis."

In the case of tropical lakes, as a result of moderately constant solar insolation, seasonal changes in water temperature are small. The annual water temperature range is only 2–3°C at the surface and even less at depths greater than 30 m.[21] Winds and precipitation, both play a vital role in mixing. Because of the large difference in rainfall between wet and dry seasons, large variation in water level is seen in some tropical lakes. Such variations have a prominent influence on dilution and nutrient supply, which, in turn, affect algal blooms, zooplankton reproduction, and fish spawning. Wind speeds are usually greater during the dry season and evaporation rates are at their highest. The subsequent heat losses, combined with the turbulence caused by wind action, stimulate the process of mixing.

As far as recreation is concerned, reservoirs are as important as natural lakes but have surplus scopes for flood control, hydropower generation, and water supply. Although both lakes and reservoirs are subjected to silt, organic, and nutrient loadings, reservoirs usually having hefty watersheds and peculiar morphometric conformations are subjected to more water quality problems. Even though lakes and reservoirs have biotic and abiotic processes in common and similar habitats, they have some significant differences. They differ in their geologic history and setting, basin morphology, and hydrologic factors.[22,23]

Non-point source of water pollution generated by expanding agricultural production is considered as a major environmental threat to some lakes. Many chemicals in common agricultural use have a strong affinity for fine soil particles. When the latter erode, these chemicals are carried with them into surface waters. The soil itself is a problem when it accumulates in great quantity in lakes. Lake Pittsfield lost nearly a quarter of its volume to sedimentation in only 24 years. Transport and deposition of eroded materials as well as substances dissolved in runoff and attached to soil particles lead to negative impacts on agricultural land and the Three Gorges Reservoir including water quality decline,[24,25] which are generally thought to be caused by land use changes of converting forest resources to agriculture in watersheds. Conversion of cropland with a slope greater than 10° into forestland meets the reduction goal.[26]

Over recent decades, the water quality of lakes and reservoirs has been deteriorating rapidly due to external and internal pollution including that from the sediment, and the eutrophication phenomenon has become a more serious global threat. Some researchers suggest applying a plan of sediment dredging to this water body. After dredging, however, a vast amount of sediment would become solid pollutants containing high concentrations of heavy metals (mainly Hg and Cd), which would certainly cause secondary pollution. This pollution should not be ignored, and a further research is therefore needed on the elaborate restoration scheme for these precious drinking-water sources.[27] Major pollution issues of lakes and reservoirs with few representative past studies are tabulated in Table 1.

TABLE 1 Major Pollution Issues of Lakes and Reservoirs

Sl. No.	Pollution Issue	Effects	Representative Case Studies
1	Eutrophication	High primary production, low oxygen concentrations and increased concentrations of hydrogen sulfide, carbon dioxide, dissolved iron, and manganese in the hypolimnion	West Twin Lake, Ohio[28]
2	Sedimentation	Volume will be decreased, macrophytes growth may be enhanced	Lake Superior at Superior Harbor, Wisconsin[29]
3	Acidification	Adverse effects on the most sensitive aquatic species	Many Scandinavian lakes[11]
4	Toxic substances and heavy metal contamination	Bioaccumulation of these toxicants poses health hazards to all members of the food chain including humans	Lake Nainital, India[30]
5	Fish depletion	As the fishery plays a vital role in the supply of animal protein, its depletion affects the food security and economy of a significant portion of humans	Lake Victoria, Uganda[11]

Sources of Lakes and Reservoir Pollution

Lakes and reservoirs tend to collect not only sediments but also most of the pollutants that are washed into them, and thus they function, in part, as environmental sinks. Eroded soil dissolves in the water and fills in lake bottoms—this activity has significantly degraded lake ecosystems across the world. The sources of pollution can be mainly classified into two: 1) point sources and 2) non-point sources. A point source of pollution is a single identifiable localized source of air, water, thermal, noise, or light pollution.[31] In earlier days, control of "point source" nutrients and toxic contaminants was the principal focus of exertions for the protection and restoration of lakes and reservoirs, but nowadays, the substantial contaminant and nutrient sources to lakes and reservoirs are "non-point" type such as agricultural runoff, erosion from urban or deforested areas, surface mining, or atmospheric depositions. Most lake water is rich in nutrients that support growth of many aquatic macrophytes and algal blooms. Besides, water is contaminated with metals like chromium (Cr), copper (Cu), iron (Fe), manganese (Mn), nickel (Ni), lead (Pb), and zinc (Zn). High concentrations of these metals are also found in sediments, but it is found that the level of metal concentrations of lake varies considerably in different seasons.

In earlier days, we believed that lakes and reservoirs were enriched with nutrients and organic matter from "point" sources such as industrial discharges and wastewater treatment plant outfalls. However, for many lakes, non-point or diffuse nutrient loading, both internal and external, is found to be momentous. These non-point sources are challenging to assess and regulate,[32] and water quality in many lakes has remained deteriorated following diversion or treatment of point sources.

The main pollutants that enter through non-point sources are various forms of phosphorus and nitrogen such as total phosphorus, phosphate phosphorus, nitrate nitrogen, nitrite nitrogen, ammoniacal nitrogen, organic nitrogen, chlorine, sodium, calcium, and suspended solids. The exports of all constituents occur mostly during rainfall-or snowmelt-generated runoff events during the spring runoff period. It has been shown that one of the major reasons for the enrichment of lakes is the conversion of previously cropped land into agricultural production while conversion to forests has very little impact on enrichment.

The runoff of nutrients from old fields depends on the nutrient status of the soils and soil water. This nutrient status reflects the soil type, fertilizer and cropping practices prior to abandonment, number of years since abandonment, and the succeeding vegetation present at any particular point. Soil data identify the reservoir of nutrients available for runoff, provide a means to relate runoff to nutrient content of that particular soil, and provide the data necessary for design of management schemes to prevent release of nutrients. Farm lands that have been abandoned for 15 to 20 years are not major non-point sources of pollution.

Water Quality Monitoring

Water quality monitoring refers to the acquisition of quantitative and representative information on the physical, chemical, and biological characteristics of a water body over time and space.[33] It is a complex task, comprising all the activities to extract information with respect to the aquatic system. A variety of contaminants, in addition to a multitude of imprudent water quality management practices and destructive land uses, are currently threatening aquatic systems on a worldwide scale. In addition, it has been shown that water of good quality is a critical component for sustainable socioeconomic development.[34] The impact and behavior of contaminants in an aquatic ecosystem are complex and may involve adsorption-desorption, precipitation-solubilization, filtration, biological uptake, excretion, and sedimentation-suspension. Besides natural processes affecting water quality, there are also anthropogenic impacts, such as man-induced point and non-point sources, xenobiotic, and alteration of water quality due to unwise water use and river engineering projects (e.g., irrigation, damming, etc.).[35]

The degradation of water resources has increased the need for determining the ambient status of water quality, in order to provide an indication of changes induced by anthropogenic activities. To understand the process dynamics of a watershed, a well-designed water quality monitoring network identifies water quality problems while establishing baseline values for short-and long-term trend analysis. The need to evaluate observed water quality conditions and their suitability for the intended uses reflects a need for cost-effective and logistically practical water quality monitoring network design methods.

Types of Monitoring

Lake Sampling

It characterizes the water quality of the lake to identify status and trends. Two main types of lake monitoring are water-column sampling and near-shore (shallow water) sampling. Water-column sampling tries to quantify the overall response of the lake to contamination. The lake's trophic status, as designated by phosphorus, turbidity, chlorophyll-a, and dissolved oxygen, is of particular concern. Some invasive species, such as spiny water flea, can also be identified by water-column sampling. During seasons when the lake is stratified, the water-column sample should be sampled in both the epilimnion (the warm upper layer) and hypolimnion (lower layer of cold water).

Near-shore refers to the depth at which rooted plants can grow. Sampling can be adequately done at the end of a dock. Sampling for pathogens and pathogen indicators is important because of contact recreation such as swimming. Near-shore monitoring allows study of the lake bottom including sediment sampling for heavy metals, macroinvertebrates, and attached or rooted invasives such as zebra mussels and Eurasian watermilfoil.

Tributary Mass Load Sampling

It determines the tributary mass loads of water contaminants entering the lake. A significant portion of the lake's water pollution is brought by the tributaries flowing into it. Determination of tributary mass loads is particularly important for management of the lake's phosphorus and sediment problems.

Tributary Water Quality Sampling

It characterizes the water quality of tributaries to identify status and trends. Tributaries may be threatened by contaminants or stresses that affect the stream health but are not significantly detrimental to the lake. The tributaries are valued for recreation and aesthetics, drinking water, irrigation, and wildlife habitat and deserve protection.

Biological Integrity Sampling

It characterizes the long-term ecological health of the lake and tributaries. Ecological sampling is useful for detecting the effects of impairments that are not present at the time of sampling, for evaluating habitat health and for determining the biological integrity of surface waters. Ecological sampling may include bioassessments of fish and benthic macroinvertebrate communities, periphyton, and single-species monitoring (trout, salmon, and freshwater mussels are often used). Biological indices, a composite of different indicators, can be developed.

Citizen Monitoring

It encourages citizen participation in the measurement of watershed quality. To the extent that people care about the watershed's lands and waters, the watershed will be protected and enhanced for generations to come. One way to encourage such stewardship is through involvement of students and other citizens in water quality monitoring. Monitoring conducted by citizen volunteers increases public awareness and knowledge about water quality and its protection.

The design of efficient water quality monitoring network is essential for effective water management. To date, many water quality monitoring networks for surface freshwaters have been rather arbitrarily designed without a consistent or logical design strategy. Moreover, design practices in recent years indicate a need for cost-effective and logistically adaptable network design approaches.[36] Furthermore, the monitoring in a water quality management program is recognized as a statistical approach, so that both the assessment and the design problems can be addressed via a statistical method. In this view, the statistical methods have been found very efficient for redesigning and assessment of water quality monitoring networks (WQM).

The International Organization for Standardization (ISO) defines water quality monitoring as: "the programmed process of sampling, measurement and subsequent recording or signaling, or both, of various water characteristics, often with the aim of assessing conformity to specified objectives." This general definition can be differentiated into three types of monitoring activities that distinguish between long-term, short-term, and continuous monitoring programs:

- Long-term observation and standardized measurement of the aquatic environment for defining the water quality status and prediction of the trend are known as monitoring.
- Intensive programs for the measurement and observation of status of the aquatic environment for a specific purpose are known as surveys. These are of limited duration.
- Continuous measurement and observation of the aquatic environment for the management of the quality of the water and other operational activities are known as surveillance.

The constituents that decide the quality of the water are transported by water from the watershed. Therefore, we have to construct a perfect water budget, as it plays a key role in identifying a lake's (reservoir's) problem. By conducting a reconnaissance survey of water from the watershed, main tributaries can be selected. Since high flows are the chief segment of the water budget and huge volume influxes are followed by high concentration, continuous gauge recording is recommended for the determination of flow in major tributaries. From a successive record of inflow and outflow in the main tributaries, an annual water budget is constructed so that estimated inflows equal outflows with a correction for lake storage.

The water budget is formulated as:[37]

$$SF_i + GW + DP + WW = SF_o + EVP + EXF + WS + \Delta STOR$$

where SF_i and SF_o are stream flow in and out, respectively; GW is groundwater in (includes deep and subsurface seepage); DP is direct precipitation on the lake surface; WW is wastewater, if any; EVP is evaporation; EXF is exfiltration; WS is removal for water supply, if any; and ASTOR is the change in lake volume.

Protective and Restorative Measures

Removal or treatment of direct input of wastewater, stormwater, or both constitutes the primary step in the restoration of water quality of eutrophic lakes and reservoirs, as these sources frequently contain comparatively high concentrations of phosphorus and nitrogen. For the realization of any long-term benefits from in-lake treatments, such external loadings should be reduced. In some cases, reduction of external loading is adequate to restore the water body (e.g., Lake Washington[38,39]), but in others, where internal loading of nutrients is significant, in-lake treatments may be indispensable to accomplish lake quality improvement (e.g., Lake Trummen[40]).

Advanced wastewater treatment (AWT) and diversions are two most commonly used techniques for the reduction of external inputs. In diversion, the treated sewage or industrial wastewater is carried away from the degraded water body to waters that are having high assimilative capacity, by the installation of interceptor lines. AWT involves the reduction of phosphorus concentration in wastewater effluent by using chemicals such as alum (aluminum sulfate), lime (calcium hydroxide), or ferric chloride. Stormwater runoff is the next dominant source of external enrichment. Even though the P concentration in stormwater is very low (2%–10%), and solubility is less than that of sewage effluent, such non-point sources can denote momentous contributions. P retention in wet detention basins and wetlands, rapid filtration through soil, and P removal in predetention basins are the principal P removal methods applied for runoff water. If internal loading of P is anticipated to hinder the recovery following primary treatments such as diversion or AWT, then supplementary in-lake treatments may be justified to accelerate reclamation. Lakes and reservoirs lose volume due to siltation. Sediment removal, together with land management and construction of device to trap silt, is an example of their restoration and protection. Some management approaches are institutional arrangement, formation and operation of lake association, sports fishery management, etc.

Eutrophication can be controlled if the phosphorus (P) concentrations in the water body are lowered to a level that will limit the growth of the algae. This can be achieved by diversion of external input, dilution, flushing, or a combination of these approaches. Where there is substantial loading reduction, comparatively augmented rate of algal flushing, and negligible recycling from sediment, in-lake phosphorus concentration can be reduced significantly and trophic state can be improved rapidly. Lake Washington is a good example for this approach.[40,41] However, for many lakes, internal phosphorus release sustained the lake's enriched trophic state and reinforced the state of continued eutrophication, in spite of the removal of a noteworthy fraction of external loading by diversion.[42,43] In those lakes, following nutrient diversion, supplementary in-lake treatment may be necessary, to avoid an extended eutrophic state.

Other methods are dilution and flushing, which are used interchangeably. Dilution involves the reduction in the concentration of the nutrients and a washout of algal cells, whereas flushing comprises only the latter. Where there is a high nutrient load, water should be diverted if possible for the low dilution rate to be most effective. This plan provides for a reduction in biomass primarily through nutrient limitation. If only moderate to high nutrient water is available, flushing may work well if the loss rate of cells is sufficiently great relative to the growth rate. Flushing rate on the order of 10%–15% per day will afford some control through washout. We can opt for the technique such as "phosphorus inactivation," which involves, usually, the application of salts of aluminum, such as aluminum sulfate (alum), sodium aluminate, etc., to precipitate phosphate as aluminum phosphate and thus to bind a significant fraction of P. Aluminum hydroxide floc, which is formed during this process, enhances the settling of the precipitate and it will continue to sorb and retain P in their molecular lattices.

Nutrient reduction mainly consists of processes such as source reduction in the watershed, issuance of fertilizer guidelines, setting up of shoreland buffer strips, and restriction of motorboats. *Biological control* involves processes such as use of bacteria for algae control, use of algae-eating fishes, biomanipulation, aquascaping, and bioscaping. *Lake aeration* comprises conventional aeration, solar powered aeration, wind-powered aeration, fountain aeration, and hypolimnetic aeration. *Addition of chemicals* includes use of barley straw, alum, buffered alum for sediment treatments, calcium compounds, liquid dyes, chlorine, and algicide[44] (Figure 3). Copper is an efficient algicide. Copper sulfate application, the typical treatment for algal problems for many decades, is often effective for short-term solution to a current algae problem, predominantly in water supply reservoirs. However, there is substantial

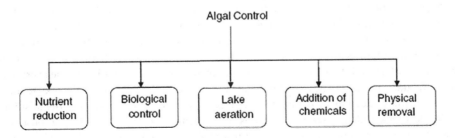

FIGURE 3 Major algal control measures.

TABLE 2 Major Restoration Measures for the Lakes and Reservoirs

Sl. No.	Restoration Measure	Advantage(s)	Drawback(s)
1	Diversion	Reduce the overall nutrient loading to the system	Presence of another water body (having high assimilative capacity) is required in the near vicinity
2	Advanced wastewater treatment	Control eutrophication by limiting nutrient concentration with suitable chemicals	Accumulation of chemicals in the system and high cost
3	Dilution and flushing	Control eutrophication by limiting the P concentration	Cannot accommodate high nutrient loading
4	Application of appropriate algicide (such as copper sulfate)	Control eutrophication by curbing algal growth	Its effects are found to be temporary, higher cost, there are major negative impacts to non-target organisms, and significant copper contamination of sediments can be possible
5	Sediment removal (dredging)	Sediment removal, together with land management can restore the lake, affected by siltation	High cost and, in some cases, may cause environmental damage

confirmation against the prolonged usage of this chemical. Its effects are found to be temporary, cost can be higher, there are major negative impacts to non-target organisms, and significant copper contamination of sediments can be possible. We can go for long-term and permanent options such as control of external and internal nutrient loading for the effective management of algal bloom. For a better water quality, and if the P is not estimated to reach an algal limiting level, an in-lake treatment for the control sediment should be established soon after external controls are in place.

Macrophytes includes all macroscopic aquatic flora, comprising macroalgae such as the stoneworts *Chara* and *Nitella*, aquatic liverworts, mosses, ferns, and flowering vascular plants. Aquatic plant management aims to curb annoyance species, to exploit the favorable features of plants in water bodies, and to reorganize plant communities. Its principal objective is the establishment of stable, diverse, aquatic plant communities comprising high percentages of desirable species. An exhaustive understanding of macrophytes biology is the foundation for evolving innovative management tactics. Continued research and development will advance our understanding of the relationship of aquatic plants to overall lake and reservoir quality and our capability to manage aquatic plant communities to preserve or improve that quality.[36,37] Table 2 shows major restoration measures for the lakes and reservoirs.

Eutrophication Model Framework

Phosphorus-loading models are often employed for the evaluation of eutrophication problems in lakes and reservoirs. In these models, phosphorus loading is linked to the average total phosphorus concentration in the lake water and to other indicators of water quality that are associated to algal bloom, such as chlorophyll and transparency. Physical and hydrologic features influence the response of the lake to phosphorus loading and thus these models take in to account various characteristics such as lake volume, average depth, flushing rate, etc. The underlying principles behind the eutrophication model are as follows: 1) phosphorus acts as the limiting nutrient for algal growth; 2) any change in the amount of phosphorus discharged into a lake over an annual or seasonal period will alter the average concentration of P in lakes and hence the extent of algal bloom; and 3) the capacity of the lake to adjust with the P loading, without causing algal blooms increases with the volume, depth, and the flushing rate of the lake. Models recapitulate these relationships in mathematical forms, based upon observed water quality responses of large numbers of lakes and reservoirs. Eutrophication models are pitched for the prediction of average status of water quality over a season or a year. Averaging is mainly done over three dimensions: 1) depth; 2) sampling stations; and 1) season.[45]

Conventions for the Protection of Lakes and Reservoirs

Lakes and Wetlands—Ramsar Convention, Iran, February 2, 1971

The Ramsar Convention (The Convention on Wetlands of International Importance) is an international treaty for the conservation and sustainable utilization of wetlands. The convention was developed and implemented by participating countries at a meeting in Ramsar, hosted by the Iranian Department of Environment, and came into force on December 21, 1975. The Ramsar List of Wetlands of International Importance presently comprises 1950 sites (known as Ramsar Sites) covering around 1,900,000 km^2 up from 1021 sites in 2000. The nation with the highest number of sites is the United Kingdom at 168; the nation with the greatest area of listed wetlands is Canada, with more than 130,000 km^2. Presently, there are 161 contracting parties, up from 119 in 1999 and from 21 initial signatory nations in 1971. Signatories meet every 3 years as the Conference of the Contracting Parties (COP); the first was held in Cagliari, Italy, in 1981. There is a standing committee, a scientific review panel, and a secretariat. The headquarters is located in Gland, Switzerland, shared with the International Union for Conservation of Nature (IUCN).[46]

UNECE Water Convention, Helsinki, March 17, 1992

The Convention on the Protection and Use of Transboundary Watercourses and International Lakes (Water Convention) is intended to strengthen national measures for the protection and ecologically sound management of transboundary surface waters and groundwater. The Parties to this convention are obliged to prevent, control, and reduce transboundary impact; use transboundary waters in a reasonable and equitable way; and ensure their sustainable management. Parties bordering the same transboundary waters shall cooperate by entering into specific agreements and establishing joint bodies. The Convention includes provisions on monitoring, research and development, consultations, warning and alarm systems, mutual assistance, and exchange of information, as well as access to information by the public.

Protocol on Water and Health, London, June 17, 1999

The Protocol on Water and Health aims to protect human health and well-being by better water management, including the protection of water ecosystems, and by preventing, controlling, and reducing water-related diseases. It is the first international agreement of its kind adopted specifically to attain an adequate supply of safe drinking water and adequate sanitation for everyone and effectively protect water used as a source of drinking water. Parties to the Protocol commit to set targets in relation to the entire water cycle.[47]

Conclusion

Many of the lakes and reservoirs are situated in developing countries or industrializing countries that only now identify the devastating economic and social impact that pollution, overfishing, and habitat degradation are having on their water resources. Once these natural systems are severely polluted, they often cannot be restored completely; they can at best only be improved to a level where they can meet basic functions, and society must bear the increased costs and risks to human health.

The reviewed case studies in this entry and elsewhere illustrate that contaminated lakes and reservoirs can be restored, at least partially. The entry presented a number of examples where this process has been undertaken effectively. From an economic and environmental point of view, it appears wise to prevent the occurrence or exacerbation of problems, rather than going for radical and expensive restorative actions. Ample evaluations of water resources should be conducted to understand the state of lakes and reservoirs, so that we can provide the essential means for the establishment of imperative water resource management goals and objectives.

All-inclusive watershed management programs that comprise significant water quality objectives precise to the hydraulic characteristics of lakes and reservoirs should be prepared. The differences between temperate and tropical lake situations need to be highlighted. During the formulation of management strategies, all exertion should be made to avert the discharge of toxic substances to lakes and reservoirs. Introduction of exotic and invasive species should be banned, unless there has been ample environmental assessment.

We should incorporate the "precautionary" approach and the "polluter pays" principle in the management of lakes and reservoirs, which can benefit from various economic tools and ample financial policies. Primary action areas for protection and restoration of lakes and reservoirs involve watershed assessment, watershed control measures, and best management practices, which comprise management of pollution from agriculture, silviculture, mining, industrial pollution, urban runoff, and other pollutant sources, predominantly for nutrients and persistent toxic pollutants. Management programs for lakes and reservoirs should be comprehensive in scope and watershed-wide in nature. If necessary, joint governance institutions, such as joint commissions, composed of high-level government officials, should be established to improve management of transboundary lake and reservoir basins. The formulation of legal instruments, treaties, and a hierarchy for making the decision are essential for the advancement of the water resources management. It is clear that the world's lake reservoirs are threatened. In many cases, lake degradation is so advanced that populations depending on them are in great danger. The scope, magnitude, and dimension of the problem demand an international cooperation as well as national exertions.

A logical and consistent design methodology that allows more efficient and effective data collection and, hence, more useful information extraction should be developed. Such an approach not only permits better water pollution control recommendations and better allocation of financial resources but also, ultimately, a better understanding of the ecosystems. To appreciate the challenges of designing water quality networks, it is crucial to clearly define objectives and identify statistically acceptable assumptions. Assumptions are an inherent part of the monitoring network design process, mainly due to the stochastic influences on water quality variables in the aquatic environment. The number and type of simplifying assumptions made or allowed are dependent upon network objectives. Furthermore, assumptions in monitoring network design should be made relative to water quality hydrologic principles, applicable statistics, information utilization, and budget constraints.

References

1. Bindler, R.; Renberg, I.; Brännvall, M.-L.; Emteryd, O.; El-Daoushy F. A whole-basin study of sediment accumulation using stable lead isotopes and flyash particles in an acidified lake, Sweden. Limnol. Oceanogr. **2001**, *46*, 178–188.
2. Available at http://www.wisconsinlakes.org/index.php/the-science-of-lakes/21-lake-types (accessed May 2012).
3. Available at http://www.unep.or.jp/ietc/publications/short_series/lakereservoirs-1/fwd.asp (accessed February 2011).
4. Available at http://ga.water.usgs.gov/edu/earthlakes.html (accessed February 2012).
5. Available at http://www.waterontheweb.org/under/lakeecology/16_trophicstatus.html (accessed February 2012).
6. Dodds, W.K. *Freshwater Ecology—Concepts and Environmental Applications, 1st Ed.;* Elsevier: New Delhi, India, *2006.*
7. Tilzer, M.M.; Bossard, P. Large lakes and their sustainable development. J. Great Lakes Res. **1992**, *18* (3), 508–517.
8. Papatheodorou, G.; Demopoulou, G.; Lambrakis, N. A long-term study of temporal hydrochemical data in a shallow lake using multivariate statistical techniques. Ecological Modell. **2005**, *193*, 759–776.

9. Zhao, Y.; Yang, Z.; Li, Y. Investigation of water pollution in Baiyangdian Lake, China. Procedia Environ. Sci. **2010**, 2, 737–748.

10. Yu, F.; Fang, G.; Ru, X. Eutrophication, health risk assessment and spatial analysis of water quality in Gucheng Lake, China. Environ. Earth Sci. **2010**, *59* (8), 1741–1748.

11. Dinar, A.; Seidl, P.; Olem, H.; Jordan, V.; Duda, A.; Johnson, R. *Restoring and Protecting the World's Lakes and Reservoirs*; World Bank Technical Paper Number 289, The World Bank: Washington, DC, 1995; 1–113.

12. Richards, R.P. The Lake Erie agricultural systems for environmental quality project: An introduction. J. Environ. Qual., **2002**, *31*, 6–16.

13. Zemenchik, R.A. Bio-available phosphorus in runoff from alfalfa, smooth bromegrass, and alfalfa-smooth bromegrass. J. Environ. Qual. **2002**, *31*, 280–286.

14. Ouyang, W.; Hao, F.H.; Wang, X.L.; Cheng H G. Nonpoint source pollution responses simulation for conversion cropland to forest in mountains by SWAT in China. Environ. Manage. **2008**, *41*, 79–89.

15. Cook, G.D.; Carlson, R.E. *Reservoir Management for Water Quality and THMPrecursor Control*; American Water Works Association Research Foundation: Denver, CO, 1989.

16. Cook, G.D.; Kennedy, R.H. Managing drinking water supplies. Lakes Reservoirs Manage. **2001**, *17*, 157–174.

17. Osgood, R.A. Lake mixes and internal phosphorous dynamics. Arch. Hydrobiol. **1988**, *113*, 629–638.

18. Larsen, D.P.; Schultz, D.W.; Malueg, K.W. Summer internal phosphorous supplies in Shagava Lake, Minnesota. Limnol. Oceanogr. **1981**, *26*, 740–753.

19. Jørgensen, S.E. *Lake and Reservoir Management*, Revised Ed.; Developments in Water Science; Elsevier, 2005; Vol. 54.

20. WRI (World Resources Institute). World Resources 1992–93. Washington, DC, 1994.

21. Bartram, J; Balance, R. *Water Quality Monitoring—A Practical Guide to the Design and Implementation of Freshwater Quality Studies and Monitoring Programmes;* Published on behalf of United Nations Environment Programme and the World Health Organization UNEP/WHO ISBN 0 419 22320 7 (Hbk) 0 419 21730 4 (Pbk), 1996.

22. Kennedy, R.H.; Thornton, K.W.; Ford, D.E. Characterization of the reservoir ecosystem. In *Microbial Processes in Reservoirs*; Gunnison, D., Ed.; Junk Publishers: The Hague, Netherlands, 1985; 27–38.

23. Kennedy, R.H. Consideration for establishing nutrient criteria for reservoirs. Lake Reservoir Manage. **2001**, *17*, 175–187.

24. Kurtz, D.A. *Long Range Transport of Pesticides;* Kurtz, Ed.; Lewis Publisher: Chelsea, MI, 1990; 440 pp.

25. Cotham, W.E; Bidleman, T.F. Estimating the atmospheric deposition of organochlorine contaminants to the Arctic. Chemosphere **1991**, 22 (1–2), 165–188.

26. Wolf, M.S; Toniolo, P.G. Environmental organochlorine exposure as potential etiologic factor in breast cancer. Environ. Health Perspect. **1995**, *103*, 141–145.

27. Wania, F.; Mackay, D. Global fractionation and cold condensation of low volatility of organochlorine compounds in Polar Regions. Ambio **1994**, 22, 10–18.

28. Cook, G.D.; Heath, R.T.; Kennedy, R.H.; MComas, M.R. The Effect of Sewage Diversion and Aluminium Sulfate Application on Two Eutrophic Lakes. USEPA-600/3–78-033, 1978.

29. Kostic, S.; Parker, G. Progradational sand-mud deltas in lakes and reservoirs. Part 1. Theory and numerical modeling. J. Hydraulic Res. **2003**, *41* (2), 127–140.

30. Ali, M.B.; Tripathi, R.D.; Rai, U.N.; Pal, A.; Singh, S.P. Physico-chemical characteristics and pollution level of lake Nainital (U.P., India): Role of macrophytes and phytoplankton in biomonitoring and peytoremediation of toxic metal ions. Chemosphere, **1999**, *39* (12), 2171–2182.

31. Liu, Y.; Islam, M.A.; Gao, J. Quantification of shallow water quality parameters by means of remote sensing. Prog. Phys. Geogr. **2003**, 27(1), 24–43.

32. Line, D.E.; Jennings, G.D.; McLaughlin, R.A.; Osmond, D.L.; Harman, W.A.; Lombardo, L.A.; Tweedy, K.L.; Spooner, J. Nonpoint sources. Water Environ. Res. **1999**, 71, 1054–1069.

33. Sanders, T.G.; Ward, R.C.; Loftis, J.C.; Steele, T.D.; Adrian, D.D.; Yevjevich, V. *Design of Networks for Monitoring Water Quality;* Water Resources Publications LLC: Highlands Ranch, CO, 1983.

34. Bartram, J.; Balance, R. *Water Quality Monitoring—A Practical Guide to the Design and Implementation of Freshwater Quality Studies and Monitoring Programmes;* Published on behalf of United Nations Environment Programme and the World Health Organization 1996, UNEP/ WHO ISBN 0 419 22320 7 (Hbk) 0 419 21730 4 (Pbk).

35. Chapman, D., Ed. *Water Quality Assessments. A Guide to the Use of Biota, Sediments and Water in Environmental Monitoring;* Chapman and Hall: London, 1996.

36. Strobl, R.O.; Robillard, P.D. Network design for water quality monitoring of surface freshwaters: A review. J. Environ. Manage. **2008**, *87*, 639–648.

37. Cooke, G.D.; Welch, E.B.; Peterson, S.A.; Nichols. S.A. *Restoration and Management of Lakes and Reservoirs*, 3rd Ed.; CRC Taylor and Francis Group, New York, 2005.

38. Edmondson, W.T. *Trophic Equilibrium of Lake Washington.* USEPA-600/3-77-087, 1978.

39. Edmondson, W.T. Sixty years of Lake Washington: A curriculum vitae. Lake Reservoir Manage. **1994**, *10*, 75–84.

40. Bjork, S. *European Lake Rehabilitation Activities.* Inst. Limnol. Rept. University of Lund: Sweden, 1974.

41. Edmondson, W.T. Phosphorus, nitrogen, and algae in Lake Washington after diversion of sewage. *Science* **1970**, *169*, 690–691.

42. Cullen, P.; Fosberg, C. Experiences with reducing point sources of phosphorous to lakes. Hydrobiologia **1988**, *170*, 321–336.

43. Scheffer, M. *Ecology of Shallow Lakes;* Chapman and Hall: New York, 1998.

44. McComas, S. *Lake and Pond Management Guidebook*; Lewis Publishers, A CRC Press Company: U.S., 2003.

45. Moore, L.; Thornton, K. *The Lake and Reservoir Restoration Guidance Manual.* EPA 440/5-88-002 First Edition. Washington, DC, 1988.

46. Available at http://www.ramsar.org/cda/ramsar/display/main/main.jsp?zn=ramsar&cp=1_4000_0_ (accessed February 2012).

47. Available at http://www.unece.org/env/water/ (accessed February 2012).

7

Mines: Acidic Drainage Water

Introduction 81
    What Is Acid Mine Drainage? • Why Is Acid Mine Drainage a
    Problem? • What Causes Acid Mine Drainage?
Mine Drainage Chemistry.. 83
Mine Drainage Mineralogy.. 83
Mine Drainage Microbiology 84
Environmental Impacts of Mine Drainage 84
Dealing with Mine Drainage... 84
    Prevention • Treatment
References.. 85

Wendy B. Gagliano
and Jerry M. Bigham

Introduction

What Is Acid Mine Drainage?

Acid mine drainage refers to metal-rich sulfuric acid solutions released from mine tunnels, open pits, and waste rock piles (Table 1). Similar solutions are produced by the drainage of some coastal wetlands, resulting in the formation of acid sulfate soils. Acid mine drainage typically yields pH values ranging from 2 to 4; however, extreme sites such as Iron Mountain, California, have produced pH values as low as –3.6.[1] Neutral to alkaline mine drainage is also common in areas where the surrounding geologic units contain carbonate rocks to buffer acidity (Table 1).

Why Is Acid Mine Drainage a Problem?

Landscapes exposed to acid mine drainage do not support vegetation and are susceptible to erosion. When acid mine drainage enters natural waterways, changes in pH and the formation of voluminous

TABLE 1 Summary of Mine Drainage Chemistry from 101 Bituminous Coal Mine Sites in Pennsylvania

	Range	Median	Mean
pH	2.7–7.3	5.2	3.6
Fe (mg/L)	0.16–512.0	43.0	58.9
Al (mg/L)	0.01–108.0	1.3	9.8
Mn (mg/L)	0.12–74.0	2.2	6.2
SO4 (mg/L)	120–2000	580.0	711.2

Unpublished data from C. Cravolta, III, 2001. USGS, Lemoyne, PA.

FIGURE 1

precipitates of metal hydroxides can devastate fish populations and other aquatic life (Figure 1). The corrosion of engineered structures such as bridges is also greatly accelerated. There may be as many as 500,000 inactive or abandoned mines in the United States, with mine drainage severely impacting approximately 19,300 km of streams and more than 72,000 ha of lakes and reservoirs.[2,3] Once initiated, mine drainage may persist for decades, making it a challenging problem to solve.

What Causes Acid Mine Drainage?

Mine drainage results from the oxidation of sulfide minerals such as pyrite (cubic FeS_2), marcasite (orthorhombic FeS_2), pyrrhotite ($Fe_{1-X}S$), chalcopyrite ($CuFeS_2$), and arsenopyrite ($FeAsS$). These minerals are commonly found in coal and ore deposits and are stable until exposed to oxygen and water. Their oxidation causes the release of metals and the production of sulfuric acid. This process can occur as a form of natural mineral weathering but is exacerbated by mining because of the sudden, large-scale exposure of unweathered rock to atmospheric conditions.

Mine Drainage Chemistry

Mine drainage is a complex biogeochemical process involving oxidation-reduction, hydrolysis, precipitation, and dissolution reactions as well as microbial catalysis.[1] The entire sequence is commonly represented by Reaction 1, which describes the overall oxidation of pyrite by oxygen in the presence of water to form iron hydroxide [$Fe(OH)_3$] and sulfuric acid.

$$FeS_{2(s)} + 3\frac{3}{4}O_{2(g)} + 3\frac{1}{2}H_2O_{(l)} \rightarrow Fe(OH)_{3(s)} + 2H_2SO_{4(aq)} \tag{1}$$

The actual oxidation process is considerably more complicated.

Pyrite and related sulfide minerals contain both Fe and S in reduced oxidation states. When exposed to oxygen and water, the sulfur moiety is oxidized first, releasing Fe^{2+} and sulfuric acid to solution (Reaction 2). The rate of oxidation is dependent on environmental factors like temperature, pH, Eh, and relative humidity, as well as mineral surface area and microbial catalysis.

$$FeS_{2(s)} + 3\frac{1}{2}O_{2(g)} + H_2O_{(l)} \rightarrow Fe^{2+}_{(aq)} + 2SO^{2-}_{4(aq)} + 2H^+_{(aq)} \tag{2}$$

Reaction 2 is most important in the initial stages of mine drainage generation and can be either strictly abiotic or mediated by contact with sulfur-oxidizing bacteria.[4] The Fe^{2+} released by pyrite decomposition is rapidly oxidized by oxygen at pH>3 as per Reaction 3.

$$Fe^{2+}_{(aq)} + \frac{1}{4}O_{2(aq)} + H^+_{(aq)} \rightarrow Fe^{3+}_{(aq)} + \frac{1}{2}H_2O_{(l)} \tag{3}$$

If acidity generated by Reaction 2 exceeds the buffering capacity of the system, the pH eventually decreases. Below pH 3, Fe^{3+} solubility increases and a second mechanism of pyrite oxidation becomes important[5] (Reaction 4).

$$FeS_{1(2)} + 14Fe^{3}_{(aq)} + 8H_2O_{(l)} \rightarrow 15Fe^{2+}_{(aq)} + 2SO^{2-}_{4(aq)} + 16H^+_{(aq)} \tag{4}$$

In this case, pyrite is oxidized by Fe^{3+} resulting in the generation of even greater acidity than when oxygen is the primary oxidant. Pyrite decomposition is thus controlled by the rate at which Fe^{2+} is converted to Fe^{3+} at low pH.[6] At pH<3, Fe^{2+} oxidation is very slow unless it is catalyzed by populations of iron-oxidizing bacteria like *Acidithiobacillus ferrooxidans* or *Leptospirillum ferrooxidans*. These acidophilic bacteria oxidize Fe^{2+} as a means of generating energy to fix carbon. In doing so, they supply soluble Fe^{3+} at a rate equal to or slightly greater than the rate of pyrite oxidation by Fe^{3+}.[5] Pyrite oxidation then regenerates Fe^{2+} (Reaction 4), creating a cyclic situation that leads to vigorous acidification of mine drainage water.

Mine Drainage Mineralogy

The hydrolysis of Fe^{3+} causes the precipitation of various iron minerals—generally represented as ($Fe[OH]_3$)—that are often the most obvious indicators of mine drainage contamination (Reaction 5).

$$Fe^{3+}_{(aq)} + 3H_2O_{(l)} \rightarrow Fe(OH)_{3(s)} + 3H^+_{(aq)} \tag{5}$$

These precipitates are yellow-to-red-to-brown in color and have long been referred to by North American miners as "yellow boy." The actual mineralogy of the precipitates is determined by solution parameters like pH, sulfate, and metal concentration and can vary both spatially and temporally. Some of the most common mine drainage minerals are goethite (α-FeOOH), ferrihydrite ($Fe_5HO_8.4H_2O$), schwertmannite ($Fe_8O_8[OH]_6SO_4$), and jarosite ($[H,K,Na]$ $Fe_3[OH]_6[SO_4]_2$).[7]

Goethite is a crystalline oxyhydroxide that occurs over a wide pH range, is relatively stable, and may represent a final transformation product of other mine drainage minerals.[8] Ferrihydrite is a poorly crystalline ferric oxide that forms in higher pH (>6.5) environments. Schwertmannite is commonly found in drainage waters with pH ranging from 2.8 to 4.5, and with moderate to high sulfate contents. It may be the dominant phase controlling major and minor element activities in most acid mine drainage. Jarosite group minerals form in more extreme environments with pH < 3, very high sulfate concentrations, and in the presence of appropriate cations like Na and K.

Mine Drainage Microbiology

The most studied bacterial species in mine drainage systems belong to the genus *Acidithiobacillus* (formerly *Thiobacillus*)[9] Species like *Acidithiobacillus thiooxidans* and *A. ferrooxidans* are important to sulfur and iron oxidation in acid drainage; however, many other microorganisms may also be involved.[10] Bacteria have been found in close association with pyrite grains and may play a direct role in mineral oxidation, but they most likely function indirectly through oxidation of dissolved Fe^{2+} as described previously. In low pH systems (<3), *A. ferrooxidans* can increase the rate of iron oxidation as much as five orders of magnitude relative to strictly abiotic rates.[6]

Iron-oxidizing bacteria are chemolithotrophic, meaning they oxidize inorganic compounds, like Fe^{2+}, to generate energy and use CO_2 as a source of carbon. Iron oxidation, however, is a very low energy yielding process. It has been estimated that the oxidation of 90.1 mol of Fe^{2+} is required to assimilate 1 mol of C into biomass.[11] Thus, large amounts of Fe^{2+} must be oxidized to achieve even modest growth.

In addition to mediating iron oxidation, bacteria may play an additional role in mineral formation. Bacteria in mine drainage systems have been shown to be partially encrusted with mineral precipitates.[12] Bacterial cell walls provide reactive sites for the sorption of metal cations, which can accumulate and subsequently develop into precipitates, using the bacterial surface (living or dead) as a template.[13,14]

Environmental Impacts of Mine Drainage

Mine drainage is primarily released from open mine shafts or from mine spoil left exposed to the atmosphere. The drainage produced can have devastating effects on the surrounding ecosystem. Chemical precipitates can obstruct water flow, dramatically increase turbidity, and ruin stream aesthetics. Dissolved metals and acidity can also affect plant and aquatic animal populations.

Besides iron, Al is the most common dissolved metal in acid mine drainage. The primary source of Al is the acid dissolution of aluminosilicates found in soil, spoils, tailings deposits, and gangue material.[7] At high concentrations, Al can be toxic to plants, and colloidal aluminum precipitates can irritate the gills of fish, causing suffocation. Aluminum occurs as a dissolved species at low pH but rapidly hydrolyzes at about pH 5 to form felsöbanyáite $[Al_4(SO_4)(OH)_{10}$ $4H_2O]$ or gibbsite $[Al(OH)_3]$.[7,15] Aluminum precipitates are white in color but are readily masked by associated iron compounds.

Elevated levels of trace elements like As, Cu, Ni, Pb, and Zn may be released during the oxidation of sulfide minerals. These elements can play a role in mineralization processes by forming coprecipitates[16,17] but occur primarily as sorbed species.[18] Mine drainage precipitates can retain both anions and cations, depending on pH. While coprecipitation and sorption function to immobilize trace elements by removing them from solution, this effect may not be permanent. Dissolution of precipitates and shifts in pH can result in the release of sorbed species, providing a latent source of pollution.[19]

Dealing with Mine Drainage

Successful control of mine drainage usually involves elements of both prevention and treatment.

Prevention

Prevention techniques include sealing mine shafts, burying or submerging spoil piles, and adding bactericides to limit the function of iron-oxidizing bacteria. These techniques often have limited success. Sealing of mines is extremely difficult due to fractures and the permeability of surrounding rocks. Covering spoil with soil material can decrease the degree of sulfide oxidation by limiting exposure to oxygen, but establishment of a vegetative cover is necessary to prevent erosion from re-exposing the spoil. Inhibition of iron-oxidizing bacteria with bactericides can decrease sulfide oxidation and reduce metal mobility; however, reapplication is necessary and adequate distribution to all affected areas is difficult. In addition, target bacteria may develop resistance, and beneficial bacteria may be harmed.[20]

Treatment

Solution pH usually underestimates the total acidity of mine drainage. Total acidity is the sum of "proton acidity" and "mineral acidity" generated upon oxidation and hydrolysis of metals like Fe^{2+}, Fe^{3+}, Mn, and Al^{3+}.[21] The traditional approach to treatment of acid mine drainage involves neutralization of total acidity by the addition of alkaline agents like caustic soda (NaOH) or hydrated lime [$Ca(OH)_2$]. This method is effective in neutralizing acidity and precipitating dissolved metals; however, it requires continuous oversight and produces large amounts of waste sludge that require disposal. Newer remediation strategies focus on low-cost, sustainable methods for treatment of drainage waters. For example, limestone drains coupled with settling ponds or compost wetlands have shown some promise as passive remediation technologies.[22] In these systems, drainage is channeled through either oxic or anoxic limestone substrates to neutralize active acidity. Dissolved metals are then allowed to hydrolyze and precipitate in ponds or wetland cells. A major difficulty is the loss of reactive surface by armoring of limestone particles with precipitates of Fe and Al that eventually obstruct flow.

Compost wetlands are designed to stimulate the development of anaerobic microbial populations, particularly sulfate- reducing bacteria. The bacteria use the compost as an organic substrate and remove sulfate from solution, either by converting it to H_2S, which is lost to the atmosphere, or by forming insoluble iron sulfides (Reactions 6 and 7).

$$2CH_3CHOHCOO^-_{(aq)} + SO^{2-}_{4(aq)} \rightarrow 2CH_3COO^-_{(aq)} + 2HCO^-_{3(aq)} + H_2S_{(g)} \tag{6}$$

$$H_2S_{(s)} + Fe^{2+}_{(aq)} \rightarrow FeS_{(s)} + 2H^+_{(aq)} \tag{7}$$

Bicarbonate is formed as a by-product of sulfate reduction and functions to buffer acidity. These systems have also shown limited success in the field. The sulfate removal rates are usually low (<10%), and pH often remains unchanged or decreases within the wetland.[23]

References

1. Nordstrom, D.K.; Alpers, C.N. Geochemistry of acid mine waters. In *The Environmental Geochemistry of Mineral Deposits*; Plumlee, G.S., Logsdon, M.J., Eds.; Society of Economic Geologists Inc: Littleton, CO, 1999; Chapt. 6, 133–160.
2. Kleinmann, R.L.P. Acid mine drainage in the United States: Controlling the impact on streams and rivers. *4th World Congress on the Conservation of Built and Natural Environments;* Toronto, Ontario, 1989; 1–10.
3. Lyon, J.S.; Hilliard, T.J.; Bethel, T.N. *Burden of Guilt;* Mineral Policy Center: Washington, DC, 1993; 68.
4. Rojas, J.; Giersig, M.; Tributsch, H. Sulfur colloids as temporary energy reservoirs for *Thiobacillus ferrooxidans* during pyrite oxidation. Arch. Microbiol. **1995**, *163*, 352–356.

5. Nordstrom, D.K. Aqueous pyrite oxidation and the consequent formation of secondary iron minerals. In *Acid Sulfate Weathering;* Kittrick, J.A., Fanning, D.S., Hossner, L.S., Eds.; Soil Science Society of America: Madison, WI, 1982; 37–56.

6. Singer, P.C.; Stumm, W. Acidic mine drainage: The rate determining step. Science **1970**, *167,* 1121–1123.

7. Bigham, J.M.; Nordstrom, D.K. Iron and aluminum hydroxy- sulfates from acid mine waters. In *Sulfate Minerals: Crystallography, Geochemistry and Environmental Significance;* Alpers, C.N., Jambor, J.L., Nordstrom, D.K., Eds.; Reviews in Mineralogy and Geochemistry; The Mineralogical Society of America: Washington, DC, 2000; Vol. 40, 351–403.

8. Gagliano, W.B.; Brill, M.R.; Bigham, J.M.; Jones, F.S.; Traina, S.J. Chemistry and mineralogy of ochreous sediments in a constructed mine drainage wetland. Geochim. Cosmochim. Acta **2004**, *68*, 2119–2128.

9. Kelly, D.P.; Wood, A.P. Reclassification of some species of *Thiobacillus* to the newly designated genera *Acidithio- bacilus* gen. nov., *Halothiobacillus* gen. nov. and *Thermi- thiobacillus* gen. nov. Int. J. Syst. Evol. Microbiol. **2000**, *50,* 511–516.

10. Gould, W.D.; Bechard, G.; Lortie, L. The nature and role of microorganisms in the tailings environment. In *The Environmental Geochemistry of Sulfide Mine-Wastes*; Blowes, D.W., Jambor, J.L., Eds.; Mineralogical Association of Canada Short Course, 1994; Vol. 22, 185–200.

11. Ehrlich, H.L. *Geomicrobiology*, 3rd Ed.; Marcel Dekker, Inc.: New York, 1996.

12. Clarke, W.A.; Konhouser, K.O.; Thomas, J.C.; Bottrell, S.H. Ferric hydroxide and ferric hydroxy-sulfate precipitation by bacteria in an acid mine drainage lagoon. FEMS Microbiol. Rev. **1997**, *20,* 351–361.

13. Schultze-Lam, S.; Fortin, D.; Davis, B.S.; Beveridge, T.J. Mineralization of bacterial surfaces. Chem. Geol. **1996**, *132,* 171–181.

14. Konhauser, K.O. Diversity of bacterial iron mineralization. Earth-Sci. Rev. **1998**, *43,* 91–121.

15. Nordstrom, D.K. Mine waters: Acid to circumneutral. Elements **2011**, *7,* 393–398.

16. Carlson, L.; Bigham, J.M.; Schwertmann, U.; Kyek, A.; Wagner, F. Scavenging of As from acid mine drainage by schw- ertmannite and ferrihydrite: A comparison with synthetic analogues. Environ. Sci. Technol. **2002**, *36,* 1712–1719.

17. Hita, R.; Torrent, J.; Bigham, J.M. Experimental oxidative dissolution of sphalerite in the Aznalcóllar sludge and other pyritic matrices. J. Environ. Qual. **2006**, *35,* 1032–1039.

18. Winland, R.L.; Traina, S.J.; Bigham, J.M. Chemical composition of ochreous precipitates from Ohio coal mine drainage. J. Environ. Qual. **1991**, *20,* 452–460.

19. Lee, G.; Faure, G.; Bigham, J.M.; Williams, D.J. Metal release from bottom sediments of Ocoee Lake No. 3, a primary catchment area for the Ducktown Mining District. J. Environ. Qual. **2008**, *37,* 344–352.

20. Ledin, M.; Pedersen, K. The environmental impact of mine wastes—Roles of microorganisms and their significance in treatment of mine wastes. Earth-Sci. Rev. **1996**, *41,* 67–108.

21. Kirby, C.S.; Cravotta, C.A., III. Net alkalinity and net acidity I: Theoretical considerations. Appl. Geochem. **2005**, *20,* 1920–1940.

22. Cravotta, C.A., III; Trahan, M.K. Limestone drains to increase pH and remove dissolved metals from acidic mine drainage. Appl. Geochem. **1999**, *14,* 581–606.

23. Mitsch, W.J.; Wise, K.M. Water quality, fate of metals, and predictive model validation of a constructed wetland treating acid mine drainage. Water Resour. **1998**, *32,* 1888–1900.

8

Rivers and Lakes: Acidification

Introduction 87
What Is Acidification, and How Is It Measured? 87
Historical Perspectives 88
Sources of Freshwater Acidification 89
    Natural Sources of Freshwater Acidification • Redox Reactions in Water–
    Sediment Systems • Biological Factors • Climatic Factors • Anthropogenic
    Acidification of Rivers and Lakes
Environmental Problems Caused by Acidification of
    Lakes and Rivers 96
Solutions to the Problem of Freshwater Acidification 96
    Environmental Legislation • Technology
Conclusion 98
References 98

Agniezka Gałuszka
and Zdzistaw M.
Migaszewski

Introduction

Acidification of rivers and lakes was one of the major environmental issues in the second half of the 20th century. As a consequence of the release of large amounts of acidifying gases (SO_2 and NO_x) to the atmosphere, mostly from coal burning and vehicle exhausts, the pH of precipitation dropped below 5.6. Acid deposition was responsible for lowering of the pH of surface waters and soils, and in many regions of the world, it resulted in forest decline and fish kills.

This entry presents the problem of acidification of rivers and lakes caused by natural and anthropogenic factors that contribute to decrease in the pH of freshwaters. In the first section, a definition of acidification and characteristics of processes causing acidification are provided with a brief description of the metrics used for acidification measurements. Historical perspectives of the problem are also briefly described. In a subsequent section, natural and anthropogenic acidification of rivers and lakes is discussed in more detail. The final part of the entry summarizes the consequences of freshwater acidification and the attempts made to solve this issue.

What Is Acidification, and How Is It Measured?

The most frequently used parameter that characterizes the acidity or alkalinity of natural waters is the pH. It is defined as a negative logarithm of hydrogen (hydronium) ion activity. Typical pH values in lakes and rivers are in the 6–9 range. The term "acidification" is usually used for freshwater with a pH of about 5.0 (and alkalinity below 0).[1] Some authors suggest that the pH threshold for acidified water should be 5.65 because this value characterizes pure water in equilibrium with atmospheric CO_2.[2] Two types of acidification of lakes and rivers can be distinguished: 1) chronic or long term in case acidifying factor is permanent (on a decade– century timescale), for example, related to bedrock or soil characteristics; and 2) episodic or short term (on an hour– day timescale), when acidifying factor is temporal, for example, acidic metal-rich effluent, snowmelt runoff, heavy rainfalls etc.

Acidification of surface waters begins when the concentration of hydrogen ions exceeds base ions present in water. The bases are produced during hydrolysis of carbonate minerals (mostly calcite) that occur in rocks and soils. The difference between the sum of the cations of strong bases and the sum of the anions of strong acids is termed acid-neutralizing capacity (ANC). This parameter is widely used in modeling of freshwater acidification and in assessment of freshwater sensitivity to acidification (critical load concept). Its components include both strong (OH^-) and weak bases (e.g., carbonate species, aluminohydroxides, organic acids). Alkalinity expresses the ability of water to neutralize acids. It is often understood as a synonym of ANC only measured in filtered water samples, whereas ANC is measured in unfiltered samples.[3] The buffering capacity of natural waters is mostly governed by a bicarbonate buffering system. Gaseous CO_2, which is a product of weathering of carbonate rocks or decay of organic remains, dissolves in water and produces carbonic acid, which subsequently dissociates to form H^+, HCO_3^-, and CO_3^{2-} ions. These reactions remain in equilibrium:

$$CO_2 + H_2O \leftrightarrow H_2CO_3 \leftrightarrow HCO_3^- + H^+ \leftrightarrow CO_3^{2-} + 2H^+$$

Bicarbonate and carbonate ions are capable of removing H^+ from water, causing the reactions to shift to the left. The bicarbonate buffering system results in a relatively constant pH and is mainly responsible for the alkalinity of natural waters. Thus, the influence of acidifying compounds on freshwater pH is more complex and it depends on ANC of the catchment area. The concept of "critical load" was introduced to find catchments that are the most vulnerable to acidification. Critical loads, defined as "an ecological threshold or intolerance to the accumulation of a pollutant in an ecosystem,"[4] can be estimated precisely on the basis of deposition rates of acidifying compounds and freshwater chemistry (empirical approach). The critical loads for lakes is a pH of 6.0 and an ANC of 0.02 meq/L.[5] Waters that have low alkalinity values (in the range of 0–10 mg/L $CaCO_3$) are highly sensitive to acidification.[6]

Alternatively, information about local geologic makeup, soil, and the land use of a study area may be employed to assess the sensitivity of freshwater to acidification.[7] The response of diatom assemblages to acid deposition and calcium levels in waters can also be used to calculate critical loads.[8] The difference between deposition of acidifying compounds and critical load values is termed the exceedance of critical load, and its values are crucial in assessment of potential acidification of freshwaters in a given area.

Paleoenvironmental data and fly-ash particle analysis in dated sediment cores are applied to the study of the history of surface water acidification and allow one to address the question regarding the origin of acidification (anthropogenic or natural). Paleolimnological pH reconstruction is based on the assumption that diatom and chrysophyte microfossils in dated sediment core intervals reflect the pH of water at a time when these organisms thrived in it. Based on the diatom techniques, it is possible 1) to evaluate models of acidification; 2) to determine critical loads of acidity; 3) to study the extent of episodic acidification; and 4) eventually to monitor lakes and rivers that are recovering from acidification.[9] The results of paleolimnological studies have indicated that the pH of some lakes located in northern Europe, eastern Canada, and northeastern United States have dropped by 0.5–1.5 from the early 1800s to recent times.[10]

Historical Perspectives

Acid deposition has been studied since the onset of the industrial revolution when Robert Smith, a creator of the term "acid rain", published the first results of his study on the rainwater chemistry in Manchester, U.K.[10] Surprisingly, it took more than 100 years to recognize the influence of acid deposition on freshwater acidification.[11,12] Although the first studies on acid deposition were carried out in Europe, the impact of precipitation chemistry on freshwater ecosystems was originally identified in North America in the late fifties by Gorham.[11] In 1968, Svante Odén showed the relationship between sulfur emissions in continental Europe and sulfate deposition affecting acidification of lakes in Scandinavia.[12]

Fleisher and others[13] have distinguished three periods of research regarding acidification of the environment: 1) an inventory period (during the 1960s and 1970s); 2) a period of mechanism studies (late 1970s and 1980s); and 3) a synthesizing period (1990s). This sequence of scientific interests has been supplemented by Herrmann[14] who suggested that a consequence of detailed knowledge on this issue should trigger a period of implementation.

Lake and river acidification became one of the major environmental issues in the 1970s and 1980s, when most of the reports from regional monitoring surveys were published. Freshwater acidification was most intensively studied in Europe (the U.K. and Scandinavian countries including Norway, Sweden, and Finland) and North America (Canada and United States).

During the 1980s, the concept of critical load was applied to quantify ecosystem vulnerability to acidification. This approach was also widely used in an assessment of freshwater sensitivity to acidification, mostly in Europe, North America, and Asia.[6] Recent research activities are focused on the recovery of lakes and rivers from long term acidification in most susceptible to acidification areas throughout the world, with some constraints to their future.[15–17] Attention is also being given to the emerging acidification of freshwaters in developing countries that experience rapid industrialization (Asia, Africa, Latin America).[18,19]

Sources of Freshwater Acidification

Factors influencing freshwater acidification may be natural (Figure 1) or anthropogenic (Figure 2) in origin. The lowest pH values are recorded in inland waters influenced by natural sources of acidifying compounds, especially of geologic origin. However, it should be stressed that natural acidification is commonly overlapped or enhanced by anthropogenic influences. Hence, separating natural acidity from anthropogenic freshwater acidification due to their different effects on biota is reasonable.[20] An important difference between anthropogenic and natural acidification is that the former one is reversible, as shown by many studies conducted after decreasing SO_2 and NO_x emissions.[21]

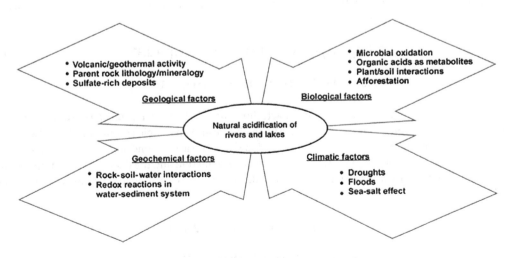

FIGURE 1 The most important natural factors influencing freshwater acidification.

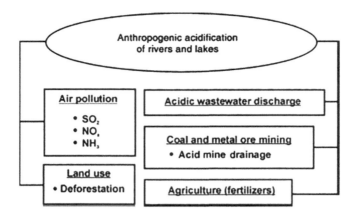

FIGURE 2   The most important anthropogenic factors influencing freshwater acidification.

Natural Sources of Freshwater Acidification

Geological Factors

Acidification of lakes and rivers as a consequence of natural processes in the environment is highly dependent on geologic factors, such as the presence of volcanic and geothermal activity and lithology/mineralogy of parent rocks. The volcanic/postvolcanic processes are responsible for formation of the world's mostly acidified surface waters; for example, the pH of caldera lake Kawah Ijen in western Java, Indonesia, is <0.3,[22] whereas the Banyupahit river, which originates from this lake, exhibits a pH in the range of about 0.7 to 3.3. Another example is the Kuril Islands (currently Russia) where many lakes and rivers are influenced by volcanic/postvolcanic activity. The main constituents of exhalations are sulfuric and hydrochloric acids. For example, the pH of Lake Usoriko (Japan) varies from 3.4 to 3.8.[23]

The extremely low pH values of inland waters are also caused by oxidation of pyrite and iron-bearing sulfide minerals (e.g., chalcopyrite, pyrrhotite, and marcasite). The mechanism of this process is natural and is known as "acid rock drainage" (ARD). It should be emphasized that ARD formation rate is usually accelerated by mining activities, and this process is called "acid mine drainage" (AMD), which will be further discussed in the next section. A series of reactions accompany the weathering of pyrite. Pyrite can undergo oxidation by two natural oxidants: 1) oxygen and 2) the even more effective ferric iron (Fe^{3+}), according to simplified reactions[24]:

$$FeS_2 \left(pyrite \right) + 3.5O_2 + H_2O \rightarrow Fe^{2+} + 2SO_4^{2-} + 2H^+ \tag{1}$$

$$FeS_2 + 14Fe^{3+} + 8H_2O \rightarrow 15Fe^{2+} + 2SO_4^{2-} + 16H^+ \tag{2}$$

The first reaction is assisted by acidophilous bacterium species *Acidithiobacillus ferrooxidans*. In case of the second reaction, Fe^{3+} rapidly oxidizes pyrite in abiotic and anaerobic conditions. These two reactions (1 and 2) bring about a substantial drop in pH and an increase in concentrations of ferrous (Fe^{2+}) and sulfate ions in water. Subsequently, the Fe^{2+} undergoes oxidation and the dissolved Fe^{3+} supports the second reaction or may hydrolyze and ferric hydroxide precipitates according to the following simplified reactions:

$$Fe^{2+} + 1/4O_2 + H^+ \rightarrow Fe^{3+} + 1/2H_2O$$

$$Fe^{3+} + 3H_2O \rightarrow Fe(OH)_3 \downarrow 3H^+$$

This acidity has further been generated by complex processes of transformation of iron oxyhydroxysulfates and oxyhydroxides into goethite (α-FeOOH).

Although geologic processes lead to the formation of extremely low pH waters, their detrimental influence on the environment is usually restricted to relatively small areas. Besides, these factors occur in the environment for a long time, creating a unique ecological niche for acid tolerant organisms. Table 1 presents examples of surface inland waters acidified mostly by geological factors.

Geochemical Factors

Rock–Soil–Water Interactions The pH and concentrations of metals in surface waters are largely controlled by the chemistry and mineralogy of bedrock and soils. In areas where bedrock consists of igneous and metamorphic rocks (granites, gneisses, porphyries) or silicoclastic rocks, soils have a low buffering capacity, resulting in a naturally low pH of lakes and rivers. The slow weathering of rocks contributes particularly to increasing freshwater acidification. This process can be accelerated by construction activities or land melioration. The extent of exposure of mineralized rock formations to weathering at the surface, the amount of scattered sulfide minerals, and the presence or absence of buffering carbonate rocks or gangue minerals are decisive factors that affect the range of acidification of surface waters. Other geochemical factors having an influence on acidification are acidic metalliferous springs, whose waters have interacted with mineralized rock formations, and unmined mineral deposits.

Other soil properties can affect the pH of surface waters, mostly through ion-exchange reactions and their influence on water buffering capacity. Lakes and rivers located in areas with natural acidic soils are more susceptible to acidification.

Acid sulfate soils are soils abundant in pyrite and other sulfide minerals. These soils, formed under waterlogged conditions, when exposed to the air (e.g., by draining, excavation, cultivation), undergo chemical reactions that are responsible for sulfuric acid generation and subsequent acidification.

TABLE 1 Examples of Surface Inland Waters Acidified Mostly by Geological Factors

Name	Type	pH	Country	Reference
Influence of volcanic/geothermal activity				
Kawah Ijen	Lake	<0.3	Indonesia	[22]
Banyupahit	River	0.7–3.3		
Usoriko	Lake	3.4–3.8	Japan	[23]
Caviahue	Lake	2.40–2.48	Argentina	[25]
Upper Rio Agrio	River	1.78		
Yugama	Lake	0.90–1.80	Japan	[26]
Popocatépetl	Lake	1.37–1.50	Mexico	[27]
Poás	Lake	0	Costa Rica	[28]
Kislyi Creek	River	2.45	Russia	[29]
Lesnaya	River	4.8		
Influence of bedrock lithology/mineralogy or acid rock drainage				
Woods Lake	Lake	4.4–5.9	United States (Adirondack Mts)	[30]
Lake Härkälampi	Lake	3.8–7.3	Finland	[31]
Langedalstjenn	Lake	4.4	Norway	[32]
Isiurqutuuq	Lake	4.5	Canada	[33]
Snake River	River	3.0	United States, Colorado	[34]
Rio Tinto	River	<3.0	Spain	[35]

Pyrite-rich soils that are not exposed to weathering pose no hazard to the environment and are called "potential acid sulfate soils" or cat-clays.

Most of the acid sulfate soils formed during the last 10,000 years following the postglacial sea level rise and the eustatic uplift of landmass. The acid sulfate soils occur mostly in lowland coastal regions (Australia, southeastern Asia, Finland), but they may also be associated with other environmental compartments favoring pyrite formation and oxidation, such as freshwater wetlands or, as mentioned above, areas with sulfide-scattered rock formations. It is estimated that acidic sulfate soils encompass an area of about 20 million ha globally.[36] The impact of acidic sulfate soils on freshwaters is highlighted by leaching of oxidation products into streams and lakes. The increased acidity of both soils and waters intensifies the mobility of many elements creating potentially toxic environments for living organisms.

Although acidic sulfate soils influence surface water acidification, there are also other soil types that may contribute to this process. Red soils developed on silicate rocks in Okinawa Island, Japan,[37] are one example. Their pH is low (4.5–5.5) and surface waters in the red-soil-dominated areas exhibit a pH within the range of 4.9–5.8.

Redox Reactions in Water–Sediment Systems

Freshwater pH can be changed by redox processes in sediments. During oxidation reactions, H^+ ions are produced, which decreases the pH of water. Nitrogen and sulfur (as inorganic and organic species) and iron are the most important elements in sediment/water redox processes. The changes in their oxidation state are usually microbially mediated. The effective acid production capacity (APC_{eff}) is a measure of acidification caused by oxidation reactions in sediment suspension. It is defined as

$$APC_{eff} = V/W.\left(\left[H^+\right]_e - \left[H^+\right]_O\right)$$

where V is the suspension volume, W is the solid mass, and $[H^+]_{e,o}$ is the H^+ concentration before and after oxidation.[38] The oxidation reactions that are responsible for acid generation in aquatic systems include 1) oxidation of sulfur from hydrogen sulfide, atomic sulfur, metal sulfides, and organic compounds (R-SH); 2) oxidation of pyrite (FeS_2) and Fe^{2+} ions; and 3) oxidation of N from NH_4^+, NO_2^- (nitrification), and organic species (R-NH_2) (ammonification).

Several important buffer reactions occur in sediments, which can consume excess hydrogen ions and contribute to an increased ANC. These reactions involve $CaCO_3$, Al_2O_3, and Fe_2O_3:

$$CaCO_3 + 2H^+ \rightarrow Ca^{2+} + H_2O + CO_2 \qquad (5)$$

$$Al_2O_3 + 6H^+ \rightarrow 2Al^{3+} + 3H_2O \qquad (6)$$

$$Fe_2O_3 + 6H^+ \rightarrow 2Fe^{3+} + 3H_2O \qquad (7)$$

Table 2 presents examples of surface inland waters acidified mostly by geochemical factors.

Biological Factors

Metabolic processes of living organisms also can contribute to the lowering of surface water pH. This drop in the pH values may be caused by numerous oxidation reactions that are mediated by microorganisms and produce H^+ as described in the previous sections.

As suggested by Rosenqvist as early as 1978,[44] expansion of coniferous forests brought about a decrease in the pH of surface waters in Norway. Afforestation of acid sensitive areas may result in freshwater acidification due to the following factors: 1) increased evapotranspiration; 2) intensive uptake of

TABLE 2   Examples of Surface Inland Waters Acidified Mostly by Geochemical Factors

Name	Type	pH	Country	Reference
Cudgen Lake	Lake	2.5	Australia	[39]
Rocky Mouth Creek	Stream	<4.5	Australia	[40]
Esse River		5.9–6.8		[41]
Purmo River	Rivers	4.8		
Kronoby River		5.0	Finland	[42]
Kovjoki River		4.7		
Larsmo Lake	Lake	4.7–5.7		
Colour Lake	Lake	3.6–4.7	Canada	[43]

cations during the tree-stand growth; 3) increased filtering of gases and aerosols in the canopy and a subsequent increase of dry deposition; and 4) influx of organic acids from litter fall decay.[45] Acidifying effects of afforestation depends on the tree-stand maturity. After reaching a state of equilibrium between the standing and dead biomass, the forest-induced acidity decreases due to the lower uptake of cations from the soil by mature trees.[46]

Organic acids produced by living organisms are also responsible for acidification of freshwater. It has been estimated that organic acids may change the pH of lake waters having ANC values within the range of 0–50 µeq/L by 0.5–2.5 pH units.[47] On the other hand, organic acids prevent surface waters from further acidification due to their role as buffering substances for acids originating from acid deposition. Dissociation of humic and fulvic acids increases H^+ concentrations in waters leading to their pH decrease. The typical pH of lakes affected by organic acids generated by *Sphagnum* mosses is 3.3–4.5.[1] Table 3 presents examples of surface inland waters acidified mostly by biological factors.

Climatic Factors

The influence of climatic factors on freshwater acidification is mostly related to precipitation and temperature variations. Several processes are involved with acidification brought about by seasonal climatic changes. Water table lowering during long drought periods accelerates oxidation reactions and accompanying production of H^+ ions, especially in wetlands.[55] Water table rising during floods causes intensive water flow through surface soil horizons, which are usually more acidic as a result of higher organic acid concentrations and higher susceptibility to anthropogenic acidification in relation to underlying horizons. During winter, the ice-covered lake waters accumulate CO2 originating from respiration, organic decay, and rock weathering. Due to this process, the pH of waters in epilimnion

TABLE 3   Examples of Surface Inland Waters Acidified Mostly by Biological Factors

Name	Type	pH	Country	Reference
Valkea-Kotinen	Lake	5.0–5.4	Finland	[48]
Rio Negro	River	4.63–5.80	Brazil	[49]
Rio Daraá	River	3.91		
Dumai	River	4.3	Indonesia	[50]
Pebbleloggitch	Lake	4.3	Canada	[51]
Yuanyang	Lake	4.8–5.5	Taiwan	[52]
Grosse Fuchskuhle	Lake	4.2–6.1	Germany	[53]
Liepsalas	Lake	4.1–4.3		
Murmasts	Lake	3.7–3.9	Latvia	[54]
Pieslaistes	Lake	4.0–4.2		

is decreased. Precipitation waters are ionically deficient and have a very low ANC. Therefore, their discharges to surface waters during spring snowmelt floods or heavy storms bring about an episodic decrease in the pH of surface waters.

Occasional acidification of lakes and rivers in coastal areas is caused by deposition of marine aerosol salts during high wind periods, changing cation-exchange equilibria in soil. Cations present in seawater (Na^+ and Mg^{2+}) undergo cation-exchange reactions with hydrogen and aluminum ions, which are released from soil to water. This phenomenon is termed "sea-salt effect."[45]

Winds that carry $CaCO_3$-rich. dusts may increase ANC and pH of surface waters by mitigating the effect of freshwater acidification, as was reported for the influx of Saharan dust to southern Alps.[56] Table 4 presents examples of surface inland waters acidified mostly by climatic factors.

Anthropogenic Acidification of Rivers and Lakes

Air Pollution with Acidifying Compounds

Serious attention to the problem of freshwater acidification was given when it became clear that there was an influence of air pollutants on the formation of acid precipitation and subsequent changes in the pH of surface waters and soils. Atmospheric deposition of acidifying compounds from anthropogenic sources has greatly contributed to the river and lake acidification. Sulfur dioxide and nitrogen oxides undergo chemical transformations in the air, producing sulfuric and nitric acids, which are then deposited on the land surface, lowering the pH of surface waters. Oxidation of SO_2 may undergo gas-phase or aqueous-phase reactions, and the sulfuric acid produced may occur as liquid aerosols, or it may dissolve in clouds or rainwater. It is noteworthy that only a small fraction of acid precipitation is deposited directly in rivers and lakes; freshwaters are mostly impacted by surface runoff.

Most of the studies on the influence of acid deposition on acidification of soils and waters have been focused on SO_2 emissions derived from fossil fuel combustion and smelting of sulfide minerals, as well as on NO_x released from fossil fuel combustion and vehicle exhausts.[67] Before reduction of SO_2 emissions, acid precipitation was mostly attributed to sulfuric acid. Since the 1990s, nitrogen compounds have been regarded as a major contributor to lowering of the precipitation pH.[68] Hydrochloric acid also causes a decrease in the precipitation pH, although to a much lesser extent than sulfur and nitrogen oxides. Anthropogenic emissions of NH_3, derived mostly from livestock farming and agriculture, neutralize acids that occur in the atmosphere. However, after deposition, $NH4^+$ ions affect freshwater pH through nitrification reactions that produce H^+ ions and via uptake of NH_4^+ ions, which in turn

TABLE 4   Examples of Surface Inland Waters Acidified Mostly by Climatic Factors

Name	Type	pH	Country	Reference
Induced by drought or flood				
Swan Lake	Lake	4.5	Canada	[57]
White pine Lake	Lake	4.1– 5.7	Canada	[58]
Hermanninlampi	Lake	4.0	Finland	[59]
Sink Beck	Pool	4.3	U.K.	[60]
Van Campens Brook	Stream	4.5	United States	[61]
Biwa	Lake	4.5– 6.5	Japan	[62]
Sea-salt effect (pH values measured during episodes)				
Svela	Stream	4.45	Norway	[63]
Lake Terjevann	Lake	4.28	Norway	[64]
Espedal	River	4.6	Norway	[65]
Allt a'Mharcaidh	Stream	<5.5	U.K.	[66]

release H^+ to soil solution. The acidification caused by strong acids contained in precipitation occurs immediately, whereas the acidification brought about by NH_3/NH_4^+ deposition is delayed in time. It is interesting to note that the acidification of ecosystems from NH_3 emissions may be equal to that from SO_2 emissions or even greater than that from NO_x emissions.[69]

Acid Mine Drainage

Acidification of freshwaters can be triggered by weathering of pyrite and other iron-bearing sulfide minerals in metal and coal mining areas. This process is termed acid mine drainage (AMD). Mine waste piles, mineral ponds, underground mines, and lignite open pits can potentially form acidic and/or metal-rich effluents that can enter local streams and rivers, which in turn may eventually reach various inland water bodies. The mechanism of acid formation resulting from pyrite oxidation was previously described in the subsection "Geological Factors."

AMD was responsible for serious environmental disasters in many areas throughout the world where metal ores or coals were mined.[70] The acid mine waters lead to lowering of the pH combined with the increase of sulfates and toxic element levels in surface water and groundwater. The most acidified waters in the world, originating from abandoned metal ore mines, were found in the Iron Mountain mining area in Shasta County, northern California.[71] In this area, several mines extracted pyrite (for sulfuric acid production) and Ag, Au, Cu, Fe, and Zn sulfide ores from the 1860s to 1962. The abandoned mines contain approximately 600,000 m^3 of strongly acidic water having a pH of about 1 and concentrations of Zn, Cu, and Cd in the range of several grams per liter. Prior to remediation, extremely high loads of dissolved Cd, Cu, and Zn (about 300 tons/yr) were discharged to the Sacramento River. Several massive fish kills were recorded as a consequence of high runoff episodes. Subsequent to the closure of the mines, more than 20 such events have occurred, with the most severe one in 1967, when at least 47,000 trout were killed during a single week.[72] Table 5 presents examples of surface inland waters acidified mostly by anthropogenic factors.

Other Sources

Aside from the natural and anthropogenic factors that were described in the previous subsections, there are also numerous sources of surface water acidification that are manifested by small-scale pH variations or are related to very specific cases.

Acidic wastewaters are also generated by metal processing industries, electroplating, oil refineries, and other chemical industries. Some wastewaters are not acidic, but eventually will become so after they undergo microbially mediated transformations (e.g., brewery wastewaters rich in organic carbon). The knowledge of wastewater properties enables us to control and avoid their detrimental influence on the environment with the help of the proper treatment technology.

Land use is sometimes responsible for acidification of surface waters. One of the examples is deforestation, which causes increased sensitivity of lakes and streams to acidification due to the removal of

TABLE 5 Examples of Surface Inland Waters Acidified Mostly by Anthripogenic Factors

Name	Type	Minimal pH Value	Country	Reference
Big Moose	Lake	4.6	United States (Adirondack Mts.)	[73]
Spring Creek	Stream	2.5	United States (Iron Mountain, California)	[74]
Lohi	Lake	4.4	Canada	[75]
ML111	Mining lake	2.54	Germany (Lusatia)	[76]
Lake Orta	Lake	3.9	Italy	[77]
Lysevatten	Lake	4.5	Sweden	[78]
Vikedal	River	5.35	Norway	[79]
Gentil Sapin	Stream	4.16	France	[80]
Čertovo	Lake	4.3	Czech Republic	[81]

cations with harvested biomass and intensification of nitrification of soils.[2] The increased acidity of lakes and rivers may also occur in agricultural lands, where acidification may come from fertilizers.

There are some additional natural factors contributing to surface water acidification, which need mentioning. Acidification of lakes depends on the watershed location because headwater lakes are more susceptible to acidification than downstream ones. Lakes located in the headwaters have a lower ANC.[82] Large amounts of sulfur dioxide released during natural fires in exposed lignite beds in Smoking Hills (Canada) caused acidification from deposition of sulfuric acid that was produced in the atmosphere.[83]

Environmental Problems Caused by Acidification of Lakes and Rivers

Inland water acidification has serious consequences for living organisms and ecosystem functioning. The direct influence of low pH on aquatic biota includes 1) the loss of species, especially fish (below pH 4.5), zooplankton, phytoplankton, benthic invertebrates, and periphyton; and 2) decreased or inhibited reproduction of fish and amphibians.[84]

Biological response to acidification may be rapid (a short-term effect) or delayed in time (a long-term effect).[85] Rapid response is caused by episodic events that result in a sudden decrease in the water pH, such as spring floods, heavy storms, or discharge of acidic wastewaters. The long-term effects on aquatic biota are highlighted mostly by changes in the ecosystem structure, with some reductions in species diversity and disappearance of acidsensitive species and dominance of acid-tolerant ones. The reduced biodiversity of freshwater systems in turn has an influence on the balance in biogeochemical cycles[86] and causes adverse changes in food chains. Acidification of lakes and rivers affects geochemical processes, including 1) higher mobility of most metals (e.g., Al, Pb, Zn, Mg, Cd); 2) increased methylmercury production; and 3) decreased phosphorus availability.

Aluminum mobilization from soil minerals to waters is responsible for increased fish mortality. The lethal aluminum toxicity to aquatic animals is most often recorded when acidic water mixes with alkaline water and $Al(OH)_3$ precipitates on the gills and filtering apparatus.[87] Some authors regard the release of free aluminum ions, which represent a bioavailable form for aquatic biota, as the most detrimental effect of freshwater acidification.[2]

Although methylmercury production is reduced at a low pH, synthesis of dimethylmercury by base-catalyzed disproportionation contributes to conversion of dimethyl mercury to monomethyl mercury by acid hydrolysis:

$$(CH_3)_2 Hg + H^+ \rightarrow CH_3Hg^+ + CH_4 \uparrow \qquad (8)$$

As a consequence of this reaction, an easily available and very toxic form of mercury is released to acidified surface waters.[86]

Bioavailability of phosphorus in acidified surface waters is diminished, as this element is adsorbed on aluminum oxides and hydroxides and may be irreversibly retained in sediments.[88] Acidic freshwaters decrease bioavailability of essential elements, such as calcium and magnesium.[89]

The response of aquatic biota to natural and anthropogenic acidification is believed to be different.[90] The mechanism of this difference is not well understood and can be explained by biogeographical histories of the region or the long-term evolutionary adaptation to natural acidification sources.[90,91]

Solutions to the Problem of Freshwater Acidification

The problem of freshwater acidification has stimulated actions to minimize its consequences, mostly through legislation and various international initiatives that allowed reduction of the levels of acidifying air pollutants. An important tool for rehabilitation of acidified waters has been offered by technology.

Every effort undertaken to reduce the risk of freshwater acidification, apart from its character, has been driven by progress in scientific research as well as by international cooperation.

Environmental Legislation

Environmental legislation, both on national and international scales, has led to substantial reduction of SO_2 and NO_x emissions in Europe and North America in recent decades and has ameliorated the effect of acid precipitation on acidification of rivers and lakes. One of the most important documents concerning the control of acidifying compounds is The Convention on Long-Range Transboundary Air Pollution (LRTAP) under the auspices of the United Nations Economic Commission for Europe.[92] This Convention has been extended by four Protocols that control emission of acidifying compounds: 1) the 1999 Protocol to Abate Acidification, Eutrophication, and Ground-Level Ozone, supplemented with the revised guidance document on ammonia; 2) the 1994 Protocol on Further Reduction of Sulphur Emissions; 3) the 1988 Protocol Pertaining to the Control of Nitrogen Oxides or Their Transboundary Fluxes; and 4) the 1985 Protocol on the Reduction of Sulphur Emissions or their Transboundary Fluxes by at least 30%. One of the initiatives directly regarding acidification of lakes and rivers is The International Cooperative Programme on Assessment and Monitoring of Acidification of Rivers and Lakes (ICP Waters), which is the action of The Working Group on Effects within LRTAP Convention. The ICP Waters initiative collects data that are essential for monitoring the effects of acid precipitation on surface waters from 18 European countries, United States, and Canada.

For European Union member states, the emission standards of SO_2 and NO_x from specific sources are regulated by adequate directives, for example, emission limit values for new plants and industrial facilities from the Large Combustion Plant Directive (88/609/EEC) or the Directives 98/70/EC and 1999/32/EC that establish the limits on the sulfur content of gas oil for stationary and mobile sources and for heavy fuel oil.[93] In Europe, present emissions of SO_2 are less than half of the 1980s levels and emissions of NOx have been reduced by about 20%.[21]

In the United States, reductions in SO_2 and NO_x emission are regulated by the Amendment to the Clean Air Act, Title IV: Acid Deposition Control, also known as the Acid Rain Program, which was implemented in 1990. This document imposed reductions in SO_2 and NO_x emissions from fossil fuel power plants. The emissions of SO_2 from power generating plants in the United States were reduced from 17.3 million tons in 1980 to 11.2 million tons in 2000 and those of NO_x from 6.0 to 5.1 million tons, respectively.[94]

Technology

Both environmental technology and engineering help to reduce the emissions of acidifying gases from industrial sources and facilitate the proper management of acidified freshwaters. The practice, which is most commonly used for rehabilitation of acidified lakes and rivers, is liming. Its purpose is to restore the water to its preacidification state, which is often reconstructed with paleolimnological techniques. Limestone powder can be dispersed on the whole surface of the lake or, alternatively, can be dosed onto the ice cover. It is suggested that liming of lakes and inlet streams should be done with terrestrial liming, which ameliorates the effect of reacidification of the littoral zone during the ice cover and the snowmelt events.[95] Liming was extensively applied in Scandinavian countries; for example, in Sweden, it was used for about 8000 lakes and 12,000 km of surface waters to achieve the following restoration goals: pH > 6 and ANC > 100 meq/L.[96] Although liming is very popular, there are some limitations in its application; for example, Wällstedt and Borg,[97] in their study on remobilization of metals from limed and nonlimed lake sediments, have shown that there is a risk of potential toxicity of Al, Cd, Mn, and Zn in lakes that have been limed. Moreover, some toxic anionic metalloids (especially arsenic), which are more mobile at a higher pH, may be released from the sediment to the water column in case of alkalization by lime.

The Swedish Forest Agency has recommended that wood ash (received from timber waste combusted for energy production) should be used to mitigate acidification and to supplement nutrient removal.[98] In this method, the wood ash is added to the soil cover. Treatment of acidified waters is also possible with the use of ecotechnologies, such as constructed wetlands or controlled eutrophication.[99] The employment of these methods has many advantages. They are cost-effective, they are based on natural processes, and they can be applied to reduce acidification originating from non-point pollution sources (e.g., AMD).

Conclusion

Lake and river acidification has been probably one of the most often discussed environmental issues in recent decades. Although natural acidic freshwaters occur throughout the world, the human-induced changes in water pH have aroused international concern, especially in countries where anthropogenic emissions of SO_2, NO_2, and NH_3 caused a substantial decrease in the freshwater pH due to their low ANC (Scandinavian countries, eastern Canada, northeastern United States). Acidic lakes and rivers have jeopardized aquatic biota, not only due to the low acid tolerance of living organisms but also because of the geochemical changes that have led to increased geo and bioavailability of toxic metals (Al, Zn, Mn, Pb, Hg, Cd) and reduction in the amount of bioavailable essential elements (Ca, Mg, P). The long-term international interest in the problem of acidification resulted in better understanding of the acidification processes and mechanisms and in well-established knowledge of the methods of monitoring and remediation and proper management of acidified rivers and lakes. As a result of international efforts in the abatement of SO_2 and NO_x emissions from industrial sources, many lakes and streams have already recovered from anthropogenic acidification. Both environmental legislation and technology have contributed to substantial improvement of the quality of surface inland waters. The lessons learned from almost 50 years of studies on acidification and its mitigation have shown that environmental pollution is a global problem, which can only be solved with cooperation of governments, scientists, and societies. Whether these lessons will protect us from similar problems in the future is an open question. Many authors claim that acidification will soon develop in rapidly developing countries of Asia, Africa, and Latin America. The experience gained from remediation of acidified lakes and rivers in the northern hemisphere may thus become invaluable for the global sustainable future.

References

1. Mattson, M.D. Acid lakes and rivers. In *Encyclopedia of Environmental Science*; Encyclopedia of Earth Sciences Series; Alexander, D.E., Fairbridge, R.W., Eds.; Kluwer Academic Publishers: Dordrecht, 1999; 6–9.
2. Norton, S.A.; Vesely J. *Acidification and Acid Rain*; Treatise on Geochemistry; Lollar B.S., Ed.; Pergamon: Oxford, 2005; 367–406.
3. Allen, D.J.; Castillo, M.M. *Stream Ecology: Structure and Function of Running Waters*, 2nd Ed.; Kluwer Academic Publishers: Boston, 2007.
4. O'Riordan, T. Critical load. In *Encyclopedia of Environmental Science*; Encyclopedia of Earth Sciences Series; Alexander, D.E., Fairbridge, R.W., Eds.; Kluwer Academic Publishers: Dordrecht, 1999; 102.
5. Elvingson, P.; Ågren, C. *Air and the Environment*; Swedish NGO Secretariat on Acid Rain: Goteborg, 2004.
6. Krzyzanowski, J.; Innes, J.L. Back to the basics—Estimating the sensitivity of freshwater to acidification using traditional approaches. J. Environ. Manage. **2010**, *91*, 1227–1236.
7. Hall, J.R.; Wright, S.M.; Sparks, T.H.; Ullyeti, J.; Allot, T.E.H.; Hornung, M. Predicting freshwater critical loads from national data on geology, soil and land use. Water Air Soil Pollut. **1995**, *85*, 2443–2448.

8. Battarbee, R.W.; Allot, T.E.H.; Juggins, S.; Kreiser, A.M.; Curtis, C.; Harriman, R. Critical loads of acidity to surface waters: An empirical diatom-based palaeolimnological model. Ambio **1995**, *25*, 366–369.

9. Battarbee, R.W.; Charles, B.F.; Bigler, C.; Cumming, B.F.; Renberg, I. Diatoms as indicators of surface-water acidity. In *The Diatoms: Aplications for the Environmental and Earth Sciences*; Smol, J., Stoermer, E., Eds.; Cambridge University Press: New York, 2010; 98–121.

10. Smith, R.A. On the air and rain of Manchester. Mem. Lit. Philos. Soc. Manchester **1852**, *10*, 207–217.

11. Gorham, E. The influence and importance of daily weather conditions in the supply of chloride, sulphate, and other ions to fresh waters from atmospheric precipitations. Philos. Trans. R. Soc. B **1958**, *247*, 147–178.

12. Odén, S. *The acidification of air and precipitation and its consequences in the natural environment*; Ecological Committee Bulletin No. 1. Swedish Natural Science Research Council, Stockholm. Translations Consultants, Ltd.: Arlington, Virginia, 1968.

13. Fleisher, S.; Andersson, G.; Brodin, Y.W.; Dickson, W.; Herrmann, J.; Muniz, I.P. Acid water research in Sweden— Knowledge for tomorrow? Ambio **1993**, *22*, 258–263.

14. Herrmann, J. Aluminum is harmful to benthic invertebrates in acidified waters, but at what threshold(s)? Water Air Soil Pollut. **2001**, *130*, 837–842.

15. Arseneau, K.M.A.; Driscoll, C.T.; Brager, L.M.; Ross K.A.; Cumming B.F. Recent evidence of biological recovery from acidification in the Adirondacks (New York, USA): A multiproxy paleolimnological investigation of Big Moose Lake. Can. J. Fish. Aquat. Sci. **2011**, *68*, 575–592.

16. Skjelkvále, B.L.; Evans, C.; Larssen, T.; Hindar, A.; Raddum, G.G. Recovery from acidification in European surface waters: A view to the future. Ambio **2003**, *32* (3), 170–175.

17. Helliwell, R.C.; Simpson, G.L. The present is the key to the past, but what does the future hold for the recovery of surface waters from acidification? Water Res. **2010**, *44* (10), 3166–3180.

18. Kuylenstierna, J.C.I.; Rodhe, H.; Cinderby, S.; Hicks, K. Acidification in developing countries: Ecosystem sensitivity and the critical load approach on a global scale. Ambio **2001**, *30* (1), 20–28.

19. Ye, X.; Hao, J.; Duan, L.; Zhou, Z. Acidification sensitivity and critical loads of acid deposition for surface waters in China. Sci. Total Environ. **2002**, *289*, 189–203.

20. Petrin, Z.; Englund, G.; Malmqvist B. Contrasting effects of anthropogenic and natural acidity in streams: A meta-analysis. Proc. R. Soc. B **2008**, *275*, 1143–1148.

21. Evans, C.D.; Cullen, J.M.; Alewell, C.; Kopácek, J.; Marchetto, A.; Moldan, F.; Prechtel, A.; Rogora, M.; Vesely, J.; Wright, R. Recovery from acidification in European surface waters. Hydrol. Earth Syst. Sci. **2001**, *5* (3), 283–297.

22. Löhr, A.J.; Bogaard, T.A.; Heikens, A.; Hendriks, M.R.; Sumarti, S.; Van Bergen, M.J.; Van Gestel, C.A.; Van Straalen, N.M.; Vroon, P.Z.; Widianarko, B. Natural pollution caused by the extremely acidic crater lake Kawah Ijen, East Java, Indonesia. Environ. Sci. Pollut. Res. Int. **2005**, *12* (2), 89–95.

23. Satake, K.; Oyagi, A.; Iwao, Y. Natural acidification of lakes and rivers in Japan: The ecosystem of Lake Usoriko (pH 3.4–3.8). Water Air Soil Pollut. **1995**, *85*, 511–516.

24. Nordstrom, D.K.; Alpers, C.N. Geochemistry of acid mine waters. In *The Environmental Geochemistry of Mineral Deposits*; Plumlee, G.S., Logsdon, M.J., Eds.; Rev. Econ. Geol. 6A: Littleton, CO, 1999; 133–160.

25. Pedrozo, F.; Kelly, L.; Diaz, M.; Temporetti, P.; Baffico, G.; Kringel, R.; Friese, K.; Mages, M.; Geller, W.; Woelfl, S. First results on the water chemistry, algae and trophic status of an Andean acidic lake system of volcanic origin in Patagonia (Lake Caviahue). Hydrobiologia **2001**, *452*, 129–137.

26. Takano, B.; Watanuki, K. Monitoring of volcanic eruptions at Yugama crater lake by aqueous sulfur oxyanions. J. Volcanol. Geotherm. Res. **1990**, *40*, 71–87.

27. Armienta, M.A.; De la Cruz-Reyna, S.; Macías, J.L. Chemical characteristics of the crater lakes of Popocatépetl, El Chichón, and Nevado de Toluca volcanoes, Mexico. J. Volcanol. Geotherm. Res. **2000**, *97*, 105–125.

28. Brantley, S.L.; Borgia, A.; Rowe, G.; Fernandez, J.F.; Reynolds, J.R. Poás volcano crater lake acts as a condenser for acid metal-rich brine. Nature **1987**, *330* (3), 470–472.

29. Chudaev, O.; Chudaeva, V.; Sugimori, K.; Kuno, A.; Matsuo, M. Geochemistry of recent hydrothermal systems of Mendeleev Volcano, Kuril Islands, Russia. J. Geochem. Explor. **2006**, *88*, 95–100.

30. April, R.; Newton, R. Influence of geology on lake acidification in the ILWAS watersheds. Water Air Soil Pollut. **1985**, *26*, 373–386.

31. Loukola-Ruskeeniemi, K.; Uutela, A.; Tenhola, M.; Paukola, T. Environmental impact of metalliferous black shales at Talvivaara in Finland, with indication of lake acidification 9000 years ago. J. Geochem. Explor. **1998**, *64* (1–3), 395–407.

32. Hindar, A.; Lydresen, E. Extreme acidification of lake in southern Norway caused by weathering of sulphide-containing bedrock. Water Air Soil Pollut. **1994**, *77*, 17–25.

33. Cameron, E.M.; Prévost, C.L.; McCurdy, M.; Hall, G.E.M.; Doidge, B. Recent (1930s) natural acidification and fish-kill in a lake that was an important food source for an Inuit community in northern Québec, Canada. J. Geochem. Explor. **1998**, *64*, 197–213.

34. Munk, L.A.; Faure, G.; Pride, D.E.; Bigham, J.M. Sorption of trace metals to an aluminum precipitate in a stream receiving acid rock-drainage; Snake River, Summit County, Colorado. Appl. Geochem. **2002**, *17* (4), 421–430.

35. Aguilera, A.; Manrubia, S.C.; Gómez, F.; Rodríguez, N.; Amils, R. Eukaryotic community distribution and its relationship to water physicochemical parameters in an extreme acidic environment, Río Tinto (Southwestern Spain). Appl. Environ. Microbiol. **2006**, *72* (8), 5325–5330.

36. Burton, E.D.; Bush, R.T.; Sullivan, L.A. Sedimentary iron geochemistry in acidic water-ways associated with coastal lowland acid sulfate soils. Geochim. Cosmochim. Acta **2006**, *70*, 5455–5468.

37. Vuai, S.A.; Ishiki, M.; Tokuyama, A. Acidification of freshwaters by red soil in a subtropical silicate rock area, Okinawa, Japan. Limnology **2003**, *4*, 63–71.

38. Calmano, W.; von der Kammer, F.; Schwartz, R. Characterization of redox conditions in soils and sediments: Heavy metals. In *Soil and Sediment Remediation: Mechanisms, Technologies and Applications,* 1st Ed.; Lens, P., Grotenhuis, T., Malina, G., Tabak, H., Eds.; IWA Publishing: London, 2005; 102–120.

39. White, I.; Melville, M.D.; Wilson, B.P; Sammut, J. Reducing acidic discharges from coastal wetlands in eastern Australia. Wetlands Ecol. Manage. **1997**, *5*, 55–72.

40. Lin, C.; Wood, M.; Haskins, P.; Ryffel, T.; Lin, J. Controls on water acidification and de-oxygenation in an estuarine waterway, eastern Australia. Estuarine, Coastal Shelf Sci. *61* (1), 55–63.

41. Roos, M.; Åström, M. Seasonal and spatial variations in major and trace elements in a regulated boreal river (Esse River) affected by acid sulphate soils. River Res. Appl.2004 *21*, 351– 361.

42. Toivonen, J.; Österholm, P. Characterization of acid sulfate soils and assessing their impact on a humic boreal lake. J. Geochem. Explor. **2011**, *110* (2), 107–117, doi:10.1016/ j.gexplo.2011.04.003.

43. Johannesson, K.H.; Lyons, W.B. Rare-earth element geochemistry of Colour Lake, an acidic freshwater lake on Axel Heiberg Island, Northwest Territories, Canada. Chem. Geol. **1995**, *119*, 209– 223.

44. Rosenqvist, I. Alternative sources of acidification of river water in Norway. Sci. Total Environ. **1978**, *10*, 39– 49.

45. Larssen, T.; Holme, J. Afforestation, seasalt episodes and acidification—A paired catchment study in western Norway. Environ. Pollut. **2006**, *139*, 440– 450.

46. Emmett, B.A.; Reynolds, B.; Stevens, P.A.; Norris, D.A.; Hughes, S.; Gorres, J.; Lubrecht, I. Nitrate leaching from afforested Welsh catchment—Interactions between stand age and nitrogen deposition. Ambio **1993**, *22*, 386– 394.

47. Lydersen, E. Humus and acidification. In *Aquatic Humic Substances. Ecology and Biogeochemistry*; Hessen D.O., Tranvik L.J., Eds.; Springer: Berlin-Heidelberg, 1998; 63– 92.

48. Keskitalo, J.; Salonen, K.; Holopainen, A.-L. Long-term fluctuations in environmental conditions, plankton and macrophytes in a humic lake, Valkea-Kotinen. Boreal Environ. Res. **1998**, *3*, 251– 262.

49. Küchler, I.L.; Miekeley, N.; Forsberg, B.R. A contribution to the chemical characterization of rivers in the Rio Negro Basin, Brazil. J. Braz. Chem. Soc. **2000**, *3*, 286– 292.

50. Alkhatib, M.; Jennerjahn, T.C. Biogeochemistry of the Dumai River estuary, Sumatra, Indonesia, a tropical blackwater River. Limnol. Oceanogr. **2007**, *52* (6), 2410– 2417.

51. Kerekes, J.; Freedman, B.; Beauchamp, S.; Tordon, R. Physical and chemical characteristics of three oligotrophic lakes and their watersheds in Kejimkujik National Park, Nova Scotia. Water Air Soil Pollut. **1989**, *46* (1– 4), 99– 117.

52. Wu, J.-T.; Chang, S.-C.; Wang, Y.-S.; Wang, Y.-F.; Hsu, M.-K. Characteristics of the acidic environment of the Yuanyang Lake (Taiwan). Bot. Bull. Acad. Sin. **2001**, *42*, 17– 22.

53. Casper, P.; Chan, O.C.; Furtado, A.L.S.; Adams, D.D. Methane in an acidic bog lake: The influence of peat in the catchment on the biogeochemistry of methane. Aquat. Sci. **2003**, *65*, 36– 46.

54. Klavins, M.; Rodinov, V.; Druvietis, I. Aquatic chemistry and humic substances in bog lakes in Latvia. Boreal Environ. Res. **2003**, *8*, 113– 123.

55. Laudon, H. Recovery from episodic acidification delayed by drought and high sea salt deposition. Hydrol. Earth Syst. Sci. **2008**, *12*, 363– 370.

56. Rogora, M.; Mosello, R.; Marchetto, A. Long-term trends in the chemistry of atmospheric deposition in NW Italy: The role of increasing Saharan dust deposition. Tellus **2004**, *56B*, 426– 434.

57. Yan, N.D.; Keller, W.; Scully, N.M., Lean, D.R.S.; Dillon, P. Increased UV-B penetration in a lake owing to droughtinduced acidification. Nature **1996**, *381*, 141– 143.

58. Tranter, M.; Davies, T.D.; Wigington P.J.; Eshleman, K.N. Episodic acidification of freshwater systems in Canada— Physical and geochemical processes. Water Air Soil Pollut.1994 *72* (1– 4), 19– 39.

59. Holopainen, I.J. The effects of low pH on planktonic communities. Case history of a small forest pond in eastern Finland. Ann. Zool. Fenn. **1992**, *28*, 95– 103.

60. Tipping, E.; Smith, E.J.; Lawlor, A.J.; Hughes S.; Stevens P.A. Predicting the release of metals from ombrotrophic peat due to drought-induced acidification. Environ. Pollut. **2003**, *123* (2), 239– 253.

61. Stansley, W.; Cooper, G. An acidic snowmelt event in a New Jersey stream: Evidence of effects on an indigenous trout population Water Air Soil Pollut. **1990**, *53* (3– 4), 227– 237.

62. Fushimi, H.; Kawamura, T.; Iida, H.; Ochiai, M.; Nakajima, T.; Azuma, Y. Internal distribution of acid materials within snow crystals. Water Air Soil Pollut. **2001**, *130* (1– 4), 1709– 1714.

63. Hindar, A.; Henriksen, A.; Kaste, Ø.; Tørseth, K. Extreme acidification in small catchments in southwestern Norway associated with a sea salt episode. Water Air Soil Pollut.1994, *85* (2), 547–552.

64. Andersen, D.O.; Seip, H.M. Effects of a rainstorm high in sea-salts on labile inorganic aluminium in drainage from the acidified catchments of Lake Terjevann, southernmost Norway. J. Hydrol. **1999**, *224*, 64–79.

65. Teien, H.-C.; Salbu, B.; Heier, L.S.; Kroglund, F.R.; Bjørn, O. Fish mortality during sea salt episodes—Catchment liming as a countermeasure. J. Environ. Monit. **2005**, *7*, 989–998.

66. Bonjean, M.C.; Hutchins, M.; Neal, C. Acid episodes in the Allt a'Mharcaidh, Scotland: An investigation based on subhourly monitoring data and climatic patterns. Hydrol. Earth Syst. Sci. **2007**, *11* (1), 340–355.

67. Gorham, E. Acid precipitation and its influence upon aquatic ecosystems—An overview. Water Air Soil Pollut. **1976**, *6*, 457–481.

68. Psenner, R. Environmental impacts on freshwaters: Acidification as a global problem. Sci. Total Environ. **1994**, *143* (1), 53–61.

69. Galloway, J.N. Acid deposition: Perspectives in time and space. Water Air Soil Pollut. **1995**, *85*, 15–24.

70. *Lottermoser, B.G.* Mine wastes: Characterization, treatment and environmental impacts; *Springer: Berlin, Germany, 2007.*

71. Nordstrom, D.K.; Alpers, C.N.; Ptacek, C.J.; Blowers, D.W. Negative pH and extremely acid mine waters from Iron Mountain Superfund site, California. Environ. Sci. Technol. **2000**, *34* (2), 254–258.

72. Nordstrom, D.K.; Alpers, C.N. Negative pH, efflorescent mineralogy, and consequences for environmental restoration at the Iron Mountain Superfund site, California. Proc. Natl. Acad. Sci. U. S. A. **1999**, *96* (7), 3455–3462.

73. Charles, D.F.; Whitehead, D.R.; Engstrom, D.R.; Fry, B.D.; Hites, R.A.; Norton, S.A.; Owen, J.S.; Roll, L.A.; Schindler, S.C.; Smol, J.P.; Uutala, A.J.; White, J.R.; Wise, R.J. Paleolimnological evidence for recent acidification of Big Moose Lake, Adirondack Mountains, N.Y. (USA). Biogeochemistry **1987**, *3*, 267–296.

74. Nordstrom, D.K.; Alpers, C.N.; Coston, J.A.; Taylor, H.E.; McCleskey, R.B.; Ball, J.W.; Davis, J.A.; Ogle, S. Geochemistry, toxicity, and sorption properties of contaminated sediments and pore waters from two reservoirs receiving acid mine drainage. In Proceedings: Charleston, S.C., U.S. Geological Survey Water-Resources Investigations Report, 99–4018-A, Morganwalp, D.W., Buxton, H.T., Eds.; USGS: Denver, CO, 1999; 289–296.

75. Yan, N.D. Effects of changes in pH on transparency and thermal regimes of Lohi Lake, near Sudbury, Ontario. Can. J. Fish. Aquat. Sci. **1983**, *40* (5), 621–626.

76. Knöller, K.; Fauville, A.; Mayer, B.; Strauch, G.; Friese, K.; Veizer, J. Sulfur cycling in an acid mining lake and its vicinity in Lusatia, Germany. Chem. Geol. **2004**, *204* (3–4), 303–323.

77. Bonacina, C. Lake Orta: The undermining of an ecosystem. J. Limnol. **2001**, *60* (1), 53–59.

78. Renberg, I.; Hultberg, H. A paleolimnological assessment of acidification and liming effects on diatom assemblages in a Swedish lake. Can. J. Fish. Aquat. Sci. **1992**, *49* (1), 65–72.

79. Hesthagen, T. Fish kills of Atlantic salmon (*Salmo salar*) and brown trout (*Salmo trutta*) in an acidified river of SW Norway. Water Air Soil Pollut. **1986**, *30* (3–4), 619–628.

80. Dangles, O.J.; Guérold, F.A. Structural and functional responses of benthic macroinvertebrates to acid precipitation in two forested headwater streams (Vosges Mountains, northeastern France). Hydrobiologia 2000, *418*, 25–31.

81. Vesely, J.; Almquist-Jacobson, H.; Miller, L.M.; Norton, S.A.; Appleby, P.; Dixit, A.S.; Smol, J.P. The history and impact of air pollution at Certovo Lake, southwestern Czech Republic. J. Paleolimnol. **1993**, *8*, 211–231.

82. Kratz, T.K.; Webster, K.E.; Bowser, C.J.; Benson, B.J. Influence of landscape position on lakes in northern Wisconsin. Freshwater Biol. **1997**, *37*, 209–217.

83. Delleur, J.W. Hydrological response to acid rain. In *Monitoring to Detect Changes in Water Quality Series*, Proceedings of the Budapest Symposium, Budapest, Hungary July 1986; International Association of Hydrological Sciences Publication no. 157, Wallingford, U.K., 1986; 175–184.

84. Driscoll, C.; Lambert, K.F.; Chen, L. Acidic deposition: Sources and ecological effects. In *Acid in the Environment: Lessons Learned and Future Prospects*; Visgilio, G.R., Whitelaw, D.M., Eds.; Springer Science + Business Media: New York, 2007; 27–58.

85. Galloway, J.N. Acidification of the world: Natural and anthropogenic. Water Air Soil Pollut. **2001**, *130*, 17–24.

86. Wood, J.M. Effects of acidification on the mobility of metals and metalloids: An overview. Environ. Health Perspect. **1985**, *63*, 115–119.

87. Havas, M.; Rosseland, B.O. Response of zooplankton, benthos, and fish to acidification: An overview. Water Air Soil Pollut. **1995**, *85*, 51–62.

88. Kopàcek, J.; Ulrich, K.; Hejzlar, J.; Borovec, J.; Stuchlík, E. Natural inactivation of phosphorus by aluminum in atmospherically acidified water bodies. Water Res. **2001**, *35*, 3783–3790.

89. Scheuhammer, A.M. Effects of acidification on the availability of toxic metals and calcium to wild birds and mammals. Environ. Pollut. **1991**, *71* (2–4), 329–375.

90. Petrin, Z.; Laudon, H.; Malmqvist, B. Does freshwater macroinvertebrate diversity along a pH gradient reflect adaptation to low pH? Freshwater Biol. **2007**, *52*, 2172–2183.

91. Dangles, O.; Malmqvist, B.; Laudon, H. Naturally acid freshwater ecosystems are diverse and functional: evidence from boreal streams. Oikos **2004**, *104*, 149–155.

92. United Nations Economic Commission for Europe. Convention on Long-range Transboundary Air Pollution; United Nations: New York and Geneva, 1999.

93. Schöpp, W.; Posch, M.; Mylona, S.; Johansson, M. Longterm development of acid deposition (1880–2030) in sensitive freshwater regions in Europe. Hydrol. Earth Syst. Sci. **2003**, *7* (4), 436–446.

94. Chestnut, L.G.; Mills, D.M. A fresh look at the benefits and costs of the US acid rain program. J. Environ. Manage. **2005**, *77*, 252–266.

95. Henrikson, L.; Hindar, A.; Abrahamsson, I. Restoring acidified lakes: An overview. In *The Lakes Handbook. Volume 2: Lake Restoration and Rehabilitation;* O'Sullivan, P., Reynolds, C.S., Eds.; Blackwell Publishing: Oxford, 2005; 483–500.

96. Norberg, M.; Bigler, C.; Renberg, I. Monitoring compared with paleolimnology: Implications for the definition of reference condition in limed lakes in Sweden. Environ. Monit. Assess. **2008**, *146*, 295–308.

97. Wällstedt, T.; Borg, H. Effects of experimental acidification on mobilisation of metals from sediments of limed and nonlimed lakes. Environ. Pollut. **2003**, *126*, 381–391.

98. Aronsson, K.A.; Ekelund, N.G.A. Limnological effects on a first order stream after wood ash application to a boreal forest catchment in Bispgården, Sweden. For. Ecol. Manage. **2008**, *255*, 245– 253.

99. Ugochukwu, C.N.C.; Nukpezah, D. Ecotechnological methods as strategies to reduce eutrophication and acidification in lakes. Environmentalist **2008**, *28*, 137– 142.

9

Rivers: Pollution

Introduction 105
Sources of River Pollution 106
    Acid Rain • Industrial Pollution • Agricultural Pollution • Radionuclides
Removal of Pollutants in River Waters 107
Conclusions 108
References 108

Bogdan Skwarzec

Introduction

More than 97% of all water on Earth is salty, and most of the remaining 3% is frozen in the polar ice caps. The atmosphere, rivers, lakes, and underground stores hold less than 1% of all freshwater and this tiny amount has to provide the freshwater needed to support the earth's population. Most of the water for human consumption originates from riverine system. The quantity of this water is very important for human life because the concentration of chemical and biological pollutants in many rivers is high.[1,2] The big rivers of Asia are the most polluted rivers in the world. They contain three times as many bacteria from human waste as the global average and 20 times more lead than rivers in industrialized countries. Water from half of the tested sections of China's and Indian's seven major rivers was found to be undrinkable because of pollution.[3–7] Almost 40% of America's rivers are too polluted for fishing, swimming, or aquatic life. The Mississippi River—which drains nearly 40% of the continental United States, including its central farmlands—carries an estimated 1.5 million metric tons of nitrogen pollution into the Gulf of Mexico each year. The resulting hypoxic coastal dead zone in the Gulf each summer is about the size of Massachusetts.[8–12] Also, other major rivers in the world (Amazon, Niger, big Russian rivers, and Australian rivers) are polluted by trace elements and heavy metals.[13–18] The King River is Australia's most polluted river, suffering from a severe acidic condition related to mining operations.[19,20] In European Union countries, rivers are polluted with sewage or fertilizer. The Sarno is the most polluted river in Europe, featuring a nasty mix of sewage, untreated agricultural waste, industrial waste, and chemicals.[21]

Apart from anthropogenic pollutants, continental material from land is also transported to the marine ecosystem through rivers. The total flux of dissolved and suspended matter transported by rivers is estimated to be 20×10^{15} g yr^{-1}, i.e., 15.5×10^{15} and 4.5×10^{15} g yr^{-1} for solid and dissolved loads, respectively.[1] The total load of pollutants in the river ecosystems varies among several regions and depends on the population density, location of industry centers, and the abundance as well as intensity of the exploitation of natural resources. The anthropogenic sources of riverine pollutants are mainly industrial wastewater, leakage from products in use and those removed from service, natural degradation of pro-products, as well as pollution from different types of land use, e.g., fertilizing and mining.[22] Most freshwater pollution is caused by the addition of organic material, which is mainly sewage but can be food waste or farm effluent. Bacteria, viruses, or parasites cause diseases such as cholera, typhoid, schistosomiasis, dysentery, and other diarrheal diseases. Moreover, other microorganisms feed on organic matter and large populations quickly develop, using up much of the oxygen dissolved in the water.[23–27]

Normally, oxygen is present in high quantities, but even a small drop in the level can have a harmful effect on the river animals. Animals can be listed according to their ability to tolerate low levels of oxygen. In the following species of animals are listed in the order of the least tolerant to the most tolerant to low oxygen levels: stone-fly nymphs, mayfly nymphs, freshwater shrimps, freshwater hog lice, blood worms, tubifex worms, and rat-tailed maggots.[23]

Sources of River Pollution

The general important sources of water pollution that work together to reduce the overall river water quality are acid rain and industrial and agricultural discharge liquid waste products. Rain, as it falls through the air or drains from urban areas and farmland, absorbs contaminants. Serious incidents resulting from spillages or discharges of toxic chemicals are pollution events that make the news.[1]

Acid Rain

Rain, when falling through polluted air, absorbs some pollutants. The main pollutant gases are sulfur dioxide (SO_2) and nitrogen oxides (N_xO_y), which are formed when fuels are burned. They react with rainwater to form sulfuric and nitric acids. On reaching the ground, the acid liquid produces a number of effects. It can release harmful substances such as aluminum and heavy metals from the soil. These substances are normally present in an inert, harmless state; however, in acid conditions, they can turn into compounds that are poisonous to plant and animal life. When washed into lakes and streams, aluminum can kill small water creatures and fish. Particularly at risk is the dipper, a river bird that actually walks underwater to catch its insect food. If acid rain kills the insects, then the dippers will disappear in turn.[28,29]

Industrial Pollution

Factories use water from rivers to power or cool down machinery. Dirty water containing chemicals flows back in the river. Water used for cooling is warmer than the river itself. Raising the temperature of the water lowers the level of dissolved oxygen and upsets the balance of life in the water. Many industrial wastes discharged into river water are mixtures of chemicals that are difficult to treat. Some industrial wastes are so toxic that they are strictly controlled, making them an expensive problem to deal with. Some companies try to cut the costs of safely dealing with waste by illegally dumping chemicals at times and in places where they think they will not be caught.[30–32] Chemical waste products from industrial processes are sometimes accidentally discharged into rivers. Examples of such pollutants include cyanide and toxic heavy metals (zinc, lead, copper, cadmium, and mercury) as well as oil. These substances may enter the water in such high concentrations that fish and other animals are killed immediately. Sometimes, the pollutants enter a food chain and accumulate until they reach toxic levels, eventually killing birds, fish, and mammals.[1,33,34]

Agricultural Pollution

Farmers use fertilizers and pesticides to enhance crop growth. However, these fertilizers and pesticides can be washed through the soil by rain, to end up in rivers. If large amounts of fertilizers or farm waste drain into a river, the concentration of nitrate and phosphate in the water increases considerably.[35] Algae use these substances to grow and multiply rapidly, turning the water green. This massive growth of algae, called eutrophication, leads to pollution. When the algae die, they are broken down by the action of bacteria, which quickly multiply, using up all the oxygen in the water, leading to the death of many animals. Also, when organic farm wastes such as silage or liquid manure (slurry) escape into rivers, the amount of oxygen in the water is reduced.[36,37] Nitrate pollution problems occur when too much chemical fertilizer is applied to the land. The excess runs off and can find its way into drinking water sources, or can trickle into rivers and lakes. Some experts believe that high levels of nitrate in drinking water may pose a threat to health. A European directive states that drinking water should not contain more than 50 mg of nitrates per liter of water. In rivers, streams, ponds, and lakes, too much nitrate can create a "pea soup" effect. The water becomes clogged with fast-growing plant life such as algae and weeds.[36–39]

Radionuclides

Radionuclides and radioactive waste are stored in riverine water ponds next to nuclear power stations. Nuclear atmospheric weapons tests and the Chernobyl accident were sources of many artificial radionuclides in oceans.[40–44] The river runoff of radioactive substances is an important source of marine radionuclide pollution, especially for closed seas, estuaries, and coastal zones. For example, the annual runoff of radiocesium, radiostrontium, polonium, uranium, and plutonium is an important source of radionuclides in southern Baltic Sea. The total runoff of these radionuclides from rivers to the southern Baltic Sea was calculated to consist of about 3.3 TBq of ^{90}Sr, 5.2 TBq of ^{137}Cs, 0.7 TBq of ^{210}Po, 4.5 TBq of $^{234+238}$U, and 1.3 GBq of $^{239+240}$Pu.[45]

Removal of Pollutants in River Waters

Many types of pollution are discharged into rivers, and purification processes remove them at different speeds. Some heavy metals, for example, are removed relatively quickly because suspended clay and

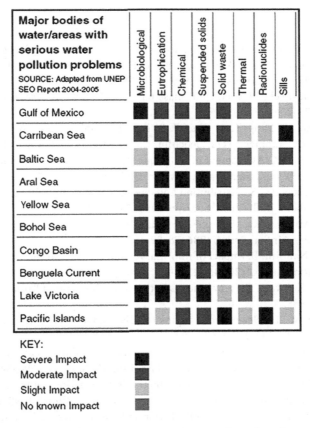

FIGURE 1 Impact of inflow of pollutants from riverine systems on ecological conditions of several coastal zones, estuaries, and closed seas.
Source: Grinning Planet, Water Pollution Facts, 26th July 2005, http://www.grinningplanet.com/2005/07-26/water-pollution-facts-article.htm.[46]

organic particles have a slight electric charge and adsorb the metal atoms. When the clay or organic particles settle out of the water, they take the metal atoms with them. Unfortunately, some pollutants are very persistent in the water and can accumulate downstream, causing great hazard.[1] Suspended solids in a moving body of water will settle out at various points or be carried within longer distances, depending on their size and the rate of the flow. The higher the amount of suspended solids is, the cloudier or more turbid the water becomes. Suspended matter can affect the amount of light entering the water, thus restricting the amount of photosynthesis that can occur, which consequently can limit the growth of plants. Small particles settling out in large amounts on the bottom of a water body can prevent some organisms from living there as well as prevent green plants from photosynthesizing.[19,21]

Conclusions

Rivers are contaminated by many natural and anthropogenic industrial pollutants. A lot of these pollutants are highly toxic to river organisms and to humans. The most important toxic chemicals in the seas are organic compounds (PCBs, DDT, PAHs, pesticides, furans, dioxins, phenol, oil), heavy metals, and radionuclides. Chemical and biological pollutants flow from big areas of river catchments to the seas and oceans. This runoff of toxic substances through river water is very dangerous, especially for coastal and estuarine zones and closed small seas. Figure 1 presents 10 of the most polluted areas and bodies of water in the world.[46]

References

1. Martin, J.-M.; Meybeck, M. The significance of the river input of chemical elements to the ocean. In *Trace Metals in Sea Water*; Wong, C.S., Boyle, E., Bruland, K.W., Burton, J.D., Goldberg, E.D., Eds.; Plenum Publishing Corporation, New York, 1983; 265–296
2. Martin, J.-M.; Meybeck, M. Elemental mass balance of material carried by major rivers. Mar. Chem. **1979**, *7*, 173–206.
3. Li, Y.-H.; Teraoka, H.; Yang, T.-S.; Chen, J.-S. The elemental composition of suspended particles from the Yellow and Yangtze rivers. Geochim. Cosmochim. Acta **1984**, *48*, 1561–1564.
4. Bhatt, S.D.; Pathak, J.K. Assessment of water quality and aspect of pollution in a stretch of river Gomati (Kumaun: Lesser Himalaya). J. Environ. Biol. **1992**, *13* (2), 113–126.
5. Tanizaki, Y.; Nagatasuka, S. Distribution and chemical behavior of 21 elements in river water in the Yakushima Island. Bull. Chem. Soc. Jpn. **1983**, *56*, 619–624.
6. Agarwal, H.C.; Mittal, P.K.; Menon, K.B.; Pillal, M.K.K. DTT residues in the River Jumuna in Delhi, India. Water Air Soil Pollut. **1986**, *28*, 89–104.
7. Bishwanath, G.; Nandini, B. Impact of informal regulation of pollution on water quality in rivers in India. J. Environ. Manage. **2004**, *73*, 117–130.
8. Newchurch, E.J.; Kahwa, I. A. Heavy metals in the Lower Mississippi River. J. Environ. Sci. Health **1984**, *19* (8), 973–988.
9. Presley, B.J.; Trefry, J.H. Heavy metal inputs to Mississippi delta sediments—A historical view. Water Air Soil Pollut. **1980**, *13* (4), 481–494.
10. Taylor, H.E.; Garbarino, J.R.; Brinton, T.I. The occurrence and distribution of trace metals in the Mississippi River and its tributaries. Sci. Total Environ. **1990**, *97/98*, 369–384.
11. Taylor, H.E.; Shiller, A.M. Mississippi River methods comparison study—Implications for water quality monitoring of dissolved trace elements. Environ. Sci. Technol. **1995**, *29*, 1313–1317.
12. Brinton, T.I.; Garbarino, J.R.; Peart, D.B.; Taylor, H.E.; Antweiler, R.C. Concentration and transport data for dissolved inorganic constituents in water collected during seven cruises on the Mississippi River and some of its tributaries, July 1987–June 1990, U.S. Geological Survey Open-File Report 94–524, 102, 1995.

13. Sholkovitz, E.R.; Price, N.B. The major element chemistry of suspended matter in the Amazon estuary. Geochim. Cosmochim. Acta **1980**, *44*, 163–171.

14. Sholkovitz, E.R. The geochemistry of rare earth elements in the Amazon River estuary. Geochim. Cosmochim. Acta **1993**, *57*, 2181–2190.

15. Vital, H.; Stattegger, K. Major and trace elements of stream sediments from the lowermost Amazon River. Chem. Geol. **2000**, *168*, 151–168.

16. Olajire, A.A.; Altenberger, R.; Kuster E.; Brack W. Chemical and ecotoxicological assessment of polycyclic aromatic hydrocarbon-contaminated sediment of the Niger Delta, Southern Nigeria. Sci. Total Environ. **2005**, *340*, 123–136.

17. Queirolo, F.; Stegen, S.; Mondaca, J.; Cortes, R.; Rojas, R.; Contreras, C.; Munoz, L.; Schwuger, M.J.; Ostapczuk, P. Total arsenic, lead, cadmium, copper and zinc in some salt rivers in the northern Andes of Antofagasta, Chile. Sci. Total Environ. **2000**, *255*, 85–95.

18. Orr, J.C.; Fabry, V.J.; Aumont, O.; Bopp, L.; Doney, S.C.; Feely, R.A.; Gnanadesikian, A.; Gruber, N.; Ishida, A.; Joos, F.; Key, R.M.; Lindsay, K.; Maier-Reimer, E.; Matear, R.; Monfray, Mouchet, A.; Najjar, R.G.; Plattner, G.K.; Rodgers, K.B.; Sabine, C.L.; Sarminto, J.L.; Schlitzer, Slater, R.D.; Totterdell, I.J.; Weirig, M.F.; Yamanaka, Y.; Yool, A., Anthropogenic ocean acidification over the twenty-first century and its impact on calcifying organisms. Nature **2005**, *437* (7059), 681–686.

19. Likens, G.E.; Keene, W.C.; Miller J.M; Galloway, J.N. Chemistry of precipitation from a remote, terrestrial site in Australia. J. Geophys. Res. **1987**, *92* (D11), 13,299–13,314.

20. Trefry, J.H.; Nelson, T.A.; Tocine, R.P.; Metz, S.; Vetter, T.W. Trace metal fluxes through the Mississippi River delta system. In *Contaminant Fluxes through the Coastal Zone;* Kullenberg, Ed.; Rapports et Proces-Verbaux des Reunions; Conseil International pour l'Exploration de la Mer, Copenhague, Denmark, 1986; Vol. 186, 277–288.

21. Smith, R.A.; Alexander, R.B.; Wolman, M.G. Water-quality trends in the Nation's rivers. Science **1987**, *235*, 1607–1615.

22. Meybeck, M. Atmospheric inputs and river transport of dissolved substances. Int. Assoc. Hydrol. Sci. Publ. **1983**, *141*, 173–192.

23. Meybeck, M.; Helmer, R. The water quality of rivers: From pristine stage to the global pollution. Palaeogeogr. Palaeoclimatol. Palaeoecol. (Global Planet. Change Sect.) **1989**, *75*, 283–309.

24. Morrison, T.E.; Fraser, R.J.; Smith, Pn.N.; Mahalingam,S.; Heise, M.T. Complement contributes to inflammatory tissue destruction in a mouse model of Ross River virus-induced disease. J. Virol. **2007**, *81* (10), 5132–43.

25. Harley, D.; Sleigh, A.; Ritchie S. Ross River virus transmission, infection, and disease: A cross-disciplinary review. Clin. Microbiol. Rev. **2001**, *14* (4), 909–932.

26. Knight, J. Meet the Herod bug. Nature **2001**, *412*, 12–14.

27. Breault, J.L. Candiru: Amazonian parasitic catfish. J. Wilderness Med. **1998**, *2* (4), 304–312.

28. Likens, G.E.; Driscoll, C.T.; Buso, D.C. Long-term effects of acid rain: Response and recovery of a forest ecosystem. Science **1996**, *272*, 244–246.

29. Galloway, J.N., Zhao, D.; Xiong, J.; Likens, G.E. Acid rain: A comparison of China, United States and a remote area. Science **1987**, *236*, 1559–1562.

30. Szefer, P. *Metals, Metalloids and Radionuclides in the Baltic Sea Ecosystem;* Elsevier: Amsterdam, the Netherlands, 2002; Vol. 5.

31. Laws, E.A. *Aquatic Pollution;* John Wiley and Sons, 2000.

32. Bruneau, L. Pollution from industries in the drainage area of the Baltic. Ambio **1980**, *9*, 145–152.

33. Billen, G.; Garnier, J.; Delinge, C.; Billen, C. Estimate of early-industrial inputs of nutrients to river system: Implication for coastal eutrophication. Sci. Total Environ. **1999**, *243/244*, 43–52.

34. Gribble, G.W. The Natural production of chlorinated compounds. Environ. Sci. Technol. **1994**, *28*, 310–319.

35. Horrigan, L.; Lawrence, R.S.; Walker, P. How sustainable agriculture can address the environmental and human health harms of industrial agriculture. Environ. Health Perspect. **2002**, *110* (5), 445–456.

36. Ambio. Special issues: Marine eutrophication, Ambio **1990**, *19*, 101–176.

37. Smith, V.H.; Tilman, G.D.; Nekola, J.C. Eutrophication: Impacts of excess nutrient inputs on freshwater, marine, and terrestrial ecosystems. Environ. Pollut. **1999**, *100* (1–3), 179–196.

38. Raimammake, A.; Pietilainen, O.P.; Rekolainen, S.; Kauppila, P.; Pitkanen, H.; Niemi, J.; Raateland, A.; Vuorenmaa, J. Trends of phosphorus, nitrogen, and chlorophyll *a* concentrations in Finnish rivers and lakes in 1975–2000. Sci. Total Environ. **2003**, *310*, 47–59.

39. Cole, J.J.; Peierls, B.L.; Caraco, N.F.; Pace, M.L. Nitrogen loading of rivers as a human-driven process. In *Humans as Components of Ecosystems,* McDonnell, Pickett, S.T.A., Eds.; Springer-Verlag: New York, 1993; 141–157.

40. Skwarzec, B. Radiochemia srodowiska i ochrona radiologiczna (in Polish), Environmental radiochemistry and radiological protection, Wydawnictwo DJ sc., Gdańsk, 2002.

41. More, W.S. Amazon and Mississippi River concentration of uranium, thorium, and radium isotopes. Earth Planet. Sci. Lett. **1967**, *2*, 231–234.

42. Spalding, R.F.; Sackett, W.M. Uranium in runoff from the Gulf of Mexico distributive province: Anomalous concentrations. Science **1972**, *175*, 629–631.

43. Sayles, F.; Livingston, H.; Panteleyev, G. The history and source of particulate ^{137}Cs and [239+240]Pu, deposition in sediment of the Ob River Delta, Siberia. Sci. Total Environ. **1997**, *202*, 25–41.

44. Johnson-Pyrtle, A.; Scott, M.R.; Laing, T.E.; Smol, J.P. [137]Cs distribution and geochemistry of Lena River (Siberia) drainage basin lake sediments. Sci. Total Environ. **2000**, *255*, 145–159.

45. Skwarzec, B. Polonium, uranium and plutonium radionuclides in aquatic environment of Poland and southern Baltic. Baltic Coastal Zone **2009**, *13*, 127–166.

46. Grinning Planet, Water Pollution Facts, 26th July 2005, http://www.grinningplanet.com/2005/07-26/waterpollution-facts-article.htm.

10

Sea: Pollution

Introduction 111
Sources of Marine Pollution 112
    Land and River Runoffs • Oil and Ship Pollution • Atmospheric
    Pollution • Deep Sea Mining • Acidification and Acid Rains • Radioactive
    Waste • Eutrophication • Plastic Debris • Noise Pollution
Conclusions 116
References 117

Bogdan Skwarzec

Introduction

Pollution is the release of harmful environmental contaminants, or the substances so released. Generally, the process needs to result from human activity to be regarded as pollution. Even relatively benign products of human activity are liable to be regarded as pollution, if they precipitate negative effects later on. The different forms of pollution are as follows: air pollution, water pollution, soil contamination, radioactive contamination, noise pollution, light pollution, and thermal pollution. Chemical pollutants can be natural substances (organic and mineral matter), or synthetic degradable or non-degradable substances (plastics, pesticides). They can be toxic or not with varying endurance. For example, in humans, polychlorinated biphenyls (PCBs) play a part in causing breast, lung, liver, and colon cancers. It can cause retardations in neurological development and growth. Persistent organic pollutants, drugs and their metabolites, and brominated organic compounds are concentrated in soft tissue of marine organisms and other filters, and they can develop typhoid, hepatitis, and other illnesses if they are ingested.[1–3]

Marine pollution occurs when harmful effects, or potentially harmful effects, can result from the entry into the ocean system of chemical, industrial, agricultural, and residential wastes, as well as invasive organisms. Marine pollution is, in the majority of cases, of land and atmospheric origin. Catastrophic accidents (shipping, platform) are also an important factor of marine pollution. Generally, sea pollutants are classified as having chemical and biological origin; however, toxic chemical substances are dominant. The most important chemical substances belong to the next group: trace and heavy toxic metals, metalo-organic compounds (tributyltin, TBT), nutrients, acid gases, radionuclides (especially artificial), radioactive waste, military toxic substances, and organic toxic substances (oil, pesticides, dioxins, furans, phenols, halogeno-organic compounds—especially DDT, PCBs, polychlorinated terphenyls [PCT], and their metabolites, as well as aromatic hydrocarbons polycyclic aromatic hydrocarbon, [PAHs]). Toxins can accumulate in the tissues of many species of aquatic life in a process called bioaccumulation.[1–3]

Many potentially toxic chemicals adhere to tiny particles that are taken up by marine organisms (plankton and benthos animals), most of which are either deposit or filter feeders (bivalvia). In this way, by bioconcentration of dissolved species, the toxins are concentrated upward within ocean food chains. When pesticides are incorporated into the marine ecosystem, they quickly become absorbed into marine food webs. Once in the food webs, these pesticides can cause mutations, as well as diseases, which can be harmful to humans as well as the entire food web. Toxic metals, metalloids, and radionuclides can also be introduced into marine food webs. These can cause a change to tissue matter, biochemistry, behavior, reproduction, and suppress growth in marine life. In addition, many animal feeds have a high fish meal or fish hydrolysate content. In this way, marine toxins can be transferred to land animals, and appear later in meat and dairy products. Pollution in marine environment is often classified as point source or non-point source pollution. Point source pollution occurs when there is a single, identifiable, and localized source of the pollution. An example is a direct discharge of sewage and industrial waste into the ocean. Pollution such as this occurs particularly in developing nations. Non-point source pollution occurs when the pollution comes from ill-defined and diffuse sources.[4,5]

Sources of Marine Pollution

There are many different ways to categorize and examine the inputs of pollution into our marine ecosystems. In principle, there are three main types of inputs of pollution into the ocean: direct discharge of waste into the oceans, runoff into the waters due to rain, and pollutants that are released from the atmosphere.[1,6]

Land and River Runoffs

Between 75% and 80% of marine pollution is caused by land, particularly agriculture, and 30% of this is from the atmosphere; however, around 12% of the pollution is caused by marine transport.[7] In South America, 98% of domestic wastewater ends up, untreated, in the sea. The countries along the Mediterranean Sea throw 50 million tons waste into it every year, and the Chinese throw 60 million tons of waste into the Yellow Sea daily.[8,9] Also, closed marine basins, such as the Baltic, Caspian, and Black Seas, are polluted every year with many million tons of chemical waste. Baltic fauna is practically decimated on half of the central Baltic's seabed as an effect of land-based pollution, especially nitrogenous.[1] In the Caspian Sea, 140 million tons of pollutants are poured every year in the northern part, especially that of the Volga delta. Also, the Black Sea is polluted by the Dniestr, Dniepr, Danube, and Don rivers. For eutrophication reason, 80% drop in fishing yield, only 6 species of fish can still be fished out of the 26 that could previously be commonly marked in this basin.[3] One important common path of entry by land and atmosphere contaminants to the sea are rivers.[9] The evaporation of water from oceans exceeds precipitation. The balance is restored by rain over the continents entering rivers and then being returned to the sea. Pollutants enter rivers and the sea directly from urban sewerage and industrial waste discharges, sometimes in the form of hazardous and toxic wastes (inorganic, radioactive, and organic). Most of the pollution is simply soil, which ends up in rivers flowing to the sea. More than 80% of all marine pollution comes from land-based activities, and many pollutants are deposited in estuarine and coastal waters. Here, the pollutants enter marine food chains, building up their concentrations until they reach toxic levels. It often takes human casualties to alert us to pollution and such was the case in Minamata Bay in Japan, when 649 people died from eating fish and shellfish contaminated with mercury and 3500 people had mercury poisoning. A factory was discharging waste containing mercury in low concentration into the sea and, as this pollutant passed through food chains, it became more concentrated in the tissues of marine organisms until it reached toxic levels.[10]

Surface water runoff contains pollutants from farming. Also, urban runoff and runoff from the construction of roads, buildings, ports, channels, and harbors can carry soil and particles laden with carbon, nitrogen, phosphorus, and minerals. This nutrient-rich water can lead to fleshy algae and phytoplankton thriving in coastal areas, known as algal blooms, which have the potential to create hypoxic conditions by using all available oxygen. Polluted runoff from roads and highways can be a significant source of water pollution in coastal areas.[11]

Oil and Ship Pollution

About 2.7 million tons of oil pollution enters our oceans each year. Of this, less than 10% is from natural seepage of oil from the ocean floor and sedimentary rock erosion. The remaining 90% comes from human activities. Offshore drilling, as a result of accident spills and other operations, accounts for only 2%. Spills from large tankers, which are reported worldwide, account for just over 5%.[3] Air pollution from cars and industry accounts for just over 13% of the total, as the hundreds of tons of hydrocarbons land in our oceans from particle fallout aided by rains, which washed the particles from air. Almost 4 times the amount of oil that comes from spills from large tankers spills, 19% is regularly released into the ocean from routine maintenance, which includes discharge as well as other ship operations. Oil is perhaps the most publicly recognized toxic pollutant. Tanker accidents between 1967 and 2007 spilled to the oceans nearly 4.5 million tons of oil. Now, from May 2010, the next million tons of oil has been delivered to the Gulf of Mexico, as a result of platform crash (BP accident). The biggest spill ever recorded took place during the 1979 drilling platform in the Gulf of Mexico as well as the 1991 Persian Gulf War, when about 2 million tons of oil was spilled. The effect of oil pollution on wildlife can be terrible. Many marine organisms (fishes, invertebrates, mammals, and birds) died.[3]

Ships can pollute waterways and oceans in many ways. Oil spills can have devastating effects. While being toxic to marine life, PAHs, the components in crude oil, are very difficult to clean up and last for years in the sediment and marine environment.[1,2] Discharge of cargo residues from bulk carriers can pollute ports, waterways, and oceans. In many instances, vessels intentionally discharge illegal waste despite foreign and domestic regulations prohibiting such actions. It has been estimated that container ships lose over 10,000 containers at sea each year (usually during storms). Ships also create noise pollution that disturbs natural wildlife, and water from ballast tanks can spread harmful algae and other invasive species. Also, ballast water taken up at sea and released in port is a major source of unwanted exotic marine life. The invasive marine species, native to the Black, Caspian, and Azov Seas, were transported to other basins (Baltic Sea, Great Lakes) via ballast water from a transoceanic vessel. Invasive species can take over once occupied areas, facilitate the spread of new diseases, introduce new genetic material, alter underwater seascapes, and jeopardize the ability of native species to obtain food.[12] Some strong toxic metalo-organic substances, especially TBT, are a source of marine pollution, as a result of commercial shipping.[13–16]

Atmospheric Pollution

Another important pathway of pollution occurs through the atmosphere.[17–21] Wind-blown dust and debris, including plastic bags, are blown seaward from landfills and other areas. Dust from the Sahara, moving around the southern periphery of the subtropical ridge, moves into the Caribbean and Florida during the warm season as the ridge builds and moves northward through the subtropical Atlantic. Dust can also be attributed to a global transport from the Gobi and Taklamakan deserts across Korea, Japan, and the Northern Pacific to the Hawaiian Islands.[22] Since 1970, dust outbreaks have worsened due to periods of drought in Africa. There is a large variability in dust transport to the Caribbean and Florida from year to year; however, the flux is greater during positive phases of the North Atlantic Oscillation.[23] Atmospheric particles containing chemical pollutants can be transported over thousands of kilometers, and airborne particles with heavy metals and radionuclides originating from anthropogenic sources can be detected even in remote areas.[24–26] The trace and heavy metals produced and emitted in Europe or North America are transported by air and deposited in the North Atlantic environment.[27,28] Also, organic toxic substances (especially pesticides) used in agriculture by farmers in warm regions (Africa and South Asia) are sublimated to air and transported to Arctic Ocean and accumulated in fish and mammals.[29]

Deep Sea Mining

Deep sea mining is a relatively new mineral retrieval process that takes place on the ocean floor; however, the complete consequences of full-scale mining operations are still unknown. Ocean mining sites are located usually around large areas of polymetallic nodules or active and extinct hydrothermal vents at about 1400–3700 m below the ocean's surface.[30] The vents create sulfide deposits, which contain precious metals such as silver, gold, copper, manganese, cobalt, and zinc.[31,32] The deposits are mined using either hydraulic pumps or bucket systems that take ore to the surface to be processed. As with all mining operations, deep sea mining raises questions about environmental damage to the surrounding areas. The removal of parts of the seafloor will result in disturbances to the benthic layer, increased toxicity of the water column, and sediment plumes from tailings. Removing parts of the seafloor disturbs the habitat of benthic organisms, possibly depending on the type of mining and location, causing permanent disturbances.[33] Among the deep sea mining products, sediment plumes could have the greatest impact. Plumes are caused when the tailings from mining (usually fine particles) are dumped back into the ocean, creating a cloud of particles floating in the water. Two types of plumes occur: near-bottom plumes and surface plumes.[30] Near-bottom plumes occur when the tailings are pumped back down to the mining site. The floating particles increase the turbidity, or cloudiness, of the water, clogging filter-feeding apparatuses used by benthic organisms. Surface plumes cause a more serious problem. Depending on the size of the particles and water currents, the plumes could spread over vast areas. The plumes could affect zooplankton and light penetration, in turn affecting the food web of the area.[30,34]

Acidification and Acid Rains

The ocean waters absorb carbon dioxide (CO_2), sulfur dioxide (SO_2) and trioxide (SO_3), and nitrogen oxides (N_xO_y) from the atmosphere. Because the atmospheric concentrations of these gaseous oxides are increasing, the oceans are becoming more acidic.[35] The potential consequences of ocean acidification are not fully understood; however, there are concerns that structures made of calcium carbonate may become vulnerable to dissolution, affecting corals and the ability of shellfish to form shells.[36,37] Oceans and coastal ecosystems play an important role in the global carbon cycle and have removed about 25% of the carbon dioxide emitted by human activities between 2000 and 2007. Rising ocean temperatures and ocean acidification means that the capacity of the ocean carbon sink will gradually become weaker. Also, the methane clathrate reservoirs, containing large amounts of the greenhouse gas methane, under sediments on the ocean floors, can potentially release the methane when oceanic water becomes warm. In 2004, the global inventory of ocean methane clathrates was estimated to occupy between 1 and 5 million cubic kilometers. This estimate corresponds to 500–2500 gigatonnes carbon (Gt C), and can be compared with the 5000 Gt C estimated for all other fossil fuel reserves.[38]

Radioactive Waste

Radionuclide and radioactive wastes are stored in marine water ponds next to nuclear power stations (electricity nuclear power and nuclear reprocessing factory) until it is considered safe for disposal. The waste is released directly into the sea about some kilometers from the coast. The ocean depth is used to dump high-level nuclear waste.[39,40] The liquid waste is sealed in glass, a process called vitrification, and stored steel canisters contain concrete. These containers are dumped in the sediment on the ocean floor. Also, nuclear atmospheric weapons test and the Chernobyl accident were sources of many artificial radionuclides in oceans.[24] Natural radionuclides in the environment can also be enhanced owing to human activity: industry, coal power, phosphate fertilizers in agriculture, and domestic and industrial sewage.[39] Moreover, coal mining is a source of huge amounts of waste containing large quantities of natural radionuclides, especially polonium, thorium, and uranium. During ashing of coal in power plants, some natural radionuclides are emitted to the atmosphere as a gas and radioactive dust, whereas the others stay as concentrated ash.[41,42] Also during production of phosphate fertilizers, about 10% of the initial of ^{226}Ra, 20% of uranium, and about 85% of ^{210}Po is found in the phosphogypsum waste.[43] Phosphogypsum wastes are often located near coasts of seas and enhance the concentration of soil groundwater and river water. For this reason, the surroundings of phosphogypsum waste are strongly polluted by natural radionuclides.[44,45] The runoff of radioactive substances are important sources of marine radionuclides pollution, especially for closed seas. For example, the total runoff of natural and artificial radionuclides from rivers to the southern Baltic Sea was calculated to comprise about 3.3 TBq of ^{90}Sr, 5.2 TBq of ^{137}Cs, 0.7 TBq of ^{210}Po, 4.5 TBq of $^{234+238}$U, and 1.3 GBq of $^{239+240}$Pu.[46] In turn, some basins of the North Sea (Irish Sea and French Channel) are strongly contaminated with artificial radionuclides (especially, radiocesium and plutonium) emitted from European nuclear reprocessing facilities in Sellafield and Cap de la Hague.[40] Radionuclides in the marine environment are strongly accumulated by biota, and the values of bioaccumulation factor for some radioactive elements (polonium, plutonium, americium) in sea algae, benthic animals, and fish are more than 5000.[47] Transuranic elements belong to the group of pollutants caused by human activity and are important from the radiological point of view owing to their high radiotoxicity, long physical lifetime, high chemical reactivity, and long residence in biological systems in the marine environment.[48]

Eutrophication

Eutrophication is an increase in chemical nutrients, typically compounds containing nitrogen or phosphorus, in an ecosystem. It can result in an increase in the ecosystem's primary productivity (excessive plant growth and decay), and further effects, including lack of oxygen and severe reductions in water quality and in fish and other animal populations.[11] The biggest culprits are rivers that empty into the ocean, and along with it many chemicals used as fertilizers in agriculture as well as waste from livestock and humans. An excess of oxygen-depleting chemicals in the water can lead to hypoxia and the creation of a dead zone.[49–51] Estuaries tend to be naturally eutrophic because land-derived nutrients are concentrated where runoff enters the marine environment in a confined channel. The World Resources Institute has identified 375 hypoxic coastal zones around the world, concentrated in coastal areas in Western Europe, the eastern and southern coasts of the United States, and East Asia, particularly in Japan. In the ocean, there are frequent red tide algae blooms that kill fish and marine mammals and cause respiratory problems in humans and some domestic animals when the blooms reach close to shore.[52–54] In addition to land runoff, atmospheric anthropogenic nitrogen can enter the open ocean. A study in 2008 found that this could account for around one-third of the ocean's external (non-recycled) nitrogen supply and up to 3% of the annual new marine biological production. It has been suggested that accumulating reactive nitrogen in the environment may have consequences as serious as putting carbon dioxide in the atmosphere.[54]

Plastic Debris

Marine debris is mainly discarded human rubbish that floats on, or is suspended in the ocean. Eighty percent of marine debris is plastic—a component that has been rapidly accumulating since the end of World War II. The mass of plastic in the oceans may be as high as 100 million metric tons. Discarded plastic bags, six-pack rings, and other forms of plastic wastes that finish up in the ocean constitute dangers to wildlife and fisheries. Aquatic life can be threatened through entanglement, suffocation, and ingestion. Fishing nets, usually made of plastic, can be left or lost in the ocean by fishermen. Known as ghost nets, they entangle fish, dolphins, sea turtles, sharks, dugongs, crocodiles, seabirds, crabs, and other creatures, restricting movement, causing starvation, laceration and infection, and, in those that need to return to the surface to breathe, suffocation. Many animals that live on or in the sea consume flotsam by mistake, as it often looks similar to their natural prey. Plastic debris, when bulky or tangled, is difficult to pass, and may become permanently lodged in the digestive tracts of these animals, blocking the passage of food and causing death through starvation or infection.[55] Plastics accumulate because they do not biodegrade in the way many other substances do. They will photodegrade on exposure to the sun, but they do so properly only under dry conditions, and water inhibits this process. In marine environments, photodegraded plastic disintegrates into ever smaller pieces while remaining polymers, even down to the molecular level. When floating plastic particles photodegrade down to zooplankton sizes, jellyfish attempt to consume them, and in this way the plastic enters the ocean food chain. Many of these long-lasting pieces end up in the stomachs of marine birds and animals, including sea turtles and black-footed albatross.[56,57] Plastic debris tends to accumulate at the center of ocean gyres. In particular, the Great Pacific Garbage Patch has a very high level of plastic particulate suspended in the upper water column. In samples taken in 1999, the mass of plastic exceeded that of zooplankton (the dominant animal life in the area) by a factor of 6.[57] Midway Atoll, in common with all the Hawaiian Islands, receives substantial amounts of debris from the garbage patch. When it comes to 90% plastic, this debris accumulates on the beaches of Midway where it becomes a hazard to the bird population of the island. Midway Atoll is home to two-thirds (1.5 million) of the global population of Laysan Albatross. Nearly all of these albatrosses have plastic in their digestive system and one-third of their chicks die.[58] Some plastic additives are known to disrupt the endocrine system when consumed; others can suppress the immune system or decrease reproductive rates. Floating debris can also absorb persistent organic pollutants from seawater, including PCBs, DDT, and PAHs.[58] Aside from toxic effects, when ingested, some of these are mistaken by the animal brain for estradiol, causing hormone disruption in the affected wildlife.[59]

Noise Pollution

Marine life can be susceptible to noise or sound pollution from sources such as passing ships, oil exploration seismic surveys, and naval low-frequency active sonar. Sound travels more rapidly and over larger distances in the sea than in the atmosphere. Marine animals, such as cetaceans, often have weak eyesight, and live in a world largely defined by acoustic information. This applies also to many deeper sea fish, which live in a world of darkness. Between 1950 and 1975, ambient noise in the ocean increased by about 10 dB (i.e., a tenfold increase).[3] Noise also makes species communicate louder, which is called the Lombard vocal response. Whale songs are longer when submarine detectors are on. If creatures do not "speak" loud enough, their voice can be masked by anthropogenic sounds. These unheard voices might be warnings, finding of prey, or preparations of net bubbling. When one species begins to speak louder, it will mask other species' voices, causing the whole ecosystem to eventually speak louder. Undersea noise pollution is like the death of a thousand cuts. Each sound in itself may not be a matter of critical concern, but taken altogether, the noise from shipping, seismic surveys, and military activity is creating a totally different environment than that existed even 50 years ago. That high level of noise is bound to have a hard, sweeping impact on life in the sea.[3,60]

Conclusions

The seas and oceans are contaminated with many natural and anthropogenic industrial pollutants. A lot of them are strongly toxic to the organisms and to humans. Apart from plastic, a very particular problem are toxins in the marine environment that do not disintegrate rapidly enough. The most important toxins in the seas are organic compounds (PCBs, DDT, PAHs, pesticides, furans, dioxins, phenol, oil), heavy metals, and radioactive waste. Heavy metals (especially, mercury, lead, nickel, and cadmium) are toxic (in the oxidized form, as ions) or poisonous at low concentrations. These toxins can accumulate in the tissues of many aquatic species. In particular, accumulation of toxins in benthic animals and fish in contaminated estuarine areas and closed seas is very dangerous, due to the high position of these organisms in the food chain and the strong toxicity of the chemicals. Moreover, surface runoff of pesticides to the marine environment can genetically alter the gender of snails and fish species, transforming male into female fish. Finally, the marine environment is an important part of the natural environment consisting of people and human activities, which in the last century have been responsible for production and deposition of many toxins in the oceans and seas.

References

1. Szefer, P. *Metals, Metalloids and Radionuclides in the Baltic Sea Ecosystem*; Elsevier: Amsterdam, the Netherlands, 2002; Vol. 5.
2. Falandysz, J.; Trzosinska, A.; Szefer, P.; Warzocha, P.; Draganic, B. The Baltic Sea, especially southern and eastern regions (Ch. 7). In *Seas in the Millennium: An Environmental Evaluation, Vol. I: Europe, The Americas and West Africa*; Sheppard, C.R.C., Ed.; Pergamon, Elsevier: Amsterdam, 2000; 99–120.
3. Laws, E.A. *Aquatic Pollution*; John Wiley and Sons: New York, 2000.
4. Dowrd, B.M.; Press, D.; Los Huertos, M. Agricultural nonpoint sources: Water pollution policy: The case of California's central coast. Agric. Ecosyst. Environ. **2008**, *128* (3), 151–161.
5. Carpenter, S.R.; Caraco, R.F.; Cornell, D.F.; Howarth, R.W.; Sharpley, A.N.; Smith V.V. Nonpoint pollution of surface waters with phosphorus and nitrogen. Ecol. Appl. **1998**, *8*, 559–568.
6. Forstner, U.; Wittmann G.T.W. *Metal Pollution in the Aquatic Environment*, 2nd Ed.; Springer-Verlag: Berlin, 1983.
7. Daoji, L.; Dag, D. Ocean pollution from land-based sources: East China Sea. Ambio **2004**, *33* (1/2), 107–113.
8. Ahn, Y.H.; Hong, G.H.; Neelamani, S.; Philip, L.; Shanmugam, P. Assessment of levels of coastal marine pollution of Chennai city, southern India. Water Resour. Manage. **2006**, *21* (7), 1187–1206.
9. Li, Y.-H.; Teraoka, H.; Yang, T.-S.; Chen J.-S. The elemental composition of suspended particles from Yellow and Yangtze Rivers. Geochim. Cosmochim. Acta **1984**, *48*, 1561–1564.
10. Mance, G. Pollution Threat of Heavy Metals in Aquatic Environments; Elsevier Applied Science: London, 1987.
11. Ambio. Special issues: Marine eutrophication, Ambio, **1990**, *19*, 101–176.
12. Pimentel, D.; Zuniga, R.; Morrison, D. Update on the environmental and economic coast associate with alieninvasive species in the United States. Ecol. Econ. **2005**, *52*, 273–288.
13. Strandenes, S.P. The second order effects on commercial shipping of restrictions on the use of TBT. Sci. Total Environ. **2000**, *258*, 111–117.
14. Tanabe, S. Butyltin concentration in marine mammals. A review. Mar. Pollut. Bull. **1999**, *39*, 62–72.
15. Alzieu, C. Environmental impact of TBT: The French experience. Sci. Total Environ. **2000**, *258*, 99–102.
16. Evans, S.M. Tributyltin pollution: The catastrophe that never happened. Mar. Pollut. Bull. **1999**, *38*, 629–636.
17. Cambray, R.S.; Jefferies, D.F.; Topping, G. The atmospheric input of trace elements to the North Sea. Mar. Sci. Commun. **1979**, *5*, 175–194.
18. Spengler, J.D.; Sexton K. Indoor air pollution: A public health perspective. Science (New Ser.) **1983**, *221* (4605), 9–17.
19. Nriagu, J.O. A global assessment of natural sources of atmospheric trace metals. Nature **1989**, *338*, 47–51.
20. Pacyna, J.M. Estimation of the atmospheric emissions of trace elements from anthropogenic sources in Europe. Atmos. Environ. **1984**, *18*, 41–50.
21. Fowler, S.W. Critical review of selected heavy metal and chlorinated hydrocarbon concentrations in the marine environment. Mar. Environ. Res. **1990**, *29*, 1–64.
22. Duce, R.A.; Unni, C.K.; Ray, B.J.; Prospero, J.M.; Merrill, J.T. Long-range atmospheric transport of soil dust from Asia to the tropical North Pacific: Temporal variability. Science **1980**, *209*, 1522–1524.

23. Prospero, J.M.; Nees, R.T. Impact of the North African drought and El Niño on mineral dust in the Barbados trade winds. Nature **1986**, *320*, 735–738.

24. Aarkrog, A. The radiological impact if the Chernobyl debris compared with than from weapons fallout. J. Environ. Radioact. **1988**, *6*, 151–162.

25. Dick, A.L. Concentrations and sources of metals in the Antarctic Peninsula aerosol. Geochim. Cosmochim. Acta **1991**, *55*, 1827–1836.

26. Boutron, C. Trace metals in remote Arctic snow: Natural or anthropogenic? Nature **1980**, *284*, 574–576.

27. Dietz, R.; Nielsen, C.O.; Hansen M.M.; Hansen C.T. Organic mercury in Greenland birds and mammals. Sci. Total Environ. **1990**, *95*, 41–51.

28. Dietz, R.; Riget, F.; Johansen P. Lead, cadmium, mercury and selenium in Greenland marine animals. Sci. Total Environ. **1996**, *186*, 67–93.

29. Donaldson, G.M.; Braune B.M.; Gaston A.J.; Noble D.G. Organochlorine and heavy metal residue in breast muscle of know-age thick-milled murres (*Uria lomvia*) from the Canadian Arctic. Arch. Environ. Contam. Toxicol. **1997**, *33*, 430–435.

30. Ahnert, A.; Borowski, C. Environmental risk assessment of anthropogenic activity in the deep sea. J. Aquat. Ecosyst. Stress Recov. **2000**, *7* (4), 299.

31. Halfar, J.; Fujita, R.M. 2007. Danger of deep-sea mining. Science **2007**, *316* (5827), 987.

32. Glasby, G.P. Lessons learned from deep-sea mining. Sci. Mag. **2000**, *289*, 551–553.

33. Nath, B.; Sharma, R. Environment and deep-sea mining: A perspective. Mar. Georesour. Geotechnol. **2000**, *18* (3), 285–294.

34. Sharma, R. Deep-sea impact experiments and their future requirements. Mar. Georesour. Geotechnol. **2005**, *23* (4), 331–338.

35. Orr, J.C.; Fabry, V.J.; Aumont, O.; Bopp, L.; Doney, S.C.; Feely, R.A.; Gnanadesikian, A.; Gruber, N.; Ishida, A.; Joos, F.; Key, R.M.; Lindsay, K.; Maier-Reimer, E.; Matear, R.; Monfray, P.; Mouchet, A.; Najjar, R.G.; Plattner, G.K.; Rodgers, K.B.; Sabine, C.L.; Sarminto, J.L.; Schlitzer, R.; Slater, R.D.; Totterdell, I.J.; Weirig, M.F.; Yamanaka, Y.; Yool, A. Anthropogenic ocean acidification over the twenty-first century and its impact on calcifying organisms. Nature **2005**, *437* (7059), 681–686.

36. Key, R.M.; Kozyr, A.; Sabine, C.L.; Lee, K.; Wanninkhof, R.; Bullister, J.; Feely, R.A.; Millero, F.; Mordy, C.; Peng, T.-H. A global ocean carbon climatology: Results from GLODAP. Global Biogeochem. Cycles **2004**, *18*, GB4031.

37. Feely, R.; Sabine, C.L.; Hernandez-Ayon, J.M.; Ianson, D.; Hales, B. Evidence for Upwelling of corrosive "acidified" seawater onto the continental shelf. Science **2008**, *10*.

38. Milkov, AV. Global estimates of hydrate-bound gas in marine sediments: how much is really out there? Earth Sci. Rev. **2004**, *66* (3–4): 183–197.

39. Skwarzec, B. Radiochemia srodowiska i ochrona radiologiczna (in Polish), Environmental radiochemistry and radiological protection, Wydawnictwo DJ sc., Gdańsk, 2002.

40. Kershaw, P.J.; McCubbin, D.M.; Leonard, K.S. Continuing contamination of the North Arctic and Arctic waters by Sellafield radionuclides. Sci. Total Environ. **1999**, *237/238*, 119–132.

41. Nakaoka, A.; Fukushima, M.; Takagi S. Environmental effect of natural radionuclides from coal-fired power plants. Health Phys. **1984**, *3*, 407–416.

42. Flues, M.; Morales, M.; Mazilli, B.P. The influence of a coal-plant operation on radionuclides in soil. J. Environ. Radioactivity **2002**, *63*, 285–294.

43. Carvalho, F.P.; Oliviera, J.M.; Lopes, I.; Batista, A. Radionuclides from post uranium mining in rivers in Portugal. J. Environ. Radioact. **2007**, *98*, 298–314.

44. Borylo, A.; Nowicki, W.; Skwarzec, B. Isotope of polonium (^{210}Po) and uranium (^{234}U and ^{238}U) in the industrialized area of Wisslinka (Northern Poland). Int. J. Environ. Anal. Chem. **2009**, *89*, 677–685.

45. Skwarzec, B.; Boryło, A.; Kosiiíska, A.; Radziejewska, S. Polonium (^{210}Po) and uranium (^{234}U and ^{238}U) in water, phosphogypsum and their bioaccumulation in plants around phosphogypsum waste heap in Wisslinka (northern Poland), Nukleonika **2010**, *55* (2), 187–193.

46. Skwarzec, B. Polonium, uranium and plutonium radionuclides in aquatic environment of Poland and southern Baltic. Baltic Coastal Zone **2009**, *13*, 127–166.

47. Skwarzec, B. Polonium, uranium and plutonium in the southern Baltic Sea. Ambio **1997**, *26* (2), 113–117.

48. Coughtrey, P.J.; Jackson, D.; Jones, C.H.; Kene, P.; Thorne, M.C. Radionuclides Distribution and Transport in Terrestrial and Aquatic Ecosystems. A Critical Review of Data; A.A. Balkema: Rotterdam, 1984.

49. Rosenberg, R. Eutrophication—The future marine coastal nuisance. Mar. Pollut. Bull. **1985**, *16*, 227–231.

50. Billen, G.; Garnier, J.; Delinge, C.; Billen C. Estimate of early-industrial inputs of nutrients to river system: implication for coastal eutrophication. Sci. Total Environ. **1999**, *243/244*, 43–52.

51. Bruneau, L. Pollution from industries in the drainage area of the Baltic. Ambio **1980**, *9*, 145–152.

52. MacLean, J.L.; White A.W. Toxic dinoflagellates bloom in Asia: A growing, concern. In *Toxic Dinoflagellates*; Anderson, D.M., White, A.W., Baden, D.G., Eds.; Elsevier: New York, 1985; 517–520.

53. Paerl, H.W. Enhancement of marine primary production by nitrogen enriched rain. Nature **1985**, *315*, 747–749.

54. Selman, M. Eutrophication: An Overview of Status, Trends, Policies, and Strategies; World Resources Institute, Washington DC, UAS, 2007.

55. Sheavly, S.B.; Register, K.M. Marine debris and plastics: Environmental concerns, sources, impacts and solutions. J. Polym. Environ. **2007**, *15* (4), 301–305.

56. Thompson, R.C. Lost at sea: Where is all the plastic? Science **2004**, *304* (5672), 843.

57. Moore, C.; Moore, S.L.; Leecaster, M.K.; Weisberg, S.B. A comparison of plastic and plankton in the North Pacific Central Gyre. Mar. Pollut. Bull. **2004**, *42* (12), 1297–1300.

58. Rios, L.M.; Moore, C.; Jones, P.R. Persistent organic pollutants carried by synthetic polymers in the ocean environment. Mar. Pollut. Bull. **2007**, *54*, 1230–1237.

59. Tanabe, S.; Watanabe, M.; Minh, T.B.; Kunisue, T.; Nakanishi, S.; Ono, H.; Tanaka, H. PCDDs, PCDFs, and coplanar PCBs in albatross from the North Pacific and Southern Oceans: Levels, patterns, and toxicological implications. Environ. Sci. Technol. **2004**, *38*, 403–413.

60. Fristrup, K.M.; Hatch, L.T.; Clark, C.W. Variation in humpback whale (*Megaptera novaeangliae*) song length in relation to low-frequency sound broadcasts. Acoust. Soc. Am. J. **2003**, *113* (6), 3411–3424.

II

COV:
Comparative
Overviews
of Important
Topics for
Environmental
Management

11

Rain Water: Harvesting

Introduction 123
Advantages of Rainwater Harvesting 123
Types of Rainwater Harvesting Systems 124
    Simple Rooftop Collection Systems • Large Systems for Educational
    Institutions, Stadiums, Airports, and Other Facilities • Rooftop Collection
    Systems for High-Rise Buildings in Urbanized Areas • Land Surface
    Catchments • Collection of Stormwater in Urbanized Catchment
Design and Maintenance of Rainwater Harvesting Systems 126
    Catchment Surface • Conveyance Systems • Storage Tanks
Conclusion 127
References 127

K.F. Andrew Lo

Introduction

Among the various alternative technologies to augment water resources, rainwater harvesting is a simple, decentralized solution and imposes insignificant impact on the environment. It is an important water source in many areas with significant rainfall but lacking any kind of conventional, centralized supply system. It is also a good option in areas where good-quality fresh surface water or groundwater is lacking. Rainwater harvesting systems have been used since ancient times and evidence of roof catchment systems dates back to early Roman times. In the Negev Desert in Israel, in Libya and Egypt, in Mexico, and in the Andes Range in South America as well as in the Arizona Desert in North America, stone dams and tanks were built to divert and store rainwater for irrigation purposes.

Advantages of Rainwater Harvesting

Rainwater harvesting systems can provide water at, or near, the point where water is needed or used. The systems can be both owner-operated and utility-operated, and owner-managed and utility-managed. Rainwater collected using existing structures (rooftops, parking lots, playgrounds, parks, ponds, and flood plains) has few negative environmental impacts compared with other water resources development technologies.[1] Rainwater is relatively clean and the quality is usually acceptable for many purposes with little or even no treatment. The physical and chemical properties of rainwater are usually superior to sources of groundwater that may have been subject to contamination.

Other advantages of rainwater harvesting include the following:

1. Rainwater harvesting can coexist with, and provide a good supplement to, other water sources and utility systems, thus relieving pressure on other water sources.
2. Rainwater harvesting provides a water supply buffer for use in times of emergency or breakdown of public water supply systems, particularly during natural disasters.

3. Rainwater harvesting can reduce storm drainage load and flooding in cities.
4. The owners who operate and manage the rainwater catchment system are more willing to exercise water conservation.
5. Rainwater harvesting technologies are flexible and can be built to meet almost any requirements.

Types of Rainwater Harvesting Systems

Collection systems can vary from simple households to large catchment systems. The categorization of rainwater harvesting systems depends on factors such as the size and nature of the catchment areas and whether the systems are in urban or rural settings.[2]

Simple Rooftop Collection Systems

The main components of a simple rooftop collection system are the cistern itself, the piping that leads to the cistern, and the appurtenances within the cistern (Figure 1). The materials and the degree of sophistication of the whole system largely depend on the initial capital investment. Some cost-effective systems involve cisterns made with ferrocement. In some cases, the harvested rainwater may be filtered or disinfected.

Large Systems for Educational Institutions, Stadiums, Airports, and Other Facilities

When the systems are larger, the overall system can become more complicated (e.g., rainwater collection from roofs and grounds of institutions, storage in underground reservoirs, and treatment and use for non-potable applications) (Figure 2).

Rooftop Collection Systems for High-Rise Buildings in Urbanized Areas

In high-rise buildings, roofs can be designed for catchment purposes and the collected roof water can be kept in separate cisterns on the roofs for non-potable uses.

Land Surface Catchments

Ground catchment techniques (Figure 3) provide more opportunity for collecting water from a larger surface area. By retaining small creek and stream flows in small storage surfaces or underground reservoirs,

FIGURE 1 A simple roof catchment system (illustrated by Chia-Ming Lin).

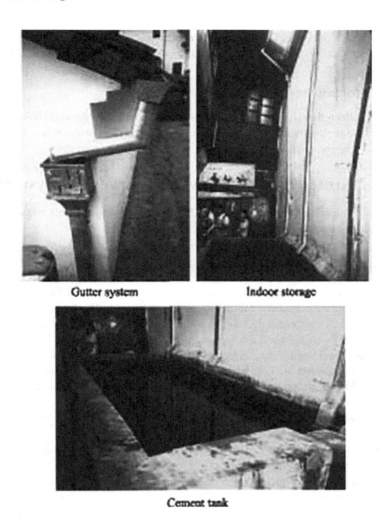

Gutter system   Indoor storage

Cement tank

FIGURE 2   An indoor storage system in a monastery in China (photographed by K. F. Andrew Lo).

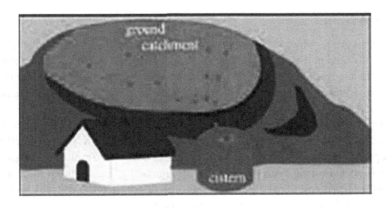

FIGURE 3   A land catchment system (illustrated by Chia-Ming Lin).

can meet water demands during dry periods. However, there is a possibility of high seepage loss to the ground. The marginal quality of the water collected is suitable for use mainly in agriculture.

Collection of Stormwater in Urbanized Catchment

The surface runoff collected in stormwater ponds/reservoirs from urban areas is subject to a wide variety of contaminants. Keeping these catchments clean is of primary importance; hence the cost of water pollution control can be considerable.

Design and Maintenance of Rainwater Harvesting Systems

Typically, a rainwater harvesting system consists of three basic elements: the collection system, the conveyance system, and the storage system.

Catchment Surface

The effective catchment area and the material used in constructing the catchment surface influence collection efficiency and water quality. Materials commonly used for roof catchment are corrugated aluminum and galvanized iron, concrete, fiberglass shingles, tiles, and slates. Mud is used primarily in rural areas. Bamboo roofs are least suitable because of possible health hazards. The catchment surface materials must be non-toxic and must not contain substances that impair water quality. Roofs with metallic paint or other coatings are not recommended because they may impart tastes or color to the collected water. Catchment surfaces and collection devices should be cleaned regularly to remove dust, leaves, and bird droppings to minimize bacterial contamination and to maintain the quality of collected water. Roofs should also be free from overhanging trees because birds and animals in the trees may defecate on the roofs.

When land surfaces are used as catchment areas, various techniques are available to increase runoff capacity: 1) clearing or altering vegetation cover; 2) increasing the land slope with artificial ground cover; and 3) reducing soil permeability by soil compaction. Specially constructed ground surfaces (concrete, paving stones, or some kind of liner) or paved runways can also be used to collect and convey rainwater to storage tanks or reservoirs. Care is required to avoid land surface damage and contamination by people and animals. If required, these surfaces should be fenced to prevent people and animal entry. Large cracks in the paved catchment because of soil movement, earthquakes, or prolonged exposure should be repaired immediately. Maintenance, typically consisting of the removal of dirt, leaves, and other accumulated materials, should take place annually before the start of the major rainfall season.

Conveyance Systems

Conveyance systems are required to transfer the rainwater collected on catchment surfaces to storage tanks. This is usually accomplished by making connections to one or more downpipes connected to collection devices. The pipes used for conveying rainwater, wherever possible, should be made of plastic, polyvinyl chloride (PVC), or other inert substance because the pH of rainwater can be acidic and may cause corrosion and mobilization of metals in metal pipes.

When it first starts to rain, dirt and debris from catchment surfaces and collection devices will be washed into the conveyance systems. Relatively clean water will only be available sometime later in the storm. The first part of each rainfall should be diverted from the storage tank. There are several possible options for selectively collecting clean water for the storage tanks. The common method is a sediment trap, which prevents debris entry into the tank. Installing a first-flush (or foul-flush) device is also useful to divert the initial batch of rainwater away from the tank.[3]

Rainwater pipes must be permanently marked in such a way that there is no risk of confusing them with drinking water pipes. Gutters and downpipes need to be periodically inspected and carefully cleaned. A good time to inspect gutters and downpipes is while it is raining, so that leaks can be easily detected.

Storage Tanks

Various types of rainwater storage facilities can be found in practice. Storage tanks should be constructed of inert material. Reinforced concrete, fiberglass, polyethylene, and stainless steel are suitable materials. Ferrocement tanks and jars made of mortar or earthen materials are commonly used. As an alternative, interconnected tanks made of pottery or polyethylene may be suitable. They are easy to clean. Bamboo-reinforced tanks are less successful because they may become infested with termites, bacteria, and fungi.

Precautions are required to prevent the entry of contaminants into storage tanks. The main sources of external contamination are pollution from debris, bird and animal droppings, and insects. A solid and secure cover is required to avoid breeding of mosquitoes, to prevent insects and rodents from entering the tank, and to keep out sunlight to prevent algae growth inside the tank.[4] A coarse inlet filter is also desirable for excluding coarse debris, dirt, leaves, and other solid materials.

All tanks need cleaning and their designs should allow for thorough scrubbing of the inner walls and floors. A sloped bottom and the provision of a pump and a drain are useful for collection and discharge of settled grit and sediment. Chlorination of the cisterns or storage tanks is necessary if the water is to be used for drinking and domestic uses. Cracks in the storage tanks can create major problems and should be repaired immediately.

The extraction system (taps/faucets, pumps) must not contaminate the stored water. Taps/faucets should be installed at least 10 cm above the base of the tank because this allows any debris entering the tank to settle on the bottom.[5] If it remains undisturbed, it will not affect the quality of the water. The handle of taps might be detachable to avoid misuse by children. Periodic maintenance should also be carried out on pumps used to lift water.

Conclusion

In the future, water scarcity in both developing and developed countries is inevitable.[6] The challenge of meeting the water demand can be largely met by appropriate understanding, study, and application of rainwater harvesting. Rainwater harvesting is about to come of age.[7] It has an appropriate image about it that meshes well with the gentler ideas of the late 20th century. Because the technique makes use of an untapped resource—precipitation that would otherwise be evaporated before it had a chance to play a useful role in feeding the human population—it looks like getting something for nothing. Making use of such a resource has certain poetry to it, particularly in a field where the resource itself can never be increased or decreased; unlike food, water cannot be grown to order, even given the right soil and the right fertilizer. But, like food, water can be harvested more efficiently. Doing so is a major priority for the 21st century.

References

1. Geiger, W.F. Integrated water management for urban areas, Proceedings of the 7th International Rainwater Catchment Systems Conference, Beijing, China, June 21–25, 1995; Liu, C.M., Ed.; Institute of Geographical Sciences and Natural Resources, Chinese Academy of Sciences: Beijing, 1995, 7-1–7–21.
2. Gould, J.; Nissen-Peterson, E. *Rainwater Catchment Systems for Domestic Supply: Design, Construction and Implementation*; Intermediate Technology Publications: London, 1999.

3. Macomber, P.S.H. *Guidelines on Rain Water Catchment Systems for Hawaii*; College of Tropical Agriculture and Human Resources, University of Hawaii at Manoa: Hawaii, 2001.
4. Cunliffe, D.A. *Guidance on the Use of Rainwater Tanks;* National Environmental Health Forum Monographs, Water Series No. 3; National Environmental Health Forum: Australia, 1998.
5. Texas Water Development Board *Texas Guide to Rainwater Harvesting*, 2nd Ed.; Texas Water Development Board: Texas, 1997.
6. Clarke, R. *Water: The International Crisis*; The MIT Press: Cambridge, 1993.
7. Pacey, A.; Cullis, A. *Rainwater Harvesting: The Collection of Rainfall and Runoff in Rural Areas*; Intermediate Technology Publications: London, 1986.

12

Water Harvesting

Introduction 129
General Description 129
Water Harvesting for Livestock Drinking Water 130
Water Harvesting for Domestic Use 130
Water Harvesting for Growing of Crops 130
Other Considerations 130
References 131

Gary W. Frasier

Introduction

Water is one of the critical elements to support life. Without water, life as we know it cannot exist. Without water, plants will die. Humans and animals will die without drinking water. Yet, the planet we call earth has more area covered with water than landmass. From space, the planet looks blue from the water in the seas and the clouds. Even with all this water, there are many places on the land, classified as arid and semiarid land, where there is not enough drinking water for man and animals to live or plants to grow. In many of these arid and semiarid places, sufficient water can be obtained for drinking-water supplies and growing of plants by a process called water harvesting. In technical terms, water harvesting is defined as "the process of collecting and storing precipitation (rain or snow) for beneficial use from an area that has been treated or modified to increase precipitation runoff."[1]

With the modern advent of water transfers via canals and deep wells for supplying water for homes and irrigation of crops, there is a tendency to consider water harvesting as a "new" water supply technique. In actuality, water harvesting is an ancient practice that has been dated back to the Edom Mountain areas of Southern Jordan 9000 years ago.[2] There is evidence that water harvesting systems were used 3000 to 4000 years ago in various places of what we now call the Middle East. Some of these early systems were located in areas having less than 200 mm annual precipitation.[3] Water harvesting systems are currently being used for water supply in various places around the world such as Israel, Egypt, Jordan, Mexico, Australia, and the United States.

General Description

There are two general classifications of water harvesting systems. One group is used for providing drinking-water supplies for human and animal consumption. The other group is used to provide water for the growing of plants (runoff farming). All water harvesting systems have two major components: an area where the precipitation is collected (catchment area) and some means for storing the collected water.

The water collection area can be any surface that has been modified to increase precipitation runoff. They can include rock outcroppings with diversion facilities at the lower edge to direct the runoff water to a central point; smoothed compacted soil, chemically treated soil surfaces; soil covered with membranes of rubber, wood, or metal; and building rooftops. The water storage facility can be a tank, pond, or cistern with an impermeable lining or, in the case of runoff farming, the soil around the plants.

Water Harvesting for Livestock Drinking Water

Providing animal, domestic, and wildlife drinking water supplies is one of the newer applications of water harvesting (past 20–30 years). The catchment area can be a hillslope, an area covered or treated to reduce water infiltration and increase surface water runoff, or even roofs of buildings. The catchment area can be modified via various techniques, frequently separated into 3 general categories: 1) topography modification; 2) soil modification; and 3) impermeable coverings or membranes. Topography modification consists of land smoothing and clearing and frequently yields 20%–35% runoff efficiency. Some typical soil modification treatments used are sodium salts mixed in the top 6–12 cm of soil and compacted with runoff yields of 50%–80%. Water repellents and paraffin wax applied to the soil surface will yield 60%–95% runoff. Asphalts or bitumen sprayed on the soil surface provide 50%–80% runoff. More expensive catchment treatments consist of various impermeable coverings. They consist of gravel-covered sheeting (75%–95% runoff); asphalt membranes (85%–95% runoff); and concrete, sheet metal, and artificial rubber sheeting (60%–95% runoff).[4,5]

Storage of the collected water is a key component of any water harvesting system, but it is also one of the most expensive components. Unlined pits or ponds are usually not suitable unless seepage losses can be stopped. There are many types, shapes, and sizes of wooden, metal, concrete, or reinforced plastic storage containers that can be used. Also, in most arid and semiarid areas where the precipitation events are erratic, it is necessary to provide some means of controlling evaporative losses from the stored-water facility. Evaporation control on sloping-sided pits or ponds is difficult because the water surface area changes with depth.[5]

Water Harvesting for Domestic Use

Most of the techniques used for water harvesting for livestock and wildlife drinking supplies can be used to provide water for domestic uses. One technique that is used in several places in the world today is the collection of water from the roofs of buildings.[6] A relatively small roof area can provide significant quantities of water if collected and stored. One millimeter of precipitation will yield 1 L of water. This technique is very effective in areas where there are frequent precipitation events but little or no surface water, because of highly permeable soils or groundwater that has high concentrations of undesirable minerals. A simple wooden or metal tank or concrete cistern stores the collected water.

Water Harvesting for Growing of Crops

One form of water harvesting used for growing crops is termed runoff farming. Runoff farming maximizes the effect of limited precipitation by collecting surface runoff from a large area and applying it to a smaller cropped area. The collected water can be applied directly to the crop area during the precipitation event, referred to as floodwater farming,[3] or stored and applied later by some form of irrigation system.[7] A modified version is termed microcatchment farming, where a small runoff area is situated directly upslope of the growing area. This technique has been used for tree crops such as pistachio, olive, and almond.[3]

Other Considerations

Water harvesting should not be considered as a cheap or inexpensive means of water supply. It can supply water in areas where other water sources are unattainable or unsuitable. Water harvesting is not a new means of water supply or source but is receiving renewed interest in many places in the world.

It became a water source of interest with the work of Evenari and his colleagues[8] in the early 1960s reconstructing ancient water harvesting systems in the Negev Desert of Israel. Following this work, several research organizations around the world conducted extensive studies, primarily exploring various low-cost means of treating the soil to reduce infiltration and increase precipitation runoff. Much of this effort has currently been completed. Many water harvesting systems are being used around the world, most notably in arid and semiarid, underdeveloped areas where there is plentiful labor. At the same time, there are many places where the systems have failed. One of the most common reasons for failure is the lack of adequate maintenance of the collection area and the water storage facility. When properly designed and maintained, a water harvesting system can last for decades.

References

1. Frasier, G.W. Water harvesting for rangeland water supplies: a historic perspective. In Proceedings of a Symposium on Environmental, Economic, and Legal Issues Related to Rangeland Water Development, Tempe, Arizona, November 13–15, 1997; The Center for the Study of Law, Science, and Technology, Arizona State University: Tempe, Arizona, 1998; 17–24.
2. Bruins, H.M.; Evenari, M.; Nessler, U. Rainwater-harvesting agriculture for food production in arid zones. The challenges of the African famine. Appl. Geogr. **1986**, *6*, 13–33.
3. Frasier, G.W. Water harvesting. In *Encyclopedia of Soil Science;* Marcel Dekker Inc.: New York, 2002; 1387–1389.
4. Frasier, G.W.; Cooley, K.R.; Griggs, J.R. Performance evaluation of water harvesting catchments. J. Range Manage. **1979**, *32* (6), 453–456.
5. Frasier, G.W.; Myers, L.E. *Handbook of Water Harvesting,* Agriculture Handbook Number 600; U.S. Department of Agriculture: Washington, D.C., 1983; 48 pp.
6. Frasier, G.W. Harvesting water for agricultural, wildlife, and domestic uses. J. Soil Water Conserv. **1980**, *35* (3), 125–128.
7. Scrimgeour, F.G.; Frasier, G.W. Runoff impoundment for supplemental irrigation: An economic assessment. Amer. J. Alter. Agric. **1991**, *6* (3), 139–145.
8. Evenari, M.; Shanan, L.; Tadmor, N.H. Runoff farming: In the desert. I. Experimental Layout. Agron. J. **1968**, *60*, 29–32.

13

Groundwater: Saltwater Intrusion

Introduction 133
Saltwater Intrusion Problems around the World 134
    United States • The Netherlands • Israel and Palestinian
    Territories • Mexico • Italy • Other Parts of the World
Mechanisms of Saltwater Intrusion 135
    Ghyben-Herzberg Relation • Up-Coning • Transition Zone
Monitoring and Exploration of Saltwater Intrusion..................................140
    Geological Investigation • Geophysical Investigation • Geochemical
    Investigation
Mathematical Modeling...143
    Constitutive Equations • Darcy's Law • Mass Balance Equation for
    Water • Mass Balance Equation for Dissolved Salt • Well-Posed Initial and
    Boundary Value Problem
Computer Models..146
Combating Saltwater Intrusion 146
Planning and Management 147
References..147

Alexander
H.-D. Cheng

Introduction

Human beings have a tendency to live near coastal areas. About 23% of the world's population lives within 100 km of the coast. This figure is likely to increase to 50% in the next 25 years. In the United States, 54% of all Americans now live in 772 coastal counties adjacent to the Atlantic and Pacific Oceans, the Gulf of Mexico, and the Great Lakes. By the year 2025, nearly 75% of all Americans are expected to live in coastal counties.

Coastal aquifers serve as major sources for freshwater supply in many countries around the world, especially in arid and semiarid zones. Many coastal areas are also heavily urbanized, a fact that makes the need for freshwater even more acute. Coastal aquifers are highly sensitive to disturbances. Inappropriate management of a coastal aquifer can lead to its destruction as a source for freshwater much earlier than aquifers that are not connected to the sea. The reason is the threat of seawater intrusion into freshwater aquifers.

The origin of saltwater intrusion into freshwater aquifers can come from natural sources such as seawater and deep formation brines or from anthropogenic sources such as deicing salt, agricultural return flow, and leachate from landfills. The most frequent occurrences are found in coastal regions where overexploitation of groundwater has caused the encroachment of seawater into freshwater aquifers. Once an aquifer is invaded, a part of the salt will adsorb onto the solid surface, making it difficult to reverse the process and restore the aquifer. The slow movement of groundwater also makes the remediation time long. Salinity in water poses a health hazard for humans and livestock, damages crops, and corrodes pipes and boilers in industrial uses. Hence, the invasion of saltwater into a freshwater aquifer means the loss of that aquifer for water sources.

Saltwater Intrusion Problems around the World

Serious saltwater intrusion problems exist in many parts of the world. In the following, we shall review some of the more prominently reported cases.

United States

In the United States, the potential for saltwater intrusion was recognized as early as 1824 in New Jersey and 1854 on Long Island, New York.[1] In modern times, serious problems of saltwater intrusion into aquifers are found in the Miami, Long Island, Hawaii, and Los Angeles areas.

Seawater intrusion into aquifers is a problem throughout coastal Florida. Particularly, in the Miami area, a network of drainage canals was constructed from 1909 through the 1930s. The resulting drainage lowered water levels about 2 m in the Everglades. This drainage, combined with coastal pumping, caused seawater to advance progressively into the Biscayne aquifer.[2] The intrusion front has reached several kilometers inland.[3] The South Florida Water Management District is one of the regional water management agencies in the state that has the regulatory power to issue or deny a well construction permit, to require the installation of monitoring wells, and to impose restrictions on groundwater extraction during drought periods based on seawater intrusion considerations.

In Los Angeles County, California, severe groundwater overdraft from the early 1900s to the late 1950s caused water levels to drop below sea level, allowing saltwater to intrude into the potable aquifers, rendering coastal wells out of service. In an effort to halt the intrusion, groundwater management agencies took three major steps from the mid-1950s to mid-1960s, including the following: 1) construction of freshwater injection wells along the coast to prevent the saltwater intrusion; 2) limiting the amount of groundwater that could be pumped annually; and 3) creation of the Water Replenishment District of Southern California to purchase artificial replenishment water to make up the annual and accumulated overdrafts.[5]

For the Oahu Island of Hawaii, fresh groundwater exists as a lens floating on top of saltwater. In the predevelopment time (pre-1880), the freshwater lens in the Pearl Harbor area was estimated to be 400 m to 500 m thick.[6] In 1990, measured salinity profiles showed the freshwater lens to be only 200 m to 300 m thick.[7] Thus, the 100 years of development has caused the freshwater lens to reduce to half its thickness.

The Netherlands

The Netherlands is located on river deltas, with much of its territory below sea level, reclaimed from the sea. The low- lying lands, known as polders, are particularly susceptible to the invasion of seawater. To keep out the brackish water, year-round recharge of aquifers using reclaimed water is needed.[8]

A great concern for the Netherlands is the threat of sea level rise, as a result of the global warming trend. The rise of sea level can inundate land and push the saltwater-freshwater equilibrium point further inland. It will also increase the hydraulic pressure of seepage flow for the low-lying areas. Possible countermeasures have been investigated.[9]

Israel and Palestinian Territories

Due to the semiarid climate, Israel and Palestinian territories (Gaza Strip and West Bank) are critically dependent on groundwater for water supply. In fact, the sandstone Quaternary aquifer that underlies coastal Israel and the Gaza Strip is being "mined," and the Mediterranean Sea intrudes along a saltwater front.[10,11] The water demands have been so much that the strategy is no longer to push the saltwater front outward to sea. The management goal is to allow the aquifer to be contaminated and to intercept the largest possible percentage of freshwater outflow to sea.

Mexico

The aquifer in northwestern Yucatan contains a freshwater lens that floats above a denser saline water wedge. It has been shown that at certain locations, the penetration of saline water can be more than 100 km inland.[12] The aquifer, which is unconfined except for a narrow band along the coast, is the sole freshwater source in northwestern Yucatan. Development of industry and agriculture and other land use changes pose a potential threat to the quantity and quality of freshwater resources in the Yucatan Peninsula.[13]

Italy

In the 1920s, Venice went through an Industrial Revolution. During this time period, water was constantly being pumped from the underlying aquifer. The lowering of the piezometric head has caused not only land subsidence but also the contamination of aquifer by saline water. Geophysical investigations have shown that the saline water may extend inshore up to 20 km from the Adriatic Sea coastline. The saltwater plume is observed from the near ground surface down to 100 m.[14,15] The combined effect of sea level rise and land subsidence can further enhance the saltwater contamination and the related soil salinization, with serious environmental and socioeconomic impacts.

Other Parts of the World

Saltwater intrusion problems are also found in the Nile Delta of Egypt,[16] Cap Bon of Tunisia,[17] China,[18] Cyprus,[19] Morocco,[20] and many other parts of the world.

Mechanisms of Saltwater Intrusion

Figure 1 gives a simplified view of seawater intrusion into an unconfined aquifer. We observe that saltwater is heavier and hence tends to stay underneath the freshwater. The freshwater, however, has a hydraulic gradient with head decreasing from inland toward the sea, thus driving a freshwater flow to the sea. This outflow momentum counteracts the density-driven seawater. Without it, seawater will continue to move inland until the entire aquifer below sea level is occupied by the heavier saltwater. However, since such a hydraulic gradient always exists due to the precipitation recharge inland, an equilibrium position of saltwater and freshwater will be established, schematically shown as the interface in Figure 1.

Ghyben-Herzberg Relation

The earliest and the simplest explanation of the saltwater intrusion mechanism was provided by Du Cummun,[21] Badon-Ghyben,[22] and Herzberg,[23] commonly known as the Ghyben-Herzberg relation. By making the following assumptions—1) a sharp interface exists, separating saltwater from freshwater; 2) the freshwater flow lines are approximately horizontal; and 3) there is no flow in the saltwater zone—a simple relation between the freshwater free surface elevation, h_f and the interface location below the sea level, η, can be developed as (see Figure 1)

$$\eta = \frac{\rho_f}{\rho_s - \rho_f} h_f \approx 40 h_f \qquad (1)$$

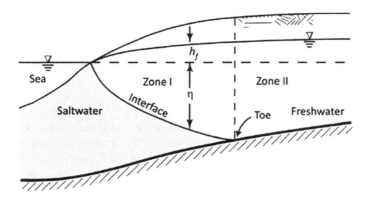

FIGURE 1 Seawater intrusion into an unconfined aquifer.

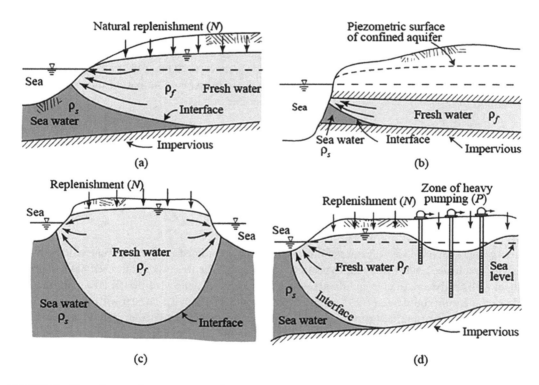

FIGURE 2 Typical vertical cross sections of seawater intrusion in coastal aquifers: (a) unconfined aquifer with replenishment; (b) confined aquifer; (c) freshwater lens on an island; (d) unconfined aquifer with pumping.

where ρ_f is freshwater density (1 g/cm³) and ρ_s is saltwater density (approximately 1.025 g/cm³). This is known as the

What the above relation says is that for every meter of freshwater head above sea level, the interface is pushed down 40 m below sea level. When the interface touches the bottom of an aquifer, a toe for the seawater wedge is located (Figure 1). Landward from the toe (zone 2), only freshwater exists; seaward from the toe (zone 1), saltwater and freshwater coexist.

Figure 2 presents some typical cross sections with interfaces in coastal aquifers under natural conditions (a–c) and with pumping (d). Like all figures that describe aquifers, these figures are highly distorted, with vertical scale much more magnified.

Up-Coning

This 40:1 ratio mentioned above based on the Ghyben–Herzberg relation may sound like good news for repelling saltwater by freshwater head. However, if pumping activity is increased in the coastal zone, as often is the case due to increasing population and economic activities, the reduced freshwater head will allow saltwater to move a large distance landward. Also, if a pumping well is situated above the interface in the saltwater-freshwater coexisting zone, the drawdown of a water table or piezometric head will cause the interface to rise up steeply to meet the well. This phenomenon is known as up-coning. Once the saltwater reaches the well, the well is generally considered to be lost.

For a pumping well situated at a distance d above an undisturbed interface (Figure 3), a formula that predicts the maximum allowable discharge, Q_{max}, before the saltwater entering the well is given by[24,25]

$$Q_{max} < 0.6\pi\Delta sKd^2 \approx 0.047Kd^2 \tag{2}$$

where K is the hydraulic conductivity, and $\Delta s = (\rho_s - \rho_f)/\rho_f \approx 0.25$. Take, for example, $K = 100$ m¹/day; then a well with its screened section situated 10 m above the interface can pump up to 470 m²/day. On the other hand, if the distance d is reduced to 5 m, then the pumping rate is restricted to 117 m³/day.

In Figure 4, we demonstrate a few saltwater intrusion scenarios with a well pumping above the interface. Figure 4a shows a confined aquifer with a well pumping at a relatively small rate. In this case, a freshwater capture zone is formed similar to a capture zone in the entirely freshwater region. Below the well, although there is an up-coning, the well is pumping freshwater.

In Figure 4b, the pumping rate is increased but not large enough to cause the interface to reach the well. At the cross section where the well is located, the (three-dimensional) capture zone touches the interface. Near the interface, there exists mixing of saltwater and freshwater due to hydrodynamic dispersion. Thus, trace salt content may be observed in the well. Figure 4c shows a similar situation as Figure 4b but for an unconfined aquifer.

Figure 4d gives a two-dimensional situation. A two dimensional pumping situation can be created either by a gallery (collector) well, that is, a perforated pipeline parallel to the coastline, or by a number of wells along the coast and close to each other, such that their influence zone overlaps. In this case, as long as the discharge per unit length of the gallery well does not exceed the freshwater outflow per unit length of coastline, the excess freshwater must flow underneath the well; hence, the interface cannot reach the well. In this way, the well can pump nearly as much as the entire freshwater outflow, not

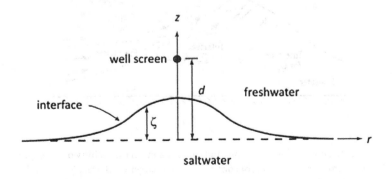

FIGURE 3   Interface up-coning below a pumping well.

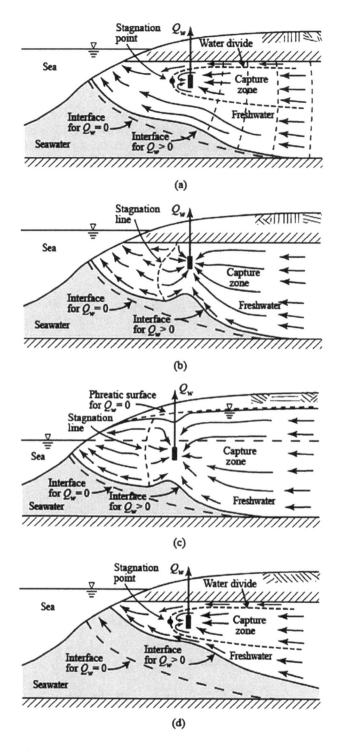

FIGURE 4 Up-coning, freshwater flow directions and capture zone in the vicinity of a well pumping above interface: (a) confined aquifer, smaller discharge; (b) confined aquifer, larger discharge; (c) unconfined aquifer; and (d) two-dimensional flow (with gallery well).

Source: Bear and Cheng.[26]

wasting any available freshwater. This type of skimming well that distributes the pumping rate over a line or over an area to avoid concentrated local up-coning beneath a point well has been practiced in Israel,[27] Palestinian territories,[28] and Pakistan.[29,30]

Transition Zone

In the above figures, saltwater and freshwater are shown to be separated by a sharp interface. This is a simplification of the actual situation, as saltwater and freshwater are miscible liquids. Actually, they constitute a single liquid phase—water—with different concentrations of total dissolved solid (TDS), such as salt. Often, the term "interface" is used for the iso-density surface that is midway between freshwater and seawater.

For the sake of simplicity, we shall continue to refer to two liquids—freshwater and seawater. The passage from the portion of the aquifer that is occupied by one liquid to the other takes the form of a transition zone, rather than a sharp interface. Depending on the extent of seawater intrusion and certain aquifer properties, this transition zone may be rather wide or narrow. The width of the transition zone is dictated by three phenomena:

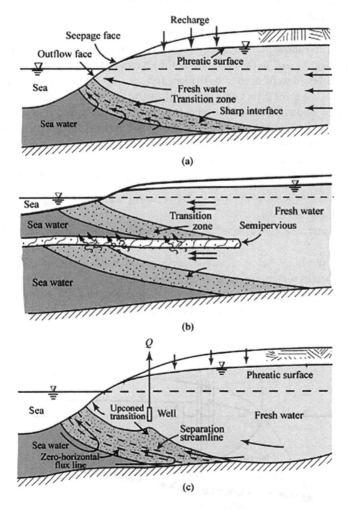

FIGURE 5 Transition zones between saltwater and freshwater regions: (a) an unconfined aquifer; (b) two aquifers separated by a semipervious layer; and (c) with a pumping well above interface.

- Hydrodynamic dispersion (dispersion and molecular diffusion)
- Advection of water—fresh and mixed—toward the sea (or, under certain conditions, landward)
- Recirculation of seawater and mixed water

Figure 5 demonstrates a few scenarios of transition zones between the saltwater and freshwater region. In Figure 5a, we observe that we can approximate a sharp interface location as the 50% relative concentration line. Figure 5b depicts a multilayer situation, where there are two transition zones. As the partition layer is semipervious, saltwater and freshwater exchange can also take place by leakage through the semipervious layer. Figure 5c shows the case with a pumping well above the interface, creating an up-coning situation.

Monitoring and Exploration of Saltwater Intrusion

Geological Investigation

A direct way of detecting saltwater intrusion is by the geological technique of drilling monitoring wells. In aquifers where a thick saltwater-freshwater transition zone exists, it is of interest to monitor the continuous change in salt concentration across the transition zone. Monitoring such information requires taking water samples at different depths. This can be achieved by the use of packers that seal sections of the well, selectively, such that each section has its own screen. Water samples taken from different sections are not mixed with each other. Obviously, such an operation is tedious, and the data obtained are limited to the well's location and to the elevation of the screened portions of the well. In order to cover a large area, a large number of wells need to be drilled, and the associated costs are usually high. The data obtained, however, are the most direct and, probably, the most accurate. Data obtained by this method are often used to calibrate data obtained by indirect methods, such as geophysical methods.

One way to reduce the effort and cost of monitoring is to utilize the Ghyben-Herzberg relation to interpret the interface location. In this case, monitoring wells need only to reach the phreatic surface, and water sampling for salt concentration is not necessary. With the Ghyben-Herzberg relation, it is possible to map the saltwater-freshwater interface using the following steps in a field investigation:

1. A network of shallow wells is deployed to observe the water table height, to give the freshwater head h_f above mean sea level.
2. The water table surface can be created using an interpolation procedure, such as kriging, expressed as $h_f(x,y)$.
3. The interface depth h(x,y) below mean sea level is represented by the same function $hf(x,y)$, except that it is amplified by a factor of 40, as indicated by Eq. 1.
4. The location of aquifer bottom $b(x,y)$ is interpolated from geological coring or map.
5. The intersection of the two surfaces h(x, y) and $b(x,y)$ is sought, which represents the toe location of the saltwater wedge.

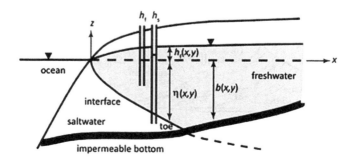

FIGURE 6   Locating saltwater toe by Ghyben–Herzberg relation.

This procedure is illustrated in Figure 6 using a vertical cross section. The above procedure is applicable for confined as well as unconfined aquifers.

The Ghyben–Herzberg relation, Eq. 1, contains the assumption that the saltwater is static. A better approximation for locating the interface that does not contain this assumption is given by[31]

$$\eta = \frac{\rho_f}{\rho_s - \rho_f} h_f - \frac{\rho_s}{\rho_s - \rho_f} h_s \qquad (3)$$

The above relation states that if we can measure at the same location both the freshwater head h_f and the saltwater head h_s, as illustrated in Figure 6, then the interface location can be better determined.

Geophysical Investigation

Geophysical methods make use of physical properties, such as the velocity of seismic waves, electrical conductivity, electromagnetic permeability, thermal conductivity, etc., of the geomaterials, such as the soil, and the water occupying the void space, with the latter varying with the concentration of dissolved matter. The typical advantages of the geophysical methods are the following:

- Measurements are taken at ground surface, although some techniques also utilize existing or specially drilled boreholes.
- Sometimes, measurements can be conducted airborne by aircraft. Usually, this means less time consumed and lower cost.
- Measurements cover a large surface area, thus making the methods more suitable for field-scale investigations.

The disadvantage of the surface-based geophysical methods, however, is that they are indirect, measuring some analogous quantities, thus often requiring calibration from direct measurements. The optimal solution to this problem is to conduct an integrated survey that combines results from geophysical techniques with data obtained from direct water sampling.

Several geophysical methods that can be used for detecting the presence of saline water in geological formations are briefly discussed below, based on Stewart:[32]

- Direct current (DC) resistivity. The method is based on the increase in electrical conductivity with increasing pore water salinity. The principal advantages of the method are its simplicity and the relatively low cost of the required equipment. In the DC method, an electrical current is introduced into the ground through electrodes driven into the soil. The resulting electrical potential (voltage) is measured between the two electrodes. The measured resistance represents the integrated resistivity over the electrically heterogeneous soil. In the more advanced equipment, multiconductor cables connect a large number of electrodes along a profile. A receiver then cycles sequentially through the electrodes, producing a series of closely spaced soundings along the profile. A computer program then performs 2-D or 3-D resistivity inversions on the field data. The inversion, however, is not unique and is dependent on the assumed "model," e.g., the number of soil layers and their thickness. This technique may not be robust enough to accurately determine a continuously varying salt concentration in a transition zone. In the case of a sharp interface, however, there is a distinct change in electrical conductivity across such interface; hence, the interface location can, generally, be determined quite accurately.[33]
- Frequency-domain electromagnetic methods (FDEMs). The FDEM involves the generation of an electromagnetic field, which induces current in the soil, which in turn causes the subsurface to create a magnetic field. By measuring this secondary magnetic field, various subsurface properties and features can be deduced. The currents in the soil, termed eddy currents, are induced by time-varying magnetic fields produced by a frequency- controlled AC (alternating current) in a transmitter coil. The transmitted electromagnetic field is called the primary field. The induced

eddy currents in the soil produce a secondary field, which is usually 90° out of phase with the primary one. The ratio of the out-of-phase component of the secondary field to the in-phase component of the primary one is an indication of the terrain's conductivity. The depth of investigation by the FDEM is primarily a function of the frequency of the primary field, with lower frequencies having greater penetration.

- Airborne electromagnetic methods. The FDEM creates eddy currents through electromagnetic induction, so that, actually, no contact with the ground is required. This means that a frequency EM system can be flown by fixed-wing aircraft or by helicopters. The typical airborne system uses several receiver-transmitter coil pairs at varying frequencies. These coil pairs are placed in a "bird," which is towed behind the aircraft at elevations of 25–50 m above ground surface. The depth of investigation is determined by the transmitter's frequencies. Common frequencies range from 200 to 56,000 Hz, yielding penetration depths from tens of meters to less than one meter, respectively. The output, as an apparent resistivity map for each frequency, is produced by an inverse method. Interpretation is normally qualitative and typically needs to be calibrated with ground surface data. The airborne method is generally applied to surveys of large areas; it has been successfully applied to the detection of saltwater intrusion in freshwater aquifers and to the exploration of freshwater lenses in saltwater environments.
- Time-domain electromagnetic methods (TDEMs). The TDEM (transient) employs a transmitter that drives an AC through a square loop of insulated electrical cable laid on the ground. The current consists of equal periods of time on and time off, with base frequencies that range from 3 to 75 Hz, producing an electromagnetic field. Similar to the FDEM, the electrical current generates a primary, time-varying electromagnetic field, which in turn creates a secondary electromagnetic field. Time-domain electromagnetic method soundings can be used to detect saltwater at depths of 5 m to several hundred meters below land surface. The TDEM has several significant advantages over DC soundings, notably depths of investigation up to twice the transmitter coil dimension and the ability to sound through a conductive, near-surface unit, such as a clay confining layer. Time-domain electromagnetic method equipment, however, is more expensive and complicated to use than DC equipment, and the interpretation of TDEM data requires sophisticated interpretation software.

Other geophysical investigation techniques that have been employed for the detection of saline water in the subsurface include the ground penetrating radar, the loop-loop electromagnetic method, the very-low-frequency electromagnetic method, and various borehole geophysical methods. More details can be found in Stewart.[32]

Geochemical Investigation

As a part of exploration of freshwater contaminated by intruding saline water, it is important to identify the origin of the latter. In coastal aquifers, seawater encroachment inland is the most common reason for the increase in salinity; however, other sources or processes can contribute to groundwater salinity. Custodio[34] lists a number of salinity sources that can contaminate freshwater supply, which are not directly related to seawater encroachment. These include entrapped fossil seawater; sea-spray accumulation; evaporite rock dissolution; displacement of old saline groundwater from underlying or adjacent aquifers or aquitards through natural or man-imposed advection or by thermal convection; leaking aquitards through fault systems; and anthropogenic pollution from various sources, including sewage effluents, industrial effluents, mine water, road-deicing salts, effluents from water softening or deionization plants, and agriculture return flows.

In general, seawater has a uniform chemistry due to the long residence time of the major constituents. Its main features are[35] predominance of Cl^- and Na^+, with a molar ratio of 0.86; an excess of Cl^- over the alkali ions (Na and K); and Mg greatly in excess of Ca^{2+} (Mg/Ca = 4.5–5.2). In contrast, continental fresh groundwater is characterized by a highly variable chemical composition, although the predominant anions are HCO_3^-, SO_3^{-4} and Cl^-. If not anthropogenetically polluted, the fundamental cations are Ca^{2+} and Mg^{2+} and, to a lesser extent, the alkali ions, Na^+ and K^+. In most cases, Ca^{2+} predominates over Mg^{2+}. Seawater solutes are specifically characterized by $Mg > SO_4 + HCO_3$, whereas meteoric waters (dilute or saline), even if dominated by resolution of marine salts, reflect $Na > Cl$. In contrast, sedimentary basin fluids can carry significant Ca and perhaps K excess over $SO_4 + HCO_3$, due to diagenetic carbonate or silicate reactions.

Geochemists generally use the following criteria to define the signature and to distinguish the sources of salinization:[35]

- Salinity: In coastal aquifers, a time series of steadily increasing chloride concentrations can indicate the early evolution of a salinity breakthrough from seawater, due to the overexploitation of groundwater and reduction of the piezometric head.
- Cl/Br ratios: The Cl/Br ratio can be used as a reliable tracer, as both Cl and Br usually behave conservatively, except in the presence of very high amounts of organic matter. Seawater (Cl/Br weight ratio = 297) is distinguished from relics of evaporated seawater (hypersaline brine Cl/Br <297), evaporite-dissolution products (over 1000), and anthropogenic sources like sewage effluents (Cl/Br ratios up to 800) or agriculture return flows (low Cl/Br ratios).
- Na/Cl ratios: Na/Cl ratios of saltwater intrusion are usually lower than the marine values (i.e., <0.86 molar ratio). On the other hand, high (>1) Na/Cl ratios typically characterize anthropogenic sources, like domestic waste waters. Thus, low Na/Cl ratios, combined with other geochemical parameters, can foretell the arrival of saltwater intrusion, even at relatively low chloride concentrations, during early stages of salinization.
- Ca/Mg, Ca/($HCO_3 + SO_4$) ratios: One of the most conspicuous features of saltwater intrusion is the enrichment of Ca over its concentration in seawater. High Ca/Mg and Ca/($HCO_3 + SO_4$) ratios (>1) are further indicators of the intrusion of seawater.
- O and H isotopes: The stable O and H isotopes can be used to describe the mixing process between saline water and freshwater. Fresh groundwater is generally depleted in both ^{18}O and ^2H (deuterium) relative to seawater. Mixing of freshwater and seawater should result in a linear correlation. Different sources with high salinity (e.g., agriculture return flows, sewage effluents) would result in different slopes due to evaporation processes that would change the isotopic composition.
- Boron isotopes: One of the processes that modify the chemistry of seawater intrusion is the adsorption of potassium, boron, and lithium onto clay minerals in the host aquifer. These elements are relatively depleted in saline water associated with seawater intrusion. Hence, the boron isotopic composition of groundwater can be used as a tool to discern the salinization sources, in particular, to distinguish seawater from anthropogenic contamination, such as domestic waste water.

As a conclusion of this section, we observe that many of the techniques discussed above for the exploration of an environment intruded by saline water are of a qualitative nature. Hence, in island or coastal regions, an exploration program usually requires the conjunctive and integrated use of two or more complementary geophysical, geological, and geochemical methods. Using several methods can increase the confidence level of the interpretation of observed data. The collected data can be used to validate the numerical model constructed for the simulation of seawater intrusion for management purposes.

Mathematical Modeling

The use of field surveys, such as geophysical and geochemical studies, can reveal the present state of saltwater intrusion and perhaps some insight into its history. However, it cannot make predictions into the future and, particularly, cannot be used for scenario building and impact assessment based on different levels of anthropogenic activities. Mathematical models are needed for these purposes.

The Ghyben–Herzberg relation is a highly simplified model. More rigorously, the dynamic movement of groundwater flow and the solute transport of salt need to be considered. Generally speaking, there does not exist a sharp division between saltwater and freshwater zones, as implied in Figure 1. The salt concentration continuously changes from that of seawater to that of freshwater. A solute transport model including advection and dispersion is needed for the modeling. In addition, the salt at higher concentration is an active solute, because it can affect the density of water and can drive the flow. Hence a density- dependent solute transport model should be used. There are occasions, however, when the predominant change of concentration from saltwater to freshwater takes place within a narrow region called the transition zone. In that case, a simplified model using the sharp interface assumption can be attempted. Furthermore, if the aquifer modeled is of regional scale, then the flow variables are often averaged in the depth direction to reduce the three-dimensional problems to two-dimensional ones.

While these different models are discussed in more detail in Bear[36] and Bear and Cheng,[26] in the following, only the governing equations for the density-dependent solute transport model are presented.

First, rather than using a piezometric head h associated with a variable density p, we shall define a reference piezometric head h', associated with the (constant) freshwater density ρ f, as[37]

$$h'(x,y,z,t)=z+\frac{p}{\rho_f g} \tag{4}$$

In the above, p is pressure, z is the elevation above datum, and g is gravity acceleration. Next, we define the mass fraction (mass of dissolved salt, or TDS, per unit mass of fluid) as

$$\omega=\frac{c}{\rho} \tag{5}$$

where c is the concentration of salt in fluid (mass of dissolved salt per unit volume of fluid). The normalized salt mass fraction is defined as

$$C=\frac{\omega-\omega_f}{\omega_s-\omega_f};0\leq C\leq 1.0 \tag{6}$$

where ω, ω_s, ω_f, are respectively the salt mass fraction in mixed (salt- and fresh-) water, unmixed seawater, and unmixed freshwater.

Constitutive Equations

Constitutive equations express the relationship between the fluid density and viscosity with the pressure and the salt mass fraction. For fluid density, we ignore the pressure effect and present only the linearized constitutive equations here:

$$\rho=\rho_f\left(1+\beta_c C\right) \tag{7}$$

where β_c is a density difference factor defined as

$$\beta_c=\beta_\omega''\left(\omega_s-\omega_f\right) \tag{8}$$

and β_ω'' is a coefficient expressing the effect of the change in salt mass fraction on the fluid density (at constant pressure), given by

$$\beta_\omega''=\frac{1}{-\rho_f}\frac{d\rho}{d\omega}. \tag{9}$$

The constitutive equation for the fluid dynamic viscosity can be expressed as[38]

$$\mu = \mu_0 \left(1 + 1.85\omega - 4.1\omega^2 + 44.5\omega^2 \right) \tag{10}$$

in which the reference viscosity μ_o corresponds to viscosity at $\omega = 0$.

Darcy's Law

Darcy's law can be written in terms of h' and C as

$$q \frac{K_f}{\mu_r} = \left(\nabla h' + \beta_c C \nabla z \right) \tag{11}$$

where $\mu_r = \mu/\mu_o$ is the relative viscosity,

$$K_f \frac{\rho_f g k}{\mu_f} \tag{12}$$

is the hydraulic conductivity with respect to the reference density and viscosity (of freshwater), and k is the intrinsic permeability.

Mass Balance Equation for Water

The mass balance of water can be expressed in terms of h' and C as

$$S_o \frac{\partial h'}{\partial t} + \phi \beta_c \frac{\rho}{\rho_f} \frac{\partial C}{\partial t} = \nabla \cdot \left[\frac{1 + \beta_c C}{\mu} K_f \cdot \left(\nabla h' + \beta_c C \nabla_z \right) + \phi \beta_c \mathbf{D} \cdot \nabla C \right]$$

$$+ \frac{\rho_R}{\rho_f} Q_R - \left(1 + \beta_c C \right) Q_p \tag{13}$$

where S_o is the aquifer's specific storativity; ϕ is porosity; $\mathbf{D}$ is the coefficient of mechanical dispersion, a second- rank tensor; Q_R and Q_p are, respectively, rate of injection (recharge) and withdrawal (pumping) of water from the aquifer; and ρ_R is the density of the recharged water.

Mass Balance Equation for Dissolved Salt

The mass balance of TDS can be expressed as

$$\frac{\partial \phi \rho C}{\partial t} = -\nabla \cdot \left(\rho C \mathbf{q} - \phi \rho \mathbf{D}_h \cdot \nabla C \right) + \rho_R C_R Q_R - \rho C Q_P \tag{14}$$

where $D_h = D + D^*$ is the coefficient of hydrodynamic dispersion, D^* is the coefficient of molecular diffusion, and C_R is the salt mass fraction of recharged water.

Well-Posed Initial and Boundary Value Problem

Eq. 11, Darcy's law, predicts the specific discharge q, which can be used to eliminate that quantity in Eqs. 13 and 14. The equations of state given in the Constitutive Equations subsection tie the properties, such as density and viscosity, to the normalized salt mass fraction C. Hence, the two mass balance equations, Eqs. 13 and 14, contain two unknowns, h' and C. Given a domain, together with a set of well-posed initial and boundary conditions (see Bear[36] or Bear and Cheng[26] for a full description of initial and boundary conditions under various physical conditions), the system can be solved for these two unknowns.

Computer Models

With the exception of some simple geometries of saltwater intrusion, for which analytical solutions are available,[39] numerical solutions are needed for practical applications. Two of the most widely used computer codes are the SHARP[40,41] for sharp interface model and the SUTRA[42,43] for the density-dependent solute transport model; both are developed by the U.S. Geological Survey (USGS). However, like many complex engineering problems, there is no single code that can be most versatile, efficient, accurate, and stable at the same time, thus dominating the rest of the codes. Depending on the availability and reliability of input data, and the limited resources dedicated to modeling, different computer codes have been developed to offer a wide range of choices. A few of these codes are listed below. A comprehensive survey can be found in Sorek and Pinder[44] and Bear and Cheng.[26]

- SHARP (a quasi-three-dimensional, numerical finite difference method that simulates freshwater and saltwater flow separated by a sharp interface in layered coastal aquifer systems)[40,41] is an implicit finite-difference code that simulates layered aquifers with Dupuit assumption.
- MOCDENSE (a two-constituent solute transport model for groundwater with variable density)[45] and MOCDENSE3D[46] are, respectively, the two-dimensional and three-dimensional rendition of MOC (computer model of two-dimensional solute transport and dispersion in groundwater)[47] and MOC3D (three dimensional method-of-characteristics groundwater flow and transport model) [48] to allow for variable- density modeling.
- SUTRA (model for 2D or 3D saturated-unsaturated, variable-density groundwater flow, with solute or energy transport)[42,43] is a Galerkin finite-element code that solves groundwater flow and transport problems under saturated and unsaturated conditions.
- SEAWAT (computer program for simulation of three dimensional variable-density groundwater flow)[49] is a USGS code that combines MODFLOW[50] and MT3DMS[51] into a single computer program for the purpose of simulating saltwater intrusion. It is a finite- difference, Eulerian-Lagrangian code, in contrast to SUTRA, which is a finite-element code.
- CODESA-3D (coupled variable-density and saturation 3D model)[52] is0 a finite-element code, similar to SUTRA.
- FEMWATER (three-dimensional finite-element model of water flow through saturated-unsaturated media).[53] The code is available from the U.S. Environmental Protection Agency.

Combating Saltwater Intrusion

One of the most effective ways of combating saltwater intrusion is to regulate pumping activities. Generally speaking, the amount of groundwater extraction should not exceed that of natural replenishment. Optimization of pumping patterns to maximize the yield and minimize the extent of intrusion is a high-priority management issue. Recharge of natural surface water or reclaimed wastewater into aquifers can increase the freshwater outflow rate to push back saltwater wedge. A recharge near the coast can build a local freshwater mound that forms a barrier to protect the water table depression inland. Extraction of saltwater in an invaded saltwater wedge can also protect the freshwater behind, if a proper way can be found to dispose of the extracted saltwater. A similar method involving pumping simultaneously in the upper freshwater zone and the lower saltwater zone to prevent up-coning, known as double pumping, has been attempted. Using collector wells (horizontal wells) to skim the thin layer of freshwater floating on top of the saltwater wedge has been effectively used in water-poor countries such as Israel. Land reclamation has the added effect of pushing saltwater to the sea. Finally, in places where large freshwater springs flowing to the sea can be identified, physical barriers, such as solid walls or slurry curtains, can be used to intercept freshwater.

Planning and Management

Groundwater resource planning and management in coastal areas is similar to the traditional water resource planning in inland areas but with the additional complication of saltwater intrusion to eventually render portions of the coastal aquifer unusable as a source of drinking water. The recommended planning elements for managing coastal aquifers include the following:[54]

1. Collect and analyze existing data: Data like piezometric head, chlorine concentration, aquifer transmissivity, formation thickness, well location, screen depth, etc., need to be collected.
2. Develop an integrated database: The data need to be organized and integrated, typically using geographic information system software.
3. Identify saltwater intrusion problem and hypothesize intrusion scenarios, such as horizontally from the sea or vertically from the bottom brackish water.
4. Develop a groundwater simulation model using one of the above-mentioned computer codes and collected data.
5. Perform field studies for model validation: The collected existing data are likely to be insufficient, and the computer simulations are likely to be unreliable, lacking field validation. Hence, field studies are needed to fill the information gap. The geological, geophysical, and geochemical methods reviewed above can be used to conduct field studies.
6. Set planning objective: Once the extent and the time scale of the threat are understood, planning objectives, such as reducing pumping, increasing recharge, etc., should be set by consulting with the stakeholders.
7. Identify solutions and actions: One or more of the saltwater intrusion combating strategies can be recommended.
8. Evaluate management alternatives: Water resource management involves complex tradeoffs. The alternative actions need to be evaluated and selected using a multicriteria decision-making process.

References

1. Back, W.; Freeze, R.A., Eds. *Chemical Hydrogeology. Benchmark Papers in Geology*; Hutchinson Ross Publ. Co.: Stroudsburg, PA, 1983; Vol. 73.
2. Cooper, H.H.J.; Kohout, F.A.; Henry, H.R.; Glover, R.E. *Sea Water in Coastal Aquifers*; U.S. Geological Survey Water-Supply Paper 1613-C; 1964.
3. Sonenshein, R.S. Delineation and Extent of Saltwater Intrusion in the Biscayne Aquifer, Eastern Dade County, Florida, U.S. Geological Survey Water-Resources Investigation Report 96–4285; 1997.
4. Konikow, L.F.; Reilly, T.E. Seawater intrusion in the United States. In Seawater Intrusion in Coastal Aquifers— Concepts, Methods and Practices; Bear, J., et al., Eds; Kluwer, 1999; 463–506.
5. Johnson, T.A.; Whitaker, R. Saltwater intrusion in the coastal aquifers of Los Angeles County, California. In *Coastal Aquifer Management-Monitoring, Modeling, and Case Studies*; Cheng, A.H.D., Ouazar, D., Eds.; Lewis Publ., 2003; 29–48.
6. Mink, J.F. State of the groundwater resources of southern Oahu. In *Board of Water Supply, City and County of Honolulu*; 1980; 83 pp.
7. Voss, C.I.; Souza, W.R. Dynamics of a Regional Freshwater-Saltwater Transition Zone in an Anisotropic Coastal Aquifer System, U.S. Geological Survey Open-File Report 98–398; 1998; 88 pp.

8. Stakelbeek, A. Movement of brackish groundwater near a deep-well infiltration system in the Netherlands. In *Seawater Intrusion into Coastal Aquifers—Concepts, Methods and Practices;* Bear, J., et al., Eds.; Kluwer, 1999; 531–541.

9. Oude Essink, G. Impact of sea level rise in the Netherlands. In *Seawater Intrusion into Coastal Aquifers—Concepts, Methods and Practices;* Bear, J., et al., Eds.; Kluwer, 1999; 507–530.

10. Melloul, A.J.; Zeitoun, D.G. A semi-empirical approach to intrusion monitoring in Israeli coastal aquifer. In Seawater Intrusion into Coastal Aquifers—Concepts, Methods and Practices; Bear, J., et al., Eds.; Kluwer, 1999; 544–558.

11. Melloul, A.J.; Collin, M.L. Sustainable groundwater management of the stressed coastal aquifer in the Gaza region. Hydrolo. Sci. J. **2000**, *45* (1), 147–159.

12. Steinich, B.; Marin, L.E. Hydrogeological investigations in northwestern Yucatan, Mexico, using resistivity surveys. Ground Water **1996**, *34* (4), 640–646.

13. Marin, L.E.; Perry, E.C.; Essaid, H.I.; Steinich, B. Hydrogeological investigations and numerical simulation of groundwater flow in the karstic aquifer of Northwestern Yucatan, Mexico. In *Coastal Aquifer Management Monitoring, Modeling, and Case Studies;* Cheng, A.H.D., Ouazar, D., Eds.; Lewis Publishers, 2003; 257–277.

14. de Franco, R.; Biella, G.; Tosi, L.; Teatini, P.; Lozej, A.; Chiozzotto, B.; Giada, M.; Rizzetto, F.; Claude, C.; Mayer, A.; Bassan, V.; Gasparetto-Stori, G. Monitoring the saltwater intrusion by time lapse electrical resistivity tomography: The Chioggia test site (Venice Lagoon, Italy). J. Appl. Geo- phys. 2009, *69* (3–4), 117–130.

15. Viezzoli, A.; Tosi, L.; Teatini, P.; Silvestri, S. Surface water-groundwater exchange in transitional coastal environments by airborne electromagnetics: The Venice Lagoon example. Geophys. Res Lett. **2010**, *37,* L01402.

16. Sherif, M. Nile Delta aquifer in Egypt. In *Seawater Intrusion in Coastal Aquifers—Concepts, Methods and Practices;* Bear, J., et al., Eds.; Kluwer, 1999; 559–590.

17. Paniconi, C.; Khlaifi, I.; Lecca, G.; Giacomelli, A.; Tar- houni, J. Modeling and analysis of seawater intrusion in the coastal aquifer of eastern Cap-Bon, Tunisia. Transp. Porous Media **2001**, *43* (1), 3–28.

18. Wu, J.C.; Meng, F.H.; Wang, X.W.; Wang, D. The development and control of the seawater intrusion in the eastern coastal of Laizhou Bay, China. Environ. Geol. **2008**, *54* (8), 1763–1770.

19. Ergil, M.E. The salination problem of the Guzelyurt aquifer, Cyprus. Water Res. **2000**, *34* (4), 1201–1214.

20. Benkabbour, B.; Toto, E.A.; Fakir, Y. Using DC resistivity method to characterize the geometry and the salinity of the Plioquaternary consolidated coastal aquifer of the Mamora plain, Morocco. Environ. Geol. **2004**, *45* (4), 518–526.

21. Du Commun, J. On the cause of freshwater springs, fountains, etc. Am. J. Sci. Arts **1828**, *14,* 174–175.

22. Badon-Ghyben, W. Nota in verband met de voorgenomen putboring nabij Amsterdam (Notes on the Probable Results of Well Drilling near AMSTERDAM); Tijdschrift van het Koninklijk Instituut van Ingenieurs: Hague, 1888; 822.

23. Herzberg, A. Die Wasserversorgung einiger Nordseeb- der (The water supply of parts of the North Sea Coast in Germany). Z. Gasbeleucht. Wasserversorg. **1901**, *44/45,* 815–819, 842–844.

24. Bear, J.; Dagan, G. Some exact solutions of interface problems by means of hodograph method. J. Geophys. Res. **1964**, *69* (8), 1563–1572.

25. Dagan, G.; Bear, J. Solving the problem of local interface upconing in a coastal aquifer by the method of small perturbation. J. Hydraul. Res. **1968**, **6**, 15–44.

26. Bear, J.; Cheng, A.H.D. *Modeling Groundwater Flow and Contaminant Transport;* Springer, 2010; 834 pp.

27. Schmorak, S.; Mercado, A. Upconing of freshwater-saltwater interface below pumping wells, field study. Water Resour. Res. **1969**, **5** (6), 1290–1311.

28. Aliewi, A.S.; Mackay, R.; Jayyousi, A.; Nasereddin, K.; Mushtaha, A.; Yaqubi, A. Numerical simulation of the movement of saltwater under skimming and scavenger pumping in the Pleistocene aquifer of Gaza and Jericho areas, Palestine. Transp. Porous Media **2001**, *43* (1), 195–212.

29. Saeed, M.M.; Ashraf, M.; Asghar, M.N. Hydraulic and hydro-salinity behavior of skimming wells under different pumping regimes. Agric. Water Manage. **2003**, *61* (3), 163–177.

30. Saeed, M.M.; Ashraf, M. Feasible design and operational guidelines for skimming wells in the Indus basin, Pakistan. Agric. Water Manage. **2005**, *74* (3), 165–188.

31. Lusczynski, N.J. Head and flow of ground water of variable density. J. Geophys. Res. **1961**, *66*, (12), 4247–4256.

32. Stewart, M.T. Geophysical investigations. In *Seawater Intrusion into Coastal Aquifers—Concepts, Methods and Practices*; Bear, J., et al., Eds.; Kluwer, 1999; 9–50.

33. Stewart, M.T. Rapid reconnaissance mapping of freshwater lenses on small oceanic islands. In *Geotechnical and Environmental Geophysics, v. II, Soc. Explor. Geophysicists, Inv. in Geophysics, n. 5*; Ward, S., Ed.; 1990; 57–66.

34. Custodio, E. Studying, monitoring and controlling seawater intrusion in coastal aquifers. Guidelines for study, monitoring and control. *FAO Water Reports* **1997**, *11*, 7–23.

35. Jones, B.F.; Vengosh, A.; Rosenthal, E.; Yechieli, Y. *Geochemical investigations*. In *Seawater Intrusion into Coastal Aquifers—Concepts, Methods and Practices*; Bear, J., et al., Eds.; Kluwer, 1999; 51–71.

36. Bear, J. Conceptual and mathematical modeling. In *Seawater Intrusion into Coastal Aquifers—Concepts, Methods and Practices*; Bear, J., et al., Eds.; Kluwer, 1999; 127161.

37. Bear, J.; Zhou, Q. Sea water intrusion into coastal aquifers. In *The Handbook of Groundwater Engineering*, 2nd Ed.; Delleur, J.W., Ed.; CRC Press, 2006.

38. Lever, D.A.; Jackson, C.P. On the Equations for the Flow of a Concentrated Salt Solution through a Porous Medium, Harwell Report AERE-R 11765; HMSO: London, 1985.

39. Cheng, A.H.D.; Ouazar, D. Analytical solutions. In *Seawater Intrusion in Coastal Aquifers—Concepts, Methods, and Practices*; Bear, J., et al., Eds.; Kluwer, 1999; 163–191.

40. Essaid, H.I. The Computer Model SHARP, a Quasi-ThreeDimensional Finite-Difference Model to Simulate Freshwater and Saltwater Flow in Layered Coastal Aquifer Systems, U.S. Geological Survey Water-Resources Investigations Report 90–4130; 1990; 181 pp.

41. Essaid, H.I. USGS SHARP model. In Seawater Intrusion in Coastal Aquifers—Concepts, Methods, and Practices; Bear, J., et al., Eds.; Kluwer, 1999; 213–247.

42. Voss, C.I. USGS SUTRA code—History, practical use, and application in Hawaii. In *Seawater Intrusion in Coastal Aquifers—Concepts, Methods and Practices*; Bear, J., et al., Eds.; Kluwer, 1999; 249–313.

43. Voss, C.I.; Provost, A.M. SUTRA, A Model for Saturated- Unsaturated Variable-Density Ground-Water Flow with Solute or Energy Transport, *U.S. Geological Survey Water- Resources Investigations Report 02–4231*; 2002.

44. Sorek, S.; Pinder, G.F. Survey of computer codes and case histories. In *Seawater Intrusion into Coastal Aquifers— Concepts, Methods and Practices*; Bear, J., et al., Eds.; Kluwer, 1999; 399–461.

45. Sanford, W.E.; Konikow, L.F. *A Two-Constituent Solute Transport Model for Ground Water Having Variable Density*, U.S. Geological Survey Water Resources Investigation Report 85–4279; 1985; 89 pp.

46. Oude Essink, G.H.P. Modeling three-dimensional density dependent groundwater flow at the Island of Texel, the Netherlands. In *Coastal Aquifer Management—Monitoring, Modeling, and Case Studies*; Cheng, A.H.D., Ouazar, D., Eds.; Lewis Publ, 2004; 77–94.

47. Konikow, L.F.; Bredehoeft, J.D. Computer model of twodimensional solute transport and dispersion in ground water. In *Techniques of Water-Resources Investigations of the U.S. Geological Survey*; 1978; 90 pp.

48. Konikow, L.F.; Goode, D.J.; Hornberger, G.Z. *A ThreeDimensional Method-of-Characteristics Solute-Transport Model (MOC3D)*, U.S. Geological Survey Water-Resources Investigations Report 96–4267; 1996; 87 pp.

49. Langevin, C.D.; Shoemaker, W.B.; Guo, W. MODFLOW-2000, the U.S. Geological Survey Modular Ground-Water Model—Documentation of the SEAWAT-2000 Version with the Variable-Density Flow Process (VDF) and the Inte- gratedMT3DMS Transport Process (IMT), U.S. Geological Survey Open-File Report 03–426; 2003; 43 pp.

50. Harbaugh, A.W. MODFLOW-2005, the U.S. Geological Survey Modular Ground-Water Model—The GroundWater Flow Process, U.S. Geological Survey Techniques and Methods 6-A16; 2005.

51. Zheng, C. MT3DMS v5.2 supplemental user's guide. In *Technical Report to the U.S. Army Engineer Research and Development Center*; Department of Geological Sciences, University of Alabama; 2006; 41 pp.

52. Gambolati, G.; Putti, M.; Paniconi, C. Three-dimensional model of coupled density-dependent flow and miscible salt transport. In *Seawater Intrusion in Coastal Aquifers— Concepts, Methods, and Practices;* Bear, J., et al., Eds.; Kluwer, 1999; 315–362.

53. Lin, H.C.; Richards, D.R.; Yeh, G.T.; Cheng, J.R.; Chang, H.P.; Jones, N.L. *FEMWATER: A Three-Dimensional Finite Element Computer Model for Simulating Density Dependent Flow and Transport,* U.S. Army Engineer Waterways Experiment Station Technical Report; 1996; 129 pp.

54. Maimone, M.; Harley, B.; Fitzgerald, R.; Moe, H.; Hos-sain, R.; Heywood, B. Coastal aquifer planning elements. In *Coastal Aquifer Management—Monitoring, Modeling, and Case Studies;* Cheng, A.H.D., Ouazar, D., Eds.; Lewis Publishers, 2003; 1–27.

14

Irrigation Systems: Water Conservation

Introduction 151
Uniformity of Water Application and Design Considerations...............151
Hydraulic Design of Drip Irrigation Systems...152
Drip Irrigation for Optimal Return, Water Conservation, and
    Environmental Protection 153
Conclusion 154
References... 154

Pai Wu, Javier
Barragan, and
Vince Bralts

Introduction

With the increasing consequence of limited water resources and the increasing need for environmental protection, drip irrigation will play an even more important role in the future. Drip irrigation systems can be used for many different types of agricultural crops, including fruit trees, vegetables, pastures, specialty crops such as sugarcane, ornamentals, golf course grasses, and high economic value crops grown in greenhouses. An understanding of drip irrigation systems, irrigation scheduling, crop response, and economic ramifications will encourage greater use of drip irrigation in future agricultural production.

Uniformity of Water Application and Design Considerations

The desired uniformity of water application and the specific crops to be grown guides the creation of drip irrigation systems. There are two types of drip irrigation uniformity: system uniformity and spatial uniformity in the field. The consistency of system distribution of water into the field describes the system uniformity. The spatial uniformity is the regularity of water distribution considering overlapping emitter flow and translocation of water in the soil. For drip irrigation systems designed for trees with large spacing, the system uniformity is equal to the water application uniformity in the field. For high-density plantings, the emitter spacing should be designed considering overlapped wetting patterns and the spatial uniformity in the field. The uniformity of a drip irrigation system depends primarily on the hydraulic design, but must also consider the manufacturer's variation, temperature effects, and potential emitter plugging. The effect of water temperature is generally negligible when using turbulent flow emitters. A combination of proper filtration and turbulent emitters can control emitter plugging. When grouping a number of emitters together as a unit, such as those designed to irrigate an individual plant's root system, the uniformity of water application with respect to the plant will improve.

Many expressions have been used to describe uniformity. The system uniformity, or emitter flow uniformity, can be expressed as the range or variation of water distribution in the field. This term was initially used for hydraulic design of drip irrigation systems given that the minimum and maximum emitter flows could be calculated and determined.[1] When more emitter flows are used or more samples are required for determining variation or spatial uniformity in the field, the Christiansen uniformity coefficient (UCC)[2] and coefficient of variation (CV), which is the ratio of standard deviation and the mean, are used. Each of the uniformity expressions are highly correlated with one other.

Hydraulic Design of Drip Irrigation Systems

Once selection of the type of drip irrigation emitter is complete, the hydraulic design can be made to achieve the expected uniformity of irrigation application.

The hydraulic design of a drip irrigation system involves designing both the submain and lateral lines. Early research in drip irrigation hydraulic design concentrated mainly on the single lateral line approach,[1,3,4] but in 1985 Bralts and Segerlind developed a method to design a submain unit. The hydraulic design is based on the energy relations in the drip tubing, the friction drop, and energy changes due to slopes in the field. Direct calculations of water pressures along a lateral line or in a submain unit are made by using an energy gradient line approach.[1] All emitter flows along a lateral line and in a submain can be determined based on their corresponding water pressures. Once the emitter flows are determined, the emitter flow variation, q_{var} is expressed by

$$q_{var} = \frac{q_{max} - q_{min}}{q_{max}} \tag{1}$$

where q_{max} is the maximum emitter flow and q_{min} is the minimum emitter flow. Based on these data, other uniformity parameters such as UCC and CV can also be determined. There is a strong correlation between any two of the three uniformity parameters in the hydraulic design of drip irrigation systems, thus any one of the uniformity parameters can be used as a design criterion. This correlation also justifies using the simple emitter flow variation q_{var} for hydraulic design. The emitter flow variation q_{var} is converted to the CV when it is combined with the manufacturer's variation of emitter flow.

The total emitter flow variation caused by both hydraulic and manufacturer's variation can be expressed by[5]

$$CV_{HM} = \sqrt{CV_H^2 + CV_M^2} \tag{2}$$

where CV_{HM} is the coefficient of variation of emitter flows caused by both hydraulic and manufacturer's variation; CV_H and CV_M are the coefficients of variation of emitter flows caused by hydraulic design and manufacturer's variation, respectively.

The design criterion for emitter flow variation q_{var} for drip irrigation design is arbitrarily set as 10.0%–20.0%, which is equivalent to a CV, from 0.033 to 0.076, or 3.0%–8.0%. Based on the research of last 30 years, the manufacturer's variation of turbulent emitters is maintained only in a range 3.0%–5.0%, expressed by CV. When this variation is combined with emitter flow variation caused by hydraulic design with a range 3.0%–8.0% in CV, the total variation determined by the equation above will be limited to a CV of less than 10.0%. This variation illustrates that the drip irrigation systems are designed to achieve high uniformity and irrigation application efficiency.

Economic return can also be the basis of design criteria for drip irrigation. A new set of design criteria for drip irrigation was developed,[6] based on achieving an expected economic return with various water resources and environmental considerations (Table 1).

TABLE 1 Design Criteria for Uniformity of Drip Irrigation System Design

Design Consideration	CV (%)	UCC (%)
Water is abundant and no environmental	30–20	75–85
pollution problems	20–10	80–90
Water is abundant but with environmental protection considerations	25–15	80–90
Limited water resources but with no environmental pollution problems	15–5	85–95
Considerations for both water conservation and environmental protection		

Drip Irrigation for Optimal Return, Water Conservation, and Environmental Protection

When the uniformity of a drip irrigation system is designed with a UCC of 70.0%, 30.0% or less in CV, the irrigation application is expressed as a straight-line distribution,[7,8] as shown in Figure 1. This figure was plotted using percent of area (PA) against a relative irrigation depth, X, which is the ratio of required irrigation depth to mean irrigation application. The straight-line distribution in the dimensionless plot can be specified by a minimum value, a, a maximum value, $(a+b)$, in the X-scale and a slope b, where b specifies the uniformity of water application.[9]

When a drip irrigation system is designed with fixed uniformity, it is possible to determine the sloped straight line with known value of a and b. A value (X) can then be selected between value a and $(a+b)$ and plotted (Figure 1). The triangle formed above the horizontal line (X) results in an irrigation deficit and yield reduction. The triangle below the horizontal line results in over-irrigation and deep seepage.

An important irrigation scheduling parameter, the relative irrigation depth, (X) indicates how much irrigation water is applied. The effectiveness of drip irrigation is shown not only by the high uniformity of the drip irrigation system, but also by the irrigation requirement and the strategy of irrigation scheduling. As illustrated in Figure 1, the irrigation scheduling parameter (X) affects the areas of overirrigation and water deficit conditions in the field and is directly related to the economic return. Practically

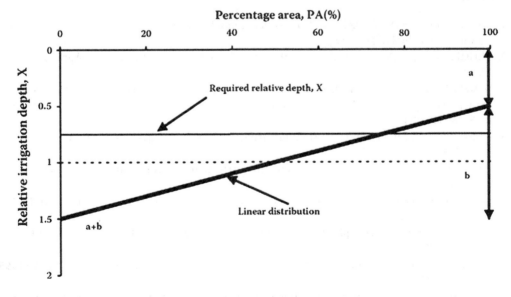

FIGURE 1 A linear water application model for drip irrigation.

speaking, the X parameter is selected in a range from a to $(a+b)$, as shown in Figure 1. Three typical irrigation schedules can be expressed by X and are as follows:

$X=a$	This schedule is a conventional irrigation schedule, which is based on the minimum emitter or minimum water application. The field is fully irrigated and whole field is over-irrigated except the point of minimum irrigation application.
$X=X_0$	For an optimal return there is a value of X for the irrigation scheduling parameter between a and (a+b).
$X=(a+b)$	This irrigation schedule is based on the maximum emitter flow or maximum irrigation application. The whole field is under deficit condition except the point of maximum water application. There is no deep percolation.

An optimal irrigation schedule for maximum economic return was determined[9] based on cost of water, price of the yield, and damage such as environmental pollution and groundwater contamination caused by over irrigation. Different irrigation strategies require different amounts of water application. Water conservation and environmental protection are realized by comparing any two of the irrigation strategies.[10]

Conclusion

Drip irrigation is an irrigation method that can distribute irrigation water uniformly and directly into the root zone of crops. It is one of the most efficient irrigation methods and can be designed and scheduled to meet the water requirement of crop and produce maximum yield in the field.

When the drip irrigation system is designed with high uniformity, the slope b of the straight line of water application function (Figure 1) can be controlled to achieve the desired variation. In this case the conventional irrigation schedule, $X=a$, optimal irrigation schedule, X_0, which is a location between a and $(a+b)$, and the irrigation schedule for environmental protection, $X=a+b$, are in close proximity. This closeness shows that the drip irrigation system can achieve optimal economic return, water conservation, and environmental protection.

References

1. Wu, I.P.; Gitlin, H.M. *Design of Drip Irrigation Lines*; HAES Technical Bulletin 96, University of Hawaii: Honolulu, HI, 1974; 29 pp.
2. Christiansen, J.E. The uniformity of application of water by sprinkler systems. Agric. Eng. **1941**, *22*, 89–92.
3. Howell, T.A.; Hiler, E.A. Trickle irrigation lateral design. Trans. ASAE **1974**, *17* (5), 902–908.
4. Bralts, V.F.; Segerlind, L.J. Finite elements analysis of drip irrigation submain unit. Trans. ASAE **1985**, *28*, 809–814.
5. Bralts, V.F.; Wu, I.P.; Gitlin, H.M. Manufacturing variation and drip irrigation uniformity. Trans. ASAE **1981**, *24* (1), 113–119.
6. Wu, I.P.; Barragan, J. Design criteria for microirrigation systems. Trans. ASAE **2000**, *43* (5), 1145–1154.
7. Seginer, I. A note on the economic significance of uniform water application. Irrig. Sci. **1978**, *1*, 19–25.
8. Wu, I.P. Linearized water application function for drip irrigation schedules. Trans. ASAE **1988**, *31* (6), 1743–1749.
9. Wu, I.P. Optimal scheduling and minimizing deep seepage in microirrigation. Trans. ASAE **1995**, *38* (5), 1385–1392.
10. Barragan, J.; Wu, I.P. Optimal scheduling of a microirrigation system under deficit irrigation. J. Agric. Eng. Res. **2001**, *80* (2), 201–208.

15

Irrigation: Erosion

Introduction 155
Importance of Irrigation 155
Unique Aspects of Irrigation Erosion 156
Surface-Irrigation Erosion 157
Sprinkler-Irrigation Erosion 159
Conclusions 161
References 161

David L. Bjorneberg

Introduction

Irrigation is vital to food production in the world. However, irrigation-induced soil erosion reduces productivity of irrigated land and can cause off-site water quality problems. Surface irrigation utilizes the soil to distribute water through the field. Water flowing over soil inherently detaches and transports sediment. Sprinkler and drip irrigation distribute water through fields in pipes, eliminating erosion from water distribution, but erosion can still occur if water is applied faster than it can infiltrate into the soil. This entry will briefly discuss the importance of irrigation to global food production and then discuss the important factors affecting soil erosion for surface-and sprinkler-irrigated land. Much of the information will focus on the United States, with international information included when possible.

Importance of Irrigation

Irrigated agriculture contributes a disproportionate amount to global food production. The most cited statistics indicate that irrigated cropland produces about one-third of the world's crop production on only 16% of the cropland that is irrigated.[1] In the United States, farms with all cropland irrigated account for only 8% of the total cropland and about half of the total irrigated land.[2] These farms produce 33% of the market value of crops and 12% of the market value of livestock. Over half of the crop value (55%) is produced on farms with some irrigated land, and these farms account for only 26% of the total cropland in the United States.[3] In some areas, irrigation provides essentially all of the water necessary for crop growth. In other areas, irrigation provides only a small portion of the total crop water requirement but reduces the potential for water stress during critical periods.

   While irrigation is critical to global food production, applying water to soil can cause erosion. This is especially true with surface irrigation, where the soil conveys and distributes water through a field by gravity. Sprinkler irrigation and microirrigation use pipes to distribute water through the field. Surface irrigation is generally thought to cause more erosion than sprinkler irrigation; however, erosion can occur any time water flows over soil. Water can be applied with sprinkler irrigation so no runoff occurs, and therefore, no erosion will occur. However, there are situations, especially with moving irrigation systems like center pivots, where water is applied faster than it can infiltrate into the soil, resulting in ponding and, possibly, runoff.

Unique Aspects of Irrigation Erosion

The factors affecting soil erosion from irrigation are the same as rainfall. Water detaches and transports sediment in both situations. However, there are some unique differences in how the factors occur with irrigation.[4] For example, rainfall occurs relatively uniformly over an entire field, whereas irrigation is seldom applied to an entire field at the same time. Irrigation is a controlled procedure where water is applied to a specific field, or portion of a field, at a specific time. This can affect the hydrology of the erosion processes on surface- and sprinkler-irrigated fields. A center pivot, for example, is essentially a moving storm that covers only 1%–2% of the field at any given time. This results in unique runoff conditions where water can do the following: 1) flow parallel to the lateral under similar conditions as rainfall; 2) flow from wet soil onto dry soil if the lateral is moving downhill; or 3) flow onto wet soil if the lateral is moving uphill.

In surface irrigation, water flow rate decreases with distance during surface irrigation as water infiltrates. Furrow flow rates also increase with time as infiltration rate decreases (Figure 1). This creates a condition where sediment can be detached on the upper end of the field and deposited on the lower end. Trout[4] documented erosion rates on the upper end of a field that were 6 to 20 times greater than the field-average erosion rates. Figure 2 shows eroded furrows on the upper end of a field after one furrow irrigation. During rainfall, raindrops wet the soil surface and detach soil particles. As runoff begins, rills form in wet soil. In contrast, irrigation furrows are formed prior to irrigation, and water flows onto initially dry soil. Furrows with initially dry soil have greater soil erosion than furrows that were prewet immediately before furrow irrigation.[5] Irrigation water flowing in furrows is not exposed to falling raindrops that can increase sediment detachment and decrease deposition.

The quality of irrigation water can vary dramatically among water sources, or even within an irrigation tract if drainage water is reused. Conversely, electrolyte concentration of rainfall is quite consistent. Electrolyte concentration in irrigation water affects erosion for both surface and sprinkler irrigation. Furrow-irrigation erosion was greater on a silt loam when irrigation water had low electrical conductivity (EC=0.7 dS m^{-1}) and high sodium adsorption ratio (SAR=9.1) compared with low EC (0.5 dS m^{-1}) and low SAR (0.9), high EC (2.1 dS m^{-1}) and low SAR (0.5), and high EC (1.7 dS m^{-1}) and high SAR (9.3).[6] Soil erosion was also greater with low-EC water in laboratory and field rainfall simulation studies.[7,8] Lower electrolyte concentrations in water cause greater dispersion of soil particles, which tends to reduce infiltration and increase soil loss.[9]

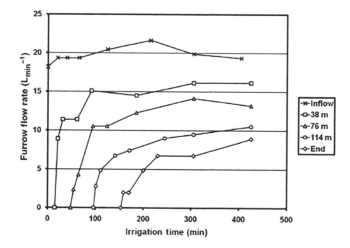

FIGURE 1 Furrow flow rate with time at five points in a 150 m–long furrow.

FIGURE 2 Eroded furrows on the upper end of a furrow-irrigated field in Idaho with approximately 1% slope.

Surface-Irrigation Erosion

Surface irrigation continues to be the most common method of irrigation in the world. The four countries with the most irrigated land are India (60.8 Mha), China (57.8 Mha), United States (22.4 Mha), and Pakistan (19.6 Mha).[10] These four countries account for 58% of the irrigated area in the world. All other countries have less than 10 Mha of irrigated land.[10] According to the country fact sheets on the Food and Agriculture Organization's Aquastat Web site,[11] surface irrigation is used on 97% of the irrigated land in India, 94% in China, 44% in the United States, and 100% in Pakistan.

Koluvek et al.[12] provided a good overview of soil erosion from irrigation in the United States. Unfortunately, this information has not been updated, and similar information is not readily available from other countries, so it is difficult to track erosion trends on irrigated lands. Some early studies documented erosion rates as great as 145 Mg ha^{-1} in 1 h[13] and 40 Mg ha^{-1} in 30 min.[14] While these rates represent extreme conditions that can occur, not typical season-long soil loss rates, these studies indicate the potential severity of the problem. One study measured annual soil losses of 1 to 141 Mg ha^{-1} from 33 fields with silt loam soils.[15] The greatest soil loss occurred on a sugar beet (*Beta vulgaris* L.) field with 4% slope. The authors noted that erosion increased sharply when field slope was greater than 1%. Close-growing crops like alfalfa *(Medicago sativa* L.) or wheat (*Triticum aestivum* L.) on fields with 1% slope had annual soil loss of less than 1 Mg ha^{-1}. A recent study in the same area documented that average soil loss from an 80,000 ha irrigated watershed decreased from 450 kg ha^{-1} in 1970 to less than 50 kg ha^{-1} in 2005.[16] This watershed was approximately 90% furrow irrigated in 1970 and 60% furrow irrigated in 2005. Another study measured daily sediment loads of 0.4 kg ha^{-1} in a watershed with no furrow irrigation compared to 19 kg ha^{-1} in a watershed with 58% of the cropland furrow irrigated.[17] Irrigation method explained 67% of the variation in soil loss measured in April and May in these nine watersheds.

The main factors affecting surface-irrigation erosion are soil type, field slope and flow rate. Soil erosion is typically not a concern where field slopes are less than 0.5% (Figure 3). However, erosion tends to increase exponentially for increasing inflow rate and field slope, with an exponent between 1 and 3 for flow rate, and between 2 and 3 for slope.[12,18,19,20] Increasing inflow rate 20% increased erosion 30% and 70% on the upper quarter of two fields.[4] Increasing inflow rate another 20% increased erosion 50% and 100%, which indicates that the exponent between erosion and flow rate was between 2 and 3.[4] Figure 4 shows soil loss from 10 furrows during a 4 h irrigation at Kimberly, Idaho, with inflow rates randomly set for each furrow.

FIGURE 3 Level furrow irrigation in Arizona. (Photo by Jeff Vanuga, USDA Natural Resources Conservation Service [NRCSAZ02037]).

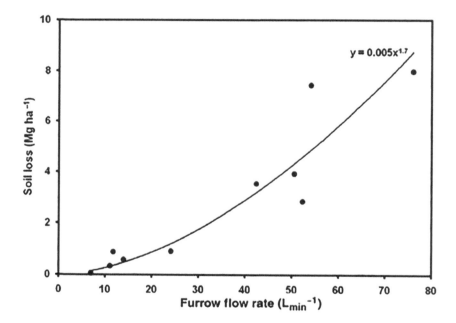

FIGURE 4 Relationship between furrow flow rate and soil loss for 10 and 30 m-long furrows with randomly set inflow rates.

Reducing field slope by grading the land is a costly practice that is not feasible in most situations compared with alternatives like installing a sprinkler-irrigation system. Reducing inflow rate is a good practice as long as the water advances down the field fast enough to uniformly irrigate the field. Slow water advance rates from low inflow rates cause overirrigation on the inflow end of the field and underirrigation on the lower end of the field due to differences in infiltration opportunity time. This results in poor distribution uniformity but little runoff. Soil loss decreases as distribution uniformity decreases.[20] An excellent practice for reducing irrigation erosion without affecting irrigation uniformity is applying small amounts of polyacrylamide (PAM) with irrigation water.[21,22] Dissolving 10 mg L^{-1} of high-molecular-weight, anionic PAM in furrow-irrigation inflow can reduce soil loss 60%–99% compared with untreated furrows. Other technologies like filter strips and sediment ponds on the lower end of the field remove sediment from the water rather than reducing erosion from occurring on the field.

Sprinkler-Irrigation Erosion

Ideally, sprinkler-irrigation systems are designed and managed to have all applied water infiltrate into the soil where it was applied. When all water infiltrates, there is no runoff or soil erosion. Solid-set (sprinklers located in the same position for the entire irrigation season) or set-move (sprinklers remaining in a location for 12 to 24 h, then moved to the next set) irrigation systems usually apply water at a low rate (e.g., 3 to 6 mm h^{-1}), so irrigation application rate does not exceed the soil infiltration rate and no soil erosion occurs. Moving irrigation systems, like center pivots, traveling guns, and lateral-move systems, often apply water faster than the infiltration rate. This occurs because the irrigation system must apply enough water as it moves across the field to meet crop water needs until the next time it irrigates that portion of the field. For example, a center pivot operating at 60 h per revolution needs to apply 20 mm per revolution to meet an 8 mm d^{-1} crop water requirement. The irrigation application rate increases with distance from the center pivot because the lateral irrigates more land as the radial distance from the pivot point increases.[23] Near the pivot point, the mean application rate could be about 4 mm h^{-1} (assuming 15 m wetting diameter). Near the end of the pivot, about 400 m, the mean application rate would be about 60 mm h^{-1}. An important fact about moving irrigation systems is that the application rate is a function of irrigation system capacity, or system flow rate. Operating the system faster decreases only the application depth, not the application rate. For example, the same center pivot operating at 48 h per revolution will apply 16 mm of water at the same application rates.

Center-pivot irrigation is the most popular type of irrigation system in the United States. According to the United States Department of Agriculture (USDA) National Agricultural Statistics Service, center-pivot irrigation was used on 47% of the irrigated land and 83% of the sprinkler-irrigated land in 2008, an increase from 25% of the irrigated land in 1988.[2] More land was irrigated by center pivots in the United States in 2008 (10.4 Mha) than all types of gravity irrigation combined (8.9 Mha). As center pivots gained popularity, researchers began to consider runoff potential, mainly to efficiently apply irrigation water. Most sprinkler-irrigation studies were not concerned with soil erosion, probably because the effects of sprinkler-irrigation erosion tend to occur within the field rather than off site.

A 1969 study evaluated center-pivot runoff from a theoretical perspective and showed the importance of modifying infiltration parameters, determined from pond infiltration tests, for the low initial application rate that occurs with moving irrigation systems.[23] Their theoretical evaluation showed that 0%–40% of the applied water could run off with typical operating conditions. A 1971 field study documented 11%–41% runoff on four center pivots operated by farmers.[24] Runoff with center-pivot irrigation became a more important issue as low-pressure sprinklers began to be used to reduce energy costs. Early types of low-pressure sprinklers applied water to a smaller area, which increased application rates and potential runoff.[25] Low-pressure sprinklers (40 and 100 Pa) averaged 69 or 70 mm of runoff compared with 8 to 10 mm of runoff for high-pressure sprinklers (170 and 345 Pa) during a 4-year field study.[26] Reducing pressure from 380 to 140 Pa increased irrigation runoff 30% for a center pivot with impact sprinklers.[27] Peak application rate at the outer end of a center pivot would be about 30 mm h^{-1} for a high-pressure impact sprinkler with 20 m wetted radius and more than 100 mm h^{-1} for a low-pressure spray sprinkler with 5 m wetted radius.[25] Figure 5 shows two sprinkler application rate curves with time and an infiltration rate curve. The volume of water applied when application rate exceeds infiltration is potential runoff. All of this water may not run off if some is ponded or stored on the soil surface.

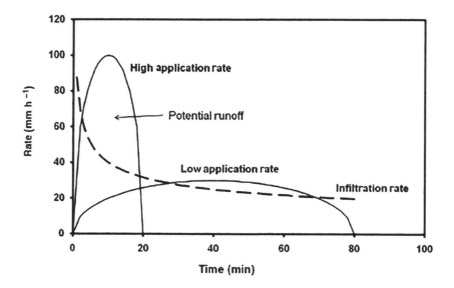

FIGURE 5 Example of soil infiltration rate and sprinkler application rates for high- and low-application-rate sprinklers. Runoff potentially occurs when sprinkler application rate exceeds the soil infiltration rate.

Many types of sprinklers are now available for center pivots. Some apply water in defined streams with a wetted diameter over 20 m with nozzle pressure of 200 Pa or less. Others distribute water evenly over the wetted area with various combinations of droplet sizes. Various sprinkler designs are the result of manufacturers trying to reduce the kinetic energy applied to the soil during irrigation, so all applied water can infiltrate. Kincaid[28] developed a model in 1996 to estimate kinetic energy per unit drop volume for common sprinkler types. Calculating area-weighted kinetic energies per unit drop volume for individual sprinklers showed that sprinklers with the smallest drop size distributions had the lowest kinetic energy. Sprinklers with smaller sized drops tend to have smaller wetted diameters because small drops cannot travel as far as large drops. Larger drops travel farther and therefore cover a greater portion of the circular wetted area for an individual sprinkler. A smaller wetted diameter also results in a higher application rate when sprinkler application patterns are overlapped like occurs on a center pivot. An alternative method for characterizing sprinkler kinetic energy is calculating the rate that energy is applied to the soil, or specific power, as a function of radial distance from the sprinkler.[29] The specific power distribution is energy per drop volume multiplied by the application rate and can be overlapped, like water application rate, to develop a composite specific power profile for a sprinkler system. A flat-plate sprinkler with small-sized drops had higher average composite specific power than two other sprinklers with larger drop sizes and larger wetted diameter.[29] Recent field research on small plots showed that soil erosion was significantly greater with the flat-plate spray sprinkler compared with the two other sprinklers with larger drop size distributions.[30] This directly contradicts previous conventional thinking that sprinklers with smaller drops caused less erosion.

The most effective way to control sprinkler-irrigation erosion is to eliminate runoff, which also increases water application efficiency. One way to control runoff is to increase water storage on the soil surface.[25] Reservoir tillage is a practice that forms small pits in the soil to store water. Each pit can hold 5 to 10 L of water.[31] This is especially important on sloping fields. Reservoir tillage reduced runoff 68% and soil erosion 92% during a 50 mm simulated irrigation on a field with 10% slope.[32] Runoff was not different when field slope was only 1%. Increasing surface residue also decreases sprinkler-irrigation runoff similar to rainfall runoff.[27] Disking corn stubble prior to planting, which left approximately 30% of the soil surface covered with crop residue, reduced runoff to 17% of applied irrigation compared with

25% runoff for moldboard plow plots in a 4 years study.[33] In addition to reducing runoff, disking also reduced soil loss about 50% compared with moldboard plowing.

Applying PAM with sprinkler irrigation can also improve infiltration, which reduces runoff and soil erosion. Field studies have shown that erosion decreased under moving sprinkler systems when 20 kg PAM ha^{-1} was applied to the soil before irrigation.[34,35] Lower PAM application rates can be effective when PAM is applied with irrigation water rather than sprayed directly on the soil surface. In laboratory studies with 2 m^2 soil boxes, applying 2 to 4 kg PAM ha^{-1} at 10 to 20 mg L^{-1} with sprinkler-irrigation water reduced soil erosion 75% compared with untreated soil, but these benefits decreased with subsequent irrigations without PAM.[36] In a similar laboratory study, applying 1 kg PAM ha^{-1} with three consecutive irrigations reduced cumulative runoff 50% compared with untreated soil, while applying 3 kg PAM ha^{-1} with one irrigation only reduced runoff by 35%.[37] Field tests in the United States showed that applying PAM with four irrigations (2 to 3 kg ha^{-1} total applied) significantly reduced soil erosion from 52 and 34 kg ha^{-1} for the control to 21 and 5 kg ha^{-1} for the PAM treatment during the 2 years of the study.[38] Soil erosion was not significantly different for a similar field study in Portugal with lower PAM application rates (0.3 kg ha^{-1}).[38]

Conclusions

Irrigation is vital to world food production, but soil erosion during irrigation threatens the long-term productivity of irrigation. Soil erosion is generally greater from surface irrigation because water flows over the soil during irrigation. Surface-irrigation management is often a tradeoff between irrigation uniformity and erosion. High flow rates can cause erosion; low flow rates can cause poor irrigation uniformity. Ideally, sprinkler irrigation should not have any runoff; however, moving irrigation systems, like center pivots, often apply water faster than it can infiltrate into the soil. Current research is attempting to quantify runoff and erosion potential for various types of center-pivot sprinklers so manufacturers can improve sprinkler designs. Irrigation can be managed to minimize erosion and maintain productivity.

References

1. Kendall, H.W.; Pimentel, D. Constraints on the expansion of the global food supply. Ambio **1994**, *23* (3), 198–205.
2. USDA National Agricultural Statistics Service. *Farm and Ranch Irrigation Survey*, available at http://www.agcensus.usda.gov (accessed December 2010).
3. Bjorneberg, D.L.; Kincaid, D.C.; Lentz, R.D.; Sojka, R.E.; Trout, T.J. Unique aspects of modeling irrigation-induced soil erosion. Int. J. Sediment Res. **1999**, *15* (2), 245–252.
4. Trout, T.J. Furrow irrigation erosion and sedimentation: Onfield distribution. Trans. ASAE **1996**, *39* (5), 1717–1723.
5. Bjorneberg, D.L.; Sojka, R.E.; Aase, J.K. Pre-wetting effect on furrow irrigation erosion: A field study. Trans. ASAE **2002**, *45* (3), 717–722.
6. Lentz, R.D.; Sojka, R.E.; Carter, D.L. Furrow irrigation water-quality effects on soil loss and infiltration. Soil Sci. Soc. Am. J. **1996**, *60* (1), 238–245.
7. Kim, K.H.; Miller, W.P. Effect of rainfall electrolyte concentration and slope on infiltration and erosion. Soil Technol. **1996**, *9* (3), 173–185.
8. Flanagan, D.C.; Norton, L.D.; Shainberg, I. Effect of water chemistry and soil amendments on a silt loam soil. Part II: Soil erosion. Trans. ASAE **1997**, *40* (6), 1555–1561.
9. Levy, G.J.; Levin, J.; Shainberg, I. Seal formation and interrill soil erosion. Soil Sci. Soc. Am. J. **1994**, *58* (1), 203–209.
10. International Commission on Irrigation and Drainage, available at http://www.icid.org/imp_data.pdf (accessed November 2010).

11. FAO Aquastat, http://www.fao.org/nr/water/aquastat/main/index.stm (accessed November 2010).

12. Koluvek, P.K.; Tanji, K.K.; Trout, T.J. Overview of soil erosion from irrigation. J. Irrig. Drain. Eng. **1993**, *119* (6), 929–946.

13. Israelson, O.W.; Clyde, G.D.; Lauritzen, C.W. *Soil erosion in small irrigation furrows.* Bull. 320. Utah Agricultural Experiment Station: Logan, UT, 1946.

14. Mech, S.J. Effect of slope and length of run on erosion under irrigation. Agric. Eng. **1949**, *30* (8), 379–383, 389.

15. Berg, R.D.; Carter, D.L. Furrow erosion and sediment losses on irrigated cropland. J. Soil Water Conserv. **1980**, *35* (6), 267–270.

16. Bjorneberg, D.L.; Westermann, D.T.; Nelson, N.O.; Kendrick, J.H. Conservation practice effectiveness in the irrigated Upper Snake River/Rock Creek watershed. J. Soil Water Conserv. **2008**, *63* (6), 487–495.

17. Ebbert, J.C.; Kim, M.H. Relation between irrigation method, sediment yields, and losses of pesticides and nitrogen. J. Environ. Qual. **1998**, *27* (2), 372–380.

18. Kemper, W.D.; Trout, T.J.; Brown, M.J.; Rosenau, R.C. Furrow erosion and water and soil management. Trans. ASAE **1985**, *28* (5), 1564–1572.

19. Mailapalli, D.R.; Raghuwanshi, N.S.; Singh, R. Sediment transport in furrow irrigation. Irrig. Sci. **2009**, *27* (6), 449–456.

20. Fernandez-Gomez, R.; Mateos, L; Giraldez, J.V. Furrow irrigation erosion and management. Irrig. Sci. **2004**, *23* (3), 123–131.

21. Lentz, R.D.; Sojka, R.E. Long-term polyacrylamide formulation effects on soil erosion, water infiltration, and yields of furrow-irrigated crops. Agron. J. **2009**, *101* (2), 305–314.

22. Lentz, R.D.; Sojka, R.E. Field results using polyacrylamide to manage furrow erosion and infiltration. Soil Sci. **1994**, *158* (4), 274–282.

23. Kincaid, D.C.; Heermann, D.F.; Kruse, E.G. Application rates and runoff in center-pivot sprinkler irrigation. Trans. ASAE **1969**, *12* (6), 790–794,797.

24. Aarstad, J.S.; Miller, D.E. Soil management to reduce runoff under center-pivot sprinkler systems. J. Soil Water Conserv. **1973**, *28* (4), 171–173.

25. Gilley, J.R. Suitability of reduced pressure center-pivots. J. Irrig. Drain. Eng. **1984**, *110* (1), 22–34.

26. DeBoer, D.W.; Beck, D.L.; Bender, A.R. A field evaluation of low, medium, and high pressure sprinklers. Trans. ASAE **1992**, *35* (4), 1185–1189.

27. Mickelson, R.H.; Schweizer, E.E. Till-plant systems for reducing runoff under low-pressure, center-pivot irrigation. J. Soil Water Conserv. **1987**, *42* (2), 107–111.

28. Kincaid, D.C. Spraydrop kinetic energy from irrigation sprinklers. Trans. ASAE **1996**, 39 *(3)*, 847–853.

29. King, B.A.; Bjorneberg, D.L. Characterizing droplet kinetic energy applied by moving spray-plate center-pivot irrigation sprinklers. Trans. ASABE **2010**, 53 *(1)*, 137–145.

30. King, B.A.; Bjorneberg, D.L. Evaluation of potential runoff and erosion of four center pivot irrigation sprinklers. Trans. ASABE **2011**, in press.

31. Oliveira, C.A.S.; Hanks, R.J.; Shani, U. Infiltration and runoff as affected by pitting, mulching and sprinkler irrigation. Irrig. Sci. **1987**, *8* (1), 49–64.

32. Kranz, W.L.; Eisenhauer, D.E. Sprinkler irrigation runoff and erosion control using interrow tillage techniques. Appl. Eng. Agric. **1990**, *6* (6), 739–744.

33. DeBoer, D.W.; Beck, D.L. Conservation tillage on a silt loam soil with reduced pressure sprinkler irrigation. Appl. Eng. Agric. **1991**, *7* (5), 557–562.

34. Levy, G.J.; Ben-Hur, M.; Agassi, M. The effect of polyacrylamide on runoff, erosion, and cotton yield from fields irrigated with moving sprinkler systems. Irrig. Sci. **1991**, *12* (2), 55–60.

35. Stern, R.; Van Der Merwe, A.J.; Laker, M.C.; Shainberg, I. Effect of soil surface treatments on runoff and wheat yields under irrigation. Agron. J. **1992**, *84* (1), 114–119.

36. Aase, J.K.; Bjorneberg, D.L.; Sojka, R.E. Sprinkler irrigation runoff and erosion control with polyacrylamide— Laboratory tests. Soil Sci. Soc. Am. J. **1998**, *62* (6), 1681–1687.

37. Bjorneberg, D.L.; Aase, J.K. Multiple polyacrylamide applications for controlling sprinkler irrigation runoff and erosion. Appl. Eng. Agric. **2000**, *16* (5), 501–504.

38. Bjorneberg, D.L.; Santos, F.L.; Castanheira, N.S.; Martins, O.C.; Reis, J.L.; Aase, J.K.; Sojka, R.E. Using polyacrylamide with sprinkler irrigation to improve infiltration. J. Soil Water Conserv. **2003**, *58* (5), 283–289.

16

Irrigation: River Flow Impact

Introduction 165
    Depletion
Hydrograph Modification 166
    Reservoir Storage • Irrigation Return Flows • Irrigation
    Methods • Irrigation Efficiencies
Basin Irrigation Efficiency 168
Environmental Concerns 169
    Water Quality Implications for Agriculture • Salt Loading
    Pick-Up • In-Stream Flow Requirements
References 170

Robert W. Hill and
Ivan A. Walter

Introduction

In hydrologic studies it is common engineering practice to quantify the impact upon the stream(s) from which the irrigation water is diverted. The impact upon the stream is actually of two kinds: 1) diversions that decrease the streamflow and 2) return flows that increase the streamflow. The engineering term used to describe the overall impact is "streamflow depletion" which means the net reduction in stream-flow resulting from diversion to irrigation uses. Actual stream depletions are a function of many factors including the amount and timing of diversions, the type of diversion structure (well vs. ditch), crops grown, soil type, depth to groundwater, irrigation method, irrigation efficiency, properties of the alluvial aquifer, area irrigated, and evapotranspiration of precipitation, groundwater, and irrigation water.

Depletion

Depletion, in this context, is the consumptive abstraction of water from the hydrologic system as a result of irrigation. It is in addition to consumptive water use that would have occurred in the unmodified natural situation. As an example, waters of the Bear River Basin of Southern Idaho, Northern Utah, and Western Wyoming, because it is an interstate system, are administered by a federally established commission under the authority of the Bear River Compact.[1] Depletion is the basis, in the compact, for allocating Bear River water use among the three states. It is defined by a "Commission Approved Procedure" which includes consideration of land use and incorporates an equation for estimating depletion based on evapotranspiration. In a study for the commission, Hill[5] defined crop depletion as

$$Dpl = Et - Smco - Pef \qquad (1)$$

where Dpl is estimated depletion for a given site or sub-basin; Et is calculated crop water use; SMco is moisture which is "carried over" from the previous non-growing season (October 1–April 30) as stored soil

165

water in the root zone available for crop water use subsequent to May 1; and Pef is an estimate of that portion of precipitation measured at an NWS station during May–September, which could be used by crops.

The carry-over soil moisture (SMco) was estimated by assuming that 67% of adjusted precipitation from October through April could be stored in the root zone. If this exceeded 75% of the available soil water-holding capacity of the average root zone in the sub-basin, the excess was considered as lost to drainage or runoff and not available for crop use. Growing season precipitation was considered to be 80% effective in contributing to crop water use. The effectiveness factor of 80% allowed for precipitation depths throughout a sub-basin that might differ from NWS rain-gage amounts. It also included a reduction for mismatches in timing between rainfall events and irrigation scheduling.

Hydrograph Modification

Diversion of significant amounts of water from rivers and streams for irrigation and subsequent return flows alters the shape and timing of downstream hydrographs. In watersheds where mountain snowmelt provides the irrigation supply, such as in the Western United States, diversion during the spring runoff attenuates the peak flow rate while later return flows extend the flow duration into late summer and early fall.

Reservoir Storage

Storage of water in reservoirs can significantly modify the natural stream hydrograph depending on the timing and quantity of the storage right. Irrigators with junior rights may only be able to store during time periods with low irrigation demand, such as during the winter, or during peak flow periods. Reductions of stream flow during the winter time may have considerable impact on downstream instream flows. Whereas, storage during periods of peak runoff may not affect minimum in-stream flow needs, but could deposit considerable amounts of sediment in the reservoir.

Irrigation Return Flows

Irrigation return flows are comprised of surface runoff and/or subsurface drainage that becomes available for subsequent rediversion from either a surface stream or a groundwater aquifer downstream (hydrologically) of the initial use. Reusable return flow can be estimated as irrigation diversion minus crop related depletions minus additional abstractions. Additional abstractions include incidental consumptive use from water surfaces as in open drains, along with non-crop vegetation. The timing of return flow varies from nearly instantaneous (recaptured tailwater) to delays of weeks and months or perhaps longer with deep percolation subsurface drainage. In a hydrologic model study of the Bear River Basin[4] delay times between diversion and subsequent appearance of the return flow at the next downstream river gage varied from 1.5 months to as long as 6 months. The delay appeared to be related to sub-basin shape and size.

Irrigation Methods

Four general irrigation methods are used: surface, subsurface, sprinkler, and trickle (also known as low flow or drip). Surface methods include wild or controlled flooding, furrow, border-strip, and ponded water (basin, paddy, or low-head bubbler). Hand move, wheel move, and center pivot are examples of sprinkler irrigation. Trickle irrigation includes point source emitters, microspray, bubbler, and line-source drip tape (above or below ground). Whereas the efficiency of surface irrigation is dependent upon the skills and experience of the irrigator, the performance of trickle and sprinkler systems is more dependent on the design. Generally, the more control that the system design (hardware) has on the irrigation system performance, the higher the application efficiency (E_a) can be. Thus, typical wheel move sprinklers have higher E_a values than surface irrigation, but lower values than for center pivots or trickle, assuming better than average management practices for each method.

The impact on river flows can be quite different among the various irrigation methods. The nature of furrow and border surface irrigation generally produces tail water runoff, which can be immediately recaptured and reused, as well as deep percolation, which may not be available for reuse until after a period of time. Tailwater is essentially eliminated and deep percolation reduced with sprinklers (Figure 1) compared to conventional surface irrigation. Whereas, with drip methods, deep percolation can be further reduced. The reduction of deep percolation implies increased salt concentration in the root zone leachate, but, perhaps significant reduction in salt pick-up potential from geologic conditions.

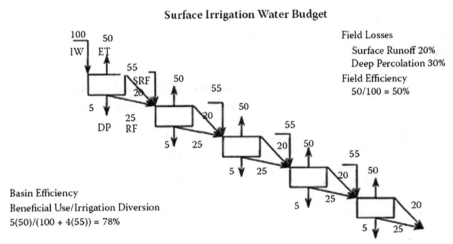

IW - Irrigation Water Supply
ET - Evapotranspiration (beneficial use) from irrigation water
DP - Deep Percolation below root zone to subsurface water
SRF - Surface Return Flow
RF - Return Flow of subsurface water
EV - Evaporation from droplets in air and wind drift losses with sprinkler

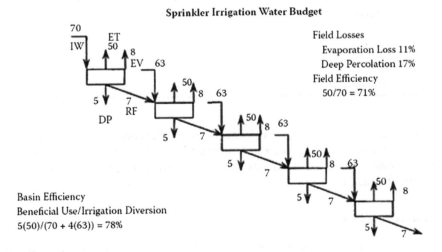

FIGURE 1 Comparison of basin efficiencies between surface and sprinkler irrigation methods with four return flow reuse cycles.

Irrigation Efficiencies

Although a full discussion of the several variations of irrigation efficiency is beyond the scope herein, two terms will be defined and discussed. More complete discussions relating to irrigation efficiencies and water requirements are given elsewhere.[6-9,13] Keller and Bliesner[9] give a particularly thorough presentation of distribution uniformity and efficiencies.

Application efficiency (E_a):

$$E_a = 100 \times \frac{\text{Volume of water stored in the root zone}(V_s)}{\text{Volume of water delivered in the farm or field}(V_f)}.$$

Distribution uniformity:

The distribution uniformity is a measure of how evenly the on-farm irrigation system distributes the water across the field. The definition of DU is:

$$DU = 100 \times \frac{\text{Average of the lowest 25\% of infiltrated water depth}}{\text{Average of all infiltrated water depths across the field}}.$$

On-farm or field application efficiencies can be affected by the distribution uniformity and vary widely for both surface and sprinkle irrigation methods. This is largely due to difference in management practices, appropriateness of design in matching the site conditions (slope, soils, and wind), and the degree of maintenance. In addition, for a given system uniformity, the higher the proportion of the field that is adequately irrigated (i.e., infiltrated water refills the soil water deficit) the lower will be the application efficiency. This is due to greater deep percolation losses in the overirrigated portions of the distribution pattern. Some values determined in recent Utah field evaluations are:

Method	Observed		
	High (%)	Low (%)	Typical (%)
Surface irrigation			
E_a	72	24	50
Tailwater	55	5	20
Deep percolation	65	20	30
Sprinkler irrigation			
E_a	84	52	70
Evaporation	45	8	12
Deep percolation	37	8	18

The E_a for a particular field may vary greatly during the season. Cultivation practices, microconsolidation of the soil surface and vegetation will alter surface irrigation efficiency both up and down from the seasonal average. Seasonal and diurnal variations in wind, humidity, and temperature will also affect sprinkle application efficiencies.

Basin Irrigation Efficiency

The actual irrigation efficiency realized for several successive downstream fields where capture and reuse of return flows is experienced is higher than the E_a of an individual field. This notion of "Basin Irrigation Efficiency"[12,13] is illustrated in Figure 1. This simple example comparison of surface and sprinkle methods assumes four reuse cycles. In each of the five "fields" Et is assumed to be 50 units. The surface runoff is captured for reuse on the next field. All of the irrigation-related evaporation is assumed "lost" as well as 5 units of deep percolation. After the fifth field, all surface and subsurface flows are lost. The basin efficiency for surface is 78%, which is the sameas for sprinkle. The surface irrigation basin efficiency increase is dependent upon the surface return flow reuse, which is 20 units in this example. However, the depletion is greater for sprinkler due to the extra evaporation. In a Colorado field study, Walter and Altenhofen[11] found a progressive increase in irrigation efficiencies from field (average E_a of 45%), to farm, to efficiency of ditch or sectors (average of 83%). This was due to the reuse of tailwater (10%–20% of delivery) and deep percolation (46% of delivery).

Environmental Concerns

The process of evapotranspiration, or crop water use, extracts pure water from the soil water reservoir, which leaves behind the dissolved solids (salts) contained in the applied irrigation water. The "evapo-concentration" of salts is an inevitable result of irrigation for crop production. As stated by Bishop and Peterson:[2]

...Other uses add something to the water, but irrigation basically takes some of the water away, concentrating the residual salts. Irrigation may also add substances by leaching natural salts or other materials from the soil or washing them from the surface. Irrigation return flow is a process by which the concentrated salts and other substances are conveyed from agricultural lands to the common stream or the underground water supply...

Water Quality Implications for Agriculture

Irrigated agriculture is dependent upon adequate, reasonably good quality water supplies. As the level of salt increases in an irrigation source, the quality of water for plant growth decreases. Since all irrigation waters contain a mixture of natural salt, irrigated soils will contain a similar mix to that in the applied water, but generally at a higher concentration. This necessitates applying extra irrigation water, or taking advantage of non-growing season precipitation, to leach the salts below the root zone.

Salt Loading Pick-Up

Water percolating below the root zone or leaking from canals and ditches may "pick-up" additional salts from mineral weathering or from salt-bearing geologic formations (such as the Mancos shale of Western Colorado and Eastern Utah). This salt pick-up will increase the salt load of return flows and consequently increase the salinity of receiving waters.

In the Colorado River Basin in the United States and Mexico salinity is a concern because of its adverse effects on agricultural, municipal, and industrial users.[10] The Salinity Control Act of 1974 (Public Law 93–320) created the Colorado River Basin Salinity Control Program to develop projects to reduce salt loading to the Colorado River. Salinity control projects include lining open canals and laterals (or replacing with pipe) and installing sprinklers in place of surface irrigation for the purpose of decreasing salt loading caused by canal leakage and irrigated crop deep percolation. Recently selenium in irrigation return flow has become a concern[3] and may also be reduced by salinity reduction projects.

In-Stream Flow Requirements

Diversions in some reaches in some Western United States streams are "dried up" immediately downstream of diversion structures during times of peak irrigation demand. This condition eliminates any use of the reach for fisheries and other uses which depend on in-stream flow. In some instances, negotiated agreements with senior water rights users have allowed for bypass of minimal amounts of water to sustain the fishery or habitat, and for control of tail-water runoff to reduce agricultural related chemicals in the receiving water.

References

1. Bear River Commission. *Amended Bear River Compact,* S.B. 255, 1979 Utah Legislature Session. 1979.
2. Bishop, A.A.; Peterson, H.B. (team leaders). Characteristics and Pollution Problems of Irrigation Return Flow. Final Report Project 14-12-408, Fed. Water Pollution Control Adm. U.S. Dept. of Interior (Ada, Oklahoma). Utah State University Foundation, Logan, Utah. May 1969.
3. Butler, D.L. *Effects of Piping Irrigation Laterals on Selenium and Salt Loads, Montrose Arroyo Basin, Western Colorado;* U.S. Geological Survey Water-Resources Investigations Report 01–4204 (in cooperation with the U.S. Bureau of Reclamation): Denver, Colorado, 2001.
4. Hill, R.W.; Israelsen, E.K.; Huber, A.L.; Riley, J.P. A *Hydrologic Model of the Bear River Basin;* PRWG72-1, Utah Water Resource Laboratory, Utah State University: Logan, Utah, 1970.
5. Hill, R.W.; Brockway, C.E.; Burman, R.D.; Allen, L.N.; Robison, C.W. *Duty of Water Under the Bear River Compact: Field Verification of Empirical Methods for Estimating Depletion. Final Report;* Utah Agriculture Experiment Station Research Report No. 125, Utah State University: Logan, Utah, January 1989.
6. Hoffman, G.J.; Howell, T.; Solomon, K. *Management of Farm Irrigation Systems;* The American Society of Agricultural Engineers: St. Joseph, MI, 1990.
7. Jensen, M.E. *Design and Operation of Farm Irrigation Systems;* ASAE Monograph, American Society of Agricultural Engineers: St. Joseph, MI, 1983.
8. Jensen, M.E.; Burman, R.D.; Allen, R.G.; Eds. *Evapotranspiration and Irrigation Water Requirements;* ASCE Manual No. 70, American Society of Civil Engineers: New York, 1990.
9. Keller, J.; Bliesner, R.D. *Sprinkle and Trickle Irrigation;* Van Nostrand Reinhold: New York, 1990.
10. U.S. Department of the Interior. *Quality of Water—Colorado River Basin: Bureau of Reclamation, Upper Colorado Region;* Progress Report no 19; Salt Lake City, Utah, 1999.
11. Walter, I.A.; Altenhofen, J. Irrigation efficiency studies—Northern Colorado. Proceedings USCID Water Management Seminar. Sacramento, CA Oct 5–7; USCID: Denver, CO, 1995.
12. Willardson, L.S.; Wagenet, R.J. *Basin-Wide Impacts of Irrigation Efficiency;* Proceedings of an ASCE Specialty Conference on Advances in Irrigation and Drainage, Jackson, WY, 1983.
13. Willardson, L.S.; Allen, R.G.; Frederiksen, H.D. *Elimination of Irrigation Efficiencies;* Proceedings 13th Technical Conference, USCID: Denver, CO, 1994.

17

Irrigation: Saline Water

Introduction 171
Irrigating with Saline Waters 171
Blending Low-Salt and Salty Waters 173
Conclusion 173
References.. 173

B.A. Stewart

Introduction

As water becomes more limited, there is increasing use of saline waters for irrigation that were previously considered unsuitable. Rhoades, Kandiah, and Marshali[1] classified saline waters as shown in Table 1. Electrical conductivity is a convenient and practical method for classifying saline waters because there is a direct relationship between the salt content of the water and the conductance of an electrical current through water containing salts. Electrical conductivity values are expressed in siemens (S) at a standard temperature of 25°C.

Most waters used for irrigation have electrical conductivities less than 2 dS m^{-1}.[1] When water higher than this level is used, there can be serious negative effects on both plants and soils. As salinity in the root zone increases, the osmotic potential of the soil solution decreases and therefore reduces the availability of water to plants. At some point, the concentration of salts in the root zone can become so great that water will actually move from the plant cells to the root zone because of the osmotic effect. Salts containing ions such as boron, chloride, and sodium can also be toxic to plants when accumulated in large quantities in the leaves. The extent that plant growth is affected by saline water is dependent on the crop species. Some plants, such as barley and cotton, are much more resistant to salt than crops like beans. Rhoades, Kandiah, and Marshali[1] list the tolerance levels of a wide range of fiber, grain, and special crops; grasses and forage crops; vegetable and fruit crops; woody crops; and ornamental shrubs, trees, and ground cover. Soils are also negatively impacted by salt, particularly sodium salts. Sodium ions tend to disperse clay particles and this has deleterious effects on infiltration rate, structure, and other soil physical properties.

Irrigating with Saline Waters

Water limitations and the need to increase food and fiber production in many parts of the world have resulted in the use of water for irrigation containing increasing levels of salts. The United States, Israel, Tunisia, India, and Egypt have been particularly active in irrigating with saline waters.[1] Rhoades, Kandiah, and Marshali[1] published an extensive paper on the use of saline waters for crop production and it is a valuable guide for anyone interested in the subject. They reported that many drainage waters, including shallow ground waters underlying irrigated lands, fall in the range of 2 dS m^{-1} to 10 dS m^{-1}

TABLE 1   Classification of Saline Waters

Water Class	Electrical Conductivity (dSm−1)	Salt Concentration (mg L−1)	Type of Water
Non-saline	<0.7	<500	Drinking and irrigation
Slightly saline	0.7–2	500–1500	Irrigation
Moderately saline	2–10	1500–7000	Primary drainage and groundwater
Highly saline	10–25	7000–15,000	Secondary drainage and groundwater
Very highly saline	25–45	15,000–35,000	Very saline groundwater
Brine	>45	>45,000	Seawater

in electrical conductivity. Such waters are in ample supply in many developed irrigated lands and have good potential even though they are often discharged to better quality surface waters or to waste outlets. These waters can be successfully used in many cases with proper management. Reuse of second-generation drainage waters with electrical conductivity values of 10 dS m^{-1} to 25 dS m^{-1} is also sometimes possible but to a much lesser degree because the crops that can be grown with these waters are atypical and much less experience exists upon which to base management recommendations.

Miller and Gardiner[2] suggest that successful irrigation with saline water requires three principles. First, the soil should be maintained near field capacity to keep the salt concentration as low as possible. Second, application techniques should avoid any wetting of the foliage. Third, salts accumulating in the soil should be periodically leached. To accomplish these objectives, Miller and Gardiner[2] recommend the following general rules:

- Apply water at or below soil surface. Sprinklers should be used only if they avoid wilting the foliage (such as sprinkling before plant emergence or below-canopy to avoid salt-burn damage).
- Keep water additions almost continuous, but at or below field capacity so that most flow is unsaturated. This maintains adequate aeration.
- Enough water should be added to keep salts moving downward, thus avoiding salt buildup in the root zone.

Miller and Gardiner[2] stress that these rules are difficult to meet and are best satisfied by some form of drip irrigation. They also state that due to the need for high water levels and because of high sodium ratios that sandy soils are more adaptable to the use of saline waters than soils containing high percentages of silt and clay particles.

Rhoades, Kandiah, and Marshali[1] also list specific management practices for producing crops with salty waters. Their list includes the following guidelines:

- Selection of crops or crop varieties that will produce satisfactory yields under the existing or predicted conditions of salinity or sodicity.
- Special planting procedures that minimize or compensate for salt accumulation in the vicinity of the seed.
- Irrigation to maintain a relatively high level of soil moisture and to achieve periodic leaching of the soil.
- Use of land preparation to increase the uniformity of water distribution and infiltration, leaching and removal of salinity.
- Special treatments (such as tillage and additions of chemical amendments, organic matter and growing green manure crops) to maintain soil permeability and tilth. The crop grown, the quality of water used for irrigation, the rainfall pattern and climate, and the soil properties determine to a large degree the kind and extent of management practices needed.

Blending Low-Salt and Salty Waters

Miller and Gardiner[2] reported that countries such as Israel have developed extensive canal and reservoir systems where both low-salt and salty waters are mixed to obtain usable water. Rhoades, Kandiah, and Marshali,[1] however, state that blending or diluting excessively saline waters with good quality water supplies should only be undertaken after consideration is given to how this affects the volumes of consumable water in the combined and separate supplies. They suggest that blending or diluting drainage waters with good quality waters in order to increase water supplies or to meet discharge standards may be inappropriate under certain situations. More crop production can usually be achieved from the total water supply by keeping the water components separated. Serious consideration should be given for keeping saline drainage waters separate from the good quality water, especially when the good quality waters are used for irrigation of salt-sensitive crops. The saline waters can be used more effectively by substituting them for good quality water to irrigate certain crops grown in the rotation after seeding establishment.

Conclusion

There is ample evidence that saline waters once considered unacceptable for irrigation can be used successfully provided that they are properly managed. There is also ample evidence, however, to show that these waters can be highly damaging to the environment and to the soil resource base when improperly managed. Therefore, saline waters should be only used for irrigation after careful study and considering as many factors as possible. Then, when the waters are used for irrigation, a careful monitoring program should be implemented of both the crops produced and of the resulting soil and environmental changes.

References

1. Rhoades, J.D.; Kandiah, A.; Marshali, A.M. The use of saline water for crop production. *FAO Irrigation and Drainage Paper 48;* Food and Agriculture Organization of the United Nations: Rome, 1992.
2. Miller, R.W.; Gardiner, D.T. *Soils in Our Environment,* 8th Ed.; Prentice Hall: Upper Saddle River, NJ, 1998.

18

Irrigation: Sewage Effluent Use

Introduction 175
Concerns of Irrigating with Sewage Effluent 175
Reuse Standards... 176
Monitoring Guidelines 176
References.. 177

B.A. Stewart

Introduction

One of the primary functions of soil is to buffer environmental change. This is the result of the biological, chemical, and physical processes that occur in soils. The soil matrix serves as an incubation chamber for decomposing organic wastes including pesticides, sewage, solid wastes, and many other wastes. Soils store, decompose, or immobilize nitrates, phosphorus, pesticides, and other substances that can become pollutants in air or water. Consequently, soil has, for centuries, been used for the application of sewage effluents. Sewage effluent provides farmers with a nutrient-enriched water supply and society with a reliable and inexpensive means of wastewater treatment and disposal. It should not, however, be assumed that irrigation is always the best solution for wastewater disposal. Disposal by irrigation should always be compared with alternative options based on environmental, social, and economic costs and benefits.

While disposal is the primary objective in many cases, the need of water for irrigation is becoming more often the driver for using sewage effluent on land. This is particularly true in areas like the Middle East where population growth is resulting in severe water shortages. The guidelines for using effluent for irrigation vary considerably among countries and other governing bodies. Cameron[1] conducted a literature review and found wide differences of guidelines for effluent irrigation projects being used throughout the world. In general, however, sustainable and environmentally sound systems can be developed in most situations provided proper management practices are followed.

Concerns of Irrigating with Sewage Effluent

In spite of the documented benefits associated with the use of sewage effluent for irrigation, there are numerous concerns. Many industrial wastewaters have been routinely dumped into municipal sewage lines. While this issue has been addressed in some jurisdictions, it has not in many others. In the United States, the Environmental Protection Agency requires that wastewaters be treated prior to disposal into municipal treatment plants or back into groundwater. Irrigating with wastewaters partially cleans water by percolation through the soil, but soluble salts and some inorganic and organic chemicals may continue to flow with the water to groundwater or surface supplies. In general, the Environmental

Protection Agency allows sewage effluents to be used for irrigation only if it does not cause: 1) extensive groundwater pollution; 2) a direct public health hazard; 3) an accumulation in the soil or water of hazardous substances that can get into the food chain; 4) an accumulation of pollutants such as odors into the atmosphere; and 5) other aesthetic losses, within the limits.[2]

Bouwer[3] has also expressed concerns about the use of sewage effluent for irrigation. He is particularly concerned with pathogens and warns that complete removal of viruses, bacteria, and protozoa and other parasites should be required before the effluent can be used to irrigate fruits/vegetables consumed raw or brought into the kitchen, or parks, playgrounds and other areas with free public access. Bouwer also stresses that long-term effects of sewage effluent irrigation on underlying groundwater should be considered in addition to the changes in nitrate and salinity. Ground water in low rainfall regions can be highly affected by percolating sewage effluent because much of the water is used by the growing crops and this greatly concentrates the chemicals in the small amounts of water that actually percolate to the groundwater. These chemicals can include disinfection byproducts, pharmaceutically active chemicals, and compounds derived from humic and fulvic acids formed by the decomposition of plant material. Bouwer claims that many of these chemicals are suspected carcinogens or toxic. Therefore, Bouwer concludes that while sewage irrigation looks good on the surface, a more extensive look reveals a potential for serious contamination of groundwater. He states that municipalities and other entities responsible for irrigation with sewage effluent should do a groundwater impact analysis to develop management protocols and be prepared for liability actions. Those who benefit are local and state institutions in water resources, environmental quality protection, public health, consultants, and operators of effluent irrigation projects.

Reuse Standards

The standards for using sewage effluent for irrigation of agricultural crops vary widely among different countries of the world. Mexico and many South American countries, e.g., use untreated wastewater for irrigation.[4] Most of these countries do not have the resources or capital to treat sewage effluents. Wastewater is utilized after little or no treatment, and health risks are minimized by crop selection. Mexico does not allow wastewater to be used to irrigate lettuce, cabbage, beets, coriander, radishes, carrots, spinach, and parsley. Acceptable crops include alfalfa, cereals, beans, chili, and green tomatoes. In contrast, Israel has very stringent water reuse requirements. Effluent water requires a high level of treatment (large soil-aquifer recharge systems with dewatering) before the water can be reused for irrigation of vegetables to be consumed raw.[5] Health guidelines for irrigation with treated wastewater developed in California indicate that effluent waters used on food crops must be disinfected, oxidized, coagulated, clarified, and filtered.[6] Total coliform counts cannot exceed a median value of 2.2/100 mL or a single sample value of 25/100 mL. Total coliforms must be monitored daily and turbidity cannot exceed 2 nephelometric turbidity units and must be monitored continuously. Less restrictive guidelines developed by Shuval et al.,[7] and adopted by most of the international agencies, suggested that effluent water reuse was relatively safe to use if it contained less than 1 helminth egg L^{-1}, and less than 1000 fecal coliforms/100 mL.

Monitoring Guidelines

Site selection is a critical and necessary step in initiating a sewage effluent irrigation system. The U.S. Environmental Protection Agency[8] published detailed information on site characterization and evaluation. Information was provided on the design of systems, site characteristics, expected quality of the effluent water after land treatment, and typical permeabilities and textural classes suitable for each land treatment process. Information was provided for designing and monitoring site characteristics for slow rate processes (sprinkler and other typical farm irrigation systems), rapid infiltration basins, and overland flow systems. Monitoring requirements will vary considerably among projects depending on the cropping patterns, soil characteristics, and specific environmental concerns. In most cases, monitoring procedures and criteria will be site specific. In all cases, however, the objectives should be to use the resources effectively, protect the land, protect the groundwater, protect the surface water, and protect the community amenity.

References

1. Cameron, D.R. *Sustainable Effluent Irrigation Phase 1: Literature Review International Perspective and Standards,* Technical Report Prepared for Irrigation Sustainability committee; Canada–Saskatchewan Agriculture Green Plan, 1996.
2. Miller, R.W.; Gardiner, D.T. *Soils in Our Environment,* 8th Ed.; Prentice-Hall Inc: Upper Saddle River, NJ, 1998.
3. Bouwer, H. *Groundwater Problems Caused by Irrigation with Sewage Effluent;* Irrigation and Water Quality Laboratory, USDA-ARS: Phoenix, AZ, 2000.
4. Strauss, M.; Blumenthal, U.J. *Human Waste Use in Agriculture and Aquaculture: Utilization Practices and Health Perspectives*; IRWCD Report No. 09/90; International Reference Centre for Waste Disposal: Deubendorf, 1990.
5. Shelef, G. The role of wastewater reuse in water resources management in Israel. Water Sci. Tech. **1990**, *23*, 2081–2089, Switzerland.
6. Ongerth, H.J.; Jopling, W.F. Water reuse in California. In *Water Renovation and Reuse;* Shuval, H.I., Ed.; Academic Press: New York, 1977.
7. Shuval, H.I.; Adin, A.; Fattal, B.; Rawitz, E.; Yekutiel, P. *Wastewater Irrigation in Developing Countries. Health Effects and Technical Solutions;* World Bank Tech. Pap.; 1986; Vol. 51, 325 pp.
8. U.S. EPA. *Process Design Manual: Land Treatment of Municipal Wastewater;* EPA 625/1–81-013; U.S. EPA Center for Environmental Research Information: Cincinnati, OH, 1981.

19

Irrigation: Soil Salinity

Introduction 179
Deleterious Effects of Salts on Plants, Soils, and Waters 179
Causes of Salination Induced by Irrigation and Drainage 180
Irrigation and Drainage Management to Control Soil Salinity 181
References 182

James D. Rhoades

Introduction

Irrigation is an ancient practice that predates recorded history. While irrigated farmland comprises only about 15% of the worlds' total farmland, it contributes about 36% of the total supply of food and fiber, and it stabilizes production against the vagaries of weather.[1] In 30 years time, irrigated agriculture is expected to have to supply 50% of the worlds' food production requirements.[1] However, over the last 20 years, irrigation growth has actually slowed to a rate that is now inadequate to keep up with the projected expanding food requirements.[1] Furthermore, irrigation has resulted in considerable salination of associated land and water. It has been estimated variably that the salinized area is as low as 20 and as high as 50% of the worlds' irrigated land.[2–4] Worldwide, about 76.6 Mha of land have become degraded by human-induced salination over the last 45–50 years.[3] It has been estimated that the world is losing at least three hectares of arable land every minute to soil salination (about 1.6 Mha per year), second only to erosion as the leading worldwide cause of soil degradation.[5–7] These data imply that the rate of salinization in developed irrigation projects now exceeds the rate of irrigation expansion.[8]

Surviving the salinity threat requires that the seriousness of the problem be recognized more widely, the processes contributing to salination of irrigated lands be understood, effective control measures be developed and implemented that will sustain the viability of irrigated agriculture, and that practical reclamation measures be implemented to rejuvenate the presently degraded lands.[9,10]

Deleterious Effects of Salts on Plants, Soils, and Waters

Salt-affected soils have reduced value for agriculture because of their content and proportions of salts, consisting mainly of sodium, magnesium, calcium, chloride, and sulfate and secondarily of potassium, bicarbonate, carbonate, nitrate, and boron. Saline soils contain excessive amounts of soluble salts for the practical and normal production of most agricultural crops. Sodic soils are those that contain excessive amounts of adsorbed sodium in proportion to calcium and magnesium, given the salinity level of the soil water. An example of a salt-affected irrigated soil is shown in Figure 1.

Soluble salts exert both general and specific effects on plants, both of which reduce crop yield.[11] Excess salinity in the seedbed hinders seedling establishment and in the crop root zone causes a general reduction in growth rate. In addition, certain salt constituents are specifically toxic to some plants. For example, boron is highly toxic to susceptible crops when present in the soil water at concentrations of

FIGURE 1   Photograph of salt-affected irrigated field.

only a few parts per million. In some woody crops sodium and chloride may accumulate in the tissue over time to toxic levels. These toxicity problems are, however, much less prevalent than is the general salinity problem.

Salts may also change soil properties that affect the suitability of the soil as a medium for plant growth.[12] The suitability of soils for cropping depends appreciably on the readiness with which they conduct water and air (permeability) and on their aggregate properties (structure), which control the friability (ease with which crumbled) of the seedbed (tilth). In contrast to saline soils, which are well aggregated and whose tillage properties and permeability to water and air are equal to or higher than those of similar nonsaline soils, sodic soils have reduced permeabilities and poor tilth. These problems are caused by the swelling and dispersion of clay minerals and by the breakdown of soil structure (slaking and crusting), which results in loss of permeability and tilth. Sodic soils are generally less extensive but more difficult to reclaim than saline soils.

Beneficial use of water in irrigation consists of transpiration and leaching for salinity control (the leaching requirement). Plant growth is directly proportional to water consumption through transpiration.[13] From the point of view of irrigated agriculture, the ultimate objective of irrigation is to increase the amount of water available to support transpiration. Salts reduce the fraction of water in a supply (or in the soil profile) that can be consumed beneficially in plant transpiration.[14] In considering the use of a saline water for irrigation and in selecting appropriate policies and practices of irrigation and drainage management, it is important to recognize that the total volume of a saline water supply cannot be consumed beneficially in crop production (i.e., transpired by the plant). A plant will not grow properly when the salt concentration in the soil water exceeds some limit specific to it under the given conditions of climate and management.[11] This is even true for halophytes.[15] Thus, the practice of blending or diluting excessively saline waters with good quality water supplies should be undertaken only after consideration is given to how it affects the volumes of consumable (usable) water in the combined and separated supplies.[14].

Causes of Salination Induced by Irrigation and Drainage

While salt-affected soils occur extensively under natural conditions, the salt problems of greatest importance to agriculture arise when previously productive soils become salinized as a result of agricultural activities (the so-called secondary salination). The extent and salt balance of salt-affected areas has been modified considerably by the redistribution of water (hence salt) through irrigation and drainage.

The development of large-scale irrigation and drainage projects, which involves diversion of rivers, construction of large reservoirs, and irrigation of large landscapes, causes large changes in the natural water and salt balances of entire geohydrologic systems. The impact of such developments can extend well beyond that of the immediate irrigated area. Excessive water diversions and applications are major causes of soil and water salination in irrigated lands. It is not unusual to find that less than 60% of the water diverted for irrigation is used in crop transpiration.[9] This implies that about 40% of the irrigation water eventually ends up as deep percolation. This drainage water contains more salt than that added with the irrigation water because of salt dissolution and mineral weathering[14] within the root zone. It often gains additional salt-load as it dissolves salts of geologic origin from the underlying substrata through which it flows in its down-gradient path. This drainage water often flows laterally to lower lying areas, eventually resulting in shallow saline groundwaters of large areas of land (waterlogging). Salination occurs in soils underlain by saline shallow groundwater through the process of "capillary rise" as groundwater (hence, salt) is driven upwards by the force of evaporation of water from the soil surface. Correspondingly, saline soils and waterlogging are closely associated problems.

Seepage from unlined or inadequately lined delivery canals occurs in many irrigation projects and is often substantial. Law, Skogerboe, and Denit[16] estimated that 20% of the total water diverted for irrigation in the United States is lost by seepage from conveyance and irrigation canals. Biswas[17] estimated that 57% of the total water diverted for irrigation in the world is lost from conveyance and distribution canals. Analogous to on-farm deep percolation resulting from irrigation, these seepage waters typically percolate through the underlying strata (often dissolving additional salts in the process), flow to lower elevation lands or waters, and add to the problems of waterlogging and salt-loading associated with on-farm irrigation there. A classic example of the rise in the water table following the development of irrigation has been documented in Pakistan and is described by Jensen, Rangeley, and Diele-man[9] and Ghassemi, Jakeman, and Nix.[2] The depth to the water table in the irrigated landscape located between three major river-tributaries rose from 20 to 30 m over a period of 80–100 years, i.e., from preirrigated time (about 1860) to the early 1960s, until it was nearly at the soil-surface. In one region, the water table rose nearly linearly from 1929 to 1950, demonstrating that deep percolation and seepage resulting from irrigation were the primary causes. Ahmad[18] concluded that about 50% of the water diverted into irrigation canals in Pakistan eventually goes to the groundwater by seepage and deep percolation.

The role of irrigated agriculture in salinizing soil systems has been well recognized for hundreds of years. It is of relatively more recent recognition that salination of water resources from agricultural activities is a major and widespread phenomenon of likely equal concern to that of soil salination. The causes of water salination are essentially the same as those of soils, only the final reservoir of the discharged salt-load is a water supply in the former case.[14] The volume of the water supply is reduced through irrigation diversions and irrigation; thus, its capacity to assimilate such received salts before reaching use-limiting levels is reduced proportionately. Only in the past 15 years has it become apparent that trace toxic constituents, such as selenium, in agricultural drainage waters can also cause serious pollution problems.[19]

Irrigation and Drainage Management to Control Soil Salinity

The key to overall salinity control is strict control that maintains a net downward movement of soil water in the root zone of irrigated fields over time while minimizing excess irrigation diversions, applications, and deep percolation.[20] The direct effect of salinity on plant growth is minimized by maintaining the soil-water content in the root zone within a narrow range at a relatively high level, while at the same time avoiding surface-ponding and oxygen deletion and minimizing deep percolation. Combined methods of pressurized, high-frequency irrigation and irrigation scheduling have been developed that permit substantially the desired control to be achieved.[21,22] These systems transfer control of water distribution and infiltration from the soil to the irrigation equipment. This results in less excess water (and hence, less salt) being applied overall to the field to meet the needs of a part of the field area having lowest intake rate, as done in the more traditional gravity irrigated systems. However, gravity irrigation systems can be designed to achieve good irrigation efficiency and salinity control, even though surface ponding still does occur. The so-called level-basin, multi-set, cablegation, surge, and tailwater-return systems are among them.[21,22] The need for irrigation and the amount required to meet evapotranspiration and leaching requirement is determined from plant stress measurements, calculations of evapotranspiration amounts, measurements of soil-water depletion, measurements of soil (or soil-water) salinity, or a combination of them.[21,22]

In addition to effective methods of irrigation scheduling and application, appropriate irrigation and salinity management also require an effective delivery system. Delivery systems have generally been designed to provide water on a regular schedule. Efficient irrigation systems require more flexible deliveries that can provide water on demand as each crop and particular field have need of it. Delivery systems can be improved by lining the canals, by containing the water within closed conduits, and by implementing techniques that increase the flexibility of delivery.

As briefly discussed earlier, irrigated agriculture is a major contributor to the salinity of many rivers and groundwaters, as well as soils. Reducing deep percolation generally lessens the salt load that is returned to rivers or groundwater and their pollution.[14] Additionally, saline drainage waters should be intercepted before being allowed to mix with water of better quality. The intercepted saline drainage water should be desalted and reused, disposed of by pond evaporation or by injection into some suitably isolated deep aquifer, or better yet it should be used for irrigation in a situation where brackish water is appropriate. Various irrigation and drainage strategies have been developed for minimizing the pollution of waters from irrigation and for using brackish waters for irrigation.[14,23] Desalination of agricultural drainage waters is not now economically feasible, but improved techniques for doing this exist and some are being implemented. However, more needs to be done in this regard.

Traditionally, the concepts of leaching requirement and salt-balance index have been used to plan and judge the appropriateness of irrigation and drainage systems, operations and practices with respect to salinity control, water use efficiency, and irrigation sustainability. However, these approaches are inadequate. The recommended method is to monitor directly the root-zone salinity levels and distributions across fields as a means to evaluate the effectiveness of salinity, irrigation, and drainage management practices, to detect problems (current and developing), to help determine the underlying causes of problems, and to determine source areas of major water and salt-load contributions to the underlying groundwater. Theory, equipment, and practical technology have been developed for these purposes.[24] More information about irrigation and drainage management to control soil and water salinity is found elsewhere.[25–27]

References

1. FAO. *World Agriculture Toward 2000:* An FAO Study; Alexandratos, N., Ed.; Bellhaven Press: London, 1988; 338 pp.
2. Ghassemi, F.; Jakeman, A.J.; Nix, H.A. *Salination of Land and Water Resources. Human Causes, Extent, Management and Case Studies;* CAB International: Wallingford, U.K., 1995; 526 pp.
3. Oldeman, L.R.; van Engelen, V.N.P.; Pulles, J.H.M. The Extent of human-induced soil degradation. In *World Map of the Status of Human-Induced Soil Degradation: An Explanatory Note;* Oldeman, L.R., Hakkeling, R.T.A., Sombroek, W.G., Eds.; International Soil Reference and Information Center (ISRIC): Wageningen, the Netherlands, 1991; 27–33.
4. Adams, W.M.; Hughes, F.M.R. Irrigation development in desert environments. In *Techniques for Desert Reclamation;* Goudie, A.S., Ed.; Wiley: New York, 1990; 135–160.
5. Buringh, P. Food production potential of the world. In *The World Food Problem: Consensus and Conflict;* Sinha, R., Ed.; Pergamon Press: Oxford, 1977; 477–485.
6. Dregne, H.; Kassas, M.; Razanov, B. A new assessment of the world status of desertification. Desertification Control Bull. **1991**, *20*, 6–18.

7. Umali, D.L. Irrigation-induced salinity. In *A Growing Problem for Development and Environment*; Technical Paper; World Bank: Washington, DC, 1993.

8. Seckler, D. *The New Era of Water Resources Management: From "Dry" to "Wet" Water Savings*; Consultative Group on International Agricultural Research: Washington, DC, 1996.

9. Jensen, M.E.; Rangeley, W.R.; Dieleman, P.J. Irrigation trends in world agriculture. In *Irrigation of Agricultural Crops*; American Society of Agronomy Monograph No. 30; ASA: Madison, WI, 1990; 31–67.

10. UNEP. *Saving Our Planet: Challenges and Hopes*; Nairobi, United Nations Environment Program: Nairobi, Kenya, 1992; 20 pp.

11. Maas, E.V. Crop salt tolerance. ASCE Manuals and Reports on Engineering No. 71. In *Agricultural Salinity Assessment and Management Manual*; Tanji, K.K., Ed.; ASCE: New York, 1990; 262–304.

12. Rhoades, J.D. Principal effects of salts on soils and plants. In *Water, Soil and Crop Management Relating to the Use of Saline Water*; Kandiah, A., Ed.; FAO (AGL) Mise. Series Publication 16/90; Food and Agriculture Organization of the United Nations: Rome, 1990; 1933.

13. Sinclair, T.R. Limits to crop yield? In *Physiology and Determination of Crop Yield*; Boone, K.J., Ed.; American Society of Agronomy: Madison, WI, 1994; 509–532.

14. Rhoades, J.D.; Kandiah, A.; Mashali, A.M. *The Use of Saline Waters for Crop Production*; FAO Irrigation and Drainage Paper 48; FAO: Rome, Italy, 1992; 133 pp.

15. Miyamoto, S.; Glenn, E.P.; Oslen, M.W. Growth, water use and salt uptake of four halophytes irrigated with highly saline water. J. Arid Environ. **1996**, *32*, 141–159.

16. Law, J.P.; Skogerboe, G.V.; Denit, J.D. The need for implementing irrigation return flow control. p. 1–17. *In Managing Irrigated Agriculture to Improve Water Quality*; Proc. Math. Conf. Manag. Irrig. Agric. Improve Water Avail., Denver, CO, May 1972; Graphics Manage. Corp.: Washington, DC, 16–18.

17. Biswas, A.K. Conservation and management of water resources. In *Techniques for Desert Reclamation*; Goudie, A.S., Ed.; Wiley: New York, 1990; 251–265.

18. Ahmad, N. Planning for Future Water Resources of Pakistan. Proceedings of Darves Bornoz Spec. Conference, National Committee of Pakistan; ICID: New Delhi, India, 1986; 279–294.

19. Letey, J.; Roberts, C.; Penberth, M.; Vasek, C. *An Agricul-turl Dilemma: Drainage Water and Toxics Disposal in the San Joaquin Valley*; Special Publication 3319; University of California: Oakland, 1986.

20. Rhoades, J.D. Soil salinity—causes and controls. In *Techniques for Desert Reclamation*; Goude, A.S., Ed.; Wiley: New York, 1990; 109–134.

21. Hoffman, G.J.; Rhoades, J.D.; Letey, J.; Sheng, F. Salinity management. In *Management of Farm Irrigation Systems*; Hoffman, G.J., Howell, T.A., Solomon, K.H., Eds.; ASCE: St. Joseph, MI, 1990; 667–715.

22. Kruse, E.G.; Willardson, L.; Ayars, J. On-farm irrigation and drainage practices. In *Agricultural Salinity Assessment and Management Manual*; Tanji, K.K., Ed.; ASCE Manuals and Reports on Engineering No. 71; ASCE: New York, 1990; 349–371.

23. Rhoades, J.D. Use of saline drainage water for irrigation. In *Agricultural Drainage*; ASA Drainage Monograph 38; Skaggs, R.W., van Schilfgaarde, J., Eds.; ASA Drainage Monograph 38; American Society of Agronomy: Madison, WI, 1999; 619–657.

24. Rhoades, J.D.; Chanduvi, F.; Lesch, S. *Soil Salinity Assessment: Methods and Interpretation of Electrical Conductivity*; FAO Irrigation and Drainage Paper 57; FAO, United Nations: Rome, Italy, 1999; 152 pp.

25. Rhoades, J.D. Use of saline and brackish waters for irrigation: implications and role in increasing food production, conserving water, sustaining irrigation and controlling soil and water degradation. Proceedings of the International Workshop on "The Use of Saline and Brackish Waters for Irrigation: Implications for the Management of Irrigation, Drainage and Crops" at the

10th Afro-Asian Conference of the International Committee on Irrigation and Drainage, Bali, Indonesia, July 23–24; Ragab, R., Pearce, G., Eds.; International Committee on Irrigation and Drainage: Bali, Indonesia, 1998; 261–304.

26. Rhoades, J.D.; Loveday, J. Salinity in irrigated agriculture. In *Irrigation of Agricultural Crops;* Stewart, B.A., Nielsen, D.R., Eds.; Agron. Monograph. No. 30; American Society of Agronomy: Madison, Wisconsin, 1990; 1089–1142.

27. Tanji, K.K. Nature and extent of agricultural salinity. In *Agricultural Salinity Assessment and Management;* Tanji, K.K., Ed.; ASCE Manuals and Reports on Engineering No. 71, ASCE: New York, 1990; 1–17.

20

Managing Water Resources and Hydrological Systems

Introduction 185
Models for Agricultural Water Consumption Estimation 186
    Hydrological Models • Dynamic Crop Models • Hybrid Models •
    Environmental Extended Multi-Region Input–Output Model
Water-Sustainable Agricultural Adaptations 188
    Field Water-Saving Irrigation Technologies • Cropping System Adaptations
Physical Water Transfer or Virtual Water Flow 189
    Freshwater Diversion Projects • Virtual Water Flows via Food Trade
Conclusions 190
References 191

Honglin Zhong,
Zhuoran Liang,
and Cheng Li

Introduction

Water scarcity has become a global issue as freshwater demand keeps soaring from the rapid increase in population and food demand under fast social-economic development, with additional pressure from altered water supply under climate change driven by anthropogenic emissions (Vörösmarty et al. 2000; Piao et al. 2010). Many regions in the world are already under severe water shortages and may suffer even worse water crisis under projected climate change (Konikow and Kendy 2005). Hotspots, such as the intensive cropped and heavy irrigated fertile major cropping zone of North China Plain (NCP), experience severe environmental problems from underground water over-exploitation for agricultural irrigation. This area is likely to experience more frequent deadly heatwave under projected future climate (Kang and Eltahir 2018). This will threaten both crop production and food security, risk farmer livelihoods, and constrain social-economic development in the future.

As the biggest freshwater consumption sector, agricultural production has achieved great success in food security with vast irrigation expansion since the "green revolution". Only 20% of the cropland with irrigation produced as much as 40% of the total agricultural production in the world (Molden 2007). However, to achieve these results, irrigated crops consume 70% of global total freshwater withdrawal (Rosegrant et al. 2009) and cause rapid groundwater depletion in many major cropping zones such as North China, Northwestern India, and US high plains (Konikow and Kendy 2005). In NCP, the fast groundwater table decline has already caused severe environmental issues of dried-up rivers and lakes, seawater intrusion, land subsidence, and ground fissures (Xue et al. 2000). Public health may also be threatened when groundwater table drops to a deep level containing toxic levels of fluoride and arsenic (Currell et al. 2012).

Groundwater irrigation-intensive crops should be abandoned, and cropland fallow should be adopted for water-sustainable development (Wang et al. 2015, 2016). Shifting double cropping system (wheat–maize rotation) to single cropping system (winter fallow–spring maize rotation) and wheat fallow subsidy are highly recommended and already adopted by local farmers and government for groundwater recovery in the NCP (Feng et al. 2007; Zhong et al. 2017; Ren et al. 2018).

In this chapter, we reviewed the models, measurements, strategies, and projects for agricultural water consumption, conservation, and adaptation at the field, regional, and global levels. With a specific focus on the NCP, where experienced a significant transformation of agricultural practice and policies from previous water-intensive high-yield production-oriented to recent sustainable agricultural water management and cropland fallow with ecological compensation. With all the existing studies covering physical water saving, freshwater transfer, "virtual water" flow, and water footprint via food trade in the semi-arid NCP, farmers, scientists, stakeholders, and government from other water-deficit regions with heavy food production will get deeper insights into the water crisis and inspire local adaptations for the sustainable social-economic development around the world.

Models for Agricultural Water Consumption Estimation

Hydrological Models

Hydrological models are effective tools to simulate local water resources, hydrological processes, and crop growth at sub-basin/watershed level. Simulating cropland water balance is possible under different crop growth and water management with evapotranspiration, surface runoff, infiltration, percolation, shallow, and deep aquifer flow (Arnold et al. 1998). Major hydrological models include the Soil and Water Assessment Tool (SWAT) (Arnold et al. 1998), Variable Infiltration Capacity model (VIC) (Wood et al. 1992), and Systeme Hydrologique European (SHE) model (Abbott et al. 1986). These models provide a better understanding of the climate variability on hydrological processes and water resources for agricultural production (Sun and Li 2013; Wang et al. 2016). Advantages of the hydrological models are applied in the watershed level to optimize crop irrigation schedules for different crops (Sun and Li 2014) and identify the groundwater depletion from cropland irrigation expansion (Zhang et al. 2016). But the relatively simplified crop growth processes compared with dynamic crop models make the crop module within the macro-scale hydrological models focus more on the change of hydrological resources under given local water management conditions, rather than design the water-saving irrigation technologies or schedules (Xiong et al. 2019). In fact, the optimal crop irrigation schedule for water saving using hydrological model is obtained in tandem with field-based experiments (Yu et al. 2006).

Dynamic Crop Models

Process-based dynamic crop models have great advantages in simulating the interaction between crop growth and the sounding environment at a daily time step. Those models are widely used to determine optimal crop water management. Major crop models include Decision Support System for Agrotechnology Transfer (DSSAT) (Jones et al. 2003), Environmental Policy Integrated Climate (EPIC) (Williams et al. 1989), Agricultural Production Systems sIMulator (APSIM) (McCown et al. 1996), and AquaCrop (Steduto et al. 2009). The DSSAT model was employed to identify the long-term relationship between crop irrigation water consumption and groundwater level depletion in the Piedmont of NCP (Yang et al. 2006). For the regional agricultural water consumption simulation, a spatial parameterization by integrating the agroecological zones with site field information of all the sites was developed for the DSSAT model upscaling (Tian et al. 2012) and applied to simulate crop water consumption under different cropping systems in the NCP (Zhong et al. 2017, 2019; Tian et al. 2018).

Several limitations of the dynamic crop models need to be mentioned: (1) Crop models require heavy inputs for model calibration, validation, and simulation, but detailed information that will affect the

TABLE 1 Models for Agricultural Water Consumption Estimation

Model Types	Model Name	Description	Model References
Hydrological Models	SWAT	The SWAT model is developed to assess the impact of management on water supplies and nonpoint source pollution, and predict the impact of management on water, sediment, and agricultural chemical yields in watersheds.	Arnold et al. (1998)
	SHE/MIKE SHE	The SHE/MIKE SHE is a differential model to simulate the surface water movement from the hydrological processes of mass, energy, and momentum in a river basin.	Abbott et al. (1986)
	VIC	The VIC model simulates land-surface hydrology from infiltration variation, with a simplified approach of estimating infiltration, evaporation, and a base flow.	Wood et al. (1992)
Dynamic crop models	DSSAT	The DSSAT model is a modular structure with separate primary scientific components (soil, crop, weather, and management) and various process-based models that can be added or replaced.	Jones et al. (2003)
	EPIC	The EPIC model is developed to estimate the soil productivity as affected by soil erosion, plant growth, and related processes.	Williams et al. (1989)
	APSIM	The APSIM model focuses more on the "system" aspect of cropping, such as crop growth, rotation, fallow, and dynamic crop management to the soil processes and climate conditions.	McCown et al. (1996)
	AquaCrop	The AquaCrop model is a water-driven process model that focuses on the differential sensitivity to water stress of key plant processes and the biomass water productivity during the crop growth cycle.	Steduto et al. (2009)

total water consumption is lacking at the regional scale, especially the field water management. As most upscaling methods depend on observations from experiment sites, agricultural water consumption will be overestimated as field management levels in those sites are much higher than the farmer's practice (Tian et al. 2012). (2) Because process-based crop models mainly focus on soil water balance in the root zone during the crop growing period, hydrological processes, such as additional water sources from the surface stream, other water bodies, and underground water flow are not included. For example, the mountain-front discharge in the Piedmont of the Hebei Plain is an important groundwater recharge source but only considered in site experiment with field observations (Oort et al. 2016). Groundwater recharge from precipitation and irrigation water deep percolation are usually simplified or ignored (Yang et al. 2015) (Table 1).

Hybrid Models

To overcome the constraint of the process-based models and hydrological models in optimizing agricultural water management, there is a growing trend of integrating/coupling modeling studies. For example, there are case studies in NCP to improve the groundwater depletion simulation by coupling the crop model (DSSAT) with hydrology model (SWAT) and groundwater flow model (MODFLOW) (Hu et al. 2010). The Catchment-based Ecohydrology model was integrated with the crop model (DSSAT) to simulate the impact of agricultural water consumption on groundwater resources (Nakayama et al. 2006). In addition, there is a great potential of applying these model-based studies to identify the sustainable water management for different crops and various cropping systems under future climate projections, which will greatly support the agricultural adaptation decisions for the local farmers, stakeholders, and government.

Environmental Extended Multi-Region Input–Output Model

In addition to the water resources and direct consumption simulated by physical models, the indirect virtual water embodied in interregional food trade is also estimated by using the environmental extended multi-region input–output model (Wiedmann 2009). The virtual water or water footprint represents both the direct and indirect volumes of freshwater consumption to produce goods and services which are consumed by social and economic activities (Hubacek et al. 2009). The agricultural virtual water, which includes irrigation water as blue water, rainwater as green water, and wastewater as gray water, was introduced to identify the embodied water import and export from interregional trade and to design alternative water management strategies (Chapagain and Hoekstra 2004). A more comprehensive understanding of the direct and indirect water consumption, especially the groundwater depletion embodied food trade, will support local agricultural adaptations and reallocation/alleviation of water stress in water-deficit regions via food and product trade (Dalin et al. 2017). For example, by tracking "virtual water" embodied in the global and regional food trade, researchers showed that a great amount of food is exported from the water-deficit arid and semi-arid regions at the expense of greater groundwater depletion, and it is not reasonable for those regions to continue their role as "breadbasket" for other regions in the world (Dalin et al. 2017; Ren et al. 2018).

Water-Sustainable Agricultural Adaptations

As the biggest water consumption sector, agriculture plays a critical role in local water management. Reducing irrigation water and increasing water-use efficiency are critical to the environment and ecosystem sustainability, especially in the major cropping regions under arid and semi-arid climate.

Field Water-Saving Irrigation Technologies

Directly reducing agricultural irrigation is the most prevalent measure to save water. Field experiments have been designed and validated to identify the optimal water management for different crops in the arid and semi-arid regions. There are two major field water-saving practices: (1) water-saving irrigation and optimal irrigation schedule, and (2) plastic/straw mulching and no tillage.

Various water-saving irrigation technologies were developed to reduce the frequency and amount of irrigation. Compared with traditional flood or sprinkler irrigation, the dripping irrigation will significantly increase the water-use efficiency by the crops with little "waste" water from surface runoff and percolation (Alcon and Burton 2011; Ayars et al. 2015). More aggressive deficit irrigation with the tradeoffs of crop yield loss was also adopted in the arid regions (Geerts et al. 2010; Al-Ghobari and Dewidar 2018). Optimizing the irrigation schedule with lower frequency and a focus on the critical crop growth stages is another effective practice (Jiang et al. 2016). Experiments showed that optimal irrigation could reduce irrigation water by about 50 mm compared with local farmer's current practices in the NCP (Sun et al. 2011). Maize irrigation amount could reduce up to 50% in the Yingke irrigation district in Northwest China under the optimal irrigation schedule (Jiang et al. 2016). However, water conservation subsidies for water-efficient irrigation may consume even more water as groundwater recharge reduced while more water was extracted to meet the demand of water-intensive crops, more comprehensive social and economic policies are needed to achieve groundwater conservation instead of solely focusing on water-saving technologies (Ward and Pulido-Velazquez 2008).

Plastic mulching during the crop seedling stage is an effective way to reduce irrigation by decreasing soil evaporation and conserve soil moisture during the dry spring in Northwest China (Ran et al. 2018). Because of the negative environmental impact of plastic film, a more environment-friendly straw residual mulching and no tillage practice was developed: winter wheat straw residuals were left after harvest and no tillage was applied for soil moisture conservation before the following summer crop sowing (Tan et al. 2017). Experiments in the Luancheng Station of the NCP showed that soil evaporation could be reduced up to 40% with higher water-use efficiency of maize (Chen et al. 2007).

Cropping System Adaptations

However, no matter how much efforts were put on reducing irrigation water, the groundwater still dropped in the NCP, even when reducing the irrigation frequency to one time per crop growth cycle (deficit irrigation at the expense of limit crop yield reduction) (Sun et al. 2015). This is because the water requirements of the existing intensive wheat–maize double cropping systems are far more than local water supply. Agricultural deintensification and water-sustainable cropping systems become a feasible alternative (Zhong et al. 2017).

Many groundwater neutral (no groundwater table drop) cropping system with less water demand crops or cropland fallow were introduced to further reduce irrigation. For example, because of the dry winter and spring in NCP, winter wheat required much more irrigation water than other crops. Many alternative cropping system with partially or completely wheat fallow were introduced to replace the dominant wheat–maize cropping system in this region, including single cropping of spring maize (Pei et al. 2015), double cropping per year of early maize and late maize (Meng et al. 2017), and triple harvests in 2 years of WM-S followed by spring maize (Meng et al. 2012), and strip–relay intercropping of wheat followed by spring maize (Gao et al. 2009). A slow adaptation of wheat planting area shifting has happened at the farmer's initiative in the NCP. A "spring maize belt" was established in the driest regions of the NCP. Farmers had already abandoned winter wheat and replaced the wheat–maize double cropping system with spring maize single cropping (Feng et al. 2007; Wang et al. 2016); and, a north–south shift of winter wheat planting area occurred where more wheat was planted in the wetter part of NCP (Wang et al. 2015). Inevitably, the regional total grain production will suffer a loss when less intensive cropping systems are adopted for groundwater recovery. The current situation of groundwater over-exploitation should come to an end before there is "no water" to use when extreme drought happens and a great food production failure under projected warmer and drier climate in NCP (Oort et al. 2016).

Fortunately, policies have been made and implemented in the groundwater over-exploited NCP, where a pilot project has started to leave cropland area fallow. Ecological compensation was also estimated for winter wheat fallow in NCP (Wang et al. 2016). Investments to improve the cropland irrigation infrastructure, such as concrete canals to reduce leakage of water when transferring, are also needed to enhance the water saving and groundwater recovery.

Physical Water Transfer or Virtual Water Flow

Freshwater Diversion Projects

The uneven distribution of freshwater resource (e.g., the arid and semi-arid NCP produces 1/2 and 1/3 of the national total wheat and maize production, respectively, while south China has much more freshwater resources) and the soaring water demand from rapid urbanization and industrialization have driven many massive surface water diversion projects (Aeschbach-Hertig and Gleeson 2012; Zhao et al. 2015). The biggest one is the South-North Water Transfer (SNWT) project, with a designed water delivery capacity of 20 billion m³ water from the western route, 9.5 billion m³ water from the central route, and 14.8 billion m³ water from the eastern route (Liu et al. 2013).

This SNWT project has also raised great controversy because of its huge investment and cost for maintenance/operation every year and the environmental impact to the downstream (Barnett et al. 2015). Moreover, studies also pointed out that even with the supplementary water from SNWT project, the groundwater table level would not stop declining in Shijiazhuang city in North China (Shu et al. 2012).

Adaptations, such as improving agricultural water-use efficiency and reducing crop irrigation, increasing urban water supply from rainwater harvesting and wastewater recycling, should play a more important role for the water sustainability in those regions (Barnett et al. 2015).

Virtual Water Flows via Food Trade

Compared with physical water transfer, the virtual water flow as embodied water in the interregional food and production trade is much bigger. Physical water flows via the major water transfer projects amounted to 4.5% of national water supply, whereas virtual water flows accounted for 35% in 2007 (Zhao et al. 2015). This suggests the demand-oriented water management should be highlighted rather than solely focusing on physical water saving, especially reducing the export of water-intensive products from the water-deficit regions. For example, the physically transferred freshwater was less than the embodied blue water export from Hebei Province to Beijing and Tianjin and had worsened the local groundwater depletion. Moreover, export from Hebei province with much less water-use efficiency to the higher water-use efficient Beijing and Tianjin was also not reasonable (Zhao et al. 2017). Case studies in the NCP indicated that reducing the export from low water-use efficiency and high-water-stress regions will be more effective than transfer freshwater via huge water diversion projects, despite the negative sides of huge investment, cost of maintain and migration, and environment impact to the downstream (Zhang 2009; He et al. 2010). Therefore, reducing the heavy-water-intensity products' export, increasing water-use efficiency, and giving up the role of "breadbasket" for Beijing and Tianjin should be the next step in Hebei Province (Ren et al. 2018). Appropriate subsidies to compensate the local farmers' income loss are critical to the success of cropland fallow in the NCP (Wang et al. 2016), and 87% of the local farmers would adopt the winter wheat fallow policy (Xie et al. 2017).

Conclusions

Soaring global water consumption from agriculture, urbanization, and industrialization, which is expected to grow up to 21% in 2050 (Rosegrant et al. 2009), and the unsustainable use of groundwater have caused severe environmental issues in many arid and semi-arid regions. The potential water crisis has raised great concern from the public and government. As the biggest water consumption sector, improving agricultural water-use efficiency and reducing crop irrigation water consumption is critical to global water sustainability.

Various models, agricultural water-saving adaptations, and water alleviation strategies were developed and tested at the field, regional, and global levels. Many water-saving irrigation technologies showed the potential of reducing crop irrigation and maintaining high level of food production. Initial plans and policies had already taken to slow or stop the drop in the groundwater drop in the over-exploited regions. On the other hand, instead of only focusing on improving field water-saving technologies, water-sustainable cropping systems adaptations, and supplementary freshwater transfer projects to alleviate the water deficit, it is also very important to increase the overall water-use efficiency and reduce the water footprint in all social-economic activities. "Virtual water" flow and economic cost-effective approaches should also be taken into consideration when deciding the local water diversion projects. In addition, a broader view of the water management should include social-economic water demand/consumption and local water resource. The Global Hydro-economic Model (Wada et al. 2016) can provide a feasible solution for the water management under different future climate projections and social-economic development scenarios.

Nevertheless, local water sustainability needs joint efforts from local farmers, government, and even global corporations. Future risks from projected climate change, growing population, and urbanization, shifting to more water-intensive lifestyle in many developing countries, still put global water shortage in an alarming situation, and continued efforts to improve the water management are required for sustainable development.

References

Abbott M., J. Bathurst, J. Cunge, P. O'Connell, and J. Rasmussen. 1986. "An Introduction to the European Hydrological System — Systeme Hydrologique Europeen, 'SHE', 1: History and Philosophy of a Physically-Based, Distributed Modelling System." *Journal of Hydrology* 87 (1): 45–59.

Aeschbach-Hertig Werner and Tom Gleeson. 2012. "Regional Strategies for the Accelerating Global Problem of Groundwater Depletion." *Nature Geoscience* 5: 853.

Alcon Francisco María Dolores de Miguel and Michael Burton. 2011. "Duration Analysis of Adoption of Drip Irrigation Technology in Southeastern Spain." *Technological Forecasting and Social Change* 78 (6): 991–1001.

Al-Ghobari Hussein M. and Ahmed Z. Dewidar. 2018. "Integrating Deficit Irrigation into Surface and Subsurface Drip Irrigation as a Strategy to Save Water in Arid Regions." *Agricultural Water Management* 209: 55–61.

Arnold J., R. Srinivasan, R.S. Muttiah, and J.R. Williams. 1998. "Large Area Hydrologic Modeling and Assessment Part I: Model Development." *JAWRA: Journal of the American Water Resources Association* 34 (1): 73–89.

Ayars J., A. Fulton, and B. Taylor. 2015. "Subsurface Drip Irrigation in California—Here to Stay?" *Agricultural Water Management* 157: 39–47.

Barnett Jon, Sarah Rogers, Michael Webber, Brian Finlayson, and Mark Wang. 2015. "Sustainability: Transfer Project Cannot Meet China's Water Needs." *Nature* 527 (7578): 295–97.

Chapagain A. and A. Hoekstra. 2004. *Water Footprints of Nations.* Value of Water Research Report Series BT – Water Footprints of Nations. Delft: Unesco-IHE Institute for Water Education.

Chen S., X. Zhang, D. Pei, H. Sun, and S. Chen. 2007. "Effects of Straw Mulching on Soil Temperature, Evaporation and Yield of Winter Wheat: Field Experiments on the North China Plain." *Annals of Applied Biology* 150 (3): 261–68.

Currell Matthew J., Dongmei Han, Zongyu Chen, and Ian Cartwright. 2012. "Sustainability of Groundwater Usage in Northern China: Dependence on Palaeowaters and Effects on Water Quality, Quantity and Ecosystem Health." *Hydrological Processes* 26 (26): 4050–66.

Dalin Carole, Yoshihide Wada, Thomas Kastner, and Michael J. Puma. 2017. "Groundwater Depletion Embedded in International Food Trade." *Nature* 543: 700.

Feng Zhiming, Dengwei Liu, and Yuehong Zhang. 2007. "Water Requirements and Irrigation Scheduling of Spring Maize Using GIS and CropWat Model in Beijing-Tianjin-Hebei Region." *Chinese Geographical Science* 17 (1): 56–63.

Gao Yang, Aiwang Duan, Jingsheng Sun, Fusheng Li, Zugui Liu, Hao Liu, and Zhandong Liu. 2009. "Crop Coefficient and Water-Use Efficiency of Winter Wheat/Spring Maize Strip Intercropping." *Field Crops Research* 111 (1): 65–73.

Geerts S., D. Raes, and M. Garcia. 2010. "Using AquaCrop to Derive Deficit Irrigation Schedules." *Agricultural Water Management* 98 (1): 213–16.

He Chansheng, Xiaoying He, and Li Fu. 2010. "China's South-to-North Water Transfer Project: Is It Needed?" *Geography Compass* 4 (9): 1312–23.

Hu Yukun, Juana Paul Moiwo, Yonghui Yang, Shumin Han, and Yanmin Yang. 2010. "Agricultural Water-Saving and Sustainable Groundwater Management in Shijiazhuang Irrigation District, North China Plain." *Journal of Hydrology* 393 (3): 219–32.

Hubacek Klaus, Dabo Guan, John Barrett, and Thomas Wiedmann. 2009. "Environmental Implications of Urbanization and Lifestyle Change in China: Ecological and Water Footprints." *Journal of Cleaner Production* 17 (14): 1241–48.

Jiang Yiwen, Lanhui Zhang, Baoqing Zhang, Chansheng He, Xin Jin, and Xiao Bai. 2016. "Modeling Irrigation Management for Water Conservation by DSSAT-Maize Model in Arid Northwestern China." *Agricultural Water Management* 177: 37–45.

Jones J., G. Hoogenboom, C. Porter, K. Boote, W. Batchelor, L. Hunt, P. Wilkens, U. Singh, A. Gijsman, and J. Ritchie. 2003. "The DSSAT Cropping System Model." *European Journal of Agronomy* 18 (3): 235–65.

Kang Suchul and Elfatih A. B. Eltahir. 2018. "North China Plain Threatened by Deadly Heatwaves Due to Climate Change and Irrigation." *Nature Communications* 9 (1): 2894.

Konikow Leonard and Eloise Kendy. 2005. "Groundwater Depletion: A Global Problem." *Hydrogeology Journal* 13 (1): 317–20.

Liu Junguo, Chuanfu Zang, Shiying Tian, Jianguo Liu, Hong Yang, Shaofeng Jia, Liangzhi You, Bo Liu, and Miao Zhang. 2013. "Water Conservancy Projects in China: Achievements, Challenges and Way Forward." *Global Environmental Change* 23 (3): 633–43.

McCown R., G. Hammer, J. Hargreaves, D. Holzworth, and D. Freebairn. 1996. "APSIM: A Novel Software System for Model Development, Model Testing and Simulation in Agricultural Systems Research." *Agricultural Systems* 50 (3): 255–71.

Meng Qingfeng, Hongfei Wang, Peng Yan, Junxiao Pan, Dianjun Lu, Zhenling Cui, Fusuo Zhang, and Xinping Chen. 2017. "Designing a New Cropping System for High Productivity and Sustainable Water Usage under Climate Change." *Scientific Reports* 7: 41587.

Meng Qingfeng, Qinping Sun, Xinping Chen, Zhenling Cui, Shanchao Yue, Fusuo Zhang, and Volker Römheld. 2012. "Alternative Cropping Systems for Sustainable Water and Nitrogen Use in the North China Plain." *Agriculture, Ecosystems & Environment* 146 (1): 93–102.

Molden David. 2007. *Water for Food, Water for Life: A Comprehensive Assessment of Water Management in Agriculture*. London, Earthscane, Colombo: International Water Management Institute.

Nakayama Tadanobu, Yonghui Yang, Masataka Watanabe, and Xiying Zhang. 2006. "Simulation of Groundwater Dynamics in the North China Plain by Coupled Hydrology and Agricultural Models." *Hydrological Processes* 20 (16): 3441–66.

Oort P.A.J van, G. Wang, J. Vos, H. Meinke, B. Li, J. Huang, and W. van der Werf. 2016. "Towards Groundwater Neutral Cropping Systems in the Alluvial Fans of the North China Plain." *Agricultural Water Management* 165: 131–40.

Pei Hongwei, Bridget R. Scanlon, Yanjun Shen, Robert C. Reedy, Di Long, and Changming Liu. 2015. "Impacts of Varying Agricultural Intensification on Crop Yield and Groundwater Resources: Comparison of the North China Plain and US High Plains." *Environmental Research Letters* 10 (4): 44013.

Piao Shilong, Philippe Ciais, Yao Huang, Zehao Shen, Shushi Peng, Junsheng Li, Liping Zhou, et al. 2010. "The Impacts of Climate Change on Water Resources and Agriculture in China." *Nature* 467: 43.

Ran Hui, Shaozhong Kang, Fusheng Li, Taisheng Du, Ling Tong, Sien Li, Risheng Ding, and Xiaotao Zhang. 2018. "Parameterization of the AquaCrop Model for Full and Deficit Irrigated Maize for Seed Production in Arid Northwest China." *Agricultural Water Management* 203: 438–50.

Ren Dandan, Yonghui Yang, Yanmin Yang, Keith Richards, and Xinyao Zhou. 2018. "Land-Water-Food Nexus and Indications of Crop Adjustment for Water Shortage Solution." *Science of The Total Environment* 626: 11–21.

Rosegrant Mark, Claudia Ringler, and Tingju Zhu. 2009. "Water for Agriculture: Maintaining Food Security under Growing Scarcity." *Annual Review of Environment and Resources* 34 (1): 205–22.

Shu Yunqiao, Karen G. Villholth, Karsten H. Jensen, Simon Stisen, and Yuping Lei. 2012. "Integrated Hydrological Modeling of the North China Plain: Options for Sustainable Groundwater Use in the Alluvial Plain of Mt. Taihang." *Journal of Hydrology* 464–465: 79–93.

Steduto Pasquale, Theodore C. Hsiao, Dirk Raes, and Elias Fereres. 2009. "AquaCrop-The FAO Crop Model to Simulate Yield Response to Water: I. Concepts and Underlying Principles." *Agronomy Journal* 101 (3): 426–37.

Sun Chen and L. Ren. 2013. "Assessment of Surface Water Resources and Evapotranspiration in the Haihe River Basin of China Using SWAT Model." *Hydrological Processes* 27 (8): 1200–222.

Sun Chen and L. Ren. 2014. "Assessing Crop Yield and Crop Water Productivity and Optimizing Irrigation Scheduling of Winter Wheat and Summer Maize in the Haihe Plain Using SWAT Model." *Hydrological Processes* 28 (4): 2478–98.

Sun Hongyong, Xiying Zhang, Enli Wang, Suying Chen, and Liwei Shao. 2015. "Quantifying the Impact of Irrigation on Groundwater Reserve and Crop Production – A Case Study in the North China Plain." *European Journal of Agronomy* 70: 48–56.

Sun Qinping, Roland Kröbel, Torsten Müller, Volker Römheld, Zhenling Cui, Fusuo Zhang, and Xinping Chen. 2011. "Optimization of Yield and Water-Use of Different Cropping Systems for Sustainable Groundwater Use in North China Plain." *Agricultural Water Management* 98 (5): 808–14.

Tan Shuai, Quanjiu Wang, Di Xu, Jihong Zhang, and Yuyang Shan. 2017. "Evaluating Effects of Four Controlling Methods in Bare Strips on Soil Temperature, Water, and Salt Accumulation under Film-Mulched Drip Irrigation." *Field Crops Research* 214: 350–58.

Tian Zhan, Honglin Zhong, Runhe Shi, Laixiang Sun, Gunther Fischer, and Zhuoran Liang. 2012. "Estimating Potential Yield of Wheat Production in China Based on Cross-Scale Data-Model Fusion." *Frontiers of Earth Science* 6 (4): 388–95.

Tian Zhan, Yilong Niu, Dongli Fan, Laixiang Sun, Gunther Ficsher, Honglin Zhong, Jia Deng, and Francesco N. Tubiello. 2018. "Maintaining Rice Production While Mitigating Methane and Nitrous Oxide Emissions from Paddy Fields in China: Evaluating Tradeoffs by Using Coupled Agricultural Systems Models." *Agricultural Systems* 159: 175–86.

Vörösmarty Charles, Pamela Green, Joseph Salisbury, and Richard B. Lammers. 2000. "Global Water Resources: Vulnerability from Climate Change and Population Growth." *Science* 289 (5477): 284–88.

Wada Y., M. Flörke, N. Hanasaki, S. Eisner, G. Fischer, S. Tramberend, Y. Satoh, M. van Vliet, P. Yillia, C. Ringler, P. Burek, and D. Wiberg. 2016. Modeling Global Water Use for the 21st Century: Water Futures and Solutions (WFaS) Initiative and Its Approaches. *Geoscientific Model Development* 9: 175–222.

Wang Ruoyu, Laura C. Bowling, and Keith A. Cherkauer. 2016a. "Estimation of the Effects of Climate Variability on Crop Yield in the Midwest USA." *Agricultural and Forest Meteorology* 216: 141–56.

Wang Xue, Xiubin Li, Günther Fischer, Laixiang Sun, Minghong Tan, Liangjie Xin, and Zhuoran Liang. 2015. "Impact of the Changing Area Sown to Winter Wheat on Crop Water Footprint in the North China Plain." *Ecological Indicators* 57: 100–109.

Wang Xue, Xiubin Li, Liangjie Xin, Minghong Tan, Shengfa Li, and Renjing Wang. 2016b. "Ecological Compensation for Winter Wheat Abandonment in Groundwater Over-Exploited Areas in the North China Plain." *Journal of Geographical Sciences* 26 (10): 1463–76.

Ward Frank A. and Manuel Pulido-Velazquez. 2008. "Water Conservation in Irrigation Can Increase Water Use." *Proceedings of the National Academy of Sciences* 105 (47): 18215–20.

Wiedmann Thomas. 2009. "A Review of Recent Multi-Region Input–Output Models Used for Consumption-Based Emission and Resource Accounting." *Ecological Economics* 69 (2): 211–22.

Williams J., C. Jones, J. Kiniry, and D. Spanel. 1989. "The EPIC Crop Growth Model." *Transactions of the ASAE* 32 (2): 497–511.

Wood E., D. Lettenmaier, and V. Zartarian. 1992. "A Land-Surface Hydrology Parameterization with Subgrid Variability for General Circulation Models." *Journal of Geophysical Research - Atmospheres* 97 (D3): 2717–28.

Xie Hualin, Lingjuan Cheng, and Tiangui Lv. 2017. "Factors Influencing Farmer Willingness to Fallow Winter Wheat and Ecological Compensation Standards in a Groundwater Funnel Area in Hengshui, Hebei Province, China". *Sustainability* 9 (5): 839.

Xiong Lvyang, Xu Xu, Dongyang Ren, Quanzhong Huang, and Guanhua Huang. 2019. "Enhancing the Capability of Hydrological Models to Simulate the Regional Agro-Hydrological Processes in Watersheds with Shallow Groundwater: Based on the SWAT Framework." *Journal of Hydrology* 572: 1–16.

Xue Yuqun, Jichun Wu, Shujun Ye, and Yongxiang Zhang. 2000. "Hydrogeological and Hydrogeochemical Studies for Salt Water Intrusion on the South Coast of Laizhou Bay, China." *Groundwater* 38 (1): 38–45.

Yang Xiaolin, Yuanquan Chen, Steven Pacenka, Wangsheng Gao, Min Zhang, Peng Sui, and Tammo S. Steenhuis. 2015. "Recharge and Groundwater Use in the North China Plain for Six Irrigated Crops for an Eleven Year Period." *PLOS ONE* 10 (1): e0115269.

Yang Yonghui, Masataka Watanabe, Xiying Zhang, Xiaohua Hao, and Jiqun Zhang. 2006. "Estimation of Groundwater Use by Crop Production Simulated by DSSAT-Wheat and DSSAT-Maize Models in the Piedmont Region of the North China Plain." *Hydrological Processes* 20 (13): 2787–2802.

Yu Q., S. Saseendran, L. Ma, G. Flerchinger, T. Green, and L. Ahuja. 2006. "Modeling a Wheat–Maize Double Cropping System in China Using Two Plant Growth Modules in RZWQM." *Agricultural Systems* 89 (2–3): 457–77.

Zhang Quanfa. 2009. "The South-to-North Water Transfer Project of China: Environmental Implications and Monitoring Strategy1." *JAWRA Journal of the American Water Resources Association* 45 (5): 1238–47.

Zhang Xueliang, Li Ren, and Xiangbin Kong. 2016. "Estimating Spatiotemporal Variability and Sustainability of Shallow Groundwater in a Well-Irrigated Plain of the Haihe River Basin Using SWAT Model." *Journal of Hydrology* 541: 1221–40.

Zhao Dandan, Yu Tang, Junguo Liu, and Martin R. Tillotson. 2017. "Water Footprint of Jing-Jin-Ji Urban Agglomeration in China." *Journal of Cleaner Production* 167: 919–28.

Zhao Xu, Junguo Liu, Qingying Liu, Martin R. Tillotson, Dabo Guan, and Klaus Hubacek. 2015. "Physical and Virtual Water Transfers for Regional Water Stress Alleviation in China." *Proceedings of the National Academy of Sciences* 112(4): 1031–35.

Zhong Honglin, Laixiang Sun, Günther Fischer, Zhan Tian, and Zhuoran Liang. 2019. "Optimizing Regional Cropping Systems with a Dynamic Adaptation Strategy for Water Sustainable Agriculture in the Hebei Plain." *Agricultural Systems* 173 (July): 94–106.

Zhong Honglin, Laixiang Sun, Gunther Fischer, Zhan Tian, Harrij van Velthuizen, and Zhuoran Liang. 2017. "Mission Impossible? Maintaining Regional Grain Production Level and Recovering Local Groundwater Table by Cropping System Adaptation across the North China Plain." *Agricultural Water Management* 193: 1–12.

21

Runoff Water

Introduction 195
Modification of the Stream of Wet Deposition following Contact
with Substances in/on the Substrate 195
Road Runoff 196
Roof Runoff..200
Throughfall...203
    Runoff from Farming Areas • Runoff from Airports
Changes in the Composition of Samples Depending on the
Moment of Sampling during Precipitation or Runoff..............................207
Summary 208
References..208

Zaneta Polkowska

Introduction

The composition of the stream of wet deposition changes significantly on contact with the substrate, be this natural or artificial. Following contact with the surfaces of artificial objects (e.g., roofs, roads, airport aprons and runways, steel and timber structures, communal and industrial landfill, cultivated fields) or natural surfaces (e.g., trees, shrubs, conifer needles, leaves, grass, vegetated fallow land), precipitation and atmospheric deposits (rain, dew, fog, melting snow, hoar frost, or rime), flow over them (runoff) or, trickling through the soil, replenish groundwaters (Figure 1). The extent of surface runoff depends on the properties of the soil and the degree to which it is covered by plants—the lusher the vegetation, the less runoff there is. Certain properties of the soil, such as minimal adsorption or good permeability, promote the vertical movement of water in the soil, enabling it to reach the groundwater; the intensity of surface runoff is therefore less. Conversely, if the properties of a surface favor runoff (e.g., heavy, impermeable soils, asphalt, or concrete surfaces), pollutants are transported across the land surface.[1,2]

Pollutants in runoff come from three main sources: dry deposition (from the atmosphere), wet deposition (precipitation and atmospheric deposits), and flushing from the surface of the material with which precipitation and deposits come into contact (e.g., roof coverings, the crowns and trunks of trees, cultivated fields, roads and railway lines, landfill sites).[3]

Modification of the Stream of Wet Deposition following Contact with Substances in/on the Substrate

The composition of atmospheric water changes radically on contact with the Earth's surface. The composition of runoff depends to a large extent on the type of surface (including that of the land) in which precipitation falls on. Runoff from fields may contain pesticides and nutrients. Urban areas may be the source of many different substances, such as aliphatic and aromatic hydrocarbons, Polycyclic aromatic hydrocarbons (PAHs), fatty acids, ketones, and heavy metals.

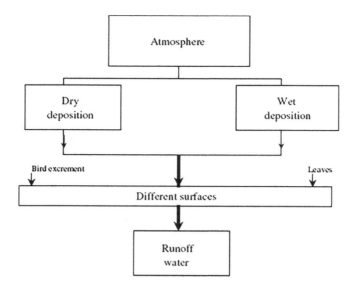

FIGURE 1 Modifications to the flow of wet deposition obstructed by different surfaces (trees, lawns, roofs, roads, motor vehicles, airport aprons and runways, steel and timber structures, communal and industrial landfill sites).

The composition of precipitation in wooded areas changes mainly under the influence of the physiological processes taking place in plants and the dissolution or flushing of atmospheric aerosols and gaseous contaminants from the surfaces of branches, bark, and leaves.[4–15]

Runoff from urban areas may also contain petroleum hydrocarbons. Road runoff contains dusts derived from the wear and tear of asphalt road surfaces and rubber tires.[16–48] The material from which a roof covering or guttering is made is a potential source of pollution taken up by water falling onto these surfaces. The compounds leached out of these materials react with the contaminants in the rainwater, changing its composition and properties. Roof runoff usually contains heavy metals and pesticides.[1,43,49–70] On the other hand, runoff from landfill sites and refuse tips are variously aggressive towards the environment, depending on the state of degradation of the refuse. On this, however, the literature supplies no data. A considerable number of papers have been published on runoff from different surfaces, although there is hardly any information on the concentrations of the constituents of runoff samples collected in Poland; what little there is refers to runoff from the crowns and trunks of trees.[5,15]

Road Runoff

Waters running off motorways contain heavy metals in two forms: suspended solid particles and dissolved forms. Figure 2 gives the results of determinations of different heavy metal ions in samples of road runoff from Europe and the United States. Similar studies were carried out in New Zealand. The results of these studies were the basis for concluding that the content of lead ions in plants (growing near roads) depends significantly on how far from the roadside they are growing.[47]

Road runoff is to a large extent contaminated by organic compounds like petroleum constituents, fuels, fats, and greases. Studies done in Madrid have shown that there is a strict relationship between the intensity of urban road traffic and the levels of PAHs and petroleum hydrocarbons in urban road runoff.[17,19] Vehicle emissions—residues of fuel and its incomplete combustion—are the chief source of these compounds in runoff.[48] Figure 3 illustrates the results of determinations of various PAHs in runoff samples collected in Europe and the United States. The concentrations were highest in the runoff samples collected along roads with a high intensity of traffic (Spain, Poland).

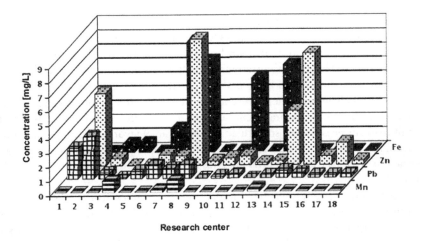

FIGURE 2   Concentrations of iron, zinc, lead, and manganese ions determined in samples of runoff water in Europe and the United States (literature data).

1 - USA [rural area - Alabama] [34]	*10 - Denmark [motorway] [71]*
2 - Switzerland [urban area - Zürich] [61]	*11 - Switzerland [motorway] [22]*
3 - England [motorway] [35]	*12 - Switzerland [1,600 vehicles / h] [41]*
4 - England [motorway] [36]	*13 - West Germany [urban area - Hamburg] [42]*
5 - England [2,000 vehicles / h] [37]	*14 - France [urban area - Paris] [43]*
6 - Norway [motorway] [21]	*15 - USA [ca 5,000 vehicles / h] [20]*
7 - USA [ca 1,000 vehicles / h] [38]	*16 - France [2,000 vehicles / day] [44]*
8 - USA [ca 7,000 vehicles / h] [39]	*17 - France [800 vehicles / day] [29]*
9 - Belgium [urban area – Brussels] [40]	*18 - France [urban area] [30]*

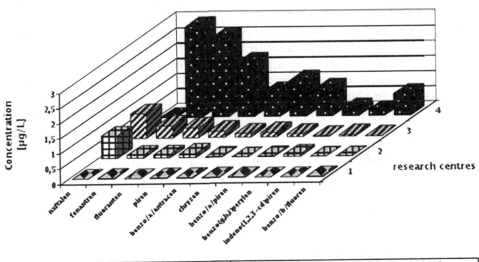

1 - England [1,600 vehicles / h] [72]	*3 - Poland [urban area – Tri-City] [73]*
2 - Hungary [urban area - Budapest] [17]	*4 - Spain 1988 [11,500 vehicles / h] [19]*

FIGURE 3   PAH content in samples of runoff water collected in England, Hungary, Poland, and Spain.

Studies regarding the presence of volatile organochlorine compounds (VOCCs) in road runoff were carried out in Poland (Tri-City: Gdansk, Sopot, Gdynia). VOCCs were found in every single sample. The most common VOCC analytes were chloroform, dichlorobromomethane, and trichloroethene. Table 1 lists the ranges of concentrations and mean concentrations of analytes in road runoff samples.[74]

Road runoff samples from Poland have also been analyzed for the presence of PAHs. The highest levels were recorded for naphthalene and phenanthrene+anthracene, while fluoranthene, pyrene, benzo(*a*)anthracene+chrysene, benzo(*b+k*)fluoranthene, and benzo(*a*)pyrene were also quite frequent. The properties of asphalt road surfaces (rather smooth and impermeable) favor runoff; thus, contaminants derived mainly from road vehicles (exhaust fumes, wear and tear of asphalt, and rubber tires) are transported across the land surface. The levels of PAH analytes rise conspicuously in the winter months (November to January).[75,76]

The organochlorine pesticides usually detected in road runoff samples from Poland include heptachlor epoxide (a product of the metabolization of heptachlor, which is synthesized in plants, the soil, and homeothermic organisms) and *o,p'*-DDE (a derivative of DDT, an insecticide that used to be widely applied). The most common of the organonitrogen and organophosphorus pesticides to be detected in road runoff are propazine and bromofos (selective weed killers). Table 2 lists the mean concentrations and ranges of concentrations of pesticides determined in road runoff samples from the Tri-City area of Poland.[74,77]

The most commonly detected petroleum hydrocarbons in runoff samples from Poland were benzene and toluene, whereby the former compound was present in every sample analyzed. Weather conditions and the street layouts of towns and cities, i.e., the speed and direction of the wind and the proximity of city buildings to one another, are important factors affecting the spread of these pollutants. Table 3 lists the maximum, minimum, and mean concentrations of analytes found in samples; the number of samples in which an analyte was detected is also given.[74,76,77]

Formaldehyde was detected in all samples of road runoff from Poland (from 0.235 to 0.433 mg/L). Figure 4 compares the levels of formaldehyde in samples of rainwater and runoff from roads in the Gdansk area and from roadside sampling points in Poland.

The formaldehyde content in road runoff is high as it is a typical pollutant produced by road vehicles. It is formed as a result of the partial oxidation of fuel in the cylinder due to the chain reactions oxidizing fuel being interrupted by the "extinguishing" effect of the fairly cool cylinder walls or to a local oxygen deficit, and is emitted with the exhaust into the atmosphere, from which it may be scavenged by dry or wet deposition. Investigations have demonstrated unequivocally that formaldehyde levels increase with increasing motor vehicle traffic in both rainwater and runoff.[78,79]

The total concentration of phenol derivatives has also been determined in road runoff samples from Poland. The usual compounds in this respect were phenol and 3,4-di- methylphenol, and the highest levels were recorded for phenol (from 91 to 120 ng/L), chlorophenol (613 ng/L), and p-nitrophenol (469 ng/L). *o*-Cresol was not detected in any of the samples. This level of pollution by phenols in runoff

TABLE 1 Ranges of Concentrations and Mean Concentrations of Analytes in Road Runoff Samples

Analyte	Concentration (µg/L)		
	Minimum	Maximum	Mean
$CHCl_3$	0.01	0.53	0.59
$CHBrCl_2 + C_2HCl_3$	n.d.	0.08	0.02
$CHBr_2Cl$	n.d.	0.01	–
CCl_4	n.d.	0.01	–

Note: n.d., not determined.
Source: Polkowska et al.[74]

TABLE 2 Mean Concentrations and Ranges of Concentrations of Pesticides Determined in Road Runoff Samples

Analyte	Concentration (µg/L)		
	Minimum	Maximum	Mean
Organo-N,-P pesticides			
Simazine	n.d.	0.48	0.27
Atrazine	n.d.	4.10	0.71
Propazine	n.d.	12.1	3.25
t-Butazine	n.d.	12.4	2.38
Bromofos	n.d.	2.86	0.27
Malathion	n.d.	0.46	0.11
Chlorfenvinfos	n.d.	0.13	0.057
Phenitrothion	n.d.	0.13	0.08
Organochlorine pesticides			
α-HCH	n.d.	5.83	3.36
γ-HCH	n.d.	61.4	31.3
Aldrin	n.d.	21.1	2.68
Heptachlor	n.d.	57.1	8.75
epoxide		127	18.1
o,p'-DDE	n.d.	8.38	3.09
p,p'-DDD	n.d.	1308	212
o,p'-DDD	n.d.	213	29.4
p,p'-DDE	n.d.	2.41	1.38
o,p'-DDT	n.d.	2.88	1.66
p,p'-DDT	n.d.	5.70	2.63
Methoxychlor	n.d.		

Note: n.d., not determined.
Source: Polkowska et al.[77]

TABLE 3 Maximum, Minimum, and Mean Concentrations of Analytes Found in Samples

Compound	Concentration of Analyte (µg/L)		
	Minimum	Maximum	Mean
Benzene	0.86	109	48.9
Toluene	0.52	54.3	32.5
Ethylbenzene	0.12	39.8	18.6
p, m-xylene	<0.03	15.3	0.96
o-xylene	<0.03	27.0	0.99
Styrene	<0.03	11.3	1.18
Cumene	0.1	2.54	0.16
m-Dichlorobenzene	<0.03	12.7	4.15
Heptane	<0.03	<0.03	-
Isooctane	<0.03	14.3	3.50
Octane	<0.03	16.7	6.23
Cyclohexanone	<0.03	<0.03	-
Decane	<0.03	14.1	4.03
Dodecane	<0.03	14.2	2.50

Source: Polkowska et al.[77]

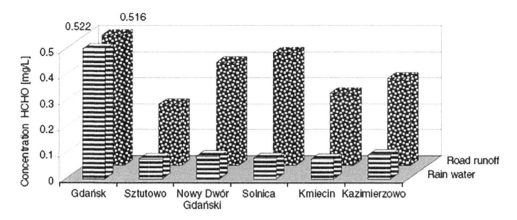

FIGURE 4 Formaldehyde (HCHO) levels determined in samples of rainwater and runoff from roads in the Gdansk area from roadside sampling points in Poland.
Source: Duncan.[78]

samples from the Tri-City area is due largely to the intensity of motor vehicle traffic and the proximity of industrial plants in the city, including the oil refinery, which discharges an average of 118 kg/yr of phenol derivatives into surface waters. The phenols present in the samples analyzed were probably formed as a result of photochemical reactions with the products of the incomplete combustion of fuel—benzene, toluene, and xylene.[80]

Roof Runoff

The most systematic investigations into the chemical composition of roof runoff have been carried out in countries where one of the sources of drinking water is water obtained from precipitation. The following factors affect the quality of water running off roofs:[81] the material from which the roof covering is made (its chemical composition, surface roughness, the area of roof covered, age, weather resistance), the physical boundary conditions (parameters) of the roof (surface area, inclination, exposure to the weather), the factors affecting precipitation (the direction and intensity of the wind, levels of pollutants in precipitation), other meteorological factors (season, type of air mass, duration of precipitation), the physicochemical properties of the pollutants (vapor pressure, partition coefficient, solubility in water, Henry's constant), and the concentration of pollutants in the atmosphere (the source and magnitude of emission, transport and spread, decomposition and degradation period, the physical form of the pollutant).

According to the available data, the pH of roof runoff is more basic than that of rainwater. In some cases, the pH may be higher, if basic soil particles have been windblown from eroded soil onto the roof surface. Roofs containing cement (concrete tiles, corrugated asbestos-cement sheet, and gravel) may cause the runoff pH to change, since they contain $CaCO_3$, which dissolves in rainwater. Table 4 presents information on the variability in runoff pH from various types of roofing materials.[54]

An in-depth study of the chemical composition of rainwater and roof runoff was carried out by German investigators. Concentrations of all the ions (except NH_4^+) were nearly twice as high in runoff waters than in rainwaters. The slight differences in the contents of ions in runoff from five types of roofing material suggest that the source of anions is more likely to be dry deposition than the roofing material.[56]

The contamination of roof runoff by heavy metals is widely described in the literature.[50,61,62,82,83] In this case, it is the roofing material that is the main factor affecting the level of these contaminants in water. Metals are the products of the corrosion of roofing and guttering materials; hence, water running off a galvanized sheet roof will contain large amounts of zinc and a certain quantity of cadmium, while roofs

TABLE 4 pH of Runoff Water from Different Types of Roofing Material

Type of Roofing Material	Median	Minimum	Maximum
Roofing felt	5.2	3.8	6.3
Cement tiles	5.4	4.6	6.5
Galvanized sheeting	6.9	6.5	7.6
Gravel roof	7.1	6.8	7.6
Concrete tiles	7.4	7.3	7.5

Source: Kennedy et al.[54]

made from copper sheeting will be a source of copper in the runoff.[64] Nonetheless, the metals present in roof runoff do not get there just as a result of leaching, since large quantities of metals have been found in runoff from roofing made from non-metallic materials, e.g., ceramic tiles, polyester, roofing felt. This does depend, however, on the location of the sampling point.[65] Heavy metal levels in roof runoff were highest in samples from the roofs of buildings situated in urban and industrial areas. Other factors affecting the amounts of heavy metals in roof runoff include the pH of rain and the angle of inclination of the roof.[64] Table 5 sets out the results of relevant research carried out in many scientific centers.

Pesticides and herbicides are frequently encountered in runoff waters. These compounds are increasingly being used in the production of roofing materials to prevent the growth on roofs of mosses and other plants. Evidence that pesticides are leached out of roofing materials was supplied by research carried out in Switzerland by a team of scientists from Eidgenössische Anstalt für Wasserversorgung, Abwasserreinigung und Gewässerschutz (EAWAG). Pesticide concentrations were lowest in runoff from ceramic tile roofs, higher in polyester roof runoff, and the highest from roofs covered with roofing felt. Particularly striking is the very high concentration of (R,S)-mecoprop in runoff from felt-covered roofs. This compound is not used on farms in Switzerland. The presence of (R,S)-mecoprop in runoff was due to the hydrolysis of Preventol® B2—a pesticide applied to root crops. It is used in large quantities as an additive to bitumen applied as a thin coating to areas on walls and roofs particularly susceptible to leaks, e.g., at the spot where the roofing felt is stuck to the wall.[51,84]

Table 6 lists the pesticides (and their concentrations) most commonly detected and determined in runoff from different types of roofing. The highest levels were recorded for heptachlor epoxide (roofing

TABLE 5 Levels of Metals in Samples of Roof Runoff with Respect to the Type of Roofing Material

Type of Roofing Material	Cu	Zn	Cd	Pb
		(μg/L)		
Ceramic tiles	1905	360	2.1	172
Polyester	6817	2076	3.1	510
Gravel	140	36	0.2	22
Gravel	1.0–56	0.05–468	0.01–0.48	0.01–2.73
Roofing felt	7.6	103.5	0.65	37.4
Cement tiles	355	53.5	0.425	38.6
Corrugated asbestos-cement sheet	10.5	22.6	0.14	24.6
Galvanized sheeting	26.4	43,500	1.23	37.6
Gravel	4.75	9,150	0.03	2.9
Ceramic tiles	71–304	10–48	0.07–0.4	13–41
Polyester	217–842	27–15	0.1–0.3	4.9–24
Gravel	18	9	0.11	2.7
Various roofing materials	14–240	582–12,357	0.2–4.5	76–2,458

Source: Quek and Forster,[50] Boller,[61] Mason et al.,[62] Zobrist et al.,[82] and Gromaire et al.[83]

TABLE 6　The Most Commonly Recorded Pesticides and Their Concentrations in Samples of Runoff Water from Roofs Depending on the Type of Roofing Material

Type of Roofing Material	Analyte (ng/L)
Roofing felt	Aldrin (6020), methoxychlor (430), heptachlor epoxide (19970), p,p'-DDD (520), o, p'-DDE (540), o,p'-DDD (340), p, p'-DDE (450), o, p'-DDT (190), p, p'-DDT (370), α-HCH (200), γ-HCH (400), propazine (1070), *tert*-butylazine (120)
Ceramic tiles	Aldrin (130), methoxychlor (430), heptachlor epoxide (430), p,p-DDD (4610), o,p'-DDE (2800), o,p'-DDD (1210), p, p'-DDE (1360), p, p'-DDT (1080), α-HCH (740), γ-HCH (600), propazine (1110), *tert*-butylazine (120), bromofos (2720)
Corrugated asbestos-cement sheeting	Aldrin (610), methoxychlor (130), heptachlor epoxide (14540), p, p'-DDD (500), o, p'-DDE (4640), o, p'-DDD (7040), p, p'-DDE (1040), o, p'-DDT (750), p, p'-DDT (8460), α-HCH (460), γ-HCH (840), atrazine (1210), *tert*-butylazine (130), bromofos (310)
Metal sheet tiles	Heptachlor epoxide (1640), o,p'-DDE (920), o,p'-DDD (1890), p,p'-DDE (680), o,p'-DDT (490), p,p'-DDD (830), α-HCH (320), γ-HCH (500), propazine (400), bromofos (4912), atrazine (2980), simazine (620), malathion (1960), fenitrothion (630), chlorfenvinfos (180)
Galvanized sheeting	Aldrin (6020), methoxychlor (430), heptachlor epoxide (19970), p,p-DDD (520), o,p'-DDE (540), o,p'-DDD (340), p,p'-DDE (450), o,p'-DDT (190), p,p'-DDT (370), α-HCH (200), γ-HCH (400), propazine (1070), *tert*-butylazine (120)
Bituminous roofing felt	Heptachlor epoxide (1680), bromofos (50)
Teflon	Bromofos (40)

Source: Grynkiewicz et al.[80] and Polkowska.[87]

felt, galvanized sheeting, corrugated asbestos-cement sheeting), p,p'-DDT (corrugated asbestos-cement sheeting), o,p'-DDD (corrugated asbestos-cement sheeting), p,p'-DDD (ceramic tiles), o,p'-DDE (corrugated asbestos-cement sheeting, ceramic tiles), aldrin (roofing felt, galvanized sheeting), bromofos (metal sheet tiles, ceramic tiles), and atrazine (metal sheet tiles, corrugated asbestos-cement sheeting). The roofing materials from which runoff samples most often contained pesticides were corrugated asbestos-cement sheeting, ceramic tiles, and metal sheet tiles.[85–87]

Table 7 summarizes the concentrations of analytes determined in samples of roof runoff from buildings in Poland. The most frequently detected VOCC analytes were $CHCl_3$ and CQ14. The presence of VOCCs in roof runoff is probably due to the use of many of these compounds as solvents and degreasing agents.

The most frequently detected petroleum hydrocarbons were benzene and toluene. Their presence in roof runoff samples points to the possible leaching of these compounds from roofing materials like roofing felt, tar, and bituminous tiles as well as to the strong adsorption of these analytes to roof surfaces.[85]

TABLE 7　Total Concentrations of Selected Analytes Determined in Samples of Roof Runoff

Total Concentrations (μg/L)	
Organochlorine Compounds	Petroleum Hydrocarbons
Station no. 1	
0.01–0.17	0.11–35
Station no. 2 (crossroads)	
0.01–0.63	0.12–33
Station no. 3 (crossroads)	
0.03–1.07	0.61–64

TABLE 8 Levels of Cations and Anions in Samples of Runoff from Roofs Covered with Different Materials

Type of Roofing Material	Cations and Anions (meq/L)
Roofing felt	F^- (0.01), Cl^- (0.22), NO_3^- (0.08), SO_4^{2-} (0.44), Na^+ (0.14), NH_4^+ (0.21), K^+ (0.04), Mg^{2+} (0.07), Ca^{2+} (0.39)
Ceramic tiles	F^- (0.02), Cl^- (3.17), NO_3^- (1.93), SO_4^{2-} (4.59), Na^+ (1.13), NH_4^+ (0.13), K^+ (0.50), Mg^{2+} (1.30), Ca^{2+} (4.28)
Corrugated asbestos-cement sheeting	Cl^- (0.16), NO_3^- (0.12), SO_4^{2-} (1.45), Na^+ (0.20), NH_4^+ (0.14), K^+ (0.05), Mg^{2+} (0.07), Ca^{2+} (1.92)
Metal sheet tiles	F^- (0.01), Cl^- (0.20), NO_3^- (0.04), SO_4^{2-} (0.17), Na^+ (0.11), NH_4 (0.15), K^+ (0.05), Mg^{2+} (0.04), Ca^{2+} (0.27)
Galvanized sheeting	F^- (0.01), Cl^- (0.12), NO_3^- (0.09), SO_4^{2-} (0.92), Na^+ (0.11), NH_4^+ (0.10), K^+ (0.07), Mg^{2+} (0.07), Ca^{2+} (1.20)
Bituminous roofing felt	Cl^- (0.05), NO_3^- (0.02), SO_4^2 (0.32), Na^+ (0.08), NH_4^+ (0.04), K^+ (0.01), Mg^{2+} (0.06), Ca^{2+} (0.59)
Teflon	Cl^- (0.02), NO_3^- (0.03), SO_4^{2-} (0.04), Na^+ (0.04), NH_4^+ (0.04), K^+ (0.03), Mg^{2+} (0.02), Ca^{2+} (0.08)

Source: Polkowska.[86]

Roof runoff samples from Poland were also analyzed for the presence of formaldehyde. The highest level of this compound was 1.37 mg/L, while its mean concentration was 0.54 mg/L and was comparable to the mean value recorded in rainwater samples.[78,79]

Table 8 lists the most frequently detected and determined ions (and their concentrations) in samples of runoff from roofs made from different materials.

Chloride and sulfate were found in all the samples analyzed. F^- ions were detected in only 4 of 68 samples analyzed. The highest concentrations of both chlorides and sulfates were found in runoff from ceramic tiled roofs. As far as cations are concerned, Na^+, K^+, Mg^{2+}, and Ca^{2+} were determined in all samples, the highest levels being registered for Na^+ and Ca^{2+}. The highest concentration of sodium ions was recorded in runoff from roofs covered with roofing felt and ceramic tiles, whereas the highest calcium levels were found in samples from roofs covered with tiles and corrugated asbestos-cement sheeting.[85,86]

Throughfall

In forest geoecosystems, precipitation does not reach the substrate directly, but as it trickles through the crowns of trees and down their trunks, it undergoes chemical changes. The chemical composition of rainwater in forest areas changes mainly under the influence of the physiological processes of plants and the dissolution and flushing from branches, bark, and leaves of aerosols and gaseous contaminants absorbed from the air. The species composition of the trees, their age, and health can also alter the properties of rainwater, giving it new physicochemical properties.

The runoff (throughfall) from tree trunks is less interesting than that from tree crowns as it makes up at most a few percent of the rainwater. In addition, trunk runoff is much less in coniferous trees, the branches of which point downwards, taking the rainwater beyond the trunk itself.[3] The proportion of throughfall in the total deposition transported to the forest floor is less than 10%. It has been found that in pine forests, 3% of precipitation reaches the ground as trunk runoff, but 83% as crown runoff.[8] The variety and numbers of different contaminants present in crown runoff are affected primarily by the type of leaves, the density of the crown, the age of the tree, the quantity of precipitation, and the wind direction.[5]

According to the results published in the literature regarding the content of cations and anions in throughfall, the concentrations of particular ions in both crown runoff and trunk runoff are considerably higher than in precipitation falling on open terrain.[88] The elevated levels of potassium, magnesium, calcium, and manganese ions in crown runoff compared to their content in rainwater sampled in open country were due to their being flushed off leaves and needles. This was particularly noticeable in the case of spruce needles, which have a much greater surface area than the leaves/needles of other tree species.

A Japanese study has shown that a high nitrogen load in atmospheric deposition (including dry deposition) may be a factor contributing to the deterioration of pine stands. This was particularly conspicuous in urban areas, where levels of nitrate and ammonium ions in runoff were high.[8,89]

Comparative studies were carried out in Japan and Spain to illustrate the effect of the location of a woodland system on the composition and numbers of contaminants entering the environment. Samples were collected at two sites: in a forest close to a road, traffic, and urban areas (exposed), and in a wood situated far from urban areas (sheltered by mountains), where the effect of pollutants is less. The respective concentrations of nitrate and sulfate ions in trunk and crown runoff were far higher in the urban area (34.4 and 50 μeq/L) than in the mountains beyond the city (4.8 and 28.8 μeq/L) and were several times higher than the levels recorded in free precipitation.[8,12] Table 9 lists the range of concentrations and the mean concentrations of contaminants (including metals from the platinum group) in samples of runoff from tree crowns.

The data on formaldehyde concentrations in throughfall have been published in paper.[79]

Runoff from Farming Areas

Good conditions for farmland runoff containing substantial loads of nutrients exist in spring. At this time of the year, the plant cover is not yet dense enough to prevent surface runoff, and the ground, still frozen after the winter, also facilitates this. Most pesticides and fertilizers are applied directly to the soil or sprayed onto arable land, plantations, and forests; they therefore gain direct entry into the environment and are the source of much contamination (aerosols containing pesticides and nitrogen and phosphorus compounds). The inappropriate storage of pesticides and mineral fertilizers (directly on the ground, without any proper protection from the elements) contributes to the leaching of active substances, which, through the mediation of surface runoff, enter surface waters, causing their eutrophication. This leads to deleterious changes in aquatic ecosystems manifested by an increase in nutrient

TABLE 9 Range of Concentrations, Mean Concentrations and Frequency of Occurrence of Analytes in Throughfall Samples

Parameter	Concentration Range (mean) [μeq/dm³]	Parameter	Concentration Range (mean) [peq/dm³]
pH	4.2–7.4 (5.6)	Fe	0.030–0.55 (0.19)
Conductivity (μS/cm)	0.020–0.050 (0.15)	Total phenols	0.20–1.6 (0.49)
Cl⁻	0.050–3.5 (0.62)	TOC	22–108 (45)
F⁻	0.0010–0.11 (0.028)	Pt	$(0.21–0.6) \times 10^{-7}$ (0.4×10^{-7})
Br⁻	–	Pd	$(1.1–2.6) \times 10^{-6}$ (1.7×10^{-6})
PO_4^{3-}	0.0070–1.0 (0.15)	Pb	0.035–0.053 (0.046)
SO_4^{2-}	0.098–1.0 (0.39)	Zn	0.0014–0.00409 (0.0022)
NO_3^-	0.0040–0.78 (0.20)	Cd	0.0039–0.014 (0.0068)
Li⁺	0.00027–0.00040 (0.00034)	Co	0.037–0.058 (0.0508)
Na⁺	0.24–1.77 (0.64)	Mn	–
NH_4^+	0.39–0.43 (0.35)	V	0.039–0.26 (0.14)
K⁺	0.21–0.14 (0.13)	Cr	0.46–1.4 (1.1)
Mg^{2+}	0.080–0.13 (0.10)	Ni	0.105–0.35 (0.21)
Ca^{2+}	0.080–0.32 (0.19)	Sn	0.0067–0.0101 (0.0084)
Formaldehyde	0.060–0.51 (0.18)	Mo	0.0081–0.015 (0.012)

levels (nitrogen and phosphorus compounds), a drop in the level of dissolved oxygen, and mass blooms of algae. These contaminants can also permeate into the shallow groundwater circulation, which serves as a source of drinking water in rural areas. Studies carried out worldwide confirm that farmland runoff carries very large amounts of nitrogen and phosphorus compounds; Cl^-, K^+, Ca^{2+}, Na^+, and Mg^{2+} ions; and pesticides.[32,90–92]

Runoff from Airports

Air transport is the most modern and the fastest developing type of transport. In 1993–2000, the number of passengers traveling by air in European Union countries increased by ca. 10% each year. Forecasts for the next 20 years put the annual increase in the global number of air passengers at around 5%.

This burgeoning development of air transport also concerns Poland. To a large extent, this is the result of the rapid expansion of the so-called budget airlines onto the Polish market and the reconstruction of many regional airports. Their operation is an unavoidable interference in the natural environment. Air transport contributes to the destruction of the ozone layer, and the harmful effects of air transport are visible on the ground as well. Precipitation runoff from airports affects surface water and groundwater to a substantial extent. Along with rainwater and snowmelt, petroleum derivatives, surfactants, de-icing agents (in winter), and other contaminants get into the environment. However, it is the de-icing agents applied when air temperatures are low that pose the greatest danger of polluting waters running off airport aprons and runways. The composition of airport runoff changes during the year and depends on a number of factors, such as the weather, the quantities of de-icing agents applied, the duration of precipitation, and the intensity of traffic at the airport in question.

Airport authorities use not only the environmentally dangerous ethylene and propylene glycols for de-icing aircraft but also the environmentally rather more friendly mixtures of acetates and formates for de-icing runways, taxiing areas, and aprons.[93,94] Airports do not usually have their own sewage treatment plants, and all the effluents, carrying petroleum derivatives, surfactants, de-icing agents, and other organic and inorganic pollutants, flow with rainwater and snowmelt into the surrounding soil and surface waters. These contaminants may significantly affect the fauna and flora of rivers, causing the death of fish and aquatic vegetation; therefore, it is crucial to monitor the runoff from such areas on a continuous basis.[95–97]

Figures 5–7 show the mean concentrations of selected analytes in samples of runoff collected at airports in Poland (Gdańsk and Warsaw) and the United States.[97,98]

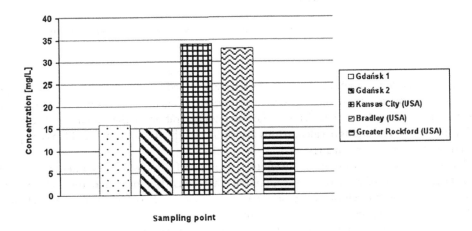

FIGURE 5 Comparison of mean concentrations of Ca in samples of runoff from airports in Gdansk and the United States.

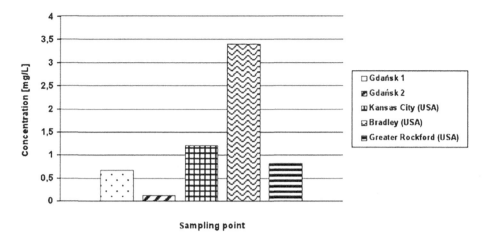

FIGURE 6   Comparison of mean concentrations of Fe in samples of runoff from airports in Gdansk and the United States.

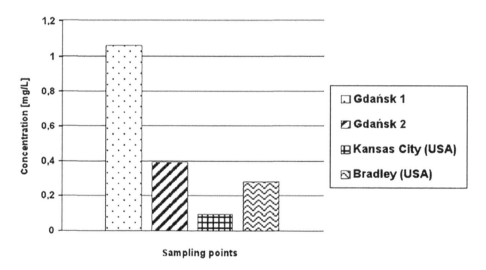

FIGURE 7   Comparison of mean concentrations of sum of phenols in samples of runoff from airports in Gdansk and the United States.

The highest levels of potassium and sodium ions were recorded in runoff from the Gdansk airport in 2009—this may have been due to the large quantities of de-icing agents (sodium and potassium formate) used during the winter. At the other airports, the mean concentrations of these analytes were very much lower. The mean concentrations of calcium ions determined in the runoff from the Gdansk airport were lower than those from American airports.

The highest level of total iron by far was found in samples from the Bradley International Airport (United States); the level of this particular contaminant at Gdansk was lower than those at U.S. airports. Comparison of the mean levels of magnesium ions shows that they have been at similar levels at all the airports, although the Mg level was lowest at Gdańsk.

The highest mean concentration of total phenols was recorded in samples taken in winter at the Gdansk airport, whereas mean Total organic carbon (TOC) levels at this time were comparable in the samples from the airports at Gdansk, Kansas City, and Bradley (United States).

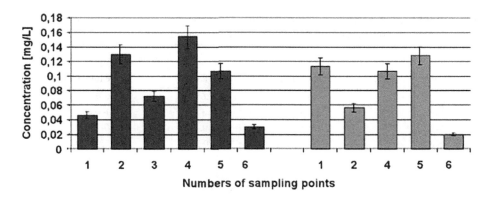

FIGURE 8 Comparison between mean concentrations of organic compounds determined in precipitation and runoff samples; 1–5, runoff samples; 6-precipitation. (March 19, 2008—black; April 8, 2008—gray).

The following organic compounds have been shown to be present in runoff from the Gdansk airport: anionic and cationic detergents, formaldehyde, phenols, PAHs, pesticides, and polychlorinated biphenyls (PCBs). Figure 8 shows the mean concentrations of cationic detergents contained in samples of precipitation and runoff.

Changes in the Composition of Samples Depending on the Moment of Sampling during Precipitation or Runoff

Literature reports indicate that atmospheric contaminants are removed during the early phases of precipitation, usually in the "first flush," and their concentrations in the precipitation drop exponentially as its volume increases. In its simplest form, this relation can be expressed by the equation

$$c = a \cdot \exp(b \cdot V) \qquad (1)$$

where c is the concentration of contaminant in the precipitation (mol/L), V is the volume of precipitation collected (L), and a and b are constants.[99]

Analysis of the results of studies carried out in Germany and Switzerland shows that concentrations of contaminants (heavy metals, pesticides, cations, anions, PAHs) are the highest in the first phase of precipitation and in the first flow of runoff.[1,56,62,87] Table 10 lists the relevant data.

From these data, it can be inferred that the presence of pesticides and inorganic compounds in the first flush of precipitation and in roof runoff may be influenced by the following: the flushing of dry deposition accumulated in the period preceding rainfall, the leaching of compounds from roofing and guttering materials and the corrosion of these materials, and the reduction in concentration of constituents with increasing amounts of precipitation, as a result of the removal by raindrops of aerosol particles and gases contained in the air.[82]

When rainfall is intense, the period of contact between the rainwater and the roofing material is short and just the contaminants accumulated on its surface are washed away; hence, in this case, the type of roofing material is practically irrelevant. However, when rainfall is less intense, the type of roofing material takes on greater significance, because contaminants are adsorbed on roofs with a rough surface like roofing felt or cement tiles, which is why the first flush of rainwater and the initial runoff do not then contain larger amounts of contaminants. These are only flushed away after a long period of rain.[54]

As far as runoff from tree crowns is concerned, there is a drop in the concentration of ions in precipitation and runoff samples with increasing intensity of rainfall.[5]

TABLE 10 The Values of Some Physicochemical Parameters and the Contents of Selected Constituents in Samples of Precipitation and Runoff

Parameter (mM)	Precipitation	Runoff	
		Initial Phase	After the Initial Phase
pH	6.9	7.0–7.6	7.0–7.6
Ca^{2+}	0.06	0.4–1.3	0.1–0.5
Mg^{2+}	0.01	0.06–0.44	0.01–0.04
Na^+	0.02	0.22–0.56	0.02–0.20
K^+	0.01	0.10–0.35	0.02–0.10
NH_4^+	0.14	0.14–0.29	0.02–0.07
Alkalinity	0.08	0.7–2.8	0.3–0.6
Cl^-	0.01	0.06–0.22	0.02–0.06
SO_4^{2-}	0.04	0.1–0.2	0.01–0.1
NO_3^-	0.06	0.29–0.64	0.02–0.20
N_{total}	–	0.5–1.2	0.07–0.30

Source: Mason et al.[62]

Summary

The atmosphere is one of the main non-point sources of pollution, and its dynamics causes a diversity of the ways that it can affect the level of contamination of soils and surface water. The composition of precipitation is closely related to a degree of pollution of air. After passing through developed areas (buildings, roads, woodland, farmland, waste dumps), wet deposition enters the environment almost exclusively as runoff.

The composition of runoff water depends to a large extent on the kind of terrain on which precipitation fell. The road runoff can contain hydrocarbons of petrochemical origin, dust, and heavy metals. The level of contamination of roof runoff depends on the roof material, which can be a source of many organic and inorganic pollutants, such as pesticides and heavy metals (copper, zinc). The dump runoff is characterized by variable aggressiveness towards the environment, dependent on the degree of processing of the refuse while the edge-of-field runoff can contain pesticides and biogenic substances. A large fraction of runoff water with the pollutants present in it enters surface water and groundwater, which can be used as a source of drinking water. Although the flux of pollutants into soil and groundwater is usually small, hazardous chemicals can undergo accumulation over a period of time.

References

1. Bucheli, T.D.; Muller, S.R.; Heberle, S.; Schwarzenbach, R.P. Occurence and behavior of pesticides in rainwater, roof runoff, and artifical stormwater infiltration. Environ. Sci. Technol. **1998**, *32*, 3457–3464.
2. Wauchope, R.D.; Graney, R.L.; Cryer, S.; Eadsforth, C.; Klein, A.W.; Racke, K.D.; Pesticide runoff: methods and interpretation of field studies. Pure Appl. Chem. **1995**, *67*, 2089–2108.
3. Yates, D.J.; Hutley, L.B.; Doley, D.; Boonsaner, A. The ecological and physiological significance of occult precipitation in an Australian rainforest, 1st Int. Conf. on Fog and Fog Collection, July 19–24. Vancouver, Canada; Schemenauer, R.S., Bridgman, H., Eds.; 1998, 81–84.
4. Balestrini, R.; Galli, L.; Tagliaferi, A.; Tartari, G. Study on throughfall deposition in two North Italian forest sites (Valtellina, Lombardy). Chemosphere **1998**, *36*, 1095–1100.

5. Walna, B.; Siepak, J. Research on the variability of physicochemical parameters characterizing acid precipitation at the Jeziory Ecological Station in The Wielkopol– ski National Park (Poland). Sci. Total Environ. **1999**, *239*, 173–187.

6. Lovett, G.M.; Traynor, M.M.; Pouyat, R.V.; Carreiro, M.M.; Zhu, W.-X.; Baxter, J.W. Atmospheric deposition to oak forests along an urbanrural gradient. Environ. Sci. Technol. **2000**, *34*, 4294–4300.

7. Forti, M.C.; Moreira-Nordemann, L.M. Rainwater and throughfall chemistry in a Terra Firme Rain Forest; Central Amazonia. J. Geophys. Res. **1991**, *96*, 7415–7421.

8. Masaaki, C.; Do Hoon, K.; Hiroshi, S. Rainfall, stemflow, and throughfall chemistry at urban- and moutain-facing sites at MT. Gokurakuj, Hiroshima, Western Japan. Water Air Soil Pollut. **2003**, *146*, 93–109.

9. Hansen, K.; Draaijers, G.P.J.; Ivens, W.P.M.F.; Gundersen, P.; van Leeumen, N.F.M. Concentration variations in rain and canopy throughfall collected sequentially during individual rain events. Atmos. Environ. **1994**, *28*, 3195–3205.

10. Houle, D.; Ouimet, R.; Paquin, R.; Laflamme, J-G. Determination of sample size for estimating ion throughfall deposition under a mixed hardwood forest at the Lake Clair Watershed (Duchesnay, Quebec). Can. J. For. Res. **1999**, *29*, 1935–1934.

11. Avila, A.; Rodrigo, A. Trace metal fluxes in bulk deposition, throughfall and stemflow at two evergreen oak stands in NE Spain subject to different exposure to the industrial environment. Atmos. Environ. **2004**, *38*, 171–180.

12. Rodrigo, A.; Avila, A.; Roda, S. The chemistry of precipitation, throughfall and stemflow in two holm oak (*Querces ilex* L.) forests under a contrasted pollution environment in NE Spain. Sci. Total Environ. **2003**, *305*, 195–205.

13. Inagaki, M.; Sakai, M.; Ohnuki, Y. The effects of organic carbon on acid rain in a temperature forest in Japan. Water Air Soil Pollut. **1995**, *85*, 2345–2350.

14. Bini, C.; Bresolin, F. Soil acidification by acid rain in forest ecosystems: A case study in northern Italy. Sci. Total Environ. **1998**, *222*, 1–15.

15. Prakasa Rao, P.S.; Momin, G.A; Safai, P.D.; Pillai, A.G.; Khemani, L.T. Rain water and throughfall chemistry in the Silent Valley in South India. Atmos. Environ. **1995**, *29*, 2025–2029.

16. Lovett, G.M.; Kinsman, J. Atmospheric deposition to high- elevation systems. Atmos. Environ. **1990**, *24*, 2767–2786.

17. Kiss, G.; Varga-Puchony, Z.; Hlavy, J. Determination of polycyclic aromatic hydrocarbons in precipitation using solid-phase extraction and column liquid chromatography. J. Chromatogr. A **1996**, *725*, 261–272.

18. Garban, B.; Blanchou, H.; Motelay-Massei, A.; Chevreuil, M.; Ollivon, D. Atmospheric bulk deposition of PAHs onto France: Trends from urban to remote sites. Atmos. Environ. **2002**, *36*, 5395–5403.

19. Bomboi, M.T.; Hernandez, A. Hydrocarbons in urban runoff: Their contribution to the wastewaters. Water Res. **1991**, *25*, 557–565.

20. Stephenson, J.B.; Zhou, W.F.; Beck, B.F.; Green, T.S. Highway stormwater runoff in karst areas—Preliminary results of baseline monitoring and design of a treatment system for a sinkhole in Knoxville, Tennessee. Eng. Geol. **1999**, *52*, 51–59.

21. Lygren, E.; Gjessing, E.; Berglind L. Pollution transport from a highway. Sci. Total Environ. **1990**, *33*, 147–159.

22. Stotz, G. Investigations of the properties of the surface water run-off from federal highways in the FRG. Sci. Total Environ. **1987**, *59*, 329–337.

23. MacKenzie, M.J.; Hunter, J.V. Sources and fates of aromatic compounds in urban stormwater runoff. Environ. Sci. Technol. **1979**, *13*, 179–183.

24. Moilleron, R.; Gonzalez, A.; Chebbo, G.; Thevenot, D.R. Determination of aliphatic hydrocarbons in urban runoff samples from the "Le Marais" experimental catchment in Paris centre. Water Res. **2002**, *36*, 1275–1285.

25. Achten, Ch.; Kolb, A.; Puttmann, W. Methyl *tert-butyl* ether (MTBE) in urban and rural precipitation in Germany. Atmos. Environ. **2001**, *35*, 6337–6345.

26. Lopes, T.J.; Fallon, J.D.; Rutherford, D.W.; Hiatt M.H. Volatile organic compounds in storm water from a parking lot. J. Environ. Eng. **2000**, *126*, 1137–1143.

27. Borden, R.C.; Black, D.C; McBlief, K.V. MTBE and aromatic hydrocarbons in North Carolina stormwater. Environ. Pollut. **2002**, *118*, 141–152.

28. Thurston, K.A. Lead and petroleum hydrocarbon changes in an urban wetland receiving stormwater runoff. Ecol. Eng. **1999**, *12*, 387–399.

29. Legret, M.; Pagotto, C. Evaluation of pollutant loadings in the runoff waters from a major rural highway. Sci. Total Environ. **1999**, *235*, 143–150.

30. Pagotto, C.; Legreti, M.; Cloirec, P.L. Comparision of the hydraulic behaviour and the quality of highway runoff water according to the type of pavement. Water Res. **2000**, *34*, 4446–4454.

31. Hvitved-Jacobsen, T.; Johansen, N.B.; Yousef, Y.A. Treatment systems for urban and highway runoff in Denmark. Sci. Total Environ. **1994**, *146/147*, 499–506.

32. Tsiouris, S.E.; Mamolos, A.P.; Kalburtji, K.L.; Alifrangis, D. The quality of runoff water collected from a wheat field margin in Greece. Agric. Ecosyst. Environ. **2002**, *89*, 117125.

33. Backstrom, M.; Nilsson, U.; Hakansson, K.; Allard, B.; Karlsson, S. Specification of heavy metals in road runoff and roadside total deposition. Water Air Soil Pollut. **2003**, *147*, 343–366.

34. Matthes, S.A.; Cramer, S.D.; Covino, B.S.; Bullard, S.J.; Holcomb, G.R. Precipitation runoff from lead, outdoor and indoor atmospheric corrosion. ASTM Spec. Tech. Publ. **2002**, 1421.

35. Maltby, L.; Boxall, B.A.; Forrow, D.M.; Calow P.; Betton, C.I. The effects of motorway runoff on freshwater ecosystems: 1. Field study. Environ. Toxicol. Chem. **1995**, *14*, 1079–1092.

36. Harrison, R.M.; Wilson, S.J. The chemical composition of highway drainage waters. I. Major ions and selected trace metals. Sci. Total Environ. **1985**, *43*, 63–77.

37. Hewitt, C.N.; Rashed, M.B. Removal rates of selected pollutants in the runoff waters from a major rural highway. Water Res. **1992**, *26*, 311–319.

38. Latimer, J.S.; Hoffman E.J.; Hoffman, G.; Fasching, J.L.; Quinn, J.G. Sources of petroleum hydrocarbons in urban runoff. Water Air Soil Pollut. **1990**, *52*, 1–21.

39. Turer, D.; Maynard, J.B.; Sansalone, J.J. Heavy metal contamination in soils of urban highways: Comparison between runoff and soil concentrations at Cincinnati, Ohio. Water Air Soil Pollut. **2001**, *132*, 293–314.

40. Singh, S.P.; Tack, F.M.; Gabriels, D.; Verloo, G. Heavy metal transport from dredged sediment derived surface soils in a labolatory rainfall simulation experiment. Water Air Soil Pollut. **2000**, *118*, 73–86.

41. Mikkelsen, P.S.; Häfliger, M.; Ochs, M.; Tjell, J.C.; Jacobsen, P.; Boller M. Experimental assessment of soil and groundwater contamination from two old infiltration systems for road run–off in Switzerland. Sci. Total Environ. **1996**, *189/190*, 341–347.

42. Dannecker, W.; Au, M.; Stechmann, H. Substance load in rainwater runoff from different streets in Hamburg. Sci. Total Environ. **1996**, *93*, 385–392.

43. Gromaire-Mertz, M.C.; Garnaud, S.; Gozalez, A.; Chebbo, G. Characterisation of urban runoff pollution in Paris. Water Sci. Technol. **1999**, *39*, 1–8.

44. Legret, M.; Colandini, V.; Le Marc, C. Effect of a porous pavement with reservoir structure of the quality of runoff water and soil. Sci. Total Environ. **1996**, *189/190*, 335–340.

45. Fernández Espinosa, A.J.; Rodriguez, M.T.; Barragán de la Rosa, F.J.; Jiménez Sánchez, J.C. A chemical speciation of trace metals for fine urban particles. Atmos. Environ. **2002**, *36*, 773–780.

46. Smith, M.E.; Kaster, J.L. Effect of rural highway runoff on stream benthic macroinvertebrates. Environ. Pollut. **1983**, *32*, 157–170.

47. Ward, N.I.; Reeves, R.D.; Brooks, R.R. Lead in soil and vegetation along a New Zealand state highway with low traffic volume. Environ. Pollut. **1975**, *9*, 243–251.

48. Bomboi, M.T.; Hernandez A. Hydrocarbons in urban runoff: Their contribution to the wastewaters. Water Res. **1991**, *5*, 557–565.
49. Yaziz, M.I.; Gunting, H.; Sapari, N.; Ghazali, A.W. Variations in rainwater quality from roof catchments. Water Res. **1989**, *23*, 761–765.
50. Quek, U.; Förster, J. Trace metals in roof runoff. Water Air Soil Pollut. **1993**, *68*, 373–389.
51. Bucheli, T.D.; Müller, S.R.; Voegelin A. Bituminous roof sealing membranes as major sources of the herbicide (*R,S*)- mecoprop in roof runoff waters: Potential contamination of groundwater and surface water. Environ. Sci. Technol. **1998**, *32*, 3465–3471.
52. Shu, P.V.; Hirner, A. Trace compounds in urban rain and roof runoff. J. High Resol. Chromatogr. **1998**, *21*, 65–68.
53. Simmons, G.; Hope, V.; Lewis, G.; Whitmore, J.; Gao, W. Contamination of potable roof-collected rainwater in Auckland in New Zealand. Water Res. **2001**, *35*, 1518–1524.
54. Kennedy, P.; Gadd, J.; Mitchell, K. Preliminary examination of the nature of urban roof runoff in New Zealand; Report, Ministry of Transport Te Manatu Waka, New Zealand, 2001.
55. Garnaud S.; Mouchel J.M.; Chebbo G.; Thevenot D.R. Heavy metal concentrations in dry and wet deposits in Paris district: Comparison with urban runoff. Sci. Total Environ. **1999**, *235*, 235–245.
56. Foster, J. Patterns of roof runoff contamination and their potential implications on practice and regulation of treatment and local infiltration. Water Sci. Technol. **1996**, *33*, 39–48.
57. Zorbist, J.; Müller, S.R.; Bucheli, T.D. Quality of roof runoff for groundwater infiltration. Water Res. **2000**, *34*, 14551462.
58. Foster, J. The influence of location and season on the concentrations of macroions and organic trace pollutants in roof runoff. Water Sci. Technol. **1998**, *38*, 83–90.
59. Davis, A.P.; Shokouhian M.; Ni, S. Loading estimates of lead, copper, cadmium and zinc in urban runoff from specific sources. Chemosphere **2001**, *44*, 997–1009.
60. Bertling, S.; Odnevall Wallinder, I.; Leygraf, C.; Berggren, D. Environmental effects of zinc runoff from roofing ma– terials—A new multidisciplinary approach. In *Outdoor Atmospheric Corrosion*; Townstend, H.E., Ed.; ASTM STP 1421; American Society for Testing and Materials International: West Conshohocken, PA, 2002.
61. Boller, M. Tracking heavy metals reveals sustainability deficits of urban drainage systems. Water Sci. Technol. **1997**, *35*, 77–87.
62. Mason, Y.; Ammann, A.A.; Ulrich, A.; Sigg, L. Behavior of heavy metals, nutrients and major components during roof runoff infiltration. Environ. Sci. Technol. **1998**, *33*, 1588–1597.
63. Heijerick, D.G.; Janssen, C.R.; Karlen, C.; Odnevall Wall- inder, I.; Leygraf, C. Bioavailability of zinc in runoff water from roofing materials. Chemosphere **2002**, *47*, 1073–1080.
64. He, W.; Odnevall Wallinder, I.; Leygraf, C. A laboratory study of copper and zinc runoff during first flush and steady- state conditions. Corros. Sci. **2001**, *43*, 127–146.
65. Heijerick, D.G.; Janssen, C.R.; Karlen, C.; Odnevall Wallinder, I.; Leygraf, C. Runoff rates and ecotoxicity of zinc induced by atmospheric corrosion. Sci. Total Environ. **2001**, *277*, 169–180.
66. Odnevall Wallinder, I.; Korpinen, T.; Soundberg, R.; Leygraf, C. Atmospheric corrosion of naturally and pre- patinated copper roofs in Singapore and Stockholm—Run- off rates and corrosion product formation. In *Outdoor Atmospheric Corrosion*; Townstend, H.E., Ed.; ASTM STP 1421; American Society for Testing and Materials International: West Conshohocken, PA, 2002.
67. Odnevall Wallinder, I.; Verbiest, P.; Leygraf, C.; He, W. Effects of exposure direction and inclination on the runoff rates of zinc and copper roofs. Corros. Sci. **2000**, *42*, 1471–1487.
68. Davis, A.P.; Burns, M. Evaluation of lead concentration in runoff from painted structures. Water Res. **1999**, *33*, 2949–2955.
69. Horstmann, M.; McLachlan, M. Concentration of polychlorinated dibenzo-p-dioxins (PCDD) and dibenzofurans (PCDF) in urban runoff and household wastewaters. Chemosphere **1995**, *31*, 2887–2896.

70. Thomas, P.; Greene, G. Rainwater quality from different roof catchments. Water Sci. Technol. **1993**, *28*, 291–299.

71. Hvitved-Jacobsen, T.; Johansen, N. B.; Yousef, Y.A. Treatment systems for urban and highway runoff in Denmark. Sci. Total Environ. **1994**, *146/147*, 499–506.

72. Hewitt, C.N.; Rashed, M.B. An integrated budget for selected pollutants for a major rural highway. Sci. Total Environ. **1990**, *93*, 375–384.

73. Grynkiewicz, M.; Polkowska, Z.; Namieśnik, J. Determination of polycyclic aromatic hydrocarbons in bulk precipitation and runoff waters in an urban region (Poland). Atmos. Environ. **2002**, *36*, 361–369.

74. Polkowska, Z.; Zabiegala, B.; Górecki, T.; Namieśnik, J. Contamination of runoff waters from roads with high traffic intensity in the urban region of Gdansk, Poland. Pol. J. Environ. Stud. **2005**, *6*, 799–807.

75. Grynkiewicz, M.; Polkowska, Z.; Namieśnik, J. Determination of polycyclic aromatic hydrocarbons in bulk precipitation and runoff waters in an urban region (Poland). Atmos Environ. **2002**, *36*, 361–369.

76. Klimaszewska, K.; Polkowska, Z.; Namieśnik, J. The influence of mobile sources on acidity of runoff waters from roads with high traffic intensity of Gdansk. Pol. J. Environ. Stud. **2007**, *16*, 883–891.

77. Polkowska, Z.; Grynkiewicz, M.; Zabiegala, B.; Namieśnik, J. Levels of pollutants in runoff water from roads with high traffic intensity in the city of Gdansk, Poland. Pol. J. Environ. Stud. **2001**, *10*, 351–363.

78. Duncan, L.C. Chemistry of rime and snow collected at a site of central Washington Cascades. Environ. Sci. Technol. **1992**, *26*, 61–66.

79. Polkowska, Z.; Skarżyńska, K.; Górecki, T.; Namieśnik, J. Formaldehyde in various forms of atmospheric precipition and deposition from highly urbanized regions. J. Atmos. Chem. **2006**, *53*, 211–236.

80. Grynkiewicz, M.; Polkowska, Z.; Kot-Wasik, A.; Namieśnik, J. Determination of phenols in runoff. Pol. J. Environ. Stud. **2002**, *11*, 85–89.

81. Förster, J. Patterns of roof runoff contamination and their potential implications on practice and local infiltration. Water Sci. Technol. **1996**, *33*, 39–48.

82. Zobrist, J.; Muller, S.R.; Ammann, A.; Bucheli, T.D.; Mottier, V.; Ochs, M.; Schoenenberger, R.; Eugster, J.; Boller, M. Quality of roof runoff for groundwater infiltration. Water Res. **2000**, *34*, 1455–1462.

83. Gromaire, M.C.; Garnaud, S.; Saad, M.; Chebbo, G. Contribution of different sources to the pollution of wet weather flows in combined sewers. Water Res. **2001**, *25*, 521–533.

84. Bucheli, T.D.; Grüebler, F.C.; Müller, S.R.; Schwarzenbach, R.P. Simultaneous determination of neutral and acidic pesticides in natural waters at the low nanogram per liter level. Anal. Chem. **1997**, *69*, 1569–1576.

85. Polkowska, Z.; Górecki, T.; Namieśnik, J. Quality of roof runoff waters from an urban region (Gdansk, Poland). Chemosphere **2002**, *49*, 1275–1283.

86. Polkowska, Z. Examining the effect of the type of roofing on pollutant content in roof runoff waters from buildings in selected districts of the city of Gdańsk. Pol. J. Environ. Stud. **2004**, *13*, 191–201.

87. Polkowska, Z.; Górecki, T.; Namieśnik, J. Occurrence of pesticides in rain and roof runoff waters from an urban region. Urban Water **2009**, *6*, 441–448.

88. Walna, B.; Siepak, J. Research on the variability of physicochemical parameters characterising acid precipitation at the Jeziory Ecological Station in the Wielkopolski National Park (Poland). Sci. Total Environ. **1999**, *239*, 173–187.

89. Rennenberg, H.; Gessler A. Consequences of N deposition to forest ecosystems—Recent results and future research needs. Water Air Soil Pollut. **1999**, *116*, 47–64.

90. Barden, Ch.J.; Geyer, W.A.; Mankin, K.R.; Ngandu, D.; Devlin, D.L.; McVay, K. Reducing runoff contaminants with riparian buffer strips. AFTA Conf. Proc. 2003, 7–14.

91. Turtola, E.; Yli-Halla, M. Fate of phosphorus applied in slurry and mineral fertilizer: Accumulation in soil and release into surface runoff water. Nutr. Cycling Agroecosyst. **1999**, *55*, 165–174.

92. Schulz, R.; Hauschild, M.; Ebeling, M.; Nanko-Drees, J.; Wogram, J.; Liess, M. A qualitative field method for monitoring pesticides in the edge-of field runoff. Chemosphere **1998**, *36*, 3071–3082.

93. Switzenbaum, M.S.; Veltman, S.; Mericas, D.; Wagoner, B.; Schoenberg, T. Best management practices for airport deicing stormwater. Chemosphere **2001**, *43*, 1051–1062.

94. Breedveld, G.D.; Roseth, R.; Sparrevik, M.; Hartnik, T.; Hem, L.J. Persistence of the de-icing additive benzotriazole at an abandoned airport. Water Air Soil Pollut. **2003**, *3*, 91–103.

95. Revitt, D.M.; Shutes, R.B.E.; Llewellyn, N.R.; Worrall, P. Experimental reedbed systems for the treatment of airport runoff. Water Sci. Technol. **1997**, *36*, 385–390.

96. Cancilla, D.A.; Baird, J.C.; Rosa, R. Detection of aircraft deicing additives in groundwater and soil samples from fairchild air force base, a small to moderate user of deicing fluids. Environ. Contam. Toxicol. **2003**, *70*, 868–875.

97. Krzemieniowski, M.; Bialowiec, A.; Zieliński, M. The effectiveness of the storm water treatment plant at Warsaw Frederick Chopin Airport. Arch. Environ. Prot. **2006**, *32*, 25–33.

98. United States Environmental Protection Agency (Amerykańska Agencja Ochrony Środowiska), Airport Deicing Operations, EPA-821-R-00–016, Waszyngton, 2000.

99. Leister, D.L.; Baker, J.E. Atmospheric deposition of organic contaminants to the Chesapeake Bay, Atmos. Environ. **1994**, *28*, 1499–1520.

22

Salt Marsh Resilience and Vulnerability to Sea-Level Rise and Other Environmental Impacts

Introduction 215
Salt Marsh Ecosystem Characteristics 215
Salt Marsh Ecosystem Services 217
Salt Marsh Sustainability 217
Conclusion 220
References 220

Daria Nikitina

Introduction

Salt marshes are coastal wetlands within the intertidal zone, characterized by highly saline soils and anoxic conditions associated with tidal flooding and vegetated by macrophytes. Salt marshes occur globally along the low-energy depositional coasts of temporal and sub-Arctic climates both in microtidal and macrotidal regimes (Allen and Pye, 1992). Commonly, salt marshes occupy broad flat areas often referred to as the marsh platform. At the landform scale, salt marshes classify based on the physical settings that include open-coast marshes, poorly developed and vulnerable to wave action (Figure 1a); back-barrier marshes that formed on the sheltered side of barrier islands or spits (Figure 1b and f); estuarine marshes that fringe estuaries and coastal lagoons where muddy sediments accumulate (Figure 1e); embayment marshes that fringe the edge of open or restricted-entrance tidal embayment (Figure 1c and d); ria/loch-head marsh, occurring mostly in Europe (Brittany, Ireland and Scotland) (Figure 1g). Salt marshes can also develop where a gently shelving coast is combined with a high concentration of suspended sediment, such as the chenier plain (Rogers and Woodroffe, 2014).

Salt Marsh Ecosystem Characteristics

While there are several types of salt marshes, in the temporal climates, salt marshes are commonly divided into low marsh vegetated by salt-tolerant *Spartina alterniflora*, high marsh dominated by *Spartina patens*, *Distichlis spicata*, *S. alterniflora* (stunted), and *Juncus roemerianus*, and the upper marsh, the transitional zone from salt marsh to freshwater upland that is commonly occupied by *Phragmites australis*, *Iva frutescens*, *Cladium jamaicense*, *Typha* spp., and *Schoenoplectus* spp. (Adams, 1963; Tiner, 1987; Stuckey and Gould, 2000; Nikitina et al., 2003) (Figure 2).

Low biodiversity of the salt marsh ecosystem is related to frequent disturbance and stressful conditions (i.e., high salinity and hypoxia), where differences in plant community are controlled by surface

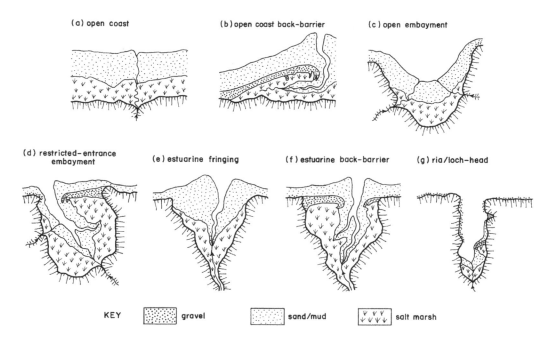

FIGURE 1 Geomorphological classification of salt marshes.
Source: Allen, 2000.

(a) Low marsh (*Sp. Alterniflora*)   (b) High marsh *(Sp. Patens/ Sp. Alterniflora)*

(c) Upper marsh (*Phragmites australis*)   (d) High marsh *(Distichlis spicata)*

(f) Upper marsh (*Iva frutescens/ Schoenoplectus americanus*)

FIGURE 2   Salt sub-environments and floral communities. (Photo credit: D. Nikitina.)

elevation, salinity regime, and duration and frequency of inundation (McKee and Patrick, 1988; Morris et al., 2002). Low salt marsh with monotypic stands of *S. alterniflora* occurs between mean tide level (MTL) and mean high water (MHW). Height variations within *S. alterniflora* are common, with interior marsh areas having lower vegetation and edges having taller vegetation. The tall (~1.5 m) herbaceous vegetation creates a dense habitat, both above and below ground, and provides habitat for fish, shellfish, and birds (Allen et al., 2018).

The high marsh lies between MHW and mean high high water (MHHW). Higher elevation zones can be more saline in drier climates, due to evaporative concentration of salts, or less saline in higher rainfall areas, due to frequent flushing of salts by fresh rainwater. In the area with hypersaline conditions, communities of succulent salt marsh plants, *Batis* and *Salicornia* spp. often develop (Allen et al., 2018).

The upper marsh occupies the brackish zone at elevations from MHHW to the highest astronomical tide (HAT) where tides mix with freshwater runoff. In climatic zones with warmer winter temperatures, temperate salt marshes naturally transition to mangrove or, in areas with lower precipitation, to salt flats. Arctic marshes are relatively narrow, species-poor, and uniform.

Salt Marsh Ecosystem Services

Salt marshes are economically important coastal landforms and productive ecosystems. Despite occupying <2% of the land, they provide critical ecosystem functions, such as production of organic material and nutrient cycling (Weinstein and Kreeger, 2000; Hopkinson et al., 2012; Howard et al., 2014). They act as nursery grounds for fisheries (Boesch and Turner, 1984; Tupper and Able, 2000); act as biological filters by sequestering pollutants from terrestrial runoff (Brin et al., 2010; Nelson and Zavaleta, 2012); buffer shorelines from storms by attenuating waves and currents that contribute to coastal inundation and damage (Costanza and Folke, 1997; De Groot et al., 2002); and provide tourism and recreation opportunities, education and research (Barbier et al., 2011).

In addition, coastal salt marshes contain 0.1% of the global sequestered terrestrial carbon and have been identified as a "Blue Carbon" ecosystem that plays a significant role in the Global Carbon System (Chung et al., 2011; Mcleod et al., 2011). Overall, "Blue Carbon" ecosystems are responsible for up to 50% of total ocean C burial and contain up to five times more C per unit area than terrestrial forest sediments and 20 times more than deep-sea sediments (Duarte et al., 2005, Cai, 2011, Ouyang and Lee, 2014).

C storage and accumulation capacity are high in salt marshes due to the continuous accumulation of organic C at a greater rate per unit area than any other "Blue Carbon" ecosystem (Chambers et al., 2001; Schlesinger and Lichter, 2001; Duarte et al., 2005; Murray et al., 2011, Chmura, 2013). Sources of organic C in a salt marsh include cordgrasses (primarily *Spartina* spp.), benthic algae, bacteria, allochthonous material including terrestrial and marine organic matter; however, the majority of C is being produced *in situ* by macrophytes (Ember et al., 1987; Radabaugh et al., 2018). Along the submerging coasts, rising sea level produces accommodation space for organic-rich sediments, thereby increasing the inventory of buried organic C through time, as long as salt marsh accretion keeps pace with the rate of rising sea (Kirwan and Blum, 2011; Kirwan and Mudd, 2012). After microbial decay, some portion of the buried C is sequestered over millennia in the marsh sediments (Mcleod et al., 2011, Chmura, 2013). If salt marshes are left undisturbed, then the C stored in the belowground sediments can remain for thousands of years due to anoxic conditions that prevent decomposition (Murray et al., 2011).

Salt Marsh Sustainability

However, despite protective measures, salt marshes are disappearing around the world at an alarming rate. It is estimated that at least 35% salt marshes have been lost globally (Duarte et al., 2008; Pendleton et al., 2012). In some countries, the loss is 70%–80% in the last 50 years (Hopkinson et al., 2012). Coastal marshes have been declining due to industrial development, aquaculture, agricultural reclamation, coastal erosion, and sea-level rise (French, 1997; Blackwell et al., 2004; Kirwan et al., 2010). Many salt marshes are at risk because of pollution by nutrient fluxes, with the largest increases in N flux occurring at coastlines with large areas of intertidal marshland in the temperate zones (Deegan et al., 2012). However, shoreline erosion and inundation are suggested to be the main mechanisms for current global salt marsh loss (Schwimmer, 2001; Mariotti and Fagherazzi, 2010; Theuerkauf et al., 2015).

In the United States, >96% of salt marsh loss is attributed to conversion of marsh to open water (Stedman and Dahl, 2008). In some places, this process could outpace C storage, especially if the depth of erosion at the seaward marsh edge is equal to or greater than the thickness of the marsh sediments (Theuerkauf et al., 2015). The potential future deterioration of marshes and the resulting conversion of their organic sediment to greenhouse gases (CO_2 and CH_4) represent significant impacts on the Global Carbon Budget contributing to global warming and will create positive feedbacks to sea-level rise and ocean acidification (Cai, 2011; Chmura et al., 2011).

The long-term sustainability of salt marsh ecosystem is dependent upon maintaining its elevation within intertidal zone by accumulating organic matter and trapping inorganic sediments (Stumpf, 1983; Morris et al., 2002). However, the processes controlling salt-marsh surface elevation are complex and operate in response to a number of factors including the rate of sea-level rise, sediment supply, tidal range, type of vegetation and primary production (Stevenson et al., 1986; Reed, 1995).

Despite multiple threats, salt marshes are resilient systems. Their ability to accrete vertically depends on concentration of suspended sediment, time of tidal submergence, and biomass density of the plants. Flooding is the main mechanism for sediment delivery to the marsh platform. During flooding, suspended sediments are transported with the tidal currents and partially deposited in distinctive spatial patterns across the marsh platform (Fagherazzi et al., 2012). While the hydrologic regime is largely determined by marsh position within the intertidal range, it is also controlled by geomorphological setting, mainly the proximity to tidal channel. Sedimentation rates across the salt marsh platform are found to decrease with increasing distance from tidal channels and from the seaward marsh edge (French et al., 1995; Leonard, 1997; Reed et al., 1999; Temmerman et al., 2003). Areas along the channel banks and marsh edge vegetated by dense canopy of *S. alterniflora* trap great amount of sediments (Christiansen et al., 2000). Interior areas, which are generally lower in elevation, are more susceptible to submergence and transition to open water, resulting in a disaggregated marsh landscape (Figure 3).

On the other hand, the lower portions of the marsh platform are flooded more frequently for a longer time period, so that more sediment is supplied and deposited. As sediments accrete vertically, sedimentation rates decrease with increasing platform elevation (Stoddart et al., 1989; Cahoon and Reed, 1995).

Hydrologic regime and marsh morphology can be directly modified by anthropogenic activity, including dredging, channelization, upstream modification of rivers and coastal engineering (Kennish, 2001; Turner, 2010). Indirect human modification of the environment including roads, pipelines, dams, oil and water wells and other structures not directly related to salt marsh may affect water and sediment flow, produce a barrier to plant and/or animal migration and contribute to habitat fragmentation, vegetation disturbance and marsh drowning (Allen et al., 2018). Dense vegetation cover favors sedimentation by decreasing tidal current velocities and turbulence once the water flows from the tidal channels into the marsh platform (Leonard and Luther, 1995; Yang, 1998; Christiansen et al., 2000).

FIGURE 3 Interior salt marsh platform recently converted to a salt pond, Slaughter beach salt marsh, Delaware, USA. (Photo credit: D. Nikitina.)

Perhaps the main threat to salt marshes globally is the recent acceleration in the relative sea-level rise (RSLR). Salt marshes have been developing under rising sea-level regime for the last 4000–7000 years by accreting vertically and keeping pace with rates of sea-level rise (Redfield and Rubin, 1962; Redfield, 1965; Stuiver and Daddario, 1963). During the 20th and early 21st centuries, rates of RSLR varied from 2.5 to 3.7 mm/yr (Kopp et al., 2014). For comparison, RSLR along the US Atlantic coast during the last 4000 years varied between 0.6 and 1.7 mm/yr and was mostly driven by glacio-isostatic adjustment of Earth's crust since the melting of the last ice sheet ~7000 years before present (Carlson et al., 2008; Engelhart et al., 2009). Geological data and the tide-gauge records show that RSLR became more rapid throughout the region since the Industrial Revolution identifying the US mid-Atlantic coast as a "hot spot" for sea-level rise (Kemp et al., 2011; Sallenger et al., 2012; Miller et al., 2013). Similar trends were observed elsewhere, reporting the global average rate of sea-level rise from 1880 to 2006 to be 1.7 ± 0.2 mm/yr, while the rate of global average sea-level rise measured from 1993 to 2013 was almost twice as high (3.2 ± 0.4 mm/yr), and this rate appears to be globally accelerating (Church and White, 2006; Jevrejeva et al., 2008; Nerem et al., 2010; Church et al., 2011; Rignot et al., 2011; Gehrels and Woodworth, 2012; Gregory et al., 2012).

Predicted marsh responses to RSLR is either retreat and conversion to tidal flat, tidal lagoon, or open water, if the future rates of RSLR exceed marshes' ability to maintain their elevation in equilibrium with sea level (Reed, 1995; Morris et al., 2002; French, 2006), keep pace with RSLR, or migrate landward. Analysis of geological data, modern observations and simulation studies suggest that salt marshes are resilient to rates of RSLR of ~7.1 mm/yr and likely to retreat when this value exceeds (Horton et al., 2018). Some landscape models using future projections of RSLR increase up to 20 mm/yr predict up to an 80% decrease in global tidal marsh area by 2100 (Spencer et al., 2016; Kopp et al., 2017).

By contrast, other studies suggest that, through biophysical feedback and inland marsh migration, marsh resilience to retreat is possible at RSLR rates in excess of 10 mm/yr (Rodger et al., 2012). Undisturbed vegetation on the marsh platform is a key factor in supporting equilibrium between marsh platform elevation and rapidly rising sea level (Morris et al., 2016). While salt marsh can rapidly subside and transition to open water, subsidence can be compensated by increased production of *S. alterniflora* regulating the elevation of sediment surface toward an equilibrium with mean sea level (Morris et al., 2002). In addition, dense vegetation canopy reduces peak flow velocities, turbulence and shear stress due to friction, enhancing the deposition of suspended sediment and ability of marsh to maintain its elevation within the tidal frame (Leonard and Luther, 1995; Christiansen et al., 2000; Neumeier and Ciavola, 2004; Bouma et al., 2005; Lightbody and Nepf, 2006; Möller et al., 2014).

Marshes also respond to RSLR by migrating into adjacent uplands where they are not restricted by topographic and anthropogenic barriers (Raabe and Stumpf, 2016). The rate of upslope marsh migration is determined by the slope and availability of suspended sediments. Computer simulations of marsh-upland system evolution under a variety of RSLR rates on a gently sloping upland, typical of Atlantic and Gulf coastal plains, suggest that marsh will contract in width under relatively low RSLR rates and low suspended sediment concentration (SSC) and expand under low-moderate-fast RSLR rates and high SSC (Kirwin et al., 2016). The rate of marsh landward transgression depends on the land use of the marsh-adjoining areas, where agricultural fields are more exposed to projected sea-level rise effects than grasslands or forested uplands (Fagherazzi et al., 2019). Construction of dikes, revetments and seawalls prevents salt marsh migration resulting in "coastal squeeze" that leads to marsh loss because erosion of the seaward salt marsh edge cannot be compensated by landward migration (Van der Wal and Pye, 2004).

Recent geological evidence suggests that salt marshes may be significantly eroded during storms and intense hurricane events (van de Plassche et al., 2006; Nikitina et al., 2014). Storm surges during hurricanes move large volumes of water mainly as fast sheet flow producing very high velocities, bottom shear stresses, and drag forces jeopardizing the stability of the vegetation and underlying sediments (Temmerman et al., 2005). Stripping of vegetation and platform erosion was widespread during hurricanes Katrina and Rita along the coast of Gulf of Mexico, USA (Morton and Barras 2011, Turner et al., 2006). While category 2–4 hurricanes are extreme and relatively rare events, salt marshes are often subject to erosion during the smaller storms because during a surge waves propagate over the marsh platform without impacting the edge of the marsh (Leonardi and Fagherazzi 2014, 2015).

Conclusion

Overall, salt marshes are highly dynamic coastal landforms vulnerable to many environmental stresses, mostly RSLR, storm erosion and anthropogenic impact. On the other hand, they are highly resilient coastal ecosystems. Their emergence and continued survival are driven by the rising sea level over millennia. Numerous feedbacks between tidal flooding, plant growth, organic matter production and sediment transport allow marshes to adapt to a wide range of RSLR rates in the vertical and lateral dimension. The global record of recent salt marsh decline correlates with anthropogenically induced acceleration in sea-level rise superimposed on other human activities (e.g., drainage and conversion to agriculture, coastal development, dredging). In many cases, the lateral expansion, the survival mechanism of salt marshes, is impossible due to the presence of topographic or anthropogenic barriers that limit salt marsh migration. However, the major risk of tidal marsh loss in the 21st century is the future rates of RSLR.

References

Adams, D.A., 1963. Factors influencing vascular plant zonation in North Carolina salt marsh. *Ecology*, 44: 445.

Allen, J.R.L., 2000. Morphodynamics of Holocene salt marshes: a review sketch from the Atlantic and Southern North Sea coasts of Europe. *Quaternary Science Reviews*, 19: 1155–1231.

Allen, J.R.L. and K. Pye, 1992. *Saltmarshes: Morphodynamics, Conservation, and Engineering Significance.* Cambridge University Press, Cambridge, 196 pp.

Allen, S.T., C.L. Stagg, J. Brenner, K.L. Goodin, D. Faber-Langendoen, C.A. Gabler, and K.W. Ames. 2018. Ecological resilience indicators for salt marsh ecosystems. In: Goodin, K.L. et al., *Ecological Resilience Indicators for Five Northern Gulf of Mexico Ecosystems*. NatureServe, Arlington, VA, 53 pp.

Barbier, E.B., S.D. Hacker, C. Kennedy, E.W. Koch, A.C. Stier, and B.R. Silliman, 2011. The value of estuarine and coastal ecosystem services. *Ecological Monographs*, 81(2): 169–193.

Blackwell, M.S.A, D.V. Hogana, and E. Maltbya, 2004. The short-term impact of managed realignment on soil environmental variables and hydrology. *Estuarine, Coastal and Shelf Science*, 59: 687–701.

Boesch, D.F. and R.E. Turner, 1984. Dependence of fishery species on salt marshes: the role of food and refuge. *Estuaries*, 7(4): 460–468.

Bouma, T.J., M.B. De Vries, E. Low, L. Kusters, P.M.J. Herman, I.C. Tanczos, A. Hesselink, S. Temmerman, P. Meire, and S. Van Regenmortel, 2005. Hydrodynamic measurements on a mudflat and in salt marsh vegetation: identifying general relationships for habitat characterisations. *Hydrobiologia*, 540: 259–274, doi:10.1007/s10750-004-7149-0.

Brin, L.D., I. Valiela, D. Goehringer, and B.L. Howes, 2010. Nitrogen interception and export by experimental salt marsh plots exposed to chronic nutrient addition. *MA Thesis*, Boston University.

Cahoon, D.R., and D.J. Reed, 1995. Relationships among marsh surface topography, hydroperiod, and soil accretion in a deteriorating Louisiana salt marsh. *Journal of Coastal Conservation*, 11: 357–369.

Cai, W.-J., 2011. Estuarine and coastal ocean carbon paradox: CO_2 sinks or sites of terrestrial carbon incineration? *Annual Review of Marine Science*, 3: 123–145.

Carlson, A.E., A.N. LeGrande, D.W. Oppo, R.E. Came, G.A. Schmidt, F.S. Anslow, J.M. Licciardi, and E.A. Obbink, 2008. Rapid early Holocene deglaciation of the Laurentide ice sheet. *Nature Geoscience*, 1: 620–624.

Chambers J.Q., N. Higuchi, E.S. Tribuzy, and S.E. Trumbore, 2001. Carbon sink for a century. *Nature*, 410: 429. doi.org/10.1038/35068624.

Chmura, G.L., 2013. What do we need to assess the sustainability of the tidal saltmarsh carbon sink? *Ocean & Coastal Management*, 83: 25–31. doi:10.1016/j.ocecoaman.2011.09.006.

Chmura, G.L., L. Kellman, and G.R. Guntenspergen, 2011. The greenhouse gas flux and potential global warming feedbacks of a northern macrotidal and microtidal salt marsh. *Environmental Research Letters*, 6(4): 1–6.

Christiansen, T., P.L. Wiberg, and T.G. Milligan, 2000. Flow and sediment transport on a tidal salt marsh surface. *Estuarine Coastal Shelf Science*, 50(3): 315–331, doi:10.1006/ecss.2000.0548.

Chung, I.K., J. Beardall, S. Mehta, D. Sahoo, and S. Stojkovic, 2011. Using marine macroalgae for carbon sequestration: a critical appraisal. *Journal of Applied Phycology*, 203(5): 877–886.

Church, J.A. and N.J. White, 2006. A 20th century acceleration in global sea-level rise. *Geophysical Research Letters*, 33, L01602, doi:10.1029/2005GL024826.

Church, J.A. and N.J. White, 2011. Sea-level rise from the late 19th to the early 21st century, *Surveys in Geophysics*, 32: 585–602, doi:10.1007/s10712-011-9119-1.

Costanza, R. and C. Folke, 1997. *Valuing Ecosystem Services with Efficiency, Fairness, and Sustainability as Goals, Nature's Services: Societal Dependence on Natural Ecosystems*. Island Press, Washington, DC, 49–70.

Deegan, L., D.S. Johnson, R.S. Warren, B.J. Peterson, J.W. Fleeger, S. Fagherazzi, and W.M. Wollheim, 2012. Coastal eutrophication as a driver of salt marsh loss. *Nature*, 490: 388–392.

De Groot, R.S., M.A. Wilson, and R.M. Boumans, 2002. A typology for the classification, description and valuation of ecosystem functions, goods and services. *Ecological Economics*, 41(3): 393–408.

Duarte, C.M., J.J. Middelburg, and N. Caraco, 2005. Major role of marine vegetation on the oceanic carbon cycle. *Biogeosciences*, 2: 1–8.

Duarte, C.M., W.C. Dennison, R.J.W. Orth, and T.J.B. Carruthers, 2008. The charisma of coastal ecosystems: addressing the imbalance. *Estuaries and Coasts*, 31: 233–238. doi:10.1007/s12237-008-9038-7.

Ember, L., D. Williams, and J. Morris, 1987. Processes that influence carbon isotope variations in salt marsh sediments. *Marine Ecology Progress Series* 36: 33–42.

Engelhart, S.E., B.P. Horton, B.C. Douglas, W.R. Peltier, and T.E. Tornqvist, 2009. Spatial variability of late Holocene and 20th century sea-level rise along the Atlantic coast of the United States. *Geology*, 37: 1115–1118.

Fagherazzi, S., S.C. Anisfeld, L.K. Blum, E.V. Long, R.A. Feagin, A. Fernandes, W.S. Kearney, and K. Williams, 2019. Sea level rise and the dynamics of the marsh-upland boundary. *Frontiers in Environmental Science*, 27, doi:10.3389/fenvs.2019.00025.

Fagherazzi, S., M.L. Kirwan, S.M. Mudd, Glenn R. Guntenspergen, S. Temmerman, A. Andrea D'Alpaos, J. van de Koppel, J.M. Rybczyk, E. Reyes, C. Craft, and J. Jonathan Clough, 2012. Numerical models of salt marsh evolution: ecological, geomorphic, and climatic factors. *Reviews of Geophysics*, 50, RG1002, doi:10.1029/2011RG000359.

French, J.R., 2006. Tidal marsh sedimentation and resilience to environmental change: exploratory modelling of tidal, sea-level and sediment supply forcing in predominantly allochthonous systems. *Marine Geology*, 235: 119–136.

French, J.R., T. Spencer, A.L. Murray, and N.S. Arnold, 1995. Geostatistical analysis of sediment deposition in two small tidal wetlands, Norfolk, United Kingdom. *Journal of Coastal Research*, 11: 308–321.

French, P.W., 1997. *Coastal and Estuarine Management* (Routledge Environmental Management Series). Routledge, London, 251.

Gehrels, W.R. and P.L. Woodworth, 2012. When did modern rates of sea-level rise start? *Global and Planetary Change*, 100: 263–277, doi:10.1016/j.gloplacha.2012.10.020.

Gregory, J.M., et al., 2012. Twentieth-century global-mean sea-level rise: is the whole greater than the sum of the parts? *Journal of Climate*, 26: 4476–4499, doi:10.1175/JCLI-D-12-00319.1.

Hopkinson, C.S., W-J. Cai, and X. Hu, 2012. Carbon sequestration in wetland dominated coastal systems – a global sink of rapidly diminishing magnitude. *Current Opinion in Environmental Sustainability*, 4: 186–194, doi.org/10.1016/j.cosust.2012.03.005.

Horton, B.P, I. Shennan, S. Bradley, N. Chahill, M. Kirwan, R.E. Kopp, and T.A. Shaw, 2018. Predicting marsh vulnerability to sea-level rise using Holocene relative sea-level data. *Nature Communications*, 9: 2687.

Howard, J., S. Hoyt, K. Isensee, E. Pidgeon, and M. Telszewski, 2014. *Coastal Blue Carbon: Methods for Assessing Carbon Stocks and Emissions Factors in Mangroves, Tidal Salt Marshes, and Seagrass Meadows*. Conservation International, Intergovernmental Oceanographic Commission of UNESCO, International Union for Conservation of Nature, Arlington, VI, USA.

Jevrejeva, S., J.S. Moore, A. Grinsted, and P.L. Woodworth, 2008. Recent global sea level acceleration started over 200 years ago? *Geophysical Research Letters*, 35, L08715, doi:10.1029/2008GL033611.

Kemp, A.C., B.P. Horton, J.P. Donnelly, M.E. Mann, M. Vermeer, and S. Rahmstorf, 2011. Climate related sea-level variations over the past two millennia. *Proceedings of the National Academy of Sciences of the United States of America*, 108: 11,017–11,022, doi:10.1073/pnas.1015619108.

Kennish, M.J., 2001. Coastal salt marsh systems in the U.S.: a review of anthropogenic impacts. *Journal of Coastal Research*, 17: 731–748.

Kirwan, M.L. and L.K. Blum, 2011. Enhanced decomposition offsets enhanced productivity and soil carbon accumulation in coastal wetlands responding to climate change. *Biogeosciences* 8(4): 987. doi:10.5194/bg-8-987-201.

Kirwan, M.L. and S.M. Mudd, 2012. Response of salt-marhs carbon accumulation to climate change. *Nature*, 489: 550–553, doi:10.1038/nature11440.

Kirwan, M.L., G.R. Guntenspergen, A. D'Alpaos, J.T. Morris, S.M. Mudd, and S. Temmerman, 2010. Limits on the adaptability of coastal marshes to rising sea level. *Geophysical Research Letters*, 37. doi:10.1029/2010GL045489.

Kirwan, M.L., S. Temmerman, E.E. Skeehan, G.R. Guntenspergen, and S. Fagherazzi, 2016. Overestimation of marsh vulnerability to sea level rise. *Nature Climate Change*, 6: 253–260.

Kopp, R.E., R.M. DeConto, D.A. Bader, C.C. Hay, R.M. Horton, S. Kulp, M. Oppenheimer, D. Pollard, and B.H. Strauss, 2017. Evolving understanding of Antarctic ice-sheet physics and ambiguity in probabilistic sea-level projections. *Earth's Future*, 5: 1217–1233, doi.org/10.1002/2017EF000663.

Kopp, R.E., R.M. Horton, C.M. Little, J.X. Mitrovica, M. Oppenheimer, D.J. Rasmussen, B.H. Strauss, and C. Tebaldi, 2014. Probalisitic 21st and 22nd century sea-level projections at a global network of tide-guage sites. *Earth's Future*, 2: 383–406.

Leonard, L.A., 1997. Controls on sediment transport and deposition in an incised mainland marsh basin, southeastern North Carolina. *Wetlands*, 17: 263–274, doi:10.1007/BF03161414.

Leonard, L.A. and M.E. Luther, 1995. Flow hydrodynamics in tidal marsh canopies. *Limnology and Oceanography*, 40(8), 1474–1484, doi:10.4319/lo.1995.40.8.1474.

Leonardi, N. and S. Fagherazzi, 2014. How waves shape salt marshes. *Geology*, 42(10): 887–890.

Leonardi, N. and S. Fagherazzi, 2015, Effect of local variability in erosional resistance on largescale morphodynamic response of salt marshes to wind waves and extreme events. *Geophysical Research Letters*, 42(14): 5872–5879.

Lightbody, A.F. and H.M. Nepf, 2006. Prediction of velocity profiles and longitudinal dispersion in emergent salt marsh vegetation. *Limnology and Oceanography*, 51: 218–228, doi:10.4319/lo.2006.51.1.0218.

Mariotti, G., and S. Fagherazzi, 2010. A numerical model for the long-term evolution of salt marshes and tidal flats. *Journal of Geophysical Research*, 115, F01004. doi:10.1029/2009JF001326.

McKee, K.L. and W.H. Patrick, 1988. The relationship of smooth cordgrass (*Spartina alterniflora*) to tidal datums: a review. *Estuaries*, 11: 143–151.

Mcleod, E., G.L. Chmura, S. Bouillon, R. Salm, M. Björk, C.M. Duarte, C.E. Lovelock, W.H. Schlesinger, and B.R. Silliman, 2011. A blueprint for blue carbon: toward an improved understanding of the role of vegetated coastal habitats in sequestering CO_2. *Frontiers in Ecology and the Environment*, 9: 552–560.

Miller, K.G., R.E. Kopp, B.P. Horton, J.V. Browning, and A.C. Kemp, 2013. A geological perspective on sea-level rise and its impacts along the U.S. mid-Atlantic coast. *Earth's Future*, doi:10.1002/2013EF000135.

Möller, I., M. Kudella , F. Rupprecht, T. Spencer, M. Paul, B.K. van Wesenbeeck, G. Wolters, K. Jensen, T.J. Bouma, M. Miranda-Lange, and S. Schimmels, 2014. Wave attenuation over coastal salt marshes under storm surge conditions. *Nature Geoscience*, 7(10): 727–731.

Morris, J.T., D.C. Barber, J.C. Callaway, R. Chambers, S.C. Hagen, C.S. Hopkinson, P. Johnson, B.J. Megonigal, S.C. Neubauer, T. Troxler, and C. Wigand, 2016. Contributions of organic and inorganic matter to sediment volume and accretion in tidal wetlands at steady state. *Earth's Future*, 4. doi:10.1002/2015EF000334.

Morris, J.T., P.V. Sundareshwar, C.T. Nietch, B. Kjerfve, and D.R. Cahoon, 2002. Responses of coastal wetlands to rising sea level. *Ecology*, 83(10): 2869–2877, doi:10.1890/0012-9658.

Morton, R.A. and J.A. Barras, 2011. Hurricane impacts on coastal wetlands: a half-century record of storm-generated features from southern Louisiana. *Journal of Coastal Research*, 27(6A): 27–43.

Murray, B.C., L. Pendleton, and S. Sifleet, 2011. *State of the Science on Coastal Blue Carbon: A Summary for Policy Makers*. Nicholas Institute for Environmental Policy Solutions Report NIR 11-06: 1–43.

Nelson, J.L. and E.S. Zavaleta, 2012, Salt marsh as a coastal filter for the oceans: changes in function with experimental increases in nitrogen loading and sea-level rise. *PloS One*, 7(8): e38558.

Nerem, R.S., D.P. Chambers, C. Choe, and G.T. Mitchum, 2010. Estimating mean sea-level change from the TOPEX and Jason Altimeter Missions, Mar. *Geodesy*, 33: 435–446, doi:10.1080/01490419.2010. 491031.

Neumeier, U. and P. Ciavola, 2004. Flow resistance and associated sedimentary processes in a Spartina maritima salt marsh. *Journal of Coastal Research*, 20: 435–447, doi:10.2112/1551-5036.

Nikitina, D.L., A.C. Kemp, B.P. Horton, C.H. Vane, O. van de Plassche, and S.E. Engelhart, 2014. Storm erosion during the past 2000 years along the north shore of Delaware Bay, USA. *Geomorphology*, 208: 160–172.

Nikitina, D.L., J.E. Pizzuto, R.E. Martin, and S.P. Hippensteel, 2003. Transgressive valley-fill stratigraphy and sea-level history of the Leipsic River, Bombay Hook National Wildlife Refuge, Delaware, U.S.A. In Olson, H.C. and Leckie, R.M. (Eds.), *Micropaleontologic Proxies for Sea-Level Changes and Stratigraphic Discontinuities*. SEPM Special Publication, Tulsa, OK, 75, 51–62.

Ouyang, X. and S.Y. Lee, 2014. Updated estimates of carbon accumulation in coastal marsh sediments. *Biogeosciences*, 11: 5057–5071.

Pendleton, L., D.C. Donato, B.C. Murray, S. Crooks, W.A. Jenkins, S. Sifleet, C. Craft, J.W. Fourqurean, J.B. Kauffman, N. Marbá, P. Megonigal, E. Pidgeon, D. Herr, D. Gordon, A. Balbera, 2012. Estimating global "Blue Carbon" emissions from conversion and degradation of vegetated coastal ecosystems. *PLoS ONE* 7(9): e43542, doi:10.1371/journal.pone.0043542.

Raabe, E.A. and R.P. Stumpf, 2016. Expansion of tidal marsh in response to sea-level rise: Gulf Coast of Florida, USA. *Estuaries Coasts*, 39: 145–157.

Radabaugh, K.R., Moyer, R.P., Chappel, A.R., Powell, C.E., Bociu, I., Clark, B.C., Smoak, J.M., 2018, Coastal Blue Carbon Assessment of Mangroves, Salt Marshes, and Salt Barrens in Tampa Bay, Florida, USA. *Estuaries and Coasts*, 41: 1496, doi:10.1007/s12237-017-0362-7.

Redfield, A.C., 1965. Ontogeny of a salt marsh estuary. *Science*, 147: 50–55, doi:10.1126/science.147.3653.50.

Redfield, A.C. and M. Rubin, 1962. The age of salt marsh peat and its relation to recent changes in sea level at Barnstable, Massachusetts. *Proceedings of the National Academy of Sciences of the United States of America*, 48(10): 1728.

Reed, D.J., 1995. The response of coastal marshes to sea-level rise: survival or submergence? *Earth Surface Processes and Landforms*, 20: 39–48. doi:10.1002/esp.3290200105.

Reed, D.J., T. Spencer, A.L. Murray, J.R. French, and L. Leonard, 1999. Marsh surface sediment deposition and the role of tidal creeks: implications for created and managed coastal marshes. *Journal of Coastal Conservation*, 5: 81–90, doi:10.1007/BF02802742.

Rignot, E., I. Velicogna, M.R. van den Broeke, A. Monaghan, and J. Lenaerts, 2011. Acceleration of the contribution of the Greenland and Antarctic ice sheets to sea-level rise. *Geophysical Research Letters*, 38, L05503, doi:10.1029/2011GL046583.

Rogers, K. and C.D. Woodroffe, 2014. Tidal Flats and Salt marshes. In G. Masselink, and R. Gehrels (Eds.), *Coastal Environments and Global Change*. Wiley & Sons, Chichester, 227–248.

Rogers, K., N. Saintilan, and C. Copeland, 2012. Modelling wetland surface elevation dynamics and its application to forecasting the effects of sea-level rise on estuarine wetlands. *Ecological Modelling* 244: 148–157.

Sallenger, A.H., Jr., K.S. Doran, and P.A. Howd, 2012. Hotspot of accelerated sea-level rise on the Atlantic Coast of North America. *Nature Climate Change*, 2: 884–888, doi:10.1038/nclimate1597.

Schlesinger, W.H. and J. Lichter, 2001. Limited carbon storage in soil and litter of experimental forest plots under increased atmospheric CO_2. *Nature*, 411: 466–469.

Schwimmer, R.A., 2001. Rates and processes of Marsh shoreline erosion in Rehoboth Bay, Delaware, U.S.A. *Journal of Coastal Research*, 17(3): 672–683.

Spencer, T., M. Schürch, R.J. Nicholls, J. Hinkel, A. Vafeidis, R. Reef, L. McFadden, et al., 2016. Global coastal wetland change under sea-level rise and related stresses: the DIVA Wetland Change Model. *Global and Planetary Change*, 139: 15–30, doi:10.1016/j.gloplacha.2015.12.018.

Stedman, S. and T.E. Dahl. 2008. Status and trends of wetlands in the coastal watersheds of the Eastern United States 1998 to 2004. National Oceanic and Atmospheric Administration, National Marine Fisheries Service and U.S. Department of the Interior, Fish and Wildlife Service. 20 pp.

Stevenson, J.C., L.G. Ward, and M.S. Kearney, 1986. Vertical accretion in marshes with varying rates of sea level rise, In D.A. Wolfe (Ed.), *Estuarine Variability*. Academic Press, San Diego, CA, 241–259.

Stoddart, D.R., D.J. Reed, and J.R. French, 1989. Understanding salt marsh accretion, Scolt Head Island, Norfolk, England. *Estuaries*, 12(4): 228–236, doi:10.2307/1351902.

Stuckey, I.H. and L.L. Gould, 2000. *Coastal Plants from Cape Cod to Cape Canaveral*. University of North Carolina Press, Chapel Hill.

Stuiver, M. and J.J. Daddario, 1963. Submergence of the New Jersey coast. *Science*, 142(3594): 951.

Stumpf, R.P., 1983. The process of sedimentation on the surface of a salt marsh, Estuarine. *Coastal and Shelf Science*, 17(5): 495–508.

Temmerman, S., T.J. Bouma, G. Govers, and D. Lauwaet, 2005. Flow paths of water and sediment in a tidal marsh: relations with marsh developmental stage and tidal inundation height. *Estuaries*, 28(3): 338–352, doi:10.1007/BF02693917.

Theuerkauf, E.J., J.D. Stephens, J.T. Ridge, F.J. Fodrie, and A.B. Rodriguez, 2015. Carbon export from fringing saltmarsh shoreline erosion overwhelms carbon storage across a critical width threshold: Estuarine. *Coastal and Shelf Science*, 164: 367–378, dx.doi.org/10.1016/j.ecss.2015.08.001.

Tiner, R.W. Jr, 1987. *Field Guide to Coastal Wetland Plants of the Northern United States*. University of Massachusetts Press, Amherst.

Tupper, M. and K. Able, 2000. Movements and food habits of striped bass (Morone saxatilis) in Delaware Bay (USA) salt marshes: comparison of a restored and a reference marsh. *Marine Biology*, 137(5–6): 1049–1058.

Turner, R.E., 2010. Beneath the salt marsh canopy: loss of soil strength with increasing nutrient loads. *Estuaries Coasts* 34: 1084–1093. doi:10.1007/s12237-010-9341-y.

Turner, R.E., J.J. Baustian, E.M. Swenson, and J.S. Spicer, 2006. Wetland sedimentation from hurricanes Katrina and Rita. *Science*, 314(5798): 449–452.

van de Plassche, O., G. Erkens, F. van Vliet, J. Brandsma, K. van der Borg, and A.F. de Jong, 2006. Salt-marsh erosion associated with hurricane landfall in southern New England in the fifteenth and seventeenth centuries. *Geology*, 34(10): 829–832.

Van der Wal, D. and K. Pye, 2004. Patterns, rates and possible causes of saltmarsh erosion in the Greater Thames area (UK). *Geomorphology*, 61: 373–391.

Weinstein, M.P. and Kreeger, D.A., 2000. *Concepts and Controversies in Tidal Marsh Ecology*, Springer, Dordrecht, doi:10.1007/0-306-47534-0.

Yang, S.L., 1998. The role of Scirpus marsh in attenuation of hydrodynamics and retention of fine sediment in the Yangtze Estuary. *Estuarine Coastal Shelf Science*, 47: 227–233, doi:10.1006/ecss.1998.0348.

23

The Evolution of Water Resources Management

Introduction 225
Water Resources Management in Pre-Modern Times 226
Colonial Times...227
The Hydraulic Paradigm...227
IWRM – Dominant Paradigm Today 228
Adaptive Management...230
Water–Energy–Food Nexus 231
Water Management in an Increasingly Urban World............................232
Conclusion 233
References..234

Francine van den
Brandeler and
Joyeeta Gupta

Introduction

Water resources are crucial for lives, livelihoods, biodiversity, ecosystems, and economic development. The adequate management of these water resources is therefore intrinsic to sustainable and inclusive development. Water problems can be broadly characterized in terms of quantity, quality, and the effects of extreme weather events (UNE 2019). Water quantity challenges can be absolute or relative (i.e., dependent on the ratio between water availability and demand), with 'chronic water shortages' defined as annual water supplies below 1000 m³ per person within a certain area drop and 'absolute water scarcity' as less than 500 m³ per person (FAO 2012). These challenges may be caused or aggravated by overabstracting water from rivers and groundwater resources. Quantitative challenges can also relate to the unequal distribution of water resources within an area and inequality of access. Qualitative water challenges relate to pollution from point sources (e.g., untreated industrial or domestic wastewater discharge) and diffuse pollution (e.g., fertilizers and pesticides from agriculture and surface runoff in urban areas) (Martinez-Santos et al. 2014; Elmqvist et al. 2013). Water-related risks linked to extreme weather events include droughts, floods, and landslides, with devastating consequences on human settlements, agriculture, and ecosystems (GRID 2018).

These challenges are affected by five driving forces that have put increased pressures on water systems around the world: (1) population dynamics, (2) urbanization, (3) technology development, (4) a constant pursuit of economic growth, and (5) climate change (UNE 2019). These, in turn, affect land use and land-use changes, such as deforestation to clear land for agriculture and urbanization; these lead to further deterioration of water quality by disturbing ecosystems and the hydrological cycle (Azzam et al. 2014; Avissar and Werth 2005). Climate variability and change affect hydrological systems by making extreme weather events more intense and more frequent (IPCC 2014).

Water resources management aims to address these challenges and underlying driving forces. Management encompasses 'the processes of decision-making, coordination, and resource deployment that occur within a given institutional setting assuming no change in rules and norms' (Hatfield-Dodds et al. 2007, 3). Although management approaches rely on scientific and technical knowledge, they are also influenced by available human and financial resources, as well as contextual factors, politics, and power relations. Indeed, "the challenge of the water crisis is first and foremost a 'crisis of governance'" (UNESCO 2006, 1), relating not only to how management systems are designed but also to how water is conceptualized, who owns water, how boundaries are delineated for management purposes, and how links to other issues are framed (Gupta et al. 2013). Water resources management approaches, in this sense, are a particular strategy of governance. They influence the institutional tools selected for allocating, regulating, and preserving water resources, the type of water (blue, black, gray, and green), infrastructure and technologies, and information systems. Water resources management approaches have evolved over time and increasingly incorporate notions of sustainability and inclusiveness, and this is reflected at global level through the adoption of a specific goal on water within the Sustainable Development Goals (SDGs) (UNGA 2015), while many other goals are directly or indirectly linked to water. Nevertheless, as water-related challenges multiply and increase in complexity, there is growing uncertainty about the most adaptive approach to manage water resources. The sections below discuss several of the main paradigms that have shaped water resources management over time, including the control of water resources to consolidate power, which started with the first ancient civilizations; water resources management as an instrument of colonization; the hydraulic paradigm; Integrated Water Resources Management (IWRM); Adaptive Management (AM); the Water–Energy–Food Nexus; and urban-centric approaches to water management.

Water Resources Management in Pre-Modern Times

Rules on managing water first developed in local customary law which later spread through the expansion of the early river basin civilizations (Gupta and Dellapenna 2009). Major rivers historically attracted communities; early states aimed at consolidating power by controlling water resources (Wittfogel 1955). Harnessing water resources was essential for the development of agriculture in Mesopotamia and Egypt, around 5700–3200 BCE, which included systems of communal canals, irrigation works, and legal frameworks to govern these (Angelakis and Zheng 2015; Kornfeld 2009). The Indus Valley civilization developed around the Indus Valley. In ancient Rome, aqueducts, fountains, and pipes on a large-scale supplied freshwater to cities and removed wastewater (Wilson 1998). Complex management systems developed first in areas where water was scarce such as the Middle East and influenced Islamic and Jewish water law and policy. The Aztecs founded Tenochtitlán (now Mexico City) in 1324 on an island surrounded by lakes within an endorheic basin. The lakes protected them from enemy invaders and provided fish and food from floating farms and through canals and irrigation infrastructures (Sosa-Rodriguez 2010). Their dependence on the lakes meant that it was essential for their survival that they preserved the hydrological balance. Exposed to frequent droughts and floods, the Aztecs also suffered from insufficient water supply, low-quality water, and inadequate sanitation. To exercise control over the basin's hydrology, they not only built drains, dams, dikes, floodgates, and aqueducts but also implemented strict rules such as the prohibition of waste disposal into water bodies. In the Netherlands, the water boards established in the 13th century to protect the lowest-lying regions from floods still exist and have been institutionalized into about 200 boards today (Lazaroms and Poos 2004). These democratically elected decentralized government authorities manage local to regional water bodies to manage water quality- and quantity-related issues and are allowed to raise resources through taxes.

The fall of civilizations has frequently been attributed to water mismanagement. The Sumerian and Indus Valley civilizations may have collapsed from soil salinization through irrigated agriculture in an arid climate (Alam et al. 2007). The Maya civilization's downfall in the early 900s may have been caused by a series of drought events that urban elites were unable to adapt to (Lucero et al. 2011). Overall, water resources management and the ability to adapt to changing conditions have been key to both the rise and fall of civilizations.

Colonial Times

During colonial times, water was central for exploration, establishing sovereignty and commercial benefits. River navigation played an important role. Portuguese colonizers in Brazil strategically used navigation to explore the newly conquered territory, map the land, control rivers, and facilitate state integration (Paganini 2008; Leite Farias 2009). In the next stage, administrating colonies required water resources management and institutions for irrigation, flood and drought prevention, and transportation (Bhattacharya 2017). After the fall of ancient civilizations, their sophisticated irrigation techniques and hydraulic works disappeared almost everywhere except in China, but these reappeared in the 19th century, as colonial powers mobilized massive corvée labor and diverted rivers to irrigate alluvial planes and deltas at great social cost (Molle et al. 2009). Colonial powers frequently cast aside indigenous water management knowledge and practices to impose theirs while possessing little understanding of local hydrology. For instance, the Spaniards transformed the basin's hydrology after they invaded Tenochtitlán. They associated the Aztec city's surrounding lakes with an unhealthy environment and a source of disease, and soon began ambitious efforts to drain the lakes, which aggravated recurrent floods (Sosa-Rodriguez 2010).

Water resources management institutions imposed as part of the colonization process often partly or largely replaced customary water regimes and asserted the colonial state's ownership over water resources (Cullet and Gupta 2009). Customary laws often continue to influence water regimes and co-exist with colonial and post-colonial institutions, leading to dual legal regimes (Nilsson and Nyanchaga 2009). In rural early America, water use was considered a community right and, as the nation consolidated, water use acquired some standing as an individual right. Conflicts between the two led to the doctrine of riparian rights, which states that water belongs to the person whose land borders a body of water (Burke 1956). With increasing competition for water resources in the 19th century, the rule of prior appropriation was introduced ('first in time, first in right'), particularly in the western states (Hanak et al. 2011).

Overall, colonialism commodified water resources while dismantling community control, causing the impoverishment of rural populations, as was the case in India where indigenous water harvesting systems were purposely destroyed (D'Souza 2006). Infrastructure and technology for drinking water provision were mainly developed in areas occupied by the colonizers as European powers were reluctant to invest in their overseas colonies unless this would be profitable, but rapidly growing cities, such as Bombay in the British Empire, experienced deterioration in urban conditions and rising tensions (Anand 2011; Nilsson and Nyanchaga 2009; Gandy 2008).

The Hydraulic Paradigm

Scientific developments, fascination with 'scientific irrigation', and ideas such as the domination of nature and the creation of new Edens in deserts cumulated into large public investments in dams and irrigation systems as of the beginning of the 20th century (Molle et al. 2009; Smythe 1905). States developed water systems as part of a 'Hydraulic mission' that aimed to increase food production, raise rural incomes, respond to growing urban water demand, strengthen state building, and legitimize governments (Hanak et al. 2011; Molle et al. 2009). In Spain, this was part of a response to the humiliating loss of colonies and the turn of colonial ambitions inward to dry and poor rural areas (Lopez-Gunn 2009). The state-led hydraulic policy served to legitimize the state, reinvent the economy, and assuage the threat of a discontented peasantry. Technological advancements, such as drilling and gasoline- and diesel-powered pumps, enhanced access to groundwater and further expanded agricultural land in dry regions such as California (Hanak et al. 2011).

The hydraulic paradigm spread worldwide, as water managers and governments strived to control the natural environment through infrastructures for reducing floods, generating hydropower and supplying cities with drinking water (Custódio 2012; Swyngedouw 1999; Lopez-Gunn 2009; Molle et al. 2009), requiring large investments and centralized management (Pahl-Wostl et al., 2007; Huitema and Meijerink, 2017). Within the United States, urban water supply systems were initially often built by small, private companies and mainly served wealthier neighborhoods (Porse 2014). However, seeking more reliable options for public health, flood control, and economic growth in urban areas, water delivery, and stormwater and sewage conveyance services were increasingly centralized and municipalized in the late 19th and early 20th centuries (Rietveld et al. 2016; Domènech 2011). As cities grew, dams and aqueducts were built to bring water from distant locations and replaced dwindling or polluted local sources (Porse 2014; Hanak et al. 2011), and the emphasis was on supply augmentation rather than demand management (Xie 2006).

The creation of national water bureaucracies in the post-World War II period and efforts to modernize customary, state, and colonial water laws soon followed to regulate many issues including large-scale engineering works (Hanak et al. 2011). The physical integration of waterworks in fragmented countries supported their integration and often became a focal point for nation-building (Huitema and Meijerink 2017). However, this process often involved the expropriation of rights from private owners to state bureaucracies in order to further national economic interests by strengthening agriculture and emerging industries. This included local and indigenous water rights under colonial and post-colonial regimes (Boelens 2009). Water management tasks were generally allocated to government agencies at provincial or national levels (Huitema and Meijerink 2017). Different tasks were the responsibility of different sectors (agriculture, domestic use, industry, environmental protection, etc.) with often little coordination between them (Xie 2006).

The Hydraulic Era gave way to different forms of water control. Large-scale, centralized infrastructure provided many societal and economic benefits but could not cover distant areas and vulnerable groups. In California, lakes dried up and rivers became wastewater canals, aquatic species declined sharply, and farmers complained about increasing salinity in their water supplies due to upstream diversions (Hanak et al. 2011). With the start of the environmental movement in the 1970s, demands for regulating pollution of water became more important. This movement spread around the world, strengthening environmental legislation and also emphasizing nature as a legitimate user of water resources (Hanak et al. 2011; Kallis and Coccossis 2002). This was institutionalized through measures such as the establishment of minimum environmental flows and maximum groundwater extractions, environmental impact assessments, and stricter regulations regarding wastewater discharges. This influenced water governance. Another feature influencing water governance was the shift from 'government to governance' and 'governance-beyond-the-state' (Huitema and Meijerink 2017; Swyngedouw 2005) and the rise of the non-state actor. This coincided with the neo-liberal demand for lean states and deregulation and led to water resources being managed by assemblages of public, private, and civil society actors with different approaches to water, shaping not only how they address problems but also how they interact with other actors (Brandeler et al. 2014). Power is transferred to both lower and higher levels. Decentralization processes have brought decision-making closer to those affected, enhancing democratic legitimacy (Dryzek 1997) and resulting in policies more likely to be supported at local levels (Kasemir et al. 2003). However, this has also led to a lack of accountability and patchy institutions that are unable to address society's needs.

Meanwhile, the recognition of global water drivers and cumulative impacts and the need for common norms have increased support for global water governance in order to address the global drivers (Gupta and Pahl-Wostl 2013; Vörösmarty et al. 2013). With the adoption of the SDGs, water governance can no longer be seen as independent of all the other goals. Some scholars see that the IWRM paradigm may have to evolve into a Nexus paradigm in order to be able to preempt the problems of tomorrow.

IWRM – Dominant Paradigm Today

The ideas behind the concept of IWRM emerged in the 1970s through practitioners, and these first appeared in official water policy circles at the global water conference in Mar del Plata in 1977, which emphasized the need for coordinating different users and authorities, participation, and legal frameworks to ensure effective water allocation (WWAP 2009). It was developed in the late 1980s and was adopted in the Dublin

Principles adopted by the International Conference on Water and the Environment in Dublin. The four principles recognized freshwater as 'a finite and vulnerable resource', integrated, participatory management at the 'lowest appropriate level', emphasis on the central role of water as well as on the economic value of water (ICWE 1992). The Global Water Partnership (GWP) defined IWRM as 'a process that promotes the coordinated development and management of water, land, and related resources, in order to maximize the resultant economic and social welfare in an equitable manner without compromising the sustainability of vital ecosystems' (GWP-TAC 2000). IWRM has since become the dominant paradigm for water resources management, having been adopted by international agencies and transnational actors such as the World Bank, the European Union (EU) (e.g., Water Framework Directive and EU Water Initiative), and the United Nations, most recently through the latter's SDGs (UNGA 2015; Abers and Keck 2013; Wallington et al. 2010; Abers 2007; Molle 2009) but also by a growing number of national governments to frame their laws and policies (Gupta and Pahl-Wostl 2013; UNEP 2012).

IWRM strives to address social equity, including the water-related needs of the poor (Hooper 2005; Rahaman and Varis 2005). It promotes the integration of human and natural systems (Dzwairo et al. 2010), sectoral responses, and viewpoints and interests (Medema et al. 2008; Jønch-Clausen and Fugl 2001). While marginalized people are given a voice, IWRM also mirrors general governance shifts towards greater private sector involvement through emphasis on cost recovery, the establishment of pricing mechanisms and cost-benefit analyses, and allocation to the most beneficial uses (Huitema and Meijerink 2017; Xie 2006). Environmental sustainability is addressed through concerns about the ecological impacts of water resources management activities, including through disruption to the water cycle, pollution control, development planning, demand management, and biodiversity conservation (Medema et al. 2008; Savenije and Van der Zaag 2008; Grigg 2008). In addition, although IWRM works through national-level legislation, policies, and institutions, it emphasizes the river basin as the ideal unit for water resources management and the creation of river basin organizations (RBOs) (Butterworth et al. 2010; Hooper 2005; Abdullah and Christensen 2004). It is considered as crucial to integrate upstream and downstream concerns, to foster bottom-up planning (Molle 2009; Xie 2006). This systems approach is often seen as the key innovation of IWRM (Huitema and Meijerink 2017).

However, there are many criticisms regarding the limitations of the IWRM concept and how to implement it in practice. River basins follow hydrological boundaries and therefore frequently cut across administrative boundaries, including national borders (Butterworth et al. 2010). This can create tensions between hydrological and institutional logics and create cooperation challenges. In addition, achieving effective participation that leads to greater inclusiveness is challenging because, even with the best intentions, it requires significant resources and time commitments from participants, as well as expert knowledge to follow and intervene in discussions (Brandeler et al. 2014; Butterworth et al. 2010). Similarly, the multiple integration goals (i.e., land/water, surface/groundwater, human/nature, upstream/downstream) are unrealistic (Biswas 2008), in particular in contexts of uncertainty, insufficient data, low political will, and limited financial resources (Agyenim 2011; Molle 2008; Watson 2004; Allan 2003). Moreover, the principle of water as an economic good can lead to it being treated as a commodity, at the expense of its nonmonetary value (Rahaman and Varis 2005). IWRM is often depoliticized, even though it is fundamentally a political process, as it involves choices such as the allocation of water between competing users (Jønch-Clausen and Fugl 2001). Despite the legitimacy and mandates that they are given, RBOs are faced with the traditional powerful line agencies, such as irrigation and agriculture authorities, which create obstacles for these organizations to assert themselves as sites of decision-making (Molle 2008; Wester and Warner 2002).

A search was conducted in ScienceDirect for IWRM and other concepts discussed below (Adaptive Water Management or AWM, Water–Energy–Food Nexus, Integrated or Sustainable Urban Water Management or SUWM) in titles, abstracts, and key words for the period 1980–2018. Figure 1 illustrates the sharp rise in publications for all these concepts since the turn of the 21st century. The growing recognition of IWRM's limitations may explain the concept's relative loss of ground to other approaches in the past decade.

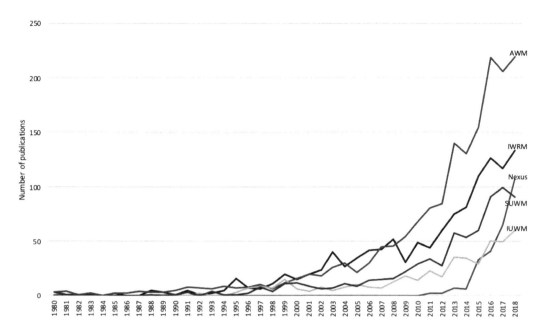

FIGURE 1   Evolution of water resources management approaches within academic literature.

Adaptive Management

AM can be traced back to the late 1970s and is based on the recognition that, although interactions between people and ecosystems can be unpredictable, it is necessary to take management action (Medema et al. 2008; Johnson 1999; Walters and Hilborn 1978; Holling 1978). It arose in a context of rising environmental challenges and public pressure to address these and aimed to support governments and policy makers through new conceptualizations (Hurlbert 2016). AM is a systematic, iterative process, centered on a learning model to not just cope with complex water-related challenges but to learn from them and improve management policies (Medema et al. 2008; Pahl-Wostl 2008). It should enable experimentation and comparison between selected policies and practices (Medema et al. 2008). As IWRM, AM therefore promotes some degree of institutional reform. However, IWRM does not clarify how implementation works within a context of uncertainty nor does it refer explicitly to adaptive capacity (Agyenim 2011; Butterworth et al. 2010). While enhancing democratic legitimacy and contextual responses, the participatory and deliberative forums of IWRM may be unequipped to deal with rapid changes (Brandeler et al. 2018).

AM is not exclusively centered around water resources and the term 'Adaptive Water Management' (AWM) is sometimes used to designate an explicit focus on these. As climate change and other global changes impact hydrological systems, the need for adaptive capacity and flexibility, AWM is increasingly recognized, as it allows for speeding up the learning cycle and faster assessments and responses (Pahl-Wostl, Craps, et al. 2007). AM encourages networked forms of governance and participatory processes that enable shared knowledge production and knowledge exchange between sectors and institutions (Pahl-Wostl, Craps, et al. 2007). This can foster social learning – understood as a process of collaborative learning through regular interaction between stakeholder groups – and thereby facilitate joint solutions, leading to more adaptive governance. Social learning processes can be strengthened by building trust and social capital and by providing space for out-of-the-box thinking and experiments

(Pahl-Wostl, Sendzimir, et al. 2007). In addition, more flexible and diverse management measures can reduce vulnerability to unpredictable shocks. AM is increasingly being complemented by adaptive governance, which includes the study of multiple social structures, interconnected resources and actors, and promotes resilience across different scales (Hurlbert and Gupta 2016). This includes measures such as scenario-building and strategic planning to prepare for a range of possible futures.

As with IWRM, AM has been described as difficult to implement and sustain in practice, especially in terms of attaining broad collaboration and the inclusion of stakeholders beyond experts and authorities (Huitema et al. 2009; Medema et al. 2008). The concept has multiple and ambiguous definitions that make it difficult for managers to apply (Medema et al. 2008). Adaptation can also be at odds with water management institutions, which tend to be path dependent and resistant to change due to institutional and technological 'lock-in' effects (Marlow et al. 2013). The large sunk costs and long life cycle of legacy infrastructure disincentivize innovative alternatives (Pahl-Wostl 2007). Another challenge of AM is that it assumes that current management processes are the cause of water resources problems and that these can be untangled, and it ignores the inherent political risks with making management 'experiments' (Medema et al. 2008).

Water–Energy–Food Nexus

In the context of water management, the Nexus approach refers to a set of interrelated activities – typically surrounding water, energy, and the environment – and their linkages as well as the boundary around them, providing a particular frame with which to address water-related problems (Muller 2015; Lofman et al. 2002). The origins of this approach can be traced back to the 1970s. The 1977 UN Water Conference in Mar del Plata highlighted the multiple linkages and interactions between water, food, and energy and the need for a coherent approach to multipurpose water resources development (Muller 2015). However, the Nexus approach gained little traction during the 1980s – seen as a lost decade for international water policy – and was further sidelined by the emergence of IWRM in the 1990s (Scheumann and Klaphake 2001). A search in ScienceDirect for the term 'Water–Energy–Food Nexus' in titles, abstracts, and/or keywords reveals that publications were virtually nonexistent before 2010 but have since grown exponentially (see Figure 1). The Nexus Conference in Bonn in 2011 became the first internationally recognized event on this theme, and it has since been promoted by the EU, the German Development Cooperation, and the United Nations in its SDGs (Benson et al. 2015). This sudden popularity has been described as a response to disillusionment with the IWRM paradigm (Muller 2015).

Through the Nexus approach, water is seen as a crosscutting issue rather than a sector or specific issue area (Gupta et al. 2013). Discussions on 'virtual water' or embodied or embedded water and the role of the food trade in water-scarce regions first highlighted the linkages between water and food (Allan 1998). The focus on the relationship between water and energy arose separately, in particular, regarding electrification in rural areas and groundwater overuse (Shah 2009). Similar to the IWRM approach, the Nexus approach aims for holistic and coordinated management between different aspects of water systems, and it promotes better resource use to allow societies to develop in an environmentally, socially, and economically sustainable way (Benson et al. 2015; Hoff 2011). This is also based on cooperation between actors and citizen participation (Hoff 2011). However, it differs from IWRM, which is a water-centric approach, by granting equal importance to different sectors (Benson et al. 2015). It considers these as inextricably interlinked and seeks to address externalities across them (e.g., energy intensity of desalination), reduce trade-offs, and increase system efficiency (Olsson 2013; Hoff 2011). In addition, rather than a focus on the river basin with an overarching centralized approach for national policy, the Nexus is implemented in 'problemsheds' where boundaries are drawn around a set of interrelated activities that take place at multiple scales (Muller 2015; Benson et al. 2015; Rouillard et al. 2014).

Criticism of the Nexus approach includes its limited emphasis on integrating water resources management with broader social and economic development activities and its lack of specific emphasis on nature, ethics, power relations, public policies, and other issues (Muller 2015; Leck et al. 2015). This may be a reflection of its focus on cross-sectoral policy coherence rather than on more normative goals, but this also depoliticizes the challenges surrounding water resources management. In addition, implementation of the Nexus approach is challenging at the institutional level, as sectoral decisions often take place at different levels (i.e., national, provincial, local) and remain decoupled (Benson et al. 2015; Scott et al. 2011). Overall, there is a burgeoning literature operationalizing this concept at multiple levels of governance, each using its own approaches and methods.

Water Management in an Increasingly Urban World

The dynamics of a growing world population is accompanied by increasing rural–urban migration with 68% of the population expected to be concentrated in cities by 2050 (UN-DESA 2018). There is a short window of opportunity as many of these cities are expected to be expanded and reach maturity in the next 20–30 years (UNE 2019). The infrastructure that is developed for water is thus critical for the future. Increasing urbanization leads to the concentration of water demand and pressure on water resources in cities' rural hinterlands (ARUP 2018). The economic development often associated with urban growth further increases per capita water use (McDonald et al. 2014). This pressure on water resources can lead to the over-exploitation of aquifers, reliance on inter-basin transfers, and conflicts between users. Urbanization causes water quality degradation due to drastic interferences in ecosystems and the hydrological cycle, including from the inadequate disposal of sewage and solid waste (Azzam et al. 2014). This aggravates regional water stress by reducing water available for use, which tends to disproportionally affect marginalized communities (Varis et al. 2006). Rapid urbanization combined with inadequate urban planning also increases vulnerability to extreme weather events, as populations move into risk-prone areas (UCLG 2016). Urban dwellers are particularly exposed to climate change as many cities are located on river banks and in coastal or flood-prone areas (Varis et al. 2006). It is expected that 40% of urban land will be in high-frequency flood zones in 2030, up from 30% in 2000 (Güneralp et al. 2015).

IWRM may not be an adequate framework to address the challenges of water in and around cities. Although it aims to coordinate between all users, it does not specifically clarify how to address urban water challenges (Brandeler et al. 2018). Cities, and in particular large metropolitan areas, are characterized by complex institutional landscapes with multiple jurisdictions and differing or even conflicting interests. This can lead to fragmented decision-making; a challenge for IWRM (i.e., integrated management and participation), Adaptive Management and Governance (i.e., social learning through collaborative processes), or the Nexus approach (i.e., dominance of urban interests over those of other sectors). Moreover, there are often important power imbalances between urban and rural areas, which affect users' ability to access water resources (Swyngedouw et al. 2002). In heavily urbanized basins where water resources are relatively scarce, competition and conflict between urban and rural users may arise due to water resources allocation regimes (Molle and Mamanpoush 2012).

City-centric frameworks, such as Integrated Urban Water Management (IUWM) and SUWM, have emerged, with roots in IWRM (Bahri 2012). IUWM promotes a holistic view of the urban water cycle, involving stakeholder participation and a flexible mode of strategic planning (Closas et al. 2012; Varis et al. 2006; Brown 2005). SUWM frames water as central to sustainability in cities and a potential starting point for urban planning (Daigger 2011). It aims to combine large and centralized infrastructure with alternative, decentralized and hybrid systems, such as green infrastructure and water reuse (Closas et al. 2012; Younos 2011; Van de Meene et al. 2011). However, such shifts are often hindered in practice by investment and technological 'lock-in' (Marlow et al. 2013). Institutional inertia, caused by factors such as overcentralization, bureaucratic inefficiencies, and a lack of sustainable finance, further limits the implementation of IUWM and SUWM, in particular as urban water challenges become increasingly large and complex (Brown and Farrelly 2007; Lee 2000). Another issue with

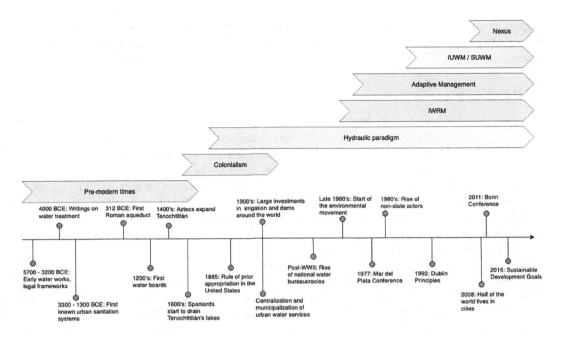

FIGURE 2 Timeline of water resources management approaches.

these frameworks is that they do not address water resources beyond urban boundaries (Brandeler et al. 2018). Ultimately, water management approaches are in constant flux (see Figure 2). As climate change and urbanization increasingly shape water-related challenges, water management approaches will need to adjust to these drivers.

Conclusion

This overview of a few of the dominant approaches to water resources management throughout history highlights society's constant struggle with harnessing the potential of water resources while mitigating its associated risks and challenges. Management approaches have had to adjust to new types of challenges, as population growth has increased water demand and competition between users and sectors and with climate change as a risk-multiplier. In the past three decades, multiple paradigms have emerged. Despite their differences, these approaches shared multiple commonalities, such as decentralization, the integration of issues and sectors, the inclusion of a larger range of actors at multiple levels, and a shift away from exclusively engineering-based solutions towards conservation, efficient water use, and green infrastructure. Around the world, water resources management actors have incorporated good governance principles within laws and policies.

However, institutional change is slow, in particular, due to path dependency and the 'lock-in' effect and as power relations often favor the status quo, leading to a lack of adherence to good governance principles in practice. To support changes in practice, it is crucial that water managers have access to reliable data and that actors across sectors and at multiple levels can effectively share knowledge and information. Decentralization has supported more inclusive and context-relevant decision-making, but effective coordination mechanisms are necessary to avoid fragmentation between actors and incoherent responses. In addition, cash-strapped agencies struggle to carry out their responsibilities, especially in the Global South. This requires not only adequate pricing mechanisms but also political will and a long-term perspective from leaders seeking fast results.

There is an urgent need to preempt possible future damage through better institutional and infrastructural design today. Although there are no panaceas to address complex water resources challenges, lessons can be learned from the advantages and limitations of past and present management approaches.

References

Abdullah, D. and B. Christensen. 2004. "Integrated River Basin Management." *Bulletin Ingenieur* 22: 21–23.

Abers, R. and M. E. Keck. 2013. "Practical Authority: Agency and Institutional Change in Brazilian Water Politics –Excerpts." In American Political Science Association 2013 Annual Meeting. https://papers.ssrn.com/sol3/papers.cfm?abstract_id=2302112.

Abers, R. N. 2007. "Organizing for Governance: Building Collaboration in Brazilian River Basins." *World Development* 35 (8): 1450–63. doi:10.1016/j.worlddev.2007.04.008.

Agyenim, J. B. 2011. "Investigating Institutional Arrangements for Integrated Water Resource Management in Developing Countries: The Case of White Volta Basin, Ghana." PhD Thesis Vrije Universiteit Amsterdam, Amsterdam.

Alam, U., P. Sahota, and P. Jeffrey. 2007. "Irrigation in the Indus Basin: A History of Unsustainability?" *Water Science and Technology: Water Supply* 7 (1): 211–18. doi:10.2166/ws.2007.024.

Allan, T. 1998. "Watersheds and Problemsheds: Explaining the Absence of Armed Conflict Over Water in the Middle East." *Middle East Review of International Affairs* 2 (1): 49–51.

Allan, T. 2003. "IWRM/IWRAM: A New Sanctioned Discourse?" https://www.soas.ac.uk/water/publications/papers/file38393.pdf.

Anand, N. 2011. "Pressure: The Polytechnics of Water Supply in Mumbai." *Cultural Anthropology*. doi:10.1111/j.1548-1360.2011.01111.x.

Angelakis, A. N. and X. Y. Zheng. 2015. "Evolution of Water Supply, Sanitation, Wastewater, and Stormwater Technologies Globally." *Water (Switzerland)* 7 (2): 455–63. doi:10.3390/w7020455.

ARUP. 2018. "Cities Alive: Water for People." Leeds, UK.

Avissar, R. and D. Werth. 2005. "Global Hydroclimatological Teleconnections Resulting from Tropical Deforestation." *Journal of Hydrometeorology* 6: 134–45. https://journals.ametsoc.org/doi/pdf/10.1175/JHM406.1.

Azzam, R., R. Strohschön, K. Baier, L. Lu, K. Wiethoff, A. Bercht, and R. Wehrhahn. 2014. "Water Quality and Socio-Ecological Vulnerability Regarding Urban Development in Selected Case Studies of Megacity Guangzhou, China." In *Megacities: Our Global Urban Future*, edited by F. Kraas, M. Aggarwal, and G. Mertins. Berlin: Springer.

Bahri, A. 2012. *Integrated Urban Water Management: Briefing Note.* Technical Committee Background Paper 16. Stockholm: Global Water Partnership.

Benson, D., A. K. Gain, and J. J. Rouillard. 2015. "Water Governance in a Comparative Perspective: From IWRM to a 'nexus' Approach?" *Water Alternatives* 8 (1). doi:10.2106/00004623-198769080-00008.

Bhattacharya, U. 2017. "From Surveys to Management: The Early Colonial State's Intervention in Water Resources of Bengal." *Indian Historical Review* 44 (2): 225–51. doi:10.1177/0376983617726471.

Biswas, A. K. 2008. "Integrated Water Resources Management: Is It Working?" *International Journal of Water Resources Development* 24 (1): 5–22. doi:10.1080/07900620701871718.

Boelens, R. 2009. "The Politics of Disciplining Water Rights." *Development and Change* 40 (2): 307–31. doi:10.1111/j.1467-7660.2009.01516.x.

Brandeler, F., J. Gupta, and M. Hordijk. 2018. "Megacities and Rivers: Scalar Mismatches between Urban Water Management and River Basin Management." *Journal of Hydrology*. doi:10.1016/j.jhydrol.2018.01.001.

Brandeler, F., M. Hordijk, K. von Schönfeld, and J. Sydenstricker-Neto. 2014. "Decentralization, Participation and Deliberation in Water Governance: A Case Study of the Implications for Guarulhos, Brazil." *Environment and Urbanization* 26 (2): 489–504. doi:10.1177/0956247814544423.

Brown, R. R. 2005. "Impediments to Integrated Urban Stormwater Management: The Need for Institutional Reform." *Environmental Management* 36 (3): 455–68. doi:10.1007/s00267-004-0217-4.

Brown, R. R. and M. Farrelly. 2007. "Barriers to Advancing Sustainable Urban Water Management: A Typology." *Rainwater and Urban Design*, 229–36. http://search.informit.com.au.ezproxy.lib. monash.edu.au/documentSummary;dn=888968667496813;res=IELENG.

Burke, W. J. 1956. "The Origin, Growth and Function of the Law of Water Use." *Wyoming Law Journal* 10 (2). http://repository.uwyo.edu/wljAvailableat:http://repository.uwyo.edu/wlj/vol10/iss2/1.

Butterworth, J., J. Warner, P. Moriarty, S. Smits, and C. Batchelor. 2010. "Finding Practical Approaches to Integrated Water Resources Management." *Water Alternatives* 3 (31): 68–81. www.water-alternatives.org.

Closas, A., M. Schuring, and D. Rodriguez. 2012. *Integrated Urban Water Management - Lessons and Recommendations from Regional Experiences in Latin America, Central Asia, and Africa.* Washington, DC: World Bank.

Cullet, P. and J. Gupta. 2009. "India: Evolution of Water Law and Policy." In *The Evolution of the Law and Politics of Water,* 157–73. Dordrecht: Springer. doi:10.1007/978-1-4020-9867-3_10.

Custódio, V. 2012. *Escassez de Água e Inundações Na Região Metropolitana de São Paulo.* São Paulo: Humanitas/Fapesp.

D'Souza, R. 2006. "Water in British India: The Making of a 'Colonial Hydrology.'" *History Compass* 4 (4): 621–28.

Daigger, G. T. 2011. "Sustainable Urban Water and Resource Management." *The Bridge* 41 (1): 13–18.

Domènech, L. 2011. "Rethinking Water Management: From Centralised to Decentralised Water Supply and Sanitation Models." *Documents d'Anàlisi Geogràfica* 57 (2): 293–310.

Dryzek, J. 1997. *The Politics of the Earth: Environmental Discourses.* Oxford: Oxford University Press.

Dzwairo, B., F. A. O. Otieno, and G. M. Ochieng. 2010. "Making a Case for Systems Thinking Approach to Integrated Water Resources Management (IWRM)." *International Journal of Water Resources and Environmental Engineering* 1 (5): 107–13. http://www.academicjournals.org/ijwree.

Elmqvist, T., M. Fragkias, J. Goodness, B. Güneralp, P. J. Marcotullio, R. I. Mcdonald, S. Parnell, et al., eds. 2013. *Urbanization, Biodiversity and Ecosystem Services: Challenges and Opportunities. A Global Assessment.* Dordrecht: Springer.

FAO. 2012. "Coping with Water Scarcity: An Action Framework for Agriculture and Food Security." Rome. http://www.fao.org.

Gandy, M. 2008. "Landscapes of Disaster: Water, Modernity, and Urban Fragmentation in Mumbai." *Environment and Planning A* 40: 108–30. doi:10.1068/a3994.

GRID. 2018. *Global Report on Internal Displacement.* Geneva, Switzerland: Internal Displacement Monitoring Centre.

Grigg, N. S. 2008. "Integrated Water Resources Management: Balancing Views and Improving Practice." *Water International* 33 (3): 1–20. doi:10.1080/02508060802272820.

Güneralp, B., İ. Güneralp, and Y. Liu. 2015. "Changing Global Patterns of Urban Exposure to Flood and Drought Hazards." *Global Environmental Change* 31 (March): 217–25. doi:10.1016/J. GLOENVCHA.2015.01.002.

Gupta, J. and J. W. Dellapenna. 2009. "The Challenges for the Twenty-First Century: A Critical Approach." In *The Evolution of the Law and Politics of Water,* edited by J. W. Dellapenna and J. Gupta, 391–410. Dordrecht: Springer. doi:10.1007/978-1-4020-9867-3_23.

Gupta, J. and C. Pahl-Wostl. 2013. "Global Water Governance in the Context of Global and Multilevel Governance: Its Need, Form, and Challenges." *Ecology and Society* 18 (4): art53. doi:10.5751/ ES-05952-180453.

Gupta, J., C. Pahl-Wostl, and R. Zondervan. 2013. "'Glocal' Water Governance: A Multi-Level Challenge in the Anthropocene." *Current Opinion in Environmental Sustainability* 5 (6): 573–80. doi:10.1016/j. cosust.2013.09.003.

GWP-TAC. 2000. "Integrated Water Resources Management." Stockholm. www.gwpforum.org.

Hanak, E., J. Lund, A. Dinar, B. Gray, R. Howitt, J. Mount, P. Moyle, and B. "Buzz" Thompson. 2011. *Managing California's Water from Connict to Reconciliation.* San Francisco, CA: Public Policy Institute of California. https://www.ppic.org/content/pubs/report/R_211EHR.pdf.

Hatfield-Dodds, S., R. Nelson, and D. C. Cook. 2007. "Adaptive Governance: An Introduction, and Implications for Public Policy." In ANZSEE Conference. Noosa Australia, 4–5 July 2007. https://pdfs.semanticscholar.org/8c29/116745551496ae166290acd780b2c988e21d.pdf.

Hoff, H. 2011. "Understanding the Nexus. Background Paper for the Bonn 2011 Nexus Conference: The Water, Energy and Food Security Nexus." https://mediamanager.sei.org/documents/Publications/SEI-Paper-Hoff-UnderstandingTheNexus-2011.pdf.

Holling, C. S., ed. 1978. *Adaptive Environmental Assessment and Management.* Chichester: Wiley.

Hooper, B. 2005. *Integrated River Basin Governance: Learning from International Experience.* Water Intelligence Online. London: IWA Publishing. doi:10.2166/9781780402970.

Huitema, D., and S. Meijerink. 2017. "The Politics of River Basin Organizations: Institutional Design Choices, Coalitions, and Consequences." *Ecology and Society* 22 (2). doi:10.5751/ES-09409-220242.

Huitema, D., E. Mostert, W. Egas, S. Moellenkamp, C. Pahl-Wostl, and R. Yalcin. 2009. "Adaptive Water Governance: Assessing the Institutional Prescriptions of Adaptive (Co-)Management from a Governance Perspective and Defining a Research Agenda." *Ecology and Society* 14 (1): 26.

Hurlbert, M. 2016. *Adaptive Governance of Disaster: Drought and Flood in Rural Areas.* Amsterdam: Springer.

Hurlbert, M., and J. Gupta. 2016. "Adaptive Governance, Uncertainty, and Risk: Policy Framing and Responses to Climate Change, Drought, and Flood." *Risk Analysis* 36 (2): 339–56. doi:10.1111/risa.12510.

ICWE. 1992. *The Dublin Statement on Water and Sustainable Development.* Dublin: International Conference on Water and the Environment. http://www.wmo.int/pages/prog/hwrp/documents/english/icwedece.html.

IPCC. 2014. "Climate Change 2014: Synthesis Report. Contribution of Working Groups I, II and III to the Fifth Assessment Report of the Intergovernmental Panel on Climate Change [Core Writing Team, RK Pachauri and LA Meyer (Eds.)]." Geneva, Switzerland.

Johnson, B. L. 1999. "The Role of Adaptive Management as an Operational Approach for Resource Management Agencies." *Conservation Ecology* 3 (2): 8. doi:10.5751/ES-00136-030208.

Jønch-Clausen, T., and J. Fugl. 2001. "Firming up the Conceptual Basis of Integrated Water Resources Management." *International Journal of Water Resources Development* 174: 501–10. doi:10.1080/07900620120094055.

Kallis, G., and H. Coccossis. 2002. "Water for the City: Lessons from Tendencies and Critical Issues in Five Advanced Metropolitan Areas." *Built Environment* 28 (2): 96–110.

Kasemir, B., J. Jäger, C. Jaeger, and M. Gardner, eds. 2003. *Public Participation in Sustainability Science: A Handbook.* Cambridge: Cambridge University Press.

Kornfeld, I. E. 2009. "Mesopotamia: A History of Water and Law." In *The Evolution of the Law and Politics of Water,* 21–36. Dordrecht: Springer. doi:10.1007/978-1-4020-9867-3_2.

Lazaroms, R., and D. Poos. 2004. "The Dutch Water Board Model." *Water Law* 15 (3/4): 137–40. http://www.fao.org/tempref/agl/emailconf/wfe2005/Art11_Lazaroms137-140.pdf.

Leck, H., D. Conway, M. Bradshaw, and J. Rees. 2015. "Tracing the Water-Energy-Food Nexus: Description, Theory and Practice." *Geography Compass* 9 (8): 445–60. doi:10.1111/gec3.12222.

Lee, T. R. 2000. "Urban Water Management for Better Urban Life in Latin America." *Urban Water* 2 (1): 71–78. www.elsevier.com/locate/urbwat.

Leite Farias, P. J. 2009. "Brazil: The Evolution of the Law and Politics of Water." In *The Evolution of the Law and Politics of Water,* edited by J. W. Dellapenna and J. Gupta, 69–86. Dordrecht: Springer.

Lofman, D., M. Petersen, and A. Bower. 2002. "Water, Energy and Environment Nexus: The California Experience." *International Journal of Water Resources Development* 18 (1): 73–85. doi:10.1080/07900620220121666.

Lopez-Gunn, E. 2009. "Agua Para Todos: A New Regionalist Hydraulic Paradigm in Spain." *Water Alternatives* 2 (3): 370–94. www.water-alternatives.org.

Lucero, L. J., J. D. Gunn, V. L. Scarborough, L. J. Lucero, J. D. Gunn, and V. L. Scarborough. 2011. "Climate Change and Classic Maya Water Management." *Water* 3 (2): 479–94. doi:10.3390/w3020479.

Marlow, D. R., M. Moglia, S. Cook, and D. J. Beale. 2013. "Towards Sustainable Urban Water Management: A Critical Reassessment." *Water Research* 47: 7150–61. doi:10.1016/j.watres.2013.07.046.

Martinez-Santos, P., M. M. Aldaya, and M. Ramón Llamas. 2014. *Integrated Water Resources Management in the 21st Century: Revisiting the Paradigm*. Leiden: CRC Press/Balkema.

McDonald, R. I., K. Weber, J. Padowski, M. Flörke, C. Schneider, P. A. Green, T. Gleeson, et al. 2014. "Water on an Urban Planet: Urbanization and the Reach of Urban Water Infrastructure." *Global Environmental Change* 27 (1). doi:10.1016/j.gloenvcha.2014.04.022.

Medema, W., B. S. Mcintosh, and P. J. Jeffrey. 2008. "From Premise to Practice: A Critical Assessment of Integrated Water Resources Management and Adaptive Management Approaches in the Water Sector." *Ecology and Society* 13 (2): 29. http://www.ecologyandsociety.org/vol13/iss2/art29/.

Meene, S. J. Van de, R. R. Brown, and M. A. Farrelly. 2011. "Towards Understanding Governance for Sustainable Urban Water Management." *Global Environmental Change* 21 (3): 1117–27. doi:10.1016/j.gloenvcha.2011.04.003.

Molle, F. 2008. "Nirvana Concepts, Narratives and Policy Models: Insights from the Water Sector." *Water Alternatives* 1 (1): 131–56.

Molle, F. 2009. "River-Basin Planning and Management: The Social Life of a Concept." *Geoforum* 40: 484–94. doi:10.1016/j.geoforum.2009.03.004.

Molle, F., and A. Mamanpoush. 2012. "Scale, Governance and the Management of River Basins: A Case Study from Central Iran." *Geoforum* 43 (2): 285–94. doi:10.1016/j.geoforum.2011.08.004.

Molle, F., P. P. Mollinga, and P. Wester. 2009. "Hydraulic Bureaucracies and the Hydraulic Mission: Flows of Water, Flows of Power." *Water Alternatives* 2 (3): 328–49. www.water-alternatives.org.

Muller, M. 2015. "The 'Nexus' As a Step Back towards a More Coherent Water Resource Management Paradigm." *Water Alternatives* 8 (1): 675–94. www.water-alternatives.org.

Nilsson, D., and E. N. Nyanchaga. 2009. "East African Water Regimes: The Case of Kenya." In *The Evolution of the Law and Politics of Water*, edited by J. W. Dellapenna and J. Gupta, 105–20. Dordrecht: Springer. doi:10.1007/978-1-4020-9867-3_7.

Olsson, G. 2013. "Water, Energy and Food Interactions—Challenges and Opportunities." *Frontiers of Environmental Science & Engineering* 7 (5): 787–93. doi:10.1007/s11783-013-0526-z.

Paganini, W. da S. 2008. *A Identidade de Um Rio de Contrastes o Tietê e Seus Múltiplos Usos*. ABES. https://books.google.nl/books/about/A_identidade_de_um_rio_de_contrastes.html?id=TF3JQwAACAAJ&redir_esc=y.

Pahl-Wostl, C. 2007. "Transitions towards Adaptive Management of Water Facing Climate and Global Change." *Water Resource Management* 21: 49–62. doi:10.1007/s11269-006-9040-4.

Pahl-Wostl, C. 2008. "Requirements for Adaptive Water Management." In *Adaptive and Integrated Water Management*, edited by C. Pahl-Wostl, P. Kabat, and J. Möltgen, 1–22. Berlin: Springer. doi:10.1007/978-3-540-75941-6_1.

Pahl-Wostl, C., M. Craps, A. Dewulf, E. Mostert, D. Tabara, and T. Taillieu. 2007. "Social Learning and Water Resources Management." *Ecology and Society* 12 (2): 5.

Pahl-Wostl, C., J. Sendzimir, P. Jeffrey, J. Aerts, G. Berkamp, and K. Cross. 2007. "Managing Change toward Adaptive Water Management through Social Learning." *Ecology and Society* 12 (2): 18. https://www.jstor.org/stable/26267877.

Porse, E. C. 2014. "Old Solutions and New Problems: On the Evolution of Urban Water Infrastructure and Environments." University of California, Davis. https://watershed.ucdavis.edu/shed/lund/students/PorseDissertation.pdf.

Rahaman, M. M., and O. Varis. 2005. "Integrated Water Resources Management: Evolution, Prospects and Future Challenges." *Sustainability: Science, Practice, & Policy* 1 (1). http://ejournal.nbii.org.

Rietveld, L. C., J. G. Siri, I. Chakravarty, A. M. Arsénio, R. Biswas, and A. Chatterjee. 2016. "Improving Health in Cities through Systems Approaches for Urban Water Management." *Environmental Health: A Global Access Science Source* 15 (Suppl 1). doi:10.1186/s12940-016-0107-2.

Rouillard, J. J., D. Benson, and A. K. Gain. 2014. "Evaluating IWRM Implementation Success: Are Water Policies in Bangladesh Enhancing Adaptive Capacity to Climate Change Impacts?" *International Journal of Water Resources Development* 30 (3): 515–27. doi:10.1080/07900627.2014.910756.

Savenije, H. H G, and P. Van der Zaag. 2008. "Integrated Water Resources Management: Concepts and Issues." *Physics and Chemistry of the Earth* 33: 290–97. doi:10.1016/j.pce.2008.02.003.

Scheumann, W., and A. Klaphake. 2001. "Freshwater Resources and Transboundary Rivers on the International Agenda: From UNCED to Rio+10." Bonn, Germany. https://www.ecolex.org/fr/details/literature/freshwater-resources-and-transboundary-rivers-on-the-international-agenda-from-unced-to-rio10-mon-067325/.

Scott, C. A., S. A. Pierce, M. J. Pasqualetti, A. L. Jones, B. E. Montz, and J. H. Hoover. 2011. "Policy and Institutional Dimensions of the Water–Energy Nexus." *Energy Policy* 39 (10): 6622–30. doi:10.1016/j.enpol.2011.08.013.

Shah, T. 2009. *Taming the Anarchy: Groundwater Governance in South Asia.* Colombo: Resources for the Future. https://cgspace.cgiar.org/handle/10568/36566.

Smythe, E. A. 1905. *The Conquest of Arid America.* 2nd ed. Seattle, WA: University of Washington Press.

Sosa-Rodriguez, F. S. 2010. "Impacts of Water-Management Decisions on the Survival of a City: From Ancient Tenochtitlan to Modern Mexico City." *International Journal of Water Resources Development* 26 (4): 675–87. doi:10.1080/07900627.2010.519503.

Swyngedouw, E. 1999. "Modernity and Hibridity: Nature, Regeneracionismo, and the Production of the Spanish Waterscape, 1890–1930." *Annals of the Association of American Geographers* 89: 443–65.

Swyngedouw, E. 2005. "Governance Innovation and the Citizen: The Janus Face of Governance-beyond-the-State." *Urban Studies* 42 (11). doi:10.1080/00420980500279869.

Swyngedouw, E., M. Kaïka, and E. Castro. 2002. "Urban Water: A Political-Ecology Perspective." *Built Environment* 28 (2): 124–37. https://www.jstor.org/stable/pdf/23288796.pdf?refreqid=excelsior%3A33d4175e87e690aec72bb0ffb516a88b.

UCLG. 2016. "Co-Creating the Urban Future: The Agenda of Metropolises, Cities and Territories." Barcelona.

UN-DESA. 2018. "World Urbanization Prospects – Population Division - United Nations." https://population.un.org/wup/Download/.

UNE. 2019. "Global Environment Outlook 6." Nairobi, Kenya.

UNEP. 2012. "The UN-Water Status Report on the Application of Integrated Approaches to Water Resources Management." Nairobi, Kenya.

UNESCO. 2006. *Water, a Shared Responsibility. The United Nations World Water Report 2.* Paris and New York: UNESCO and Berghahn Books.

UNGA, (United Nations General Assembly). 2015. "2015 Time for Global Action for People and Planet." New York. https://sustainabledevleopment.un.org/%0Apost2015.

Varis, O., A. K. Biswas, C. Tortajada, and J. Lundqvist. 2006. "Megacities and Water Management." *International Journal of Water Resources Development* 22 (2): 377–94. doi:10.1080/07900620600684550.

Vörösmarty, C. J., S. E. Bunn, and R. Lawford. 2013. "Global Water, the Anthropocene and the Transformation of a Science." *Current Opinion in Environmental Sustainability* 5 (6): 539–50. doi:10.1016/J.COSUST.2013.10.005.

Wallington, T. J., K. Maclean, T. Darbas, and C. J. Robinson. 2010. "Knowledge-Action Systems for Integrated Water Management: National and International Experiences, and Implications for South East Queensland." http://www.griffith.edu.au/.

Walters, C. J., and R. Hilborn. 1978. "Ecological Optimization and Adaptive Management." *Annual Review of Ecology and Systematics* 9 (1): 157–88. doi:10.1146/annurev.es.09.110178.001105.

Watson, N. 2004. "Integrated River Basin Management: A Case for Collaboration." *International Journal of River Basin Management* 2 (4): 243–57. doi:10.1080/15715124.2004.9635235.

Wester, P., and J. Warner. 2002. "River Basin Management Reconsidered." In *Hydropolitics in the Developing World: A Southern African Perspective*, edited by A. Turton and R. Henwood, 61–71. African Water Issues Research Unit, Centre for International Political Studies, University of Pretoria. https://library.wur.nl/WebQuery/wurpubs/123299.

Wilson, A. 1998. "Water Supply in Ancient Carthage." In *Carthage Papers: The Early Colony's Economy, Water Supply, a Public Bath and the Mobilization of State Olive Oil*, edited by J. T. Pena, J. J. Rossiter, A. I. Wilson, C. Wells, M. Carroll, J. Freed and D. Godden, 65–102. Portsmouth: Rhode Island.

Wittfogel, K. A. 1955. "Developmental Aspects of Hydraulic States." In *Irrigation Civilizations: A Comparative Study*. Washington, DC: Pan-American Union, Social Science Monographs.

WWAP. 2009. "Third World Water Development Report, Preface Xii."

Xie, M. 2006. "Integrated Water Resources Management (IWRM)-Introduction to Principles and Practices 1." http://pacificwater.org/userfiles/file/IWRM/Toolboxes/introductiontoiwrm/IWRM Introduction.pdf.

Younos, T. 2011. "Paradigm Shift: Holistic Approach for Water Management in Urban Environments." *Frontiers of Earth Science*. doi:10.1007/s11707-011-0209-7.

24

Wastewater and Water Utilities

Introduction 241
Water 241

    History • Standards and Monitoring • Sources and Transmission •
    Treatment • Hydraulics

Wastewater 247

    History • Standards and Monitoring • Sources and
    Collection • Treatment • Liquid and Solids Disposal •
    Industrial Waste • Issues

Conclusion 250
References...250

Rudolf Marloth

Introduction

Municipalities are normally responsible for providing water and disposing of wastewater, either by privately or publicly owned utilities. A water utility is responsible for the source, transmission, treatment, and distribution of potable water. Water usage is metered at the user. A wastewater treatment utility is responsible for collection, treatment, and disposal. Industrial wastewater is often metered. Residential wastewater charges are based on water usage.

Municipalities are also responsible for the collection and disposal of runoff, which consists of stormwater, misplaced or excessive irrigation, domestic car washing, etc. Runoff collects in gutters or holding ponds; then goes to storm sewers; and is dumped into a nearby river or, in coastal cities, directly into the ocean. This water carries with it a variety of pollutants: fertilizer, animal waste, oils, tire dust, etc. For municipalities that do not have combined storm and sanitary sewers, this runoff is dumped without treatment. Worse, in severe storms, combined sewer systems are likely to dump runoff combined with untreated sewage. Furthermore, runoff is not metered, so the municipality bears the cost of installing and maintaining the infrastructure. Possible approaches to reducing pollution of natural waters are (a) enactment of ordinances to make property owners responsible for reducing runoff and its pollutants, (b) diverting light flow or the first flow in a storm to a wastewater treatment plant (WWTP), and (c) installing treatment facilities to make runoff suitable for reuse. (The city of Santa Monica, California recently demonstrated its SMURRF (Santa Monica Urban Runoff Recycling Facility), the first of its kind in the nation.)

Water

The primary objective of water treatment is to make it safe for human consumption at a reasonable cost. It is possible to produce safe water that has objectionable taste, odor, or color, so a secondary goal is to make the water appealing to the consumer. Turbidity and color are qualities apparent to the naked eye.

The former is caused by particles in suspension and is measured in nephalometric turbidity units (NTU) by passing light through a sample. These particles can be removed by settling or filtering. Colloidal particles will not settle in a reasonable amount of time and must be removed by other physical processes. Dissolved substances are removed or transformed by chemical treatment. Color may be caused either by materials in solution (true color) or in suspension (apparent color).[1]

History

Early water treatment focused on what was apparent to the senses: appearance, taste, and odor. These qualities were improved by removing turbidity through filtration or precipitation. It was found much later that particles in the water harbored pathogens, which were largely removed while clarifying the water.

A slow sand filter (SSF) designed by James Simpson was commissioned in 1829, but it was some time before its full importance was realized. This simple device is essentially a tank filled with sand with water introduced at the top and removed at the bottom. Sometime after the filter is started up (a few days to a few weeks), the upper layer of sand becomes coated with a gelatinous biological layer called the schmutzedecke, made up of algae, bacteria, protozoa, and small invertebrates. This sticky layer is biologically active and converts organic matter in the water to water, carbon dioxide, and harmless salts (i.e., it is mineralized). Later, research showed the importance of biological removal and also showed that the SSF was very effective in that respect. In particular, the SSF is effective in removing *Giardia* and *Cryptosporidium* oocysts, which are nearly unaffected by chlorination.[1,2]

In the famous Broad Street pump episode of 1854, an outbreak of cholera in the Soho district of London killed more than 600 people. Dr. John Snow, who had theorized that the disease was spread by contaminated water, traced it to water from the Broad Street pump. The likely cause was from a leaking and cholera-infested cesspool located only three feet from the Broad Street well. In fact, cesspits lay under many of the houses in the district.[3] Acceptance of Snow's theory was slow, as was conversion to a sewer system that conveyed wastewater to central plants.

Recently, the Centers for Disease Control and Prevention and the National Academy of Engineering named water treatment as one of the most significant public health advancements of the 20th century.[4] This is rightly so, for in the first half of the 20th century, life expectancy in the United States increased dramatically, primarily because of water treatment, which has greatly reduced the incidence of waterborne bacterial infections such as cholera, typhoid fever, and dysentery. Even so, waterborne disease does occur in this country, the most well-known instance of which is the cryptosporidiosis outbreak of 1973 in Milwaukee. Most episodes are due to contamination of raw or treated water, inadequate treatment, and cross-contamination between sewers and water mains.

Standards and Monitoring

The Safe Drinking Water Act of 1974 and its amendments of 1986 and 1996 are the primary pieces of federal legislation protecting drinking water supplied by public water systems. Primary regulations under the act are for the protection of public health; secondary regulations are for regulations pertaining to taste, odor, and appearance.

The Surface Water Treatment Rule mandates that surface water or groundwater under the influence of surface water must be treated to remove or inactivate 99.9% of *Giardia lamblia* cysts and 99.99% of enteric viruses. These requirements are commonly stated as "log removals," where n-log removal is removal of a 1–10^{-n} fraction of the pollutant. Treatment processes using filtration are judged to comply by providing an adequate concentration/ contact time (Ct) product, where C is the concentration of the disinfectant in mg/L and t is the time in minutes.[5]

The Enhanced Surface Water Treatment Rule requires a 2-log removal of *Cryptosporidium.* Systems using filtration are granted credit if they meet certain turbidity criteria. The Filter Backwash Recycling

Rule requires treatment plants to recycle filter backwash water through the entire process cycle. The Disinfectants/Disinfection Byproduct Rule establishes maximum contaminant levels (MCLs) on total trihalomethanes, haloacetic acids, bromate ion, and chlorite ion.[5]

The most common way of testing the quality of drinking water is the coliform test, as specified by the U.S. Environmental Protection Agency. Coliforms are bacteria that are gram-negative, aerobic and facultative anaerobic, nonspore-forming rods, which ferment lactose with gas formation in 48 h at 35°C. When a sample tests positive for coliforms, it must be tested for fecal coliforms. Fecal or thermotolerant coliforms include all coliforms that can ferment lactose at 44.5°C. It is common to identify *Escherichia coli* uniquely with fecal coliforms.[5]

Sources and Transmission

Large-scale sources are primarily surface water and groundwater. Desalinated ocean water is not yet a major source in the United States, although Spain is a major producer and user of desalinated water. In the past 30 years, the energy required for desalination has fallen from 12 to 3 or 4 kWh/m^3 using reverse osmosis.[6] In a few areas, recycled water (treated wastewater) is used for irrigation. Water is transmitted by way of natural water courses, lined open channels, or pressure pipe. Los Angeles, for example, receives water from the Sacramento Delta via a concrete-lined open channel, from snowmelt in the Sierra via a natural watercourse that is diverted into an iron pipe aqueduct. New York reservoirs supply water to treatment plants through a series of underground tunnels. In some cases, water treatment plants are located right at the site where water is drawn from a river or lake.

Treatment

Treatment is tailored somewhat to the characteristics of the influent and the effluent limits, but in general, large particles are removed first by screening and then by settling. Next are particles that will float or settle with some assistance, such as air floatation or mixing with coagulants. Colloids and dissolved materials are removed last. The water should be nearly clear before disinfection so that contaminants may not hide in turbidity.

Standard water treatment (Figure 1) consists of coagulation, flocculation (aggregation into a wooly mass), sedimentation, filtration, and disinfection. The size of the units and major equipment are determined by the hydraulic loading. The examples below are based on a plant with an average flow of 24 mgd (million gal per day). Plants are commonly designed for a maximum daily flow that is 1.5 times the average daily flow. Plants whose main supply is groundwater will have a somewhat different process chain because the water will contain less settleable and suspended solids but probably more dissolved metals. Plants that draw from a river should have at their head a coarse screen for tree branches, etc., and a grit chamber for sand and silt, because this kind of grit will cause a serious maintenance problem with pumps.

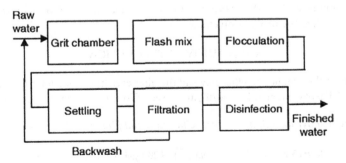

FIGURE 1   Water treatment.

The coagulation operation consists of the addition and flash mixing of chemicals meant to remove the charge on colloids and suspended solids so that they will aggregate in the flocculation step. Depending on the type of chemical used, flash mixing should occur in about one to five seconds. The addition of more than the optimum amount of chemicals can compensate for less-than-optimum mixing. A potential problem in this step is the clogging of feed lines. Various methods of flash mixing are available, the best of which are diffusion mixing by water jets and inline static mixing. Proper operation at this step requires choosing the right chemicals and quantities in response to the raw water quality and flow rate. Chemicals used in this step include aluminum sulfate (alum), polyaluminum chloride, various iron compounds, polymers, and bentonite.[5]

Flocculation is slow mixing that increases the rate of collisions between particles whose electrostatic repulsive forces have been neutralized in the coagulation step. Now the particles will stick together into sizes that will settle. Flocculation mixing is performed by mechanical mixers, baffles, or other methods. Paddle reels with a horizontal axle perform well. In a reasonable design for two parallel, 12-mgd floc tanks, each of three chambers in a tank might be about 13.5 ft^2 in cross section so that a 13-ft horizontal paddle reel would nearly fill a chamber. Perforated baffle walls separate the chambers to promote good mixing. Proper operation of this step requires continual monitoring of and adjusting for the floc size, and removing scum from the surface of the water, sludge from the bottom of the tank, and algae from the vertical surfaces (walls and baffles). Transfer of the flocculated water to the sedimentation tank must be done at low velocity to avoid breaking up the floc.

Four progressive stages of sedimentation are distinguished. Type I sediment consists of separate, destabilized particles. Type II is made of larger groups of flocculated particles. In Type III, the particles have formed a blanket that initiates hindered settling. Type IV settling is compression of the sludge blanket at the bottom of the tank.

A typical filter bed is made up of a layer of sand and a layer of coal, charcoal, or granular activated carbon. The water from the sedimentation tank is introduced at the top of the filter and moves by gravity down through the media to the underdrain. It is most desirable to have the filter backwashed once per day. Thus, the design depends upon the quality of the raw water, the required throughput, the local climate, and the skill level of the operators. The backwash water is required to be recirculated to the head of the plant.

Filter efficiency is determined by the unit filter run volume (UFRV), which is the ratio of the amount of water processed during a filter cycle to the amount that could be processed if no backwashing were necessary. The effective filtration rate, R_e, is

$$R_e = (UFRV - UBWV)/T$$

where UBWV (unit backwash volume), gal/ft^2; T, filter cycle time, min; and

$$UFRV = V_f/A \quad UBWV = V_b/A$$

where V_f, volume filtered per filter cycle, gal; A, area of filter, ft^2; V_f, volume of backwash water, gal.

The design filtration rate, R_d, is the maximum filtration rate, which can be achieved only if no backwash were necessary. Then the production efficiency is

$$R_e/R_d = (UFRV - UBWV)/UFRBV$$

Example 1. Find the production efficiency and the filter cycle time for a filter with UFRV equal to 7500 gal/ft^2, UBWV equal to 200 gal/ft^2, and design filtration rate equal to 5 gpm/ft^2.

Solution. *The effective filtration rate, filter efficiency, and filter cycle time are*

$$R_e = \left(5 \text{ gpm/ft}^2\right)\left[\left(7500 \text{ gal/ft}^2\right) - \left(200 \text{ gal/ft}^2\right)\right] / \left(7500 \text{ gal/ft}^2\right)$$
$$= 4.87 \text{ gpm/ft}^2$$

$$= R_e / R_d \left[\left(7500 \ \text{gal/ft}^2 \right) - \left(200 \ \text{gal/ft}^2 \right) \right] / \left(7500 \ \text{gal/ft}^2 \right)$$

$$= 0.973$$

and

$$T = \left[\left(7500 \ \text{gal/ft}^2 \right) - \left(200 \ \text{gal/ft}^2 \right) \right] / \left(4.87 \text{gpm/ft}^2 \right)$$

$$= 1499 \ \text{min}$$

Note that the filter cycle time is very close to one day (1440 min). Also, increasing the filtration rate will not necessarily increase the amount of water filtered per day. Increasing the rate increases the amount of deposition on the filter media, reducing the filter run time.

Most of the water treatment plants in this country disinfect with chlorine. Common alternatives are ozone, chloramine, and ultraviolet light. The disinfectant is sometimes added at the head of the plant to give an adequate concentration-contact time product. When chlorine is used, disinfection byproducts can be formed—most notably, tri- halomethanes. To suppress this, ammonia might be added at the end of sedimentation to form chloramines. Then the water is lightly rechlorinated in the clearwell to suppress regrowth of pollutants in the distribution system.

In small municipalities across the heartland, treated water is pumped into elevated tanks emblazoned with the name of the town. These towers serve to keep the water clean, meet surges in demand, and supply even pressure at the tap. Large cities tend to keep the water in open reservoirs, an unfortunate but perhaps necessary practice that leaves the water subject to the reintroduction of various undesirable substances.

Hydraulics

Flow through the plant is described by Bernoulli's equation, where all the components are expressed in terms of head in feet,

$$z_1 + P_1 / \gamma + V_1^2 / 2\text{g} = z_2 + P_2 / \gamma + V_2^2 / 2\text{g} + H_L$$

and z_i, distance of water level above datum; P_i / γ, pressure head at surface; $V_i^2 / 2\text{g}$, velocity head; H_L, head loss; $i = 1$, upstream; $i = 2$, downstream.

The head-loss term is made up of entrance and exit losses, pipe friction losses, and minor losses. Hydraulic calculations are best made starting at the clearwell and working upstream.[1]

It is preferable to have the water flow through the plant by gravity. The head losses through the A particle with settling velocity satisfying $\text{HRT} = D / V_s$ will reach the bottom of the tank before being swept up and out (as will particles with V_s greater than this). Then processes are approximately (a) rapid mixing, 1 ft; (b) flocculation, 2 ft; (c) sedimentation, 2 ft; (d) and filtration, 10 ft, for a total of 15 ft. If the water comes into the plant at the level of the clearwell, this head and the friction losses in the lift pipe must be provided by a pump. At the pump discharge, the water horsepower is HP_w; the pump input is the motor brake horsepower, HP_b; and the power required to drive the motor is P, where

$$HP_w = QH / C_1 \qquad HP_b = QH / C_1 e_P$$
$$P = C_2 HP_b / e_m$$

and Q, pump flow rate, gal/min; H, pump discharge head, ft; C_1, constant, (550 ft-lb/s-hp) (60 s/min)/ (8.34 lb/gal) 3960 ft-gal/min-hp; e_p, pump efficiency; C_2, constant, 0.746 kW/hp; e_m, motor and drive efficiency.

Example 2. Size the motor, and find the input power required to provide 15 ft of head for a 24-mgd water treatment plant. Ignore plumbing losses. Take the pump efficiency to be 70% and the motor/drive efficiency to be 90%.

Solution. *The flow rate is $(24 \times 10^6 \text{ gal/da})/[(24 \text{ h/da})(60 \text{ min/h})] = 16{,}667 \text{ gal/min}$. Thus*

$$HP_b QH/C_1 e_p = (16{,}667 \text{ gal/min})(15 \text{ ft})/\left[(3960 \text{ ft} - \text{gal/min} - \text{hp})(0.70)\right] = 90 \text{hp}$$

and

$$P = C_2 HP_b/e_m = (0.76 \text{kW/hp})(90 \text{hp})/(0.90) = 75 \text{kW}$$

Because a motor is just as efficient at 75% load as at full load, a 125-hp motor should be installed.

The important hydraulic parameters of a sedimentation tank (Figure 2) are the hydraulic retention time (HRT), the horizontal velocity of the water through the tank (V_h), and the surface overflow rate (SOR). For a tank with dimensions L, W, D for length, width, and depth, the volume $\forall$, is LWD; the surface area, A_s, is LW; and the vertical crosssectional area is $A_c = WD$. Therefore, the horizontal velocity and the SOR are

$$V_h = Q/A_c = Q/DW \text{ and } SOR = Q/A_s$$

The HRT is

$$\text{HRT} = L/V_h = L/(Q/DW) = \forall/Q$$

A particle with settling velocity satisfying $\text{HRT} = D/V_s$ will reach the bottom of the tank before being swept up and out (as will particles with V_s greater than this). Then

$$V/Q = D/V_s$$

$$\text{or } V_s = DQ/V = Q/LW = SOR$$

That is, for a particle to settle, the settling velocity must be equal to the SOR (or greater).

The power (P) required for mixing or flocculation with an impeller in a tank is dependent upon the average velocity gradient (G) in the fluid, dynamic viscosity (μ) of the fluid, and volume ($\forall$) of the tank: $P = G^2 \mu \forall$ For wastewater treatment, the average velocity gradient for rapid mixing is about 500–1500/s and for flocculation is about 50–100/s.[5]

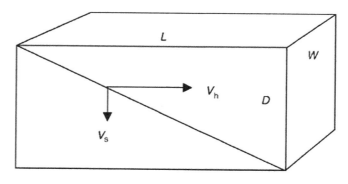

FIGURE 2   Sedimentation parameters.

Example 3. Find the power (P) required to achieve an average velocity gradient (G) of 100/s in a 1 million-gal flocculation tank whose contents are at a temperature (T) of 40°F.

Solution. *The dynamic viscosity of water at 40°F is $\mu = 2.359 \times 10^{-5}$ lb-s/ft². Therefore,*

$$P = (100/s)^2 \left(2.359 \times 10^{-5} \mathrm{lb} - \mathrm{s/ft}^2\right)\left(10^6 \ \mathrm{gal}\right)$$

$$\times (0.746 \ \mathrm{kW/hp}) / \left[\left(550 \ \mathrm{ft} - \mathrm{1b/s} - \mathrm{hp}\right)\left(7.48 \ \mathrm{gal/ft}^3\right)\right]$$

$$= 42.8 \ \mathrm{kW}$$

Example 4. Consider a plant with an input flow rate of $Q = 24$ mgd $= 24 \times 10^6$ gal/da. The two sedimentation tanks are 300 ft long, 40 ft wide, and 13 ft deep. Find the HRT, the velocity of the water through the tank, and the SOR.

Solution. *The flow rate through each tank is $Q = (24/2 \times 10^6$ gal/da)/[(24 h/da)(60 min/h)] = 8333 gal/min = (8333 gal/ min)/[(7.48 gal/ft³) (1440 min/da)] = 1114 ft³/min. Therefore, the velocity in each tank is $V = (1114$ ft³/min)/[(40 ft) (13 ft)] = 2.14 ft/min, and the surface overflow rate is SOR = (8333 gal/min)/[(40 ft) (300 ft)] = 0.694 gal/min-ft².*

Wastewater

Some authors make the following distinction: wastewater is water that has been used for domestic, industrial, or commercial purposes, whereas sewage is more inclusive in that it can include water that has not been used, such as rain runoff. In the past 50 years, sanitation agencies have made a great effort to confine runoff to storm sewers and out of treatment plants, so the term WWTP is appropriate. These facilities are also called publicly owned treatment works (POTWs).

History

The history of wastewater treatment is a sordid one of determined ignorance and apathy. Until 1965, for example, Salt Lake City was dumping raw sewage into a 9-mi open canal that emptied into the Great Salt Lake.[7] In other parts of the world, this kind of practice continues to the present day.

Dr. John Snow in London convincingly linked cholera with the consumption of contaminated waters. This most famous episode in the history of both epidemiology and water treatment occurred in the late summer of 1854. Repeated outbreaks of cholera had occurred between 1831 and 1854 in the industrial cities of England, with little being done to prevent or contain it. The particularly sudden and violent episode in and around Broad Street in September gave Dr. Snow the opportunity to verify his belief that the cause was in contaminated water. When he persuaded the authorities to remove the handle from the Broad Street pump, the spread of the disease was halted. Later, Snow established that wastes from a single infected individual had been dumped into a leaking cesspit near the Broad Street well. After some time, people accepted the fact that fecal contamination of drinking water was a major cause of disease.[3]

With better water supplies and sewer systems, there was a sharp decrease in the incidence of waterborne diseases, even before the agents were identified. After half a century of research, the concept of waterborne disease was established. The cause was known to be microorganisms in the digestive tract and the associated health hazards had been proven. Then work proceeded on two fronts: analytical methods for the detection of fecal pollution and the development of treatment methods and facilities. Research led to publication of the first edition of *Standard Methods* in 1901,[8] and the SSF was an early and very effective method for treatment. (In fact, the newer and faster conventional rapid sand filter does not remove dissolved constituents as effectively.)

Standards and Monitoring

The Clean Water Act of 1972 and subsequent legislation placed increased emphasis on the importance of reducing the discharge of pollutants to natural waters. The minimum national standards now for secondary treatment are the "30/30" rule: a 30-day average of no more than 30 mg/L of BOD_5 (5-day biochemical oxygen demand) and TSS (total suspended solids), as well as pH to be between 6.0 and 9.0 at all times. Unfortunately, meeting these standards does not guarantee the absence of disease-causing agents—notably, *Giardia lamblia* and *Cryptosporidium parvum*. Increased sophistication of monitoring techniques is leading to better treatment techniques and stricter standards.[9]

Sources and Collection

Sanitary sewers receive some groundwater infiltration and stormwater. Otherwise, 90% or more of the intended influent is of residential or commercial origin. Industrial users are either direct dischargers dumping into a waterway or indirect dischargers dumping into a POTW. (Some industrial liquid waste may also be hauled off.)

Influent is collected in closed pipe, mostly by gravity flow. When pump lift is necessary, the flow is into short runs of pressure pipe, eventually returning to gravity flow lines. Because plants are often located next to natural waters, they are typically at the low points of terrain, which keeps pumping to a minimum.

Treatment

Several levels of increasing care are defined. Preliminary treatment is the removal of large items, sand and grit, floatables, and grease. Wastewater typically contains floatable materials—particularly fats, oils, and grease (FOG)—whereas (fresh) water does not. Removal is accomplished purely by physical processes such as screening and gravity, and is intended to protect the plant equipment. Primary treatment is the removal of suspended solids and organic matter, often by the addition of chemicals. Secondary treatment is a biological process that removes organics and suspended solids (and sometimes nitrogen and phosphorus), followed by disinfection. Tertiary treatment removes remaining suspended solids by fine filtering, and may include disinfection and nutrient removal. The standard today for wastewater is full secondary treatment, meaning that all the influent to a plant is given secondary treatment. When a plant cannot handle the flow, partial secondary treatment means that all the influent is given preliminary and primary treatment, but only part of it is given secondary treatment.

A typical process chain for secondary treatment is the activated sludge process shown in Figure 3.[9] Large items are screened out. Dense noncontaminants are removed purely by gravity in the grit chamber. Primary treatment is a chemical/physical process, whereby small particles agglomerate (flocculate)

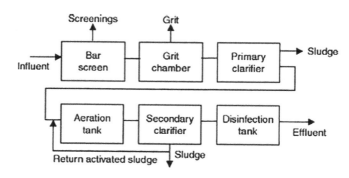

FIGURE 3   Activated sludge wastewater treatment.

and gravitate out. Secondary treatment is biological, in which microbes consume the dissolved and suspended organic matter. Disinfection kills most of the remaining contaminants.

Preliminary and primary treatment for wastewater is much like that for water. Preliminary screening of wastewater is necessary because large objects sometimes find their way into the sewer, an unfortunate example being construction debris illegally dumped into a manhole (alternatively, "maintenance hole"). (Note that a manhole cover is round for at least three reasons: (a) It will not fall through the hole, no matter how oriented; (b) it is easy to move by rolling; and (c) it need not be rotated to fit.) Primary treatment for both is nominally flash mix, flocculation, and sedimentation. Wastewater undergoes secondary treatment, in which microbes remove biological pollutants.

Both organic and inorganic particulates may be removed by settling, flotation, or filtering, depending on particle size and density. Carbon filtration is preferable to vaporizing, because the latter merely moves the substance from the water to the atmosphere.

Most reactions in waste treatment are of first order—that is, the rate of reaction is proportional to the concentration of the pollutant.

$$dC/dt = -kC$$

where C is concentration of pollutant (mg/L); t is time (min); k is reaction constant (mg/L-min).

Reactors are of three types: complete mix (or batch), plug flow, and dispersed flow. In a complete mix reactor, the reactor is filled; the reaction is allowed to take place; and then the reactor is emptied. The concentration of pollutant is equal throughout the tank. Complete mix reactors are approximately cubical and can be set up in sequence to provide an increasing proportion of removal. In a plug flow reactor, the flow moves through the reactor with the reaction taking place so that the concentration of pollutant is less at the outlet than at the inlet. These tanks are long in proportion to their length and width. Turbulence should be minimized to retain the form of the plug. Most reactors are of the dispersed flow type, intermediate between the other 2.[9]

Liquid and Solids Disposal

By the principle of conservation of mass, treatment does not make the contaminants disappear; it merely separates them from the water that bears them. When most of the contaminants have been removed and the water is sufficiently clean, it may be discharged to a natural waterway, such as a river, lake, or ocean. Most of the processes in Figure 3 produce residuals. These impurities—such as sediment, sludge, waste washwater, and brine—are left behind to be treated and disposed of in other ways. Large items from screening and grit from the grit chamber are sent to landfill. Sludges from primary and secondary treatment are sent to a digester, which itself produces waste. Waste washwater from the filters is recycled, but because it adds to the throughput volume, it should be kept to a minimum.

Industrial Waste

For both direct dischargers and indirect dischargers, the content of industrial wastewater is regulated. Industrial users are allowed to send the first fraction of runoff from rainstorms to the sanitary sewer, for the purpose of keeping pollutants out of the storm sewers and, ultimately, out of natural waters after which it must be diverted to storm sewers.

Because POTWs are set up to treat organic waste, much industrial waste is incompatible with public treatment systems, so is usually subject to pretreatment. The objectives of such treatment are to prevent interference with the process in the POTW, to prevent pass-through of pollutants to the receiving waters, and to make possible reuse of the effluent and sludge from the POTW.[10]

Some treatment strategies are flow equalization to prevent shock loading to the POTW; solids removal by straining or settling; removal of FOG by dissolved air floatation or centrifuging; neutralization of high-pH or low-pH solutions; and hydroxide precipitation of heavy metals. A notable exception to the last is the removal of Cr^{+6}, which will not respond to hydroxide precipitation. Instead, it is converted by chemical reduction at low pH to Cr^{+3}, which can be removed by hydroxide precipitation. Dissolved inorganics may be removed by hydroxide precipitation, ion exchange, or membrane filtering.[10] It should be noted that diluting industrial wastewater to reduce the concentration of pollutants is not acceptable, and that "dilution of pollution is not a solution, and can lead to prosecution." The most important principle is segregation: keeping the pollutants separated so that they can be treated individually.

Issues

Wastewater treatment plants are designed to treat organic wastes. Other substances (nutrients in fertilizer, pharmaceuticals, etc.), can pass through and cause problems in receiving waters (e.g., algal growth and abnormal growth in fauna). Yet others, such as heavy metals and toxic chemicals, can cause interference (also called upset)—disruption of the process in the biological reactor.

Failed equipment can cause raw or partially treated wastewater to flow into storm drains and then into natural waters. Runoff from heavy storms can flow into sanitary sewers and overwhelm treatment plants. Cities with combined sewers are especially subject to this problem. Inadequately sized treatment plants will discharge partially treated wastewater in times of heavy flow.

Everything removed from wastewater must be disposed of. Sometimes, objections are raised to the release of volatiles into the atmosphere. Although the creators believe that digested sludge is a fertilizer rich in nutrients, others are not convinced.[11]

More radically, some have questioned the wisdom of the whole process of fouling great quantities of cleaned water and then cleaning it again.[12]

Conclusion

A basic requirement of human existence is an adequate supply of clean water. Today, very few have access to clean, untreated water. Wealthy societies obtain clean water by treating it, while poor ones often rely on polluted sources and suffer from the resulting waterborne diseases.

The idea that many municipalities draw from surface waters that are used for the disposal of wastewater is sobering, if not chilling. In the United States, the Safe Drinking Water Act mandates water treatment standards, and the Clean Water Act mandates wastewater treatment standards. Observance of these standards has made the practice of having a common source and sink acceptable.

The conventional water treatment process described in this entry has been very effective in removing bacterial pathogens. Dechlorination to control disinfection byproducts created by chlorination was instituted as a result of the Safe Drinking Water Act. The most recent major issue is the resistance of *Cryptosporidium* oocysts to chlorination. Membrane filtration is an effective way to remove these and other very small suspended pathogens. New water treatment plants are likely to be based on this technique because of its effectiveness and ease of operation.[5]

The most important developments required in wastewater treatment are (a) building or expanding facilities to provide the capacity to subject all flow to complete secondary treatment, (b) repairing and maintaining the collection system to ensure that all wastewater reaches the treatment plant, and (c) finding practicable ways to dispose of the residual solids.

References

1. Reichenberger, J. Unpublished notes. Loyala Marymount University: Los Angeles, 2005.
2. Jesperson, K. *Search for Clean Water Continues.* Available at: http://www.nesc.wvu.edu/ndwc (accessed November 2006).

3. Judith Summers. *Broad Street Pump Outbreak*. Available at: http://www.ph.ucla.edu/epi/snow/broadstreetpump.html (accessed November 2006).

4. United States Environmental Protection Agency, Office of Water. *The History of Drinking Water Treatment*. Available at: http://epa.gov/safewater/consumer/pdf/hist.pdf (accessed November 2006).

5. Kawamura, S. *Integrated Operation and Design of Water Treatment Facilities,* 2nd Ed.; Wiley: New York, 2000.

6. Graber, C. *Desalination in Spain.* Available at: http://www.technologyreview.com/spain/water (accessed November 2006).

7. Salt Lake City Department of Public Utilities. *Conduits of Civilization.* Available at: http://www.ci.slc.ut.us/utili-ties/NewsEvents/news1998/news9281998-1.htm (accessed November 2006).

8. Greenberg, Arnold, et al., eds. *Standard Methods for the Examination of Water and Wastewater,* 21st Ed.; American Public Health Association (APHA), American Water Works Association (AWWA), and Water Environment Federation (WEF): Washington, D.C., 2005.

9. Tchobanoglous, G.; Burton, F.L.; Stensel, H.D. *Wastewater Engineering, Treatment and Reuse,* 4th Ed.; Metcalf and Eddy Inc., McGraw-Hill: New York, 2003.

10. Water Environment Federation Pretreatment of Industrial Wastes In *Manual of Practice FD-3;* Water Environment Federation: Alexandria, VA, 1994.

11. Voters seek to block sludge, *Los Angeles Times* 2006, January 2. L.A. Fights Kern County Sludge Ban. *Los Angeles Times* 2006, August 16.

12. Stauber, J.C.; Sheldon, R.S. *A Brief History of Slime.* Available at: http://www.prwatch.org/prwissues/1995Q3/slime.html (accessed November 2006).

25

Wastewater: Municipal

Municipal Wastewater 253
Combinations of Methods for the Treatment of Municipal
    Wastewater (Reduction of Bod₅) 254
Methods for the Treatment of Municipal Wastewater
    (Reduction of the Phosphorus Concentration) 257
Methods for the Treatment of Municipal Wastewater
    (Reduction of Nitrogen Concentration) 260
Recycling of Municipal Wastewater 263
References 264

Sven Erik Jørgensen

Municipal Wastewater

The composition of domestic sewage varies surprisingly little from place to place, although to a certain extent, it reflects the economic status of the society.

A typical composition including biochemical oxygen demand (BOD_5), chemical oxygen demand (COD), suspended matter, and nutrients is shown in Table 1.

Most industrialized countries require treatment of municipal wastewater. The standards vary for the European Union, the United States, Canada, Australia, New Zealand, and Japan, but it is generally required to reduce BOD_5 to about 10 mg/L, nitrogen to about 10 mg/L, and the phosphorus concentration to 1–2 mg/L. To obtain the required reductions, a combination of the methods presented in the entry "Waste Water Treatment: Overview of Conventional Methods" can be applied. The combinations of methods applied for the treatment of municipal wastewater are presented in the next sections, and

TABLE 1 Typical Composition of Municipal Wastewater (in milligrams per liter)

Constituent	Soluble	Particulate	Total
BOD_5	100–200	50–100	150–300
COD	200–500	100–200	200–700
Ammonium-N	20–40	0	20–40
Nitrate-N	5–20	0	5–20
Organic nitrogen	0	5–20	25–60
Suspended matter	–	40–80	40–80
Carbohydrates	20–40	10–15	30–55
Amino acids	10–15	15–25	25–40
Fatty acids	0	50–80	50–80
Surfactants	10–20	5–10	15–30
Creatinine	3–5	0	3–5
Phosphorus	2–4	4–10	6–14

Source: Jørgensen[1] and Jørgensen.[2]

it will be demonstrated that it is possible by a suitable combination of methods to meet all realistic effluent standards. Municipal wastewater can, however, also be treated by ecotechnological methods, for instance, by waste stabilization ponds and by constructed or natural wetlands. These methods are presented in the entries *Pollution: Point and Non-Point Source Low Cost Treatment* (p. 2174), *Waste: Stabilization Ponds* (p. 2652), *Wetlands* (p. 2846), and *Wetlands: Treatment System Use* (p. 2885).

Combinations of Methods for the Treatment of Municipal Wastewater (Reduction of Bod₅)

Models are used increasingly to design and optimize wastewater treatment methods. For details about the applied models, see Jørgensen's *Handbook of Ecological Models Used in Ecosystem and Environmental Management*,[1] which has a comprehensive entry devoted to models of wastewater treatment systems.

There are a number of different possible designs of biological treatment plants.[2]

The use of mechanical-biological wastewater treatment is a classical method, or rather the combination of the two methods. Figure 1 shows the flowchart of a classical mechanical and activated sludge plan as it is generally used all over the world. The processes involve screening, a separation of grease and sand in the grit chamber, a primary settling, and an aeration step followed by sedimentation. Chlorine is often added before discharge to the receiving water, particularly when it is used for swimming and bathing. The sludge is most often digested anaerobically, which produces biogas. In the activated sludge plant, a rapid adsorption and flocculation of suspended matter take place. Organic matter is oxidized

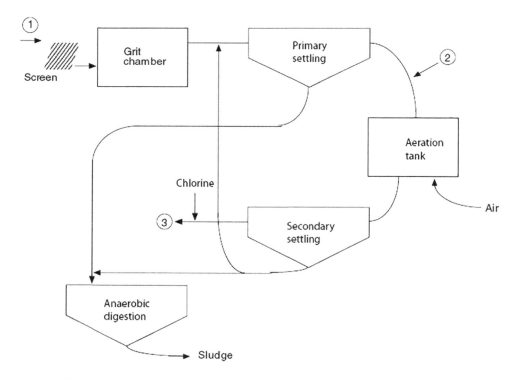

FIGURE 1 A flow diagram of a conventional activated sludge plant. It is widely used to treat municipal wastewater. Number 1 is untreated municipal wastewater that would have a BOD₅ of about 150–300 mg/L, total nitrogen of about 25–45 mg/L, and total phosphorus of about 6–12 mg/L. The mechanical treatment (sample at point 2) will reduce BOD₅ and total N by about 25%–40%, while the reduction of total phosphorus will be minor. The totally treated wastewater (point 3) will have a BOD₅ of about 10–15 mg/L, while the nitrogen is only reduced 35%–50% and the phosphorus concentration only about 10%–20%.

and decomposed. Sludge particles are dispersed and settled by the secondary settling. The processes are conceptualized in Figure 2.

Oxidation ditches (see Figure 3) can replace the activated sludge plant and the secondary settling. The rotor provides the aeration to oxidize the organic matter. The influent and the rotor are usually stopped during the night, when the inflow is low anyhow, to allow settling. The supernatant is withdrawn through an effluent launder. The retention time is usually 2–5 days.

A trickling filter is a bed packed with rocks, although recently, plastic media (celite pellets[3]) and bio-blocks with a high surface area due to the high porosity are also applied. They require substantially less space than the stone-packed trickling filter. They have usually a specific surface of $100\,m^2/m^3$ or even more. It is at least 2–3 times the specific surface for a rock trickling filter. The medium is covered by a slimy microbiological film. The wastewater is passed through the bed and oxygen and organic matter diffuse into the film, where oxidation occurs. Recirculation improves the removal efficiency. A flowchart of a treatment plant combining a trickling filter and an activated sludge plant is shown in Figure 4. The alternative ecotechnological methods used for BOD_5 reduction, lagoons, and waste stabilization ponds are presented in the entry "Waste Stabilization Ponds." Various solutions to reduce BOD_5 based on natural and constructed wetlands are covered by several entries, all with wetlands as the key word in the title.

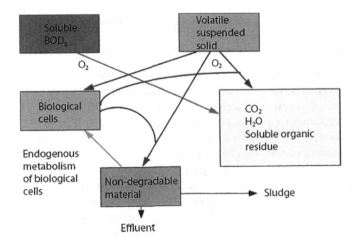

FIGURE 2 Processes characteristic for the biological treatment.

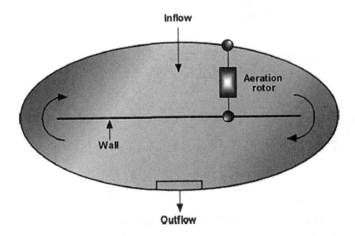

FIGURE 3 Oxidation ditch.

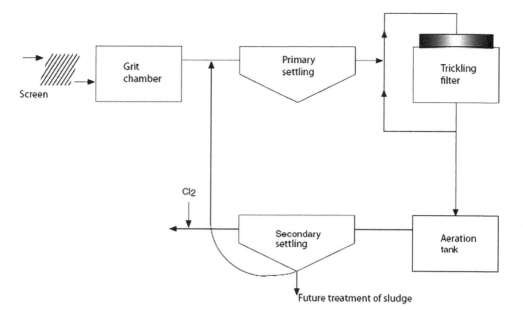

FIGURE 4 Treatment of municipal wastewater, combining trickling filter and activated sludge treatment.

Physical–chemical methods have been proposed to replace mechanical–biological treatment, for instance, the so-called AWT system, which is based on the application of a combination of mechanical treatment, precipitation, adsorption on activated carbon, and sand filtration (see Figure 5).

It has been considered that recovering proteins and grease could at least partially cover the costs of the treatment of wastewater in slaughterhouses, fish industries, starch factories, and other foodstuff industries. Figure 6 shows a method that has been applied in several, but still relatively few, cases.

If the wastewater has high concentration of grease, oil, and fat, it is possible to apply a flotation unit, offering an alternative to sedimentation. A portion of water is pressurized by 3–10 atm, and when this water is returned to the normal atmospheric pressure in the flotation unit, air bubbles are created. The air bubbles attach themselves to particles, and the air-solute mixture rises to the surface where it can be skimmed off, while the clarified water is removed from the bottom of the flotation unit. Figure 7 shows

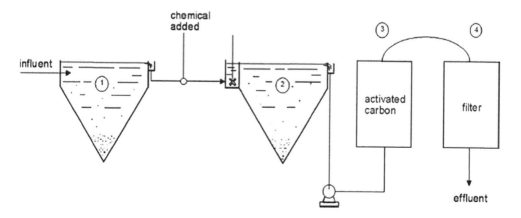

FIGURE 5 The AWT system consisting of mechanical treatment, chemical precipitation, settling, adsorption on activated carbon, and filtration. The method is not used very much. It has the advantage that it is easier to control but it is also more expensive with respect to both investment and operation.

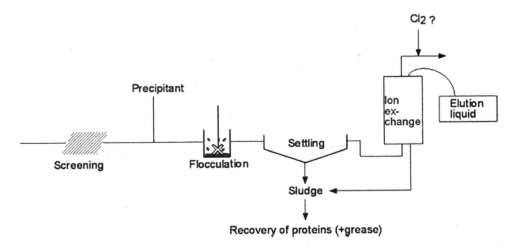

FIGURE 6 Recovery of proteins and grease from the wastewater discharge from the foodstuff industries is possible by a combination of precipitation and ion exchange. The recovery pays partially for the treatment of the wastewater by the process.

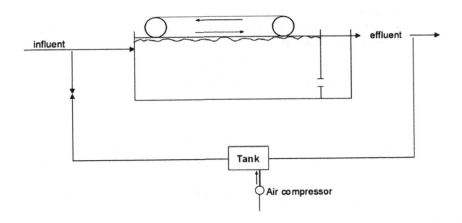

FIGURE 7 Flotation unit.

a flotation unit. Usually, the retention time in a flotation unit is 3–6 times less than for a settling unit, which means that a significant volume reduction is obtained. Flotation has therefore most frequently replaced sedimentation in slaughterhouses, fish industries, and oil industries.

Methods for the Treatment of Municipal Wastewater (Reduction of the Phosphorus Concentration)

Nutrient removal is most frequently carried out by chemical precipitation, often combined with mechanical–biological treatment. Chemical precipitation can be applied at three different points in the mechanical–biological plant as shown in Figure 8. Sometimes, both direct precipitation and posttreatment is applied, which together with sand filtration makes it possible to obtain 0.1 mg P/L or less in the effluent. In many cases where such a low phosphorus concentration is needed for particular sensitive receiving waters (mainly lakes), this double precipitation is a very attractive method, because it is

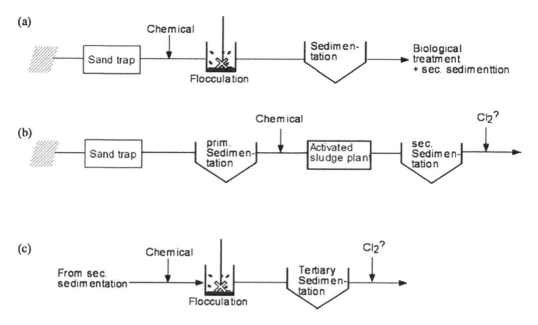

FIGURE 8 Precipitation by aluminum sulfate or other aluminum compounds, iron(III) chloride, and calcium hydroxide is able to reduce the phosphorus concentration in wastewater significantly. The precipitation is applied often in combination with mechanical biological treatment and can be carried out after the sandtrap: (a) direct precipitation, before the activated sludge plan: (b) simultaneous precipitation or after the mechanical–biological treatment: (c) posttreatment. The arrow+"Chemical" indicates the three different possibilities for addition of the precipitation chemical and "chlorine?" indicates where chlorine eventually can be added.

relatively cost moderate compared with other alternatives. Without the sand filtration and only a single precipitation, it is usually relatively easy to obtain a concentration of phosphorus in the effluent between 1.0 and 1.5 mg/L.

Aluminum sulfate, various polyaluminates, calcium hydroxide, and iron(III) chloride can be applied for the precipitation. The amount of hydrated lime or calcium hydroxide needed for the precipitation is usually 2.5–6 times higher than for the aluminum and iron compounds, because a high efficiency of the precipitation requires a pH of 10.0 or higher, which, dependent on the hardness of the water, is not possible without the indicated amount of calcium hydroxide. For a hardness of 15–30 hardness degrees, the amount is 100–480 mg calcium hydroxide per liter. It would usually give an efficiency of 90%–95% precipitation of the phosphorous compounds, which is slightly better than for most precipitations with aluminum and iron compounds. The disadvantage of precipitation with lime is the high pH, which makes adjustment of the pH after the precipitation and settling necessary. Carbon dioxide produced by incineration of the sludge or solid waste can be used for this purpose. If the sludge is incinerated, it is possible to partially recycle the calcium hydroxide and thereby reduce the costs of precipitation chemicals. Recycling 3–5 times is possible, and afterwards it can be applied as fertilizer, as it has a relatively high phosphorus concentration. A flowchart with direct precipitation and recycling of calcium hydroxide is shown in Figure 9.

If the sludge after calcium hydroxide precipitation cannot be incinerated, an adjustment of pH is needed before anaerobic digestion or aerobic sludge treatment. Heavy metals are removed more effectively by the use of calcium hydroxide than by aluminum and iron compounds. Lead, copper, and chromium are removed by a very high efficiency by all the mentioned precipitation chemicals, while only calcium hydroxide would give a high removal efficiency for cadmium and zinc.

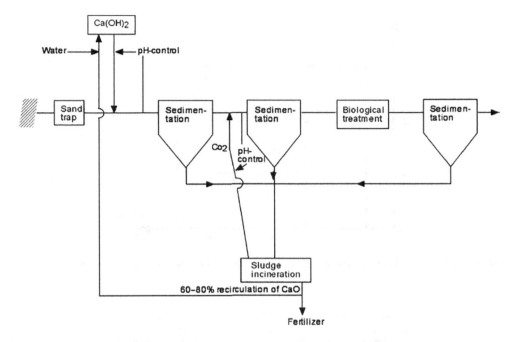

FIGURE 9 Chemical precipitation with partial recirculation of calcium hydroxide and use of carbon dioxide from incineration of the sludge for pH adjustment.
Source: Jørgensen.[1]

The amount of aluminum and iron compounds can be found in Figures 10 and 11 or by the use of the following equation based on Freundlich adsorption isotherms:

$$(C_0 - C)/n = a^* C^b$$

where C_0 is the initial concentration of phosphorus (mg P/L), C is the final concentration, n is the dose of chemical expressed as milligrams of Al per liter or milligrams of Fe per liter, and a and b are characteristic constants that can be found in Table 2.

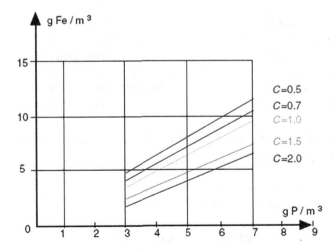

FIGURE 10 Addition of iron (III) as f(P-concentrations) at different P-concentrations in the effluent.

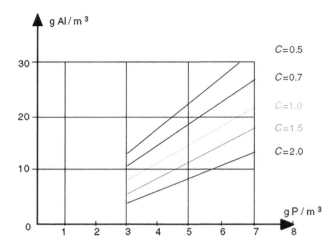

FIGURE 11　Addition of aluminum salts as f(P-concentrations) at different P-concentrations in the effluent.

TABLE 2　TThe Constants in Freundlich's Adsorption Isotherms for Aluminum Sulfate and Iron (III) Chloride

Precipitation with	a	b
Aluminum sulfate	0.63	0.2
Iron(III) c]hloride	0.26	0.4

Source: Jørgensen.[2]
Note: With good approximation, the constants for aluminum sulfate can also be applied for other aluminum compounds.

A more rapid flocculation, precipitation, and settling can be obtained by the addition of synthetic organic polymeric flocculants. They may be cationic polyelectrolytes, anionic polyelectrolytes, or non-ionic polymers. It is hardly possible to indicate which one of the polymeric flocculants would give the best result, as the ionic characteristics of municipal wastewater vary significantly. It is recommendable to test the various types of polyflocculants from case to case. The optimum design of a flocculator, before the settling of the precipitated material, should be based on a mathematical model (see Dharmappa et al.[4] and Thomas et al.[5]).

It is possible to remove phosphorus by biological treatment. Activated sludge systems with anaerobic and aerobic zones in sequence have been developed to achieve a higher phosphorus removal. The system is called EDPR—enhanced biological phosphorus removal. The shift between aerobic and anaerobic conditions activates the microorganisms to take considerably more phosphorus than under aerobic conditions, particularly if the wastewater contains relatively high concentrations of easily biodegradable organic matter. With a P/BOD_5 ratio of more than 20, a phosphorus removal of 80%–90% can be obtained.

Methods for the Treatment of Municipal Wastewater (Reduction of Nitrogen Concentration)

A combination of biological *nitrification and denitrification* can significantly reduce the nitrogen concentration in the effluent. The applied chemical processes are

$$NH_4^+ + 2O_2 \rightarrow NO_3^- + H_2O + 2H^+$$
$$4NO_3^- + 5C + H_2O \rightarrow 5HCO_3^- + 2N_2 + H^+$$

The ammonium is oxidized to nitrate and the nitrate is used to oxidize organic matter by anaerobic conditions—here indicated just as C. Thereby, the nitrate is reduced to dinitrogen, which is released to the atmosphere, where there is about 78% dinitrogen, and a minor addition of dinitrogen is therefore harmless.

Effective nitrification occurs when the sludge age is greater than the reciprocal rate of the constant of the nitrifying microorganisms.[6] The sludge age is defined as $X/\Delta X$, where X is the mass of sludge in the system and ΔX is the sludge yield per unit of time. Usually, the time unit applied is 24 hours. The relationship between nitrification efficiency in percentage and the sludge age is shown in Figure 12.

The nitrification process is a two-step biological process. Firstly, ammonia is oxidized to nitrite by *Nitrosomonas*. Secondly, nitrite is oxidized to nitrate by *Nitrobacter*. The optimum pH range for the nitrification is 6.7–7.0.[7] The oxygen concentration for the nitrification has to be at least 2 mg/L. Heavy metal ions are toxic to the nitrification at rather low concentrations. Toxic levels of about 0.2 mg/L are reported for chromium, nickel, and zinc. The nitrification process is sensitive to the temperature and an Arrhenius expression can be applied:

$$\text{Kn}\left(\text{rate constant }1/24\text{h}\right)=0.18*1.128^{(\text{temperature}-15)}$$

The denitrification takes place by a number of heterotrophic bacteria present in activated sludge. Anaerobic conditions are absolutely required. The optimum pH for this process is about 7.0. About 3 times as much BOD_5 as nitrate-N (both expressed in milligrams per liter) should be applied to ensure adequate denitrification. The process can be realized even at high concentrations of nitrate or at high salinities.[2] Addition of organic carbon—for instance, methanol—has been proposed to ensure proper denitrification, but it is of course not an attractive method to apply. Alternatives are as follows:

1. To switch between aerobic and anaerobic conditions. It is called alternative operation (for more details, see Diab et al.[8] and Halling-Sørensen and Jørgensen[9]).
2. To recycle a patty of the wastewater containing nitrate after the treatment is completed.

Both methods work properly. Other possibilities are the use of zeolite ion exchange material to obtain simultaneous nitrification and denitrification,[9] the use of electrolysis for denitrification,[10] and the use of activated carbon to enhance the denitrification.[11]

Ammonia can be removed by blowing air through the wastewater. The process is called *stripping,* and the equipment shown in Figure 13 can be used. It is of course required that the pH is sufficiently high

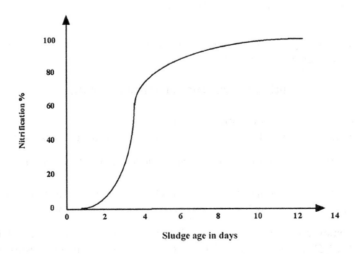

FIGURE 12 Nitrification efficiency as function of sludge age.

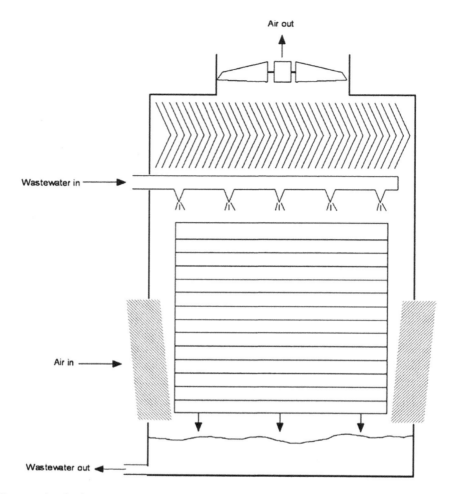

FIGURE 13   A sketch of a stripping tower is shown. The efficiency of the process is dependent on 1) pH; 2) temperature; 3) the amount of air relatively to the amount of water; and 4) the stripping tower height. Figures. 14 and 15 show how these factors influence the efficiency. As the air volume-to-water volume ratio is very high, there is a significant evaporation in the tower, which will cool the water.

enough to ensure that the gas ammonia is present. The equilibrium between ammonium and ammonia can be expressed by the following equation:

$$pH = pK + \log [\text{ammonia}] / [\text{ammonium}],$$

where pK is 9.3 for water with low concentrations of ions.

The cost of stripping is relatively low, but the process has three crucial limitations:

1. It is practically impossible to work at temperatures lower than 7°C due to the cooling effect of the evaporation.
2. Deposition of calcium carbonate can reduce the efficiency or even block the tower.
3. After the ammonia removal, pH is high and must be adjusted to 8.0 or below.

Ammonia can be removed by *chlorination* and *adsorption on activated carbon*. Chlorine can oxidize ammonia to NH_2Cl, $NHCl_2$, and NCl_3, and the activated carbon is able to adsorb the chloramines formed by these oxidation processes. By the adsorption process, it is most likely that dinitrogen and

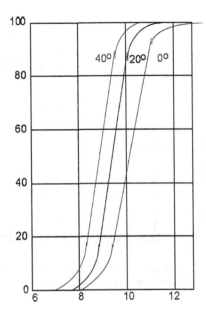

FIGURE 14 Stripping efficiency as function of temperature and pH.

chloride ions are formed. In the order of 10 parts by weight of chlorine are required for each part of ammonium-N. As the wastewater may typically contain 30 mg ammonium-N/L, 300 mg/L of chlorine is required. It would give a cost of $0.10 per cubic meter to cover only the costs of chlorine. When the capital costs and the costs of the recovery of activated carbon are added, the total costs per cubic meter may reach about $0.30. The method has therefore not found a wide application due to the high costs. The method has however a high removal efficiency even if the ammonium concentration is low, provided that sufficient chlorine is applied. Chlorination followed by active carbon adsorption can therefore be used as the last treatment for the removal of ammonium, where very low concentrations are required in the effluent.

Ion exchange by the use of the natural clay material clinoptilolite can be used for the removal of ammonium ions.[9] Clinoptilolite has a high selectivity for ammonium ions and about one-third to one-half of the ion exchange capacity, which is about 1.5 eq/L, and can, in many practical cases, be applied for the uptake of ammonium ions. It means that 1 L of clinoptilolite will be able to remove at least 1.5*14/3 g of ammonium-N = 7 g ammonium-N.

Ion exchange has furthermore been applied to remove nitrate from drinking water. A general anion exchanger is applied. It has an ion exchange capacity of about 2.5–3.0 eq/L, but the selectivity is not very high. It is however not necessary to remove the nitrate ions completely but to reduce the concentrations to the regional standards for nitrate in drinking water.

For further details about the ion exchange of wastewater and drinking water, see the entry denoted "Ion Exchange."

Recycling of Municipal Wastewater

By a combination of the available treatment methods, it is possible to recycle municipal wastewater, i.e., to produce drinking water from municipal wastewater. The recycling takes about 8–10 days from toilets back to the water taps. All drinking water is in principle recycled. It takes normally nature a couple of thousands years to recycle water on average, which of course ensures a much better water quality than a recycling with a duration of 8–10 days.

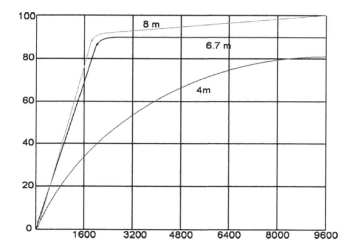

FIGURE 15 Efficiency of stripping process versus air/water ratio and tower height. Efficiency versus m³ air/m³ water for three different tower depths. The three tower depths are 4 m (insufficient), 6.7 m and 8 m. Tower height of 7–8 m is recommended.

Recycling of wastewater has been applied in Pretoria, South Africa, and in Windhoek, Namibia, due to an insufficient supply of natural water for production of potable water. In Pretoria, a treatment consisting of the following steps is applied: mechanical-biological treatment, aeration, lime precipitation, ammonia stripping, sand filtration, chlorination, adsorption on activated carbon, and a second chlorination. The plant in Windhoek has the same steps except ammonia stripping.

References

1. Jørgensen, S.E., Ed. *Handbook of Ecological Models Used in Ecosystem and Environmental Management;* CRC Press, 2011; 620 pp.
2. Jørgensen, S.E. *Principles of Pollution Abatement;* Elsevier: Amsterdam, 2000; 520 pp.
3. Sorial, G.A. et al. Evaluation of trickle-bed air biofilter performance for styrene removal. Water Res. **1998**, *32*, 1593–1603.
4. Dharmappa, H.B. et al. Optimal design of a flocculator. Water Res. **1993**, *27*, 513–519.
5. Thomas, D.N. et al. Flocculation modelling: A review. Water Res. *33*, 1579–1592.
6. Bernhardt, E.L. 1975. Nitrification in industrial treatment works. Second International Congress on Industrial Waste Water and WASTES, Stockholm, February 1975.
7. Groeneweg, J. et al. Ammonia oxidation by *Nitrosomonas* at ammonia concentrations near K_m: Effects of pH and temperature. Water Res. **1994**, *28*, 2561–2566.
8. Diab, S. et al. Nitrification pattern in a fluctuating anaerobic-aerobic pond environment. Water Res. **1993**, *27*, 1469–1475.
9. Halling-Sørensen, B.; Jørgensen, S.E. *The Removal of Nitrogen Compounds from Waste Water;* Elsevier: Amsterdam, 1993; 444 pp.
10. Islam, S.; Suidan, M.T. Electrolytic denitrification. Water Res. **1998**, *32*, 528–536.
11. Sison, N.F. et al. Denitrification with external carbon source utilizing adsorption and desorption capability of activated carbon. Water Res. **1996**, *30*, 217–227.

26

Water Quality and Quantity: Globalization

Introduction 265
Globalization and Water Resources 266
Political Control and Global Water Management: Policies and
    Political Structure 267
Global Surface Water and Groundwater Concerns and Issues:
    Introduction 268
Surface Water and Groundwater Contamination 269
Groundwater and Aquifer Depletion 271
Global Water Resource Policy: Toward a Potential Framework 273
Conclusion 274
References 274
Bibliography 276

Kristi Denise
Caravella and
Jocilyn Danise
Martinez

Introduction

Water is one of our most precious natural resources, covering nearly 70% of the Earth's surface, but only 2.5% of the Earth's water is fresh and suitable for consumption.[1] Thus, it is quite an understatement to say that freshwater is a scarce resource for the nearly 7 billion people living on Earth today.[2] Many nations around the world are currently ill equipped to deal with the demand for water. Nearly 1 billion people lack access to safe water.[3] The health and economic impacts of inaccessibility to water are staggering.

It is predicted that the problem will worsen with the increase in human population and the continued growth of the global economy. By 2050, the world population is expected to grow to more than 9 billion people, placing a strain on water utility systems around the world to meet the growing demand for the human consumption of water that is necessary for survival coupled with the need to utilize water as a tool of industry to maintain a competitive edge in the global marketplace.[3] Nations around the world must develop a coherent plan of action to deal with this problem. This entry discusses management issues related to water quality and quantity concerns that have occurred as a result of increased globalization. The purpose of discussing these issues is to mitigate water quality and quantity concerns that can have an adverse effect on human health and the environment on a global scale.

The results of globalization have created an international environment whereby economic integration is essential. Since World War II, the global trend toward greater interdependence among nation-states in the exchange of goods, services, and investments has accelerated. Much of this growth can be attributed to transnational corporations who have used technology to capitalize on expanding trade partnerships and exchanges. This increased international trade has led to an increase in global production, which has placed a strain on the environment. With regard to global water resources, urbanization led to heavy exploitation due to major advances in geological knowledge, well drilling, pump technology, and rural electrification. Given these circumstances, it is becoming increasingly important to put issues such as water quality and quantity in a global context.

The environment, water, and food production are closely interrelated at the international, local, and regional levels. Water is the major limiting factor for world agricultural production and implicitly tied to global food production.[4] As the human population continues to swell, food production will hinge on the ability to manage, preserve, and enhance global water resources.

Throughout the world, freshwater is used for drinking, industrial production, irrigation, transportation, recreation, waste processing, hydroelectric power, and a habitat for aquatic life.[5] Therefore, issues of contamination and local water depletion are intimately connected to the global economy. Increased global environmental trade, which increases the demand for "virtual" water, or water used to produce crops traded in international markets, can result in declines in water quality and quantity.[6] This entry discusses these important water resource topics from a global environmental management perspective.

Globalization and Water Resources

As globalization has become a more pervasive force, nations around the world have created development plans that prioritize industry over the protection of natural resources such as water.[7] This has become an increasingly notable trend in poorer nations. Many leaders in these nations see the growth of industry as the only realistic way to grow national industry in order to become a competitive force in the global economy and to increase quality of life standards for their citizens. These developing nations have been industrializing at a much faster pace than ever before, which can place an extreme burden on the environment. Some Asian countries (i.e., China) are industrializing in 20 years or less whereas the United States and Europe spent more than 100 years industrializing. However, following in the footsteps at the rate of developing nations such as Europe, Japan, and the United States can lead to a severe degradation of the environment.[7] The Millennium Ecosystem Assessment conducted by the United Nations (UN) finds that although there have been substantial gains in human well-being over the past 50 years due to industrial growth, broad ecosystems around the world are now degraded and under pressure.[8]

With rapidly increasing water demands, the competition among household, industrial, environmental, and agricultural water uses has been escalating in many regions as they are already limited by the amount and quality of available water. Human impacts on the quality and quantity of freshwater threaten economic prosperity, social stability, and the resilience of ecosystem services and natural capital.[5] According to Meyer and Turner,[9] human production and consumption is "unprecedented in its magnitude and rate and has truly reached a global scale" (p. 39).

Not only is human consumption of water an important aspect of water management, the environmental impacts of the changing hydrologic cycle can have an adverse effect on our accessibility to water. It is estimated that "less than 3 percent of all water is fresh enough to drink or to irrigate crops, and of that total, more than two-thirds is locked in glaciers and ice caps"[5] (p. 1029). In the coming century, global climate change and a growing imbalance among freshwater supply, consumption, and population will alter the water cycle dramatically.

Environmental changes that involve the "flow of materials and energy through the chain of extraction, production, consumption, and disposal of modem industrial society" or the "alteration of the land surface (i.e. wetlands, rainforests) and its biotic cover" become global either by "affecting a globally fluid system (the atmosphere, world climate, sea level) or by occurring in a localized or patchwork fashion in enough places to sum up to a globally significant total"[9] (p. 39). It is these two types of environmental change that can affect water resources at a global level.

Political Control and Global Water Management: Policies and Political Structure

With the growing demands on freshwater resources globally, there is an urgent need to link water policy research with improved water management strategies. Water resources sustainability means using water wisely and protecting complex ecosystems with future generations in mind.[10] Better monitoring, assessment, and forecasting of water resources will be needed to allocate water more efficiently among competing needs and to achieve water resource sustainability.[5] This will require coordination from a broad, ideally a global, perspective.

On an international level, the United Nations (UN) plays a critical role in water policy discussions. The UN defines public water policy as "the legislation and regulations that underpin water management"[11] (p. 7). The UN supports the goal of sustainable water management that "holistically addresses equity, economy, and the environment in a way that maintains the supply and quality of water for a variety of needs over the long-term, and ensures meaningful participation by all water resource management stakeholders"[11] (p. 7). The UN supports four dimensions of sustainability that include social sustainability (ensuring that the public has access to adequate and affordable water services), environmental sustainability (protecting against threats to the environment), economic sustainability (balancing costs), and institutional sustainability (establishing institutional frameworks that can function over a long-term period).

Much of the trouble in creating a framework that provides for international consensus, however, is that policymaking at the international level is decentralized and fragmented with multiple stakeholders. The primary stakeholders that serve as official and unofficial actors in issues of global water policy include government agencies at multiple levels, regulatory agencies, private sector firms, consultants, scientists, trade union groups, civil society, and religious groups.[12] Each of these stakeholders seeks to pursue their own individual agendas. While this statement is easily applicable to a variety of global policy issues, it is particularly well suited to global water policy issues because of the disparity in terms of how water policy is treated among nations. Essentially, public water policymaking occurs at all levels of government. The primary legislative framework is developed at the national level, whereas management and operational aspects are implemented at the local level. For example, American water policy involves a combination of state, local, and federal laws dealing with surface water and groundwater allocation and quality protection. Surface water and groundwater allocation is almost exclusively the province of state law, whereas surface water and groundwater quality protection is a mixture of state, local, and federal laws.[13] This fragmented decision-making structure leads to an increase in the number of stakeholders that influence water policymaking within and, inevitably, among nations.

There is fragmentation on water policy not only at the international level but also at the national level. Another aspect of fragmented water policy occurs when discussing specific hydrologic cycle issues and ecological water budgets. Precise data about regional hydrologic cycle issues and area-specific estimates of water supply and demand are difficult to achieve at the national level.[14] Because water policy decision making occurs at the national level in most nations, region-specific issues are the responsibility of state and local policymakers and not applicable to other regions. For example, it is often the case that hydrological knowledge and methodologies obtained for one specific geographic region cannot necessarily be adapted to other regions. This is especially problematic in large nations that have a multitude of differing geographical conditions within its political boundaries, such as the United States and the European Union. Policy fragmentation also occurs as a result in a changing conceptualization of the hydrologic cycle. Classical thinking about the hydrologic cycle has focused on the large-scale (or macro) aspects of the cycle, while largely ignoring the smaller microcycles that occur within it. These microcycles are an important factor in water resource recovery that can help to restore the natural functions of aquatic systems around the world to help achieve sustainability within watersheds. Restructuring these thoughts about water policymaking to include the smaller microcycles that occur in the classic hydrologic cycle model can be a positive step toward the successful regeneration of these ecological systems, but does little to achieve cohesion in national water policymaking.

A negative side effect of this fragmented policymaking is the lack of collaboration and equity in the context of water development and management with nations. When policy fragmentation occurs in this way, the design of policies, interventions, and programs aimed at the equitable distribution of benefits from water resources becomes impossible. It also leads to deficiency and irregularity in data collection and interpretation, which makes cohesive policymaking efforts next to impossible.[14]

Balance sheets of micro-, meso-, and macro-watersheds are another challenge in the policy environment of water policymaking. Evolving a robust and rigorous methodology for assessing the impact of watershed (or river basin) projects at the micro, meso, and macro levels is a major challenge facing academicians, policymakers, and practitioners.[15] Watershed development projects are a continuous process, rather than time-bound projects with a well-defined set of activities and clearly earmarked funds, which creates methodological problems for studying them. This leads to difficulties in creating policy that incorporates these resources. In addition, there has been disagreement on the indicators for success. Overall, studies at the watershed scale have been inconclusive with regard to positive effects on soil conservation, sustainability, or environmental regeneration.[15] Despite limitations, watershed projects have led to at least two positive outcomes, "creation of a large number of structures for soil water conservation; and … setting up of participatory institutions for management of the local resources"[15] (p. 1). Recent watershed studies include studies in India, the Indo-German Watershed Project, Latin America and the Caribbean, the Nile Basin (Africa), the Thukela river basin of South Africa, and the Pangani river basin of Tanzania.

In order for watershed projects to have significance in the development of water policy, Shah et al.[15] state the following issues must be addressed: "technology, physical treatments and economic returns; (creating) institutions that ensure equity and efficiency in resource use; and (creating) market linkages essential for enabling the project to sustain beyond the initial funding, and for spreading out to larger number of beneficiaries as well as areas" (p. 5). At the micro level, watershed projects will inherently require negotiation among territories. At the meso and macro levels, watershed projects will require extensive (often international), integrated collaboration among participatory institutions that can be very costly.[16] Finally, water policy that seeks to address issues at the watershed scale must address socio-economic issues that affect utilization of these resources.

There is a large disparity between water policymaking when looking at developed and developing countries.[17] There is also disparity when looking at water policymaking among different developed nations. In many developed nations, issues of water quality and quantity are managed by a well-established regulatory framework that monitors the public water supply. Many developing nations do not have such a political framework in place at the national level, leaving most of the policymaking decisions for local leaders. This can lead to duplication in implementation and inaction at the local level. Ultimately, this can surmount to even greater disparity within nations, which further limits abilities to develop a cohesive national water policy agenda and regulatory framework.

Global Surface Water and Groundwater Concerns and Issues: Introduction

William Jury and Henry Vaux, at a colloquium held in 2004 to address emerging global water problems, stated that the "optimum management of global water resources presents one of the most crucial challenges of the 21st century" (p. 15,715). Of all the water resource problems, scarcity is considered to be the most serious. According to Jury and Vaux,[18] in much of the world, existing water supplies are already insufficient to meet all of the urban, industrial, agricultural, and environmental demands.

Globalization also has led to an increase in the contamination of water resources through urbanization, agricultural intensification, and land degradation. The lack of a global water quality standard has led to varying levels of water quality around the world. This is significant as water resource contamination can have global effects on human health and the environment through increased international trade and resource sharing.

Global water concerns can be divided into two main areas: water contamination and groundwater, or aquifer, depletion. The next two sections outline the major causes of these two concerns and current issues that are being explored to rectify these issues from a global perspective.

Surface Water and Groundwater Contamination

Quality surface water and groundwater in abundance is needed for human health, agriculture, industry, and recreation. Our surface water resources have a long history of being adversely affected by point sources such as industrial and municipal waste disposal, animal waste, recreational vehicles, hazardous product spills, and air deposition of particulates. Surface water is typically heavily contaminated and is not available for usage in the water supply. According to the Food and Agriculture Organization of the UN, more than 5 million people die annually from waterborne illnesses from both surface water and groundwater contamination. Roads, railways, pipelines, hydroelectric corridors, runoff from mines and mine wastes, quarries, well sites, a large variety of recreational land uses including ski resorts, boating and marinas, campgrounds, and parks, leachates, and gases (as a result of irrigation) contribute to surface water and, eventually, groundwater contamination.

Groundwater is water that is present below the surface of the Earth in underground sources (i.e., streams and aquifers). It is one of the most important natural resources in any nation and serves as a major source of water in communities around the world for households, industries, and agricultural purposes. In comparison to surface water, it is generally considered to be much less susceptible to contamination. Hence, groundwater becomes the main source for potable water in many parts of the world.

Groundwater contamination is one of the most serious problems that can have an impact on water quality. In the past 50 years, groundwater contamination has become one of the most serious environmental problems in the world because once polluted, the remediation of aquifers may be very difficult and even impossible.[19] A groundwater contaminant is any substance that impacts the quality of groundwater. Groundwater contaminants include substances found in storage tanks, septic systems, hazardous waste sites, landfills, road salts, fertilizers, pesticides, and other chemicals.

Problems with water quality have increased primarily due to contaminations associated with human and land-use activities including agricultural activities, leaking municipal sewage treatment facilities, poor septic management, uncontrolled cattle access to water bodies, effluents from industries, and runoff from urban areas.[19] This is a particularly dangerous problem when, as noted, groundwater sources constitute the main drinking water supply in a community, as is the case in most nations. Groundwater contamination also can occur as a result of contamination in surface waters. If contaminants are present in other parts of the hydrologic cycle (i.e., surface water), these contaminants can also be transferred into groundwater supplies.

Typically, groundwater contamination does not occur as a result of a lack of legislation that establishes rules and guidelines regarding groundwater contamination. In other words, much of the problem is not that anticontamination laws do not exist, but that these laws are not necessarily enforced. In China, the major culprit of worsening groundwater pollution is industrial wastes.[20] Many industrial plants discharge waste without proper treatment because environmental protection laws are not enforced. This lack of enforcement coupled with the increase in human activities, largely due to population growth and increasing industrialization, has led to an increase in groundwater contamination. In India, the primary sources of groundwater contamination include untreated wastewater (which can lead to death and disease from sewage-borne infection), large quantities of industrial effluents being discharged, and agricultural runoff filled with hazardous chemicals found in fertilizers and pesticides that taints groundwater sources.[21] These factors are largely attributed to the fact that contamination laws are not enforced. Lack of enforcement can lead not only to decreasing water quality but also to human fatalities. For example, high levels of arsenic found in the surficial aquifer Simav Plain in Kutahya, Turkey, have been linked to an increase in deaths related to gastrointestinal cancers from 1995 to 2005.[22]

Because groundwater serves as the primary source of potable water in many communities, water quality is a critical aspect of the international water crisis. Part of the problem in regulating these systems on a global level is the large amount of diversity observed in the provision of water and policy approaches to contamination among nations around the world and the different regulatory models adopted in different countries. National programs of water quality monitoring throughout the world range from limited to non-existent. Few developing countries include water quality within a meaningful national water policy context. Water supply is considered to be a national issue while contamination is often considered to be a local problem. This can be problematic as national policy frameworks develop that fail to take into consideration the differences in regions.

Many of the national programs that do exist are dysfunctional due to years of neglect, chronic underfunding, and lack of focus. In some countries, water utilities are natural monopolies whereby public or private owners of the system are able to exercise considerable control over pricing and service quality. This places users of that system in a vulnerable situation whereby they may have no other options for water access. In some developing countries, there may be more competition among water utility vendors. This gives users greater options in terms of accessing quality water.

Yet, the regulation of water systems is critical for economic and social development of a nation. Regulatory agencies have been created and established in many different countries to help police utility systems and ensure that water quality guidelines are followed. A recent study conducted by Marques[17] identified 136 water service regulatory authorities worldwide (in January 2008). These regulators are spread across 57 countries, 12 of which are in Africa, 5 in Asia, 16 in Europe (2 in Euro-Asia), 2 in Oceania, and 22 in America. About 25% of countries have regulators for the water sector. Marques[17] suggests that the trend is moving toward nations creating new agencies to provide for the regulation of their water utilities.

In terms of water quality issues, many national governments in developed countries have created a system of rules and guidelines that establish minimum criteria that utility systems must follow in the provision of drinking water to users of that system. Although the implementation of such rules can provide heavy cost burdens on those utilities, the framework provides a uniform standardization mechanism that can assure the public that the water they are consuming is safe.

However, even among developed countries, there is little consistency among the treatment of water quality issues at the national level. Some have well-developed regulatory frameworks, while others do not. The United States passed the Safe Water Drinking Act of 1974 to regulate the nation's public drinking water supply in an effort to protect public health. The Act authorizes the United States Environmental Protection Agency (USEPA) to set national health-based standards for drinking water to protect against both naturally occurring and man-made contaminants that may be found in drinking water. USEPA, individual states, and water systems then work together to make sure that these standards are met. The European Union established a Drinking Water Directive in 1983 that protects human health by establishing healthiness and purity requirements for water. The directive set quality standards for drinking water at the tap and obliges member states to regularly monitor drinking water quality.[23] The Japanese national government has not only established Environmental Water Quality Standards and water pollution control measures but also established uniform water quality regulations for public water and groundwater that operators of utility systems must implement.[24]

Surprisingly, even some larger developed countries lack an established framework. In Canada, there is no national strategy in place to address water issues and no federal leadership to conserve, protect, and regulate Canada's water. The Federal Water Policy is more than 30 years old and needs to be updated to establish more rigid criteria for water quality monitoring. It is ironic that Canada's economy is built on the myth of an abundance of freshwater, yet only 1% of Canada's freshwater is considered renewable [25] Often associated with increased regulations is a shift toward computerized water quality monitoring. Data-driven approaches to water quality can be costly and difficult to implement in developing countries.[26] Over the last several decades, there has been a large shift toward the purchase of techno-logical upgrades that help monitor water quality and collect data in developing countries. Water quality monitoring, as developed in Western countries, is based on the premise that with enough data, a well-designed program can answer most types of water quality management issues. Traditionally, developing countries are data-poor environments. This poses a major challenge for environmental management and decision making because nations that lack water monitoring technology are not able to obtain data about the water quality in order to make appropriate surface water and groundwater contamination policy.[26] In light of the lack of enforceable regulatory systems and lack of knowledge about the actual quality of water in many nations, surface water and groundwater contamination are critical concerns for many nations around the world in a national context. Because of the diverse nature of water quality monitoring in individual nations, forming a cohesive policy regarding groundwater contamination at an international level presents considerable challenges.

Groundwater and Aquifer Depletion

By far, the most abundant and available source of freshwater is underground water supplies or well-springs known as *aquifers*. Groundwater is the primary source of water for drinking and irrigation. At least one-fourth of the world's population draws its water from underground aquifers and approxi-mately 99% of all liquid freshwater is found in underground aquifers.[5] Serious difficulties exist in fairly allocating the world's freshwater resources between and within countries.[27]

Conflicts are escalating among new industrial, agricultural, and urban sectors. According to the Global Environment Facility, worldwide such conflicts have increased from an average of 5 per year in the 1980s to 22 in 2000.[27] In 23 countries for which data are available, the cost of conflicts related to the agricultural use of water was an estimated $55 billion between 1990 and 1997. At least 20 nations obtain more than half their water from rivers that cross national boundaries,[28] and 14 countries receive 70% or more of their surface water resources from rivers that are outside their borders.[29,30] One way that governments have sought to deal with this imbalance has been to pass laws that set limits on water allocations.

Some of the major causes of groundwater depletion include population growth and increased demand for food supply, increased trade, increased agricultural irrigation, and environmental changes including changes in rainfall, temperature, evaporation rates, soil quality, vegetation type, and water runoff. This section reviews these causes and discusses remedies where available.

In the next 30 years, the Earth's population is projected to rise by approximately one-third (or 3 billion people).[5] This growth in global population is projected to be at least 3 times greater than acces-sible freshwater runoff. The growth in global population and water consumption will place additional pressure on freshwater resources. People use freshwater for many purposes including irrigation for crop production, industrial and commercial activities, and residential activities. Some water that is used by people can be reused; herein lays the difference between water consumption and water use. Water con-sumption occurs when there is no ability to reuse the water.[27] For example, when humans use water for sanitation, it can be reused and is therefore not consumed.

To make matters worse, most of the world's population growth will occur in developing countries where water is already critically short and many of the residents are impoverished. It is estimated that, today, more than 1 billion people do not have access to safe and affordable drinking water and perhaps twice that many lack adequate sanitation services.[18] Particularly in cities and towns around the world where there is population density, there is just not large enough a recharge area to support the needs of their inhabitants on a sustainable basis.[31] For example, in Beijing, China, a highly populated city, prohibitions have been placed on farmers from using local water for irrigation. Also in the Gulf Coast countries (Bahrain, Kuwait, Oman, Qatar, Saudi Arabia, and the United Arab Emirates) where there is arid climate, these countries are experiencing a severe water shortage problem that "threatens the sustainable development and hinders the national plans for human, industrial and agricultural develop-ment" throughout the region[32] (p. 59).

While floodwater is one way to replenish global water resources, floodwater requires large-scale management and coordination of water systems including reservoirs and dams. Water for irrigation, industry, and household uses must be delivered in controlled quantities at specific times, making floodwater an impractical solution to renewing water resources without the necessary management in place.

Also, as the human population has increased, so has trade among nations and more water is used to produce these goods and services. As a result of this increase in environmental trade and urbanization, countries have tried to overproduce in an effort to maintain a competitive edge in the global marketplace. This has placed a burden on the groundwater sources in many nations that have over-pumped aquifers in an effort to increase production and trade. This is most notable in the agricultural community. For example, the three main grain producers, China, United States, and India, have all felt the effects of overpumping, or pumping that exceeds the long-term average rate of replenishment. Ultimately, overpumping can lead to a depletion of aquifers and consumption of aquifer reserves, which may lead to cutbacks in grain harvests. Evidence of substantial and widespread drawdown of the piezometric surface beneath many Asian cities, as a result of heavy exploitation of aquifers, also has been accumulating since the early 1970s.

The increase in agricultural production, in both crops and livestock, is significant as much of the water that is used for production cannot be recovered or reused; in essence, it is consumed. According to the United Nations Educational, Scientific and Cultural Organization (UNESCO), world agriculture consumes 70% of the freshwater withdrawn per year.[27]

Increased crop production and irrigation negatively affects soil quality and water runoff. Irrigation causes soil erosion, which diminishes the soil quality by reducing soil depth and soil nutrients. Other problems and failures of irrigated agriculture include groundwater overdraft, water quality reduction, water logging, and salinization.[33] While these various forms of soil degradation are proven to affect water resources, on a global level, they are very hard to track.[9]

Increased livestock production directly and indirectly affects the depletion of water resources in that not only do livestock directly consume freshwater but water is also consumed in the production of grains that are fed to the livestock. According to the U.S. Department of Agriculture (USDA), increased crop and livestock production during the next five to seven decades will significantly increase the demand on all water resources, especially in the western, southern, and central United States and in many regions of the world with low rainfall.[27] Strategies such as crop rotation, replacement of trees and shrubs, soil maintenance, and the use of organic mulch promote water conservation by limiting the damage done by crop production.[27]

Appropriate water pricing is important for improved water demand and conservation of water.[27] The relatively high cost of treating and delivering water has led many world governments to subsidize water for agricultural and household use. For example, some U.S. farmers pay as little as $0.01 to $0.05 per 1000 L used in irrigation, while the public pays $0.30 to $0.80 per 1000 L treated water for personal use.[34] According to the World Bank,[35] the objectives of fair water pricing are 1) to seek revenue to pay for the operations and maintenance of water availability; 2) to improve water use efficiency; and 3) to recover the full costs of water pumping and treatment.[36]

Finally, there is a growing scientific consensus that the buildup of greenhouse gases in the atmosphere is warming the Earth.[5] The continued loss of forests and other vegetation and the accumulation of carbon dioxide, methane gas, and nitrous oxide in the atmosphere are projected to lead to global climate change or global warming. Over time, such changes may alter precipitation and temperature patterns throughout the world.[27] These temperature changes might increase soil erosion and deforestation, which could result in greater reductions in water availability worldwide. Salati and Vose[37] claim that regional forest clearance would have severe climatic consequences. Because of the high proportion of water recycled by the rainforest, deforestation would significantly lessen rainfall and increase temperatures.

Global Water Resource Policy: Toward a Potential Framework

As noted throughout this entry, there are many efforts at the national level to conserve water resources, but what is lacking is an integrated worldwide water resource management system. A well-coordinated international plan for managing the diverse and growing pressures on freshwater systems and for establishing goals and research priorities for cross-cutting water issues is required. At the most basic level, water resource management must address how to "secure water and related services, avoid degradation of water and land resources and of ecosystem integrity, and foresee changes (climate, population, diet preferences etc.)"[38] (p. 14).

This plan will have to incorporate the many national, transnational, and private organizations that already exist into global water resources conglomerations or networks. Collaboration among this myriad of agencies will inevitably require a change in the culture of water management where water agencies have been known to be proprietary in nature. In addition, a redefinition of water policy from a political, economic, cultural, and social perspective is needed.

Foster and Chilton[39] note that nations and stakeholders need to "appreciate their social and economic dependency on groundwater, and to invest in strengthening institutional provisions and building institutional capacity for its improved management" (p. 1970). This can be achieved by creating global communication strategies to fuse the dialogue between the many water management stakeholders including scientists, policymakers, water managers, and citizens.[18]

Foster and Chilton[39] also state that local field agencies might have to change their roles to aid in the development of a broader global water management system. Jury and Vaux[18] agree that many existing water management institutions were "created in times and eras when the problems of developing and managing water resources were very different from what they are today" and would therefore require substantial changes to their institutional arrangements. Currently, water management institutions tend to focus on a narrow set of interests, but with the global nature of water resources management today, these institutional arrangements will no longer be beneficial to solving world water problems.

For example, Benvenisti[40] argues that there is enough freshwater in the world to meet the existing and future needs of the world's population but that this water is poorly distributed, making the management of global water resources predominately about redistribution. Shah et al.[31] agrees that redistribution and "the co-existence of regions with undeveloped (water) resource and those with overdeveloped (water) resource" is central to the future of water management policy.

Redistribution policy management inevitably will bring into play the competing priorities of different uses and users, and since most water resources traverse political boundaries, these competing priorities often become regional conflicts between states. Policies aimed at redistribution across political boundaries must then be of "collective action despite the varied self-interest" of these regions[40] (p. 385). Benvenisti recommends an international law policy framework that borrows aspects of game theory, economics, contract legal theory, international relations theory, and international negotiations theory to encourage the cooperation needed for the redistribution of global water resources.

In addition, to move toward a global policy framework, much more precise and spatially congruent data are needed. Meyer and Turner[9] note that the reluctance of national and international institutions to fund the collection of social data is an obstacle that must be addressed. Foster and Chilton[39] note that, globally, current data used to assess the status of aquifer degradation are of questionable reliability, inadequate coverage, and poor compilation. Since data on aquifer depletion and water contamination are collected by many different agencies that might focus on different aspects of water management including engineering sciences, hydrology, climatology, and geology, institutional arrangements that can bring together these interdisciplinary concepts into a comprehensive data management system will have to be created.[18]

Finally, while there is a general consensus on some of the future challenges in a global water management system, there are still many unknown factors that may arise. Jackson et al.[5] state that "uncertainty will be the most important feature of freshwater forecasts" and that by "evaluating uncertainties, forecasters can help decision makers to anticipate the range of possible outcomes and to design flexible responses" (p. 1038). Planning for uncertainties requires "adaptive management" or the "process of designing management interventions to decrease the variance of future forecasts and recommend alternative management options" (p. 1038). Future global management of freshwater systems must, therefore, incorporate adaptive management strategies.

Conclusion

The major water concerns discussed in this entry are associated with the prevailing cultural, political, economic, scientific, and social perceptions about how humans should interact with the environment. A redefinition of water policy may lead to changes in these perceptions. On a larger scale, the global water discussion requires a redefinition of the relationships between industry and environment. A redefinition of water policy may lead to changes in human behavior that can help address some of the concerns discussed in this entry. For decades, the primary engine that has driven the growth of the global market is a continued focus of economic gain among nations. This tendency toward capitalist values has continually placed a greater value on industry at the expense of protecting the environment. Strategic thinking at the international level should emphasize sustainable development. There must be some consensus among policymakers on water policy goals. A global water policy agenda that focuses on sustainable development is necessary to meet the goals of protecting the Earth's water while meeting human water demand.

References

1. National Aeronautics and Space Administration, 2011, available at available at http://earthobservatory.nasa.gov/Features/WeighingWater/ (accessed September 7, 2011).
2. United Nations Educational, Scientific, and Cultural Organization (UNESCO), 2011, available at http://www.unesco.org/new/en/unesco/ (accessed September 7, 2011).
3. World Health Organization (WHO), 2011, available at http://www.who.int/en/ (accessed September 12, 2011).
4. Kendall, H.W.; Pimentel, D. Constraints on the expansion of the global food supply. Ambio **1994**, *23* (3), 198–205.
5. Jackson, R.B.; Carpenter, S.R.; Dahm, C.N.; McKnight, D.M.; Naiman, R.J., Postel, S.L.; Running, S.W. Water in a changing world. Ecol. Appl. **2001**, *11* (4), 1027–1045.
6. Wichelns, D. Virtual water: A helpful perspective, but not a sufficient policy criterion. Water Resour. Manage. **2010**, *24*, 2203–2219.
7. Steiner, G.; Steiner, J. *Business, Society, and Government*; McGraw-Hill: New York, 2005.
8. United Nations. Millennium Ecosystem Assessment, 2011, available at http://www.maweb.org/en/About (accessed September 10, 2001).
9. Meyer, W.B.; Turner, II, B.L. Human population growth and global land-use/cover change. Annu. Rev. Ecol. Syst. **1992**, *23*, 39–61.
10. Kinoti Mutiga, J.; Mavengano, S.T.; Zhongbo, S.; Woldai, T.; Becht, R. Water allocation as a planning tool to minimize water use conflicts in the Upper Ewaso Ng'iro North Basin, Kenya. Water Resour. Manage. **2010**, *24*, 3939–3959.
11. United Nations. United Nations Report on Guide to Responsible Business Engagement with Water Policy Report, 2010, available at http://www.unglobalcompact.org/docs/issues_doc/Environment/ceo_water_mandate/Water_Policy_Engagement_Guide_Public_Consultation_Draft.pdf (accessed September 7, 2011).
12. Ainuson, K. An advocacy coalition approach to water policy change in Ghana: A look at belief systems and policy oriented learning. J. Afr. Stud. Dev. **2009**, *1* (2), 016–027.

13. Tarlock, Dan A. Water Policy Adrift. Forum for Applied Research and Public Policy, Spring 2001, 63–70.
14. Deason, J.; Schad, T.; Sherk, G.W. Water policy in the United States: A perspective. Water Policy **2001**, *3*, 175–192.
15. Shah, A.; Devlal, R.; Joshi, H.; Desai, J.; Shenoy, R. *Benchmark Survey for Impact Assessment of Participatory Watershed Development Projects in India;* Gujarat Institute of Development Research: Ahmadabad, 2004.
16. San Martin, O. *Water Resources in Latin America and the Caribbean: Issues and Options;* Inter-American Development Bank Sustainable Development Department: Brazil, 2002.
17. Jury, W.A.; Vaux, H., Jr. The role of science in solving the world's emerging water problems. Proc. Natl. Acad. Sci. U. S. A. **2005**, *102* (44), 15715–15720.
18. Marques, R.C. *Regulation of Water and Wastewater Services;* IWA Publishing: Lisbon, 2010.
19. Shah, T.; Molden, D.; Sakthivadivel, R.; Seckler, D. *The Global Groundwater Situation: Overview of Opportunities and Challenges;* International Water Management Institute: Colombo, Sri Lanka, 2000.
20. Samake, M.; Tang, Z.; Hlaing, W.; Ndoh Mbue, I.; Kasereka, K.; Balogun, W. Groundwater vulnerability assessment in shallow aquifer in Linfen Basin, Shanxi Province, China using DRASTIC model. J. Sustainable Dev. **2011**, *4* (1), 53–71.
21. Aguilar, D. Groundwater reform in India: An equity and sustainability dilemma. Texas Int. Law J. **2011**, *46* (3), 623–653.
22. Gunduz, O.; Simsek, C.; Hasozbek, A. Arsenic pollution in the groundwater of Simav Plain, Turkey: Its impact on water quality and human health. Water, Air, Soil Pollut. **2010**, *205*, 43–62.
23. European Union Water Drinking Directive, 2011, available at http://ec.europa.eu/environment/water/water-drink/index_en.html (accessed September 7, 2011).
24. World Bank. International Trends in Water Pricing and Use, 2000, available at http://lnweb18.worldbank.org (accessed September 7, 2010).
25. The Council of Canadians, 2011, available at http://www.canadians.org/water/issues/policy/index.html (accessed September 12, 2011).
26. Ongley, E.D. Water quality management: Design, financing and sustainability considerations. Proceedings of the African Water Resources Policy Conference, Nairobi, May 26–28, 1999.
27. Pimentel, D.; Berger, B.; Filiberto, D.; Newton, M.; Wolfe, B.; Karabinakis, E.; Clark, S.; Poon, E.; Abbett, E.; Nandagopal, S. Water resources: Agricultural and environmental issues. BioScience **2004**, *54* (10), 909–918.
28. Gleick, P.H., Ed. *Water in Crisis: A Guide to the World's Fresh Water Resources;* Oxford University Press: New York, 1993.
29. Alavian, V. Shared Waters: Catalyst for Cooperation, 2003, available at http://ucowr.siu.edu/updates/pdf/V115_A2.pdf (accessed September 7, 2011).
30. Cech, T.V. *Principles of Water Resources: History, Development, Management and Policy;* John Wiley and Sons: New York, 2003.
31. Shah, T.; Molden, D.; Sakthivadivel, R.; Seckler, D. Global groundwater situation: Opportunities and challenges. Econ. Polit. Wkly. **2001**, *36* (43), 4142–4150.
32. Al-Rahsed, M.F.; Sherif, M.M. Water resources in the GCC countries: An overview. Water Resour. Manage. **2000**, *14*, 59–75.
33. Rosegrant, W.; Cai, X. Water constraints and environmental impacts of agricultural growth. Am. J. Agric. Econ. **2002**, *84* (3), 832–838.
34. Gleick, P.H. *The World's Water;* Island Press: Washington, DC, 2000.
35. World Bank. Water Resources Management in Japan. World Bank Analytical and Advisory Assistance Program, 2006, available at http://siteresources.worldbank.org (accessed September 2011).

36. Schalch, K. Providing safe drinking water to developing nations. National Public Radio, 2003, available at http://www.unesco.org/water/wwap/news/national_public_radio.shtml (accessed September 2011).

37. Salati, E.; Vose, P.B. Amazon basin: A system in equilibrium. Science **1984**, *225*, 129–138.

38. Falkenmark, M. Shift in thinking to address the 21st century hunger gap: Moving focus from blue to green water management. Water Resour. Manage. **2007**, *2*, 3–18.

39. Foster, S.S.D.; Chilton, P.J. Groundwater: The processes and global significance of aquifer degradation. Philos. Trans. R. Soc., B **2003**, *358* (1440), 1957–1972.

40. Benvenisti, E. Collective action in the utilization of shared freshwater: The challenges of international water resources. Am. J. Int. Law **1996**, *90* (3), 384–415.

Bibliography

1. Ahmad, S.; Prashar, D. Evaluating municipal water conservation policies using a dynamic simulation model. Water Resour. Manage. **2010**, *24*, 3371–3395.

2. Cobb, R.W.; Elder, C.D. *Participation in American Politics: The Dynamics of Agenda-Building,* 2nd Ed.; Allyn and Bacon: Boston, 1983.

3. Ellison, B. New thinking about water management. Public Adm. Rev. **2007**, *67* (5), 946–950.

4. EnvironmentalProtectionAgency, 2011, available at http://water.epa.gov/lawsregs/rulesregs/sdwa/index.cfm (accessed September 2011).

5. Food and Agriculture Organization of the United Nations. Control of Water Pollution from Agriculture. FAO Corporate Document Repository (accessed March 2012).

6. Gisser, M. Groundwater: Focusing on the real issue. J. Polit. Econ. **1983**, *91* (6), 1001–1027.

7. Gleick, P.H. Soft water paths. Nature **2002**, *418*, 373.

8. Hassan, R.; Scholes, R.; Ash, N. *Ecosystems and Well-Being: Current State and Trends;* Island Press: Washington, DC, 2006.

9. Ives, G.; Messerli, B. *The Himalayan Dilemma: Reconciling Development and Conservation;* Routledge: London and New York, 1989.

10. Kamel, N.H.; Sayyah, A.M.; Abdel-aal, A. Ground water in certain sites in Egypt and its treatments using a new modified ion exchange resin. J. Environ. Prot. **2011**, *2*, 435–444.

11. Leung, K.S. Public–private partnerships solve groundwater contamination issues. Am. Water Works Assoc. **2011**, *103* (1), 26–29.

12. López, E.; Schuhmacher, M.; Domingo, J. Human health risks of petroleum-contaminated groundwater. Environ. Sci. Pollut. Res. Int. **2008**, *15* (3), 278–288.

13. Shankar, B.S.; Balasubramanya, N.; Maruthesha, R. Impact of industrialization on groundwater quality—A case study of Peenya industrial area, Bangalore, India. Environ. Monit. Assess. **2008**, *142* (1–3), 263–268.

14. Wang, M. Optimal environmental management strategy and implementation for groundwater contamination prevention and restoration. Environ. Manage. **2006**, *37* (4), 553–565.

15. Water demand creates policy challenges. *Oxford Analytica Daily Brief Service,* p. 1, 2011.

16. Widdowson, M.; Mendez, III, E.; Chapelle, F. Estimating cleanup times for groundwater contamination remediation strategies. Am. Water Works Assoc. **2007**, *99* (3), 40–45.

17. Zaller, J.R. *The Nature and Origin of Mass Opinion;* Cambridge University Press: New York, 1992.

27

Water: Cost

Introduction 277
    Competitive Markets and Water Prices
Optimizing Model: Simulating a Competitive Water Market 279
    Shadow Prices and Scarcity Rents • The Role of Shadow Prices
Structuring a Price (Tariff) of Water 283
    Substitutability and Environmental Considerations • Economic
    Efficiency • Equity and Fairness • Simplicity • Sustainability • Stability
    and Quality • Water Tariffs and Pricing Strategies
Conclusions 289
References 290

Atif Kubursi and
Matthew Agarwala

Introduction

Is water really different from other commodities? While water is versatile and has many uses, it is generally considered as an asset, a consumer good, and a factor of production. If it is a resource, is it a renewable or a "non-renewable resource?" Is its value infinite and "thicker than blood?" Can a price for water be determined much like for any other commodity? Can water or water permits be traded? What constitutes an equitable distribution of this scarce resource? What is a fair price for water? What is the full-cost price of water that incorporates its many environmental functions?

These complex questions are provocative and can incite passionate debate, yet they are central to water resource management. In this note, we show that economics sheds useful light on how to address these questions. This is not to suggest that economics alone can answer all of them. There are complex social, environmental, moral, and strategic issues that need to be factored into such an analysis. These add complications to the general structure within which water prices will be considered but do not change in any fundamental way the conclusions reached.

Before going any further in answering some of these questions, it is useful to outline some of the specific attributes of water that distinguish it from other commodities. These attributes will also influence the way water prices are set and optimal allocations of this scarce resource may be affected.

- Water is a fugitive, reusable, and stochastically supplied resource whose production can be subject to economies of scale, which give rise to natural monopoly situations. In this respect, water has many of the characteristics of a common property resource and a quasi-public good.
- Water is typically a non-traded commodity that is rarely sold in a competitive market. There are few overt water markets where suppliers and demanders exchange water. Markets in water rights have emerged in several parts of the world; the most notable examples are in Australia, Colorado, California, and Argentina.[1–3] However, most of these markets are within national entities, are heavily regulated, and often represent simulated market solutions. There are few international examples of water trade, but it is not difficult to conceive schemes that would involve this trade.

- Water scarcity is not only about physical scarcity, but it is also complicated by economic scarcity where actual prices for water are typically only a fraction of the true scarcity price. When prices are below scarcity prices, waste and overuse are quickly observed.
- Water values generally differ from the price that would obtain in a free and competitive market. The distinction between price and value is crucial. A family's visit to a local park may have a price of $0, but the value of such a visit may be much greater. When prices and values diverge, market forces produce inefficient outcomes. Water often has social value that is above what private users are willing to pay for it.[4] The allocation of water often reflects national and social policies and priorities toward agriculture, the environment, food security, and national security that go beyond serving the interests of private farmers. Social and policy considerations apart, the diversion of actual prices from their scarcity values imposes social costs on the domestic economy as well as on neighboring countries. Water is part of the tragedy of the commons.[5]
- Water is not only a desirable commodity, but its availability is also critical for life. At subsistence levels, it has no substitute. Furthermore, that every person is entitled to a minimum quantity that is considered consistent with human dignity is now enshrined in international human rights law.[6,7]
- The secure supply of water in many regions of the world is quite low. Security of supply is defined here as the probability of its average flow availability for 9 out of 10 years. In Canada, a country rich in water, this probability is less than 30%, whereas it is less than 5% in the Middle East.[8]
- While the total water supply may be limited and few, if any, substitutes exist for it, there exist substantial possibilities for intersectoral and interregional substitutions. As well, there are a number of technologies and conservation programs that rationalize demand and raise the efficiency of its use.
- Part of the water scarcity crisis is the fact that agriculture on average uses about 70% of the total global supply.[9] It is typically the case that other needs are suppressed, but this leaves a wide room for intersectoral reallocations and shuffling water between users and uses.
- While the quantity of water is in short supply, concern for preserving its quality is perhaps more pressing. Pollution and saline intrusion of the aquifers are being increasingly recognized as critical factors in planning for the future.
- Water is the "universal solvent."[10] It performs numerous ecosystem services including the absorption, transportation, and dilution of pollutants. Water is indispensable to the environment's pollution abatement capacity.[11]

Competitive Markets and Water Prices

The scarcity of water suggests that it should have a price that reflects this scarcity. Since economics is particularly suited for dealing with how scarce resources are, or should be, allocated to various uses and users, it can shed useful light on its use and value.

Indeed, water has a value as does any scarce resource, input, or asset. There is a monetary equivalent to water, although this is not the way people typically speak about water. This value need not emerge from competitive markets. However, it has or could be constructed to have many of the characteristics that are associated with competitive prices. The price should reflect the scarcity of water and disseminate information about this relative scarcity. It should also invite, if not provoke, the correct responses (incentives) that prices in general are expected to do. Higher prices persuade consumers to economize on its use and suppliers to raise their quantities offered. In competitive markets, the market price is what buyers are willing to spend for an additional unit of the commodity and the extra cost of producing it (marginal cost). If the price is larger than the marginal cost of the last unit of water, this provides a signal that more is needed of the commodity and more should be produced.[12] Conversely, if the price is below marginal cost, additional units will not be produced as society will have to sacrifice more than what people would value the additional units.

It is generally accepted, however, that water is not bought and sold in competitive markets. This is because in the case of water, at least five of the basic properties of competitive markets are generally absent.[4] These five properties include the following:

- Free markets lead to an efficient allocation of scarce resources if these markets are characterized by competitive structures, that is, these markets include a large number of independent small sellers and a similarly large number of independent small buyers that no single supplier or buyer is significant enough to influence the price. Each and every buyer and seller in this market is a price taker.
- Competitive markets require freedom of entry and exit and that no barriers that preclude easy entrance or exit exist.
- The product must be homogeneous enough that each unit is quite similar to any other unit.
- For a free market to lead to an efficient allocation, externalities must be absent. In economics, an externality or spillover of an economic transaction is an impact on a party that is not directly involved in the transaction. An efficient allocation can emerge from a free market when social costs coincide with private costs. Water production, however, involves many "externalities."[4] In particular, extraction of water in one place reduces the amount available in another. Further, pumping water from an aquifer in one location can affect the cost of pumping elsewhere. Such externalities do not typically enter the private calculations of individual producers and drive a wedge between private and social costs.
- In a free market that efficiently allocates scarce resources, social benefits must coincide with private ones. If not, then (as in the case of cost externalities) the pursuit of private ends will not lead to socially optimal results. In the case of water, many uses have social benefits that exceed the private ones. The use of water in agriculture may result in benefits that exceed the private returns to farmers. Among these are food security, border security, environmental objectives, and national interest.

These conditions are often violated in the case of water, where water sources are relatively few; barriers to entry are real and high (high cost of infrastructure); a large gap exists between private and social costs; and benefits and water units are not homogeneous, with a large spectrum of different qualities observed. This, perhaps, is why it is often the case that water production facilities are owned by the state. In many respects, water is not a private good; it has what we alluded to above, many of the characteristics of quasi-public goods.

Although water is not bought or sold in competitive markets, it is possible to develop prices that reflect its value.[13–15] In practice, the full value of water is not revealed in competitive markets. It is possible, however, to build specific models of water allocations that simulate competitive conditions and where the optimal nature of markets is restored. The model can explicitly optimize the benefits to be obtained from water, taking into account the five points made above. It is also true that now we have a rich body of literature that informs the setting of water prices and tariffs to overcome some of the five points listed above. The prices that emerge from formal optimization models not only maximize profits for the producers (producer surplus) but also maximize the utility of the consumer (consumer surplus). Furthermore, as is shown later, a multi-objective framework can be developed to design prices that satisfy goals beyond pure economic efficiency. These constructed prices permit the optimal allocation of water to its best uses as would emerge under competitive conditions. In many respects, the constructed prices are designed to serve as guides to consumers and producers in much the same way as competitive prices.

Optimizing Model: Simulating a Competitive Water Market

The underlying economic theory of the "optimizing model" is simple and compelling.[14,16,17] (This section draws heavily on Fisher[4] and Fisher and Huber-Lee.[17]) In Figure 1, we show an individual household's demand curve for water, with the amounts of water (on the horizontal axis) that the household

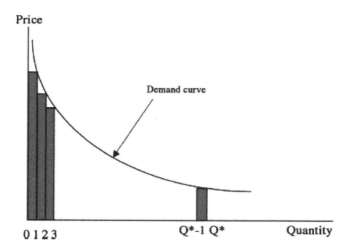

FIGURE 1   Demand curve for water.

will buy at various prices (on the vertical axis). The curve slopes down, representing the fact that the first few units of water are very valuable, while later units will be used for purposes less essential than drinking and cooking.

Now consider how much it will be worth to the household in question to have a quantity of water, Q^*, as depicted in Figure 1. We begin by asking how much the household would be willing to pay for the first small unit of water. The price that would be paid is given by a point on the curve above the interval on the horizontal axis from 0 to 1. Thus, the amount that would be paid is (approximately) the area of the leftmost vertical strip in Figure 1 (one unit of water times the price in question). Similarly, the amount that would be paid for a second unit can be approximated by the area of the second-to-left vertical strip, and so on until we reach Q^*. It is easy to see that if we make the size of the units of water smaller and smaller, the total amount that the household would be willing to pay to get Q^* approaches the area under the demand curve to the left of Q^*.

It is quite simple to reinterpret Figure 1 so that it represents not the demand curve of an individual household but the aggregate demand curve of all households in a given region. The gross (private) benefits from the water flow Q^* can thus be represented as the total area under the demand curve to the left of Q^*. These benefits are gross benefits as the cost of providing this water is not subtracted. To derive the net benefits from Q^*, we must subtract the costs of providing Q^*.

In Figure 2, the line labeled "marginal cost" shows the cost of providing an additional unit of water. That cost increases as more expensive sources of water are used. The area under the marginal-cost curve to the left of Q^* is the total cost of providing the supply, Q^*, to the households involved. Thus, the net benefit from providing Q^* to these households is the shaded area in the diagram, the area between the demand curve and the marginal-cost curve.

The amount of water that should be delivered so as to maximize the net benefits (the sum of consumer and producer surpluses) from water is Q^*, where the two curves intersect. This is the largest area between the two curves and represents the sum of the consumer surplus, defined as the difference of what households would be willing to pay and what they actually pay (the price that rules in the market at the intersection between supply and demand), and the producer surplus, defined as the amount of profits producers realize as the price of water exceeds its marginal cost. Both of these areas are defined in Figure 2.

Social value of water can exceed its private value as society may value this water beyond the benefits individuals would derive from it. A national policy to subsidize water for agriculture by say, 10 cents per cubic meter, at all quantities simply means that water to agriculture is worth 10 cents per cubic meter more to society than farmers are willing to pay for it.[17] This is represented in Figure 3. The lower

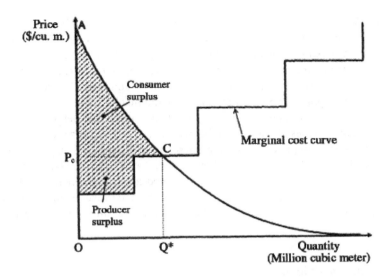

FIGURE 2   Net benefits from water.

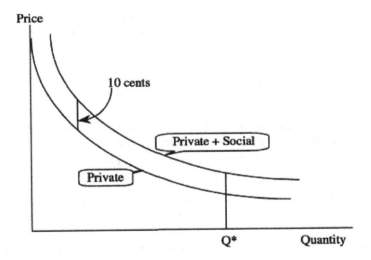

FIGURE 3   Social value of water as revealed by a subsidy.

demand curve represents the private value of water to agriculture; the upper demand curve also includes the additional public value as reflected in the policy, an additional value of 10 cents per cubic meter. As this illustrates, any consistent water policy can be represented as a change in the demand curve for water. Once such a policy has been included in the demand curves, the methods used above can be employed to measure net benefits.

Shadow Prices and Scarcity Rents

(This section too draws heavily on Fisher.[4])

As already discussed, purely private markets and the prices they generate cannot be expected to optimize net social benefits of water. Nevertheless, prices in an optimizing model can play a role very similar to that which they play in a system of competitive markets. Dr. Franklin M. Fisher[4] and his team developed such a model, called "WAS" (for "Water Allocation System") at Harvard University.[18] The model allocates water so as to maximize the net benefits (consumer surplus plus producer surplus) obtained from this allocation. The maximization of net benefits is done subject to constraints. An example of such constraints includes that at each location, the amount of water consumed cannot exceed the amount produced there plus net imports (imports minus exports) into that location.

Corresponding to the optimum quantities, there is a dual solution that defines the associated prices at the optimized level of activities. It is a general theorem that when maximization involves one or more constraints, there is a system of prices involved in the solution. These prices are called "shadow prices" or "Lagrange multipliers." Each shadow price shows the rate at which the quantity being maximized (here, net benefits from water) would increase if the associated constraint were relaxed by one unit at the optimal solution. In effect, the shadow price is the amount the maximizer would be just willing to pay (in terms of the quantity being maximized) to obtain an additional unit of the associated constrained quantity.[19]

The Role of Shadow Prices

Since water does not generally command a market price, economists have constructed models that estimate the scarcity (shadow) price of water that reflects many of its competitive equilibrium characteristics. The shadow price associated with any particular constraint shows the extent by which the net benefits from water would increase if that constraint were loosened by one unit. For example, where a pipeline is limited in capacity, the associated shadow price shows the amount by which benefits would increase per unit of pipeline capacity if that capacity were slightly increased. This is the amount that those benefiting would just be willing to pay for more capacity.

The shadow price of water at a given location is the amount by which the benefits to water users (consumers and producers) would increase were there an additional cubic meter per year available at that location. It is also the price that the buyers at that location who value additional water the most would just be willing to pay to obtain an additional cubic meter per year.

There are a number of defining characteristics of the shadow prices that need further explanation.[4] These include the following:

- Shadow prices are not necessarily the prices that water consumers are charged. That would be true in a purely private, free market system. However, in the WAS model, as in reality, the prices charged to some or all consumers can (and often will) be a matter of social or national policy. When such policy-driven prices are charged, the shadow prices of water will reflect the net benefits of additional water given the policies adopted.
- Shadow prices are outputs of the model solution, not inputs specified a priori. They depend on the policies and constraints put in by the user of the model.
- The shadow price of water in a given location does not generally equal the marginal cost of providing it there. If demand for water from a source is sufficiently high, the shadow price of that water will not be zero even if it costs nothing to produce it; benefits to water users would be increased if the capacity of the source were greater. Equivalently, buyers will be willing to pay a nonzero price for water in short supply, even though its direct marginal costs are zero.
- When demand at the source exceeds capacity, it is not costless to provide a particular user with an additional unit of water. That water can be provided only by depriving some other user of its benefits; that loss of benefits represents an opportunity cost. In other words, scarce resources have positive values and positive prices even if their direct marginal cost of production is zero. Such a positive value, the shadow price of the water in situ, is called a "scarcity rent."
- The shadow price of water used in any location equals the direct marginal cost plus the scarcity rent. For water in situ, the shadow price is the scarcity rent.

- Water will be produced at a given location only if the shadow price of water at that location exceeds the marginal cost of production. Equivalently, water will be produced only from sources whose scarcity rents are nonnegative.
- If water can be transported from location A to location B, then the shadow price of water at B can never exceed the shadow price at A by more than the cost of transporting water from A to B. If water is transported from A to B, then the shadow price at B will equal the shadow price at A plus the transportation costs. Equivalently, if water is transported from A to B, then the scarcity rent of that water should be the same in both locations.

Shadow prices generalize the role of market prices,[4] as can be seen from the following propositions:

- Where there are only private values involved, at each location, the shadow value of water is the price at which buyers of water would be just willing to buy and sellers of water just willing to sell an additional unit of water.
- Where social values do not coincide with private values, this need not hold. In particular, the shadow price of water at a given location is the price at which the user of water would be just willing to buy or sell an additional unit of water. That payment is calculated in terms of net benefits measured according to the user's own standards and values.
- Water in situ should be valued at its scarcity rent. That value is the price at which additional water is valued at any location at which it is used, less the direct marginal costs involved in transporting it there.

Structuring a Price (Tariff) of Water

Water utilities routinely design water tariffs and prices. Their choice of the price is motivated by considerations other than pure economics. They simply choose the price at a level that would help cover their operating costs plus a capital charge in addition to several other objectives, particularly environmental and sustainability objectives. Few utilities, if any, seek solely to generate sufficient revenue to cover their private costs. The generation of revenue is rarely the only purpose of a tariff, nor is it the sole consideration.[20] It has been generally recognized that water tariffs are powerful management tools with a number of complex and important functions. The tariff could create incentives for the efficient production and use of water. In this way, the tariff promotes environmental sustainability. As well, tariffs can be set to recover the full cost (including externalities, both economic and environmental) of an activity or structured to subsidize poorer customers.

Rogers, Bhatia, and Huber[21] make an interesting distinction between economic and environmental externalities. Economic externalities are more readily quantifiable and involve costs such as those imposed on downstream users by upstream diverters and polluters. Environmental externalities include the value of ecosystem degradation; the loss of biodiversity; and the health effects on human, plant, and animal populations. Each objective would lead to a different tariff design. The "best" design is the "one that strikes the most desirable balance among the different objectives."[20] However, is the "best" tariff the optimal one? Standard optimization theory tells us that net social benefits are maximized where marginal social benefit equals marginal cost. However, there is more than one cost perspective to consider when structuring an optimal tariff or price for water. Given the many different definitions for full supply cost, full economic cost, and full cost, the question remains as to which marginal cost should be equated with the marginal benefit. Full-cost pricing requires a quantitative assessment of water's intrinsic and cultural values, as well as the net of environmental externalities. The legitimacy of any such quantification is, however, severely limited by the subjectivity and unavailability of relevant data.[11,21,22] However, optimality requires quantification, and this has become a major issue in setting the optimal water tariff. The different definitions and their relevance for designing the optimal water tariff are presented in the table below.

Opportunity Cost

The opportunity cost of a commodity or service is the value of its next-best alternative use. If good X has only two possible uses, A and B, the opportunity cost of using good X for use A is the foregone benefit of employing it in use B. Similarly, the opportunity cost of using X for B is A. That is, it is the value of the foregone best alternative use. As Rogers, Bhatia, and Huber[21] argue, neglecting the value of water's opportunity costs "undervalues water, leads to failures to invest, and causes serious misallocation of the resource between users."

Full Supply Cost

This is the sum of costs directly involved in delivering water to the end user. This may include pumping, purification, storage, fixed infrastructure costs, capital investment costs, government taxes, and connection and metering costs. Full supply costs exclude all externalities and opportunity costs.[21] Capital and infrastructure costs are particularly heterogeneous vectors that vary widely depending on the types of delivery systems involved. Wells, house connections, pipes of various flow capacities, desalination projects, and recycled and treated water can lead to vastly different nominal values for full supply cost.[23]

Full Economic Cost

This is an economic concept that refers to the full supply cost plus the opportunity costs and net of economic externalities.

Full Cost

This is the full economic cost plus environmental externalities (these are more diffused and indirect and not easily quantifiable; Figure 4).

Substitutability and Environmental Considerations

Economic theory predicts that as the price of natural resource inputs rises, agents will substitute produced capital to take their place. The feasibility of the substitution of produced capital for natural resources is mixed. In some areas, the argument offers a viable solution. For example, the scarcity of whale oil, which resulted from overfishing, was made nearly irrelevant by the introduction of the electric light bulb. In telecommunications, fiber optics has been heavily substituted for copper. High energy costs promote the substitution of higher-quality insulation for fuel.[24–26] It is important to recognize and

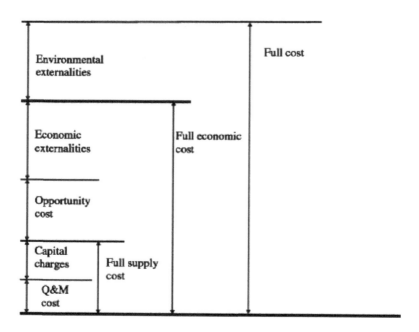

FIGURE 4   Full-cost prices.
Source: Rogers, Bhatia, and Huber.[21]

exploit the ability to substitute produced capital for natural resource inputs, but it is also important to recognize that these abilities are not universal and that they are not infinite.

This is particularly the case when it comes to water where the possibilities of substitution are more limited. As prices rise due to the introduction of full-cost pricing, human capital may be substituted for water in the form of genetically engineered seeds that yield a better harvest index. Water-scarce regions may shift production away from water-intensive activities to reflect relative scarcities (from producing wheat to producing barley) or import embodied water (virtual water, particularly agricultural products such as fruits and vegetables). Irrigation technologies that minimize evaporation and leakages will become more attractive in the face of rising water prices. Despite these possibilities, the inimitability of water and its centrality to life severely limit the potential scope and scale of substituting produced capital for water.

Given the fact that water prices are not determined in competitive markets, there exists an opportunity for economic policy makers to develop a pricing structure that incorporates broader goals. The multiple identities of water as an asset of society, a factor of production, and a consumer good are complementary and interdependent while simultaneously being rivalrous and independent. This presents a unique set of challenges for policy makers interested in setting the "right" price for water. Such a price must balance these competing identities while facilitating cooperation where they overlap. (We have abstracted in this entry from issues of intertemporal allocations of water. The models we have used are steady-state models. Treating water as an asset carries serious implications for intergenerational allocations. Suffice it to say here that sustainability requires that annual water extractions must not exceed the replenishment rates. A price that does not reflect this intertemporal conservation will misallocate resources over time.) Water as a factor of production and as an asset of society is inextricably linked by the concept of sustainable development. Similarly, the asset of society and the consumer goods are connected by the cultural and social identities associated with water use. Finally, the consumer goods and the factor of production are related by the externalities they impose upon each other. This analysis is not limited to the national level, and it is equally applicable in bilateral, multilateral, and global contexts. Given these relationships, the following principles and objectives have been identified as necessary ingredients in determining a price for water. Each of these is broadly defined and adopted in varying degrees with different weights placed on each from case to case. In Figure 5, we sketch a multiple-criteria framework for choosing prices that deal with complex objectives that balance economic, environmental, and sociocultural issues.

Economic Efficiency

Here, efficiency also incorporates multiple goals. Fees for water services must be sufficient to ensure that the utility recovers at least the full supply cost of delivery to the end user, a concept known as revenue sufficiency. Any fraction less than this fails to send the signal that water is a scarce resource; withholds information from consumers, the utility, and the regulatory agency; and prevents the proper inter-agent allocation of water (Figure 5).

It is worth noting that the standard approach to economic efficiency includes the concept of "full cost recovery," which the United Nations World Water Development Report defines as a situation in which "users pay the full cost of obtaining, collecting, treating, and distributing water, as well as collecting, treating, and disposing of wastewater."[27] In practice, as well as in theory, full cost recovery and revenue sufficiency refer only to the full supply cost.[20,21,28–30] We argue below that this is undesirable. This particular characterization is, however, of interest because it directly addresses the management of wastewater. That the importance of the economics of wastewater management has been largely ignored is demonstrated by the vast differences in progress toward the achievement of the Millennium Development Goals (MDG), and the new Sustainable Development Goals (SDGs) for water and sanitation.[27]

In stark contrast to convention, we argue that pricing at the full cost of supply does not support true economic efficiency. Supply-cost pricing brings us closer to the outcome that would exist if water were

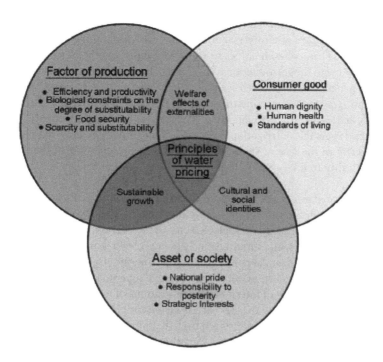

FIGURE 5   Multiple-criteria framework of prices.

traded in a perfectly competitive market. This price for water would fail to properly incorporate the value of opportunity costs, economic externalities, and environmental externalities.[21,31] Any such price prevents socially optimal allocations by promoting overuse, undervaluation, and waste. For this reason, we argue that wherever possible, full cost or full economic cost, as they are defined above, be used in the development of water tariffs.

Equity and Fairness

The principles of equity and fairness are contentious and have evoked controversy. However, given the increased recognition that a minimum amount of water is a basic human right, fairness imposes a constraint on water tariff setting. There are two dimensions to equity. On the one hand, equity has been defined as "treating equals equally" in the sense that those who purchase water that has identical costs should face identical prices.[20] On the other hand, equity and fairness are also grounded in the ideas of social justice, and water as a human right suggests that treating equals equally may satisfy only one dimension of equity. A balanced view of equity would also require treating unequals unequally. In areas of high-income inequality, poor users may be unable to afford even the minimum amount of water necessary to sustain basic human health and dignity. In such cases, some form of subsidy is required. This can take the form of internal cross-subsidization embedded in the pricing structure or of government transfers and income redistribution from other areas of the economy. The complexity of equity and fairness is hard to overstate, and policy makers are often forced to make difficult choices. The relative merits of different objectives for water subsidies are discussed in detail below.

Simplicity

Prices serve multiple objectives; one of the most important is to synthesize and convey information. Complex pricing structures confuse this information, making it difficult for producers, consumers, and regulators to respond appropriately to price signals. Simplicity and transparency also limit the potential for corruption. Insofar as simplicity facilitates the gathering and dissemination of information and economic signals, it promotes conservation, socially optimal allocations, and economic efficiency.

Sustainability

The concept of sustainability is central to each of water's roles depicted in Figure 5. Prices need to reflect this by encouraging conservation and substitution where possible.[22] A fundamental feature of the economics of sustainability is that assets are maintained over time. Water is a crucial component of natural capital. Depletion, diversion, and degradation of water assets represent an erosion of this natural capital asset, and therefore a depletion of society's wealth.

Stability and Quality

Here, stability refers to price stability, service stability, and revenue stability, all of which are interdependent. Stable prices convey stable information, avoid confusion, and promote smooth consumption. Revenue stability promotes and attracts investment while protecting service providers from shocks. Service stability reduces waste by eliminating the need to "hoard," eliminates time and travel costs of finding alternative sources of supply, promotes investment in connection to the system (which in turn increases the customer base, decreasing costs per user via economies of scale), and has the potential to raise revenue for producers as well as increase consumer surplus.

Evidence has shown that even in poor areas of developing countries, willingness to pay for improved service stability is well above current water prices. There are several reasons for this. First, the poor incur significant costs resulting from low-quality service, including travel and time costs, purification costs, reduced labor productivity due to illness and workdays lost, investment in individual water storage tanks, and purchasing expensive bottled water from private vendors.[29,32,33] A study conducted in Kathmandu, Nepal, showed that nearly 80% of respondents with a piped connection would be willing to pay as much as four times their current monthly water bill for improved water quality, stability, and more transparent billing.[32]

Water Tariffs and Pricing Strategies

The water tariff is the primary tool with which a policy maker can combine the various objectives and considerations into a socially optimal price regime. The tariff is central to the utility's ability to attract capital, creates incentives for efficiency in production and consumption, and distributes costs across consumers and time.[20]

Traditionally, decreasing block tariff (DBT) pricing strategies have been employed in water delivery systems for several reasons.[34] The DBT, as shown in Figure 6, involves prices based on quantity consumed.

The initial block, or quantity of water, costs the most, and prices decrease as consumption rises. DBTs score well on stability grounds as revenue is insulated from climatic shocks (to both supply and demand) and because the largest share of revenue comes from the initial units of water consumed. They exploit the economies of scale that arise from falling average costs as quantities supplied rise.[34] The problem, however, is that on grounds of efficiency, equity, and sustainability, this structure has less to offer than an increasing block tariff (IBT) scheme. In terms of efficiency, DBTs send exactly the wrong message to consumers. They fail to convey the fact that water is a scarce resource that must be conserved; they promote waste and fail to encourage socially optimal interagent, intersectoral, and intertemporal allocations. Furthermore, as the utility reaches capacity, the marginal cost of supply it faces rises; however, under the DBT regime, the marginal cost faced by consumers is falling. In terms of equity and fairness, the DBT fails to treat equals equally because consumers of water with identical costs of supply face

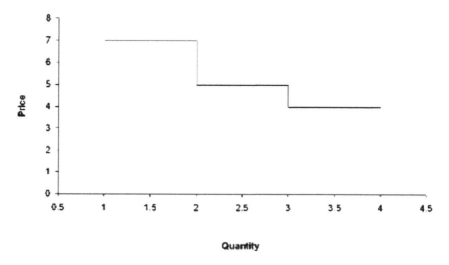

FIGURE 6   Decreasing block tariffs.

different prices. By charging higher prices for the initial units of water, the DBT also ensures that the poor who consume the least water face the highest prices (Figure 6).

In recent years, DBTs have largely been replaced by IBTs, which have become the strategy of choice in the developing world. This is because they have gained preference by donor institutions; management, engineering, and financial advisors; and water experts as being the most effective way of synthesizing the goals and principles of water supply and demand.[29] The IBT (the pricing scheme depicted in Figure 2) also features different prices based on quantities consumed; however, in this case, prices rise with quantity consumed.

In terms of economic efficiency, this is a significant improvement over the DBT. Here, tariffs are consistent with marginal-cost pricing and create an incentive structure that conveys the necessary signals to encourage socially optimal use. IBTs score well on fairness grounds because income and water demand are positively correlated and the prices faced by consumers rise with consumption. That is, IBTs are progressive, so wealthier customers that consume more water face higher prices. Furthermore, IBTs can exhibit "internal subsidization" or cross-subsidization in which large consumers pay sufficiently high prices that the first units of water consumed can be sold below cost. This feature enables the utility to provide water to the poor below the full cost of supply without violating the principle of revenue sufficiency.

The practical implementation of IBTs has yielded mixed results. If blocks are not properly structured in quantities that are relevant for the consumers in question, the benefits of IBTs are lost.[20,28,30,31,35] Specifically, the compatibility of fairness with the other principles of water pricing requires that the first block be sufficient to cover the basic requirements of human health and dignity but not more. Much attention must be paid to determining the relative block sizes and their prices, and finding the appropriate values for both requires data that are frequently unavailable.

It is possible that in some cases, low initial block prices may yield a result in which metering costs may exceed the revenue collected from low-income users. In such situations, the utility provider incurs a loss not only from the provision of water below full supply cost but also on the metering operations.[30] Alternatively, if tariffs faced by large-scale consumers such as heavy industry and agriculture are sufficiently high, these consumers may opt to disconnect from the utility and seek alternative sources of water, such as in-house direct extraction.[29,30] Even where this is illegal, government capacity to enforce such laws is frequently limited. The effect of this would be to remove the highest-paying customers, which in turn would eliminate the feasibility of internal subsidization, increase average costs, and distribute fixed-capital costs across a smaller customer base. That is, the benefits of IBTs are highly dependent upon the participation of the highest-paying customers; however, these are exactly the customers who have the greatest incentive to search for alternative sources of water.

Even where subsidies are absolutely necessary in order for the poor to meet their basic needs, it may not be ideal to provide these subsidies through the tariff structure. Chile subsidizes water consumption for the poor via transfer payments from municipal and national budgets. Many have claimed that this approach enhances equity and economic efficiency because the tariff received by the utility treats equals as equals.[30] Although we accept Yepes'[30] claim in principle, we do so with some trepidation. For the argument to hold, it must be shown that the welfare loss incurred by using the tariff structure to internally subsidize water for the poor is greater than that incurred by appropriating funds from elsewhere in the economy to fund government transfers. This reservation may be especially relevant in developing countries, where the tax base is small, government capacity is limited, and corruption is significant.

An interesting extension to the principle of IBTs involves the concept of two-tariff design. The price that end users face for water can be divided into two components: volumetric usage charges and fixed-rate charges, commonly referred to as connection fees. The two-tariff system provides an economic incentive for end users to conserve water through its volumetric component while providing an element of revenue security for the supplier through its fixed-rate component. This structure has evolved into the structure of choice in most Organization for Economic Cooperation and Development (OECD) countries.[31,34]

A two-tariff system can be carried out in the context of an IBT scheme so that a fixed cost is paid for "connection," but low levels of water can still be provided at a rate that is subsidized by larger consumers. In many areas, connection fees are sufficiently high that they exclude the poorest consumers. Since the cross-subsidization inherent in the IBT structure is beneficial only to those with connections, an argument can be made for higher subsidies on connection costs combined with lower subsidies on volumetric consumption. This approach incorporates the poorest strata of society while continuing to expand the customer base, which decreases average costs for all.

This is especially important in situations where land tenure rights are insecure. In such cases, the incentives and capacity for household investment in connection to the water utility are extremely low. It may be the case that these areas realize their greatest economic surplus by providing shared public taps that face a fixed rate rather than an IBT. There is already some evidence to support such a conclusion. Poor neighborhoods that face IBT prices often share a single connection between multiple households. These meters read high volumes of consumption, so the absolute poorest face higher prices due to the structure of the IBT.

Raising the price of water to its full cost would enhance economic efficiency and sustainability. Furthermore, Rogers, de Silva, and Bhatia[31] argue that this can be done without sacrificing equity and fairness. Situations exist in which Pareto improvements (improvements that leave at least some agents better off without making others any worse off) are possible even in cases where relative equity decreases.[29]

Conclusions

Water is not traded in competitive markets. High infrastructure costs raise barriers to entry, ruling out competitive conditions, whereas economies of scale, limited substitutes, human rights issues, and environmental externalities combine to complicate the design of a fair, efficient, and sufficient price for water.

The multiple roles that water plays in the economy, society, and the environment increase the complexity of structuring prices that capture scarcity and the multiple values of water. Economics alone cannot deal with this complex commodity, resource, and input. However, optimal allocations of this scarce resource among competing users and uses are basically an economic issue. The optimality criteria could be salvaged and applied where net social benefits are maximized within an idealized model capable of simulating competitive conditions. The shadow prices for water that emerge from the model solutions can be used to guide allocation of this scarce resource to optimize net social and environmental benefits. It is also feasible to construct a price-designing exercise where it is possible to capture the full cost of producing and delivering water and to preserve and balance multiple objectives. Many utilities do that now on a routine basis.

Water is valuable, but it is not thicker than blood, and an economic price can be set that embodies many of the competitive efficiency criteria in addition to other valuable societal and environmental objectives. Of particular importance is the sustainability consideration where the present generation must limit its water consumption to preserve quality and sufficient quantity for future generations. The emphasis must be particularly on quality, as increasingly societies face major issues of degradation of water quality and desertification.

References

1. Grafton, R.Q.; Libecap, G.D.; Edwards, E.C.; O'Brien, R.J.; Landry, C. *Water Scarcity and Water Markets: A Comparison of Institutions and Practices in the Murray Darling Basin of Australia and the Western US*; International Center for Economic Research: Washington, DC; 2010, Working Paper No. 28.
2. Newlin, B.D.; Jenkins, M.W.; Lund, J.R.; Howitt, R.E. Southern California water markets: Potential and limitations. *J. Water Resour. Plann. Manage.* 2002, *128* (1), 21–32.
3. Saliba, B.C.; Bush, D.B.; *Water Markets in Theory and Practice: Market Transfers, Water Values, and Public Policy*; Westview Press: Boulder, CO, 1987.
4. Fisher, F.M. Towards cooperation in water: The Middle East Water project. *Rev. Region Dev.* 2000, *12*, 143–165.
5. Hardin, G. The tragedy of the commons. *Science* 1968, *162*, 1243–1248.
6. United Nations General Assembly. *The Human Right to Water and Sanitation*, A/Res/64/292; United Nations General Assembly: New York, 2010; 1–3.
7. Gleick, P.H. The human right to water. *Water Policy* 1998, *1* (5), 487–503.
8. Lonergan, S.C.; Brooks, D.B. *Watershed: The Role of Fresh Water in the Israeli-Palestinian Conflict*; IDRC: Ottawa, ON, 1994.
9. Anand, P.B. *Scarcity, Entitlements and the Economics of Water in Developing Countries*; Edward Elgar Publishing Ltd.: Northampton, MA, 2007.
10. Young, R.A. Water, economics, and the nature of water policy issues. In *Determining the Economic Value of Water: Concepts and Methods*; Resources for the Future: Washington, DC, 2005; 3–16.
11. Conca, C. *Governing Water: Contentious Transnational Politics and Global Institution Building*; MIT Press: Cambridge, MA, 2006.
12. Mansfield, E.; Yohe, G. *Microeconomics: Theory and Applications*; W. W. Norton and Company: New York, 2003; Vol. 11, 764 pp.
13. Eckstein, O. *Water Resources Development: The Economics of Project Evaluation*; Harvard University Press: Cambridge, MA, 1958.
14. Maass, A.; Hufschmidt, M.M.; Dorfman, R.; Thomas, H.A.; Marglin, S.A. *Design of Water Resource Systems*; Harvard University Press: Cambridge, MA, 1962.
15. Hirshleifer, J.; DeHaven, J.C.; Milliman, J.W. *Water Supply: Economics, Technology, and Policy*; University of Chicago Press: Chicago, IL, 1960.
16. Fisher, F.M.; Huber-Lee, A. Economics, water management, and conflict resolution in the Middle East and beyond. *Environment* 2006, *48* (3), 26–41.
17. Amir, I.; Fisher, F.M. Analyzing agricultural demand for water with an optimizing model. *Agric. Syst.* 1999, *61*, 45–56.

18. Fisher, F.M.; Huber-Lee, A.T. WAS-guided cooperation in water management: Coalitions and gains. In *Game Theory and Policymaking in Natural Resources and the Environment*; Dinar, A., Albiac, J., Sánchez-Soriano, J., Eds.; Routledge: New York, 2008.

19. Dorfman, R.; Samuelson, P.; Solow, R. *Linear Programming and Economic Analysis*, The Rand Series; McGraw-Hill: Torontos, 1958.

20. Boland, J. Pricing urban water: Principles and compromises. *Water Resour. Update: Univ. Counc. Water Resour.* 1993, *92* (Summer), 7–10.

21. Rogers, P.; Bhatia, R.; Huber, A. *Water as a Social and Economic Good: How to Put the Principle into Practice*; Global Water Partnership/Swedish International Development Agency: Stockholm, 1998.

22. National Research Council. The meaning of value and use of economic valuation in the environmental policy decision-making process. In *Valuing Ecosystem Services: Toward Better Environmental Decision-Making*; The National Academies Press: Washington, DC, 2005.

23. Perks, A.R.; Kealey, T. International price of water. In *Environmental Economics and Investment Assessment*; Aravossis, K., Brebbia, C.A., Kakaras, E., Kungolos, A.G., Eds.; WIT Press: Boston, MA, 2006.

24. Krautkraemer, J. *Economics of Natural Resource Scarcity: The State of the Debate*; Resources for the Future: Washington, DC, 2005; Discussion Paper 05-14.

25. Turner, R.K.; Pearce, D.; Bateman, I. *Environmental Economics*; The Johns Hopkins University Press: Baltimore, MD, 1993.

26. Solow, R. An almost practical step toward sustainability. In *Assigning Economic Value to Natural Resources, Proceedings of the Workshop on Valuing Natural Capital for Sustainable Development*, Woods Hole, MA, July 1–3, 1993; National Academy Press: Washington, DC, 1994; 19–30.

27. World Health Organization and UNICEF. *Progress on Sanitation and Drinking Water: 2010 Update*; WHO Press: France, 2010.

28. Boland, J.J.; Whittington, D. The political economy of increasing block tariffs in developing countries. In *The Political Economy of Water Pricing Reforms*; Dinar, A., Ed.; Oxford University Press: Oxford, 2000; 215–235.

29. Whittington, D. Municipal water pricing and tariff design: A reform agenda for South Asia. *Water Policy* 2003, *5* (1), 61–76.

30. Yepes, G. *Do Cross-Subsidies Help the Poor to Benefit from Water and Wastewater Services? Lessons from Guayaquil*; UNDP-World Bank Water and Sanitation Program: Washington, DC, 1999; 1–9.

31. Rogers, P.; de Silva, R.; Bhatia, R. Water is an economic good: How to use prices to promote equity, efficiency, and sustainability. *Water Policy* 2002, *4* (1), 1–17.

32. Whittington, D.; Pattanayak, S.K.; Yang, J.C.; Kumar, K.C. Do households want improved piped water services? Evidence from Nepal. *Water Policy* 2002, *4* (6), 531–556.

33. Zerah, M.H. *Water: Unreliable Supply in Delhi*; Manohar, Centre de Sciences Humaines: New Delhi, 2000.

34. Griffin, R. Water pricing. In *Water Resource Economics: The Analysis of Scarcity, Policies, and Projects*; MIT Press: Cambridge, MA, 2006.

35. Olmstead, S.M.; Stavins, R.N. *Managing Water Demand: Price vs. Non-Price Conservation Programs, 39*; Pioneer Institute: Cambridge, MA, 2007, 1–40.

28

Wetlands: Methane Emission

Introduction 293
Methane Production and Consumption 293
Environmental Controls on Methane Emissions 294
    Water Table Position
Soil Temperature...295
    Methane Transport: Diffusion, Bubble Ebullition, and Transport through
    Plants • Vascular Plants and Substrate for Methane Production
References...296

Anna Ekberg
and Tørben Røjle
Christensen

Introduction

Methane (CH_4) is a radiatively active trace gas that plays an important role for atmospheric chemistry and the energy balance of the Earth. At present, its contribution to the greenhouse effect is about 22%, which can be compared to carbon dioxide (CO_2) that contributes approximately 65% to the climate forcing of all long-lived greenhouse gases (excluding water vapor).[1] In addition to the direct warming effects, where infrared radiation is absorbed and returned to the Earth's surface, chemical and photochemical reactions with CH_4 in the troposphere and stratosphere indirectly cause greenhouse warming. The preindustrial atmospheric concentration of CH_4 was about 0.75ppmv (parts per million by volume) and since then it has more than doubled to approximately 1.73ppmv.[1] Anthropogenic sources of CH_4 include cattle (about 15% of the annual CH_4 release), rice paddies (20%), coal mining and oil production (14%), biomass burning (10%), natural gas leaks, and landfills and sewage disposal, while the major natural sources include wetlands (20%–25%) and termites (5%).[2] Because of the substantial contribution from natural wetlands, knowledge about the dynamics leading to CH_4 formation and emission is of great importance to understand how these ecosystems interact with the climate and how they would respond to climate change.

Methane Production and Consumption

Methane is produced by strictly anaerobic archaebacteria that are limited to the use of only a few simple substrates for biosynthesis and energy production.[3] If available carbon compounds are not directly supplied, they depend on other groups of micro-organisms for the initial breakdown of more complex organic structures into simpler molecules. The reduction of CO_2 with hydrogen is a common pathway to CH_4 formation, but acetate and formate are also important precursors of CH_4 in natural environments. Growth of methanogenic bacteria further requires a very low redox potential (E_h below –400mV) and the ideal environmental conditions are often met in permanently waterlogged wetlands. However, CH_4 efflux from wetlands to the atmosphere depends not only on the rate of production (methanogenesis), but also on the extent of CH_4 consumption (methanotrophy) that may occur in oxic surface layers and in the close vicinity of plant roots.[4–6] In this process, CH_4 is oxidized to CO_2, and the rate of CH_4 emission that can be measured at the surface is hence the net result of two counteracting processes—methanogenesis and methanotrophy.

Environmental Controls on Methane Emissions

Water Table Position

Water table position in relation to the surface is considered to be the most important factor controlling CH_4 flux, because it indicates the boundary between anaerobic CH_4 production and aerobic CH_4 consumption (Figure 1). Negative correlations between water table depth and rates of CH_4 emission have frequently been reported.[7–9] It has also been found that the lowering of the water table may result in increased CH_4 emissions when episodic releases of CH_4 trapped in pore water occur and the diffusivity

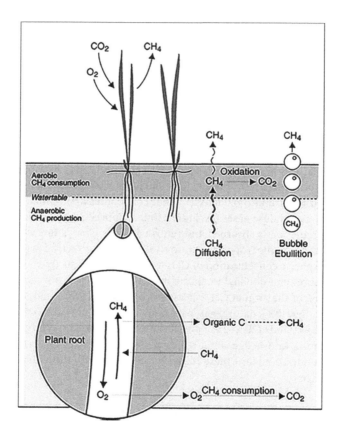

FIGURE 1 The depth of the water table is important in determining net CH_4 emissions from natural wetlands because it indicates the boundary between anaerobic CH_4 production and aerobic CH_4 consumption. Methane may be transported from the soil to the atmosphere via diffusion through the soil profile, by bubble ebullition or by plant-mediated transport. In the first case, CH_4 is subjected to methanotrophic oxidation as it passes through the oxic surface layer and this acts to decrease net CH_4 emissions. Bubbles are quite stable, meaning that little oxidation takes place, and the same is also true when CH_4 is transported through vascular plants. Vascular plants are sources of methanogenic substrate because they release labile carbon compounds into the soil via root exudation and root turnover. Leakage of O_2 from roots may lead to inhibition of methanogenesis and methanotrophic CH_4 consumption in the otherwise anoxic soil.

of gases in air-filled pore spaces increases.[10,11] Field experiments in peat-forming wetlands reveal a high spatial variability of CH_4 emissions caused by distinct microtopographical features on the wetland surface.[12] Interactions between vascular plants and mosses create differences in elevation, and therefore distance to the water table within a radius of less than a meter.[13] Scaling up of spot CH_4 emission measurements to larger areas without careful consideration of experimental plot location may therefore be misleading. Laboratory incubations have shown that there is a potential for CH_4 production both above and below the water table,[14] but the actual production in the field peaks at approximately 5–15 cm below the water table.[8] For methanotrophic CH_4 consumption to occur, a simultaneous supply of both CH_4 and oxygen is necessary. The abundance and activity of methanotrophic bacteria are therefore highest at the interface between anoxic and oxic conditions, i.e., at the depth of the water table.[5,14] Watson et al.[5] used a model to show a high capacity for CH_4 oxidation in peat, where methanotrophy accounted for 85% of the oxygen uptake potential when both oxygen and CH_4 were present in excess. In the absence of CH_4, maximum oxygen uptake potential was near the surface.

Soil Temperature

Correlations between soil temperature and CH_4 production and emission have been found at scales ranging from laboratory incubations and monolith experiments[8,11,15] to field investigations at the plot[16] and landscape scale.[9] There is a direct effect of temperature on the metabolic rate of CH_4-producing micro-organisms, but the sensitivity to temperature change is different for methanogens and methanotrophs.[11,17] In a study carried out with peat slurries under laboratory conditions, Dunfield et al.[17] found that CH_4 production in peat samples from temperate and subarctic areas was more sensitive to changes in temperature as compared to CH_4 consumption in the same samples. Increasing or decreasing temperatures also interact with other parameters with potential to control CH_4 production and emission to the atmosphere. One such interactive effect would be the relationship between temperature and water table position, where increasing temperatures lead to enhanced evaporation and plant transpiration, with lowering of the water table as a consequence.[18] It has also been suggested that the temperature dependence of CH_4 production is constrained by substrate limitation and that temperature has an effect on methanogenesis mainly via its influence on substrate availability.[19] In peat samples from a Swedish acid mire, CH_4 production was stimulated by increased temperatures only when substrate (glucose) would be the relationship between temperature and increased.[15] It has further been found that decomposition of labile material is more strongly controlled by temperature than decomposition of more recalcitrant components.[18] From the same study, it was also concluded that substrate availability lagged behind rapid temperature changes and that the thermal/hydrological history of the soil was important to determine the rate of CH_4 production.

Methane Transport: Diffusion, Bubble Ebullition, and Transport through Plants

The solubility of CH_4 in water is low, and it escapes through waterlogged soil to the atmosphere by diffusion, bubble ebullition, or by transport through vascular plants (Figure 1). The rate of CH_4 emission is largely controlled by the mode of transport because the different transportation pathways are associated with more or less extensive CH_4 consumption in the soil profile. Methane diffusing through oxic environments is subjected to methanotrophic oxidation to CO_2, and a water table depth of only a few centimeters may cause considerable reductions of CH_4 emissions to the atmosphere. Bubbles form when the rate of CH_4 production causes the sum of the partial pressures of dissolved gases to exceed the value of the hydrostatic pressure in the soil. However, the concentration of CH_4 in bubbles will be at equlibrium with dissolved pore water CH_4. At low rates of methanogenesis and with no vascular plants present, the main release of CH_4 to the atmosphere is likely to be by diffusion, but bubbles begin to form when the production rate exceeds the capacity for diffusive loss.[20] Once formed, bubbles are quite stable, meaning that little methanotrophic CH_4 oxidation occurs even as they pass through oxic environments. Studies conducted in rice paddies have shown that bubble formation tends to decrease as the plants mature because the progressively better developed rooting system acts as gas conduits and efficiently transports gases out of the soil.[21] However, a close relationship between vascular plant production and the availability of methanogenic substrate (discussed in the next section) may also speed up the rate of methanogenesis to such an extent that bubble formation is promoted. In order to supply oxygen for respiration to submerged structures, certain vascular plant species develop lacunae in stems, roots, and rhizomes.[22] Depending on the species, the ventilation of growing roots is carried out either by pressurized bulk flow or by simple diffusion that follows concentration gradients between the atmosphere and the soil. Leakage of oxygen may give rise to CH_4 oxidation and inhibition of methanogenesis in the close vicinity of the roots.[5,6] However, net CH_4 emission is generally enhanced by the presence of vascular plants with a deep rooting system because of a "chimney effect," where CH_4 transported from the soil to the atmosphere within plant lacunae is withdrawn from consumption in oxic conditions.[12,23,24]

Vascular Plants and Substrate for Methane Production

Vascular plants are sources of substrates for CH_4 production because they easily release degradable carbon compounds into the soil via root exudation and root turnover (Figure 1).[20,24,25] In peat-forming wetlands, the organic matter becomes increasingly recalcitrant with depth[26] and several studies have pointed out that plant-derived inputs of labile carbon compounds could significantly contribute to increased CH_4 production.[20,23,27] This "loading" of substrates into the soil may further be correlated to the photosynthetic rate of the vascular plants in the ecosystem,[24,28] because the amount of carbon allocated to belowground plant structures and eventually released to the soil is likely proportional to the CO_2 fixation rate.[24] The overall primary productivity in the ecosystem may therefore be of great importance for the CH_4 emission rates from a wetland area.

References

1. Lelieveld, J.; Crutzen, P.J.; Dentener, F.J. Changing concentration, lifetime and climate forcing of atmospheric methane. Tellus **1998**, *50B*, 128–150.
2. Chappelaz, J.A.; Fung, I.Y.; Thompson, A.M. The atmospheric CH_4 increase since the last glacial maximum. 1. Source estimates. Tellus **1993**, *45B*, 228–241.
3. Oremland, R.S. Biogeochemistry of methanogenic bacteria. In *Biology of Anaerobic Microorganisms*; Zehnder, A.J.B., Ed.; John Wiley: New York, 1988; 641–703.
4. King, G.M. Ecological aspects of methane oxidation, a key determinant of global methane dynamics. Adv. Microb. Ecol. **1998**, *12*, 431–468.
5. Watson, A.; Stephen, K.D.; Nedwell, D.B.; Arah, J.R.M. Oxidation of methane in peat: kinetics of CH_4 And O_2 removal and the role of plant roots. Soil. Biol. Biochem. **1997**, *29* (8), 1257–1267.
6. Popp, T.J.; Chanton, J.P.; Whiting, G.J.; Grant, N. Evaluation of methane oxidation in the rhizosphere of a Carex dominated fen in north central Alberta, Canada. Biogeochemistry **2000**, *51*, 259–281.
7. Christensen, T.R.; Friborg, T.; Sommerkorn, M.; Kaplan, J.; Illeris, L.; Soegaard, H.; Nordstroem, C.; Jonasson, S. Trace gas exchange in a high-arctic valley. 1. Variations in CO_2 and CH_4 flux between tundra vegetation. Glob. Bio- geochem. Cycles **2000**, *14* (3), 701–713.
8. Daulat, W.E.; Clymo, R.S. Effects of temperature and water table on the efflux of methane from peatland surface cores. Atmos. Environ. **1998**, *32* (19), 3207–3218.
9. Hargreaves, K.J.; Fowler, D. Quantifying the effects of water table and soil temperature on the emission of methane from peat wetland at the field scale. Atmos. Environ. **1998**, *32* (19), 3275–3282.

10. Kettunen, A.; Kaitala, V.; Alm, J.; Silvola, J.; Nykänen, H.; Martikainen, P.J. Cross-correlation analysis of the dynamics of methane emissions from a boreal peatland. Glob. Bio- geochem. Cycles **1996**, *10* (3), 457–471.

11. Moore, T.R.; Dalva, M. The influence of temperature and water table position on carbon dioxide and methane emissions from laboratory columns of peatland soils. J. Soil Sci. **1993**, *44*, 651–664.

12. Frenzel, P.; Karofeld, E. CH4 emission from a hollowridge complex in a raised bog: the role of CH_4 production and oxidation. Biogeochemistry **2000**, *51*, 91–112.

13. Waddington, J.M.; Roulet, N.T. Carbon balance of a boreal patterned peatland. Glob. Change Biol. **2000**, *6*, 87–97.

14. Moore, T.R.; Dalva, M. Methane and carbon dioxide exchange potentials of peat soils in aerobic and anaerobic laboratory incubations. Soil. Biol. Biochem. **1997**, *29* (8), 1157–1164.

15. Bergman, I.; Svensson, B.H.; Nilsson, M. Regulation of methane production in a Swedish acid mire by pH, temperature and substrate. Soil Biol. Biochem. **1998**, *30* (6), 729–741.

16. Verville, J.H.; Hobbie, S.E.; Chapin, F.S.; Hooper, D.U. Response of tundra CH_4 and CO_2 flux to manipulation of temperature and vegetation. Biogeochemistry **1998**, *41*, 215–235.

17. Dunfield, P.; Knowles, R.; Dumont, R.; Moore, T.R. Methane production and consumption in temperate and subarctic peat soils: response to temperature and pH. Soil Biol. Biochem. **1993**, *25* (3), 321–326.

18. Updegraff, K.; Bridgham, S.D.; Pastor, J.; Weishampel, P. Hysteresis in the temperature response of carbon dioxide and methane production in peat soils. Biogeochemistry **1998**, *43*, 253–272.

19. Valentine, D.W.; Holland, E.A.; Schimel, D.S. Ecosystem and physiological controls over methane production in northern wetlands. J. Geophys. Res. **1994**, *99* (D1), 1563–1571.

20. Chanton, J.P.; Whiting, G.J. Trace gas exchange in freshwater and coastal marine environments: ebullition and transport by plants. In *Biogenic Trace Gases:Measuring Emissions from Soil and Water*; Matson, P.A., Harriss, R.C., Eds.; Blackwell Science: Oxford, 1995; 98–125.

21. Holzapfel-Pschorn, A.; Conrad, R.; Seiler, W. Effects of vegetation on the emission of methane from submerged paddy soils. Plant Soil **1986**, *92*, 223–233.

22. Armstrong, W.; Justin, S.H.F.W.; Beckett, P.M.; Lythe, S. Root adaptation to soil waterlogging. Aquat. Bot. **1991**, *39*, 57–73.

23. Greenup, A.L.; Bradford, M.A.; McNamara, N.P.; Ineson, P.; Lee, J.A. The role of *Eriophorum Vaginatum* in CH_4 flux from an ombrotrophic peatland. Plant Soil **2000**, *227*, 265–272.

24. Joabsson, A.; Christensen, T.R.; Wallén, B. Vascular plant controls on methane emissions from northern peatforming wetlands. Trends Ecol. Evol. **1999**, *14*, 385–388.

25. Jones, D.L. Organic acids in the rhizosphere—a critical review. Plant Soil **1998**, *205*, 25–44.

26. Christensen, T.R.; Jonasson, S.; Callaghan, T.V.; Havström, M. On the potential CO_2 releases from tundra soils in a changing climate. Appl. Soil Ecol. **1999**, *11*, 127–134.

27. Joabsson, A.; Christensen, T.R. Methane emissions from wetlands and their relationship with vascular plants: an Arctic example. Glob. Change Biol. **2001**, *7* (8), 919–932.

28. Whiting, G.J.; Chanton, J.P. Plant-dependent CH_4 emission in a subarctic Canadian fen. Glob. Biogeochem. Cycles **1992**, *6* (3), 225–231.

III

CSS: Case Studies of Environmental Management

Alexandria Lake Maryut: Integrated Environmental Management

Introduction 301
Current State of Lake Maryut 304
    Hydraulic and Hydrological Functioning • Water Quality and
    Ecosystems • Social Considerations and Governance
Integrated Environmental Management 307
    Integrated Plan Concept • Scenario Identification • Model
    Development • Hydraulic Model • Ecological Model • Proposed Actions
Conclusions 313
Acknowledgments.. 313
References.. 313

Lindsay Beevers

Introduction

Lake Maryut is located on the Mediterranean coast of Egypt, in the delta of the River Nile and defines the southern boundary of the city of Alexandria.[1] The lake extends for about 80 km along the northwest coast of Alexandria and 30 km south and has a wetted surface area of 65 km² (Figure 1). It is a shallow water lake[2] with a water depth of approximately 1.5 m across the different basins and, unlike any of the other Nile deltaic lakes, is not directly connected to the Mediterranean Sea. The water level of the lake is kept below mean sea level, and freshwater is supplied, through irrigation canals, from the Rosetta branch of the Nile.[3] Throughout literature, the lake is referred to variously as Mariout, Mariut, or Maryut Lake. For this entry, the lake will be identified as Lake Maryut.

The lake has been in existence for more than 6000 years, as part of the Nile deltaic formation. In an extensive study of the evolution of the lake,[4] several drivers for the current form of the lake were established, namely, eustatic sea level, climate oscillations, compaction, sediment transport, and more recent anthropogenic influences such as land reclamation, irrigation, and agricultural practices. It is a mixture of these anthropogenic influences that has shaped the development of the lake in more recent years, along with its connection to the city of Alexandria.

During the Greco-Roman period, Alexandria was an active port, serving as a key navigational route from the Mediterranean, through Lake Maryut, up the Nile to Cairo. Canals linked the lake with the sea to the north and the Nile to the east. However, the siltation of the canopic mouth of the Nile in the 12th century[5,6] severed the freshwater influx to the lake and with it the navigational links. During this period, the city of Alexandria went into a phase of decline, and the lake went through several phases, which included coastal connection and influx and coastal disconnection and drying out.

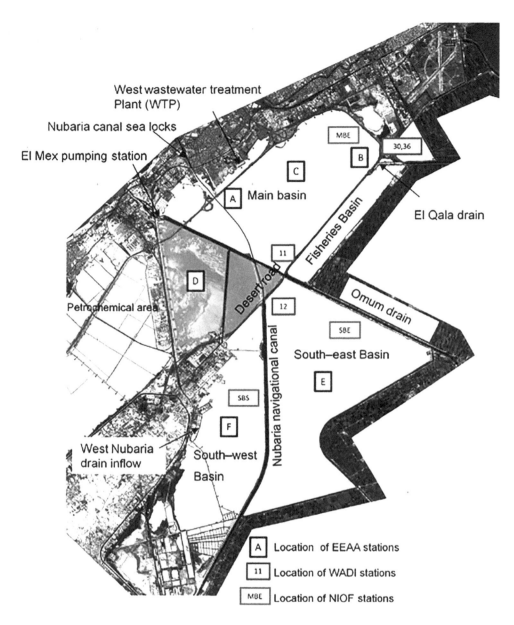

FIGURE 1   Lake Maryut: location map.

In 1892, the irrigation system of the Beheirah district was established, which restored a flow of freshwater to the lake. The lake remained disconnected from the coast, with excess water pumped to the bay through El Mex station (Figure 1), and navigation permitted through sea locks.[7] This is similar to the form the lake takes today, where the principal drainage inflows arise from agricultural irrigation channels fed by the main River Nile.

Over the years, Lake Maryut has provided Alexandria and Egypt with a source of fish, and currently, part of the lake is dedicated to aquaculture (Fisheries Basin, Figure 1). Over 7000 fishermen have rights to fish in the lake, and recent pollution levels and increased vegetation growth affect their catch.[8] *Tilapia* species are the predominant catch, representing approximately 90% of the total.[9] In addition to fisheries, the lake acts as a source of water for irrigation and for raising animals and provides areas for dwelling beside its shores.

In its current form, canals divide the lake water body into several basins. The earth embankments along the canals have several breaches that allow water to flow from the canals into the basins and vice versa, creating interaction between water bodies. There are three main inflows to the Lake Maryut: the Qala drain located to the northeast part of the lake, the Omum drain located at the east of the lake, and the Nubaria navigational canal located at the south of the lake (Figure 1). In addition to this, there are three minor inflows: one from the West Nubaria drain, one from the petrochemical industrial area, and the third from the West Wastewater Treatment Plant (WWWTP) (see Figure 1). The main outflow from the Lake Maryut is El Mex pumping station, which consists of two buildings, each housing six pumps with nominal capacities of 12.5 m³/sec.[7] Small flows are lost to the system through the navigational locks as well as through evaporation.

As a consequence of high nutrient loadings due to the agricultural and industrial activities upstream and in its surrounds, Lake Maryut has become a highly polluted and eutrophic lake. Eutrophication of lakes is a natural process that can be accelerated by human activities (Figure 2) that introduce an excess of nutrients together with other pollutants.[10] The main sources of nutrients and pollutants to Lake Maryut are human sewage, industrial waste, farm, and urban runoff.[11,12]

Eutrophic lakes often experience an excessive growth of algae and larger aquatic plants. Such growth consumes dissolved oxygen (DO), vital for the fish and other animal life.[10] This growth and subsequent proliferation of vegetation occurs in Lake Maryut due to the high nutrient loadings that has entered the lake for a number of years. Currently, 60% of the surface area of the lake is covered by vegetation (*Phragmites australis* and *Eichornia crassipes*), which reduces the DO concentrations in the lake, especially in the main basin. Significant discharge of domestic sewage (with basic primary treatment) from the Qala drain and the East Wastewater Treatment Plant (EWWTP) enter the lake. The Omum and Nubaria Canals are less polluted than the Qala drain; however, they also contribute nutrient loadings to the lake, albeit to a lesser extent. Finally, nonpoint sources such as agricultural runoff containing pesticides and fertilizers contribute to the deterioration of the environmental quality of the lake.

The lake also suffers from other anthropogenic pressures including urbanization, unplanned settlements, and land reclamation (which has reduced the surface area of the lake gradually). An analysis of the stakeholders directly responsible for managing the lake was undertaken in 2007.[13] This showed that Alex Company for Sanitary and Drainage, the Fishing Authority, and The Ministries of Industry, Water Resources, and Environment are the main authorities responsible for the lake. The coordination

FIGURE 2   Fishing activities on the lake.

and communication regarding the management of the lake between these authorities are known to be poor, with each developing management plans in isolation.

Between 2007 and 2009, a European Union (EU)-funded SMAP III project was set up to analyze the current functioning of the lake in more detail and to develop an integrated action plan for the environmental management of the lake. This entry will cover the development of this plan, specifically focusing on the scientific models set up to support the plan development and to investigate the results of different interventions.

Current State of Lake Maryut

The current functioning of the lake has been investigated through a mixture of studies commissioned to gather specific information and ongoing monitoring surveys. Hydraulic information was gathered through a United States Agency for International Development (USAID) project in 1996,[7] hydrologic information was sourced from the local airport weather station (Nouzha airport), and a longer data series was obtained from the weather station at Port Alexandria.[14] Bathymetric information of the lake came from a survey completed in 2008 by the National Institute of Oceanography and Fisheries (NIOF).[15] The ecosystem functioning and health of the basin were constructed using a number of surveys, principally the long-term monitoring strategy undertaken annually by the Egyptian Environmental Affairs Authority (EEAA),[16] a comprehensive sampling campaign by NIOF[15] in 2008 and the EU-funded Water Demand Integration (WADI) project.[8]

Using these data, a comprehensive picture of the lake can be assembled.

Hydraulic and Hydrological Functioning

Table 1 shows the monthly average meteorological data for the lake. It is clear that, generally, precipitation in the region is very low, with the lowest values recorded between April and October. During the same period, along with significant sunlight hours recorded, evapotranspiration is highest.

Table 2 summarizes the main inflows to the lake (Figure 1) over the year. These flows are known to remain reasonably steady throughout the year, though rising slightly in the months of August to February due to the connection with the main river (Nile) and agricultural requirements. The Omum drain carries the greatest flow to the basin. Direct inputs to the lake from the WWWTP, West Nubaria drain, and the petrochemical area (Figure 1) are detailed in Table 3. These flows are considered to be

TABLE 1 Average Monthly Meteorological Data

Month	Precipitation (mm/day)[14]	Evapotranspiration	Wind Speed (m³/sec)[15]	Solar Radiation (J/m²/day)[20]
January	1.73	1.4	3.23	10,000,000
February	0.92	2.2	3.97	12,000,000
March	0.44	1.7	3.57	17,000,000
April	0.12	4.0	4.17	22,000,000
May	0.04	4.0	3.60	24,000,000
June	0.00	5.6	3.54	26,000,000
July	0.00	6.4	3.32	25,000,000
August	0.01	7.7	3.04	23,000,000
September	0.04	9.3	2.15	20,000,000
October	0.25	5.5	2.58	16,000,000
November	1.15	3.4	2.92	12,000,000
December	1.82	1.3	3.77	10,000,000
Average		1.4	3.32	18,083,333

TABLE 2   Monthly Flow Discharges

Month	Discharges (m³/day)		
	Qala	Omum	Nubaria
January	529,920	3,368,640	959,040
February	466,560	2,478,640	1,270,080
March	475,200	3,248,640	462,240
April	532,224	3,318,640	666,144
May	671,328	3,680,640	691,200
June	414,720	3,453,038	243,792
July	673,920	3,136,320	285,120
August	734,400	3,412,800	838,030
September	676,500	4,275,690	492,480
October	873,500	4,813,000	559,800
November	853,500	5,353,000	559,800
December	693,500	4,103,000	559,800

Source: USAID.[7]

TABLE 3   Constant Flow Discharges

Discharges (m³/day)		
WWWTP[17]	West Nubaria Drain	Petrochemical Area[16]
410,325	259,200	47,398

constant inflows to the lake. El Mex pumping station manages the outflow of the lake. Twelve pumps with nominal capacities of 12.5 m³/ sec[7] maintains the water level in the lake to –2.8 m below sea level [7,17] In addition, small flows are lost from the lake through the navigational locks and of course through evaporation. Water depths vary throughout the lake, with the deepest occurring in the canals due to the navigational requirements. Depths in the basins vary between 0.5 and 1.2 m,[15] and in places, this is maintained by dredging and vegetation removal.

The flow direction in the lake is generally concentrated down the canals. Interaction of the canal water with the basin water occurs through the breached bunds; however, velocities and, hence, mixing in these sections are low. The point where the greatest interaction of canal and basin flow can be observed is at the junction between the Omum and Nubaria Canals in the main basin.

Water Quality and Ecosystems

Table 4 shows the time period available for each of the data sources, and Figure 1 shows the location of the sampling points. As a whole, the data of these sources were consistent and a similar magnitude was observed between sources. Table 5 presents the seasonal values available from the EEAA[16] data. Typical effluent concentrations from treatment plants with primary treatment can be found in literature.[18,19]

TABLE 4   Water Quality Data Sources

Source	Sampling Period	Number of Measurements	Number of Stations
EEAA[16]	2004–2008	2–3 per year	6
NIOF[15]	March–May 2008	5	10
WADI[8]	March 2007	1	31

TABLE 5 Water Quality Parameters from Selected Basins

EEAA stations	Year	Water Quality Parameters (EEAA 2004–2008) Yearly Averages														
		DO (mg/L)	DO %	BOD	COD	NH$_3$-N (mg/L)	NO$_2$-N (mg/L)	NO$_3$-N (mg/L)	Total N (mgN/L)	PO$_4$-P (mg/L)	Temperature (°C)	Conductivity (mS/cm)	PH	TDS	TS	TSS
Main basin in front of Qala drain (B)	2004	3.8	43.7	38.0	110.0			3.6	3.6	5.1	21.5		8.0	2454.0	2479.0	25.0
	2005	2.0	22.5	30.0	98.0	12.6	0.8	1.2	14.2		19.4	3.0	7.5	1828.0		
	2006	2.6	29.2	40.0	137.0	21.0	0.1	1.8	22.8	4.5	21.0	3.6	7.2	1825.0	1852.0	27.0
	2007					20.0		1.3	21.4	4.8						
	2008	1.0			133.0	27.8		1.5	29.3	4.8	24.0		7.8	1923.0	1958.0	35.0
	Ave	2.4	31.8	36.0	119.5	20.3	0.4	1.9	18.2		21.5	3.3	7.6	2007.5	2096.3	29.0
Middle of south basin (E)	2004	7.2	83.5	2.0	11.8	0.3	0.2	4.0	4.3	0.5	23.0	6.5	7.7	5134.0	5139.3	5.3
	2005	6.7	76.9	1.0	29.3	0.3	0.2	2.2	2.7	0.8	21.2	7.5	8.1	5266.3	5349.0	6.5
	2006	7.4	82.5	4.3	34.8	0.6	0.1	1.9	2.1	0.4	22.0	8.0	7.8	4289.8	4307.5	17.8
	2007					0.1	0.3	1.0	1.2	0.7						
	2008	9.2		2.0	55.0	0.4	0.0	2.9	3.3		27.5		8.1	5141.0	5153.5	12.5
	Ave	7.6	81.0	2.3	32.7	0.3	0.2	2.4	2.7	0.6	23.4	7.3	7.9	4957.8	4987.3	10.5
Middle of west basin (F)	2004	5.4	63.8	6.0	99.0	0.8	0.2	5.5	5.9	3.8	23.5	8.1	7.5	6953.3	6965.3	12.0
	2005	7.3	84.9	2.0	49.7	0.4	0.3	2.6	2.4	0.3	21.0	8.5	8.0	5085.0	5339.0	20.0
	2006	6.7	73.9	4.3	84.5	0.3	0.1	2.4	2.1	0.8	22.2	9.4	7.8	5885.8	5907.0	21.3
	2007					0.1	0.1	0.9	1.1	1.2						
	2008	7.9		5.0	81.5	0.4	0.1	1.9	2.4		27.0		8.2	7254.0	7274.0	20.0
	Ave	6.8	74.2	4.3	78.7	0.4	0.2	2.7	2.8	1.6	23.4	8.7	7.9	6294.5	6371.3	18.3
Middle of main basin (C)	2004	4.5	52.3	21.0	97.0	6.8		1.2	1.2	4.2	24.2	3.6	8.0	2415.0	2415.0	29.0
	2005	3.3	36.3	44.0	99.0	15.6	0.1	1.0	7.9	3.0	19.0	3.1	7.9	1961.0	1965.3	17.0
	2006	3.4	38.3	29.7	136.7	9.0	0.1	4.9	18.8		23.7	4.6	7.8	1935.7		29.7
	2007						0.1	3.9	13.0	4.7						
	2008	7.5			67.0	16.5		0.2	16.7		25.0		8.0	1941.0	1973.0	32.0
	Ave	4.7	42.3	31.6	99.9	12.0	0.1	2.2	11.5	4.0	23.0	3.8	7.9	2063.2	2117.8	26.9

The data clearly show that the most polluted basin is the main basin, and the source of that pollution comes from the Qala drain, which carries with it effluent from the EWWTP. The highest. Biological oxygen demand (BOD) values and the lowest DO% values are found at station B (Figure 1) at the entrance of the Qala drain. Similarly, the highest nitrogen loading enters the main basin through the Qala drain, along with the highest phosphorus loading. Station C shows generally lower values of the N and P as the water from the drain enters the lake and becomes mixed.

In general, the proportion of total N arising from ammonia was higher than that from nitrates, particularly in the main basin. It is thought that this is due in part to the source of pollution arising from the Qala drain and the low DO values recorded. In the West Basin, the proportion of ammonia is lower than that in the main basin, which is corroborated by the findings of El Rayis.[21] Since the levels of DO are also higher in this basin, it is believed that the nitrification process is more evident here.

This poor water quality has led to a drastic reduction in biodiversity.[8] The eutrophic nature of the lake results in a significant dominance of *P. australis,* while at the outfalls from the WWTPs, significant growths of the lead tolerant species *E. crassipes* is found. Periodic cutting of the *Phragmites* is practiced to increase the nitrogen removal from the lake. This intervention is necessary to prevent the lake area being dominated by vegetation, which in turn reduces the area available for fish habitat.

It is clear that in its present state, Lake Maryut is severely polluted with reduced biodiversity. A continual deterioration in the status of the basin will lead to a significant reduction in economic and industrial development potential.

Social Considerations and Governance

There are no specific studies that survey those living by the lake. However, the Integrated Action Plan research[22] identifies three main groups, namely, fishermen, poorer communities, and a scattered population, totaling approximately 30,000. It is estimated that 62% of the population is aged between 15 and 55, which is the average employment age in Egypt. There are high unemployment rates (up to 15%–20%) within this group, leading to low annual incomes. High illiteracy, poor health services, high mortality rates, high crime levels, and a tendency to marry young are also characteristics of the demographic.

A significant proportion of the population are living in settlements and accommodation that are temporary or informal. In these scattered settlements, some common issues arise: 1) lack of, or poor infrastructure coverage, particularly water supply and sanitation/wastewater networks and paved roads; 2) inadequate local services, especially health care, education, and youth facilities; 3) poor housing conditions; 4) lack of secure land tenure in the squatter settlements; and 5) high unemployment.

An analysis of the stakeholders directly responsible for managing the lake was undertaken in 2007.[13] This showed that the Alex Company for Sanitary and Drainage, the Fishing Authority, and The Ministries of Industry, Water Resources, and Environment are the main authorities responsible for the lake. The coordination and communication regarding the management of the lake between these authorities are known to be poor. Each authority makes strategic plans for the lake in isolation, with little discussion; hence, an integrated vision of the lake did not exist.

Integrated Environmental Management

Integrated Plan Concept

The integrated action plan for Lake Maryut draws on the principles of Integrated Coastal Zone Management (ICZM) to develop integrated actions for the sustainable development of Lake Maryut. Central to the ICZM approach is participation from stakeholders in the development of a plan that integrates strategic actions to address the environmental, economic, administrative, social, and urban issues of the lake.[22] The outcomes of this process defined four specific objectives of the action plan:

- Improve the ecological and chemical status of Lake Maryut.
- Enhance levels of economic, urban, and social development in a sustainable manner.
- Increase the potential for industrial activities around the lake and their environmental management.
- Adapt the governance system of the lake to execute the plan.

Strategic actions were proposed for each of the objectives, following analysis of the current state of the lake, and where feasible, these actions were tested. This entry will focus on the process completed for the first objective in detail and will present a short summary of the actions identified for the other three objectives.

Scenario Identification

The geographic scope of the plan was taken as the general boundary of the lake (Figure 1) and the natural extension to these boundaries according to the interaction of the lake. To the north, the plan includes the navigational locks and the El Mex pumping stations (necessary connections to the sea). To the east, the limit is bounded by the main road to Alexandria from Cairo, plus the position of the EWWTP and the Qala drain. To the west, the limit was the industrial area and a 500 m perimeter to the lake, and finally, the 500 m perimeter was extended to the south of the lake as the limit to the plan.

Through a participatory process, which included four training events and six participation workshops, a series of scenarios and potential actions for the improvement of the water ecosystem were defined by key stakeholders. These can be summarized as follows:

- Redevelopment of the waterfront and potential development of islands in the lake for urban regeneration projects.
- Improve lake mixing.
- Removal or improvement of the WWTP discharges into the lake.
- Vegetation management and sediment management.
- Industrial effluent improvement.
- Installation of wetland areas for nutrient removal.

Model Development

To assess the impact of certain actions, tools were developed to predict the hydrodynamic functioning and ecosystem functioning of the lake. Given the level of information available to create these tools and the complex nature of the processes in the lake, a mixture of approaches was proposed. To investigate the hydrodynamics and mixing of Lake Maryut, a two-dimensional (2D) hydrodynamic model was developed. To investigate the ecological improvements as a result of intervention, a point model of ecosystem functioning of different basins was set up.[23]

Hydraulic Model

Using the data set out in Tables 1–3, a 2D hydraulic model of the lake was built using MIKE21 software.[24] The computational mesh is shown in Figure 3, which defines the principal flow routes through the canal. A roughness map was developed to represent vegetation growth in the basins.[25] The model was calibrated for water levels but not velocities due to lack of data. A comprehensive sensitivity study was undertaken, which showed that the main control on the lake's hydraulic system is the pumping station at El Mex.[26] Figure 4 shows the general circulation patterns predicted by the model. The interactions are mainly driven by the drain inflows and clear areas of the lake where no vegetation is found.

To investigate the impact of particular actions on the hydraulics, a number of simulations were undertaken (Figure 5). These included the land reclamation from the lake and addition of islands to

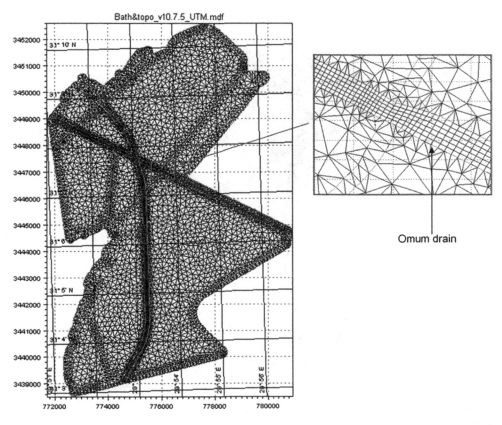

FIGURE 3 Computational mesh: Hydraulic model.
Source: Duran and Beevers.[26]

promote urban regeneration of the lakeshore, as well as the design of different entry configurations for the Qala drain to promote mixing. Figure 6 shows the magnitude of velocities predicted in the basin following these interventions. The change in velocities is highest in the main basin where most interventions are proposed; however, the magnitude of change is only in the order of a 9% increase.[26] While areas in the main basin become better mixed, areas of stagnation occur, most notably around the islands. From these results, it becomes clear that increasing mixing significantly would be difficult unless an increase to the inflow to the lake could be facilitated.

Ecological Model

Using the data set out in Tables 1–5, a dynamic 0-dimension ecological model was built for the different basins of Lake Maryut. The software PCLAKE describes the dominant ecological interaction in a shallow lake ecosystem.[23] This model was chosen for its ability to model the complex processes and allow a number of different proposed actions to be modeled, including vegetation management on a seasonal basis, dredging, wetland installation, and improvements to discharges. It should be noted that the model assumes mixing across a basin, which, for Lake Maryut, is a simplification; however, the model is detailed enough to give an indication of the potential change caused by an action.

The model was built using the seasonal data available from the EEAA monitoring for the inflows, and this was complemented by the WADI and NIOF data. Calibration was possible, comparing monitoring data available in the lake itself.

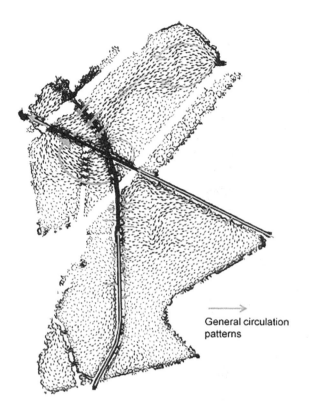

FIGURE 4   General circulation patterns: Hydraulic model.
Source: Duran and Beevers.[26]

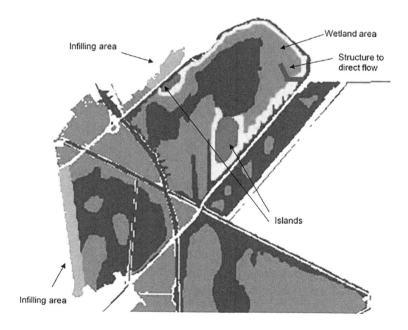

FIGURE 5   Scenario modifications: Hydraulic model.
Source: Duran and Beevers.[26]

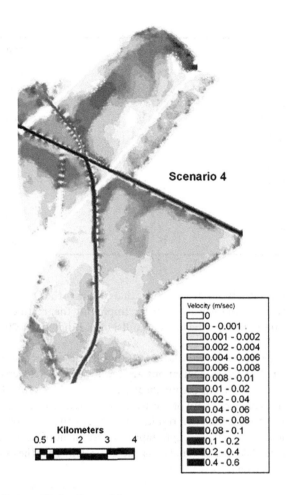

FIGURE 6   Velocity predictions: Hydraulic model.
Source: Duran and Beevers.[26]

A significant number of actions were tested in the ecological model. Available literature was used to represent the actions in the lake:

- Secondary treatment of effluent from the WWTPs.[27]
- Addition of aluminum to the WWTP process.[18]
- Installation and removal efficiencies of constructed wetlands (for the Qala drain).[28–31]
- Vegetation and sediment management.
- Table 6 shows the predicted improvement to different parameters if the following actions are taken:
- Removal of the WWWTP from the lake (outfall to the sea) after upgrade to secondary treatment.
- Upgrade of EWWTP to secondary treatment plus additional treatment using aluminum.
- Upgrade of industrial effluent through improved environmental management.
- Dredge sediments in the Qala drain and installation of a constructed wetland.
- Aquatic vegetation management.

The oxygen consumption in Lake Maryut is mainly attributed to the respiration of aquatic organisms including plankton and aerobic bacteria and the oxidation of organic matter.[32] High values of organic carbon have been reported in sediments close to the treatment plants.[21] The oxidation of these

TABLE 6   Ecological Model Results

Parameter	Unit	Baseline Value (average)	Improvement with Stated Actions	
			Increase %	Decrease %
DO	mg/L	1.3	303	
NH_4-N	mg N/L	13.4		47
NO_3-N	mg N/L	0.8	120	
Organic N	mg N/L	6.4		83
N total	mg N/L	20.5		52
P total	mg P/L	4.9		72
Chlorophyll-a	mg/m₃	93.4		85
Phyto-biomass	mg DW/L	5.9		85
Zoo biomass	g DW/m₂	0.53	39	
Detritus	mg/L	83.5		82
Secchi depth	m	0.1	377	

Source: Alvarez-Mieles and Beevers.[36]

sediments, rich in organic matter, is one of the factors that increases the DO consumption and therefore decreases the DO concentrations in the water column. Improving the treatment processes for the treatment plants has a significant impact on the model results, indicating up to a 300% improvement to DO levels in the main basin.

Ammonium is generated by heterotrophic bacteria as the primary nitrogenous end product of decomposition of organic matter. NH_4-N concentrations are usually low in oxygenated waters of oligotrophic lakes since it is utilized by plants in the nitrification process. However, at low DO concentrations, nitrification ceases, the absorptive capacity of the sediments is reduced, and an increase of the release of NH4-N from the sediments occurs. As a result, the NH_4-N concentration of shallow lakes would increase.[33] Increasing the levels of DO through improved treatment of wastewater and a reduction in pollutant loads arriving to the lake as a result of constructed wetlands has the potential to reduce overall ammonia levels in the main basin by up to 47%.

Chlorophyll-a is a primary productivity indicator and a very good estimate for monitoring and assessing the eutrophication status of lakes.[34] The supply of P and N is considered to be one of the main factors that determine the magnitude of the primary production.[35] In cases where pollution is caused by domestic wastewater with a high nutrient content, the algae production in the recipient watercourse increases considerably.[34] With the improvement to N and P and the increase in the DO predicted, a decrease in chlorophyll-a of up to 85% could be achieved for the lake.

Finally, the Secchi depth is a measure of the transparency of a water column. Secchi readings between 0.1 and 0.4 m were reported in the main basin,[32] which are directly related to the high production rate of phytoplankton. Eutrophic lakes present low values of Secchi depth due to the high concentrations of algae and detritus that increase turbidity and therefore decrease transparency. With the proposed measures and reduction in detritus, an increase in the transparency of the lake is predicted to be greater than 370%.[36]

Proposed Actions

The modeling component tested a number of actions for the improvement of the water ecosystem and showed that a significant improvement could be achieved. In addition to these, actions were proposed to address the other strategic objectives of the plan.[22] The following is a summary of such objectives:

- Prepare a Lake Maryut land and water plan.
- Promote urban regeneration around the lake shoreline and create ports and markets in the main basin.

- Promote ecotourism and encourage new ecological economic activities.
- Upgrade informal settlement areas and integrate social housing into urban development projects.
- Plan and develop fisheries activities.
- Prepare online monitoring systems along with water quantity and quality maps.
- Regulate specific limits for industrial water discharge into Lake Maryut.
- Establish a Lake Maryut Authority and set up a management unit with a clear structure and purpose.
- Establish a robust monitoring system for water quality.

The plan aims to address all four strategic objectives defined at the outset. To monitor the effect of any potential implemented action as part of the plan, a monitoring program is suggested. The implementation of such a monitoring plan would allow the incorporation of adaptive management techniques should unexpected impacts become evident.

Conclusions

An integrated action plan for Lake Maryut is proposed, using a participatory process and supported by a technical modeling study that was used to predict the impact of potential actions. It is clear from the study that significant improvement to the lake can be achieved, thus encouraging economic development and urban regeneration. However, this improvement requires substantial investment. The plan suggests that to achieve this, a financial strategy that links the urban and industrial future development potential with the environmental interventions is required to improve the lake, thereby linking potential future economic gains to the required lake renovation.

Since the completion of the proposed plan in 2009, there have been follow-up activities that move towards the adoption of a sustainable development strategy for the lake. A lake management unit is established and receives technical input from Centre for Environment and Development for the Arab Region and Europe (CEDARE) for Geographical Information System (GIS) support and NARSS (National Authority for Remote Sensing and Space Sciences of Egypt) for model development. In addition, the Ministry for Housing has proposed to use the outcomes of the study for the development of a new urban territorial plan for Alexandria that includes an integrated sustainable development vision for the Lake Maryut zone.

The study has shown that, with a coordinated and integrated approach to environmental management, the current deteriorating trend of Lake Maryut could be reversed. However, to achieve this goal, actions addressing environmental, social, and governance issues must be implemented.

Acknowledgments

The author would like to acknowledge the official partners of the ALAMIM project: MEDCITIES, Barcelona Metropolitan Area, CEDARE, Alexandria Governorate, the EEAA, NARSS, City of Marseilles, Environment and Housing Ministry of the Catalan Government, UNESCO-IHE, and the EUCC—The Coastal Union. In particular, the author is grateful to Mr. Joan Parpal, the project leader, and Ms. Roxan Duran and Ms. Gabriela Alvarez-Mieles.

References

1. Saad, M. Distribution of phosphates in Lake Maryut, a heavily polluted lake in Egypt. Water, Air, Soil Pollut. **1973**, *2*, 515–522.
2. Mesnage, V.; Picot, B. The distribution of phosphate in sediments and its relation with eutrophication of a Mediterranean coastal lagoon. Hydrobiologia **1995**, *297* (1), 29–41, DOI: 10.1007/BF00033499.

3. Hughes, R.; Hughes, S. *A Directory of African Wetlands;* IUCN: Gland, Switzerland and Cambridge, UK/UNEP: Nairobi, Kenya/WCMC: Cambridge, U.K., 1992; 820 pp., ISBN 2–88032-949–3.

4. Warne, A.; Stanley, D. Late quaternary evolutions of the Northwestern Nile Delta and adjacent coast in the Alexandria region, Egypt. J. Coastal Res. **1993**, *9* (1), 26–64.

5. Stanley, J.; Warne, A.; Schnepp, G. Geoarchaeological interpretation of the Canopic, largest of the relict Nile Delta distributaries, Egypt. J. Coastal Res. **2004**, *20* (3), 920–930, doi: 10.2112/1551–5036(2004)20[920:GI0TCL]2.0.C0;2.

6. Stanley, D.; Warne, A. Nile Delta: Recent geological evolution and human impact. Science **1993**, *260* (5108), 628–634 DOI: 10.1126/science.260.5108.628.

7. USAID. Hydrologic Studies of Lake Mariout Technical Memorandum N° 8. 1996 USAID Project N° 263–0100.

8. Mateo, M. Lake Mariut: An Ecological Assessment. Final report of the Water Demand Integration (WADI) project: INCO-C-2005—15226, 2009. Available for download at http://www.medcities. org/ocuments section (accessed April 29, 2011).

9. Bakhoum Shnoudy, A. Comparative study on length weight relationship and condition factor of the genus *Oreochromis* in polluted and non-polluted parts of Lake Mariut Egypt. Bull. Natl. Inst. Oceanogr. Fish. **1994**, *20* (1), 201–282.

10. Khan F.A.; Ansari, A.A. Eutrophication: An ecological vision. The Botanical Review. **2005**, *71* (4), 449–482. doi: 10.1663/00068101(2005)071[0449:EAEV]2.0.C0;2

11. Wahby, S.; Kinawy, S.; El-Tabbach, T.; Abdel Moneim, M. Chemical characteristics of Lake Maryût, a polluted lake south of Alexandria, Egypt. Estuarine Coastal Mar. Sci. **1978**, *7* (1), 17–28.

12. Amr, H.; El-Tawila, M.; Ramadam, M. Assessment of pollution in fish and water of main basin, Lake Maryut. J. Egypt. Public-Health Assoc. **2005**, *80* (1–2).

13. Kafafi, A. *Stakeholder analysis report,* 2007. Final report for the ALAMIM project available for download at http://www.medcities.org/ documents section (accessed April 29, 2011).

14. World Climate. Port Alexandria, Egypt Weather History and Climate Data, http://www.worldclimate.com/cgi-bin/datapl?ref=N31E029+2100+62315W (accessed April 2009).

15. National Institute of Oceanography and Fisheries (NIOF). *Lake Maryut Data Acquisition.* Final report for the ALAMIM project available for download at http://www.medcities.org/ documents section (accessed April 29, 2011).

16. Egyptian Environmental Assessment Authority (EEAA). *Monitoring Data Lake Maryut,* 2004–2008 reports.

17. CEDARE. *Alexandria Lake Mariout Integrated Management—Stocktaking Analysis report,* 2007. ALAMIM report available for download at http://www.medcities.org/ documents section (accessed April 29, 2011).

18. El-Bestawy, E.; Hussein, H. Baghdadi, H.; El-Saka, M. Comparison between biological and chemical treatment of wastewater containing nitrogen and phosphorous. J. Industrial Microbiol. Biotechnol. **2005**, *32*, 195–203.

19. Metcalf, A.; Eddy, G.; Tchobanoglous, F.; Burton, F.; Stensel, H. *Wastewater Engineering, Treatment and Reuse*, 4th Ed.; McGraw Hill Education: New York, 2003; 1329, ISBN: 0070418780.

20. Trabea, A.; Salem, I. Empirical relationship for ultraviolet solar radiation over Egypt. Egypt J. Sol. Energy **2001**, *24* (1), 123–132.

21. El Rayis, O. Impact of man's activities on a closed fishing lake, Lake Maryout in Egypt, as a case study. Mitigation Adapt. Strategies Global Change **2005**, *10*, 145–157.

22. El-Refaie, M.; Rague, X. *ALAMIM Integrated Action Plan,* 2009. ALAMIM report available for download at http://www.medcities.org/ documents section (accessed April 29, 2011).

23. Janse, J. Model studies on the eutrophication of shallow lakes and ditches, Ph.D. Thesis, Wageningen University: Wageningen, the Netherlands, 2005.

24. Warren, I.; Back, H. MIKE21: A modelling system for estuarine, coastal waters and seas. Environ. Software **1992**, *7* (4), 229–240.

25. Wang, C.; Zhu, P.; Wang, P.; Zhang, W. Effects of aquatic vegetation on flow in the Nansi Lake and its flow velocity modeling. J. Hydrodyn. **2006**, *18* (6), 640–648.

26. Duran, R.; Beevers, L. *Alexandria Lake Mariout Integrated Management: Hydrodynamic Model Report,* 2009. ALAMIM report available for download at http://www.medcities.org/ documents section.

27. Knight, R.; Rubles, R.; Kadlec, R.; Reed, S. Wetlands for wastewater treatment performance data base. In *Constructed Wetlands for Water Quality Improvement*; Moshiri, G., Ed.; CRC Press; Boca Raton, FL, 1993; 35–49.

28. Vymazal, J. Removal of nutrients in various types of constructed wetlands. Sci. Total Environ. **2007**, *380,* 48–65.

29. Moortel van de, A.; Rousseau, D.; Tack, F.; De Pauw, N. A comparative study of surface and subsurface flow constructed wetlands for treatment of combined sewer overflows: A greenhouse experiment. Ecol. Eng. **2009**, *35*, 175–183.

30. Verhoeven, J.; Meuleman, A. Wetlands for wastewater treatment: Opportunities and limitations. Ecol. Eng. **1999**, *12,* 5–12.

31. Lu, S.; Wu, F.; Lu, Y.; Xiang, C.; Zhang, P.; Jin, C. Phosphorous removal from agricultural run off by constructed wetland. Ecol. Eng. **2009**, *35*, 402–409.

32. Anwar, A.; Samaan, A. Productivity of Lake Mariut, Egypt. Part I. Physical and chemical aspects. Int. Rev. Hydrobiol. **1969**, *54* (3), 313–355.

33. Quiros, R. The relationship between nitrate and ammonia concentrations in the pelagic zone of lakes. Limnetica **2003**, *22* (1–2), 37–50.

34. Heinonen, P.; Ziglio, G.; Beken van der, A. *Hydrological and Limnological Aspects of Lake Monitoring*; Wiley: Chichester, England, 2000, ISBN 0–471–89988-7.

35. Fathi, A.; Abdelzaher, H.; Flower, R.; Ramdani, M.; Kraiem, M. Phytoplankton communities of North African wetland lakes: The Cassarina Project. Aquat. Ecol. **2001**, *35,* 303–318.

36. Alvarez-Mieles, G.; Beevers, L. *Alexandria Lake Mariout Integrated Management: Ecological Modelling for Lake Maryut,* 2009. ALAMIM report available for download at http://www. medcities.org/ documents section (accessed May 29, 2011).

30

Aral Sea Disaster

Introduction 317
Aral Sea Basin 317
Irrigation and Cotton...319
Muynak and Aralsk..320
Environmental Problems..320
Human Tragedy 321
Conclusion 321
Bibliography 322

Guy Fipps

Introduction

The cause is attributed to a vast expansion of irrigation in the Central Asian Republics beginning in the 1950s, which greatly reduced inflows to the Sea. The diversion of water for massive irrigation development was done deliberately by Soviet Union officials, unconcerned about the consequences of their actions.

The environmental, social, and economic damage has been immense. Winds pick up dust from the dry seabed and deposit it over a large populated area. The dust likely contains pesticide and chemical residues that are blamed for the serious rise in mortality and health problems in the region. The Sea, and the now exposed dry seabed, may also be contaminated by runoff from a former Soviet military base and a biological weapons lab. The ecosystem of the Aral Sea has collapsed, and climate changes in the Aral Sea Basin have been documented. Hundreds of agreements have been signed since 1980s on programs designed to address the "Aral Sea Problem" which, to date, have not been effective at preventing the continuing shrinking of the sea.

Aral Sea Basin

The Aral Sea is located in Central Asia and lies between Uzbekistan and Kazakhstan in a vast geological depression, the Turan lowlands, in the Kyzylkum and Karakum Deserts. In the 1950s, the sea covered 66,000 km², contained about 1090 km³ of water, and had a maximum depth of about 70 m. The Aral Sea supported vast fisheries and shipping industries. At that time the sea was fed by two rivers, the Amu Darya (2540 km) and the Syr Darya (2200 km), which originate in the mountain ranges of central Asia and flow through the five republics of Uzbekistan, Kazakhstan, Kyrgyzstan, Tajikistan, and Turkmenistan.

The two rivers provide most of the fresh water used in Central Asia. In the last 50 years, about 20 dams and reservoirs and 60 major irrigation schemes have been constructed. About 82% of river diversions are for agricultural use and 14% is for municipal and industrial use (Table 1).

Water demand due to population growth and industrial expansion continues to increase (Table 2). Since 1960, the population of the Central Asian republics has increased 140% and totals over 50

TABLE 1   Average Water Supply and Demand in the Aral Sea Basin

Total Water Available	km³	%
Amu Darya Basin	84.3	64
Syr Darya Basin	47.8	36
Total	132.1	100
Water demand		
Agriculture		
Amu Darya Basin	44.8	81.6
Syr Darya Basin	34.6	
Municipal Water		
Amu Darya	3	6.5
Syr Darya	3.3	
Industry		
Amu Darya	3	8.2
Syr Darya	5	
Livestock		
Amu Darya	0.2	0.2
Syr Darya	0	
Fishery		
Amu Darya	2.6	3.5
Syr Darya	0.8	
Total	97.3	100

TABLE 2   General Statistics of the Aral Sea Basin Countries in 1995

	Kazakhstan	Uzbekistan	Turkmenistan	Kyrgyzstan	Tajikistan
Area, km²	2,717,300	447,400	488,100	198,500	143,100
Irrigated land, km²	23,080	41,500	12,450	10,320	6.940
Population	17,376,615	23,089,261	4,075,316	4,769,877	6,155,474
Population growth rate, %	0.62	2.08	2.5	1.5	2.6

million. Likewise, industrial production using large amounts of water has also increased. Examples include steel production which rose 200%, cement production by 170%, and electricity generation by a factor of 12.

The total inflows to the Aral Sea began decreasing rapidly in the 1960s, and by 1990 the storage volume of the sea has decreased by 600 km³ (Table 3). As the water level fell, salinity levels have tripled, rising from about 1000 ppm to just under 3000 ppm today. By the 1980s, as the Aral Sea problem became well known in the Soviet Union, government officials proposed ambitious projects to divert water from other rivers, including ones in South Russia and Siberia, to be transported to the Aral Sea in massive canals. However, these plans died with the breakup of the Soviet Union.

The decrease in sea level has now split the Aral Sea into two separate water bodies: the Small and Large Aral Seas (Maloe More and Bol'shiye More) each separately fed by the Syr Darya and the Amu Darya, respectively. The once vast Amu Darya delta which once covered 550,000 ha has now shrunk to less than 20,000 ha.

TABLE 3 Decline of the Aral Sea during the 1980s and Total Estimated Inflows from the Amu Darya and Syr Darya Rivers

		Aral Sea	
Year	Inflows (km³)	Volume (km³)	Surface Area (km²)
1911–1960	56.0	1064	66,100
1981	6.0	618	50,500
1982	0.04	583	49,300
1983	2.3	539	47,700
1984	7.9	501	46,100
1986	0.0	424	41,100
1987	9.0		
1988	23.0		41,000
1989		300	30,000

Irrigation and Cotton

For thousands of years, Central Asian farmers diverted water from the Amu Darya and Syr Darya Rivers, transforming desert into green oases and supporting great civilizations. Historically, irrigation water use was conducted at a sustainable level. The creation of the Soviet Union and the collectivization of farmlands resulted in the end for traditional agricultural practices. Beginning as early as 1918, Soviet leaders began expanding irrigated land in Central Asia for export and hard currency. Cotton was known as "white gold." The USSR became a net exporter of cotton by the 1930s, and by the 1980s, was ranked fourth in the world in cotton production.

The policy of emphasizing cotton production was accelerated in the 1950s as Central Asia's irrigated agriculture was expanded and mechanized. In 1956, the Kara Kum Canal was opened, diverting one-third of the flow in the Amu Darya to new cultivated areas in the deserts of Uzbekistan and Turkmenistan. The year 1960 represents the critical junction when the Aral Sea began to drop. Irrigated cotton production and water diversions continued to be expanded until the break-up of the Soviet Union (Table 4).

Estimates are that upwards of 80% of the workforce is employed in agriculture. The main agricultural crops in the basin are cotton (6.4 million ha), forage (1.7 million ha), rice (0.4 million ha), and tree crops (0.4 million ha).

Some Central Asian irrigation experts estimate that only 20%–25% of the water diverted from the rivers is actually used by the crops, the rest being lost in the canals that transport the water to the fields and due to inefficient irrigation practices used on-farm. It is believed that over the past decade, adequate maintenance, repair, and renovation of the irrigation infrastructure were not performed at a meaningful level, and water losses from deteriorating canals, gates, and other facilities have increased.

Most land is under furrow irrigation, with drip irrigation accounting for about 5% of the irrigated cropland (used primarily on orchard crops), and sprinkler irrigation accounts for about 3%. Even though the

TABLE 4 Cultivated Land along the Amu Darya and Syr Darya Rivers

	Before 1917	1960	1980	1992
Millions of hectares	5.2	10	15	18.3

water saving benefits of gated pipe are well known in the region, less than one-sixth of the farms use this technology. Reasons may include costs and product availability. Most farms follow the centuries' old practice of cutting earthen canals with shovels in order to divert water into the field. The volume of water available at these farm ditches is not sufficient to provide an even distribution of water over the field. As a result, water logging and soil salinity now affects about 40% of all the cultivated land in the region.

Muynak and Aralsk

Of all the villages affected by the drying of the Aral Sea, Muynak is the best known. Historically, Muynak was located on an island of the vast Aral Sea delta at the convergence of the Amu Darya River in Karakalpakstan (a semi-autonomist republic in Uzbekistan). In 1962, the island became a peninsula. By 1970 the former seaport was 10 km from the sea. The retreat of the sea accelerated and the town was 40 km from the sea by 1980, 70 km in 1995, and close to 100 km today.

Over 3000 fishermen once worked the abundant waters around Muynak which supported 22 different commercial species of fish. In 1957, Muynak fishermen harvested 26,000 tons of fish, about half of the total catch that year taken from the Aral Sea. Muynak also produced 1.1 million farmed muskrat skins which were used to produce coats and hats.

The Kazakhstan city of Aralsk, was once located on the northern edge of the Aral Sea, and like Muynak, had major fisheries and commerce industries. A major shipping and transport industry existed between these two cities. As the Aral Sea skunk, Aralsk found itself farther and farther from the shore which had retreated nearly 129 km by the 1980s. In the early 1990s, a dam was built just to the south of the mouth of the Syr Darya, to protect the northern part of the Aral Sea, letting the southern portion of the Aral Sea evaporate. Although only 10% of the water in the Syr Darya River reaches the northern part of the Aral Sea, the Little Aral has risen 3 m since the construction of the dam, and the shoreline has crept to within 16 km of the town.

Environmental Problems

The Aral Sea is an unfortunate example of an old Uzbek proverb: "at the beginning you drink water, at the end you drink poison." As the rivers flow through cultivated areas, they pick up fertilizers, pesticides, and salts from runoff, drainage water and groundwater flow. In the 1960s, it was common for about 550 kg ha^{-1} of chemicals to be applied to cotton fields in Central Asia, compared to an average of 25 kg used for other crops in the Soviet Union. Residues of these chemicals are now found on the dry seabed. Estimates are that millions of tons of dust are picked up from the seabed and distributed over the Aral Sea region.

The Sea may have been contaminated from runoff from by two former USSR military installations in the area. A chemical weapons testing facility was located on the Ust-Jurt Plateau (north shore), and was closed in the mid-1980s. Renaissance Island (Vozrozhdeniya Island), located in the central Aral Sea, was the site of the former USSR Government's Microbiological Warfare Group which produced the deadly Anthrax virus. Some scientists believe that some containers holding the virus were not properly stored or destroyed. As the Aral Sea continues to dry and water levels recede, the ever-expanding island will soon connect to the surrounding land. Scientists fear that reptiles, including snakes that have been exposed to the various viruses, will move onto the surrounding land and possibly infect the humans living around the shores of the Aral Sea.

The Aral Sea once supported a complex ecosystem, an oasis in the vast desert. Over 20 species of fish are now extinct. Karakalpakstan scientists believe that a total of about 100 species of fish and animals that once flourished in the region are now extinct, as are many unique plants.

Residents believe that there is a direct correlation between the drying of the sea and changes in climate of the Aral Sea Basin. The moderating effect of the sea has diminished and temperatures are now about 2.5°C higher in the summer and lower in the winter. Rainfall in the already arid basin has decreased by about 20 mm.

Human Tragedy

Over the last 50 years, there has been a large increase in mortality, illnesses, and poor health in the region. Some estimate that 70%–90% of the population of Karakalpakstan suffer some an environmentally induced malady. Tuberculosis is rampant. Hardest hit are women and children. Common health problems include kidney diseases, thyroid dysfunctions, anemia, bronchitis, and cancers.

Conclusion

Some accounts are that since 1984, hundreds of international agreements have been signed to address Aral Sea problems. The early agreements had the goals of first stabilizing the Sea, then slowly increasing flows to restore its ecosystem. In 1992, the Interstate Commission for Water Coordination was formed by the five central Asian republics, which also accepted, in principle, to adhere to the limits on water diversions as set during the Soviet era in 1984 and 1987. To date, however, no progress has been made on stabilizing or reversing the declining inflows. With no water reaching the Aral Sea from the Amu Darya, scientists predict that this portion of the sea (the Large Aral Sea) will disappear by 2020 (Figure 1).

FIGURE 1 This NASA photograph (STS085–503-119) was taken in August 1997 and looks toward the southeast. The Amu Darya River is visible to the right and the Syr Darya on the left. The Aral Sea is now separated into the Small Aral to the north and the Large Aral to the south. Shown are the approximate extent of the Aral Sea in 1957 before a massive expansion of irrigation diversions from the rivers.

Bibliography

1. The Aral Sea Homepage: Available at http://www.dfd.dlr.de/app/land/aralsee/
2. Requiem for a dying sea: Available at http://www.oneworld.org/patp/pap_aral.html/
3. Disappearance of the Aral Sea: Available at http://www.grida.no/aral/maps/aral.htm
4. Earth from Space: Available at http://earth.jsc.nasa.gov/categories.html

31

Chesapeake Bay

Introduction 323
Discussion 323
Conclusions 325
References 326

Sean M. Smith

Introduction

The Chesapeake Bay is the largest estuary within the U.S.A.[1] Fresh water flows to the Chesapeake Bay from a watershed that covers an estimated 166,709 km², including portions of Delaware, Maryland, New York, Pennsylvania, Virginia, Washington, DC, and West Virginia (Figure 1). The ecological productivity of the estuary has made it an important resource for Native Americans, European immigrants, and current residents in the region.[2] The population in the contributing watershed in recent years has swelled to over 15 million people, resulting in extensive direct and indirect impacts that are now a focus of a large-scale restoration effort led by the US Environmental Protection Agency.[1]

Discussion

Located along the Mid-Atlantic coast of the U.S.A. within the limits of Maryland and Virginia, the Bay is approximately 304-km long, has an estimated surface area of 11,603 km², a width that ranges from 5.5 to 56 km, and an average depth of approximately 6.4 m.[1] Salinity in the tidal portions of the estuary transition from "fresh" conditions (i.e., 0–5 parts per thousand (ppt) salt concentration) at the northern-most end to "marine" conditions (30–35 ppt salt concentration) at the southern boundary with the Atlantic Ocean.

Evidence indicates that the modern Chesapeake Bay began forming approximately 35 million years ago with a meteorite impact in the proximity of what is now the confluence of the Bay with the Atlantic Ocean.[2,3] The impact created a topographic depression that influenced the location and alignments of several large river valleys, including those associated with the present day Susquehanna, Rappahannock, and James Rivers. Since then, the river valleys have been periodically exposed and flooded in response to cycles of global glaciation and associated fluctuations in sea level. The most recent, the Wisconsin glaciation, began retreating approximately 18,000 years ago. The retreat resulted in a rise in sea level by almost ninety meters, drowning the river valleys and forming the current Bay.

Eleven large rivers drain the Bay's watershed, the Susquehanna River from Pennsylvania and New York providing the largest contribution with an average of 98 million m³/day flowing into the northern end of the estuary. The rivers drain one or more of five different physiographic provinces within the watershed, including the Appalachian Plateau, Ridge and Valley, Blue Ridge, Piedmont, and Coastal Plain (Figure 1).[4] Each province's geologic composition and history creates dramatically different landscape settings from the western to eastern sides of the drainage basin. The Appalachian, Ridge and Valley, and Blue Ridge are characterized by mountainous terrain, a dominance of sandstone along ridge tops, and several carbonate valleys. The Piedmont has less relief, is dominated by metamorphic rocks, and is characterized by a surface that has been dissected by dendritic stream channel networks. Further to the east, the Coastal Plain is characterized by thick layers of unconsolidated geologic materials over-lying bedrock deep beneath the surface. Waterways that flow from the Piedmont into the Coastal Plain traverse the "Fall Zone," a region that is easily distinguished by waterfalls coincident with an abrupt drop in the underlying bedrock elevations. Major ports and cities were developed along the Fall Zone, including Washington, DC, Baltimore, Maryland, and Richmond, Virginia, because of their locations at the upstream terminus of navigation from tidal waters and proximity to hydropower sources.

The Chesapeake Bay estuary is naturally dynamic and characterized by physical conditions that can be stressful to aquatic organisms. The salinity gradient broadly governs the spatial distribution of aquatic habitat types. Alterations in currents, wind, and freshwater inputs can cause salinity conditions to vary over time. The shallow depths also cause colder winter and warmer summer water temperatures compared to the open ocean. These spatial and temporal fluctuations can create physiologically chal-lenging conditions. However, many organisms have adapted and use the abundant nutrients and physi-cal habitat in different portions of the estuary for specific periods of their life cycles or seasons of the year. As a result, the Bay supports an estimated 3600 species of plants, fish, and animals, including 348 finfish and 173 shellfish species. Some of the most notable of these include striped bass, American shad, blueback herring, blue crab, and the American oyster.[5] The name "Chesapeake" itself was coined from the Algonquin American-Indian word "Chesepiooc" meaning "great shellfish bay."[5]

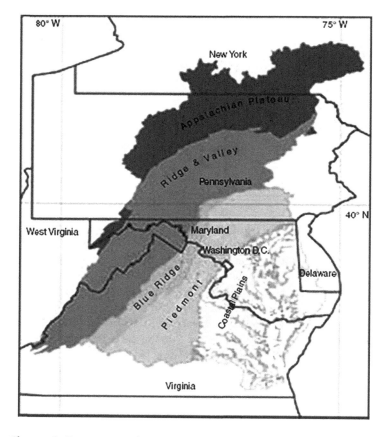

FIGURE 1 The Chesapeake Bay estuary and its watershed, including physiographic provinces and state boundar-ies. (Courtesy of M. Herrmann, Maryland Department of Natural Resources.)

Archaeologists estimate that Native American inhabitants first arrived in the Bay region from the south or west approximately 12,000 years ago as the ice sheets associated with the Wisconsin glaciation began to retreat and temperatures increased.[2] The first inhabitants are presumed to have been nomadic; however, archeological evidence suggests that selective food production started as early as 5000 years ago and settled towns began to be formed approximately 1300 years ago as the population density in the region increased. Recovered artifacts provide evidence of the extensive use of the Bay by the early inhabitants for travel, communication, tools, and food.

The first recorded European contact with the Chesapeake Bay region was by the Italian captain, Giovanni da Verrazano in 1524.[2] The English established one of the most well known early settlements at Jamestown, Virginia in 1607. English colonization expanded through expeditions to the north in the Bay, partly led by the famed Captain John Smith. Immigration to the region increased throughout the 1600s and much of the area was settled by the mid-1700s. The colonists made extensive use of the resources provided by the estuary, its wetlands, and tributaries. Shellfish, including oysters, blue crabs, and hard and soft clams, were harvested from shallow water areas.[2] The numerous piles of oyster shells that can be found near Coastal Plain tidal areas provide support for written claims of the extensive oyster beds that existed in the Bay when the European colonists arrived. Traps and nets were used to harvest finfish, including herring, striped bass, and shad. Migratory waterfowl, such as ducks and geese, were also plentiful food sources.

The rapid growth in the human population since European colonization of the Chesapeake Bay region dramatically increased the harvest of finfish, shellfish, waterfowl, and mammals naturally supported by the estuary. Extensive landscape alterations also caused direct and indirect physical changes to the Bay and its tributaries. The combination of overharvesting, pollution, and physical alterations has severely impacted the ecosystem and many of the species that historically flourished in the estuary.[1] Dramatic declines have been documented by the harvest records of popular commercial fisheries such as shad and striped bass. Records indicate a decline in the catch of blue crabs per unit of effort since the 1940s. The oyster harvest is currently at less than 1% of historic levels, although this reduction is partly attributed to disease. Many other species not harvested commercially have also been affected by the alterations in the Bay ecology that have accompanied European settlement and population growth.

One of the most important impacts to the Chesapeake Bay has been the increased erosion rates and downstream sedimentation caused by extensive deforestation of the watershed.[6] The influx of sediment into the tidal estuary has reduced the water depths in many embayments that once served as navigable ports.[7] Elevated suspended sediment inputs during storm events also increase turbidity in the tidal water column.[6] The resulting decrease in water clarity, which has been exacerbated by algal blooms associated with nutrient runoff pollution, reduces submerged aquatic vegetation (SAV) growth in shallow areas (i.e., depths less than 2 m). SAV coverage on the Bay bottom is estimated to have declined from approximately 80,900 hectares in 1937 to 15,400 hectares in 1984.[1] The loss has negative implications for a variety of species that use the vegetation for habitat, including blue crabs and juvenile finfish.

An extensive effort to restore the Bay has been undertaken by the US federal government in coordination with states in the watershed.[1,8] A large part of this effort has been focused on the recovery of the historic SAV distributions, as well as reversal of the abnormally low oxygen levels that now occur in the main stem of the Bay and its major tidal tributaries during summer months.[1,9] As with the water clarity problems, the low dissolved oxygen is related to excess nutrient inputs, mainly nitrogen and phosphorous, which stimulate algal production. The oxygen depletion occurs because of algal decomposition, resulting in, estuarine habitat degradation. Substantial reductions in nutrients from watershed runoff have been concluded to be necessary to achieve restoration goals related to both SAV and low dissolved oxygen.[1,9]

Conclusions

The Chesapeake Bay is a large and historically productive estuary on the Mid-Atlantic coast of the U.S.A., with extensive fisheries and wildlife resources. The Bay ecosystem has been impaired by watershed alterations and overharvesting accompanying human population growth in the region, thereby inspiring an extensive government-supported restoration effort. A large part of the restoration focuses on sediment and nutrient pollution associated with runoff from the contributing watershed.

References

1. http://chesapeakebay.net (accessed July 11, 2005).
2. Grumet, R.S. *Bay, Plain, and Piedmont: A landscape history of the Chesapeake Heartland From 1.3 Billion Years Ago to 2000;* The Chesapeake Bay Heritage Context Project; U.S. Department of the Interior: Annapolis, MD, 2000; 183 pp.
3. Powars, D.S.; Bruce, T.S. *The Effects of the Chesapeake Bay Impact Crater on the Geological Framework and Correlation of Hydrogeologic Units of the Lower York-James Peninsula, Virginia;* U.S. Geological Survey Professional Paper 1612; United States Geological Survey: Reston, VA, 1999.
4. Smith, S.; Gutuierrez, L.; Gagnon, A. *Streams of MarylandTake a Closer Look*; Maryland Department of Natural Resources: Annapolis, MD, 2003; 65 pp.
5. White, C.P. *Chesapeake Bay: Nature of the Estuary, A Field Guide;* Tidewater Publishers: Centreville, MD, 1989.
6. Langland, M.; Cronin, T. *A Summary Report of Sediment Processes in Chesapeake Bay and Watershed*; United States Geological Survey Water Resources Investigations Report 03–4123; United States Geological Survey: Reston, VA, 2003.
7. Gottschalk, L.C. Effects of soil erosion on navigation in upper Chesapeake Bay. Geogr. Rev. **1945**, 35, 219–238.
8. Ernst, H.R. *Chesapeake Bay Blues*; Rowman and Littleford Publishers, Inc.: Lanham, MD, 2003.
9. Macalaster, E.G., Barker, D.A., Kasper, M., Eds.; *Chesapeake Bay Program Technical Studies: A Synthesis*; United States Environmental Protection Agency: Washington, DC, 1983.

32

Giant Reed (*Arundo donax*): Streams and Water Resources

Introduction 327
Biology and Ecology 327
Effects on Streams and Water Resources 329
    Flooding • Water Use • Wildfire • Biodiversity and Wildlife • Control
    Methods, Restoration, and Revegetation
Conclusions 333
References 333

Gretchen C.
Coffman

Introduction

Both natural and anthropogenic disturbances along rivers in Mediterranean-type climates are thought to promote the spread of invasive plant species in natural as well as altered riparian ecosystems.[1–5] A large bamboo-like member of the grass family (Poaceae), giant reed (*Arundo donax* L.), is one of the greatest invasive species threats to streams in arid and Mediterranean-type climate regions. *A. donax* forms extensive monotypic stands, successfully invading rivers in these regions. Infestations of *A. donax* are known to increase risks of flooding, compete with indigenous riparian species for scarce water resources, create unnatural fire hazards, and reduce the value of riparian habitat for most wildlife. Although many organizations are actively removing small areas of *A. donax* from streams in California, larger strategic watershed-scale removal is necessary to prevent further invasion and impacts to water resources.

Biology and Ecology

A. donax is one of the most successful weedy invaders in the highly dynamic, disturbance-defined rivers and riparian ecosystems of arid and Mediterranean-type climate regions[6,7] (Figure 1). *A. donax* is a tall, erect perennial grass species with culms 1–4 cm in diameter and 2–10 m in height, with two ranked leaves 2–6 cm wide at the base tapering to a point and up to 70 cm in length or more, and tough fibrous roots that penetrate more than 8 ft deep emanating from large horizontal creeping rhizomes. Although the seeds of this reed-like grass species are mostly sterile outside its native range,[8,9] *A. donax* colonizes readily via vegetative propagation; it is dispersed downstream when small pieces of its culm or rhizome break off during flooding and land on bare, moist substrates.[1,6,7,10,11] Fragments of the rhizome or culm as small as 2 cm² sprout under most soil types, depths, and soil moisture conditions.[7,10,11] Growing at an extremely rapid rate (up to 6.25 cm per day under optimal conditions), *A. donax* quickly establishes on exposed or sparsely vegetated soil and grows to more than 8 m in height after only a few months.[12,13]

FIGURE 1   *A. donax* infestation along the Santa Clara River in Ventura County, California.

Once established, *A. donax* clumps expand outward by clonal propagation (large rhizomes), crowding and displacing indigenous shrubs, herbs and grasses, and trees, especially under elevated soil moisture, nutrient, and light conditions.[8] In this manner, *A. donax* forms extensive stands, or monocultures, along floodplains and terraces of river and stream systems. *A. donax* is able to persist in ecosystems that experience seasonal drought by storing carbohydrate reserves in large rhizomes and sharing water resources through clonal integration.[14,15] When stressful conditions subside, shoots emerge.

   A. donax is thought to be indigenous to freshwaters of Eurasia,[16] extending from Southeast Asia to the Mediterranean Basin, although the precise extent of its native distribution is unclear. Herbivore diversity suggests that it is Mediterranean in origin, but its native range may extend much farther. Several thousand years ago, *A. donax* was thought to have been spread around the Mediterranean Basin for use in erosion control; production of reeds for wind instruments; and construction of roofs, ceilings, fences, and baskets.[17] It has been introduced to most tropical and warm, temperate regions worldwide, including North and South America, Southern Africa, and Australia, and thrives below 350 m in elevation.[6,18] In North America, *A. donax* has become especially devastating to riparian habitats in California's Mediterranean climate region, creating significant impacts to natural river functioning and sustainability.[19] In Southern California, *A. donax* was originally planted along irrigation canals for erosion control and used as building materials and windbreaks.[20] Carried by floodwaters, *A. donax* eventually made its way to adjacent streams and rivers, and by the 1820s, patches were commonly found along floodplains of many streams, including the Los Angeles River.[20] However, it appears that *A. donax* has only recently succeeded in invading natural riparian ecosystems in Southern California (i.e., replacing native riparian vegetation).

   Because of its clonal growth strategy, ability to colonize rapidly after disturbance, use of available resources, tolerance of stress, and high growth rate, *A. donax* is one of the most successful riparian weedy invaders in arid and Mediterranean-type climates.[12] Following an era of human alterations to river systems in Southern California, it was widely dispersed throughout riparian ecosystems in the floods of 1969, established in terrace and floodplain locations, and is now thriving in riparian ecosystems throughout this region.[21] Factors such as water, nutrients, light, and fire that are abundant in highly modified riparian ecosystems of arid and Mediterranean-climate regions increase the competitive ability of *A. donax*.[21] Although *A. donax* grows primarily in floodplains and terraces of low-gradient river and stream systems,[21,22] it may be found on beaches, around homes, in higher elevations, and next to hot springs where planted. *A. donax* forms huge infestations in open floodplains with high soil moisture and excess nutrients and in areas susceptible to wildfire.[21] *A. donax* successfully invades areas consisting of any soil type and once established can grow well in many soil moisture regimes.[7,21,23] Established stands recover readily after aboveground biomass is removed by wildfire, floods, frost, or mechanical means. In fact, the natural flood and wildfire regime characteristic of Mediterranean-type climates promotes growth and invasion of *A. donax*.[21]

From the time of early human settlement in arid and Mediterranean-type climates, humans have dammed, channelized, mined, diverted, and developed along rivers.[24,25] These alterations have magnified the susceptibility of streams in these regions to plant invasions by weedy species.[26,27] Human alterations associated with urbanization of watersheds in California in addition to the natural flood and wildfire processes have created ideal conditions for *A. donax* invasion. Increased water, nutrients, and light availability, as well as occurrence of fire in riparian ecosystems in Southern California, are thought to promote *A. donax* invasion.[6,21,28,29] Ever-expanding residential and agricultural development in coastal Southern California has led to increased water availability and nutrient loading of riparian ecosystems. The once vast low-lying areas of riparian forest continue to be removed to make room for agriculture, golf courses, and residential and commercial development. Consequently, open areas along floodplains formed by floods and clearing of terraces for development create an ideal location for *A. donax* to establish and invade riparian ecosystems. Furthermore, fire is more frequent in riparian corridors owing to anthropogenic ignition during the dry summer and fall months when *A. donax*-infested areas provide a large amount of dry fuel.[21,30] Because of higher post-fire growth rate and immediate growth response compared to native plant species, fire appears to contribute to the *A. donax* invasion process, especially in riparian terraces located next to fire-prone shrublands.[21]

Effects on Streams and Water Resources

Infestations of *A. donax* have created serious physical and biological problems along streams in arid and Mediterranean-type climates.[6,13,22,29] Where it grows extensively along floodplains, *A. donax* physically obstructs natural water flow, thereby increasing the risk of flooding. *A. donax* uses more water than native plant species, outcompetes native riparian species, thus reducing the value of riparian habitats for wildlife, and creating unnatural fire hazards.[21]

Flooding

Large infestations of *A. donax* within the active floodplains increase stream roughness during moderate to large flood events, forcing flood stages to higher levels and flooding adjacent property. During very high winter flows in California, *A. donax* is removed from the floodplain, floats downstream, and creates debris dams at bridges and culverts.[22] In addition, *A. donax* biomass collects in large piles along beaches after large flood events (Figure 2). Although originally planted to assist in erosion control, it now acts as an agent of erosion in California streams. The shallow rhizomes of the large, top-heavy plants growing along stream banks are undercut by high flows, causing bank erosion and instability. Economic losses due to effects of *A. donax* invasion include costs associated with repair of flood damage to property and bridges, beach cleanup, and bank stabilization repair.

Water Use

Water loss due to high evapotranspiration (ET) of *A. donax* infestations is of increasing concern in arid and Mediterranean-type climates where water resources are scarce and the plant continues to spread.

FIGURE 2   *A. donax* debris litters beaches in California after winter storms.

Using transpiration rates of rice (another C_3 species thought to have similar transpiration rates), estimates of *A. donax* water use suggest that it uses three times more water than native riparian species.[31] Other studies using a variety of methods indicate that ET of *A. donax* (1.2–7.5 m/yr) may be much higher than that of native riparian vegetation such as *Salix* spp., *Populus* spp. (1.0–3.3 m/yr), and mixed riparian communities of arid and Mediterranean-type climates (0.11–1.6 m/yr).[32–35] On the Santa Ana River alone, *A. donax* was estimated to transpire 37,500 acre-feet more water per year than native plants worth approximately $12 million at drinking water costs.[31] However, comprehensive studies are needed that compare water-use efficiency (WUE) of *A. donax* to various native species under different environmental conditions to determine exactly how much water is lost due to this invasive species. Excess water used in *A. donax* transpiration could be salvaged for groundwater recharge, drinking water supply, agricultural irrigation, and augmentation of in-stream flow for native vegetation and wildlife.[36] Furthermore, saturated soils that persist after flood events provide a lasting legacy to sustain plants over longer timescales. Using oxygen ($\delta^{18}O$) and hydrogen (δ^2H) stable isotopes analyses, Moore et al.[37] found that groundwater was not the dominant source for *A. donax* when soil moisture was sufficient for plant uptake following a flood.

Recent studies focused on water use, photosynthetic performance, and growth of *A. donax* under varying levels of soil moisture, carbon dioxide, and drought conditions. Watts and Moore[38] found that stands of *A. donax* used approximately 8.8 ± 0.9 mm of water per day during the peak of the growing season on the Lower Rio Grande in South Texas, the high end of the spectrum for plants. They also found that transpiration and leaf area index declined as water availability decreased, suggesting a sensitivity to drought and declining water tables. In another study, *A. donax* plants were subjected to a gradual drought stress for 3 weeks, after which they were returned to fully hydrated soil conditions for 1 week.[39] Overall, plant dry weight and key growth parameters were not significantly affected; however, photosynthetic performance was impaired. Physiological functions were restored after rewatering, reflecting the environmental plasticity of *A. donax*. Both atmospheric enrichment of carbon dioxide (CO_2) and increased drought are predicted to occur in regions where *A. donax* grows in arid and Mediterranean regions of the United States. Nackley et al.[40] found that increased CO_2 affected growth and water use of *A. donax*, reducing transpiration rates significantly. Reduced transpiration delayed drought responses, increased WUEs, and extended periods of assimilation but could not prevent desiccation and photosynthetic decline during extreme drought.

Wildfire

Wildfires ignited by humans at unnatural and dangerous times of the year burn rapidly through riparian corridors infested with *A. donax* and may help spread fires across watersheds and along riparian corridors.[21] Historically, dense biomass that accumulated over a period of 30–50 years or more in chaparral communities of California and shrublands in other Mediterranean-type climate regions caused fires to ignite.[41–44] Although fire was once a natural part of shrubland ecosystems in many Mediterranean-type regions, large riparian ecosystems provided natural firebreaks because native vegetation retained foliar water that resisted ignition.[13] Lightning was the primary cause of wildfires, especially during July and August under dry, low-humidity conditions and would commonly burn slowly for months.[42] Currently, however, most wildfires in these areas are anthropogenic in origin, occur much more frequently, and during strong Santa Ana wind conditions starting in September. For example, all 14 concurrent fires in October 2003 (739,597 acres burned) resulted from human activities.[41]

Invasion of annual grass species has been linked to altered fire regimes in rangelands, deserts, and wild lands of California and the Western United States.[41,47–49] However, *A. donax* may be an even bigger problem in riparian ecosystems of altered Southern California fire regimes because of its perennial growth form (the large volume of biomass produced) and rapid recovery after fire (Figure 3a and b).[21] Several accounts suggest that infestations of giant reed have increased fuel load as well as fire frequency and intensity along riparian corridors.[12,13,30,50] Thus, *A. donax* invasion appears to have created a positive feedback cycle or an invasive plant-fire regime[21] similar to those presented by other grass species.[45,46]

Biodiversity and Wildlife

A. donax has little habitat or food value for wildlife because of its dense growth structure and high content of noxious chemicals.[6,12,13] The federally endangered least Bell's vireo (*Vireo bellii pusillus*) and other riparian birds require structural diversity provided by riparian scrub and mature forest communities for breeding.[6,13,51] When naturally diverse riparian vegetation types are replaced by thick stands of *A. donax*, bird species abundance and other native wildlife have been found to decline.[6,13,45,46] Movement of medium to large mammals is most likely impaired by dense *A. donax* infestations. Herrera and Dudley[52] showed that arthropod abundance and diversity associated with native riparian vegetation was twice that associated with *A. donax* infestations. In addition, fish and aquatic invertebrates may be affected by increased stream temperature owing to lack of shading where *A. donax* has replaced mature riparian forests.[6]

Control Methods, Restoration, and Revegetation

Tens of millions of dollars ($7–$25 thousand per acre) have been spent in efforts to remove *A. donax* from riparian ecosystems in the Central Valley and coastal California. Although most attempts have been successful in removing small infestations on riparian terraces, *A. donax* continues to thrive in floodplains. An understanding of the ecological conditions that promote continued growth and invasion of *A. donax* is needed for its effective control. Management strategies for the control and removal of *A. donax* should be based on the location and size of the infestation. Priority should be given to removal of *A. donax* from riparian terrace habitats where infested areas are easily accessible and require less maintenance than along floodplains, especially infestations located adjacent to fire-prone shrubland plant communities.[53] Removal of large *A. donax* infestations on riparian terraces with high soil moisture and nutrient availability will be most difficult but is essential in removing the largest source of propagules to prevent future reinfestation. Active revegetation with native plants after *A. donax* removal

FIGURE 3   Three weeks (a) and 6 months (b) after Verdale–Simi Fire, *A. donax* invades riparian terrace along Santa Clara River in Ventura County.

is recommended to prevent reinfestation of *A. donax* or other weeds and restore functional riparian ecosystems. Unless *A. donax* is removed from floodplains on a watershed-scale working from the head-waters downstream, it is likely to recolonize removal areas after flood events. Watershed removal planning is underway in several large streams in Southern California to eradicate *A. donax* from floodplains, including the Santa Clara and Santa Ana Rivers.

Both mechanical and hand clearing techniques may be used to remove *A. donax*. Mechanical clearing methods include mulching or total excavation of all aboveground and belowground biomass. Hand clearing methods include either painting of *A. donax* stumps with herbicide after cutting or foliar applications of herbicide (glyphosate).[54] Research on biocontrol agents for *A. donax* is underway on the Santa Clara River, California and in Weslaco, Texas (Dudley, *Personal Communication*, 2006).

Conclusions

One of the biggest threats to streams and water resources in Mediterranean-type climates is invasion of *A. donax*. Forming large monocultures under ideal resource conditions along streams, *A. donax* increases flooding, promotes the spread of wildfire, outcompetes natives for water resources, and decreases wildlife value of riparian habitat. Although millions of dollars are spent every year to remove *A. donax* in California, many rivers and streams are still heavily infested. Effective removal and control strategies must be based on an ecological understanding of the invasion process and removal areas prioritized based on gaining the greatest ecological benefit for the least effort. Control of *A. donax* from watersheds in Mediterranean-type climates is an important initial step in restoration and long-term sustainability of riparian ecosystems.

References

1. Else, J.A.; Zedler, P. 1996 Annual Combined Meeting, Providence, RI, 10–14 August 1996, *Bull. Ecol. Soc. Am.* **1996**, *77*, 129.
2. Gregory, S.V.; Swanson, F.J.; McKee, W.A.; Cummins, K.W. An ecosystem perspective of riparian zones, *BioScience* **1991**, *41*, 540–551.
3. Pysek, P.; Prach, K. How important are rivers for supporting plant invasions? In *Ecology and Management of Invasive Riverside Plants*; Waal, L.C.d., Child, L.E., Wade, P.M., Brock, J.H., Eds.; John Wiley & Sons: New York, 1994; 19–26.
4. Drake, J.A.; Mooney, H.A.; diCastri, F.; Groves, R.; Kruger, F.J.; Rejmanek, M.; Williamson, M. *Biological Invasions: A Global Perspective*; John Wiley & Sons: Chichester, 1989.
5. Dudley, T. Exotic plant invasions in California riparian areas and wetlands. *Fremontia* **1998**, *26*, 24–29.
6. Bell, G.P. Ecology and management of Arundo donax, and approaches to riparian habitat restoration in Southern California. In *Plant Invasions: Studies from North America and Europe*; Brock, J.H., Wade, M., Pysek, P., Green, D., Eds.; Blackhuys Publishers: Leiden, 1997; 103–113.
7. Boose, A.B.; Holt, J.S. Environmental effects on asexual reproduction in *Arundo donax*. *Weed Res.* **1999**, *39*, 117–127.
8. Decruyenaere, J.G.; Holt, J.S. Seasonality of clonal propagation in giant reed. *Weed Sci.* **2001**, *49*, 760–767.
9. Khudamrongsawat, J.; Tayyar, R.; Holt, J.S. Genetic diversity of giant reed in the Santa Ana River, California. *Weed Sci.* **2004**, *32*, 395–405.
10. Else, J.A. San Diego State University: San Diego, 1996; 74 pp.
11. Wijte, A.H.B.M.; Mizutani, T.; Motamed, E.R.; Merryfield, M.L.; Miller, D.E.; Alexander, D.E. Temperature and endogenous factors cause seasonal patterns in rooting by stem fragments of the invasive giant reed, *Arundo donax* (Poaceae). *Int. J. Plant Sci.* **2005**, *166*, 507–517.
12. Rieger, J.P.; Kreager, D.A. *Proceedings of the California Riparian Systems Conference: Protection, Management, and Restoration for the 1990s*; Abell, D.L., Ed.; USDA Forest Service Gen. Tech. Rep. PSW-110; Berkeley, CA, 1989; 222–225.
13. Bell, G.P. *Arundo donax* Workshop Proceedings, Nov, 19, 1993; Jackson, N.E., Frandsen, P., Douthit, S., Eds.; Ontario, CA, 1994; 1–6.
14. Mann, J.J.; Barney, J.N.; Kyser, G.B.; DiTomaso, J.M. *Miscanthus x giganteus* and *Arundo donax* shoot and rhizome tolerance of extreme moisture stress. *GCB Bioenergy* **2013**, *5*, 693–700.
15. Kui, L.; Li, F.; Moore, G.; West, J. *Can the riparian invader, Arundo donax, benefit from clonal integration*. Weed Res. **2013**, *53*, 370–377.

16. Polunin, O.; Huxley, A. *Flowers of the Mediterranean;* Hogarth Press: London, 1987.
17. Perdue, R.E. *Arundo donax*—source of muscial reeds and industrial cellulose. *Econ. Botany* **1958**, *12*, 157–172.
18. Brossard, C.C.; Randall, J.M.; Hoshovsky, M.C. *Invasive Plants of California's Wildlands;* University of California Press: Berkeley, CA, 2000.
19. Rundel, P.W. Alien species in the flora and vegetation of the Santa Monica Mountains, CA: Patterns, processes, and management implications. In *2nd Interface* between *Ecology and Land Development in California*; Keeley, J.E., Baer-Keeley, M., Fotheringham, C.J., Eds.; U.S. Geological Survey Open-File Report 00-62, 2000; 145–152.
20. Robbins, W.W.; Bellue, M.K.; Ball, W.S. *Weeds of California;* California Department of Agriculture: Sacramento, 1951.
21. Coffman, G.C. *Factors Influencing Invasion of Giant Reed (Arundo donax) in Riparian Ecosystems of Mediterranean-type Climates*; University of California: Los Angeles.
22. DiTomaso, J.M. *Proceedings of the Arundo and Saltceder: The Deadly Duo Workshop*, Ontario, CA, June, 17, 1998; 1–5.
23. Singh, C.; Kumar, V.; Pacholi, R.K. Growth performance of *Arundo donax* (reed grass) under difficult site conditions of Doon Valley for erosion control. *Indian Forester* **1997**, *123*, 73–76.
24. Mount, J.F. *California Rivers and Streams: The Conflict Between Fluvial Process and Land Use*; University of California Press: Berkeley, 1995.
25. Palmer, T. *California's Threatened Environment: Restoring the Dream*; Island Press: Washington, DC, 1993.
26. Rundel, P.W. Landscape disturbance in Mediterranean-type ecosystems: An overview. In *Landscape Disturbance and Biodiversity in Mediterranean-Type Ecosystems. Ecological Studies 136*; Rundel, P.W., Montenegro, G., Jaksic, F.M., Eds.; Springer-Verlag: Berlin, 1998; 3–22.
27. Randall, J.M.; Rejmanek, M.; Hunter, J.C. Characteristics of the exotic flora of California. *Fremontia* **1998**, *26*, 3–12.
28. Wang, A. Environmental Science: Policy and Practice. In *Proceedings, Senior Research Seminar, Environmental Sciences Group Major-UGIS*; Dudley, T., Kennedy, K., Eds.; University of California: Berkeley, 1998; 720 pp.
29. Rundel, P.W. Invasive species. In *Southern California Environmental Report Card 2003*; Carlson, A.E., Winer, A.M., Eds.; UCLA Institute of the Environment: Los Angeles, CA, 2003; 4–11.
30. Scott, G. *Arundo donax* Workshop Proceedings; Jackson, N.E., Frandsen, P., Douthit, S., Eds.; Ontario, CA, 1994; 17–18.
31. Iverson, M.E. *Arundo donax* Workshop Proceedings, Ontario, CA. Nov, 1993; Jackson, N.E., Frandsen, P., Douthit, S., Eds.; 1994; 19–25.
32. Shafroth, P.B.; Cleverly, J.R.; Dudley, T.L.; Taylor, J.P.; Van Riper, C.; Weeks, E.P.; Stuart, J.N. Control of *Tamarisk* in the western United States: Implications for water salvage, wildlife use and riparian restoration. *Environ. Manage.* **2005**, *33* (3), 231–246.
33. Zimmerman, P. *Proceedings of the California Exotic Pest Plant Council*; California Exotic Pest Plant Council: Sacramento, CA, 1999.
34. Hendrickson, D.; McGaugh, S. http://desertfishes.org/cuatroc/organisms/non-native/arundo/Arundo.html, 2005; 17 pp.
35. Abichandani, S.L. *The Potential Impact of the Invasive Species Arundo donax on Water Resources along the Santa Clara River: Seasonal and Diurnal Transpiration.* University of California: Los Angeles. 2007.
36. The Nature Conservancy. Enhancing Water Supply through Invasive Plant Removal: A Literature Review of Evapotranspiration Studies on Arundo donax. file:///C:/Users/13106/Documents/Research/Arundo%20Research/TNC_Arundo_ET_Literature_Review_Feb2019.pdf, 2019; 6 pp.

37. Moore, G.; Li, F.; Kui, L.; West, J. Flood water legacy as a persistent source for riparian vegetation during prolonged drought: an isotopic study of *Arundo donax* on the Rio Grande. *Ecohydrology* **2016**, *9*, 909–917.

38. Watts, D.A.; Moore G.W. Water-use dynamics of an invasive reed, Arundo donax, from leaf to stand. *Wetlands*, **2011**, *31*, 725–734.

39. Pompeiano, A.; Remorini, D.; Vita, F., Guglielminetti, L.; Miele, S.; Morini, S. *Growth and physiological response of Arundo donax L. to controlled drought stress and recovery. Plant Biosyst.* **2017**, *151* (5), 906–914.

40. Nackley, L.L.; Vogt, K.A.; Kim, S.H. *Arundo donax* water use and photosynthetic response to drought and elevated CO_2. *Agric. Water Manage.* **2014**, *136*, 13–22.

41. Keeley, J.E.; Fotheringham, C.J. Lessons learned from the wildfires of October 2003. In *Fire, Chaparral, and Survival in Southern California*; Halsey, R.W., Ed.; Sunbelt Publications: San Diego, CA, 2005; 112–122.

42. Keeley, J.E.; Fotheringham, C.J. Historic fire regime in Southern California shrublands. *Conserv. Biol.* **2001**, *13*, 1536–1548.

43. Minnich, R.A. Fire mosaics in Southern California and northern Baja California. *Science* **1983**, *219*, 1287–1294.

44. Keeley, J.E.; Fotheringham, C.J.; Morais, M. Reexamining fire suppression impacts on brushland fire regimes. *Science* **1999**, *284*, 1829–1931.

45. D'Antonio, C.M.; Vitousek, P.M. Biological invasions by exotic grasses, the grass/fire cycle and global change. *Ann. Rev. Ecol. Syst.* **1992**, *23*, 63–87.

46. Brooks, M.L.; D'Antonio, C.M.; Richardson, D.M.; Grace, J.B.; Keeley, J.E.; DiTomaso, J.M.; Hobbs, R.J.; Pellant, M.; Pyke, D. Effects of invasive plants on fire regimes. *BioScience* **2004**, *34*, 677–688.

47. Keeley, J.E. Invasive plants and fire management in California Mediterranean-climate ecosystems. In *10th International Conference on Mediterranean Climate Ecosystems (MEDE- COS)*; Arianoutsou, M., Papanastasis, V.P., Eds.; Millpress: Rhodes, Greece, 2004; (full text on cd).

48. D'Antonio, C.M. Fire, plant invasions, and global changes. In *Invasive Species in a Changing World*; Mooney, H.A., Hobbs, R.J., Eds.; Island Press: Washington, DC, 2000; 65–93.

49. Brooks, M.L. Peak fire temperatures and effects on annual plants in the Mojave Desert. *Ecol. Appl.* **2002**, *12*, 1088–1102.

50. Dukes, J.S.; Mooney, H.A. Disruption of ecosystem processes in western North America by invasive species. *Rev. Chil. Hist. Nat.* **2004**, *77*, 411–437.

51. Zembal, R. Riparian habitat and breeding birds along the Santa Margarita and Santa Ana Rivers of southern California. In *Endangered Plant Communities of Southern California*; Schoenherr, A.A., Ed.; Southern California Botanists: Claremont, CA, 1990; 98–114.

52. Herrera, A.M.; Dudley, T.L. Reduction of riparian arthropod abundance and diversity as a consequence of giant reed (*Arundo donax*) invasion. *Biol. Inv.* **2003**, *5*, 167–177.

53. Coffman, G.C.; Ambrose, R.F.; Rundel, P.W. *10th International Conference on Mediterranean Climate Ecosystems*; Arianoutsou, M., Papanastasis, V.P., Eds.; Millpress: Rhodes, Greece, 2004; (full text on cd).

54. Cornwall, C.; Dale, R.; Newhonser, M. 1999. *Arundo donax: A Landowner Handbook*. Sonoma Ecology Center and California State University, Sacramento Media Services: Long Beach, CA.

33

Inland Seas and Lakes: Central Asia Case Study

Introduction 337
Degradation of Lakes in the World...338
The Aral Sea 340
Kara-Bogaz-Gol Bay 343
Conclusions..345
Acknowledgments 346
References.. 346

Andrey G.
Kostianoy

Introduction

Shallowing, desiccation, and degradation of freshwater and salt lakes and inland seas are among the major environmental problems at the beginning of the 21st century. All over the globe, water is being diverted for industrial, agricultural, and household uses, and many lakes are suffering from the resulting lack of inflow. There are clear indications that the growth of human population and the increasing use of natural resources, especially water, combined with climate changes, exert a considerable stress on closed or semi-enclosed seas and lakes. In many regions of the world, marine and lacustrine hydrosystems are or have been the objects of severe or fatal alterations ranging from changes in regional hydrological regimes and/or modifications of the quantity or quality of water resources, deterioration of geochemical balances (increased salinity, oxygen depletion, etc.), to mutations of ecosystems (eutrophication, decrease in biological diversity, etc.), to socioeconomic perturbations, which have been the consequences or may soon be in the near future.[1]

About 97.5% of earth's water is saltwater in oceans and seas.[1] Around 1% of that is brackish groundwater. The remaining 2.5% is freshwater. Nearly 70% of the world's freshwater is frozen in the Antarctic and Greenland ice sheets, glaciers, and permanent snow cover and ice. About 30% of all freshwater is groundwater. Lakes and rivers contain only about 0.25% of all freshwater. In 2002, the United Nations Environment Program (UNEP) announced that about 3 billion people would face severe water shortages by 2025, if the present consumption rates persist. Global and regional climate change, global and regional warming, and droughts amplify the freshwater problem. According to NASA Goddard Institute for Space Studies, the year 2018 was the fourth warmest since 1880. It was ranked behind those of 2016, 2017, and 2015. The past 5 years were the warmest years in the modern record.[2]

The goal of this chapter is to pay attention to the degradation of inland seas and lakes all over the world. They are very often called "dying" or "critical," meaning that the lake or inland sea is facing either a severe anthropogenic pressure in some form or a rapid change of its physical conditions owing to regional/global climate change.[3–6] In many cases, both mechanisms work together, and it is really very difficult to discriminate between natural and anthropogenic impact and to quantitatively assess them.

In this chapter, we address the most known examples of the degradation of the lakes all over the world, and then we focus on two Central Asia case studies—the Aral Sea and the Kara-Bogaz-Gol Bay. Conclusions summarize this chapter and briefly touch on future prospects related to the state of the Central Asia water bodies.

Degradation of Lakes in the World

A significant fall in the water level and/or increase in salinity of many large saline lakes and inland seas have taken place worldwide during the past century.[3-6] Examples include the following: in North America—Great Salt Lake (Utah), Walker and Pyramid lakes (Nevada), Lake Mead (Nevada, Arizona), Owens and Mono lakes and the Salton Sea (California), and Deadmoose Lake (Canada); in South America—Llancanelo (Argentina); in the Middle East—the Dead Sea (Israel/Jordan/West Bank) and Lake Van (Turkey); in Central Asia—the Aral and Caspian seas and Kara-Bogaz-Gol Bay (Turkmenistan), Lake Balkhash (Kazakhstan), and Lake Issyk-Kul (Kyrgyzstan); in China—Lop Nor and Qinghai Hu lakes; in Africa—Lake Chad and Lake Elmenteita (Kenya); in Japan—Lake Biwa; and in Australia—Keilambete, Eyre, Corangamite, Gnotuk, and Bullenmerri lakes.[5]

The most striking examples include (1) the Lop Nor Lake in China, which completely dried up by 1972; (2) the Aral Sea, which is following such a fate; (3) the Dead Sea, whose level dropped 23 m from 1970 to 2006; and (4) Lake Chad, which at the same time shrunk to about 5% of its size in 1963.

Northern and northwestern China has been experiencing a desiccation process since the 1950s owing to a decrease in precipitation by at least 30%.[3] As a result, Lake Lop Nor vanished completely in 1972 and became a nuclear testing site; the depth of Lake Ohlin at the head of the Yellow River dropped by more than 2 cm annually; the Qinghai Hu Lake water level decreased by an average of 10 cm/yr between 1959 and 1982 owing to a decrease in rainfall, groundwater supply, and the unsustainable use of the water for irrigation (the total drop since 1908 reached 11.7 m; the lake salinity increased from 5.6 to 12 g/L since 1950). Because of rapid population growth, the surface of Ebi Nor, the largest salt lake in northwest China's Xinjiang Uygur Autonomous Region, has shrunk to 530 km^2 in the past five decades (its surface was 1200 km^2 in the 1950s). As a result, many plant and animal species living in and around the lake have been extirpated.

The Dead Sea is a 378 m deep salty terminal lake at the border between Israel, Jordan, and the West Bank. The Dead Sea waters are among the saltiest (340 g/L) and densest (1.237 g/cm^3) lake waters/sea-waters in the world.[7] Its level is determined by the balance between river runoff, precipitation, and evaporation. Since the 1960s, the hydrological regime of the Dead Sea has been strongly influenced by use of its watershed by Israel, Syria, and Jordan. Moreover, Israel and Jordan use the seawaters for mineral production at salt evaporation ponds located south of the sea, responsible for 25%–30% of the total Dead Sea evaporation. As a result, evaporation exceeded freshwater inflow to the sea. Since 1977, the length of the sea decreased from 80 to 50 km and the level has dropped by 23 m (1970–2006) at rates of 0.6–1.0 m/yr.[7] From 2000 to 2018, the sea level has dropped with a rate of 1.1 m/yr. The Dead Sea level drop has been followed by a groundwater level drop, causing brines in the underground layers near the shoreline to be flushed out by freshwater. This is the cause of the recent appearance of large sinkholes along the western shore. Plans have been developed to construct a pipeline to bring water from the Red Sea into the Dead Sea to stabilize or raise the water level. The pipeline, referred to as the "Peace Conduit" project, would bring about 450×10^6 m^3 of Red Sea water into the Dead Sea annually.[4] In May 2009 at the World Economic Forum, Jordan announced its plans to construct the "Jordan National Red Sea Development Project" (JRSP). This is a plan to convey seawater from the Red Sea near the Gulf of Aqaba to the Dead Sea. Water would be desalinated along the route to provide freshwater to Jordan, with the brine discharge sent to the Dead Sea for replenishment. In 2016, Ministry of Water and Irrigation of Jordan announced that the project will be completed by 2021.

Global warming (during the 20th century, the mean land surface temperature in Africa has increased by 0.9°C) and withdrawal and/or diversion of water from inflowing rivers are the reasons for the water level drop in several African lakes.[3-5] Lake Chad, once one of the largest in Africa, has been a source of freshwater for irrigation projects in Chad, Niger, Nigeria, and Cameroon. Since 1963, the lake has shrunk to nearly 1/20th of its original size (from 25,000 to 1,350 km² in 2000), owing to both climatic changes (including a 50% decline in rainfall) and high agricultural water (between 1983 and 1994, irrigation water use increased fourfold). The UNEP says that about half of the lake's decrease is attributable to human water use such as inefficient damming and irrigation methods. The other half of the shrinkage is due to regional climate change. There are considerations that the lake will shrink further and perhaps even disappear in the course of the 21st century.

Lake Victoria, shared by Kenya, Uganda, and Tanzania, with a surface area of 68,000 km², is the world's second largest and the largest African body of freshwater in terms of surface area.[3] It is of great socioeconomic importance for 20 million people living in the basin. Over the past few decades, Lake Victoria has been subjected to drastic ecological and water quality changes owing to pollution (including sewage discharges and agricultural runoff), sediments resulting from soil erosion in the catchment area because of deforestation and overgrazing and industrial pollution from many local industries. All these factors have resulted in the eutrophication of the lake because of the increase in nutrient supply to the lake, algal blooms, and massive fish deaths. Moreover, the introduction of exotic fish species such as the Nile perch has altered the freshwater ecosystem of the lake and driven several hundred species of native cichlids to extinction or near extinction. The water hyacinth *Eichhornia crassipes*, a native of the tropical Americas, was introduced by colonists to Rwanda and then advanced by natural means to Lake Victoria where it was first sighted in 1988. Without any natural enemies, it became an ecological plague. By forming thick mats of vegetation, it causes difficulties to transportation, fishing, hydroelectric power generation, and drinking water supply.

A number of lakes in North America have also experienced desiccation in the past century.[3,5] For instance, the Pyramid Lake in Nevada experienced a 21 m level drop since 1910, accompanied by a salinity increase from 3.8 to 5.5 g/L between 1933 and 1980.

In California, Owens Lake (once 280 km² large and 7–15 m deep) dried completely by 1926 due to irrigation diversions and withdrawals. The dry bed of Owens Lake has produced enormous amounts of windblown dust since the desiccation of the lake. The lake bed is probably the largest single source of PM10 dust (aerosol particles smaller than 10 microns in aerodynamic diameter) in the United States (by one estimate, 900,000–8,000,000 metric tons per year).[8] Unusually fine-grained, alkaline dust storms from the dry lake bed are a significant health hazard to residents of Owens Valley and nearby areas and impact air quality in a large region (40 km) around the lake bed.

In California, the Mono Lake's level dropped 17 m from 1919 to 1982 owing to diversions of their tributary system to the Los Angeles Aqueduct system. By 1982, the lake was reduced to 153 km², having lost 31% of its 1941 surface area. Before 1941, average salinity was approximately 50 g/L, but in January 1982, when the lake reached its lowest level (1942 m), the salinity had nearly doubled to 99 g/L. In 2002, it was measured at 78 g/L and is expected to stabilize at an average of 69 g/L as the lake replenishes (in August 2019, its level was at 1946 m).

Even the world's largest freshwater system, the North American Great Lakes, may be shrinking. They contain about 95% of the fresh surface water supply for the United States and 20% for the world. In 2002, the aggregate level of the five Great Lakes was at the lowest in more than 30 years. Since 1997, Lakes Huron, Michigan, and Erie have dropped over 1 m, and an additional drop of 0.5–1 m has been predicted. These changes are attributed to decrease in precipitation, enhanced evapotranspiration, and reduced ice cover because of higher temperatures and irreversible loss of water for urban and industrial uses (e.g., Chicago sends its used water taken from the lakes to the Mississippi basin after treatment, instead of back to the Great Lakes).

From the late 1990s until 2005, a multiyear drought in the Upper Colorado River Basin caused water level drop in Lake Powell. As people's demands on the river overcame the rate of recharge, the volume of the lake dropped to only 33% of capacity in 2005.[9]

In August 2010, Lake Mead reached its lowest level since 1956. The largest reservoir in the United States was straining from 12 years of persistent drought and increasing human demand after decades of population growth in the American Southwest. In August 2010, Lake Mead held just 37% of the lake's capacity (35 km³). The lake level dropped 39 m since August 1985. Refilling the lake depends almost entirely on snowmelt from the Rocky Mountains and how much of that water is released from dams and reservoirs in the upper basin of the Colorado River (Wyoming, Colorado, Utah, and New Mexico).[10]

The Aral Sea

Before 1960, the Aral Sea (Figure 1) was a water-abundant sea lake in Central Asia that was the fourth largest in the world list of lakes after the Caspian Sea (USSR, Iran), Great Lakes (United States, Canada), and Victoria Lake (Africa).[11–18] The Aral Sea was about the size of the Netherlands and Belgium taken together. It is located in the largest deserts—Karakum and Kyzylkum. Navigation and the fishery (yearly catches of 44,000 tons) were developed here. The deltas of the Amu Darya, the major river of Central Asia, and Syr Darya bringing their waters into the Aral Sea were famous for their biodiversity, fishery, muskrat rearing, and reed production. The local population found occupation in the spheres related to the water infrastructure. This was a natural and stable period of the Aral Sea evolution that since 1960 was followed by the anthropogenic one, which continues until the present day.

The Aral Sea existed thanks to runoff from the two main rivers of Central Asia—Amudarya and Syrdarya. However, since 1960, riverine water resources have been irrationally used for increasing irrigation of agricultural lands and creation of artificial water reservoirs. As a result, the Aral Sea water balance was disrupted and irreversible alterations in the sea regime appeared that later escalated into one of the "largest ecological disasters of the 20th century." During the last 50 years, we have observed a progressive degradation of the Aral Sea and its environment. During this time period, the sea shrunk in size from 66,100 km² (in 1961) to 10,400 km² (in 2008), its volume decreased from 1066 to 110 km³, the sea level dropped by 24 m (maximum depth of 69 m was observed in 1961), and its salinity (mineralization) rose from 10 to 116 g/L in the Western Large Aral Sea and about 210 g/L in the Eastern Large Aral

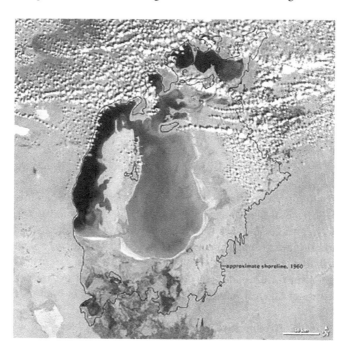

FIGURE 1 Satellite image (MODIS-Terra, 250 m resolution) of the Aral Sea on August 19, 2000, with a black line indicating the 1960 shoreline (http://earthobservatory.nasa.gov/Features/WorldOfChange/aral_sea.php, accessed on October 15, 2010).

Sea (Figure 1).[15,18,19] Today, the Aral Sea represents two isolated water bodies—a narrow and relatively deep the Western Large Aral Sea and the Small Aral Sea located in the north-eastern part of the satellite image presented in Figure 2. The former bottom of the Large Aral Sea is now a desert which is often called the Aralkum. Recent measurements (May 2019) of physical and chemical characteristics in the Western Large Aral Sea, performed by P.P. Shirshov Institute of Oceanology, Russian Academy of Sciences yearly since 2002, have recorded that the sea level has dropped by 30 m since 1961, and salinity is of 121 g/L (Personal communication by P.O. Zavialov). The ongoing desiccation, shallowing, and salinization of the Aral Sea have resulted in profound changes in its shape and its physical, chemical, and biological regime. The Aral Sea lost its economic importance, and the aftermath of its degradation represents a serious threat to the local population due to a lack of freshwater, water quality loss, salinization of soils, dust and salt storms, climate deterioration, various diseases, etc.

Until the 1960s, the deltas of the Amudarya and Syrdarya rivers, which received a lot of water and sediments, were among the most dynamic in the world, were notable for their high biodiversity and biological productivity, and resisted well against the deserts of Central Asia. As a result of dramatic man-induced reduction of river flow and a drop of the Aral Sea level, the deltas of the Amudarya and Syrdarya rivers, their hydrographic network, and landscapes have undergone severe degradation.

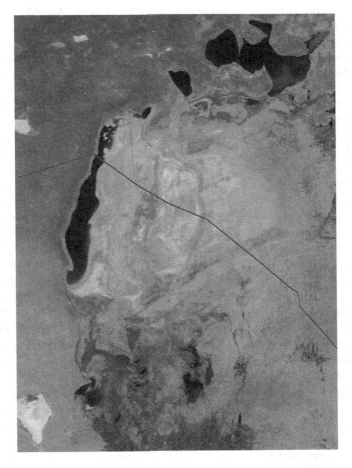

FIGURE 2 Satellite image (MODIS-Terra, 250 m resolution) of the Aral Sea on August 9, 2019 (https://wvs.earthdata.nasa.gov/?LAYERS=MODIS_Terra_CorrectedReflectance_TrueColor,Reference_Features&CRS=EPSG:4326&TIME=2019-08-09&COORDINATES=42.967532,57.612308,46.834720,61.596683&FORMAT=image/jpeg&AUTOSCALE=TRUE&RESOLUTION=250m, accessed on August 11, 2019).

The ongoing desiccation of the Aral Sea has resulted in significant changes in the distributions of dissolved gases in the residual water body. In particular, the once fully oxygenized sea developed anoxic conditions and intermittent hydrogen sulfide contamination in the bottom layers. The sulfide content depends on the density stratification and is mainly controlled by the physical regime of the sea. However, H_2S is a variable rather than a permanent feature of the present Aral Sea.[15,17]

In the first half of the 20th century in the Aral Sea, 33 species of fishes, 61 species of bottom invertebrates, 49 species of zooplankton, 306 species of phytoplankton, and 37 species of macrophytes were found. The Aral Sea accounted for 7% of the total internal waters fishery of the USSR. The main trade species were roach, sazan, and bream. From 1960 until now, the fauna of the Aral Sea evolved from mainly freshwater to hyperhalinic with the increase in water salinity. During the first decade of salinization, more than 70% of the species of fish and invertebrates vanished. By 2004, when mineralization had reached 90 ppt in the Large Aral Sea, fish fauna of the Large Aral had progressively disappeared. In 2008, zooplankton was represented by only one hypersaline species, *Artemia parthenogenetica*, introduced into the Aral Sea in 1996.[17]

It is generally accepted that the main reason for desiccation of the Aral Sea has been irrational use of Amudarya and Syrdarya waters for development of irrigation of agricultural lands and the filling of artificial water reservoirs, but regional climate change (rise of air temperature and decrease of atmospheric precipitation) also plays an important role in this process. According to estimates of the Intergovernmental Panel on Climate Change (IPCC), the trend of the mean annual air temperature in the Aral region in 1901–2005 was 1.1°C–1.7°C per 100 years, and only in 1979–2005, it was 0.3°C–0.7°C per 100 years. The results of these estimates presented in the *Fourth Assessment Report* of the IPCC have indicated that by the late 21st century, the air temperature in the Aral region depending on the emission scenarios may become 2°C–7°C higher compared to 1981–2000.[20] Our estimates of the amount of water precipitated from the atmosphere over the catchment areas of the Amudarya and Syrdarya rivers for the period 1979–2001 revealed a marked decreasing trend for the Amudarya catchment area from 7–8 to 4–5 km³/mo on average.[21] Thus, both the effects of regional climate change significantly influenced the water balance of the Aral Sea in the past 40 years leading to its supplementary desiccation.

By the mid-1980s, the Aral crisis had been acknowledged by the whole world and became one of the most significant environment protection issues.[11–16] The Aral Sea crisis redoubled when, after disintegration of the USSR in 1990–1991, the investigations and monitoring of the sea practically ceased. Almost all hydrometeorological stations have been closed, sea expeditions have been stopped, and assessments of ongoing quick changes in the Aral Sea environment have ceased. At the same time, the interest in the Aral Sea problem has risen sharply on the international level. Many United Nations (UN) organizations (UN University, UNDP, UNESCO, UNEP, UNIDO, FAO, WMO, UN High Commissioner for Refugees, and the International Labor Organization), financial organizations (World Bank, Asian Development Bank, European Bank for Reconstruction and Development, International Monetary Fund, and Global Environment Facility), European Union Programs (TACIS, INTAS, INCO-Copernicus, OSCE, and TEMPUS), international nongovernmental organizations ("Doctors Without Borders"), regional organizations (International Fund for Saving of the Aral Sea, Interstate Coordination Water Commission, Commission on Sustainable Development, and Central Asian Economic Community), and bilateral organizations [U.S. Agency for International Development, Soros Foundation (United States), Konrad Adenauer Foundation, Friedrich Ebert Foundation, Germany Agency for Technical Cooperation (Germany), NOVIB (the Netherlands), NATO Program "Science for Peace," JAIKA, Global Infrastructure Fund Research Foundation (Japan), and others] were involved in the implementation of several hundreds of projects.[16,17]

Many of the abovementioned projects produced a lot of interesting scientific results, which traced in detail the development of the environmental crisis, but unfortunately did not result in real measures promoting salvation of the Aral Sea. Scientists, researchers, designers, and politicians could not come to a consensus on a strategy for the preservation and restoration of the Aral Sea. By the end of 2019, we can state that the main progress made towards saving of the Aral Sea occurred only in Kazakhstan with the construction of the Kokaral dam between the Small Aral Sea and Eastern Large Aral Sea in August 2005. Thus, the Small Aral Sea is now slowly reviving, while the Large Aral Sea continues to disappear progressively (Figure 2).

Kara-Bogaz-Gol Bay

Kara-Bogaz-Gol is a bay that incises deeply into the mainland and forms a highly saline bay on the eastern side of the Caspian Sea (Figure 3).[22–24] Kara-Bogaz-Gol is a Turkmen name that means "Black Trap Bay." Ancient legends say that in the bay, there was a very deep hollow that trapped the Caspian waters and even the ships that took a risk to call at this vast lagoon. This is the Caspian's largest salt-generating lagoon separated from the sea with two sand spits extending meridionally for more than 90 km (Figure 4). These sand spits form a strait 7–9 km long, 120–800 m wide, and 3–6 m deep. Due to the difference of water levels in the Caspian Sea and the bay, the waters from the sea rush at a speed of 50–100 cm/s along the strait to the bay where they evaporate completely (at a rate of 800–1000 mm/yr, on the average). Therefore, with the average annual atmospheric precipitations in this region being no more than 110 mm, Kara-Bogaz-Gol represents an enormous natural evaporation basin of seawater. Due to high evaporation, the bay is filled with brine, the salinity of which reaches 270‰–300‰ and even more. This brine is a concentrated solution of salts such as chlorides of sodium, magnesium, and potassium; magnesium sulfate; and small quantities of rare earth elements. The Kara-Bogaz-Gol Bay is the largest salt deposit where up to 20 salt minerals are found. The salt deposits of the bay accumulated a dozen billion tons of various salts that make the most valuable raw material for development of the chemical industry, agriculture, nonferrous metallurgy, medicine, and other branches of the economics.[23,24]

In the first decades of the 20th century, the level difference between the sea and the bay was about 0.5 m, and they were hydraulically linked. The level in Kara-Bogaz-Gol was close to −26.5 m relative to the world ocean. The area of the bay (in the beginning of the 20th century and after 1996) was greater than 18,300 km², the volume of water was 130 km³, and the prevailing depths were 8–10 m (maximum depth was 13 m).

Starting from the 1930s, the nature of Kara-Bogaz-Gol Bay has experienced significant changes related to the sharp fall in the level of the Caspian Sea.[22] In 1939, the supply of seawater to the bay decreased to 6 km³/yr (from 20 to 25 km³/yr in 1929–1930) and the water balance was distorted. The maximum depth of the bay decreased from 12 m in the 1920s–1930s to 4.5 m in the early 1950s. The decrease in the water volume of Kara-Bogaz-Gol Bay, accompanied by the long-term sea-level fall and related growth in the salt concentration in the brine and changes in its chemical composition, negatively affected the conditions of sulfate mining.

In the 1960s, the difference between the levels of the sea and the bay became larger and the world's only sea waterfall about 3.7 m high (1970) was formed in the strait mouth. The maximum flow velocity in this waterfall was more than 2.5 m/s. In the 1960s to early 1970s with the water flow being 9–10 km³, the bay area varied around 10,000–11,500 km², while the volume of water was 25–29 km³ and the maximum depth of the bay decreased to 3.5 m. In the late 1970s, the Caspian water flux into the bay reduced to 5–7 km³/yr, the water level in the lagoon dropped to −32 m, the area decreased to 10,000 km², and the volume of water decreased to 20–22 km³, while the brine concentration increased to 270‰–300‰.[24] In order to reduce the Caspian Sea water use and retard the fall of the water level in the Caspian Sea that in 1977 reached its absolute minimum for the last 400 years (−29 m), in March 1980, the strait was dammed, thus stopping the inflow of seawater into the bay.

Construction of a dam led in late 1983 to complete desiccation of the bay and considerable reduction of chemical products manufactured here (sodium sulfate, etc.). By the end of 1983, the area of the bay was only 1000 km², its volume was 0.2 km³, and its depth was 0.1–0.3 m. Brine salinity reached 330‰–380‰. By late August 1984, no common brine surface remained. In September 1984, tubes were installed in the dam body to direct the seawater into the bay in the amount of 1.6–1.8 km³/yr. Such restricted supply of seawater lasted for nearly a decade and did not result in noticeable improvement of the hydrological

FIGURE 3 Satellite image of the Caspian Sea and Kara-Bogaz-Gol Bay obtained from MODIS-Terra on July 30, 2019 (MODIS Land Rapid Response Team, NASA GSFC, https://wvs.earthdata.nasa.gov/api/v1/sna pshot?REQUEST=GetSnapshot&LAYERS=MODIS_Terra_CorrectedReflectance_TrueColor,Reference_ Features&CRS=EPSG:4326&TIME=2019-07-30&BBOX=36.027841,46.596679,47.629403,55.327149&FORMAT=im age/jpeg&WIDTH=1987&HEIGHT=2640&AUTOSCALE=TRUE&ts=1565516768981, accessed on August 11, 2019).

and hydrochemical situation in the bay. In June 1992, the Turkmenistan president ordered to dismantle the dam, and by 1996, the bay was filled with Caspian water until its equilibrium state (Figure 4). At the high water level in the Caspian Sea in 1993–1995, the annual water flow reached 37–52 km³/yr and the difference between levels in the sea and bay decreased from 6.9 m in June 1992 to 0.2–0.6 m in 1996.[24] The situation became stable, the behavior of the bay returned back to its natural conditions related to the Caspian Sea, and the salt production increased.

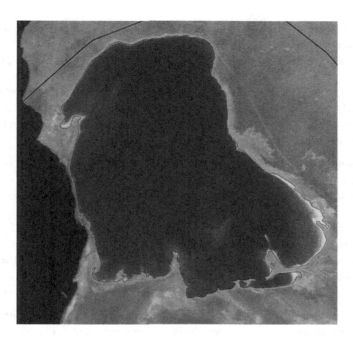

FIGURE 4   Satellite image of Kara-Bogaz-Gol Bay obtained from MODIS-Terra on August 5, 2019 (MODIS Rapid Response System, NASA GSFC, https://wvs.earthdata.nasa.gov/api/v1/snapshot?REQUEST=GetSnapshot&LAY ERS=MODIS_Terra_CorrectedReflectance_TrueColor,Reference_Features&CRS=EPSG:4326&TIME=2019-08-05&BBOX=40.465395,52.395631,42.230532,54.867555&FORMAT=image/jpeg&WIDTH=1125&HEIGHT=803&AU TOSCALE=TRUE&ts=1565517161424, accessed on August 11, 2019).

Conclusions

The history of Kara-Bogaz-Gol Bay has been instructive and showed that without multidisciplinary and comprehensive research of the ecological, economic, and social aftereffects, one should not alter natural equilibriums reached over thousands of years, and human intervention into the complicated environmental processes may lead to catastrophic results.

Within a life span of only two generations, the Aral Sea as a single natural water body practically ceased to exist, and the main reason for this should be sought in man's economic activities. Since 1960, riverine water resources have been irrationally used for increasing irrigation of agricultural lands and creation of artificial water reservoirs. From the other side, regional climate change significantly influenced the water balance of the Aral Sea in the past 30 years leading to its supplementary desiccation. Thus, both effects have led the Aral Sea to one of the "largest ecological disasters of the 20th century."

Desiccation of the Aral Sea in the so-called anthropogenic period (since 1961) led not only to considerable changes in its morphometric, physical, chemical, biological, and other parameters but also to the disappearance of infrastructure in the coastal zone, including meteorological and sea-level gauge stations, ships, ports, villages, and roads. The current lack of reliable in situ measurements and time series for sea surface temperature, sea level, ice cover, and morphometric characteristics since the mid-1980s may be successfully replaced by using corresponding satellite information available directly from satellites or through the world databases.[19,21,25,26] Images from the Advanced Very High-Resolution Radiometers (AVHRR onboard National Oceanic and Atmospheric Administration (NOAA) satellites) and Moderate Resolution Imaging Spectroradiometers (MODIS onboard Terra and Aqua satellites) provide a possibility to follow the changes in the sea's coastline and observe interesting phenomena in the water, in the atmosphere, and on the dried parts of the Aral Sea.[19]

Regular measurements of the sea/lake level are practically absent in many regions such as Central Asia and Africa. Monitoring of the evolution of these and other water bodies worldwide may be accomplished by satellite altimetry from the TOPEX/Poseidon, Jason-1/-2/-3, ERS-1, ERS-2, GFO, and Envisat satellites.[26–31]

In 1986, the International Lake Environment Committee and UNEP started a project called "Survey of the State of World Lakes," aimed at collecting and analyzing environmental data on >500 lakes from 73 countries.[32,33] The results have indicated that environmental problems, common for the lakes in all continents, may be classified in the following categories:[32]

1. Lake shallowing and salinization owing to overuse of water from lakes and/or tributary rivers, resulting in a degradation of water quality and lake ecosystems.
2. Accelerated sedimentation in lakes and reservoirs resulting from anthropogenic or natural soil erosion.
3. Lake water acidification resulting from acid precipitation, which may result in the extinction of ecosystems and contamination of water with toxic agricultural and/or industrial chemicals.
4. Eutrophication owing to inflow of nitrogen and phosphorus compounds or other nutrients in the discharged water or wastewater inflows, strongly affecting biodiversity.
5. In extreme cases, a complete collapse of aquatic ecosystems and desiccation of lakes.

Increased freshwater consumption for agricultural, industrial, and urban uses and uncontrolled irrigation pose a serious threat to inland seas, lakes, rivers, and wetlands as well. In addition, shallowing and desiccation of inland seas often lead to nonhydrologic consequences, such as air pollution from dust/salt storms caused by wind erosion of exposed lake beds.

Global warming is expected to lead to large air and water temperature anomalies and make droughts and desertification more severe in the future. Climate change is also the reason for the water level drop. Shallowing or desiccation of lakes has caused local climate changes in numerous drainage basins, and if the present climatic trends persist, global climate changes could trigger almost untenable environmental effects for people and aquatic ecosystems. One of the principal tasks for future research is the delimitation of the anthropogenic and natural climate change impacts on lakes and inland seas. Degradation of many inland water bodies is a continuing global environmental problem, and its social and economic implications have attracted the growing attention of many individuals, organizations, and governments, resulting in many national and international research projects.

Acknowledgments

This research was partially supported by the Russian Science Foundation Project N 19-77-20060 "Assessing ecological variability of the Caspian Sea in the current century using satellite remote sensing data" (2019–2022).

References

1. Water, our thirsty world. *Natl. Geogr.* **2010**, *217* (special issue), N4.
2. Global Climate Change. 2018 fourth warmest year in continued warming trend, according to NASA, NOAA. Available at https://climate.nasa.gov/news/2841/2018-fourth-warmest-year-in-continued-warming-trend-according-to-nasa-noaa/ (accessed 11 August 2019).
3. Kostianoy, A.G.; Zavialov, P.O.; Lebedev, S.A. What do we know about dead, dying and endangered lakes and seas? In Dying and Dead Seas. Climatic versus Anthropic Causes; Nihoul, J.C.J., Zavialov, P.O., Micklin, Ph.P., Eds.; NATO ARW/ASI Series, Kluwer Acad. Publ.: Dordrecht, 2004; 1–48.
4. Nihoul, J.C.J.; Zavialov, P.O.; Micklin, Ph.P., Eds. *Dying and Dead Seas. Climatic versus Anthropic Causes*; NATO ARW/ASI Series, Kluwer Acad. Publ.: Dordrecht, 2004.

5. Kostianoy, A.G. Dead and dying seas. In Encyclopedia of Water Science; Taylor & Francis: London, 2007, doi:10.1081/E-EWS-120042068.

6. Lerman, A., Ed. Saline lakes and global change. *Aquat. Geochem.* **2009**, *15* (special issue), 1–5.

7. Gertman, I.; Hecht, A. The Dead Sea hydrography from 1992 to 2000. *J. Mar. Syst.* **2002**, *35*, 169–181.

8. Gill, T.E.; Gillette, D.A. Owens Lake: A natural laboratory for aridification, playa desiccation and desert dust. *Geol. Soc. Am. Abstr. Programs* **1991**, *23* (5), 462.

9. NASA Earth Observatory. April 2010 Water Level in Lake Powell. Available at http://earthobservatory.nasa.gov/IOTD/view.php?id=43933 (accessed October 2010).

10. NASA Earth Observatory. Water Level Changes in Lake Mead. Available at http://earthobservatory.nasa.gov/IOTD/view.php?id=45945 (accessed October 2010).

11. Letolle, R.; Mainguet, M. *Aral*; Springer Verlag: Paris, 1993.

12. Glantz, M.H., Ed. *Creeping Environmental Problems and Sustainable Development in the Aral Sea Basin*; Cambridge University Press: Cambridge, 1995.

13. Nihoul, J.C.J.; Kosarev, A.N.; Kostianoy, A.G.; Zonn, I.S., Eds. *The Aral Sea: Selected Bibliography*; Noosphere: Moscow, 2002.

14. Kostianoy, A.G.; Wiseman, W., Eds. The dying Aral Sea. *J. Mar. Syst.* **2004**, *47* (1–4) (special issue), 1–152.

15. Zavialov, P.O. *Physical Oceanography of the Dying Aral Sea*; Springer and Praxis Publishing: Chichester, 2005.

16. Zonn, I.S.; Glantz, M.; Kostianoy, A.G.; Kosarev, A.N. *The Aral Sea Encyclopedia*; Springer-Verlag: New York, 2009.

17. Kostianoy, A.G.; Kosarev, A.N., Eds. *The Aral Sea Environment*; The Handbook of Environmental Chemistry; Springer-Verlag: New York, 2010.

18. Kosarev, A.N.; Kostianoy, A.G. The Aral Sea under natural conditions (till 1960). In The Aral Sea Environment; Kostianoy, A.G., Kosarev, A.N., Eds.; The Handbook of Environmental Chemistry; Springer-Verlag: New York, 2010; Vol. 7, 45–63.

19. Ginzburg, A.I.; Kostianoy, A.G.; Sheremet, N.A.; Kravtsova, V.I. Satellite monitoring of the Aral Sea region. In The Aral Sea Environment; Kostianoy, A.G., Kosarev, A.N., Eds.; The Handbook of Environmental Chemistry; Springer-Verlag: New York, 2010; Vol. 7, 147–179.

20. IPCC. Climate change 2007: The physical science basis. In Contribution of Working Group I to the Fourth Assessment Report of the Intergovernmental Panel on Climate Change; Solomon, S., Qin, D., Manning, M., Chen, Z., Marquis, M., Averyt, K.B., Tignor, M., Miller, H.L., Eds.; Cambridge University Press: Cambridge, 2007.

21. Nezlin, N.P.; Kostianoy, A.G.; Lebedev, S.A. Interannual variability of the discharge of Amu Darya and Syr Darya estimated from global atmospheric precipitation. *J. Mar. Syst.* 2004, 47 (1–4), 67–75.

22. Kostianoy, A.G.; Kosarev, A.N., Eds. *The Caspian Sea Environment*; The Handbook of Environmental Chemistry; Springer-Verlag: New York, 2005.

23. Kosarev, A.N.; Kostianoy, A.G. Kara-Bogaz-Gol Bay. In The Caspian Sea Environment; Kostianoy, A.G., Kosarev, A.N., Eds.; Springer-Verlag: New York, 2005; 211–221.

24. Kosarev, A.N.; Kostianoy, A.G.; Zonn, I.S. Kara-Bogaz-Gol Bay: Physical and chemical evolution. *Aquat. Geochem.* 2009, 15 (1–2), Special Issue: Saline Lakes and Global Change, 223–236, doi:10.1007/s10498-008-9054-z.

25. Nezlin, N.P.; Kostianoy, A.G.; Li, B.-L. Interannual variability and interaction of remote-sensed vegetation and atmospheric precipitation in the Aral Sea Region. *J. Arid Environ.* **2005**, *62* (4), 677–700.

26. Kouraev, A.V.; Kostianoy, A.G.; Lebedev, S.A. Recent changes of sea level and ice cover in the Aral Sea derived from satellite data (1992–2006). *J. Mar. Syst.* **2009**, *76* (3), 272–286. doi:10.1016/j.jmarsys.2008.03.016.

27. Vignudelli, S.; Kostianoy, A.G.; Cipollini, P.; Benveniste, J., Eds. *Coastal Altimetry*; Springer-Verlag: Berlin, 2011.

28. Birkett, C.; Reynolds, C.; Beckley, B.; Doorn, B. From research to operations: The USDA Global Reservoir and Lake Monitor. In Coastal Altimetry; Vignudelli, S., Kostianoy, A.G., Cipollini, P., Benveniste, J., Eds.; Springer-Verlag: Berlin, 2011; 19–50.

29. Crétaux, J.-F.; Calmant, S.; Abarca del Rio, R.; Kouraev, A.; Bergé-Nguyen, M.; Maisongrande, P. Lakes studies from satellite altimetry. In Coastal Altimetry; Vignudelli, S., Kostianoy, A.G., Cipollini, P., Benveniste, J., Eds.; Springer-Verlag: Berlin, 2011; 509–534.

30. Hydroweb. LEGOS. Available at http://ctoh.legos.obs-mip.fr/data/hydroweb/ (accessed August 11, 2019).

31. USDA. The USDA Global Reservoir and Lake Monitor. https://ipad.fas.usda.gov/cropexplorer/global_reservoir/ (accessed August 11, 2019).

32. Kira, T. Survey of the state of world lakes. In Guidelines of Lake Management. Vol.8: The World's Lakes in Crisis; Jorgensen, S.E., Matsui, S., Eds.; ILEC Foundation and UNEP: Kusatsu, 1997; 147–155.

33. World Lake Database. Available at http://wldb.ilec.or.jp/ (accessed August 11, 2019).

34

Oil Pollution: The Baltic Sea

Introduction 349
Main Sources of Oil Pollution in the Baltic Sea 350
Oil Spill Observations, Statistics, and Tendencies 352
Operational Satellite Monitoring of Oil Pollution 355
Numerical Modeling of Oil Spill Drift 359
Problems and Solutions 364
Conclusions 365
References 366

Andrey G.
Kostianoy

Introduction

Detection of oil pollution is among the most important goals of monitoring a coastal zone. Public interest in the problem of oil pollution arises mainly during dramatic tanker and oil platform catastrophes such as those that involved the following: *Amoco Cadiz* (France, 1978), *Ixtoc I* (Gulf of Mexico, 1979–1980), *Exxon Valdez* (Alaska, 1989), *The Sea Empress* (Wales, 1996), *Erica* (France, 1999), *Prestige* (Spain, 2002), and *Deepwater Horizon* (Gulf of Mexico, 2010). However, tanker and oil platform catastrophes are only one among the many causes of oil pollution. Oil and oil product spillages at sea take place all the time, and it would be a delusion to consider tanker accidents as the main environmental danger. According to the International Tanker Owners Pollution Federation (ITOPF), over the period of 1970–2017, spillages resulting from collisions, groundings, tanker holes, and fires amounted to 52% of total leakages during tanker loading/unloading and bunkering operations.[1] In the category of 7–700 tons, some 38% of spills occurred during routine operations, most especially loading or discharging (31%). Accidents were the main cause of large spills (>700 tons), with groundings and collisions accounting for 61% of the total during the period 1970–2017.[1] Other significant causes included hull failures and fire/explosion. Discharge of wastewater containing oil products is another important source, by pollutant volume comparable to offshore oil extraction and damaged underwater pipelines. The greatest but hardest-to-estimate oil inputs come from domestic and industrial discharges, direct or via rivers, and natural hydrocarbon seeps. The long-term effects of this chronic pollution are arguably more harmful to the coastal environment than a single, large-scale accident.

Each year, ships and industries damage delicate coastal ecosystems in many parts of the world by releasing oil or pollutants into ocean, coastal waters, and rivers. Offshore environments are polluted by mineral oil mainly due to tanker accidents, illegal oil discharges by ships, and natural oil seepage. Shipping activities in European coastal seas, including oil transport and oil handled in harbors, have a number of negative impacts on the marine environment and coastal zone. Oil discharges from ships represent a significant threat to marine ecosystems. Oil spills cause the contamination of seawater, sediments, shores, and beaches, which may persist for several months and even years and represent a threat to marine resources.[1,2]

As highlighted by Oceana in its report *The Other Side of Oil Slicks*, chronic hydrocarbon contamination from washing out tanks and dumping bilge water and other oily waste represents a danger at least three times higher than that posed by the oil slicks resulting from oil tanker accidents.[3,4] For example, in the North Sea, the volume of illegal hydrocarbon dumping is estimated at 15,000–60,000 tons/year, added to which are another 10,000–20,000 tons of authorized dumping. Oil and gas platforms account for 75% of the oil pollution in the North Sea via seepage and the intentional release of oil-based drilling muds.[5] In the Mediterranean Sea, it is estimated at 400,000–1,000,000 tons/year. Of this, about 50% comes from routine ship operations and the remaining 50% comes from land-based sources via surface runoff.[5] In the Baltic Sea, this volume is estimated at 1750–5000 tons/year.[3,4] In 2004, the Finnish Environment Institute[6] estimated the total annual number of oil spills in the Baltic Sea to be 10,000 and the total amount of oil running into the sea to be as much as 10,000 tons, which is considerably more than the amount of oil pouring into the sea in accidents. In 2014, an analysis of different sources of information on oil pollution gave annual values of 20–20,000 tons for the Baltic Sea.[7] It is impressive to note that the total amount of oil spilled annually worldwide from all sources to the ocean is estimated as 1.7–8.8 million tons (the more realistic value was about 3.2 million tons) in the 1970s, 0.47–8.3 (1.3) million tons in the 1990s, and 2.6–4.8 million tons in the 2000s, which is about 0.05%–0.1% from the world oil production (4.76 billion tons in 2011).[7]

One of the most important goals of the ecological monitoring of European seas is monitoring and detection of oil pollution.[8–11] After a tanker accident or illegal oil discharge, the biggest problem is to obtain an overall view of the phenomenon, getting a clear idea of the extent of the slick and predicting the way it will move. For natural and man-made oil spills, it is necessary to operate regular and operational monitoring. Oil pollution monitoring in the Mediterranean Sea, North Sea, and Baltic Sea is normally carried out by aircrafts or ships. This is expensive and is constrained by the limited availability of resources. Aerial surveys over large areas of the seas to check for the presence of oil are limited to the daylight hours, good weather conditions, and maritime boundaries between countries.

Satellite imagery can help greatly in identifying probable spills over very large areas and in guiding aerial surveys for precise observation of specific locations. The synthetic aperture radar (SAR) instrument, which can collect data independently of weather and light conditions, is an excellent tool to monitor and detect oil on water surfaces.[7–11] This instrument offers the most effective means of monitoring oil pollution: oil slicks appear as dark patches on SAR images because of the damping effect of the oil on the backscattered signals from the radar instrument. This type of instrument is currently on board the European Space Agency's (ESA's) *Sentinel-1A and -1B* satellites, the Canadian Space Agency's *RADARSAT-2* satellite, the German Earth observation satellite *TerraSAR-X*, and other spacecraft. ASAR/SAR (advanced synthetic aperture radar) instrument is used for mapping sea ice and oil slick monitoring, measurements of ocean surface features (currents, fronts, eddies, internal waves), ship detection, oil and gas exploration, etc. Users of remotely sensed data for oil spill applications include the Coast Guard, national environmental protection agencies and departments, oil companies, shipping, insurance and fishing industries, national departments of fisheries and oceans, and other organizations.

In this entry, we will focus on oil pollution in the Baltic Sea, briefly describing main sources of oil pollution, the number of oil spills observed yearly, the results of operational satellite monitoring and numerical modeling of oil spills, and the remaining problems.

Main Sources of Oil Pollution in the Baltic Sea

Crude oil and petroleum products account for about 40% of the total exports of Russia. Russian Federation stands as one of the leading operators in the international oil business, being the largest oil exporter after Saudi Arabia. In 2000, Russia exported approximately 145 million tons of crude oil and

50 million tons of petroleum products. Since 2000, exports of petroleum and petroleum products began to grow and virtually doubled for the period from 1996 to 2005. According to the Federal Customs Service of Russia, in 2008, Russia exported 221.6 million tons, and in 2017, 253 million tons of oil.

The ports on the Baltic Sea play a huge role in the export of oil from Russia. The main oil terminals here used to be the Latvian port of Ventspils and the Port of Tallinn, Estonia. In the last 20 years, a number of new oil terminals have been built in the Baltic Sea area, resulting in increased transport of oil by ships and, consequently, an increased risk of accidents and increased risk of pollution of the marine environment. Today, in the Gulf of Finland, there are more than 20 oil terminals in Russia, Finland, and Estonia. The following are the major existing and projected oil terminals in the Gulf of Finland from Russia: Primorsk, Vysotsk, Big Port of St. Petersburg, Ust-Luga, Batareinaya, Vistino, Gorki, and Lomonosov.[12,13]

Primorsk is the largest Baltic oil terminal located on Russian territory. In 2008, 75.6 million tons of oil products were exported from Primorsk, 13.6 million tons from Vysotsk, and 14.4 million tons from the St. Petersburg oil terminal. In 2018, 53.5 million tons of oil were exported from Primorsk, maximum export possibility of the Primorsk terminal is estimated at 120 million tons, while that of Vysotsk is at 20.5 million tons. In November 2000, Lukoil has opened an oil terminal in Kaliningrad. In 2001, the company built another terminal in Kaliningrad with a declared capacity of 2.5 million tons. These terminals can overload up to 3–5 million tons of oil annually.

According to estimates of the Centre for Maritime Studies at the University of Turku (Finland), in 2007, 263 million tons of cargo were transported through the Gulf of Finland, among which the share of oil is 56%.[13] Russian ports handled 60% of goods, Finnish ports handled 23%, and Estonian ports handled 17%. The share of imports was 22%, that of export was 76%, and that of local transportation was 2%. Russian ports held 68.6% of the total turnover of petroleum products, Estonian ports held 17.2%, and Finnish ports held 14.2%.[13] The major ports were the following: Primorsk (74.2 million tons), St. Petersburg (59.5 million tons), Tallinn (35.9 million tons), Skoldvik (19.8 million tons), Vysotsk (16.5 million tons), and Helsinki (13.4 million tons). In 2007, the ports of the Gulf of Finland carried out about 53,600 ship calls, most of which were in St. Petersburg (14,651), Helsinki (11,727), and Tallinn (10,614). In 2009, vessels entered or left the Baltic Sea via Skaw 62,743 times; this is 20% more than in 2006. Approximately 21% of those ships were tankers, 46% were other cargo ships, and 4.5% were passenger ships.[14]

The growth of oil and other cargo through the terminals and the Baltic ports inevitably leads to an increase in the number of tankers and other types of vessels, which then leads to an increase in chronic sea pollution and a higher probability of ship accidents. According to statistics, shipping accounts for 45% of oil pollution in the ocean, while oil production at the shelf accounts for only 2%. In the Baltic Sea, about 2000 large ships and tankers are at sea every day; thus, shipping, including oil transport, has a major negative impact on the marine environment and coastal zone. Illegal discharges of oil and petroleum products from ships, ship accidents, collisions, and groundings represent a significant threat to the Baltic Sea.[8]

According to Global Marine Oil Pollution Information Gateway,[15] major oil spills in the region in 1970–2007 resulted from the following ship accidents: *Othello* (1970, Tralhavet Bay, Sweden, spill of 60,000 tons), *Tsesis* (1977, off Nynäshamn, Sweden, spill of 1000 tons), *Antonio Gramsci* (1979, off Ventspils, Latvia, spill of 5500 tons; another incident in 1985, off Porvoo, Finland, spill of 580 tons), *Jose Marti* (1981, off Dalarö, Sweden, spill of 1000 tons), *Globe Asimi* (1982, off Klaipeda, Lithuania, spill of 16,000 tons), *Sivona* (1984, in The Sound, Sweden, spill of 800 tons), *Volgoneft* (1990, off Karlskrona, Sweden, spill of 1000 tons), *Baltic Carrier* (2001, international waters between Denmark and Germany, spill of 2700 tons), *Fu Shan Hai* (2003, between the Danish island of Bornholm and coast of Sweden, spill of 1200 tons), *Haaga* (2003, St. Petersburg port, Russia, spill of 1300 tons), and *Golden Sky* (2007, off Ventspils, Latvia, spill of 25,000 tons).

According to HELCOM data (http://www.helcom.fi), the total number of accidents on ships in 2000–2004 is 374, of which 29 resulted in the pollution of marine waters. The number of accidents has risen since 2006, which can be linked to the 20% increase in ship traffic. Now, there are 120–140 shipping accidents yearly in the Baltic Sea area.[14] The majority of accidents are groundings and collisions. The share of groundings in the total number of accidents is higher for the Baltic Sea than for other European waters. On average, 7% of the shipping accidents in the Baltic Sea result in some kind of pollution, usually containing not more than 0.1–1 ton of oil. For the last 10 years, no major accidental oil spill has happened in the Baltic Sea.[14]

As far as oil exploitation at sea and on the coast is concerned, offshore operations have been taking place for some years in Polish waters (two jack-up rigs); Germany operated two platforms very close to the coast; in March 2004, Russia started oil production at Lukoil D-6 platform in the waters between the Kaliningrad area (Russian Federation) and Lithuania, and Latvia plans to drill for oil in the waters between them and Lithuania.[15]

Oil Spill Observations, Statistics, and Tendencies

Every ship entering the Baltic Sea must comply with the antipollution regulations of the Helsinki Convention and MARPOL (marine pollution) Convention. Even though strict controls over ships' discharges have been established by the Baltic Sea countries, illegal spills and discharges continue to happen.[14] Fortunately, the number of illegal oil spills detected by aerial surveillance has been reduced significantly over the last 30 years, from 763 spills in 1989 to 52 spills in 2017 (Table 1 and Figure 1). Also, the volume of the spills has been decreasing—most are between 1 and 0.1 m^3 today.[14,16]

A decreasing trend in the number of observed illegal oil discharges despite rapidly growing density of shipping, increased frequency of surveillance flights (Figure 1 and Table 2), and usage of satellite imagery, provided by the CleanSeaNet satellite service of the European Maritime Safety Agency (EMSA) (http://cleanseanet.emsa.europa.eu/), illustrates the positive results of the complex set of measures known as the Baltic Strategy, implemented by the Contracting Parties to the Helsinki Convention.[16]

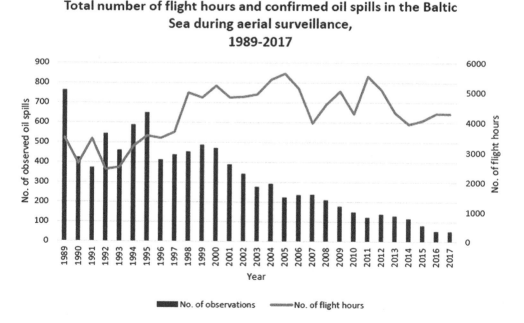

FIGURE 1 Total number of flight hours and observed oil spills in the HELCOM area during aerial surveillance in 1989–2017. (Annual Report on discharges observed during aerial surveillance in the Baltic Sea in 2017.[16])

TABLE 1 Country-Wise Data on the Number of Illegal Oil Discharges Observed in National Waters in the Baltic Sea in 1988–2017

	Denmark	Estonia	Finland	Germany	Lithuania	Latvia	Poland	Russia	Sweden	Total
1988	129			90			40	82	168	509
1989	159			139			69	184	212	763
1990	34			45		73	88		184	424
1991	46			85	8	20	14	3	197	373
1992	18	18		76	34	15	92	13	278	544
1993	17	7		43	28	6	110		250	461
1994	30	4		75			104		375	588
1995	48	3	26	55			72		445	649
1996	36		42	44			50		241	413
1997	38	3	104	34			25		234	438
1998	53	10	53	23		33	33		249	454
1999	87	33	63	72		18	18		197	488
2000	68	38	89	51		17	51		158	472
2001	93	11	107	51	0	6	24		98	390
2002	54	8	75	44		21	25		117	344
2003	37	4	40	60		14	39		84	278
2004	30	19	36	42	0	13	10		143	293
2005	28	24	32	34	0	5	5	2	94	224
2006	41	31	29	22	0	0	3		110	236
2007	43	58	29	30		2	15		61	238
2008	41	46	28	24		5	22		44	210
2009	34	20	16	15		1	27		65	178
2010	33	25	15	22	1	0	14	0	39	149
2011	18	14	16	13	0	0	5	0	56	122
2012	19	8	24	25	0	0	5	0	58	139
2013	14	8	9	7	1	0	27	0	64	130
2014	25	9	11	16	0	0	10		46	117
2015	3	12	17	12	0	0	9		29	82
2016	3	5	17	4	0	0	7		17	53
2017	4	2	8	7	0	0	9		22	52

Source: Annual Report on discharges observed during aerial surveillance in the Baltic Sea in 2017.[16]

In order to obtain the geographical distribution of oil spills, we put the spills observed in 1988–2002 on the same map (Figure 2). Analysis of Tables 1 and 2 and Figure 2 shows the following:

1. HELCOM data seem to be underestimated in figures in comparison with other estimates.[3,4,6,7]
2. Since 1993, Russia does not carry out aerial surveillance in the Southeastern Baltic Sea and in the Gulf of Finland.
3. Since 1994, Lithuania seems to have no regular and effective aerial surveillance, because no oil spills have been detected.
4. Since 2005, Latvia seems to have no effective aerial surveillance, because 0–5 oil spills yearly are not realistic figures.
5. Traces of oil spills in Figure 2 show the main ship routes in the Baltic Sea, as well as the approaches to major sea ports and oil terminals.
6. Figure 2 proves that ships are primarily responsible for the oil pollution in the Baltic Sea.

TABLE 2 Number of Aerial Surveillance Flight Hours Performed by the HELCOM Countries in 1989–2017

	Denmark	Estonia	Finland	Germany	Lithuania	Latvia	Poland	Russia	Sweden	Total
1989				142			131	1618	1600	3491
1990	292			168		400	164		1600	2624
1991	199			129	348	408	140	629	1600	3453
1992	172			267	78	127	62	32	1700	2438
1993	153	40		201	133	24	49		1900	2500
1994	253	420		290		18	179		2038	3198
1995	225	420	355	291		8	301		1953	3553
1996	275	305	400	313	65	8	345		1763	3474
1997	209	284	355	288		64	291		2189	3680
1998	325	236	649	206		577	465		2544	5002
1999	416	268	603	286		320	375		2565	4833
2000	497=	212	660	439	250	436	362		2374	5230
2001	463	161	567	466	300	412	187		2281	4837
2002	412	153	605	469		387	320		2518	4864
2003	510	201	615	446		414	228		2532	4946
2004	265	198	644	491	100	365	239		3231	5534
2005	251	178	625	549	54	384	141		3455	5638
2006	290	471	517	504	64	311	131		2842	5128
2007	271	410	529	598	41	343	380		1397	3969
2008	246	503	438	650		298	406		2063	4603
2009	240	371	351	638	66	61	561		2758	5046
2010	156	266	605	558		48	421	10	2215	4279
2011	188	315	645	648	3	18	499		3225	5541
2012	227	220	631	769		4	318		2921	5090
2013	207	327	625	470	0	19	387		2283	4317
2014	239	362	505	596	5	12	393		1823	3935
2015	254	356	490	700	0	8	259		1995	4062
2016	271	267	484	726	0	0	290		2256	4295
2017	324	268	428	891	0	0	321		2098	4331

Source: Annual report on discharges observed during aerial surveillance in the Baltic Sea in 2017.[16]

The Baltic Sea States aerial surveillance fleet today consists of more than 20 airplanes and helicopters, most of which are equipped with up-to-date remote-sensing equipment—side-looking airborne radar, SAR, infrared (IR) and ultraviolet scanner, microwave radiometer, laser fluorosensor (lidar), and forward-looking infrared (FLIR) high-resolution camera. The Baltic Sea States have to conduct aerial surveillance for detecting oil pollution and suspected ships at a minimum of twice per week over regular traffic zones including approaches to major sea ports as well as in regions with regular offshore activities. Other regions with sporadic traffic and fishing activities should be covered once per week.[16] Also, the Coordinated Extended Pollution Control Flights, which constitutes continuous surveillance of specific areas in the Baltic Sea for 24 hours or more, should be carried out twice a year.[16] Although the number of observations of illegal oil discharges shows a decreasing trend over 30 years, it should be noted that for some areas and countries, aerial surveillance is not evenly and regularly carried out and therefore there are no reliable figures for these areas.[16]

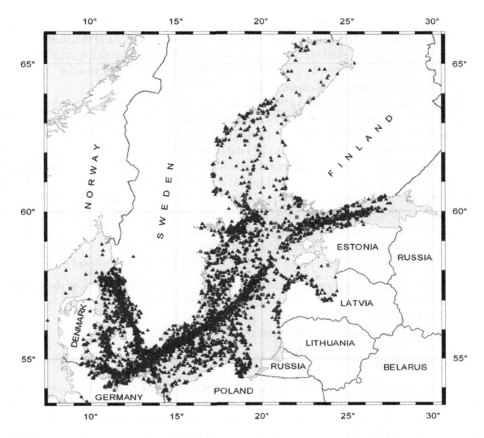

FIGURE 2 Oil spills detected in the Baltic Sea by aerial survey in 1989–2002 based on HELCOM data. (Kostianoy et al.[19])

Since 2007, aerial surveillance in the Baltic Sea was supported by the satellite remote-sensing technique for oil spill detection. CleanSeaNet is a near-real-time satellite-based oil spill and vessel-monitoring service.[8] It entered into operation on April 16, 2007.[16] The service is continually being expanded and improved and provides a range of different products to the Commission and European Union member states. The legal basis for the CleanSeaNet service is Directive 2005/35/EC on ship-source pollution and on the introduction of penalties, including criminal penalties, for pollution offenses (as amended by Directive 2009/123/EC). EMSA has been tasked to "work with the Member States in developing technical solutions and providing technical assistance in relation to the implementation of this Directive, in actions such as tracing discharges by satellite monitoring and surveillance."[16]

Operational Satellite Monitoring of Oil Pollution

A satellite-based remote-sensing system is capable of ensuring a relatively low-cost, high-standard observational system for oil pollution monitoring. SAR is the best instrument for the detection of oil slicks on the sea surface from space because slicks modify seawater viscosity and damp short waves measured by SAR. SAR images can be acquired regardless of cloud cover and light conditions. Along with 300–400 km wide swath, this is the main advantage of SAR in comparison with aerial surveillance. However, oil spill detection by SAR has a problem of distinguishing oil slicks from look-alikes, such as sea areas covered by organic films, algal bloom, sea ice, wind shadows, rain cells, and upwelling zones. Therefore, reliable automatic detection of oil spills on the basis of SAR data is not yet achieved and there is a risk of false alarms. This problem can be significantly reduced by a new approach, which consists in the combined use of all available quasi-concurrent satellite, oceanographic, and meteorological information, along with numerical modeling of oil spill transport. This operational system was specially elaborated in the beginning of 2004 for monitoring oil pollution in the vicinity of the Lukoil D-6 oil platform in the Southeastern Baltic, Russian Federation.[8,17–21]

Since 1993, regular aerial surveillance of the oil spills in the Russian sector of the Southeastern Baltic Sea and in the Gulf of Finland has stopped (Tables 1 and 2, Figure 2). In June 2004, we organized daily service for monitoring of oil spills in the Southeastern Baltic Sea based on the operational receiving and analysis of ASAR *ENVISAT* and SAR *RADARSAT-1* data as well as of other satellite IR and optical (VIS) data, meteorological information, and numerical modeling of currents required for the identification of the slick nature in the sea and forecast of the oil spill drift.[8,17–21] This work was initiated and financed by Lukoil-Kaliningradmorneft (Kaliningrad, Russia) in connection with the start of oil production from the continental shelf of Russia in March 2004. The principal differences from the existing projects and satellite services were (1) an operational monitoring regime of 24 hours/day, 7 days/week for 18 months and (2) a complex approach to the oil spill detection and forecast of their drift.

The general goals of the satellite oil pollution monitoring in the Baltic Sea were as follows:

1. Correct detection of oil spills in the vicinity of the D-6 oil platform as well as in the large area of the Southeastern Baltic Sea between 54°20′–58°N and 18°–22°E
2. Identification of sources of oil pollution
3. Forecast of the oil spill drift by different methods
4. Data systematization and archiving
5. Cooperation with authorities.

Operational monitoring of oil pollution in the sea was based on the processing and analysis of ASAR *ENVISAT* (every pass over the Southeastern Baltic Sea, frame of 400×400 km, 75 m/pixel spatial resolution) and SAR *RADARSAT-1* (300×300 km, 25 m/pixel resolution) images received from KSAT Station (Kongsberg Satellite Services, Tromsø, Norway) in operational regime (1–2 hour after the satellite's overpass). For interpretation of ASAR *ENVISAT* imagery and forecast of the oil spill drift, IR and VIS AVHRR (the Advanced Very High-Resolution Radiometer aboard the National Oceanic and Atmospheric Administration (NOAA) satellites) and Moderate Resolution Imaging Spectroradiometer (MODIS) (*Terra* and *Aqua*) images were received, processed, and analyzed, as well as the *QuikSCAT* scatterometer and the *Jason-1* altimeter data.[18–21] The total area covered by the monitoring was equal to about 60,000 km², which is almost one-sixth of the Baltic Sea total surface.

The satellite receiving station at the Marine Hydrophysical Institute (MHI) in Sevastopol was used for operational (24 hours/day, 7 days/week) receiving of the AVHRR NOAA data for the construction of the sea surface temperature (SST), optical characteristics of seawater, and currents maps. SST variability and intensive algae bloom (high concentration of blue–green algae on the sea surface in the summertime) allow one to highlight meso- and small-scale water dynamics in the Baltic Sea and to follow movements of currents, eddies, dipoles, jets, filaments, river plumes, and outflows from the Vistula and Curonian bays. Sequence of daily MODIS IR and VIS imagery allows reconstruction of a real field of surface currents (direction and velocity) with 0.25–1 km resolution, which is very important for a forecast of a direction and velocity of a potential pollution drift including oil spills. The combination of ASAR *ENVISAT* images with high-resolution VIS and IR MODIS images allows understanding of the observed form of the detected oil spills and prediction of their transport by currents.[18–21]

Sea wind speed fields were derived from scatterometer data from every path of the *QuikSCAT* satellite over the Baltic Sea (twice a day). These data were combined with data from coastal meteorological stations in Russia, Lithuania, Latvia, Estonia, Finland, Sweden, Denmark, Germany, and Poland and numerical weather models. Altimetry data from every track of the *Jason-1* satellite over the Baltic Sea were used for compilation of sea wave height charts, which include the results of the FNMOC (Fleet Numerical Meteorology and Oceanography Center, the United States) World War III Model. Both data were used for the analysis of the ASAR *ENVISAT* imagery and estimates of the oil spill drift direction and velocity.[18–21]

In total, 274 oil spills were detected in 230 ASAR *ENVISAT* images and 17 SAR *RADARSAT-1* images received during 18 months (June 2004 to November 2005).[8,20] One example from the oil spill gallery is shown in Figure 3 where an illegal release of oil from three ships was detected on August 25, 2005.

The interactive numerical model Seatrack Web of the Swedish Meteorological and Hydrological Institute (SMHI) was used for a forecast of the drift of satellite-detected oil spills and ecological risk assessment of the Lukoil D-6 oil platform.[18–20]

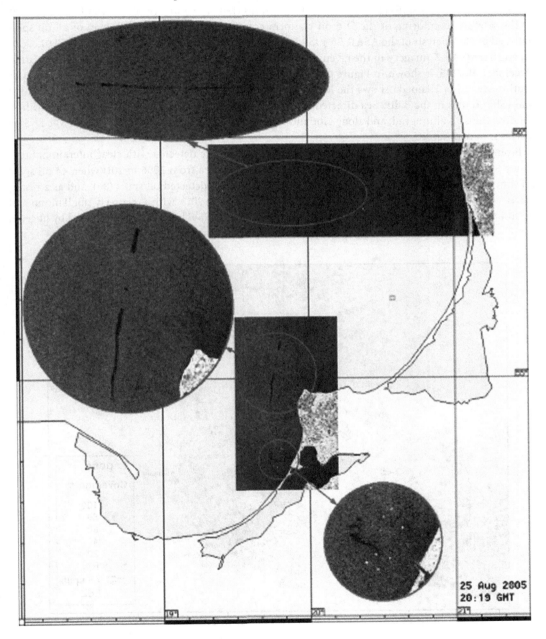

FIGURE 3 A release of oil from three ships on August 25, 2005 (ASAR *ENVISAT*). The length of the spill in front of Klaipeda is 33.6 km, surface—8.6 km². The length of another long spill is 22 km.

Since 2006, the satellite monitoring of the D-6 oil platform was transformed, reduced in size to about 24,000 km², and unfortunately, it lost its main peculiarity—a complex approach to the oil spill detection and forecast of their drift. In 2006–2009, 638 oil spills have been identified on 804 ASAR images, from which 319 spills were detected in this reduced area.[21] Combined analysis of the location and shape of the detected spills with location of the ships thanks to AIS (Automatic Identification System for ships) clearly indicates that the major source of sea pollution is shipping. In the Southeastern Baltic Sea, an area with no HELCOM statistical data, we also observe a decreasing trend in oil spill number and their total surface. In 2006, the total number of oil spills (and area of oil pollution) amounted to 114 (371.7 km²); in 2007, 94 spills (213.7 km²); in 2008, 67 spills (198.7 km²); and in 2009, 44 spills (81.7 km²).[21]

The satellite monitoring of the D-6 oil platform continues till present.[21–28] A map of all oil spills detected by the analysis of the ASAR *ENVISAT, RADARSAT-1, RADARSAT-2, Cosmo-SkyMED-1, -2, -3, -4,* and *TerraSAR-X* imagery in the given area of the Southeastern Baltic Sea from June 12, 2004, until December 31, 2015, is shown in Figure 4. A real form and dimension of oil spills are shown. A square southwestward of Klaipeda shows the location of the D-6 oil platform. Oil spills clearly revealed the main ship routes in the Baltic Sea directed to ports of Ventspils, Liepaja, Klaipeda (routes from different directions), Kaliningrad, and along Gotland Island (Figure 4). No spills originated from the D-6 oil platform were observed.

From June 12, 2004, to December 31, 2015, 1232 oil spills were detected with clear interannual tendency to reduction of oil pollution in the Southeastern Baltic Sea from 2006 to 2011 when 44 oil spills (147 km²) were recorded. In 2012, an increase in the number of detected oil spills (86), and as a result, increase of a total area of oil pollution (470 km²) was observed. In 2013, with a relatively small number of oil spills (52), we have observed a high total area of oil pollution (341 km²) what is explained by huge oil

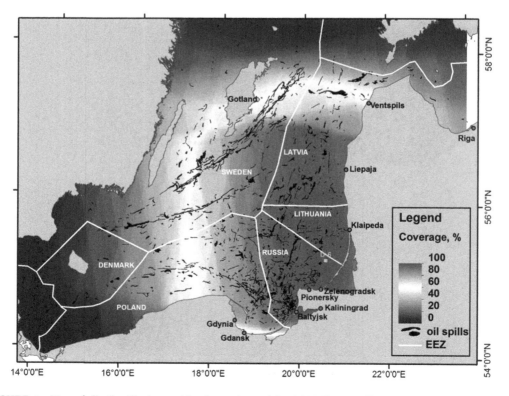

FIGURE 4 Map of all oil spills detected by the analysis of the ASAR/SAR satellite imagery from June 2004 to December 2015.

spills detected on September 13, 2013, and September 20, 2013. In 2014 and 2015, the number of oil spills has reduced to 47 and 39 with a total surface of about 188 km² for both years.[26]

A significant seasonal variability in oil spill detection is observed. During autumn and winter, we detect oil spills four times less than in spring and summer.[27] This huge difference is explained by limitations of the SAR method to detect oil spills when the wind is stronger than 10 m/s, which is very often during the cold season in the Baltic Sea. In addition, strong wind-wave mixing contributes to more rapid formation of emulsions ("water in oil" and "oil in water"), thus preventing formation of the oil slicks on the sea surface.

Image comparison of the number of detected oil spills in the morning (about 11:00 local time) and in the evening (about 22:00 local time) for 2006–2009 showed that the probability of finding oil pollution in the morning is about 40% higher than that in the afternoon and evening.[21] This fact indicates that the illegal discharge of oil from vessels occurs more often at night, when it is impossible for patrol aircraft or ships to record this fact by photo and video camera. This once again confirms the advantages of satellite radar imagery for monitoring of oil pollution.

Based on the number of oil spills detected in 2014 and 2015 (about 40 spills yearly on the area of about 24,000 km²), we can estimate the total number of oil spills for the Baltic Sea (377,000 km²) as about 630 yearly and the total surface of oil pollution as about 3000 km² yearly. These values are six to seven times higher than reported by HELCOM for the same years (Table 1). These values may double and even triple if we take into account the following: (1) SAR satellites pass over a specific area in the Baltic Sea every 2 days in average, (2) significant reduction of oil spill observation in autumn and winter due to unfavorable weather conditions, and (3) spatial resolution of SAR/ASAR imagery of 25–75 m/pixel.

Numerical Modeling of Oil Spill Drift

The abovementioned satellite monitoring of the Southeastern Baltic Sea was coupled with numerical modeling of oil spill transport. The interactive numerical model Seatrack Web SMHI was used for a forecast of the drift of (1) all large oil spills detected by ASAR *ENVISAT* in the Southeastern Baltic Sea and (2) virtual (simulated) oil spills from the Lukoil D-6 platform. The latter was done daily for the operational correction of the action plan for accident elimination at the D-6 and ecological risk assessment (oil pollution of the sea and the Curonian Spit).

This version of a numerical model on the Internet platform has been developed at the SMHI in close cooperation with the Danish Maritime Safety Administration, Bundesamt fur Seeschifffahrt und Hafen, and the Finnish Environment Institute.[29] The first version of Seatrack Web was introduced in 1995, and since then, Seatrack Web has been used successfully for oil spill drift forecast. It was developed to be a friendly tool for authorities responsible for oil spill response in the Baltic Sea region. Seatrack Web's main purpose is to calculate the drift and transformation of oil spills in the Gulf of Bothnia, the Gulf of Finland, the Baltic Sea Proper, the Sounds, the Kattegat, the Skagerrak, and part of the North Sea (to 3°E).[29] The program can also be used for substances other than oil, such as chemicals, algae, and floating objects. In addition to an oil drift forecast, it is possible to make a backward calculation. Then, calculation starts at the position where a substance was found and the program calculates the drift backward in time and traces the origin of the substance or an object.

Today, the Seatrack Web is hosted by SMHI and developed together by SMHI in close cooperation with the FCOO (Defence Centre for Operational Oceanography) in Denmark, the BSH (Federal Maritime and Hydrographic Agency) in Germany, and the FMI (Finnish Meteorological Institute) in Finland. Current fields are modeled with the NEMO-Nordic model (Nucleus of European Modeling of the Ocean), which is a three-dimensional circulation model covering the Gulf of Bothnia, the Gulf of Finland, the Baltic Sea, the Sounds, the Kattegat, the Skagerack, the North Sea, and the English Channel. The meteorological model used in Seatrack Web is ECMWF (the European Centre for Medium-Range Weather Forecasts), forecast for 4 days ahead and 6 days backward. The Seatrack Web provides oil spill drift with a time step of 15 min and 2 n.m. spatial resolution.

The oil spreading calculation is added to the currents, as well as oil evaporation, emulsification, sinking, stranding, and dispersion.[30]

The AIS functionality in Seatrack Web is a tool to identify the ship that causes an oil spill. The AIS function gives the ship tracks in the area where the spill was detected and backward to its probable origin. By using the ship tracks simultaneously with the oil tracks, the probability of identifying a suspected ship increases. This powerful system was recommended by HELCOM for operational use in the Baltic countries.[30]

The Seatrack Web model was very useful for ecological risk assessment related to exploitation of the Lukoil D-6 oil platform, which is installed 22.5 km from the Curonian Spit, a UNESCO World Heritage Site. Virtual (simulated) oil spills of 10 m³ were released daily from the platform for 6 months in order to calculate the shape, direction, distance, and velocity of their drift. Then, all 180 oil tracks were accumulated on one map showing the potential impact of oil pollution in case of an accident at the platform. Statistics, based on daily forecast of the oil spill drift in July–December 2004, shows potential probability (%) of the appearance of an oil spill in any point of the area during 48 hours after an accidental release of 10 m³ of oil (Figure 5). The probability of the oil spill drift directed to the Curonian Spit (150° sector from D-6) is equal to 67%, but only in half of these cases did oil spills reach the coast due to a coastal current.[18,20] This new technology allowed a quantitative assessment of ecological risks in the whole Baltic Sea.[31]

Later, the same methodology was applied to the risk assessment of the Nord Stream gas pipeline construction and assessment of the impact of oil pollution along the ship routes on the Baltic Sea Marine Protected Areas (BSPAs).[24,31] Figures 6 and 7 show two examples of oil spill drift modeling and a probability of the drift for specific points along main ship routes in the Gulf of Finland and southward of Gotland Island for July and August 2007. Figure 6a shows a drift of a virtual oil spill of 10 m³ during 48 hours, which

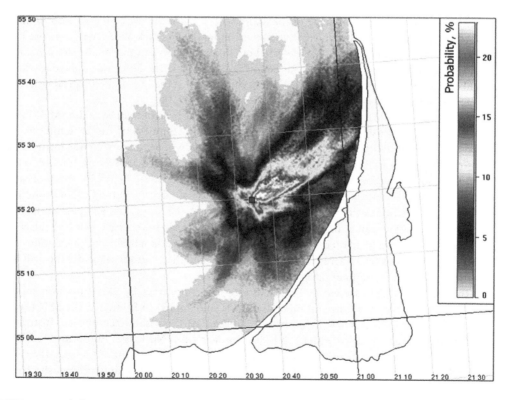

FIGURE 5 Probability of observation of potential oil pollution from the D-6 platform during the first 48 hours after an accidental release of 10 m³ of oil (based on 6 months daily release of oil spill from the oil platform).

was released on July 23, 2007, at a specific point (red square) of the ship route passing through the Gulf of Finland. The same numerical experiment was performed daily from July 1 to August 31, 2007. Thus, based on the compilation of 62 maps of oil spill drifts, we could construct Figure 6b and calculate a probability of oil spill drift. Figure 6b shows that for this time period, there is no impact of possible oil spill drift on the surrounding BSPAs along the coasts of Finland and Estonia, which are marked by blue and rose colors. This is explained by low wind speed and weak currents observed during July and August 2007. These weather conditions differ from those observed from July 26 to August 15, 2006, when a virtual oil spill could drift 33.5 nautical miles during 2 days with a velocity up to 50 cm/s (Figure 6c). Thus, potential releases of oil spills from the ships may represent a threat to seven protected areas located along the coasts of Finland and Estonia, as well as to the islands and coasts of these Baltic countries. The same dot lies on the trajectory of the Nord Stream gas pipeline (upper line in Figure 6c), which was under construction in the Gulf of Finland since May 2010. Risk assessment for the construction of the gas pipeline was performed using the same methodology in 2006 for seven key points of the pipeline.[32]

Figure 7a shows a simultaneous release of oil from a long part of the ship route located southward of Gotland Island. Figure 7b shows a significant impact of oil pollution produced by this virtual oil release. Both BSPAs were subjected to oil pollution, but at different degrees, so that a quantitative estimation was possible with the help of the Seatrack Web model (Figure 7c and d). For instance, about 60% of the first BSPA and 95% of the second one will be polluted with the indicated probability. In this case, even the coastal zone of Gotland Island was potentially threatened by oil pollution with a clearly calculated probability.[32]

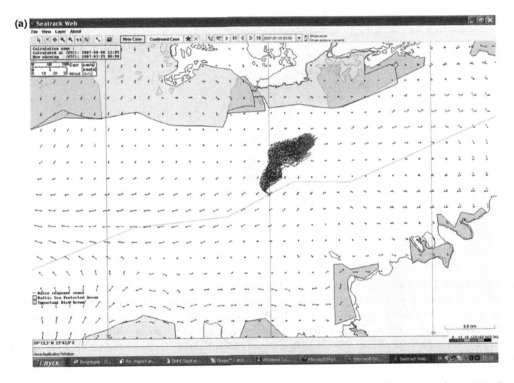

FIGURE 6 Modeling of oil spill drift in the Gulf of Finland. Panel (a) shows oil spill drift on July 23, 2007. Panel (b) shows probability (%) of oil spill drift calculated on the basis of daily modeling at this point for real wind and current conditions in July–August 2007. Panel (c) shows probability (%) of oil spill drift calculated for July 16 to August 15, 2006. BSPAs along the coastlines are shown in light grey color; important bird areas are in grey colors. The lines in panel (a) and (b) show MARIS response zones. The upper line in panel (c) shows approximate position of the Nord Stream gas pipeline, while the lower line is the MARIS response zones.

(Continued)

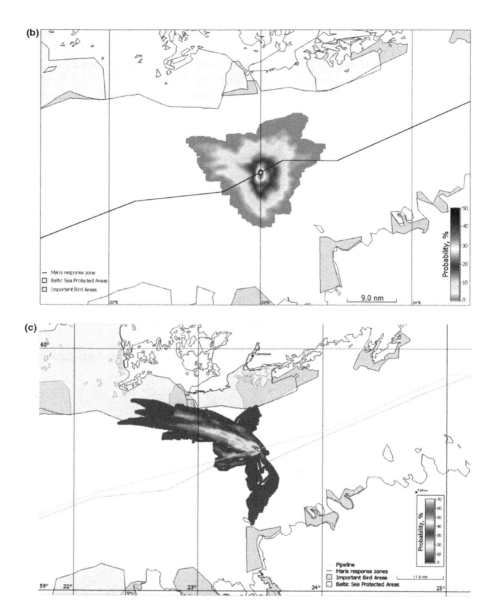

FIGURE 6 (CONTINUED)   Modeling of oil spill drift in the Gulf of Finland. Panel (a) shows oil spill drift on July 23, 2007. Panel (b) shows probability (%) of oil spill drift calculated on the basis of daily modeling at this point for real wind and current conditions in July–August 2007. Panel (c) shows probability (%) of oil spill drift calculated for July 16 to August 15, 2006. BSPAs along the coastlines are shown in light grey color; important bird areas are in grey colors. The lines in panel (a) and (b) show MARIS response zones. The upper line in panel (c) shows approximate position of the Nord Stream gas pipeline, while the lower line is the MARIS response zones.

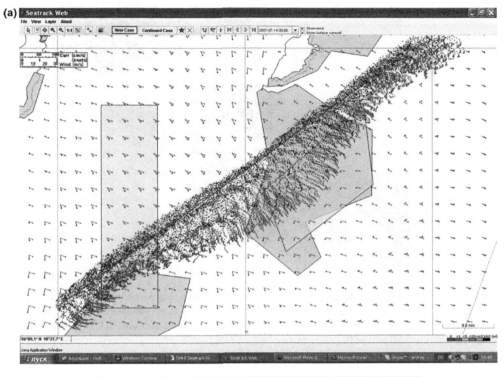

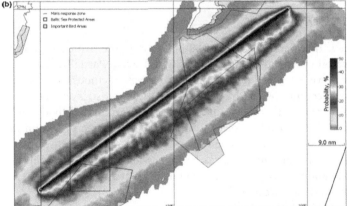

FIGURE 7 Modeling of oil spill drift released from a long part of the ship route located southward of Gotland. Panel (a) shows oil spill drift on July 12, 2007. Panel (b) shows probability (%) of oil spill drift calculated on the basis of daily modeling at this line for real wind and current conditions in July–August 2007. Panels (c) and (d) show the impact of this part of the ship route on both BSPAs.

(Continued)

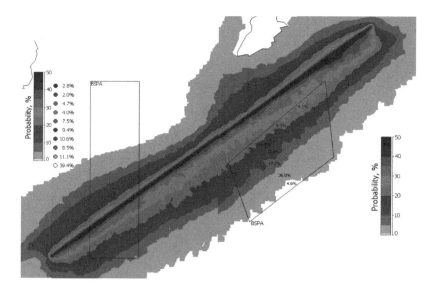

FIGURE 7 (CONTINUED)   Modeling of oil spill drift released from a long part of the ship route located south-ward of Gotland. Panel (a) shows oil spill drift on July 12, 2007. Panel (b) shows probability (%) of oil spill drift cal-culated on the basis of daily modeling at this line for real wind and current conditions in July–August 2007. Panels (c) shows the impact of this part of the ship route on both BSPAs.

Problems and Solutions

The main challenge is that the real degree of oil pollution in the Baltic Sea is unknown, because the number of observed oil spills and estimates differ significantly. Partially, this is explained by known differences in aerial surveillance and satellite monitoring. Both methods have their own advantages and limitations and should complement each other. It is clear that statistics on oil spills is not complete and not comparable in different parts of the sea due to different efforts and methods applied for oil pollution monitoring (different number of oil patrol ships, aircraft, and helicopters per country and per unit of the sea area; the number of flight hours per country and unit of the sea area; use of satellites; a number of ASAR/SAR images acquired and analyzed yearly; application of complex satellite monitoring based on the multisensor and multiplatform approach along with the analysis of metocean data; local peculiari-ties of the water area; and numerical modeling).

For example, since 1993, Russia does not carry out aerial surveillance in the Gulf of Finland and the Southeastern Baltic Sea. The existing satellite monitoring is performed on a regular basis only in the Southeastern Baltic Sea and by a private company, Lukoil-Kaliningradmorneft. According to HELCOM data, since 1994 and 2005, respectively, Lithuania and Latvia seem to have had no regular aerial surveillance of oil pollution. Well-equipped regular aerial surveillance is very expensive, and it is clear that countries in economic recession reduce their aerial and in situ monitoring. Satellite monitoring may partially solve this problem, because satellites cover simultaneously very large areas of the Baltic Sea.

Organization of the Baltic International Satellite Monitoring Center in HELCOM could solve many problems in the operational monitoring of oil pollution in the Baltic Sea.[33] It will:

1. Ensure full and uniform coverage of the Baltic Sea area by remote-sensing control
2. Reinforce aerial surveillance and improve the oil pollution monitoring
3. Establish satellite monitoring for the countries where it is not yet applied

4. Remove duplication of satellite monitoring for the same area performed by neighboring countries
5. Significantly reduce the total cost of operational satellite monitoring for all countries
6. Provide data to all the Baltic Sea States in the same format
7. Solve the problem regarding different technologies, methods, and algorithms used for the analysis of satellite data in different countries
8. Solve the problem of the "night" oil spill pollution that is getting more and more acute
9. Stimulate exchange of data and cooperation between countries
10. Contribute to early warning in case of transboundary oil spill drift
11. Improve the ecological state of the Baltic Sea, coastal zones, and shores of the Baltic Sea States
12. Stimulate organization of analogous operational monitoring centers for the seas with a high density of shipping and/or oil/gas exploration/production industry, i.e., the North Sea, the Mediterranean Sea, the Black Sea, the Caspian Sea, the Gulf of Mexico, etc.

Wide usage of the SMHI Seatrack Web model for oil spill drift forecast is required. Originally, the model was not devoted to the ecological risk assessment, but we found it very useful for this purpose as well. [31] The ecological risk assessment for all ports, oil terminals and platforms, subsurface oil pipelines, ship routes, the Baltic Sea Protected Areas, and any part of the 8000 km long coastline of the Baltic Sea can be performed based on the methodology we elaborated in 2004 and successfully used for Lukoil D-6 oil platform and Nord Stream gas pipeline construction.[18,20,24,31,32] This will quantitatively and precisely reveal the hot spots in the marine area, islands, and coastline of the Baltic Sea that are vulnerable to the impact of the shipping oil pollution. Such a general map of the Baltic Sea with calculated probability for any point of the sea and the coastline to be polluted may serve as a guideline for the Baltic Sea States to improve their monitoring systems.

Conclusions

Oil is a major threat to the Baltic Sea ecosystems. In the last decade, maritime transportation in the Baltic Sea region has been growing steadily, reflecting the intensified trade and oil export. An increase in the number of ships also means that we can expect a larger number of illegal oil discharges. Both oil tankers and all types of ships are responsible for oil pollution of the Baltic Sea. Any discharge into the Baltic Sea of oil, or diluted mixtures containing oil in any form including crude oil, fuel oil, oil sludge, or refined products, is prohibited. This applies to oily water from the machinery spaces of any ship, as well as from ballast or cargo tanks from oil tankers. Every ship entering the Baltic Sea must comply with the antipollution regulations of the Helsinki Convention and MARPOL Convention. Even though strict controls over ships' discharges were established by the Baltic Sea countries, illegal spills and discharges still happen. The number of illegal oil spills has been reduced significantly over the last 30 years, from 763 spills in 1989 to 52 spills in 2017, and this is an evident and positive tendency, resulted from the long-term efforts of HELCOM.[16]

However, the actual total number of oil spills and their volume seem to be unknown because these values contradict significantly (10–20 times) with estimates of different organizations[3,4,6,8] and with the results of complex operational satellite monitoring performed in 2004–2017 in the Southeastern Baltic Sea.[8,21–28] Although the number of observations of illegal oil discharges shows a decreasing trend over the years, it should be kept in mind that for some areas and countries, aerial and satellite surveillance is not evenly and regularly carried out and, therefore there are no reliable figures. We have to add to the uncertainties in the oil pollution statistics considerable seasonal variability in observations of oil spills on the sea surface and predominance of the "night" discharge of oil spills from the ships used to avoid any direct visible evidence of pollution and responsibility for this fact.

So far, as the Baltic Sea ecosystem undergoes increasing human-induced impacts, especially associated with intensifying oil transport and production, further research on the link between physical, chemical, and biological parameters of the ecosystem; the complex monitoring of the Baltic Sea States;

and, especially, oil spill monitoring are of great importance. Oil spill behavior, modeling, prevention, effects, control, and cleanup techniques require supplementary information about a large number of complex physical, chemical, and biological processes and phenomena.

ASAR/SAR satellite imagery provides effective opportunities to monitor oil spills, particularly in the Baltic Sea, as well as in other European seas.[8–11] Combined with satellite remote sensing (AVHRR NOAA, MODIS-*Terra* and -*Aqua*, QuikSCAT, *Jason-1*, etc.), of the SST, sea level, chlorophyll and suspended matter concentration, meso- and small-scale dynamics, and wind and waves, this observational system represents a powerful method for long-term monitoring of the ecological state of semi-enclosed seas especially vulnerable to oil pollution. Our experience in the operational oil pollution monitoring in the Baltic Sea could be easily applied to the Caspian, Black, Mediterranean, and other European seas.

Since 2004, we have elaborated several operational satellite monitoring systems for oil and gas companies in Russia and performed complex satellite monitoring of the ecological state of coastal waters in the Baltic, Black, Caspian, and Mediterranean seas.[24,34] The accident on the BP oil platform "Deepwater Horizon" on April 20, 2010, in the Gulf of Mexico showed that the absence of such a permanent complex satellite monitoring system makes all efforts related to cleaning operations at sea and on the shore during the first weeks after the accident less effective.[35]

A large number of discharges of hydrocarbons that annually take place in European waters, the vast quantity of waste generated by the sea traffic in Europe, the lack of adequate port installations for waste management, and the toxicity of compounds thrown into the sea make solving the chronic hydrocarbon pollution problem a priority for improving the environmental quality of European seas. The growing availability of satellite and sea observation data should encourage interest, involvement, and investment into the complex operational monitoring systems from the side of the state authorities responsible for the environment, pollution control, meteorology, coastal protection, transport, fisheries, and hazard management, as well as from the side of private companies operating in the sea and coastal zone.

References

1. ITOPF (International Tanker Owners Pollution Federation Limited), Handbook 2018/2019, https://www.itopf.org/fileadmin/data/Documents/Company_Lit/ITOPF_Handbook_2018.pdf (accessed April 24, 2019).
2. European Environment Agency. *The European Environment—State and Outlook 2005*; EEA: Copenhagen, 2005.
3. Oceana. *The Other Side of Oil Slicks*; The Dumping of Hydrocarbons from Ships into the Seas and Oceans of Europe; 2003, http://www.oceana.org/north-america/publications/reports/the-other-side-of-oil-slicks (accessed May 22, 2012).
4. Oceana. *The EU Fleet and Chronic Hydrocarbon Contamination of the Oceans*, 2004, http://www.oceana.org/north-america/publications/reports/the-eu-fleet-and-chronic-hydrocarbon-contamination (accessed May 22, 2012).
5. UNESCO. *The Integrated, Strategic Design Plan for the Coastal Ocean Observations Module of the Global Ocean Observing System*, GOOS Report N 125, IOC Information Documents Series N 1183; UNESCO: Paris, 2003.
6. Finnish Environment Institute, 2004, http://www.ymparisto.fi (accessed March 15, 2005).
7. Kostianoy, A.G.; Lavrova, O.Yu. Introduction. In: *Oil Pollution in the Baltic Sea*. A.G. Kostianoy and O.Yu. Lavrova (Eds.). Springer-Verlag: Berlin, Heidelberg, New York, 2014. Vol. 27, pp. 1–13.
8. Kostianoy, A.G.; Lavrova, O.Yu. (Eds.) *Oil pollution in the Baltic Sea. The Handbook of Environmental Chemistry*. Springer-Verlag: Berlin, Heidelberg, New York, 2014, Vol. 27, 268pp.
9. Carpenter, A. (Ed.) *Oil Pollution in the North Sea*. Springer International Publishing AG: Cham, Switzerland, 2016, 312pp.

10. Carpenter, A.; Kostianoy, A.G. (Eds.) *Oil Pollution in the Mediterranean Sea: Part I – The International Context*. Springer International Publishing AG: Cham, Switzerland, 2018, 350pp.

11. Carpenter, A.; Kostianoy, A.G. (Eds.) *Oil Pollution in the Mediterranean Sea: Part II – National Case Studies*. Springer International Publishing AG: Cham, Switzerland, 2018, 291pp.

12. Hanninen, S.; Rytkonen, J. *Oil transportation and terminal development in the Gulf of Finland*; VIT publication N 547; VIT Technical Research Center of Finland, 2004, 141pp.

13. Kuronen, J.; Helminen, R.; Lehikoinen, A.; Tapaninen, U. *Maritime transportation in the Gulf of Finland in 2007 and in 2015*; Publications from the Centre for Maritime Studies, University of Turku, 2008, N A-45, 114pp.

14. HELCOM. Maritime activities in the Baltic Sea—An integrated thematic assessment on maritime activities and response to pollution at sea in the Baltic Sea Region. *Baltic Sea Environment* **2010**, *123*, 68.

15. Global Marine Oil Pollution Information Gateway, 2004, http://oils.gpa.unep.org/framework/region-2next.htm (accessed March 15, 2005).

16. Annual Report on discharges observed during aerial surveillance in the Baltic Sea in 2017. HELCOM, 2018, http://www.helcom.fi/Lists/Publications/HELCOM%20Annual%20report%20on%20discharges%20observed%20during%20aerial%20surveillance%20in%20the%20Baltic%20Sea%202017.pdf (accessed August 13, 2019).

17. Kostianoy, A.G.; Lebedev, S.A.; Litovchenko, K.Ts.; Stanichny, S.V.; Pichuzhkina, O.E. Satellite remote sensing of oil spill pollution in the southeastern Baltic Sea. *Gayana* **2004**, *68* (2), 327–332.

18. Kostianoy, A.G.; Lebedev, S.A.; Soloviev, D.M.; Pichuzhkina, O.E. *Satellite monitoring of the Southeastern Baltic Sea. Annual Report 2004*; Lukoil-Kaliningradmorneft: Kaliningrad, 2005.

19. Kostianoy, A.G.; Lebedev, S.A.; Litovchenko, K.Ts.; Stanichny, S.V.; Pichuzhkina, O.E. Oil spill monitoring in the Southeastern Baltic Sea. *Environmental Research, Engineering and Management* **2005**, *3* (33), 73–79.

20. Kostianoy, A.G.; Litovchenko, K.Ts.; Lavrova, O.Y.; Mityagina, M.I.; Bocharova, T.Y.; Lebedev, S.A.; Stanichny, S.V.; Soloviev, D.M.; Sirota, A.M.; Pichuzhkina, O.E. Operational satellite monitoring of oil spill pollution in the southeastern Baltic Sea: 18 months experience. *Environmental Research, Engineering and Management* **2006**, *43* (38), 70–77.

21. Bulycheva, E.V.; Kostianoy, A.G. Results of the satellite monitoring of oil pollution in the Southeastern Baltic Sea in 2006–2009. *Current Problems of Remote Sensing of the Earth from Space*, **2011**, *8* (2), 74–83 (in Russian).

22. Bulycheva, E.V.; Kostianoy, A.G. Results of satellite monitoring of the sea surface oil pollution in the Southeastern Baltic Sea in 2004–2013. *Current Problems in Remote Sensing of the Earth from Space*, **2014**, *11* (4), 111–126 (in Russian).

23. Lavrova, O.Yu.; Mityagina, M.I.; Kostianoy, A.G.; Semenov, A.V. Oil pollution in the southeastern Baltic Sea in 2009–2011. *Transport and Telecommunication*, **2014**, *15* (4), 322–331.

24. Kostianoy, A.G.; Bulycheva, E.V.; Semenov, A.V.; Krainyukov, A.V. Satellite monitoring systems for shipping, and offshore oil and gas industry in the Baltic Sea. *Transport and Telecommunication*, **2015**, *16* (2), 117–126.

25. Bulycheva, E.V.; Krek, A.V.; Kostianoy, A.G.; Semenov, A.V.; Joksimovich, A. Oil pollution of the Southeastern Baltic Sea by satellite remote sensing data and in-situ measurements. *Transport and Telecommunication*, **2015**, *16* (4), 296–304.

26. Bulycheva, E.V.; Krek, A.V.; Kostianoy, A.G.; Semenov, A.V.; Joksimovich, A. Oil pollution in the Southeastern Baltic Sea by satellite remote sensing data in 2004–2015. *Transport and Telecommunication*, **2016**, *17* (2), 155–163.

27. Krek, E.V.; Krek, A.V.; Kostianoy, A.G. Seasonal variability of oil pollution in the Southeastern Baltic Sea. *Current Problems in Remote Sensing of the Earth from Space*, **2018**, *15* (3), 171–182 (in Russian).

28. Krek, E.; Kostianoy, A.; Krek, A.; Semenov, A.V. Spatial distribution of oil spills at the sea surface in the Southeastern Baltic Sea according to satellite SAR data. *Transport and Telecommunication,* **2018**, *19* (4), 294–300.

29. Ambjorn, C. Seatrack Web, forecast of oil spills, a new version. *Environmental Research, Engineering and Management,* **2007**, *3* (41), 60–66.

30. SeaTrackWeb User Guide, 2019, https://stw.smhi.se/player/help/classic/?domain=helcom (accessed on August 13, 2019).

31. Kostianoy, A.G.; Ambjörn, C.; Solovyov, D.M. Seatrack web – A numerical tool for environmental risk assessment in the Baltic Sea. In: *Oil Pollution in the Baltic Sea,* A.G. Kostianoy and O.Yu. Lavrova (Eds.), Springer-Verlag: Berlin, Heidelberg, New York, 2014, Vol. 27, pp. 185–220.

32. Kostianoy, A.; Ermakov, P.; Soloviev, D. Complex satellite monitoring of the Nord Stream gas pipeline construction. *Proceedings of the US/EU Baltic 2008 International Symposium "Ocean Observations, Ecosystem-Based Management and Forecasting",* May 27–29, 2008, Tallinn, Estonia, 2008.

33. Kostianoy, A.G. Oil pollution: Baltic Sea. In: *Encyclopedia of Environmental Management,* S.E. Jorgensen (Ed.), Taylor & Francis: New York, 2013, Vol. III, pp. 1851–1866.

34. Kostianoy, A.G.; Solovyov, D.M. Operational satellite monitoring systems for marine oil and gas industry. *Proceedings of the 2010 Taiwan Water Industry Conference,* October 28–29, 2010, Tainan, Taiwan, 2010, pp. B173–B185.

35. Lavrova, O.Yu.; Kostianoy, A.G. A catastrophic oil spill in the Gulf of Mexico in April–May 2010. *Russian Journal of Remote Sensing (Issledovanie Zemli iz Kosmosa)* **2010**, *6,* 67–72 (in Russian).

35

Status of Groundwater Arsenic Contamination in the GMB Plain

Abhijit Das, Antara Das, Meenakshi Mukherjee, Bhaskar Das, Subhas Chandra Mukherjee, Shyamapada Pati, Rathindra Nath Dutta, Quazi Quamruzzaman, Khitish Chandra Saha, Mohammad Mahmudur Rahman, Dipankar Chakraborti, and Tarit Roychowdhury

Introduction 369
Background 369
Groundwater Arsenic Contamination in West Bengal, India 370
Groundwater Arsenic Contamination in Kolkata, India 373
Groundwater Arsenic Contamination in Bihar, India............................374
Groundwater Arsenic Contamination in Uttar Pradesh, India..............374
Groundwater Arsenic Contamination in Jharkhand, India 375
Groundwater Arsenic Contamination in Assam, India 375
Groundwater Arsenic Contamination in Manipur, India.......................375
Groundwater Arsenic Contamination in Haryana, India.......................375
Groundwater Arsenic Contamination in Punjab, India.........................376
Groundwater Arsenic Contamination in Bangladesh 376
Conclusion 379
References...379

Introduction

During the past few decades with the discovery of newer arsenic sites around the world, high arsenic contamination in drinking water has emerged as a major public health problem. Asian countries, especially the Ganga–Meghna–Brahmaputra (GMB) plain of India and Bangladesh, are the worst affected among the world arsenic scenario. This chapter gives a brief update of the global arsenic contamination situation with a detailed update of arsenic contamination in the GMB plain. It appears that a good portion of all states and countries in the GMB plain, comprising an area of more than 569,749 km² and a population of more than 500 million, may be at risk from groundwater arsenic contamination

Background

The natural arsenic content in soil is found at an average value of 5 mg/kg in the range of 1–40 mg/kg.[1,2] Agricultural soils have been detected to contain mean As concentrations in the range of 50–60 mg/kg.[3] Large variations of arsenic content can be found in sediments of lakes, rivers, and streams.[4] In surface water and groundwater not polluted by arsenic, its concentration is 10 µg/L, and in open ocean water, its concentration is typically 1–2 µg/L.[5] Arsenic can be found in inorganic forms of As (III) and As (V). Arsenite [As (III)] is the dominant form under reducing conditions, while arsenate [As (V)] is generally

369

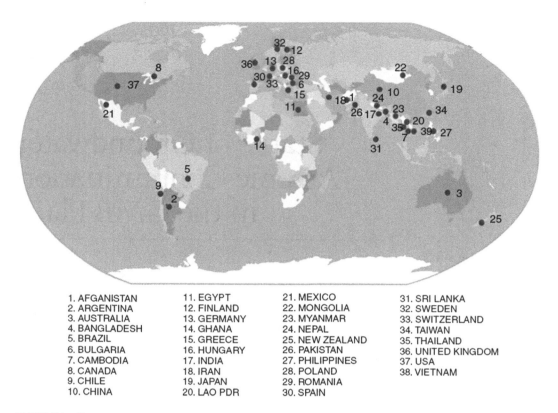

1. AFGANISTAN	11. EGYPT	21. MEXICO	31. SRI LANKA
2. ARGENTINA	12. FINLAND	22. MONGOLIA	32. SWEDEN
3. AUSTRALIA	13. GERMANY	23. MYANMAR	33. SWITZERLAND
4. BANGLADESH	14. GHANA	24. NEPAL	34. TAIWAN
5. BRAZIL	15. GREECE	25. NEW ZEALAND	35. THAILAND
6. BULGARIA	16. HUNGARY	26. PAKISTAN	36. UNITED KINGDOM
7. CAMBODIA	17. INDIA	27. PHILIPPINES	37. USA
8. CANADA	18. IRAN	28. POLAND	38. VIETNAM
9. CHILE	19. JAPAN	29. ROMANIA	
10. CHINA	20. LAO PDR	30. SPAIN	

FIGURE 1   Current arsenic contamination scenario around the world.

the stable form of arsenic in oxygenated environments. Presently, there have been 42 incidents of arsenic contamination reported from 38 countries in the world. Figure 1 illustrates the update of these incidents.

However, the magnitude of arsenic contamination is considered highest in the following four Asian countries: Bangladesh, India, China, and Taiwan. Based on the survey carried out over the last 23 years, it can be said that the groundwater of some parts of all the states (Uttar Pradesh, Bihar, Jharkhand, West Bengal, Arunachal Pradesh, and Assam) in GMB plain in India and six out of seven of the northeastern states (all except Mizoram) will be arsenic affected. By the year 2014, the number of states in the pockets of GMB plain where groundwater is reported to be As polluted has increased from seven to ten; Haryana, Punjab, and Karnataka have been added to the abovementioned list of states.[6] Figure 2 shows the arsenic-affected states and countries in the GMB plain. The scenario of groundwater As contamination in the affected states in India is given in detail in Table 1.

Groundwater Arsenic Contamination in West Bengal, India

The severity of Arsenic pollution in groundwater of the districts was categorized into severely affected (>300 μg/L), mildly affected (10–50 μg/L) and unaffected (<10 μg/L).[6] The districts of Maldah, Murshidabad, Nadia, North 24 Parganas, South 24 Parganas, Bardhaman, Howrah, Hugli, and Kolkata are the nine affected districts. During the last 20 years, SOES analyzed 140,150 hand tube-well water samples for the presence of arsenic above the WHO's recommended level from the aforementioned nine districts of West Bengal.[7] From the findings, it was observed that 49.7% of these tube wells had arsenic >10 μg/L, 24.7% had arsenic >50 μg/L, and 3.4% of the analyzed tube wells had arsenic concentrations >300 μg/L (the concentration predicting overt arsenical skin lesions).[8] A maximum arsenic concentration of 3700 μg/L was found in the Ramnagar village of Gram Panchayat (GP) Ramnagar II,

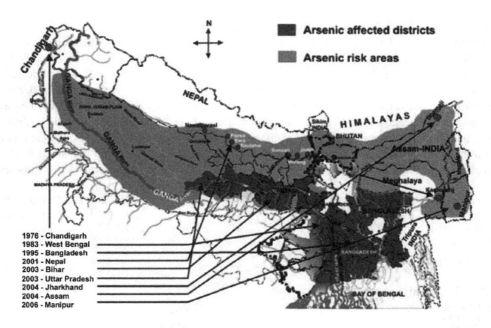

1976 - Chandigarh
1983 - West Bengal
1995 - Bangladesh
2001 - Nepal
2003 - Bihar
2003 - Uttar Pradesh
2004 - Jharkhand
2004 - Assam
2006 - Manipur

FIGURE 2 Groundwater arsenic contamination situation in India and Bangladesh in the GMB plain.

TABLE 1 Groundwater Arsenic Contamination in India[24,25]

S. No.	Name of the States	Name of the Districts	Level of As Contamination (µg/L)
1	West Bengal	Maldah, Murshidabad, Nadia, North 24 Parganas, South 24 Parganas, Bardhaman, Howrah, Hugli, and Kolkata	50–3700
2	Bihar	Darbhanga, Katihar, Khagaria, Kishanganj, Lakhisarai, Munger, Patna, Purnea, Samastipur, Saran, Vaishali, Begusarai, Bhagalpur, Bhojpur, Buxar	>50
3	Uttar Pradesh	Bahraich, Balia, Balrampur, Bareilly, Basti, Bijnor, Chandauli, Ghazipur, Gonda, Gorakhpur, Lakhimpur Kheri, Meerut, Mirzapur, Moradabad, Raebareli, Sant Kabir Nagar, Shahjahanpur, Siddharthnagar, Sant Ravidas Nagar, Unnao	>50
4	Jharkhand	Sahibgunj	>50
5	Assam	Sivasagar, Jorhat, Golaghat, Sonitpur, Lakhimpur, Dhemaji, Hailakandi, Karimganj, Cachar, Barpeta, Bongaigaon, Goalpara, Dhubri, Nalbari, Nagaon, Morigaon, Darrang, and Baksha	50–657
6	Chhattisgarh	Rajnandgaon	52–88
7	Haryana	Ambala, Bhiwani, Faridabad, Fatehabad, Hissar, Jhajjar, Jind, Karnal, Panipat, Rohtak, Sirsa, Sonepat, Yamunanagar	>50
8	Karnataka	Raichur and Yadgir	>50
9	Punjab	Mansa, Amritsar, Gurdaspur, Hoshiarpur, Kapurthala, Ropar etc.	>50
10	Manipur	Bishnupur and Thoubal	798–986
11	Tripura	North Tripura, Dhalai, West Tripura	65–444
12	Nagaland	Mokokchung and Mon	>50

Baruipur block in South 24 Parganas district. Groundwater arsenic contamination in all the districts of West Bengal is shown in Figure 3. The United Nations Children's Fund (UNICEF) collaborated with the Public Health Engineering Department (PHED), Government of West Bengal. They tested 132,262 government-installed hand tube wells and private tube wells for arsenic from eight arsenic-affected districts of West Bengal.[9] Those tube wells that have water with an arsenic concentration >10 µg/L account for 57.9% of the tested hand tube wells; similarly, 25.5% of the tested hand tube wells contained water that has arsenic concentration >50 µg/L. It is estimated that about 9.5 million people could be drinking water contaminated with an arsenic level of 10 µg/L. A recent study on Nadia, West Bengal reported that all the 17 blocks are affected with arsenic in groundwater. It has been found that about 51.4% and 17.3% of the tube wells had As >10 and 50 µg/L, respectively. The maximum observed level of arsenic in

FIGURE 3 Present groundwater arsenic contamination situation in the state of West Bengal in India.

groundwater is found to be 3200 µg/L.[10] Chakdah is one of the worst affected blocks in Nadia where the shallow aquifers as well as the agricultural land show a high magnitude of arsenic content. According to a survey carried out in the nine districts, out of 135,555 samples of groundwater pumped up by tube well, As concentration of above 10 µg/L was reported in 67,306 (49.7%) samples and above 50 µg/L in 33,470 (24.7%) samples. In affected areas, the demographic survey showed that 13.85 million people were under the threat of As concentration above 10 µg/L of which 6.96 million people are exposed to a concentration above 50 µg/L.[11]

Groundwater Arsenic Contamination in Kolkata, India

The present demand of drinking water in Kolkata per day is around 1262 million liters per day (MLD) which is supplied through the pipelines laid down by Metropolitan authority. Out of that, 1096 MLD is treated surface water and the rest is groundwater. A total of 3626 samples of hand tube-well water was collected from 100 out of 141 administrative wards in Kolkata. Tube wells in 65 wards in Kolkata have arsenic concentration levels >10 µg/L, and in another 30 wards, arsenic concentration was found to be >50 µg/L. Based on water samples from 1057 wells, dug deeper than 100 m, 220 samples (20.8%) have shown arsenic concentrations >10 µg/L, 113 samples (10.7%) have arsenic concentrations >50 µg/L, and 72 samples (6.8%) have arsenic concentrations >100 µg/L. Water samples were collected from 977 wells that were 91–100 m deep. Of 977 samples, 149 (15.3%) have arsenic concentrations >10 µg/L and 13 (1.3%) have arsenic concentrations >100 µg/L. However, the tube wells that were deeper than 300 m have been detected with arsenic concentrations >50 µg/L. Maximum arsenic concentration in this region was found to be at 800 µg/L at a depth of 20 m. An analysis of water from these Kolkata Municipal Corporation (KMC) wells shows that out of 734 wells, 121 (16.5%) have arsenic concentrations >10 µg/L, 26 (3.6%) have arsenic concentrations of 50 µg/L, and only 2 (0.2%) have arsenic concentrations >300 µg/L. It has also been found that the southern part of the city is more contaminated by arsenic than the northern and central parts (Figure 4).

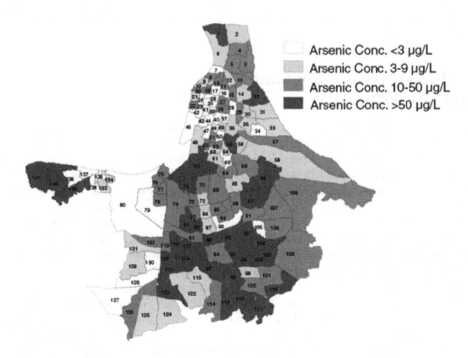

Arsenic Conc. <3 µg/L
Arsenic Conc. 3-9 µg/L
Arsenic Conc. 10-50 µg/L
Arsenic Conc. >50 µg/L

FIGURE 4 Present groundwater arsenic contamination situation in all the 141 wards of KMC, West Bengal.

At present in Kolkata, 70% of KMC area, i.e., 100 out of 144 wards, has been reported with As concentration above permissible limit in shallow unconfined aquifers with 51 wards (i.e. 35.4%) having As concentration >50 µg/L and remaining 49 wards having As concentration between 11 and 50 µg/L.[12] During the past 23 years, 14.2% of samples throughout 77 wards in KMC, exceeded the permissible limit of As in drinking water and around 5.2% samples in 37 wards showed five times higher value of the permissible limit of As in drinking water.[13]

Groundwater Arsenic Contamination in Bihar, India

Water samples from 19,961 tube wells were analyzed, and arsenic content in water samples was found to be >50 µg/L in 12 districts of Bihar. SOES's analysis of water samples from these tube wells revealed that 313 villages have water containing arsenic concentrations >10 µg/L and 240 villages have water containing arsenic concentrations >50 µg/L. Analytical results show that 32.7% of tube wells in that region have concentrations >10 µg/L, 17.75% tube wells have arsenic concentrations >50 µg/L, and 4.55% have arsenic concentrations >300 µg/L. The maximum arsenic concentration was found in the Chakani village of Brahampur GP belonging to the Barahara block of Buxar district. The level of arsenic concentration in its tube well's water is 2182 µg/L. In the two districts of Bihar, Vaishali, and Bhagalpur were reported to have arsenic contamination in groundwater where the maximum arsenic concentration was found to be 20 and 143 µg/L, respectively; 73% of the groundwater samples showed As (III)-dominant species whereas 27% showed As (V). Katihar being the most vulnerable district because of the socioeconomic and biophysical conditions needs a serious concern followed by Vaishali, Samastipur, Khagaria, and Purnia. They investigated 17 districts out of 37 and a total of 87 of 532 community blocks for groundwater arsenic contamination where 72 community blocks are reported with arsenic beyond permissible limit in drinking water.[14] On a further analysis by SOES, Jadavpur, it has been observed that on analysis of 4704 tube-well water samples from all 88 villages of Shahpur block of the district Bhojpur of Bihar, India, 21.1% of water samples is contaminated with arsenic >50 µg/L according to BIS, 2009, whereas nearly 40.3% of water samples is found containing arsenic >10 µg/L (WHO, 2004). The mean and maximum arsenic concentration is found to be 43.5 and 1805 µg/L, respectively.

Groundwater Arsenic Contamination in Uttar Pradesh, India

It was perceived in 2003 that UP is arsenic contaminated.[15] The rural population of the Allahabad district and the urban population of Shuklaganj-Kanpur of the Unnao district in the Allahabad-Kanpur track of the upper Ganga plain first reported cases of arsenic contamination in 49 districts in January 2009.[16] A recent report[11] showed that in a survey done in 2003 in 25 villages of Ballia district, Ghazipur, and Varanasi were added making a total of 20 districts affected by As where the people mostly depended on water from hand pumps which pumped the groundwater from shallow aquifers (20 m deep). Furthermore, it was reported[17] that only 3 out of 72 samples had As concentration <10 µg/L, while 95.83% had As concentration above the permissible limit. The maximum As concentration in the samples collected during pre-monsoon and monsoon was reported to be 75.6 and 74.46 µg/L, respectively. A total of 136 groundwater samples were collected and analyzed from 12 villages on the floodplain of the Rapti River. About 42% of 103 samples collected from nine villages situated on the left bank were reported to have As concentration >10 µg/L with maximum concentration of 399 µg/L while none of the samples collected from the three villages situated on the right bank had As concentration above the permissible limit.[18] On analyzing 5044 water samples from five districts of UP, SOES found 43.1% of those samples to have arsenic concentrations >10 µg/L and 27.5% to have arsenic concentrations >50 µg/L.

Groundwater Arsenic Contamination in Jharkhand, India

In this state, there are only three districts affected by arsenic concentration.[19] It was observed that the groundwater As concentration in Sahibganj district during three seasons was highest in post-monsoon ranging between 1 and 133 µg/L compared to (2–98 µg/L) in monsoon and (7–115 µg/L) in pre-monsoon with the trend of As concentration increasing with depth.[20] SOES has conducted several surveys in the nine blocks of the Sahibganj district (in the middle of Ganga plain) from December 2003 till now, where they have analyzed 3832 hand tube-well water samples for arsenic concentration. The findings suggest that 32.28% of the water samples had arsenic concentrations >10 µg/L, 13.44% contained arsenic concentrations >50 µg/L, and 2.61% had arsenic concentrations >300 µg/L. The highest arsenic concentration in water has been detected in the Hajipur Vitta village of Hajipur Porsun GP, Sahibganj block, in the Sahibganj district of Jharkhand. The amount of arsenic concentration found in the water sample from that region is 1018 µg/L.

Groundwater Arsenic Contamination in Assam, India

Dhemaji and Karimganj in Assam have been surveyed for groundwater arsenic contamination, and both have been found arsenic contaminated in January 2004. So far, water samples from 241 hand tube wells have been analyzed for arsenic contamination from these two districts, and it has been found that 42.3% of these samples have arsenic concentrations >10 µg/L and 19.1% of these samples have arsenic concentrations >50 µg/L.

A total of 27 villages have been identified from these two districts where groundwater arsenic concentrations are above 50 µg/L. Initially, As concentration >50 µg/L was found in a few samples of groundwater collected from Karimganj, Dhubri, and Dhemaji districts. Eighteen out of twenty three districts covering 76 blocks, and a population of 603 are reported to be affected by As toxicity according to a new report by UNICEF.[11] A work in 2014 reported that the mean As concentration in the groundwater of Titabor, Jorhat, Assam is 210 µg/L.[21] It was also noticed that among the six blocks of Golaghat district, Assam, the most severely affected area was Gamariguri block, where 100% of the samples had an As concentration >10 µg/L. The other blocks were also badly affected showing arsenic contamination in groundwater by 87.5%, 71.9%, 42.4%, 42.1%, and 36.4% in the blocks Podumoni, Kathalguri, Dergaon, Morangi, and Kakodonga, respectively.[22]

Groundwater Arsenic Contamination in Manipur, India

Since the preliminary study in Manipur in May 2006, SOES confirmed groundwater arsenic contamination in two districts of Manipur. They are the districts of Thoubal and Imphal. Findings suggest that 41.27% of the collected and analyzed samples have arsenic concentrations >50 µg/L and 64.72% of the samples have arsenic concentrations >10 µg/L. After the preliminary study by SOES in 2006, again 628 water samples of tube wells were analyzed in the Manipur valley which consists of four districts, where 63.3% sample holds arsenic concentration in groundwater >10 µg/L. Among the four districts, Thoubal is the most affected one where 77.6% groundwater contains arsenic >10 µg/L.[23]

Apart from Assam and Manipur, the other north-eastern states like Arunachal Pradesh, Nagaland, and Tripura are also reported with high range of arsenic contamination in groundwater (50–986 µg/L).

Groundwater Arsenic Contamination in Haryana, India

The districts on the banks of Yamuna, originating from Himalaya, and its tributaries are covered by alluvial aquifers resulting in groundwater contamination due to mobilization of As from As-containing minerals. The alluvial sediments are the main source of As in the sporadic region. Under Aquifer Mapping Project, 2013, groundwater As concentration >50 µg/L has been reported in Chandigarh.[11]

Groundwater Arsenic Contamination in Punjab, India

In 2004, 261 shallow groundwater samples were analyzed to find As concentration more than 10 µg/L in 12 districts including Ropar, Amritsar, Hoshiarpur, Gurdaspur, and Kapurthala. In July 2010, out of 105 samples collected from the district of Mansa at a depth ranging from 13 to 35 m, six locations were reported with As concentration more than 10 µg/L.[11]

Groundwater Arsenic Contamination in Bangladesh

In International Conference on Arsenic, Dhaka, held in 1998, it was reported that 66% of water from hand tube wells from Bangladesh contain arsenic concentrations >10 µg/L and 51% have arsenic concentration levels >50 µg/L. The researchers also said that the arsenic detected in the groundwater of 52 districts is >10 µg/L and that in 41 districts is >50 µg/L.[26] In surveys carried out by the British Geological Survey (BGS) along with the Department of Public Health Engineering (DPHE), it was further reported that water samples from 46% of 3534 analyzed tube wells exceeded arsenic concentrations of 10 µg/L and 27% of the samples exceeded arsenic concentration levels of 50 µg/L. This survey had taken samples from tube wells from all over Bangladesh, leaving out the Chittagong Hill tract. Based on their study, BGS-DPHE reported that 57 million and 35 million people could be drinking As-contaminated water >10 µg/L and 50 µg/L, respectively.[27] Between 2000 and 2001, half of 6500 tube wells were sampled for testing by Van Geen in the Araihazar Police Station, which is known as Thana/Upazila. It was found that half of these water samples contain arsenic with concentrations >10 µg/L and one-quarter of these samples contain arsenic concentrations >50 µg/L.[28] It can be found from this database that 1.4 million out of 4.8 million water samples, which is 30% of the water samples in Bangladesh, contain arsenic concentrations >50 µg/L.[29]

The important incidents related to groundwater arsenic contamination in the GMB plain are shown in Table 2.

Bangladesh can be divided into four existing geomorphological regions of which the hill tract and the tableland are free from arsenic contamination. The highly contaminated regions of Bangladesh fall within the deltaic region (Figure 5).

Arsenic content in water in the tableland area (detected in 17 out of 9755 analyzed hand tube wells, which is 0.2%) is >50 µg/L. In another 204 out of the total of 9755, which is 2.1%, arsenic concentration was found between 10 and 50 µg/L. In the floodplain region of Bangladesh, there are 229 thanas in total. Out of these, 19,845 water samples were collected from 158 thanas and were analyzed which suggest that 50.8% of these samples have arsenic concentrations greater than 10 µg/L and another 34.9% of the samples have arsenic concentrations >50 µg/L. Another survey by BGS-DPHE in 1999 has brought out data that indicate that 4587 out of 16,513 (27.8%) of the water samples analyzed have shown concentrations of arsenic >50 µg/L. This survey was also held in the same region. A total of 22,113 water samples from hand tube wells in the deltaic region including the coastal belt were taken for analysis. These samples were collected from 97 of the 139 thanas in that region. On analysis, it was found that 54.7% of the samples have arsenic concentrations >10 µg/L and 32.8% >50 µg/L. In 1999, BGS-DPHE also reported about 12,245 water samples from 126 thanas of this region. A total of 3344 samples (27.5%) have arsenic concentrations >50 µg/L. The study had only collected samples from Chittagong and Cox's Bazar districts. The percentage of samples that have arsenic concentrations >1000 µg/L in the floodplains is 1.14% (227 samples), which is greater than % found in the deltaic region, which is 0.33% (74 samples). Shallow tube wells varying in depth from 10 to 70 m are highly contaminated compared to deep aquifers leading to Arsenic contamination in 61 out of 65 districts of Bangladesh.[30] The maximum arsenic concentration found in the three regions are 134, 4730, and 2190 µg/L, respectively (n = 9755, 19,845 and 22,113, respectively). Groundwater As contamination in different areas of Bangladesh reported so far has been given in Table 3.

TABLE 2 Groundwater Arsenic Contamination in India and Bangladesh (September 2010)

Parameters	West Bengal	Bihar	UP	Jharkhand (Sahibganj)	Assam	Manipur	Chhattisgarh (Rajnandgaon)	Bangladesh
Area (km²)	88,750	94,163	238,000	1,600	78,438	22,327	6,396	147,570
Population (in millions)	80.2	83	166	1	26.6	2.29	1.5	124.3
Arsenic-affected area (km²)	38,861	21,271	17,919	725	8,822	2,238	6,396	–
Total arsenic-affected districts (As > 50 µg/L)	12	12	5	1	3	4	1	50
Total arsenic-affected blocks/PS (As >50 µg/L)	111	36	9	3	9	–	–	197
No. of villages where groundwater arsenic >50 µg/L	3417	235	70	68	–	–	146	–
Total hand tube-well water samples analyzed	140,231	19,961	5,044	3,354	1,448	628	146	52,202
% of samples having arsenic >10 µg/L	48.1	32.70	46.1	36.1	39.0	63.3	25.34	43.0
% of samples having arsenic >50 µg/L	23.8	17.75	26.9	15.4	13.81	23.2	8.22	27.2
Total number of biological samples analyzed (hair, nail, urine)	39,624	1,833	258	367		57 (urine)		11,298
Total people screened by SOES-JU	96,000	3,012	989	522	–	–	a	18,991
People registered with arsenical skin lesions	9,356	457	154	71	–	–	80	3,762
People drinking arsenic-contaminated water >10 µg/L (in millions)	9.5	3.1	1.1	0.15				36.6
People drinking arsenic-contaminated water >50 µg/L (in millions)	4.6	1.7	0.6	0.06	1.2			22.7
Population potentially at risk from arsenic contamination >10 µg/L (in millions)	26	9	3	0.4		1		

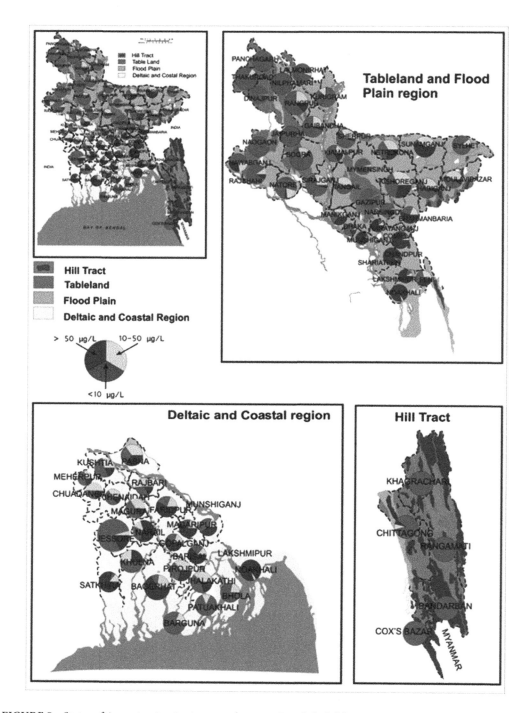

FIGURE 5   Status of As contamination in groundwater in Bangladesh.[26]

Bangladesh and West Bengal are considered the two worst arsenic-affected areas in the world. In Bangladesh, 42.1%, 27.2%, and 7.5% of the hand tube wells had As 10, 50, and 300 µg/L, respectively, whereas in West Bengal, the corresponding values were 48.1%, 23.8%, and 3.3%. SOES has detected 301—(0.6%) hand tube-well water samples with arsenic concentrations >1000 µg/L in Bangladesh, whereas 187 (0.1%) water samples had arsenic concentrations >1000 µg/L in West Bengal.

TABLE 3 Arsenic Contamination Scenario in Groundwater of Bangladesh

S. No.	Cases of As Contamination Reported	Reference
1	Groundwater of 47 districts was reported to be As contaminated on the basis of 22,003 tube-well water samples collected over 64 districts and 6000 samples which were collected over 25 km^2 area and reported to have As concentration exceeding the Bangladesh drinking water standard of 50 µg/L	[31]
2	1,00,000 tube wells in Chandpur	[32]
3	Groundwater As concentration varies from 6 to 934 µg/L with an average of 347 µg/L in Meghna river delta	[33]
4	As concentration >50 µg/L found in 27.2% of 5202 water samples collected from 64 districts	[34]
5	60% tube wells are severely affected in southern and eastern Bangladesh with a maximum concentration of 1000 µg/L	[35–37]

Conclusion

The situation of As contamination has been deteriorating ever since the first evidence of As in Asian countries such as India, Bangladesh, China, Nepal, Taiwan, and Vietnam as well as Mexico, Argentina, etc. Besides the geogenic cause, excessive withdrawal of groundwater to satisfy the increasing demand of the growing population has played an important role in groundwater As pollution. In spite of several mitigation strategies already undertaken, the global situation is even worse than before due to lack of awareness and insufficient steps taken to deal with the problem in domestic level. Since the toxicity of As is directly dependent on the nutritional status as well as economic condition of the people, efforts should be made to improve the condition at ground level.

References

1. Beyer, W.N.; Cromartie, E.J. A survey of Pb, Cu, Zn, Cd, Cr, As and Se in earthworms and soil from diverse sites. *Environmental Monitoring and Assessment* **1987**, *8*(1), 27–36.
2. Bowen, H.J.M. *Elemental Chemistry of the Elements*; Academic Press: London, 1979.
3. Sanok, W.J.; Ebel, J.G.; Manzell, K.L.; Gutenmann, W.H.; Lisk, D.J. Residues of arsenic and lead in potato soils on Long Island. *Chemosphere* **1995**, *30*(4), 803–806.
4. Welch, A.H.; Lico, M.S.; Hughes, J.L. Arsenic in groundwater of the western United States. *Ground Water*, **1988**, *26*(3), 333–347.
5. Smith, R.A.; Alexander, R.B.; Wolman, M.G. Water quality trends in the nation's rivers. *Science* **1987**, *235*, 1607–1615.
6. Bhattacharya, A.K. An analysis of arsenic contamination in the groundwater of India, Bangladesh and Nepal with a special focus on the stabilisation of Arsenic Laden Sludge from arsenic filters. *Electronic Journal of Geotechnical Engineering* **2019** *24*(1).
7. Chakraborti, D.; Das, B.; Rahman, M.M.; Chowdhury, U.K.; Biswas; Goswami, A.B.; Nayak, B.; Pal, A.; Sengupta, M.K.; Ahamed, S.; Hossain, M.A.; Basu, G.; Roychowdhury, T.; Das, D. Status of Groundwater arsenic contamination in the state of West Bengal, India: A 20-year study report. *Molecular Nutrition & Food Research* **2009**, *53*, 542–551.
8. Rahman, M.M.; Chowdhury, U.K.; Mukherjee, S.C.; Mondal, B.K.; Paul, K.; Lodh, D.; Chanda, C.R.; Basu, G.K.; Biswas, B.K.; Saha, K.C.; Roy, S.; Das, R.; Palit, S.K.; Quamruzzaman, Q.; Chakraborti, D. Chronic arsenic toxicity in Bangladesh and West Bengal-India. *Journal of Toxicology: Clinical Toxicology* **2001**, *39*(7), 683–700.
9. Nickson, R.; Sengupta, C.; Mitra, P.; Dave, S.N.; Banerjee, A.K.; Bhattacharya, A.; Basu, S.; Kakoti, S.; Moorthy, N.S.; Wasuja, M.; Kumar, M.; Mishra, D.S.; Ghosh, A.; Vaish, D.P.; Srivastava, A.K.; Tripathi, R.M.; Singh, S.N.; Prasad, R.; Bhattacharya, S.; Deverill, P. Current knowledge on arsenic in groundwater in five states of India. *Journal of Environmental Science and Health Part A* **2007**, *42*, 1707–1718.

10. Rahman, M. M.; Mondal, D.; Das, B.; Sengupta, M. K.; Ahamed, S.; Hossain, M. A.; Samal, A.C.; Saha, K.C.; Mukherjee, S.C.; Dutta, R.N.; Chakraborti, D. Status of groundwater arsenic contamination in all 17 blocks of Nadia district in the state of West Bengal, India: A 23-year study report. *Journal of Hydrology* **2014**, *518*, 363–372.

11. Bhattacharya, A. K.; Lodh, R. Arsenic contamination in the groundwater of India with a special focus on the stabilization of Arsenic-Laden Sludge from arsenic filters. *Electronic Journal of Geotechnical Engineering* **2018**, *23*(1), 575–600.

12. Malakar, A.; Islam, S.; Ali, M. A.; Ray, S. Rapid decadal evolution in the groundwater arsenic content of Kolkata, India and its correlation with the practices of her dwellers. *Environmental Monitoring and Assessment* **2016**, *188*(10), 584.

13. Chakraborti, D.; Das, B.; Rahman, M. M.; Nayak, B.; Pal, A.; Sengupta, M. K.; Ahamed, S.; Hossain, M.A.; Chowdhury, U.K.; Biswas, B.K.; Saha, K. C. Arsenic in groundwater of the Kolkata Municipal Corporation (KMC), India: Critical review and modes of mitigation. *Chemosphere* **2017**, *180*, 437–447.

14. Singh, S. K. Groundwater arsenic contamination in the Middle-Gangetic Plain, Bihar (India): the danger arrived. *International Research Journal of Environmental Science* **2015**, *4*(2), 70–76.

15. Chakraborti, D.; Mukherjee, S.C.; Pati, S.; Sengupta, M.K.; Rahman, M.M.; Chowdhury, U.K.; Lodh, D.; Chanda, C.R.; Chakraborti, A.K.; Basu, G.K. Arsenic groundwater contamination in middle Ganga Plain, Bihar, India: A future danger. *Environmental Health Perspectives* **2003**, *111*, 1194–1201.

16. Chakraborti, D.; Ghorai, S.; Das, B.; Pal, A.; Nayak, B.; Shah, B.A. Arsenic exposure through groundwater to the rural and urban population in the Allahabad-Kanpur track in the upper Ganga plain. *Journal of Environmental Monitoring*, **2009**, *11*, 1455–1459.

17. Singh, A. L.; Singh, V. K. Assessment of groundwater quality of Ballia district, Uttar Pradesh, India, with reference to arsenic contamination using multivariate statistical analysis. *Applied Water Science* **2018**, *8*(3), 95.

18. Singh, C. K.; Kumar, A.; Bindal, S. Arsenic contamination in Rapti River Basin, Terai region of India. *Journal of Geochemical Exploration* **2018**, *192*, 120–131.

19. Nayak, B.; Das, B.; Mukherjee, S.C.; Pal, A.; Ahamed, A.; Hossain, M.A.; Maity, P.; Dutta, R.N.; Dutta, S.; Chakraborti, D. Groundwater arsenic contamination in Sahibganj district of Jharkhand state, India in middle Ganga plain and adverse health effects. *Toxicology & Environmental Chemistry* **2007**, *90* (4), 673–694.

20. Alam, M. O.; Shaikh, W. A.; Chakraborty, S.; Avishek, K.; Bhattacharya, T. Groundwater arsenic contamination and potential health risk assessment of Gangetic Plains of Jharkhand, India. *Exposure and Health* **2016**, *8*(1), 125–142.

21. Sailo, L.; Mahanta, C. Arsenic mobilization in the Brahmaputra plains of Assam: groundwater and sedimentary controls. *Environmental Monitoring and Assessment* **2014**, *186*(10), 6805–6820.

22. Chetia, M.; Chatterjee, S.; Banerjee, S.; Nath, M. J.; Singh, L.; Srivastava, R. B.; Sarma, H. P. Groundwater arsenic contamination in Brahmaputra river basin: a water quality assessment in Golaghat (Assam), India. *Environmental Monitoring and Assessment* **2011**, *173*(1–4), 371–385.

23. Chakraborti, D.; Singh, E. J.; Das, B.; Shah, B. A.; Hossain, M. A.; Nayak, B.; Ahamed, S. & Singh, N. R. Groundwater arsenic contamination in Manipur, one of the seven North-Eastern Hill states of India: a future danger. *Environmental Geology* **2008**, *56*(2), 381–390.

24. Sanyal, S. K. Arsenic toxicity in soil-plant-human continuum and remedial options. In: *Dialogue- Science, Scientists, and Society.* Ed. M. Dadlani, 2018, 1–17. doi:10.29195/DSSS.01.01.0006

25. Kumar, C. P. Status and mitigation of arsenic contamination in groundwater in India. *International Journal of Earth & Environmental Sciences* **2015**, *1*(1), 1–10.

26. Chakraborti, D.; Rahman, M.M.; Das, B.; Murrill, M.; Dey, S.P.; Mukherjee, S.C.; Dhar, R.K.; Biswas, B.K.; Chowdhury, U.K.; Roy, S.; Sorif, S.; Selim, M.; Rahman, M.; Quamruzzaman, Q. Status of groundwater arseniccontamination in Bangladesh: A 14-year study report. *Water Research* **2010**, *44*(19), 5789–5802.

27. BGS-DPHE. Arsenic Contamination of Groundwater in Bangladesh, BGS Technical Report WC/00/19; British Geological Survey: Keyworth, UK, 2001.

28. Van Geen, A.; Cheng, Z.; Jia, Q.; Seddique, A.A.; Rahman, M.W.; Rahman, M.M.; Ahmed, K.M. Monitoring 51 community wells in Araihazar, Bangladesh, for up to 5 years: Implications for arsenic mitigation. *Journal of Environmental Science & Health* **2007**, *42*, 1729–1740.

29. Johnston, R.B.; Sarker, M.H. Arsenic mitigation in Bangladesh: National screening data and case studies in three upazilas. *Journal of Environmental Science & Health* **2007**, *42*, 1889–1896.

30. BGS (British Geological Survey). Executive Summary of the Main Report of Phase I, Groundwater Studies of As Contamination in Bangladesh. British Geological Survey and Mott MacDonald (UK) for the Government of Bangladesh, Ministry of Local Government, Rural Development and Cooperatives DPHE and DFID (UK), 2000. Available at http://www.engconsult.com/arsenic/article/DFID-sum.html (accessed in February 2014)

31. Van Geen, A.; Zheng, Y.; Versteeg, R.; Stute, M.; Horneman, A.; Dhar, R.; Steckler, M., Gelman, A.; Small, C.; Ahsan, H. and Graziano, J.H. Spatial variability of arsenic in 6000 tube wells in a 25 km^2 area of Bangladesh. *Water Resources Research* **2003**, *39*(5).

32. Islam, A. B. M. Arsenic pollution and the probable bioremediation (Doctoral dissertation, PhD Thesis. Kanazawa University, Japan), 2003.

33. Bibi, M. H.; Ahmed, F.; & Ishiga, H. Geochemical study of arsenic concentrations in groundwater of the Meghna River Delta, Bangladesh. *Journal of Geochemical Exploration* **2008**, *97*(2–3), 43–58.

34. Chakraborti, D.; Rahman, M. M.; Das, B.; Murrill, M.; Dey, S.; Mukherjee, S. C.; Dhar, R. K.; Biswas, B.K., Chowdhury, U.K., Roy, S., Sorif, S. Status of groundwater arsenic contamination in Bangladesh: a 14-year study report. *Water Research* **2010**, *44*(19), 5789–5802.

35. Escobar, M. O.; Hue, N. V.; Cutler, W. G. Recent developments on arsenic: contamination and remediation. *Recent Research Developments in Bioenergetics* **2006**, *4*, 1–32.

36. Kinniburgh, D. G.; Kosmus, W. Arsenic contamination in groundwater: some analytical considerations. *Talanta* **2002**, *58*(1), 165–180.

37. Uddin, R.; Huda, N. H. Arsenic poisoning in Bangladesh. *Oman Medical Journal* **2011**, *26*(3), 207–208.

36

Yellow River

Zixi Zhu, Ynuzhang
Wang, and Yifei Zhu

Introduction 383
Discussion 383
Conclusion 385
References...385

Introduction

Since 1949, China has made unremitting efforts to harness the Yellow River. A number of water control projects were undertaken, hydropower stations had been built, and the dykes in the lower reaches had been strengthened and heightened. The eroded land on the Loess Plateau had been harnessed and water and soil loss was restricted. Soil and water conservation improved both agricultural production conditions and the ecological environment. Since 1949, the Yellow River has never been breached, thus ensuring the safety of its people and property and promoting the development of economy and society.

Discussion

The Yellow River originates from the Yueguzonglie Basin, which has an elevation of 4500 m and is located at the northern slope of the Bayankera Mountain in the Qinghai–Tibet Plateau. From west to east, it flows through nine provinces/autonomous regions: Qinghai, Sichuan, Gansu, Ningxia, Inner Mongolia, Shanxi, Shaanxi, Henan, and Shandong, and empties into the Bohai Sea in Kenli County, Shandong Province. The trunk is 5464 km in length and the basin area is 795,000 km². Within the river basin, the population is 110.08 million, accounting for 8.7% of the total population in China; the cultivated land is 13.1 million ha.[1] The Yellow River basin is the cradle of Chinese nationality and the birthplace of ancient Chinese civilization. As early as 1 million years ago, the "Lantian Man" had been living in the Yellow River Basin. In this very long historical period, the basin was the center of politics, economy, and culture in China.

The terrain of the Yellow River basin is high in the west and low in the east, thus showing great disparity. The west section of the basin belongs to the Qinghai–Tibet Plateau with an elevation of 3000–5000 m. The middle section is mostly located in the Loess Plateau and reaches the east side of Taihang Mountain, with an elevation of 1000–2000 m. The east section starts from east of Taihang Mountain and ends at the Bohai Sea, lying in the Huanghuaihai Plain, with an elevation of lower than 100 m.[1] Based on the geographical position and the feature of the river, the trunk stream of the Yellow River can be divided into upper, middle, and lower reaches. The upper reaches are from the source of the river to Hekouzhen (Tuoketuo county) in Inner Mongolia, with length of 3472 km and area of 428,000 km². The middle reaches are from Hekouzhen, then extend to Taohuayu (Zhengzhou) of Henan province, comprising 1206 km in length and 344,000 km² in area. Starting from Taohuayu, the lower reaches run through nearly 800 km, occupy 23,000 km² of area, and end at the mouth of the river.[1] Because of the height difference, the climate in the Yellow River basin is very distinct, ranging from arid climate in the west through semiarid to semihumid climate in the east. Average annual temperature is −4°C at the source, 1–8°C in the upper reaches, 8–12°C in the middle reaches, and 12–14°C in the lower reaches. The average annual precipitation of the basin is 452 mm. The maximum is in the southeast part of the basin, reaching 800–1000 mm; whereas the minimum annual precipitation is less than 200 mm in the northwest part of the basin, including Ningxia and Inner Mongolia, which features the inland climate.[1] The Yellow River basin is rich in mineral and energy resources that are of great importance in China. Out of the 45 proved major mineral resources in China, 37 are found in the Yellow River basin.[1] The water energy in the upper-middle reaches of Yellow River, the coal in the middle reaches, and the oil and natural gas in the middle-lower reaches are all quite rich in deposits. Thus the Yellow River basin is called the "energy resources basin," playing an important role in China.[1]

The Yellow River provides the major water resource for northwest and north China, but the amount is comparatively poor. The average annual runoff in the basin is 58 billion m³, making up only 2% of that in China, whereas the area of the Yellow River basin accounts for 8.3% of the land area of the country.[1] The distributions of runoff in different areas are different. Annual runoff in the upper, middle, and lower reaches contributes 55.6%, 40.8%, and 4.6% of the total annual runoff, respectively. It also varies with seasons. More than 60% of the annual runoff happens during the period from July to October, while less than 40% occurs from November to June.[1] The Yellow River basin is an important production area in agriculture. The main crops are wheat, cotton, oil-bearing plants, tobacco, etc. The irrigation area along the Yellow River basin is about 7.3 million ha,[2] most of which is in the Ningxia–Inner Mongolia Plain at the upper reaches, Fen-Wei basin at the middle reaches, and the irrigation area drawing water from the Yellow River at the lower reaches. Agricultural irrigation consumes 28.4 billion m³ of water from the Yellow River annually, accounting for 92% of the overall annual water consumption of the river.[2] With high-speed development of the economy and continued population growth, water consumption has increased rapidly and the competition between water supply and demand has become more acute. As a result, zero flow in the lower reaches has resulted in recent years. From 1972 through 1998, zero flow occurred in 21 years with accumulated duration of 1051 days. The worst was in 1997, when Lijin, near the river mouth, had zero flow for 226 days; the zero-flow section extended upstream even to Kaifeng of Henan Province.[2]

The Yellow River flows through the Loess Plateau, where the surface is characterized by loose soil and sparse vegetation, and the climate is dominated by dry weather and heavy storms concentrated mostly in the summer. Therefore, it is the largest area affected by water and soil loss and the strongest intensity in erosion in China. Based on the data collected by remote sensor in 1990, the area experiencing water loss and soil erosion is up to 454,000 km², which makes up 70.9% of the Loess Plateau's area. The area of water erosion with the annual erosion mean exceeding 8000 Mg/km² is 85,000 km², accounting for 64% of the congener area in China. The severe water degradation area with annual erosion mean exceeding 15,000 Mg/km² is 36,700 km², about 89% of the congener area.[1] Average annual amount of sediment and sand washed into the river is about 1.6 billion Mg with sand content of 35 kg/m³. The maximum sand content is 933 kg/m³, which was measured at Longmen on July 18, 1966.[3] In the lower reaches of the Yellow River, a large amount of sediment is left behind to raise the riverbed. Nearly 9.2 billion Mg of sediment were deposited in the lower reaches of the Yellow River from 1950 through 1998. Thus, the riverbed is 4–6 m higher than the ground outside the river on an average. At some places, this number could even reach 10 m or more.[2] For this reason, the lower reaches of the Yellow River are called the "suspended river." Because of rainstorms occurring in the middle reaches and the channel in the lower reaches being wider in the upper part and narrower in the lower part, the lower reach area frequently suffers from heavy floods. Records indicated that from 602 BC to 1949 AD, the Yellow River was breached 1590 times and changed its route 26 times, i.e., on the average, "breach twice every three years and changing its route every century."[4] Heavy floods affected a large area from Tianjin City in the north to Huaihe River in the south, crossing 250,000 km². On the other hand, drought is another disaster that occurs frequently in the Yellow River basin. In 582 years from 1368 to 1949, severe drought occurred in 107 years, once every 5.4 years.[5] Records show that an extraordinarily serious drought lasted 4 years, from 1875 to 1878, through the whole basin of the Yellow River.

Harnessing of the Yellow River has always been a major issue concerning China's prosperity and the people's peaceful life. Since 1949, China has made unremitting efforts to harness the Yellow River and, therefore, ensured the safety of its people and property, promoted the development of economy and society, and improved the ecological environment. On the trunk of the Yellow River, 15 key water control projects and hydropower stations have been built or are being built, providing a total water capacity of 56.6 billion m³, a total installed capacity of 11.13 million kW, and an average annual power supply of 40.1 billion kWh.[2] By the end of 2000, 1400 km of dykes along the lower reaches had been strengthened and heightened four times. Large-scale channel improvement had been performed. Reservoirs exceeding 10,000 in number and in different sizes have been built, with a total storage capacity of 72 billion m³. Among them, 22 of the largest reservoirs are able to hold 61.7 billion m³ of water.[2] Other projects for irrigation and water supply had also been accomplished and the groundwater had begun to be developed. As a result, the irrigation area has increased from 0.8 million ha in 1950 to 7.3 million ha now, including 2.4 million ha that are located outside of the basin.[2] Until 2000, the area of 180,000 km² of eroded land on the Loess Plateau, which is one-third of the land with soil erosion, had been harnessed. To some extent, water and soil loss and desertification have been restricted. The amount of sediment and sand being washed into the Yellow River each year had decreased by about 300 million Mg.[2] Soil and water conservation improved the agricultural production condition and ecological environment. By taking these measures, the average annual grain yield has increased by more than 5 billion kg, which can provide enough food and clothing for more than 10 million people.[2] Since 1949, the Yellow River has never been breached, even when heavy flood with flow of 22,300 m³/sec occurred in July of 1958.[3]

Conclusion

Compared with other rivers in China, the Yellow River is characterized by shortage of water, high content of sand, and serious loss of water and soil. Average annual runoff is only 58 billion m³, taking up only 2% of that in China, and annual amount of sediment and sand washed into the river is about 1.6 billion Mg with sand content of 35 kg/m³. The riverbed in lower reaches is 4–6 m higher than the ground outside the river on average. Floods and droughts are two major disasters in the Yellow River basin. Breach occurred frequently in its history. The Chinese government has made great efforts to harness the river. A number of water control projects have been undertaken, hydropower stations have been built, and dykes in the lower reaches had been strengthened and heightened. To a certain extent, the eroded land on the Loess Plateau had been harnessed and water and soil loss was restricted. The implementation of the great-development-of-the-west strategy of China will further prompt the harnessing and exploitation of the Yellow River along with the development of economy and society.

References

1. Yellow River Conservancy Commission. A survey of the yellow river basin (Ch. 1). *The Short-Term Program on Harnessing and Exploiting the Yellow River*; The Yellow River Conservancy Publishing House: Zhengzhou, China, 2002; 1–17.
2. Yellow River Conservancy Commission. Current situation on harnessing and exploiting the yellow river and the main problems (Ch. 2). *The Short-Term Program on Harnessing and Exploiting the Yellow River*; The Yellow River Conservancy Publishing House: Zhengzhou, China, 2002; 18–32.

3. Lanqin, Z. Section of natural geography. *300 Questions on Yellow River*; The Yellow River Conservancy Publishing House: Zhengzhou, China, 1998; 3–33.
4. Xioufeng, M. Introduction. *Floods and Droughts in Yellow River Basin*; The Yellow River Conservancy Publishing House: Zhengzhou, China, 1996; 1–9.
5. Xioufeng, M. Drought in agriculture (Ch. 10). *Floods and Droughts in Yellow River Basin*; The Yellow River Conservancy Publishing House: Zhengzhou, China, 1996; 270 pp.

IV

DIA: Diagnostic Tools: Monitoring, Ecological Modeling, Ecological Indicators, and Ecological Services

37

Groundwater: Modeling

Introduction 389
What Can Be Modeled and What For? 390
    Modeled Phenomena • What Are Models Built For?
How Are Models Built: The Modeling Process 393
    Conceptualization • Discretization • Calibration and Error
    Analysis • Model Selection • Predictions and Uncertainty
Conclusion 397
References 398

Jesus Carrera

Introduction

A model is an entity built to reproduce some aspect of the behavior of a natural system. In the context of groundwater, aspects to be reproduced may include: groundwater flow (heads, water velocities, etc.); solute transport (concentrations, solute fluxes, etc.); reactive transport (concentrations of chemical species reacting among themselves and with the solid matrix, minerals dissolving or precipitating, etc.); multiphase flow (fractions of water, air, non-aqueous phase liquids, etc.); energy (soil temperature, surface radiation, etc.); and so forth.

Depending on the type of description of reality that one is seeking (qualitative or quantitative), models can be classified as conceptual or mathematical. A conceptual model is a qualitative description of "some aspect of the behavior of a natural system." This description is usually verbal, but may also be accompanied by figures and graphs. In the groundwater flow context, a conceptual model involves defining the origin of water (areas and processes of recharge) and the way it flows through and exits the aquifer. In contrast, a mathematical model is an abstract description (abstract in the sense that it is based on variables, equations, and the like) of "some aspect of the behavior of a natural system." However, the motivation of mathematical models is not abstraction, but rather quantification. For example, a groundwater flow mathematical model should yield the time evolution of heads and fluxes (water movements) at every point in the aquifer.

Both conceptual and mathematical models seek understanding. Some would argue that understanding is not possible without quantification. In the reverse, one cannot even think of writing equations without some sort of qualitative understanding. The methods of conceptual modeling are those of conventional hydrogeology (study geology, measure heads and hydraulic parameters, hydrochemistry, etc.). On the other hand, the methods of mathematical modeling (discretization, calibration, etc.) are more specific. Yet, it should be clear from the outset that conceptualization is the first step in modeling and that mathematical modeling helps in building firm conceptual models.

Depending on the manner in which equations are solved, models can be classified as analog, analytical, and numerical. Analog models are based on a physical simulation of a phenomenon governed by the same equation(s) as that of our natural system. For example, because of the equivalence between electrostatics and steady state flow, one may use conductive paper subject to an electrical current to solve the flow equation (a parallelism can be established between electric potential and hydraulic head). This kind of application, however, is restricted mainly to teaching. Boxes of resistances and condensators were used in the 1950s and 1960s as analog aquifer models, but they have become inefficient compared to computers. As a result, analog models are no longer used in practice.

Analytical models are based on closed-form solutions to the groundwater flow and transport equations. They are convenient in the sense that they are easy to evaluate and intuitive (visual inspection of the equation may yield an idea of the phenomenon). As a result, they are used very frequently. Examples include solutions of problems in well hydraulics, tracer movement, etc.

Numerical models are based on discretizing the partial differential equations governing flow and transport. This leads to linear systems of equations that can only be solved with the aid of computers. The advantage of numerical models lies in their generality. Analytical models are constrained to homogeneous domains and very simple geometry and boundary conditions. Numerical models, on the other hand, can handle spatially and temporally variable properties, arbitrary geometry and boundary conditions, and complex processes. The price to pay is methodological singularity. Analytical models are easy to use. Numerical models can be complex and, often, difficult.

Because of the methodological singularity mentioned above, this entry concentrates on mathematical numerical models. Analytical solutions are not discussed. In addition, conceptual modeling will be discussed as the first step in modeling, but not by itself.

What Can Be Modeled and What For?

Modeled Phenomena

The most basic phenomenon is groundwater flow (Figure 1) because of its intrinsic importance and because it is needed for subsequent processes. In essence, the flow equation expresses two things. First, groundwater moves according to Darcy's law. Second, a mass balance must be satisfied in the whole aquifer and in each of its parts. Therefore, the main output from flow models is a mass balance: classified inflows, outflows, and storage variations. The output also includes where water flows through the aquifer (water fluxes) and heads (water levels in the aquifer). In essence, input data are a thorough description of hydraulic conductivity (and/or transmissivity), storativity, recharge/discharge throughout the model domain, as well as conditions at the model boundaries. Obviously, these data are never available, and the modeler has to use a good deal of ingenuity to generate them. This is where the conceptual model becomes important.

Specific cases of flow phenomena are unsaturated and multiphase flow. In the first case, one models water flow in the vadose zone or, in general, in areas where water does not fill all the pores.[1] Therefore, besides heads and fluxes, one must work with water contents (volume of water per unit volume of aquifer), capillary pressures and suctions (difference between water and air pressure). From the input viewpoint, the main singularity of unsaturated flow is the need to specify the retention curve (water content vs. suction) and relative permeability (permeability vs. water content). The multiphase flow case is similar, but includes several fluids (phases). It is used to represent the flow of air or mixtures of liquids, singularly non-aqueous phase liquids (NAPLs), which have been the subject of much research in recent years.[2]

Conservative transport refers to the movement of inert substances dissolved in water. Solutes are affected by advection (displacement of the solute as linked to flowing water) and dispersion (dilution of contaminated water with clean water, which causes the size of the contaminated area to grow while reducing peak concentrations). The main input to a solute transport model is the output of a flow model (water fluxes). Additionally, porosity and dispersivity need to be specified (Figure 1). The output is the time evolution and spatial distribution of concentrations. While the amount of data needed for solute

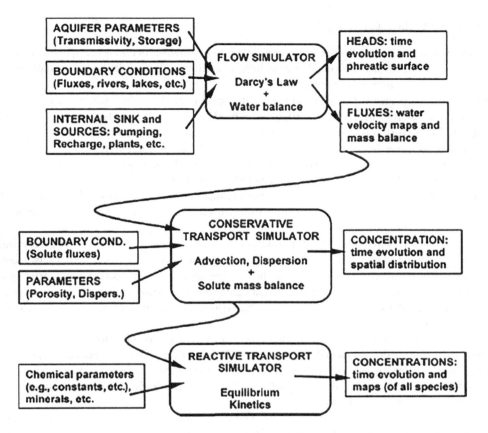

FIGURE 1 A groundwater flow model involves using a flow simulator to take aquifer parameters, boundary conditions, and internal sink and sources as inputs and obtain heads and water fluxes as output. Water fluxes are used in conservative transport models, together with porosity, diffusivity, and solute mass inflows, to yield time evolution and spatial distribution of inert tracers.

transport modeling is relatively small, it must be stressed that solute transport is extremely sensitive to variability and errors in water fluxes. A flow model may be good enough for flow results (heads and water balances) but insufficiently detailed to yield water fluxes good enough for solute transport. Therefore, modeling solute transport ends up being rather difficult.

Reactive transport refers to the movement of solutes that react among themselves and with the soil phase. Reactions can be of many kinds, ranging from sorption of a contaminant onto a solid surface to redox phenomena controlling the degradation of an organic pollutant. Input for reactive transport modeling includes not only the output of flow and conservative transport models but also the equilibrium constants of the reactions (usually available from chemistry databases) and the parameters controlling reaction kinetics. However, the most difficult input is the proper identification of relevant chemical processes. Model output includes the concentrations of all chemical species, the reaction rates, etc.

Coupled models refer to models in which different phenomena are affected reciprocally. Density dependent flow is a typical example. Variations in density affect groundwater flow (e.g., dense sea water sinks under light fresh water), which in turn affects solute transport and, hence, density distribution. Other coupled phenomena are the non-isothermal flow of water (coupling flow and energy transport) and the mechanically driven flow of water (coupling flow and mechanical deformation equations).

What Are Models Built For?

While discussing the usage of models, it is convenient to distinguish between site-specific models and generic models. The former are aimed at describing a specific aquifer while the latter emphasize processes, regardless of where they take place.

Groundwater management is the ideal use of site- specific models. Management involves deciding where to extract and/or inject water to satisfy water needs while ensuring water quality and other constraints. In this context, it is important to point out that a model is essentially a system for accounting water fluxes and stores (Figure 2) in the same way that the accounting system of a company keeps track of money fluxes and reserves. No one would imagine a well-managed company without a proper accounting system. Aquifers will not be managed accurately until they have a model running on real time. Unfortunately, at present, this is still a dream. Because of the difficulties in building and maintaining models and because of legal and practical difficulties to manage aquifers in real time, models are rarely, if ever, used in this fashion.

Instead, models are often used as decision support tools. Building an accurate model is very difficult and time consuming. As a result, one can rarely expect models to yield exact predictions. However, approximate models are much easier to build. These do not result in precise forecasts but normally allow reasonable assessments of the outcome of different management alternatives, i.e., the relative advantages and disadvantages of each alternative can be evaluated and the options ranked. This is usually all one needs for decision making.

This type of use is very frequent in aquifer rehabilitation, where one has to choose among several alternatives, including the option of doing nothing.[3] Models are also used for supporting aquifer exploration policies, i.e., for answering questions such as "how much water can be extracted?," "where should one pump to minimize environmental impact?," etc. In fact, a large body of literature is devoted to this kind of questions in an optimal fashion.[4]

Site-specific models are most frequently used, however, as a tool to support aquifer characterization efforts. This is somewhat ironic because a model is an essentially quantitative tool while site characterization is rather qualitative. Yet, experience dictates that modeling is the only way to consistently integrate the kind of data available in site characterization. These data are very diverse and range from geologic maps to isotope concentrations. One can use vastly different models to verbally explain all observations. Quantitative consistency is not so easy to check and requires the use of a model. Because of the difficulties in fully describing all data, this kind of model use is rarely described in the scientific literature.

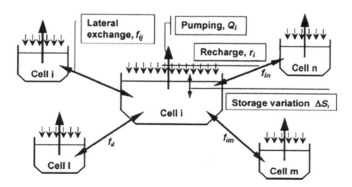

FIGURE 2 A groundwater model is the accounting system of an aquifer. It keeps track of the balance of each section (cells or compartments in the groundwater language) by evaluating exchanges with the outside (pumping Q_i, recharge, r_i, etc.) and with the adjacent sections (f_{ij}). The difference between inflow and outflow is equal to the variation in reserves (storage variation, ΔS_i). A well-managed company needs an accounting system, and so does an aquifer.

Models can also be used in generic fashion as teaching or research tools to gain understanding on physicochemical phenomena. In these cases, they do not aim at representing a specific aquifer, but at evaluating the role of some processes under idealized conditions. A classical example of this type of use is the analysis of flow on regional basins.[5] Models are used in this fashion to explain geological processes.[6,7] Much emphasis has been placed in recent years on the evaluation of the effects of spatial variability. This involves issues such as upscaling, i.e., finding the relationship between large-scale effective parameters and small-scale measurements;[8] or analysis of hydraulic tests.[9]

How Are Models Built: The Modeling Process

The procedure to build a model is outlined in Figure 3. First, one defines a conceptual model (i.e., zones of recharge, boundaries of aquifers, etc.). Second, one discretizes the model domain into a finite element or finite difference grid. This can be entered as input data for a simulation code. Unfortunately, output data will rarely fit the observed aquifer heads and concentrations. This is what motivates calibration, i.e., the modification of model parameters to ensure that model output is indeed similar to what has been observed in reality. The model thus calibrated can be considered a "representation of the natural system" and can be used for management or simulation purposes.

The above procedure is formally described in Figure 4. This section is devoted to discussing in detail the modeling steps as previously described.[10–12] In practice, the effort behind each of these tasks may be very sensitive to the objectives of the studies and model. For the time being, we will assume that one is building a model aimed at describing reality in detail for the purpose making predictions.

Conceptualization

Modeling starts by defining which processes are important and how they are represented in the model. Definition of the relevant processes is termed "process identification" and it is needed for several reasons. First, the number of processes that may affect flow and transport is very large. For practical reasons, the modeler is forced to select those that affect the phenomenon under study, most significantly. Second, not all processes are well understood and they have to be treated in a simplified manner. In short, process identification involves simplifications, both in the choice of the processes and in the way they are implemented in the model.

Model structure identification refers to the definition of parameter variability, boundary conditions, etc. In a somewhat narrower but more systematic sense, model structure identification implies expressing the model in terms of a finite number of unknowns called model parameters. Parameters controlling the above processes are variable in space. In some cases, they also vary in time or depend on heads and/or concentrations. As discussed earlier, data are scarce so that such variability cannot be expressed accurately. Therefore, the modeler is also forced to make numerous simplifications to express the patterns of parameter variations, boundary conditions, etc. These assumptions are reflected on what is denoted as model structure.

The conceptualization step of any modeling effort is somewhat subjective and dependent on the modeler's ingenuity, experience, scientific background, and way of looking at the data. Selection of the physicochemical processes to be included in the model is only rarely the most difficult issue. The most important processes affecting the movement of water and solutes underground (advection, dispersion, sorption, etc.) are relatively well known. Ignoring a relevant process will only be caused by misjudgments and should be pointed out by reviewers, which illustrates why reviewing by others is important. Difficulties arise when trying to characterize those processes and, more specifically, the spatial variability of controlling parameters.

In spite of the large amount of data usually available, their qualitative nature prevents a detailed definition of the conceptual model. Thus, more than one description of the system may result from the conceptualization step. Selecting one conceptual model among several alternatives is sometimes performed during calibration, as discussed later.

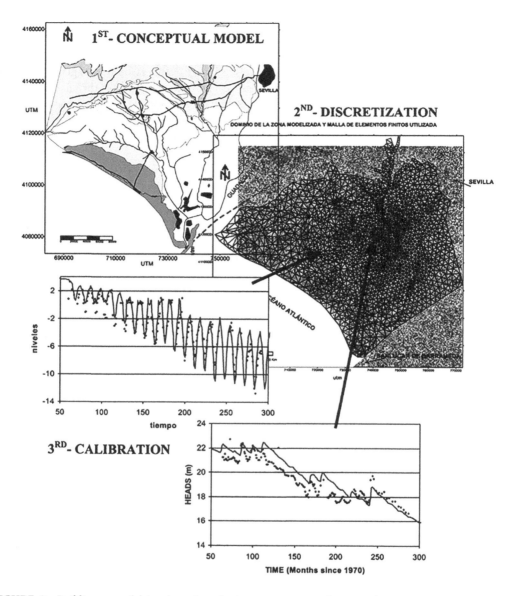

FIGURE 3 Building a model involves three basic steps: conceptualization, discretization, and calibration. Example from the Almonte-Marismas aquifer.

Discretization

Strictly speaking, discretization consists of substituting a continuum by a discrete system. However, we are extending this term here to describe the whole process of going from mathematical equations, derived from the conceptual model, to numerical expressions that can be solved by a computer. Closely related is the issue of verification, which refers to ensuring that a code accurately solves the equations that it is claimed to solve. As such, verification is a code-dependent concept. However, using a verified code is not sufficient for mathematical correctness. One should also make sure that time and space discretization is adequate for the problem being addressed. Moreover, numerical implementation of a conceptual model is not always straightforward international code comparison projects; INTRANCOIN

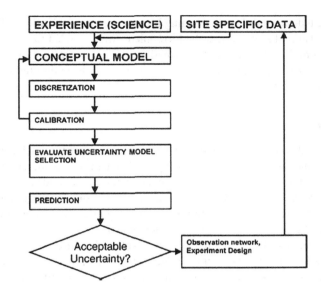

FIGURE 4   A formal description of the modeling process. Modeling starts with an understanding of the natural system (conceptual model), which is based on experience about such kind of systems (science) and on data from the site. Writing the conceptual model in a manner adequate for computer solution requires discretization. The resulting model is still dependent on many parameters that are uncertain. During calibration, these parameters are adjusted so that model outputs are close to measurements (recall Figure 3). Model predictions may be uncertain because so are the fitted parameters or because different models are consistent with observations. If uncertainty is unacceptably high, one should perform additional measurements or experiments and redo the whole process.

and HYDROCOIN have shown the need for sound conceptual models and independent checks of calculation results. Even well-posed mathematical problems lead to widely different solutions when solved by different people, because of slight variations in the solution methodology or misinterpretations in the formulation.[13] The reasons behind these differences and ways to solve them only become apparent after discussions among them.

The main concern during discretization is accuracy. In this sense, it is not conceptually difficult, although it can be complex. Accuracy is not only restricted to numerical errors (differences between numerical and exact solutions of the involved equations) but also refers to the precision with which the structure of spatial variability reproduces the natural system.

Calibration and Error Analysis

The choice of numerical values for model parameters is made during calibration, which consists of finding those values that grant a good reproduction of head and concentration data (Figure 3) and are consistent with prior independent information.

Calibration is rarely straightforward. Data come from various sources, with varying degrees of accuracy and levels of representativeness. Some parameters can be measured directly in the field, but such measurements are usually scarce and prone to error. Furthermore, since measurements are most often performed on scales and under conditions different from those required for modeling purposes, they tend to be both numerically and conceptually different from model parameters. The most dramatic example of this is dispersivity, whose representative value increases with the scale of measurement so that dispersivities derived from tracer tests cannot be used directly in a large-scale model. As a result, model parameters are calibrated by ensuring that simulated heads and concentrations are close to the corresponding field measurements.

Calibration can be tedious and time consuming because many combinations of parameters have to be evaluated, which also makes it prone to be incomplete. This, coupled to difficulties in taking into account the reliability of different pieces of information, makes it very hard to evaluate the quality of results. Therefore, it is not surprising that significant efforts have been devoted to the development of automatic calibration methods.[14–16]

Model Selection

The first step in any modeling effort involves constructing a conceptual model, describing it by means of appropriate governing equations, and translating the latter into a computer code. Model selection involves the process of choosing between alternative model forms. Methods for model selection can be classified into three broad categories. The first category is based on a comparative analysis of residuals (differences between measured and computed system responses) using objective as well as subjective criteria. The second category is denoted parameter assessment and involves evaluating whether or not computed parameters can be considered as "reasonable." The third category relies on theoretical measures of model validity known as "identification criteria." In practice, all three categories will be needed: residual analysis and parameter assessment suggest ways to modify an existing model and the resulting improvement in model performance is evaluated on the basis of identification criteria. If the modified model is judged an improvement over the previous model, the former is accepted and the latter discarded.

The most widely used tool of model identification is residual analysis. In the groundwater context, the spatial and frequency distributions of head and concentration residuals are very useful in pointing towards aspects of the model that need to be modified. For example, a long tail in the breakthrough curve not properly simulated by a single porosity model may point to a need for incorporating matrix diffusion or a similar mechanism. These modifications should, whenever possible, be guided by independent information. Qualitative data such as lithology, geological structure, geomorphology, and hydrochemistry are often useful for this purpose. A particular behavioral pattern of the residuals may be the result of varied causes that are often difficult to isolate. Spatial and/or temporal correlation among residuals may be a consequence of not only improper conceptualization, but also measurement or numerical errors. Simplifications in simulating the stresses exerted over the system are always made and they lead to correlation among residuals. Distinguishing between correlations caused by improper conceptualization and measurement errors is not an easy matter. This makes analysis of residuals a limited tool for model selection.

An expedite way of evaluating a model concept is based on assessing whether or not the parameters representing physicochemical properties can be considered "reasonable;" i.e., whether or not their values make sense and/or are consistent with those obtained elsewhere. Meaningless parameters can be a consequence of either poor conceptualization or instability. If a relevant process is ignored during conceptualization, the effect of such process may be reproduced by some other parameter. For example, the effect of sorption is to keep part of the solute attached to the solid phase, hence retarding the movement of the solute mass; in linear instantaneous sorption, this effect cannot be distinguished from standard storage in the pores. Therefore, if one needs an absent porosity (e.g., larger than one) to fit observation, one should consider the possibility of including sorption in the model. However, despite this example, parameter assessment tends to be more useful for ruling out some model concepts than for giving a hint on how to modify an inadequate model. Residual analysis is usually more helpful for this purpose.

Instability may also lead to unreasonable parameter estimates during automatic calibration, despite the validity of the conceptual model. When the number of data or their information content is low, small perturbations in the measurement or deviations in the model may lead to drastically different parameter estimates. When this happens, the model may obtain equally good fits with widely different parameter sets. Thus, one may converge to a senseless parameter set while missing other perfectly meaningful sets. This type of behavior can be easily identified by means of a thorough error analysis and corrected by fixing the values of one or several parameters.[14]

Predictions and Uncertainty

Formulation of predictions involves a conceptualization of its own. Quite often, the stresses, whose response is to be predicted, lead to significant changes in the natural system, so that the structure used for calibration is no longer valid for prediction. Changes in the hydrochemical conditions or in the flow geometry may have to be incorporated into the model. While numerical models can be used for network design or as investigation tools, most models are built in order to study the response of the medium to various scenario alternatives. Therefore, uncertainties on future natural and man-induced stresses also cause model predictions to be uncertain. Finally, even if future conditions and conceptual model are exactly known, errors in model parameters will still cause errors in the predictions. In summary, three types of prediction uncertainties can be identified: conceptual model uncertainties; stresses uncertainties; and parameters uncertainties.

The first group includes two types of problems. One is related to model selection during calibration. That is, more than one conceptual model may have been properly calibrated and data may not suffice to distinguish which one is the closest to reality. It is clear that such indetermination should be carried into the prediction stage because both models may lead to widely different results under future conditions. The second type of problems arises from improper extension of calibration to prediction conditions, i.e., from not taking into account changes in the natural system or in the scale of the problem. The only way we think about dealing with this problem consists of evaluating carefully whether or not the assumptions in which the calibration was based are still valid under future conditions. Indeed, model uncertainties can be very large.

We do not think that, strictly speaking, the second type of uncertainties, those associated with future stresses, falls in the realm of modeling. While future stresses may affect the validity of the model, they are external to it. In any case, this type of uncertainty is evaluated by carrying out simulations under a number of alternative scenarios, whose definition is an important subject in itself.

The last set of prediction uncertainties is the one associated with parameter uncertainties, which can be quantified quite well.

Conclusion

Groundwater modeling involves so many subjective decisions that it can be considered as an art. This is somewhat contrary to the widely accepted perception of models as something objective. The fact is that numerous assumptions need to be made both about the selection of relevant processes and about the manner of representing them in the computer. All these assumptions are specified in the conceptual model.

The result relies so heavily on conceptualization that models ought to be viewed as theories about the behavior of natural systems. Model predictions should rarely be viewed as firm statements about the future evolution of aquifers. Rather, they should be considered references against which actual data has to be compared. Codes do exist for modeling most processes affecting groundwater (flow, transport, reactions, thermomechanics, etc.). It is lack of understanding and lack of data what limits the actual application of those codes.

Having specified a conceptual model, the remaining steps (discretization, calibration, uncertainty analysis, prediction) are relatively objective, in the sense that systematic procedures can be followed. This explains why conceptualization is so important. It also explains why modeling is the best way of integrating widely different data. Uncertain as it is, it may represent unambiguously the overall knowledge of the aquifer.

Models represent the water balance (or solute balance, or energy balance) at the overall aquifer and at each of its parts. Therefore, they can also be viewed as accounting systems. It is argued that well-managed aquifers need real time models to help decision making, the same way that well managed companies need financial accounting systems. This is the challenge modelers must meet in the near future.

References

1. Neuman, S.P. Saturated–unsaturated seepage by finite elements. ASCE, J. Hydraulics Div. 1973, *99* (12), 2233–2250.
2. Abriola, L.M.; Pinder, G.F. A multiphase approach to the modeling of porous media contamination by organic compounds. Water Resour. Res. 1985, *21* (1), 11–26.
3. Konikow, L.F.; Bredehoeft, J.D. Groundwater models cannot be validated. Adv. Water Resour. 1992, *15* (1), 75–83.
4. Gorelick, S. A review of distributed parameter groundwater management modeling methods. Water Resour. Res. 1983, *19*, 305–319.
5. Freeze, R.A.; Witherspoon, P.A. Theoretical analysis of regional groundwater flow. II, effect of water table configuration and subsurface permeability variations. Water Resour. Res. 1966, *3*, 623–634.
6. Garven, G.; Freeze, R.A. Theoretical analysis of the role of groundwater flow in the genesis of strat-abound ore deposits. II, quantitative results. Am. J. Sci. 1984, *284*, 1125–1174.
7. Ayora, C.; Tabener, C.; Saaltink, M.W.; Carrera, J. The genesis of dedolomites: a discussion based on reactive transport modeling. J. Hydrol. 1998, *209*, 346–365.
8. Sánchez-Vila, X.; Girardi, J.; Carrera, J. A synthesis of approaches to upscaling of hydraulic conductivities. Water Resour. Res. 1995, *31* (4), 867–882.
9. Meier, P.; Carrera, J.; Sánchez-Vila, X. An evaluation of Jacob's method work for the interpretation of pumping tests in heterogeneous formations. Water Resour. Res. 1998, *34* (5), 1011–1025.
10. Carrera, J.; Mousavi, S.F.; Usunoff, E.; Sánchez-Vila, X.; Galarza, G. A discussion on validation of hydrogeological models. Reliab. Eng. Syst. Saf. 1993, *42*, 201–216.
11. Anderson, M.P.; Woessner, W.W. *Applied Groundwater Modeling;* Academic Press: San Diego, 1992.
12. Konikow, L.F.; Bredehoeft, J.D. *Computer Model of TwoDimensional Solute Transport and Dispersion in Groundwater: Reston,* VA; U.S. Geo. Surv. TWRI, Book 7, 1, 1978; 40 pp, Chap. C2.
13. NEA-SKI. *The International HYDROCOIN Project, Level 2: Model Validation;* OECD: Paris, France, 1990; 194 pp.
14. Carrera, J.; Neuman, S.P. Estimation of aquifer parameters under steady state and transient conditions: I through III. Water Resour. Res. **1986**, *22* (2), 199–242.
15. Loaiciga, H.A.; Marino, M.A. The inverse problem for confined aquifer flow: identification and estimation with extensions. Water Resour. Res. 1973, *23* (1), 92–104.
16. Hill, M.C. A Computer Program (MODFLOWP) for Estimating Parameters of a Transient, Three Dimensional, Groundwater Flow Model Using Nonlinear Regression, USGS Open-File Report, 1992; 91–484.

38

Groundwater: Numerical Method Modeling

A Generic Numerical Method for Solving Groundwater Flow..............399
Finite Differences (FD)... 401
Integrated Finite Differences (IFD) 402
Finite Element Method...402
Boundary Element Method 403
Simulating Solute Transport 403
Conclusion 404
References...405

Jesus Carrera

A Generic Numerical Method for Solving Groundwater Flow

As mentioned earlier, all methods require, first, discretizing and, second, approximating the physical phenomenon. For the generic method we are going to present here, discretization will be performed as shown in Figure 1. That is, the continuum aquifer domain will be substituted by a discrete number of cells. Furthermore, the continuum aquifer heads, $h(x,y)$, are substituted by a discrete number of model heads, h_i.

The second step, approximation, can be made in different manners. For the purpose of this section, it is sufficient to bear in mind that the flow equation is nothing but a mass balance. Therefore, we will express the mass balance in cell i as change in storage equals inflows minus outflows.

$$\Delta S_i = f_{ij} + f_{il} + f_{im} + f_{in} + g_i \qquad (1)$$

where ΔS_i is the rate of change in storage during one time step (say, between time t^k and time t^{k+1}); f_{ij} is the inflow into cell i from cell j (and the same for f_{il}, f_{im}, and f_{in}); and g_i are external inflows into cell i (for example, recharge, minus pumping, minus evaporation, minus river outflow, etc.). Each of the terms in Eq. 1 is relatively easy to approximate. Storage variation can be derived from the definition of storage coefficient, (S is the change in volume of water stored per unit surface area of aquifer and per unit change in head):

$$\Delta S_i = SA_i \frac{\left(h_i^{k+1} - h_i^k\right)}{\Delta t} \qquad (2)$$

where A_i is the surface area of cell i, h_i^k is the head in node i at time k and $\Delta t = t^{k+1} - t^k$ is the time step. Darcy's law gives lateral inflows

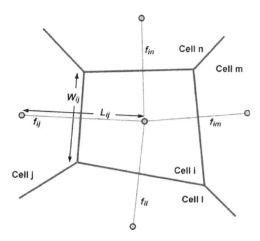

FIGURE 1 The system of equations for a generic groundwater model is obtained by establishing a water balance (inputs minus outputs equal storage variations) at each cell. Inputs and outputs include water exchanges with the outside (pumping Q_i, recharge, r_i, etc.) and with adjacent cells f_{ij}). The latter are expressed, using Darcy's law, as $f_{ij} = T_{ij}w_{ij}(h_j - h_i)/L_{ij}$.

$$f_{ij} = -Tw_{ij}\frac{h_i - h_j}{L_{ji}} = a_{ij}\left(h_i - h_j\right) \tag{3}$$

where T is transmissivity; w_{ij} is the width of the connection between nodes i and j; L_{ji} is the length of such connection; and a_{ij} is implicitly defined as $-Tw_{ij}/L_{ij}$. The remaining inflow terms, f_{il}, f_{im}, and f_{in} are defined likewise. Changing these terms to the left-hand side of Eq. 1 and rearranging terms yield:

$$SA_i \frac{\left(h_i^{k+1} - h_i^k\right)}{\Delta t} + a_{ii}h_i + a_{ij}h_j + a_{il}h_l + a_{im}h_m + a_{in}h_n = g_i \tag{4}$$

where $a_{ii} = -a_{ij} - a_{il} - a_{im} - a_{in}$. If an equation like Eq. 4 is written for all cells from $i = 1$ through N, N being the number of cells (nodes), the resulting system of equations can be rewritten in matrix form as:

$$\mathbf{D}\frac{\left(\mathbf{h}^{k+1} - \mathbf{h}^k\right)}{\Delta t} + \mathbf{A}\mathbf{h} = \mathbf{g} \tag{5}$$

where $\mathbf{D}$ is a diagonal matrix whose i-th diagonal term is precisely SA_i. This matrix is often called storage matrix. $\mathbf{A}$ is the conductance matrix, a square symmetric matrix whose components are a_{ij}. Finally, $\mathbf{g}$ is the source vector.

All the numerical methods to be outlined in subsequent sections lead to equations analogous to Eq. 5. Moreover, the meaning of the terms in such equations is always similar to that in Eq. 5. Namely, the system represents the mass balance at each of the N nodes (cells); specifically the i-th equation represents the mass balance at the i-th node. The first term, $\mathbf{D}(\mathbf{h}^{k+1} - \mathbf{h}^k)/\Delta t$, always represents storage variations. The second term, $\mathbf{A}\mathbf{h}$, represents outflows from minus inflows into the i-th cell from the adjacent cells. Finally, term $\mathbf{g}$ represents external inflows minus outflows (recharge, pumping, etc.) at all i.

Equation 5 needs to be integrated in time. For this purpose, let us assume that $\mathbf{A}\mathbf{h}$ is evaluated at time $k + 1$ ($\mathbf{A}\mathbf{h}^{k+1}$). Then, Eq. 5 can be rewritten as:

$$\left(A + \frac{D}{\Delta t}\right)h^{k+1} = g + \frac{D}{\Delta t}h^k \tag{6}$$

This is simply a linear system, which can be solved using conventional methods.

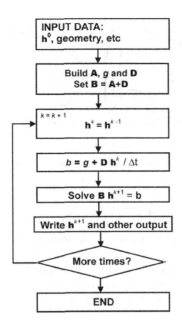

FIGURE 2 Basic steps involved in simulating groundwater flow. Heads $\mathbf{h}^{k+1}$ are computed by solving equation $\mathbf{B}h^{k+1} = \mathbf{b}$. They may be written for later drawings. They are used as initial head for the next time increment. These steps (time loop) are repeated sequentially until the last time is reached.

$$Bh^{k+1} = \mathbf{b} \tag{7}$$

where $\mathbf{B} = \mathbf{A} + \mathbf{D}/\Delta t$ and $\mathbf{b} = \mathbf{g} + \mathbf{D}h^k/\Delta t$. This system is solved sequentially in time.

That is, most codes solve Eq. 7 using the following steps (Figure 2):

1. Input all data. Set $k = 0$
2. Compute $\mathbf{g}$, $\mathbf{A}$ (Eq. 3); $\mathbf{D}$ (Eq. 5) and $\mathbf{B}$ (Eq. 7)
3. Set $k = k + 1$
4. Build $\mathbf{b}$ (Eq. 7)
5. Solve $\mathbf{B}h^{k+1} = \mathbf{b}$
6. If $k = k_{\max}$ (maximum number of time steps), end. Otherwise, return to step 3

Most codes follow a structure such as this, although each method displays specific features. Some of these are outlined below.

Finite Differences (FD)

As mentioned at the beginning, numerical methods differ in the way in which the domain is discretized and in the way in which the partial differential equation is transformed into a linear system of equations. In finite differences, the problem domain is discretized in a regular grid (Figure 3a), usually rectangular (equilateral triangles or hexagons are possible, but very rare). The grid may be centered at the corners (nodes are located at the vertices of the squares) or at the cells (nodes are located at the center of the squares, such as in Figure 3a).

Regarding the approximation of the partial differential equations, several alternatives are possible. The most intuitive consists of substituting all derivates by an incremental ratio. That is, the derivative between adjacent nodes i and j is approximated as:

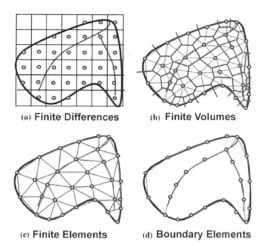

(a) **Finite Differences**   (b) **Finite Volumes**

(c) **Finite Elements**   (d) **Boundary Elements**

FIGURE 3 The most widely spread methods of discretization are Finite Differences, which consists of subdividing the model domain into regular rectangles, and Finite Elements, which is based on dividing the aquifer region into elements of arbitrary shape (often triangles). Finite Volumes, also called Integrated Finite Differences, divides the region into polygons. The Boundary Element Method is very convenient, when applicable, because it only requires discretizing boundaries (both internal and external).

$$\frac{\partial h}{\partial x} = \frac{h_j - h_i}{\Delta x} \tag{8}$$

where h_i and h_j are heads at nodes i and j, respectively, and Δx is the distance between them. Approximating all derivatives by means of equations analogous to Eq. 8 leads to a system identical to Eq. 5. In fact, the finite differences method is often introduced using a mass balance approach such as the one in the section "A Generic Numerical Method for Solving Groundwater Flow," only using a regular instead of a generic grid. This is the method used in MOD- FLOW,[1] HST3D,[2] and their children.

Integrated Finite Differences (IFD)

The basic philosophy of this method is very similar to that of the generic method introduced in the generic numerical section. Basically, the domain is discretized in a number of cells centered around arbitrarily located nodes. Frequently, the cells are the Thiessen polygons of the set of nodes. This allows adapting the node density to the problem (e.g., increasing nodes density where accuracy is needed most).

Model equations can be derived using a mass balance approach, such as in the generic method section. Integrating the flow equation over each cell and applying Green's identity to transform volume integrals in boundary fluxes can also yield model equations. This type of approach is the basis of the finite volume method, which is widely used nowadays.

Finite Element Method

Finite element method (FEM) discretization consists of elements and nodes. Elements are generalized polygons (normally triangles or curvilinear quadrilaterals). Nodes are points located at the vertices and, sometimes, at the sides or the middle of the element. Unlike FD or IFD, cells around the nodes are not defined. Still, in many cases, one may write the equations in such a way that the mass balance formulation of the generic method section is still valid. However, the most singular feature of the FEM is the way the solution is interpolated, so that it becomes defined at every point. That is, head (or concentrations) is approximated as:

$$h(x) \cong \hat{h}(x) = \sum_i h_i N_i(x) \qquad (9)$$

where h_i are nodal heads and N_i are interpolation functions. Since is not the exact solution, it would yield a residual if substituted in the flow equation. Minimizing this residual, which requires somewhat sophisticated maths, leads to a system similar to Eq. 5.

Boundary Element Method

The idea behind the boundary element method (BEM) is similar to that of the FEM. The main difference stems from the choice of interpolation functions, which are taken as the fundamental solutions of the flow equation (or whatever equation is to be solved). As a result, when the corresponding $\hat{h}$ is substituted in the flow equation, the residuals are zero. Since the equation is satisfied exactly in the model domain, one is only left with boundary conditions. In fact, as shown in Figure 3d, discretization is only required at the boundaries, where boundary heads are defined so as to satisfy approximately the boundary conditions. This method is extremely accurate, but its applicability is limited by the need of finding the fundamental solutions. This constrains the BEM to flow problems in relatively homogeneous domains.

Simulating Solute Transport

All the methods discussed above can be used for simulating solute transport. They are called Eulerian methods because they are based on a fixed (as opposed to moving) grid and all derivatives are based on a fixed coordinate system. They work fine when dispersion is dominant. Otherwise, they may lead to numerical problems (Figure 4). Two dimensionless numbers are used to anticipate numerical difficulties. Specifically, discretization must satisfy the following conditions.

Peclet number

$$Pe = \frac{v\Delta x}{D} \cong \frac{\Delta x}{\alpha} < \frac{1}{2} \qquad (10)$$

Courant number

$$Co = \frac{v\Delta t}{\Delta x} < 1 \qquad (11)$$

where v is the solute velocity, α is dispersivity, Δx is the distance between nodes and we have assumed that $D \cong \alpha v$. The condition on the Courant number implies that the solute at one node will not move

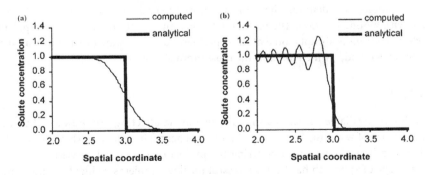

FIGURE 4 Difficulties typically associated to numerical simulation of solute transport: a) front smearing: the concentration front is more dispersed than it should; b) instability oscillations: too high and/or too low (even negative) concentrations.

beyond the following node downstream during one time step. This condition is easy to meet because usually groundwater moves slowly and also reducing Δt to satisfy Eq. 11 is not difficult. The condition on the Peclet number, Eq. 10, implies that Δx is smaller than $\alpha/2$, which may require very small elements, leading to a huge computational burden. Because of this, conventional Eulerian methods are not applicable to many groundwater problems, which have motivated the search of alternative methods.

Alternative methods can be Eulerian or Lagrangian. Among the former, the most popular is upstream weighting, introduced by Heinrich et al.[3] in the FEM, but with a huge number of papers thereafter. It consists of slightly modifying Eulerian equations so as to ensure stability. The problem is that, in doing so, it introduces numerical dispersion. As a result, the wiggles at the solute front in Figure 4 are substituted by an artificially smeared front. However, the vast majority of alternative formulations for solute transport are Lagrangian in the sense that time variations are written in terms of the material derivative, which expresses the rate of change in concentration of a particle that moves with the water. In this way, the advective term, which is the cause of problems in Eulerian formulations, disappears.

The number of Lagrangian methods is very large and many researchers have devoted much effort to find one, which is universal. The fact that so many methods have survived to date suggests that the effort has not been fully successful. Still, in practical problems, one can usually find a suitable method. Following is an outline of some of the most popular methods, with a discussion of their advantages and disadvantages and early references. The interested reader should seek further.

The most natural Lagrangian method is to write the equations on a moving grid, that is on a grid whose nodes move with water. This method is highly accurate, but expensive because the grid has to be updated every time step. Moreover, the grid can become highly deformed over time.

To avoid problems with moving grids it is frequent to work with particles. Displacing the particles with the moving water represents advection, while dispersion can be represented with a variety of methods. One such possibility is to add a random component to each particle basis displacement. This is statistically equivalent to each particle basis dispersion and is the basis of the "random walk" method.[4] The method requires careful implementation, but its main drawback is the fact that the solution is given in terms of number of particles per cell. If one is interested in spatial distributions of the solute, a huge number of particles may be needed.

The method of characteristics (MOC) overcomes the above difficulty by assigning concentrations to particles and interpolating them onto a fixed grid, where dispersion and, possibly, other transport processes are modeled. Concentrations are then interpolated back onto the particles. The method has become very popular in groundwater because of the USGS MOC Code.[5] The method is very practical, but the interpolation back-and-forth between particles and grid may introduce numerical dispersion and mass balance errors.

The modified method of characteristics[6] tries to overcome the problems associated to interpolating particle concentrations by redefining them in each time interval so that at the end of the time step they coincide with a node location. The method is very accurate, although some interpolation errors still occur when the front is abrupt. Some of these problems are overcome by the Eulerian–Lagrangian Localized Adjoint Method,[7] which looks as the most promising method.

Conclusion

Computer codes are available for simulating all phenomena affecting groundwater. In essence, they represent the balance of water (or salt, contaminants, or energy) in a manner that can be solved by the computer. This is achieved by, first, discretizing the problem and, second, rewriting it as a system of equations. This type of approach has been successful for flow problems. Solute transport, on the other hand, remains elusive. No single method is universally successful. Instead, one must seek the appropriate code in each case.

References

1. McDonald, M.G.; Harbaugh, A.W. A modular threedimensional finite difference groundwater flow model. *Techniques of Water Resources Inv.*, of the USGS, Book 6, 1988, Ch. A1.
2. Kipp, K.L. HST3D, A Computer Code for Simulation of Heat and Solute Transport in Three Dimensional Groundwater Flow Systems. USGS Water Recourses Inv. Rep. 86–4095; 1987.
3. Heinrich, J.C.; Huyakorn, P.S.; Mitchell, A.R.; Zienkie- wicz, O.C. An upwind finite element scheme for twodimensional transport equation. Int. J. Numer. Methods Eng. **1977**, *11*, 131–143.
4. Prickett, T.A.; Naymik, T.G.; Lonnquist, C.G. A "Random Walk" solute transport model for selected groundwater quality evaluations. Illinois State Water Survey Bull. **1981**, *55*, 62.
5. Konikow, L.F.; Bredehoeft, J.D. *Computer Model of TwoDimensional Solute Transport and Dispersion in Ground Water*, U.S. Geo. Surv. TWRI: Reston, Va., 1978; Book 7, 40 pp. Chap. C2.
6. Neuman, S.P. A Eulerian–lagrangian numerical scheme for the dispersion–convection equation using conjugate spacetime grids. J. Comput. Phys. **1981**, *41*, 270–294.
7. Celia, M.A.; Russell, T.F.; Herrera, I.; Ewing, R.E. An Eulerian–lagrangian localized adjoin method for the advection–diffusion equation. Adv. Water Resour. **1990**, *13*, 187–206.

39

Nitrogen (Nitrate Leaching) Index

Introduction: Importance of Nitrogen 407
Nitrogen Leaching Contributes to Environmental Problems............... 408
New Ecotechnology: Quick Tools/Indicators... 409
Conclusion 410
References...411

Jorge A. Delgado

Introduction: Importance of Nitrogen

The United Nations recently reported that a changing climate will be one of the biggest challenges that humanity will be confronting this century, especially with respect to the potential impacts of climate change on food security.[1] One of the problems of climate change is the potential increase in extreme events.[2] Adapting to the occurrence of these extreme events will be critical since the potential for higher soil erosion is also significant.[3] Increased precipitation or higher amounts of irrigation increase the potential for nitrate leaching losses.[4–9] However, there is potential to use conservation management practices to reduce losses of nitrate leaching and adapt to a changing climate.[10] Humanity will need to implement the best management policies for soil and water conservation, including nutrient management, if it is to be able to adapt to a changing climate to help ensure that agricultural production is increased over the next decades and food security is achieved.[10,11]

With the projected global population increase of over two billion by 2050 and the challenges humanity faces as far as a changing climate, reduction of water resources, and more frequent occurrence of extreme events, there is a need to continue developing new management practices that help to increase nitrogen use efficiencies. This is because increases in grain and other agricultural products will require additional inputs of nitrogen, yet we do not want to increase the pressure in the metaphorical nitrogen pipes of these intensive agricultural systems, which could then potentially increase nitrogen leakages from these systems. Conservationists and nutrient managers across the world will be asking themselves questions about how to increase yields and nitrogen inputs while simultaneously minimizing the losses of reactive nitrogen to the environment, especially leaching losses of nitrate nitrogen, which can quickly move to water bodies in cases of shallow aquifers, tile systems, and surface water bodies. Increases in agricultural production will be necessary to supply grain for a larger world population, and larger quantities of grain and forage will be needed to meet the additional demand for meat and dairy products as large populations in growing economies such as China, and other regions increase in affluence.

Nitrogen management, including management of inorganic fertilizer inputs, management of animal manure and other by-products, and management of nitrogen across the environment from agricultural fields to non-agricultural areas across the watershed, will certainly be at the center of the world's emerging sustainability challenges in the coming decades. Independent of what policies are implemented, nitrogen management will be at the core of agricultural systems since most of the world's systems depend on agricultural inputs to maximize agricultural productivity.[10] As we move closer to the middle of the 21st century and we have more limited soil and water resources, we will need a global agricultural system that is both more productive and more sustainable. Best nitrogen management practices will be an essential component of efforts to increase productivity and sustainability.

Nitrogen Leaching Contributes to Environmental Problems

Although agricultural production has increased with the use of nitrogen fertilizer, nitrogen inputs also have contributed to increased losses of reactive nitrogen from agricultural systems that have contributed to global environmental impacts.[12–16] The US Environmental Protection Agency (EPA) has reported that drinking water with nitrate concentrations above 10 mg NO_3–N L^{-1} can negatively impact humans, and is thus unfit for human consumption.[17] It is important that we reduce the transport of reactive nitrogen from agricultural systems to reduce the negative impacts of reactive nitrogen on the environment and human health.[18,19] Several scientists have reported that nitrogen contributes to the hypoxic zone in the Gulf of Mexico.[20–25] There are hypoxic zones across the USA and around the world, and another example of an area impacted is Lake Erie, with its hypoxic zone and high microcystin levels.[26] Microcystin-LR concentrations above 1.0 µg/L in treated drinking water have been reported to be unsafe by the World Health Organization,[27] while the USEPA reported a much lower concentration of 0.3 µg/L to be unsafe for children less than six years old.[28] Microcystin concentrations above 0.3 µg/L can negatively impact children less than six years old.[28] Health effects from exposure could include gastroenteritis, and liver and kidney damage.[28]

It has been widely reported across the scientific literature that nitrogen losses could impact surface and groundwater quality.[29,30] Nitrate leaching is a major pathway for nitrogen loss that contributes to poorer water quality of aquifers across the world; examples of affected areas include regions of the North China Plain,[31] Europe,[15,32] South America,[33] and the USA.[29,30,34] The Midwest region of the USA has been reported to contribute large amounts of nitrate which is transported to the Gulf of Mexico, and tile systems have been reported as a major pathway for this nitrate mass that moves from agricultural fields to coastal systems.[20–23,34,25] Domestic wells that are sources of drinkable water across the USA have also been impacted by nitrate leaching as reported by Dubrovsky et al.[25] in his extensive study that found the nitrate levels were higher than the 10 mg NO_3–N L^{-1} safe limit established by the USEPA for a significant percentage of the wells.

Applying higher-than-needed N rates does not necessarily mean that all the pathways for nitrogen losses will be affected at the same magnitude, since site-specific properties affect the main mechanism for nitrogen losses. For example, in our adaptation of the metaphorical pipe concept, a higher-than-needed N rate for a coarser-textured soil or a soil system with tile drainage will have a bigger metaphorical pipe for nitrate leaching losses since coarser soils or tile drainage systems will have higher nitrate leaching potential. In other words, applying more nitrogen than needed to a coarser-textured soil or a tile system will most likely increase the pressure in the nitrate leaching pipe and the potential for NO_3–N leaching losses in this system. Another point to consider is that leaching is driven by water movement, so arid systems are less susceptible to leaching than humid systems with higher precipitation. Similarly, irrigated systems are also more susceptible to leaching, and water management practices (e.g., management of type of irrigation) can be used to reduce the leaching potential in irrigated systems. An example of a very sensitive system would be shallow-rooted crops grown in coarse-textured soils in flood-irrigated systems, which results in high potential for nitrate leaching. The good news is that we could use management practices (e.g., the inclusion of cover crops in a rotation with irrigated, shallow-rooted crops grown in coarse-textured soils) to try to reduce the pressure on the NO_3–N leaching pipe and the losses of nitrogen from the system. Randall et al., Delgado et al., and Delgado and Follett discussed in detail how best management practices could be used to reduce nitrate leaching to the environment.[5,35,36]

As previously discussed, nitrogen application rates that are higher than needed increase the flow through the metaphorical pipes to the environment, increasing the potential for nitrogen losses that will potentially impact air and water quality. However, best nitrogen management practices can be used to reduce the pressure in the pipes and minimize the leakage. Practices that will increase denitrification, contributing to conversion of NO_3–N as it is transported across water bodies to N_2 (e.g., wood chips, wetlands), will minimize the environmental impact of reactive nitrogen on the environment. Although this is a beneficial approach environmentally, there is still an economic loss due to the losses of NO_3–N from field applications, and increasing nitrogen use efficiencies at a field level while reducing NO_3–N leaching losses from the field will be a way to reduce the environmental impact and reduce field-level economic losses. Aside from this consideration, best management practices such as denitrification traps will be good practices to reduce the transport of NO_3–N to the environment.

For global sustainability, we could use many different management practices to increase nitrogen use efficiencies and yields while reducing losses of reactive nitrogen to the environment across global agricultural systems. Quick nitrogen tools that can use some general principles of soil hydrology and soil chemical and physical properties while accounting for irrigation and precipitation balances can be used to assess practices for different cropping systems on an annual basis. New, advanced tools such as the Nitrogen Index are among the new ecotechnologies that can be applied.[35–37]

New Ecotechnology: Quick Tools/Indicators

There are a variety of indexes that have been used to assess the risk of nitrate leaching.[38] Shaffer and Delgado[39] proposed a nitrate leaching index that considers the effects of management (e.g., manure, crops, fertilizer, irrigation), soils, climate, and off-site factors; that is easy to use; and that can conduct a quick but strong assessment of the risk of nitrate leaching. They discussed the different indexes that have been developed in the past to assess the risk of nitrate leaching and the advantages and disadvantages of these indexes. Delgado et al. [35,40–44] developed such an index considering hydrological and ecological factors. This new ecotechnology has been used for about the last decade by different US agencies and other national and international institutions to assess the risk of nitrate leaching for a given set of management practices under a given set of weather, soil, and off-site conditions. The tool is available for download at the website of the National Agricultural Library, and the code of the Nitrogen Index is available to download as well (https://data.nal.usda.gov/dataset/cce-nitrogen-index-tool/resource/ec68e0b2-f3e6-4adc-88f4-aee430f6603d).

In addition to the Nitrogen Index, there are other indexes that have been used to assess nitrogen use efficiencies. Shaffer and Delgado[39] and Buczko et al.[38,45] discussed the advantages and disadvantages of several nitrogen indexes approaches. Assessment of the risk of nitrate leaching potential has been conducted in New York with the New York Nitrate Leaching Index[46] and in California with the Nitrate Leaching Hazard Index.[47] However, none of these indexes conduct a quick assessment of the mass of nitrate leaching at different root depths, and the new California Nitrogen Index and new generic Nitrogen Index can be used to conduct a quick assessment of the mass of nitrate leaching at different root depths.[35,40] One of the indexes used to assess nitrate leaching is the water Leaching Index (LI) by Williams and Kissel[48] that assesses the potential for water leaching and extrapolates that risk to nitrate leaching. However, since there is no assessment of the mass of nitrate leaching or the risk of nitrate leaching, this water leaching index is at a disadvantage in assessing the risk since it does not consider N inputs and N dynamics.[39]

Delgado[49–51] studied the movement of nitrate across the soil profile to assess the nitrate leaching potential of cropping systems with different root depths and found a significant correlation between the root depths of cropping systems and nitrate leaching, with a higher nitrate leaching potential for crops with more shallow root systems. Similarly, he assessed the nitrate leaching potential across different soil textures and nitrate leaching spatially across fields and found a significant correlation between soil texture and nitrate leaching potential when fields were managed similarly, with a higher nitrate leaching potential for the areas of the field with coarser-textured soils.[52]

Sharpley et al.[53–55] and Lemunyon and Gilbert et al.[56] reported about the need to join the Phosphorus Index with a nitrogen index to conduct a multi-nutrient assessment of the risk of losses of nitrogen and phosphorus to the environment. Heathwaite et al.[57] developed a joint nitrogen and phosphorus index. OMAFRA[58] also proposed a joint index. Similarly, Delgado et al.[35] proposed to join the new Nitrogen Index to currently available P indexes. Bolster et al.[59] developed a nitrogen and a phosphorus index analysis for USDA-NRCS in Kentucky.

De Paz[15] tested the new Nitrogen Index approach joined with GIS capabilities across one of the regions of Europe most impacted by nitrate leaching.[15,32] De Paz et al.[15] found a correlation between the nitrate leaching and water leaching predicted by the Nitrogen Index, and measured nitrate leaching and water leaching. They also found that the nitrate leaching predicted by the Nitrogen Index was significantly correlated with the measured nitrate leaching across this Mediterranean region and that the areas with higher groundwater nitrate concentrations were correlated with the areas estimated to have higher nitrate leaching potential by the Nitrogen Index.

These results from De Paz et al. are another example indicating that the hole-in-the-pipe model proposed by Firestone and Davidson[60] describing the losses via emissions of greenhouse gases from the soil in the nitrogen cycle can be applied to the whole nitrogen cycle in agricultural systems, as described in the present work, and that the application of higher nitrogen rates than needed increases the pressure in the system, resulting in increased leaking from the system, and thus higher concentrations of nitrate, impacting groundwater quality.

Figueroa et al.[61] found that in intensive manure systems in Mexico that received large manure applications, the risk of nitrate leaching was higher. The tool was used to assess the effects of manure management practices on residual soil nitrate in the profile.[61] The Instituto Nacional de Investigaciones Forestales Agricolas y Pecuaris (INIFAP), a research institution in Mexico, conducted studies in Mexico and reported that the Nitrogen Index was a tool that can be used to assess the nitrate leaching risk in forage systems in this country.[42] Saynes et al.[16] were also able to use the Nitrogen Index to assess the effects of management practices in Mexico on emissions of N_2O from these fertilized systems. This is another example that this simple tool can be used to validate the hole-in-the-pipe model proposed by Firestone and Davidson[60] as adapted in the present work in international systems since the emissions of N_2O increased with higher fertilizer rates.[16]

Conclusion

The Nitrogen Index has been tested and calibrated using data from North America,[4,35,43,44,59,62] different regions in Mexico,[16,41,42,61,63,64] the Caribbean[65] region, different regions of South America[35,66,67] (Andean region and Pampas region), a Mediterranean region,[15] and the North China Plain,[35] where personnel from various national and international agencies, universities, and/or research centers have used the tool. Results of these efforts show that the Nitrogen Index is able to evaluate the effect of management on residual soil nitrate, water leaching, nitrate leaching, emissions of nitrous oxide, crop nitrogen use efficiencies, and crop nitrogen uptake. Additionally, measured groundwater nitrate concentrations have also been correlated with nitrate leaching predicted by the Nitrogen Index, validating the potential of using the tool to assess the risk of nitrate leaching to the environment.

Although it is recommended that users get familiar with the capabilities and limitations of the Nitrogen Index, the correlation between predicted and observed values for these different losses of reactive nitrogen suggests that this is a robust tool capable of being used to assess the risk of management practices on potential nitrate leaching and losses via other pathways. These results also support using the Nitrogen Index as part of an ecological engineering approach to soil and water conservation planning.[35,43] There is potential to use this tool to assess the effects of nitrogen inputs from manure, fertilizer and other organic sources, and management practices, and to rank the risk of losses of reactive nitrogen, including the nitrate leaching pathways, as very low, low, medium, high, or very high risk.

The Nitrogen Index can potentially be used as an indicator to assess the risk of nitrate leaching to the environment, helping conservation practitioners and nutrient managers implement best management practices to increase nutrient use efficiencies while reducing the nitrate leaching risk and losses of reactive nitrogen through different pathways. This tool is a new approach and is available for download, along with the code for the tool. The Nitrogen Index shows promise for being used as a quick and effective assessment tool for global sustainability.

References

1. IPCC (Intergovernmental Panel on Climate Change). IPCC Special Report on Global Warming of 1.5: Summary for Policymakers. 2018. https://www.ipcc.ch/2018/10/08/summary-for-policymakers-of-ipcc-special-report-on-global-warming-of-1-5c-approved-by-governments/.
2. Lal, R.; Delgado, J.A.; Gulliford, J.; Nielsen, D.; Rice, C.W.; Van Pelt, R.S. Adapting agriculture to drought and extreme events. *J. Soil Water Conserv.* **2012**, *67* (6), 162A–166A.
3. Nearing, M.A.; Pruski, F.F.; O'Neal, M.R. Expected climate change impacts on soil erosion rates: A review. *J. Soil Water Conserv.* **2004**, *59* (1), 43–50.
4. Marchi, E.C.S.; Zotarelli, Lincoln; Delgado, J.A.; Rowland, D.L. Use of the nitrogen index to assess nitrate leaching and water drainage from plastic-mulched horticultural cropping systems of Florida. *Int. Soil Water Conserv. Res.* **2016**, *4* (4), 237–244.
5. Randall, G.W.; Delgado, J.A.; Schepers, J.S. Nitrogen management to protect water resources. In *Nitrogen in Agricultural Systems*; Schepers, J.S., Raun, W.R., Follett, R.F., Fox, R.H., Randall, G.W., Eds.; Agronomy Monograph 49; American Society of Agronomy: Madison, WI, 2008; 911–946.
6. Randall, G.W.; Goss, M.J.; Fausey, N.R. Nitrogen and drainage management to reduce nitrate losses to subsurface drainage. In *Advances in Nitrogen Management for Water Quality*; Delgado, J.A., Follett, R.F., Eds.; Soil and Water Conservation Society: Ankeny, IA, 2010; 61–93.
7. Pierce, F.J.; Shaffer, M.J.; Halvorson, A.D. Screening procedure for estimating potentially leachable nitrate-nitrogen below the root zone. In *Managing Nitrogen for Groundwater Quality and Farm Profitability*; Follett, R.F., Keeny, D.R., Cruse, R.M., Eds.; Soil Science Society of America: Madison, WI, 1991; 259–283.
8. Van Es, H.M.; Delgado, J.A. Nitrate leaching index. In *Encyclopedia of Soil Science*; Lal, R., Ed.; Markel and Decker: New York, 2006; 1119–1121.
9. Delgado, J.A. Quantifying the loss mechanisms of nitrogen. *J. Soil Water Conserv.* **2002**, *57*, 389–398.
10. Delgado, J.A.; Groffman, P.M.; Nearing, M.A.; Goddard, T.; Reicosky, D.; Lal, R.; Kitchen, N.; Rice, C.; Towery, D.; Salon, P. Conservation practices to mitigate and adapt to climate change. *J. Soil Water Conserv.* **2011**, *66*, 118A–129A.
11. Delgado, J.A. The Nanchang Communication about the potential for implementation of conservation practices for climate change mitigation and adaptation to achieve food security in the 21st century. *Int. Soil Water Conserv. Res.* **2016**, *4* (2), 148–150.
12. Delgado, J.A.; Gagliardi, P.; Shaffer, M.J; Cover, H.; Hesketh, E.; Ascough, J.C.; Daniel, B.M. New tools to assess nitrogen management for conservation of our biosphere. In *Advances in Nitrogen Management for Water Quality*; Delgado, J.A., Follett, R.F., Eds.; SWCS: Ankeny, IA, 2010; 373–409.
13. Cowling, E.; Galloway, J.; Furiness, C.; Erisman, J.W., et al. *Optimizing nitrogen management and energy production and environmental protection: Report from the Second International Nitrogen Conference*, Bolger Center, Potomac, MD, Oct 14–18, 2001; 2002, available at http://www.initrogen.org/fileadmin/user_upload/Second_N_Conf_Report.pdf (accessed May 2010).
14. Galloway, J.N.; Aber, J.D.; Erisman, J.W.; Seitzinger, S.P.; Howarth, R.W.; Cowling, E.B.; Cosby, B.J. The nitrogen cascade. *BioScience* **2003**, *53* (4), 341–356.

15. De Paz, J.M.; Delgado, J.A.; Ramos, C.; Shaffer, M.; Bar-barick, K. Use of a new nitrogen index-GIS assessment for evaluation of nitrate leaching across a Mediterranean region. *J. Hydrol.* **2009**, *365,* 183–194.

16. Saynes, V.; Delgado, J.A.; Tebbe, C.; Etchevers, J.D., Lapidus, Otero-Arnaiz, A. Use of the new Nitrogen Index tier zero to assess the effects of nitrogen fertilizer on N2O emissions from cropping systems in Mexico. *Ecol. Eng.* **2014**, *73,* 778–785.

17. USEPA (U.S. Environmental Protection Agency). National Primary Drinking Water Regulations. n.d. https://www.epa.gov/ground-water-and-drinking-water/ national-primary-drinking-water-regulations.

18. Follett, J.R.; Follett, R.F.; Herz, W.C. Environmental and human impacts of reactive nitrogen. In *Advances in Nitrogen Management for Water Quality*; Delgado, J.A., Follett, R.F., Eds.; Soil and Water Conservation Society: Ankeny, IA, 2010; 1–37.

19. Follett, R.F.; Walker, D.J. Groundwater quality concerns about nitrogen. In *Nitrogen Management and Groundwater Protection*; Follett, R.F., Ed.; Elsevier Sci. Pub.: Amsterdam, 1989; 1–22.

20. Turner, R.E.; Rabalais, N.N. Linking landscape and water quality in the Mississippi River Basin for 200 years. *BioScience* **2003**, *53* (6), 563–572.

21. Mitsch, W.J.; Day, J.W. Restoration of wetlands in the Mississippi–Ohio–Missouri (MOM) River Basin: Experience and needed research. *Ecol. Eng.* **2006**, *26,* 55–69.

22. Rabalais, N.N.; Turner, R.E.; Wiseman, W.J., Jr. Gulf of Mexico hypoxia, a.k.a. "The Dead Zone." *Annu. Rev. Ecol. Syst.* **2002**, *33,* 235–263.

23. Rabalais, N.N.; Turner, R.E.; Scavia, D. Beyond science into policy: Gulf of Mexico hypoxia and the Mississippi River. *BioScience* **2002**, *52* (2), 129–142.

24. Goolsby, D.A.; Battaglin, W.A.; Aulenbach, B.T.; Hooper, R.P. Nitrogen input to the Gulf of Mexico. *J. Environ. Qual.* **2001**, *30,* 329–336.

25. Dubrovsky, N.M.; Burow, K.R.; Clark, G.M.; Gronberg, J.A.M.; Hamilton, P.A.; Hitt, K.J.; Mueller, D.K.; Munn, M.D.; Puckett, L.J.; Nolan, B.T.; Rupert, M.G.; Short, T.M.; Spahr, N.E.; Sprague, L.A.; Wilbur, W.G. *Nutrients in the Nation's Streams and Groundwater, 1992–2004, Circular 1350*; U.S. Geological Survey: Reston, VA, 2010.

26. Smith, D.R.; Wilson, R.S.; King, K.W.; Zwonitzer, M.; McGrath, J.M.; Harmel, R.D.; Haney, R.L.; Johnson, L.T. Lake Erie, phosphorus, and microcystin: Is it really the farmer's fault?. *J. Soil Water Conserv.* **2018**, *73,* 48–57.

27. World Health Organization (WHO) (2011.) *Guidelines for Drinking-Water Quality.* 4th Ed. WHO Press: Geneva, Switzerland.

28. USEPA (United States Environmental Protection Agency). 2015 Drinking Water Health Advisories for Two Cyanobacterial Toxins. Office of Water 820F15003, June 2015, Washington, DC. https:// www.epa.gov/sites/production/files/2017-06/documents/cyanotoxins-fact_sheet-2015.pdf.

29. Follett, R.F.; Delgado, J.A. Nitrogen fate and transport in agricultural systems. *J. Soil Water Conserv.* **2002**, *57* (6), 402–408.

30. Hatfield, J.L.; Follett, R.F. *Nitrogen Management in the Environment. Sources Problems, and Management,* 2nd Ed.; Elsevier Science Publ.: Amsterdam, 2008.

31. Li, X.; Hu, C.; Delgado, J.A.; Zhang, Y.; Ouyang, Z. Increased nitrogen use efficiencies as a key mitigation alternative to reduce nitrate leaching in North China Plain. *Agric. Water Manage.* **2007**, *89,* 137–147.

32. Juergens-Gschwind, S. Ground water nitrates in other developed countries (Europe)—Relations to land use patterns. In *Nitrogen Management and Groundwater Protection*; Follett, R.F., Ed., Elsevier Science Publishers: New York, 1989; 75–138.

33. Lavado, R.S.; de Paz, J.M.; Delgado J.A.; Rimski-Korsakov, H. Evaluation of best nitrogen management practices across regions of Argentina and Spain. In *Advances in Nitrogen Management for Water Quality*; Delgado, J.A., Follett, R.F., Eds.; Soil and Water Conservation Society: Ankeny, IA, **2010**; 313–342.

34. Rupert, M.G. Decadal-scale changes of nitrate in ground water of the United States, 1988–2004. *J. Environ. Qual.* **2008**, *37*, S240–S248.

35. Delgado, J.A.; Shaffer, M.; Hu, C.; Lavado, R.; Cueto Wong, J.A.; Joosse, P.; Sotomayor, D.; Colon, W.; Follett, R; Del Grosso, S.; Li, X.; Rimski-Korsakov, H. An index approach to assess nitrogen losses to the environment. *Ecol. Eng.* **2008**, *32*, 108–120.

36. Delgado, J.A.; Follett, R.F., Eds. *Advances in Nitrogen Management for Water Quality*; SWCS: Ankeny, IA, 2010.

37. Delgado, J.A.; Shaffer, M.J.; Lal, H.; McKinney, S.; Gross, C.M.; Cover, H. Assessment of nitrogen losses to the environment with a Nitrogen Trading Tool (NTT). *Comput. Electron. Agric.* **2008**, *63*, 193–206.

38. Buczko, U.; Kuchenbuch, R.O.; Lennartz, B. Assessment of the predictive quality of simple indicator approaches for nitrate leaching from agricultural fields. *J. Environ. Manage.* **2010**, *91*, 1305–1315.

39. Shaffer, M.J.; Delgado, J.A. Essentials of a national nitrate leaching index assessment tool. *J. Soil Water Conserv.* **2002**, *57*, 327–335.

40. Delgado, J.A.; Shaffer, M.; Hu, C.; Lavado, R.S.; Cueto Wong, J.; Joosse, P.; Li, X.; Rimski-Korsakov, H.; Follett, R.; Colon; W.; Sotomayor, D. A decade of change in nutrient management requires a new tool: A new nitrogen index. *J. Soil Water Conserv.* **2006**, *61*, 62A–71A.

41. Figueroa Viramontes, U., Delgado, J.A., Cueto Wong, José. Índice de Nitrógeno Ver. 4.4 Adaptado para la Producción de Forrajes en México. Manual del Usuario. (In Spanish.) Mexico Nitrogen Index User Manual. Instituto Nacional de Investigaciones Forestales, Agrícolas, y Pecuarias (INIFAP) and Secretaría de Agricultura, Ganadería, Desarrollo Rural, Pesca y Alimentación (SAGARPA), Matamoros, Coahuila, Mexico, 2011.

42. Instituto Nacional de Investigaciones Forestales Agricolas y Pecuaris (INIFAP). Reporte Anual 2009 Ciencia y Tecnología para el Campo Mexicano Instituto Nacional de Investigaciones Forestales, Agrícolas y Pecuarias. Reporte Anual 2009 México, D. F. Publicación Especial Núm. 5; ISBN 978-607-425-316-0: Mexico, 2010; 31 pp.

43. Gross, C.M.; Delgado, J.A.; Shaffer, M.J.; Gasseling, D.; Bunch, T.; Fry, R. A tiered approach to nitrogen management: A USDA perspective. In *Advances in Nitrogen Management for Water Quality*; Delgado, J.A., Follett, R.F., Eds.; Soil and Water Conservation Society: Ankeny, IA, 2010; 410–424.

44. USDA. National NRCS Nutrient Management Tools. https://www.nrcs.usda.gov/wps/portal/nrcs/detail/national/nwmc/partners/?cid=nrcs143_015091.

45. Buczko, U.; Kuchenbuch, R.O. Environmental indicators to assess the risk of diffuse nitrogen losses from agriculture. *Environ. Manage.* **2010**, *45*, 1201–1222.

46. Van Es, H.M.; Czymmek, K.J.; Ketterings, Q.M. Management effects on nitrogen leaching and guidelines for a nitrogen leaching index in New York. *J. Soil Water Conserv.* **2002**, *57*, 499–504.

47. Wu, L.; Letey, J.; French, C.; Wood, Y.; Bikie, D. Nitrate leaching hazard index developed for irrigated agriculture. *J. Soil Water Conserv.* **2005**, *60*, 90A–95A.

48. Williams, J.R.; Kissel, D.E. Water percolation: An indicator of nitrogen-leaching potential. In *Managing Nitrogen for Groundwater Quality and Farm Profitability*; Follett, R.F., Keeney, D.R., Cruse, R.M., Eds.; Soil Science Society of America: Madison, WI, 1991; 59–83.

49. Delgado, J.A. Sequential NLEAP simulations to examine effect of early and late planted winter cover crops on nitrogen dynamics. *J. Soil Water Conserv.* **1998**, *53*, 241–244.

50. Delgado, J.A. Use of simulations for evaluation of best management practices on irrigated cropping systems. In *Modeling Carbon and Nitrogen Dynamics for Soil Management*; Shaffer, M.J., Ma, L., Hansen, S., Eds.; Lewis Publishers: Boca Raton, FL, 2001; 355–381.

51. Delgado, J.A.; Shaffer, M.; Brodahl; M.K. New NLEAP for shallow and deep rooted rotations: Irrigated agriculture in the San Luis Valley of south central Colorado. *J. Soil Water Conserv.* **1998**, *53* (4), 338–340.

52. Delgado, J.A.; Bausch, W.C. Potential use of precision conservation techniques to reduce nitrate leaching in irrigated crops. *J. Soil Water Conserv.* **2005**, *60*, 379–387.

53. Sharpley, A.N.; Daniel, T.; Sims, T; Lemunyon, J.; Stevens, R.; Parry, R. *Agricultural Phosphorus and Eutrophication*, U.S. Department of Agriculture Agricultural Research Service No. ARS-149; 1999.

54. Sharpley, A.N.; Kleinman, P.; McDowell, R. Innovative management of agricultural phosphorous to protect soil and water resources. *Commun. Soil Sci. Plant Anal.* **2001**, *32* (7 & 8), 1071–1100.

55. Sharpley, A.N.; Weld, J.L.; Beegle, D.B.; Kleinman, P.J.A.; Gburuk, W.J.; Moore, P.A., Jr.; Mullins, G. Development of phosphorus indices for nutrient management planning strategies in the United States. *J. Soil Water Conserv.* **2003**, *53*, 137–151.

56. Lemunyon, J.L.; Gilbert, R.G. The concept and need for a phosphorus assessment tool. *J. Prod. Agric.* **1993**, *6*, 483–486.

57. Heathwaite, L.; Sharpley, A.; Gburek, W. A conceptual approach for integrating phosphorous and nitrogen management at watershed scales. *J. Environ. Qual.* **2000**, *29*, 158–166.

58. Ontario Ministry of Agriculture, Food and Affairs (OMAFRA). Fundamentals of Nutrient Management Training Course; Queen's Printer of Ontario: Ontario, Canada, 2005. Available at http://www.omafra.gov.on.ca/english/nm/nman/default.htm (accessed May 2011).

59. Bolster, C.H.; Horvath, T.; Lee, B.D.; Mehlhope, S.; Higgins, S.; Delogado, J.A. Development and testing of a new phosphorus index for Kentucky. *J. Soil Water Conserv.* **2014**, *69* (3), 183–196.

60. Firestone, M.K.; Davidson, E.A. Microbiological basis of NO and N2O production and consumption in soil. In *Exchange of Trace Gases between Terrestrial Eco systems and the Atmosphere;* Andreae, M.O.; Schimel, D.S., Eds. John Wiley & Sons: New York, 1989; 7–21.

61. Figueroa, V.U.; Núñez, H.G.; Delgado, J.A.; Cueto, J.A.; Flores, M.J.P. Estimación de la producción de estiércol y de la excreción de nitrógeno, fósforo y potasio por bovino lechero en la Comarca Lagunera. In *Agricultura Orgánica*, 2nd Ed.; Orona et al., Eds.; Sociedad Mexicana de la Ciencia del Suelo, Consejo Nacional de Ciencia y Tecnología, and Universidad Autónoma del Estado de Durango-Facultad de Agricultura y Zootecnia: Gomez Palacio, 2009; 128–151, 425.

62. Delgado, J.A.; Gagliardi, P.M.; Rau, E.J.; Fry, R.; Figueroa, U.; Gross, C.; Cueto-Wong, J.; Shaffer, M.; Kowalski, K.; Neer, D.; Sotomayor-Ramirez, D.; Alwang, J.; Monar, C.; Escudero, L. and Saavedra-Rivera. A.K. Nitrogen Index 4.4 User Manual. USDA-ARS-SPNR, Fort Collins, CO. 2011 (USDA-ARS-SPNR & NRCS User Manual; available at official USDA ARS SPNR. http://www.ars.usda.gov/npa/spnr/nitrogentools).

63. Figueroa-Viramontes, U.; Delgado, J.A.; Cueto-Wong, J.A.; Núñez-Hernández, G.; Reta-Sánchez, D.G.; Barbarick, K.A. A new Nitrogen Index to evaluate nitrogen losses in intensive forage systems in Mexico. *Agric. Ecosyst. Environ.* **2011**, *142*, 352–364.

64. Figueroa-Viramontes, U.; Delgado, J.A.; Sánchez-Duarte, J.I.; Ochoa-Martínez, E.; Núñez-Hernández, G. A nitrogen index for improving nutrient management within commercial Mexican dairy operations. *Int. Soil Water Conserv. Res.* **2016**, *4*, 1–5.

65. Oliveras-Berrocales, M.; Delgado, J.A.; Pérez-Alegría, L.R. The new Caribbean Nitrogen Index to assess nitrogen dynamics in vegetable production systems in southwestern Puerto Rico. *Int. Soil Water Conserv. Res.*, **2017**, *5* (1), 69–75.

66. Escudero, L.; Delgado, J.A.; Monar, C.; Valverde, F.; Barrera, V.; Alwang, J. A new nitrogen index for assessment of nitrogen management of Andean mountain cropping systems of Ecuador. *Soil Sci.* **2014**, *179*, 130–140.

67. Saavedra, A.K.; Delgado, J.A.; Botello, R.; Mamani, P.; Alwang, J. A new N Index to assess nitrogen dynamics in potato (*Solanum tuberosum* L.) production systems of Bolivia. *Agrociencia*, **2014**, *48*, 667–678.

40

Nitrogen (Nutrient) Trading Tool

Importance of Nitrogen 415
Need for Nitrogen Loss Measurements 416
Water Quality Trading Programs 417
Payments for Environmental Services Programs 418
Conclusion 420
Disclaimer 420
References 420

Jorge A. Delgado

Importance of Nitrogen

Nitrogen is an essential nutrient with very important functions for life; it is a component of nucleic acids, the building blocks of genetic information encoded in DNA, and protein, the building blocks of biological tissues such as muscles or hair. As evidenced by its contribution to increased yields of cropping systems during the 20th century, nitrogen is also important for food security and therefore global stability and peace. Almost all global cropping systems respond to nitrogen inputs with higher yields and economic returns. When its management is combined with conservation practices, the results are very positive; for example, just recently Delgado et al. reported that in systems implementing conservation agriculture in the Andean region of South America, nitrogen inputs increased profits for farmers by 22%, and that these improved best management practices of fertilizer application and conservation agriculture could increase the profits for 200,000 small farmers.[1]

The ever-increasing global population creates additional pressure to increase agricultural intensification to achieve the needed increase in agricultural production needed to feed 2 billion additional people by 2050. The success of the green revolution has contributed to increased global production of grains and other crops such as potatoes, as well as increased yields per area during the last 100 years. [2] Similarly, the total use of nitrogen and nutrients worldwide has also been increasing as the human population and fertilizer needed to increase total global production also increase.[3] It is expected that the use of nitrogen and other nutrients will continue to increase during the 21st century; thus, management of nitrogen and other nutrients will be key to reducing the environmental impacts of adding additional nutrients to intensive agricultural systems to avoid repeating the errors of the 20th century that contributed to losses of reactive nitrogen to the environment in large quantities.

Environmental systems around the world are under pressure due to different factors such as severe droughts, extreme events, reduction of water sources, deforestation, erosion, and impacts of nutrients on aquatic and other natural systems, among others. The projected additional use of nitrogen and other nutrients at a global scale will continue to put additional pressure on these already impacted environmental systems. Nitrogen will continue to be key during the 21st century and important for food security, global stability and peace. The projected increase in the use of nutrients, the effects of extreme events and a changing climate will all put additional pressure on agricultural systems and increase the potential for losses of nutrients such as nitrogen. Adding more reactive nitrogen to the global nitrogen cycle will also increase pressure on these systems since nitrogen cycles through the environment in a phenomenon known as the nitrogen cascade,[4,5] creating a cascade of effects that accompanies this cycling.

Ribaudo et al. discussed the effects of reactive nitrogen from agricultural sources across the environment and concluded that removing nitrate from drinking water costs about 1.7 billion dollars annually in the United States.[6] Follett et al. reported on the effects of reactive nitrogen on human health and the importance of reducing losses of nitrogen to the environment.[7] In a recent paper, Temkin et al. calculated that exposure to nitrate could be connected to close to 3,000 cases of low birth weight and about 2,300–12,500 cancer cases annually in the United States, with annual economic impacts ranging from 250 million to 1.5 billion US dollars, as well as potentially 1.3–6.5 billion US dollars in lost productivity.[8] These impacts will be much higher on a global scale.[8,9] In addition to the impact of nitrate on groundwater and surface water (which will need to be removed for safety if the concentration is above 10 ppm NO_3–N),[10] surface water with hypoxic zones and algae blooms around the world could potentially develop high microcystin levels.[11] Microcystin-LR concentrations above 1.0 µg/L have been reported to be unsafe by the World Health Organization,[12] while the US Environmental Protection Agency (USEPA) reports that safe concentration limits of microcystin are 0.3 µg/L for children less than 6 years old[13] and that microcystin could contribute to gastroenteritis, and liver and kidney damage.[13] Since current projections suggest that use of nitrogen and other nutrients will continue to increase around the world to produce the additional food needed to feed the growing population, the use of conservation practices and management to reduce losses of nutrients will be of great importance.[14,15] Use of nitrogen trading tools that could assess the positive effects of best management practices will be needed so that savings (reductions in nitrogen loss) could be traded in ecosystem markets during the 21st century. Nitrogen is important for global food security, sustainability and peace; the question is how to manage it in the 21st century in the face of a changing climate, which could contribute to higher losses and make management of nitrogen more complex, and how to avoid the errors that were made in the 20th century, which contributed to a stable food supply but resulted in greater losses of nitrogen to the atmosphere, groundwater and surface water.

Need for Nitrogen Loss Measurements

As climate change continues, the occurrence of extreme events such as droughts and floods, desertification, shrinking water resources and other challenges are beginning to put additional stress on agricultural systems throughout the world. These challenges increase the pressure on these systems to maintain productivity and keep up with the growth in food demand that follows human population growth. Humans are significantly impacting the global nitrogen cycle; these impacts are reflected in, for example, increased concentrations of N_2O in the atmosphere, increased nitrate (NO_3) levels in groundwater or surface water resources and increased ammonia (NH_3) deposition in natural areas. Not only can these reactive nitrogen losses impact human health, as has been previously reported, they can also impact ecosystem health and ecosystem balances.[7–9] Over 20 years ago, the USEPA established that the safe limit for drinking water was 10 mg NO_3–N/L[10]; however, a recent paper has reported a positive correlation of NO_3–N in drinking water as low as 0.14 mg/L nitrate with a colorectal cancer risk of one in a million and greater risk at higher concentrations.[8]

The correlation of food security with nitrogen application is strong, and there is a delicate balancing act to be made: on the one hand, nitrogen applications provide food production benefits, but on the other hand, there are environmental impacts from nitrogen applications. This presents a quandary for humanity as far as how to continue to maximize productivity across agricultural systems, which requires nitrogen inputs, while minimizing nitrogen losses to the environment. Nitrogen trading tools and quantification of savings in nitrogen losses that could be traded in water and air quality markets can help provide some answers to this quandary since farmers that improve management of nitrogen could be compensated by trading these savings in air and water quality markets.[6,16,17] Users of the NTT should look at the effects of improved management in the context of the nitrogen cycle; this will help make it possible to sort out the effects of the practices on losses of reactive nitrogen via emissions of N_2O to the atmosphere (i.e., direct emissions of N_2O) and the losses of nitrogen (e.g. NO_3) through surface transport or leaching, which could impact groundwater or surface waters. How management practices interact with crop rotations depends on the previous and current crops, previous history at the site due to previous manure applications, and the site-specific factors (weather, crops, hydrology and landscape) of a given field, and this can be assessed by these types of tools to reduce nitrogen losses.[6,16,18]

Water Quality Trading Programs

Several researchers have proposed the concept of using payments as part of an ecosystem services payment system in an environmental trading/water quality trading program to reduce the negative effects of reactive nitrogen in the environment.[19–23] Hey and Hey et al.[19,21] recommended the strategic placement of wetlands across the Mississippi watershed that could be used as filters to remove nitrate as it enters these wetlands to reduce the transport of nitrogen to the Gulf of Mexico and have farmers receive environmental payments for developing wetlands for their farms to remove the nitrate. This approach is one of a series of precision conservation approaches that could be used with other conservation practices that could be set up strategically in fields and the landscape to increase the efficiency of conservation practices such as sediment traps, riparian buffers and denitrification traps to remove nitrate, phosphorus and sediment from water traveling to the Mississippi River and water bodies.[24–28] These precision conservation approaches together with precision regulation can contribute to increased adoption of conservation practices that will reduce losses of nitrogen and other nutrients to the environment. These practices can be among the key tools in the 21st-century toolbox to adapt to a changing climate and avoid some of the errors made in the 20th century.[26,28] The concept of Sustainable Precision Agriculture and Environment (SPAE)[29] is similar to the 7 R's approach to reduce the transport of nitrogen to the environment and the transport of other nutrients and sediment.[26,28] There are tools such as the Nutrient Tracking Tool and COMET-VR that can be used to assess the effect of management practices and quantify the reductions of nitrogen losses, nutrient losses and sediment transport across the environment so these reductions can potentially be traded in water quality and air quality markets.[30,31]

A review of potential trading markets by Lal et al.[32] found that precision conservation approaches have been considered by the Connecticut Department of Environmental Protection (DEP), Pennsylvania, Ohio and Oregon, and that these markets could benefit from the use of an NTT approach. The state of Pennsylvania's DEP has a nutrient trading program and additional information could be found at the program's website (https://www.dep.pa.gov/Business/Water/CleanWater/NutrientTrading/Pages/default.aspx). The nutrient trading program in Pennsylvania has developed Excel files used in calculations to assess practices such as cover crops, wetlands restoration, riparian buffers and grass buffers (https://www.dep.pa.gov/Business/Water/CleanWater/NutrientTrading/Pages/TradingResources.aspx) and their reduction of nitrogen losses.

The World Resources Institute (WRI) and the Chesapeake Bay Foundation have reported that these nutrient trading approaches in the states that are part of the Chesapeake Bay area can contribute to reduced environmental impact and could provide income for a farmer that installs a grass buffer along a stream to filter out fertilizer runoff and trades these credits with municipal wastewater treatment plants or stormwater management units that are having trouble reducing their pollution (https://www.wri.org/blog/2017/03/podcast-how-we-can-clean-chesapeake-and-save-money-process).

The concept of nutrient trading for water quality has been supported by the USEPA (https://www.epa.gov/nutrient-policy-data/collaborative-approaches-reducing-excess-nutrients#creating) with policies and regulations. Figure 1 shows nutrient trading for water quality efforts in the United States up to 2016. These nutrient trading efforts have contributed to localities and regions achieving compliance with the regulatory requirements of the Clean Water Act. The USEPA has implemented efforts to leverage

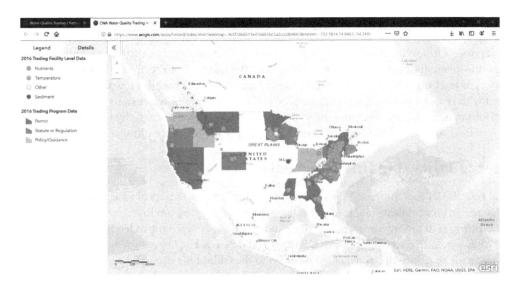

FIGURE 1   A map of water quality trading in the United States in 2016 (USEPA webpage).

emerging technologies and facilitate broader adoption of market-based programs. Customers of the water quality trading effort under the Clean Water Act can use this option for compliance with water quality-based effluent limitations with a National Pollutant Discharge Elimination System (NPDES) permit. Ribaudo and Gottlie[33] and Ribaudo et al.[34,35] reviewed different trading markets that can be used to reduce environmental effects from agriculture and reported that tools like the NTT could be used to reduce the uncertainty and provide information that could be used for these market systems.

Payments for Environmental Services Programs

Various reviews have been conducted about worldwide water quality trading programs or other types of trading programs. Selman et al. reported in 2009 that water quality trading programs were gaining traction and that there were 57 water quality trading programs worldwide, with 26 active programs, 21 under consideration for development and 10 inactive, with the majority of the programs in the United States and others in Australia, New Zealand and Canada. Stanton et al.[36] reported that there were 288 watershed payment programs. They reported that in 2008 there were 14 active water quality trading programs and 113 payments for watershed service programs with a value of transactions of 9.3 billion dollars, protecting 50.1 billion hectares.[36] They reported that in 2008 there were programs across 24 different countries including the United States, Brazil, Canada, France, China and New Zealand.

Forest Trends has a map of current active ecosystem markets that displays current watershed projects worldwide (https://www.forest-trends.org/project-list/#s). These projects have been identified as projects where people or organizations pay for conservation practices. They identified 466 watershed projects worldwide covering over 330,000,000 hectares. The water quality trading in the United States in 2016 can be found at the USEPA website (https://www.epa.gov/nutrient-policy-data/collaborative-approaches-reducing-excess-nutrients#creating), with a large number of programs concentrated around the northeast and in the Midwest. At this webpage, the user could click on each identified point on the map and find specific information about the program (Figure 2).

The Nature Conservancy identified 391 carbon sequestration projects worldwide covering over 52,000,000 hectares. These carbon projects can be used to produce carbon offsets (reductions in CO_2 emissions) that are sold in carbon markets. These sold offsets in CO_2 can be used to offset greenhouse gas emissions elsewhere. The Environmental Defense Fund has a website describing 19 multinational, national,

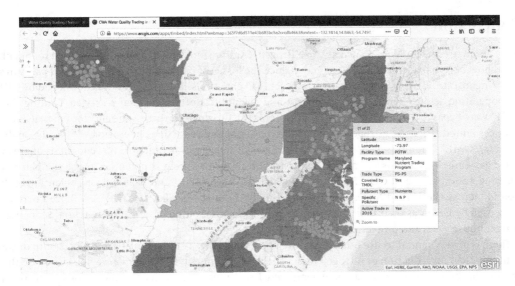

FIGURE 2   A map of water quality trading in the United States in 2016 (USEPA webpage). The user could click on each identified point on the map and find specific information about the program.

regional and local emissions trading systems (https://www.edf.org/worlds-carbon-markets). They reported that there are over 50 jurisdictions around the world with emission trading programs (or carbon markets).

The NTT has been reported as a potential tool that could be used to assess the effects of management practices on reductions of greenhouse gases (GHG).[6,16,30,31] Delgado et al. reported that since nitrogen fertilizers increase N_2O emissions and we could use the NTT to assess the effects of practices that reduce emissions of N_2O and that reduce NH_3 volatilization and NO_3 leaching losses that contribute to indirect emissions of N_2O, the NTT could be used to trade reductions of direct and indirect N_2O emissions.[6,16] Recent advances with tools such as the Nutrient Tracking Tool and COMET-VR also put these tools in a strong position to be used to assess the effects of management practices on carbon sequestration and reductions of direct and indirect N_2O emissions.[30,31] Both of these tools are being used by NRCS.[30,31]

The Nutrient Tracking Tool (NTT) is a tool being used by NRCS and is a free, online, user-friendly tool that quantitatively calculates the nitrogen, phosphorus and sediment losses from crop and pasture lands, and these savings could potentially be used for trade in water and air quality markets. The COMET-Farm tool is the official greenhouse gas quantification tool of USDA, and it could be used to the assess effects of management practices on reductions of GHG (https://www.nrcs.usda.gov/wps/portal/nrcs/detail/national/programs/?cid=stelprdb1261363).

The American Carbon Registry is a nonprofit carbon market in the United States that is a voluntary greenhouse gas registry (https://americancarbonregistry.org/). Among the different factors that this voluntary registry can keep track of is carbon sequestration in soils as well as potential reductions in N_2O emissions from agricultural fields. This registry lists accepted methodologies for quantifying reductions in greenhouse gas emissions resulting from changes in fertilizer management, such as improvement in nitrogen fertilizer management practices, which could contribute to reductions in N_2O emissions (https://americancarbonregistry.org/carbon-accounting/standards-methodologies/emissions-reductions-through-changes-in-fertilizer-management). Examples like this show that water and air (carbon) quality markets are connected, in that improved nitrogen management practices that increase nitrogen use efficiencies and reduce losses of nitrate, generate savings that can be traded in water quality markets, and increase efficiency, simultaneously reducing emissions of N_2O and generating savings that can be traded in air (carbon) trading markets. The 21st century will continue to bring opportunities to farmers to improve their nitrogen management practices and trade these savings in nitrate leaching losses and direct and indirect emissions of N_2O in water and air (carbon) trading markets.

Conclusion

There are new ecotechnologies such as the Nitrogen Trading Tool (NTT) that could be used to quickly assess how best management practices may reduce the losses of reactive nitrogen via several pathways, including nitrate leaching and direct and indirect N_2O emissions. The first NTT was developed by ARS in cooperation with NRCS.[6,16,37] The concept and capabilities of the original Nitrogen Trading Tool (NTT) were expanded in the Nutrient Tracking Tool (NTT), which can assess the effects of management practices on reductions of nitrogen and phosphorus losses and reductions in erosion.[18,30] Additionally, the new COMET-VR model can also assess the effects of conservation practices on reductions of greenhouse gas emissions and assess carbon sequestration in farms. There is potential to use the Nutrient Tracking Tool (NTT) and COMET-VR tools, which are both robust and have a strong scientific background, across the United States and internationally. Both the NTT and COMET-VR are currently being used by USDA as their tools to assess the effects of conservation practices on potential reductions in greenhouse gases and protection of water quality.

There is also potential to use other tools and approaches that could quantify the effects of management practices on reduction of nitrogen losses to the environment. For example, various water quality trading markets have developed coefficients and Excel files to quantify the reductions in nitrogen losses when a given conservation practice is applied (www.dep.pa.gov/Business/Water/CleanWater/NutrientTrading/Pages/default.aspx; https://www.wri.org/blog/2017/03/podcast-how-we-can-clean-chesapeake-and-save-money-process). Webpages developed by the EPA and Forest Trends are examples of online tools that track the development of active water and air quality markets where there are opportunities for tools like the NTT be to applied (https://www.epa.gov/nutrient-policy-data/collaborative-approaches-reducing-excess-nutrients#creating; https://www.forest-trends.org/project-list/#s). Independent of what tools are being used to assess the benefits of management practices in reducing losses of reactive nitrogen to the environment, application of conservation practices across watersheds will provide opportunities to reduce losses of nitrogen that can potentially impact water quality and air (carbon) quality. The use of new technologies and tools like the Nutrient Tracking Tool and COMET-VR that can assess temporal and spatial variability will allow us to assess hot-spot sources of nitrate leaching and greenhouse gas emissions across the environment and effectively apply conservation practices by considering spatial and temporal variability with precision conservation management and precision regulation.[27-31] The concept of the NTT and new ecotechnologies such as COMET-VR and the Nutrient Tracking Tool will help nutrient managers, conservation practitioners, farmers and other users quickly identify conservation practices that maximize nitrogen use efficiency while reducing nitrogen losses to the environment, providing an opportunity to trade the saved nitrogen in water and air (carbon) markets and contributing to conservation of the biosphere.[6,16,18,17,29-31]

Disclaimer

Trade and manufacturer's names are necessary to report factually on available data; however, the USDA neither guarantees nor warrants the standard of the product, and the use of the name by USDA implies no approval of the product to the exclusion of others that may also be suitable.

References

1. Delgado, J.A.; Barrera Mosquera, V.H.; Escudero López, L.O.; Cartagena Ayala, Y.E.; Alwang, J.R.; Stehouwer, R.C.; Arévalo Tenelema, J.C.; D Adamo, R.E.; Domínguez Andrade, J.M.; Valverde, F.; Alvarado Ochoa, S.P.; Conservation Agriculture increases profits in an Andean region of South America. *Agrosystems, Geosciences & Environ.* **2019** doi:10.2134/age2018.10.0050.

2. Max Roser, M.; Ritchie, H.; *Yields and Land Use in Agriculture*. University of Oxford. Global Change Data Laboratory. Published online at OurWorldInData.org. Retrieved from https://our-worldindata.org/yields-and-land-use-in-agriculture [Online Resource].

3. Smil, V.; Global population and the nitrogen cycle. *Scientific American* **1997**, *227*, 77–81.

4. Cowling, E.; Galloway, J.; Furiness, C.; Erisman, J.W. Optimizing nitrogen management in food and energy production and environmental protection. *Report* from *the* Second *International Nitrogen Conference*, October 14–18, 2001; Bolger Center: Potomac, MD, 2002.

5. Galloway, J.N.; Aber, J.D.; Erisman, J.W.; Seitzinger, S.P.; Howarth, R.W.; Cowling, E.B.; Cosby, B.J. The nitrogen cascade. *BioScience* **2003**, *53* (4), 341–356.

6. Delgado, J.A.; Shaffer, M.J.; Lal, H.; McKinney, S.; Gross, C.M.; Cover, H. Assessment of nitrogen losses to the environment with a Nitrogen Trading Tool. *Comput. Electron. Agric.* **2008**, *63*, 193–206.

7. Follett, J.R.; Follett, R.F.; Herz, W.C. Environmental and human impacts of reactive nitrogen. In *Advances in Nitrogen Management for Water Quality*; Delgado, J.A., Follett, R.F., Eds.; Soil and Water Conservation Society: Ankeny, IA, 2010; 1–37.

8. Temkin, A.; Evans, S.; Manidis, T.; Campbell, C.; Naidenko, O.V.; Exposure-based assessment and economic valuation of adverse birth outcomes and cancer risk due to nitrate in United States drinking water. *Environ. Res.* **2019**, doi:10.1016/j.envres.2019.04.009.

9. Ribaudo, M.; Delgado, J.; Hansen, L.; Livingston, M.; Mosheim, R.; Williamson, J. *Nitrogen in Agricultural Systems: Implications for Conservation Policy*; ERS: Economy Research Report: Washington, DC, 2010.

10. USEPA (U.S. Environmental Protection Agency). National Primary Drinking Water Regulations. n.d. https://www.epa.gov/ground-water-and-drinking-water/national-primary-drinking-water-regulations.

11. Smith, D.R.; Wilson, R.S.; King, K.W.; Zwonitzer, M.; McGrath, J.M.; Harmel, R.D.; Haney, R.L.; Johnson, L.T. Lake Erie, phosphorus, and microcystin: Is it really the farmer's fault? *J. Soil Water Conserv.* **2018**, *73*, 48–57.

12. World Health Organization (WHO). *Guidelines for Drinking-water Quality*. 4th Edn. WHO Press: Geneva, 2011.

13. USEPA (United States Environmental Protection Agency). Drinking Water Health Advisories for Two Cyanobacterial Toxins. 2015, Office of Water 820F15003 June 2015; Washington D.C. https://www.epa.gov/sites/production/files/2017-06/documents/cyanotoxins-fact_sheet-2015.pdf.

14. Delgado, J.A.; Groffman, P.M.; Nearing, M.A.; Goddard, T.; Reicosky, D.; Lal, R.; Kitchen, N.; Rice, C.; Towery, D.; Salon, P. Conservation practices to mitigate and adapt to climate change. *J. Soil Water Conserv.* **2011**, *66*, 118A–129A.

15. Lal, R.; Delgado, J.A.; Groffman, P.M.; Millar, N.; Dell, C.; Rotz, A.; Management to mitigate and adapt to climate change. *J. Soil Water Conserv.* **2011**, *66*, 276–285.

16. Delgado, J.A.; Gross, C.M.; Lal, H.; Cover, H.; Gagliardi, P.; McKinney, S.P.; Hesketh, E.; Shaffer, M.J. A new GIS nitrogen trading tool concept for conservation and reduction of reactive nitrogen losses to the environment. *Adv. Agron.* **2010**, *105*, 117–171.

17. Delgado, J.A., Follett, R.F., Eds. *Advances in Nitrogen Management for Water Quality*; Soil and Water Conservation Society: Ankeny, IA, 2010.

18. Saleh, A.; Gallego, O.; Osei, E.; Lal, H.; Gross, C.; McKinney, S.; Cover, H. Nutrient Tracking Tool—A user-friendly tool for calculating nutrient reductions for water quality trading. *J. Soil Water Conserv.* **2011**, *66*, 400–410.

19. Hey, D.L. Nitrogen farming: Harvesting a different crop. *Restor. Ecol.* **2002**, *10*, 1–10.

20. Greenhalch, S.; Sauer, A. *Awakening the Dead Zone: An Investment for Agriculture, Water Quality, and Climate Change*; World Resources Institute: Washington, DC, 2003.

21. Hey, D.L.; Urban, L.S.; Kostel, J.A. Nutrient farming: The business of environmental management. *Ecol. Eng.* **2005**, *24*, 279–287.

22. Ribaudo, M.O.; Heimlich, R.; Peters, M. Nitrogen sources and Gulf hypoxia: Potential for environmental credit trading. *Ecol. Econ.* **2005**, *52*, 159–168.

23. Glebe, T.W. The environmental impact of European farming: How legitimate are agri-environmental payments? *Rev. Agric. Econ.* **2006**, *29*, 87–102.

24. Berry, J.R.; Delgado, J.A.; Khosla, R.; Pierce, F.J. Precision conservation for environmental sustainability. *J. Soil Water Conserv.* **2003**, *58*, 332–339.

25. Delgado, J.A.; Berry, J.K. Advances in precision conservation. *Adv. Agron.* **2008**, *98*, 1–44.

26. Delgado, J. A.; 4 Rs are not enough. We need 7 Rs for nutrient management and conservation to increase nutrient use efficiency and reduce off-site transport of nutrients. In *Soil Specific Farming: Precision Agriculture;* Lal, R.; Stewart, B. A., Eds. Advances in Soil Science Series; CRC Press: Boca Raton, FL, 2016; 89–126.

27. Delgado, J. A., Sassenrath, G., Mueller, T., Eds. *Precision Conservation: Geospatial Techniques for Agricultural and Natural Resources Conservation.* Agronomy Monograph 59. American Society of Agronomy, Crop Science Society of America, and Soil Science Society of America: Madison, WI, 2018.

28. Sassenrath, G. F., Delgado, J.A. Precision Conservation and Precision Regulation Delgado, J. A., Sassenrath, G., Mueller, T., Eds. *Precision Conservation: Geospatial Techniques for Agricultural and Natural Resources Conservation.* Agronomy Monograph 59. American Society of Agronomy, Crop Science Society of America, and Soil Science Society of America: Madison, WI, 2018.

29. Delgado, J.A.; Short Jr.; N.M.; Roberts, D.P.; Vandenberg, B. Big Data Analysis for Sustainable Agriculture on a Geospatial Cloud Framework. *Front. Sust. Food Syst.* **2019**; doi:10.3389/fsufs.2019.00054.

30. Saleh, A., Osei, E., Precision conservation and water quality markets. In *Precision Conservation: Geospatial Techniques for Agricultural and Natural Resources Conservation*; Delgado, J. A., Sassenrath, G., Mueller, T., Eds. Agronomy Monograph 59. American Society of Agronomy, Crop Science Society of America, and Soil Science Society of America: Madison, WI, 2018; 313–340.

31. Paustian, K., Easter, M., Brown, K., Chambers, A., Eve, M., Huber, A., Marx, E., Layer, M., Sterner, M., Sutton, B., Swan, A., Toureene, C., Verlayudhan, S., Williams, S., Field- and farm-scale assessment of soil greenhouse gas mitigation using COMET-Farm. In *Precision Conservation: Geospatial Techniques for Agricultural and Natural Resources Conservation*; Delgado, J. A., Sassenrath, G., Mueller, T., Eds. Agronomy Monograph 59. American Society of Agronomy, Crop Science Society of America, and Soil Science Society of America: Madison, WI, 2018; 341–360.

32. Lal, H.; Delgado, J.A.; Gross, C.M.; Hesketh, E.; McKinney, S.P.; Cover, H.; Shaffer, M. Market-based approaches and tools for improving water and air quality. *Environ. Sci. Policy* **2009**, *12*, 1028–1039.

33. Ribaudo, M.; Gottlie, J. Point-nonpoint trading—Can it work? *J. Am. Water Resour. Assoc.* **2011**, *47*, 5–14.

34. Ribaudo, M.; Delgado, J.; Hansen, L.; Livingston, M.; Mosheim, R.; Williamson, J. *Nitrogen in Agricultural Systems: Implications for Conservation Policy*; ERS (Economy Research Report): Washington, DC, 2010.

35. Ribaudo, M.; Greene, C.; Hansen, L.; Hellerstein, D. Ecosystem services from agriculture: Steps for expanding markets. *Ecol. Econ.* **2010**, *69*, 2085–2092.

36. Stanton, T.; Echavarria, M.; Hamilton, K.; Ott, C. State of Watershed Payments: An Emerging Marketplace. Ecosystem Marketplace 2010, available at https://www.forest-trends.org/publications/state-of-watershed-payments/.

37. Gross, C.; Delgado, J.A.; McKinney, S.; Lal, H.; Cover, H.; Shaffer, M. Nitrogen Trading Tool (NTT) to facilitate water quality credit trading. *J. Soil Water Conserv.* **2008**, *63*, 44A–45A.

41

The Accounting Framework of Energy–Water Nexus in Socioeconomic Systems

Saige Wang and
Bin Chen

Introduction 423
Accounting Framework 424
References...425

Introduction

Water use and energy consumption are highly interwoven in economic networks due to their interconnections in both the production and consumption of primary, intermediate, and final goods and services (Chen 2016; Chen and Lu 2015). At all stages of the production chain, water and energy resources are used and "embodied" in the products. Nexus analysis highlights the interconnections and interdependencies for all kinds of economic activities (Dai et al. 2018; Hoff 2011). The tracking of energy and water flows among regions and the quantification of their interdependencies are fundamental for coordinated management of regional resources (Liu et al. 2018).

Most of the recent research on energy–water nexus has explored the energy or water sector alone or impacts of specific technologies on sectors with energy minimization and water use options (Kenway et al. 2011; Wang and Chen 2016; Wang et al. 2017a). However, there are few works published that conceptualize the role of the nexus in reconfiguring the interactions between energy- and water-related sectors in a socioeconomic system (Chen et al. 2018; Owen et al. 2018). Some studies have explored the direct and indirect interdependencies between energy and water in socioeconomic systems (Duan and Chen 2016; Fang and Chen 2017; Liu et al. 2019). By tracking sectoral economic flows through a supply chain, input–output analysis (IOA) can be used to investigate these interdependencies, which are a common focus in nexus studies (Wang et al. 2019; Wang and Chen 2016). Based on these sectoral interactions and exchanges with other economies through the supply chain, IOA-based approaches can assess both the direct and indirect energy (or water) consumption required to produce goods and services in a region (Wang et al. 2017b; Wang and Chen 2018). Compared with conventional nexus studies, which prioritize specific interdependencies (e.g., by defining that specific processes require certain resources), the system-wide approach of IOA presents a more comprehensive picture that includes direct and indirect use profiles and hotspots. Pioneering works have used IOA case studies to focus on the interplay involved in the water–energy nexus (Cao et al. 2018; Chen and Chen 2016; Wang et al. 2019). For example, Wang et al. (2017b) proposed a modified IOA that provided a unified framework to analyze the tradeoffs between urban energy and water systems. Fang and Chen (2017) used IOA and linkage analysis to detect synergetic effects of water and energy consumption and their interactions among economic sectors. Marsh suggested various IOA techniques to address multiple dimensions of the nexus (linkage, dependency, multiplier, and scenario analyses) (Marsh 2008). Kahrl and Roland-Holst built relevant metrics to quantify the nexus from physical, monetary, and distributive perspectives (Kahrl and Roland-Holst 2008).

Ecological network analysis (ENA), a system-based approach, has its unique strength in examining the structure and function of system from a system perspective (Fath 2007, 2015). The control and dependence analysis of ENA may also provide insights into the interwoven relationships among nodes due to the direct and indirect flows, which can be used to identify the regulating pathways for the system. Also, Wang and Chen (2016) established a multiregional nexus network based on multiregional IOA and ENA to explore the structural properties and sectoral interactions within urban agglomerations. However, nexus issues in energy and water policy are rarely studied from a systems perspective, an area that urgently requires investigation. Particularly in energy development planning, the role of water stress issues is insufficiently considered as a part of nexus management (Font et al. 2017; Williams et al. 2014).

Water is directly or indirectly required in all types of power generation technologies for cooling purposes, steam generation, and infrastructure manufacturing. There is a great diversity of water intensities among different energy types. The water intensity gap among different electricity types provides us the opportunity to reduce the power generation's vulnerability to water shortages. In this study, we will propose an accounting framework to extend the energy development scenarios into nexus to study interwoven connections of energy consumption and water use in economic supply chain based on IOA and ENA.

Accounting Framework

In our accounting framework, we will first quantify the energy-related water and water-related energy from consumption and production perspectives combining a bottom-up method and a life cycle analysis. Then, energy and water flows among regions and sectors will be calculated based on IOA. The direct and indirect energy and water-related energy were then combined as embodied energy to build the nexus network model. By merging the energy-related water and direct water, we will build a regional energy–water nexus network to investigate its system properties. The nexus point is built to form the energy and water flows in the same nexus network. The tools from ENA were used to explore the energy and water control and dependence relationship among sectors under the nexus impact background. Flow analysis was used to identify critical pathways for nexus management. We investigated the sectoral nexus impact for energy system by comparing energy and water-related energy network to identify critical sectors for energy-side nexus management. The sectoral nexus impact under different scenarios was compared and analyzed based on ENA. Finally, network impact through direct and indirect pathways is used to reveal the effect of energy–water nexus on all sectors in a more explicit and predictable way and the tradeoffs among energy, water, and carbon using nexus accounting and the scenario features can be discussed.

Thus, we can capture both intensities and structures of regional nexus and how they alter the local consumption. Integrated analysis and modeling of the water–energy nexus require the simulation of many human and natural systems and their complex interactions and dynamics (Wang et al. 2018, 2019). By doing this, we are trying to reveal energy–water nexus impacts of pathways and sectors on the system and track energy and water flows among sectors. System properties, such as cycling index and system robustness, can help us to identify the risks in the energy–water nexus and leverage points to promote sustainable regional energy and water management.

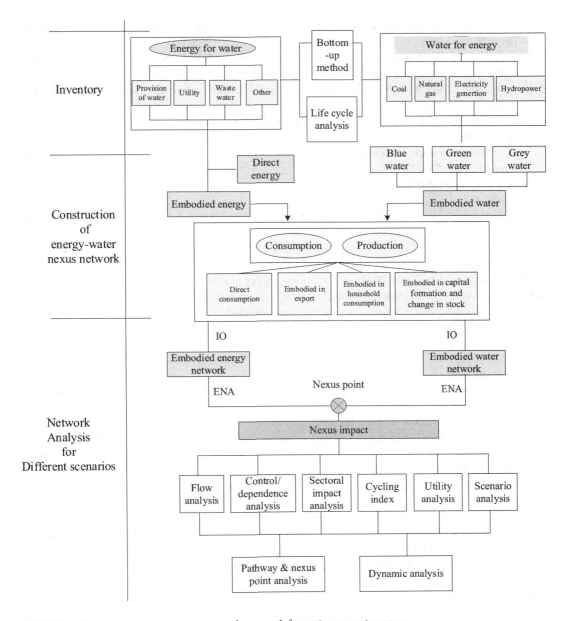

FIGURE 1   Energy–water nexus accounting framework for socioeconomic system.

References

Cao, T., Wang, S., and Chen, B. 2018. Priority in Energy-water Nexus Based on Input-output Analysis: A Case Study of China. DEStech Transactions on Environment, Energy and Earth Sciences, (iceee).

Chen, B. 2016. Energy, ecology and environment: A nexus Perspective. *Energy Ecology Environment* 1(1): 1–2.

Chen, B., and Lu, Y. 2015. Urban nexus: A new paradigm for urban studies. *Ecological Modelling* 318: 5–7.

Chen, P. C., Alvarado, V., Hsu, S. C. 2018. Water energy nexus in city and hinterlands: Multi-regional physical input-output analysis for Hong Kong and South China. *Applied Energy* 225: 986–97.

Chen, S.Q., and Chen, B. 2016. Urban energy–water nexus: A network perspective. *Applied Energy* 184: 905–14.

Dai, J., Wu, S., Han, G., Weinberg, J., Xie, X., Wu, X., … Yang, Q. 2018. Water-energy nexus: A review of methods and tools for macro-assessment. *Applied Energy*, 210: 393–408.

Duan, C.C., and Chen, B. 2016. Energy–water nexus of international energy trade of China. *Applied Energy* 194:725–34.

Fang, D.L., and Chen, B. 2017. Linkage analysis for the water–energy nexus of city. *Applied Energy* 189: 770–9.

Fath, B.D. 2007. Network Mutualism: Positive community level relations in ecosystems. *Ecological Modelling* 208: 56–67.

Fath, B.D. 2015. Quantifying economic and ecological sustainability. *Ocean Coast Management* 108: 13–9.

Font V., D., Wang, R., Hertwich, E. 2017. Nexus strength: A novel metric for assessing the global resource nexus. *Journal of Industrial Ecology* 22(6): 1473–86.

Hoff, H. 2011. Understanding the nexus. *Background Paper for the Bonn 2011 Conference: The Water, Energy and Food Security Nexus*. Stockholm Environment Institute, Stockholm.

Kahrl, F., and Roland-Holst, D. 2008. China's water–energy nexus. *Water Policy* 10(S1):51–65.

Kenway, S. J., Lant, P. A., Priestley, A., and Daniels, P. 2011. The connection between water and energy in cities: A review. *Water Science and Technology* 63(9): 1983–90.

Liu, J., Hull, V., Godfray, H. C. J., Tilman, D., Gleick, P., Hoff, H., … Li, S. 2018. Nexus approaches to global sustainable development. *Nature Sustainability*, 1(9): 466.

Liu, Y., Wang, S., Chen, B. 2019. Water–land nexus in food trade based on ecological network analysis. *Ecological Indicators* 97: 466–475.

Marsh, D.M. 2008. The water-energy nexus: A comprehensive analysis in the context of New South Wales. *Doctoral Dissertation*.

Owen, A., Scott, K., Barrett, J. 2018. Identifying critical supply chains and final products: An input-output approach to exploring the energy-water-food nexus. *Applied Energy* 210: 632–42.

Wang, S.G., Cao, T., and Chen, B. 2017a. Water–energy Nexus in China's Electric Power System. *Energy Procedia*, 105, 3972–977.

Wang, S.G., Cao, T., and Chen, B. 2017b. Urban energy–water nexus based on modified input–output analysis. *Applied Energy* 196: 208–21.

Wang, S.G., and Chen, B. 2016. Energy–water nexus of urban agglomeration based on multiregional input–output tables and ecological network analysis: A case study of the Beijing–Tianjin–Hebei region. *Applied Energy* 178: 773–83.

Wang, S. G., and Chen, B. 2018. Three-Tier carbon accounting model for cities. *Applied Energy* 229: 163–75.

Wang, S.G., Fath, B., and Chen, B. 2019. Energy–water nexus under energy mix scenarios using input–output and ecological network analyses. *Applied Energy* 233, 827–39.

Wang, S.G., Liu, Y. T., Chen, B. 2018. Multiregional input–output and ecological network analyses for regional energy–water nexus within China. *Applied Energy* 227: 353–64.

Williams, J., Bouzarovski, S., Swyngedouw, E. 2014. Politicising the nexus: Nexus technologies, urban circulation, and the coproduction of water–energy. Nexus Network Think Piece Series, Paper 001, November.

42

Water Quality: Modeling

Introduction 427
Classification of Water Quality Models..427
Uses of Water Quality Models 428
    Risk Assessment of Pesticides • Evaluation of Best Management
    Practices (BMPs) • Evaluation of Sources and/or Impacts of
    Pollutants • Explanation of Large-Scale Systems Behavior

Richard Lowrance

References...429

Introduction

Water quality models are based on some representation of hydrology and may include movement of surface water, groundwater, and mixing of water in lakes and water bodies. Based on the hydrology, water quality models then simulate some combination of sediment, nutrients, heavy metals, and xenobiotics such as pesticides. Some water quality models, especially those that deal with nutrients, may contain substantial detail related to biological processes including algal growth, nutrient transformations, and respiration. Most water quality models that portray the movement of water within a landscape or landscape components (e.g., fields, forests, streams) portray the interaction of water with soil in a variety of ways. Newer water quality models and add-ons to older water quality models are able to portray the effects of water quality parameters on the biota of lakes and streams or incorporate stream bank, riparian zone, and/or channel functions to understand the effects of these areas on chemical and sediment transport. Other water quality models are used to simulate the effects of critical inputs on the biological communities of lakes and rivers. These aquatic ecosystem models may or may not be tied to watershed models that provide simulated loading to the aquatic ecosystem under varying land use and management.

Classification of Water Quality Models

Water quality models are either built on hydrologic models, are used in conjunction with hydrologic models, or use empirical hydrologic data. Although water quality models can be physical representations of the real world such as channels and ditches built to scale, mathematical or formal models are more common.[1] Mathematical water quality models are quantitative expressions of processes or phenomena that are known to occur in the real-world. The expressions are simplifications of real-world systems through a series of equations governed by conservation of mass. Mathematical water quality models are often a combination of theoretical and empirical representations of the real-world system. Empirical models use water quality observations to provide estimates of water quality parameters through regression analysis. Process based or theoretical representations use physical, chemical, and biological causal relationships to describe the workings of a conceptual system.

Although the real world is subject to random occurrences of weather and management that drive hydrology and water quality, many models ignore the randomness of inputs and spatially distributed attributes and assume that there is a known value for all model parameters. Conversely, stochastic (or random) models use probability distributions of parameters in time or space and can provide outputs based on the distribution. Most water quality models are deterministic models in the sense that one set of inputs will provide only one set of outputs. The difference in a stochastic and deterministic model can be illustrated by how models deal with something simple like how fast water moves in a soil. A deterministic model would use one value for each soil while a stochastic model would vary the movement rate based on the range and distribution of measured water movement rates. Deterministic models are often used with a range of key parameters in order to produce a range of outputs that would better represent real world conditions. Another critical distinction among water quality models is whether they provide continuous or event-based simulations. Continuous simulation models generally provide at least some representation of groundwater/surface water interactions, while event-based models are more likely to provide only representations of hydrologic processes that take place during rainfall events.

A final distinction among models is whether they are lumped or distributed parameter models. A lumped parameter model contains little or no spatial realism and represents landscape units as homogeneous with respect to the parameters and inputs that drive the model. A distributed parameter model represents certain aspects of the landscape structure, typically by representing areas that are homogeneous with respect to soils, vegetation, and/or land use. Each of these discrete areas is modeled separately and then outputs from all the discrete areas are put together and routed through the system. Because most water quality models are tied to hydrologic models, the water quality outputs from source areas in the model are typically routed through either surface flow pathways, subsurface flow pathways, or both. Models that deal only with events are typically routed through surface flow pathways. Models that simulate continuous or daily water quality in a watershed or field generally must deal with both subsurface or groundwater routing and surface water routing.[2]

Uses of Water Quality Models

Risk Assessment of Pesticides

Knowledge of fate and transport of pesticides in the environment is essential to the assessment of risk due to dietary and drinking water exposure. The passage of the Food Quality Protection Act (FQPA) lead to a pressing need to quantitatively predict ranges and magnitudes of expected environmental pesticide concentrations in drinking water. Health-based safety standards mandated by FQPA require USEPA to consider drinking water exposures of humans to pesticides during the risk assessment process. Some state agencies and USEPA use screening models to estimate pesticide concentrations in groundwater and surface water to identify those food-use pesticides that are not expected to contribute enough exposure via drinking water to result in unacceptable levels of aggregate risk.[3] The models are used to guide regulatory agencies such as USEPA to identify where more detailed field data are needed.

Evaluation of Best Management Practices (BMPs)

Water quality improvement from extensive land uses such as agriculture and forestry depends largely on the use of BMPs. Agricultural water quality modeling attempts to adequately represent the differences among various management practices in order to compare and choose which BMPs lead to the least transport of pollutants. These models are typically structured to represent homogeneous landscape units such as fields or portions of fields in order to compare management features such as tillage, fertilizer sources, manure use, and pesticide use and predict the relative impacts on local transport of pollutants such as sediment, nitrogen, phosphorus, and pesticides. Existing models may be used to test the application of BMPs to areas for which no water quality data are available or to determine the effects of BMPs that are similar to those for which water quality effects have been quantified.

Evaluation of Sources and/or Impacts of Pollutants

Both process based and empirical models have been used successfully to examine the sources of pollutants in watersheds and the impact of pollutants or non-pollutants on aquatic ecosystems. The need to quantify the non-point source contributions for watersheds and small basins is largely driven by total maximum daily load (TMDL) assessments and implementation plans mandated by the federal Clean Water Act.[4] The TMDL assessments are done with a water quality accounting approach that typically uses water quality models to estimate non-point source pollution. The non-point and point sources of a pollutant that are causing the water quality impairment are then combined and compared to observations in the water body. If the water quality is impaired due to the direct presence of a pollutant, then the model estimates of non-point source pollution are used to design a plan for reducing non-point sources or trading point sources for non-point. If pollutants are tied indirectly to the impairment, for instance nutrient enrichment that causes low dissolved oxygen, then the behavior of the pollutant in the water-body is modeled in order to determine the necessary pollutant load reduction.

Explanation of Large-Scale Systems Behavior

As the behavior of large-scale systems becomes more of an issue and as water quality monitoring data become more available, attempts have been made to combine monitoring and modeling to predict the transport of water-borne pollutants on large scales—river basins and continents. Regression models are used to relate measured pollutant transport in streams to spatially referenced descriptors of pollutant sources, land surface characteristics, and stream channel characteristics.[5] Although mechanisms of pollutant transport are not modeled directly, coefficients that serve as surrogates for processes are used to achieve substantial explanatory power for observed water quality data.

References

1. Tim, U.S. Emerging issues in hydrologic and water quality modeling research. In *Water Quality Modeling*; Heatwole, C., Ed.; American Society of Agricultural Engineers: St. Joseph, MI, 1995; 358–373.
2. Overton, D.E.; Meadows, M.E. *Stormwater Modeling*; Academic Press: New York, 1976.
3. USEPA, OPP. 1999. Estimating the Drinking Water Component of a Dietary Exposure Assessment.
4. *NAS*. Watershed Management for Potable Water Supply: Assessing the New York City Strategy; *National Academy of Sciences: Washington, DC, 2000; 549 pp.*
5. Smith, R.A.; Schwarz, G.E.; Alexander, R.B. Regional interpretation of water quality monitoring data. Water Resources Research **1997**, *33*, 2781–2798.

V

ELE: Focuses on the Use of Legislation or Policy to Address Environmental Problems

43

Drainage: Hydrological Impacts Downstream

Mark Robinson
and D.W. Rycroft

Introduction 433
Hydrologic Impacts 433
Conclusions 436
References 436

Introduction

Land drainage is the practice of removing excess water from the land, and it is one of the most important land management tools for improving crop production in many parts of the world. Drainage systems may be broadly divided into surface drainage (comprising land grading and open ditches), shallow drainage (such as subsoiling to mechanically loosen the upper layer of soil), subsurface or groundwater drainage (buried perforated pipes or deep ditches), and the main drainage systems (commonly open channels) used to convey the drain water away.[1] Drainage will inevitably affect the pattern of water flows from the land and into the receiving watercourses. It is these downstream impacts of farmland drainage on the timing and magnitude of peak flows that are considered here, using the results of experimental studies and computer simulations, to present a coherent picture, and to answer most of the apparent anomalies and conflicts.

Hydrologic Impacts

Concern about the possible downstream effects of drainage is shown by many published papers worldwide, in North America,[2,3] Great Britain,[4,5] and continental Europe, including France,[6] Netherlands,[7] Ireland,[8,9] Fin- land,[10] and Germany.[11] The role of drainage has often been highlighted by serious flood events—for example, in the Midwest of the United States in 1993, and across Europe in 1997—which reawakened concerns that drainage could aggravate flooding downstream.

There has been a debate about the effects of drainage on streamflow for well over a century, but until recently, due to the lack of appropriate data, the debate has been largely speculation. Too often, the absence of evidence has erroneously been taken as evidence of an absence of effect. The earliest published account[12] was a report of a 4-day meeting held at the Institution of Civil Engineers in London in 1861. Many of the arguments and opinions expressed have resonance today, but due to the absence of objective measurements, the participants were unable to reach any conclusions and the meeting was inconclusive.

These conflicting opinions resulted from differences in the emphasis given to the two processes of water storage and routing. Considering the former, it may be argued that because drainage lowers the water table, the available storage capacity in the soil is enlarged and able to absorb more storm rainfall, thereby reducing peak flow rates. In contrast, according to the routing argument, the purpose of

433

drainage is to "remove water from the land more quickly" than under natural conditions, so peak out-flows must necessarily increase.

Probably more work has been carried out in Britain upon the effects of agricultural drainage upon streamflow than in any other country. Britain was the originator of modern field drainage[13] and so became the first country where concern arose about its downstream effects; it is also one of the most extensively drained countries in the world.

It is only in the last few years that it has been possible to obtain a coherent picture based on observations of field processes and supported and extended by computer modeling. This has shown that general statements that drainage "causes" or "reduces" flood risk downstream are oversimplifications of the complex processes involved, and that any consideration of the impact of drainage on streamflow must identify the point of interest, whether at the outfall from the field, along the main channel, or a combination of both at the catchment scale.

Experimental studies indicate that the provision of surface drainage will result in higher peak flows downstream. This was shown by a long-term experiment at Sandusky in Northern Ohio[14] and is a result of the reduction/ elimination of surface storage capacity, as well as the provision of more efficient faster flow routes. This has been demonstrated conclusively both by experimental studies and by computer simulations.

In contrast, there seems to be general agreement from experimental studies that subsurface drainage of waterlogged, poorly permeable clay soils reduces peak out- flows.[15–17] Since this is one of the most common situations where artificial drainage is used, it might be considered to represent the most general result of field drainage.

There are, however, instances where even on heavy soils this result may not apply. Due to their low hydraulic conductivity, most water movement in clay soils is confined to flow through macropores, such as cracks. As a result of clay shrinkage and cracking in warm, dry summers, rapid macropore flow can result in larger peak flows from the drained land than from the undrained land. The role of macropores on the seasonality of peak flows from drained land was demonstrated in detail.[18]

More permeable, drier soils may also be drained where there is an economic justification—for example, drainage of land producing high-value crops. In contrast to clay soils, relatively few scientific field studies have investigated the impact of draining lighter, more permeable soils. This may be partly due to the emphasis on draining clay soils, but also, no doubt, results from the greater practical difficulty encountered in plot definition where the soils are more permeable. Nevertheless, data from several drainage experiments on permeable soils are available. At Withernwick,[19] flow peaks were increased in the first year after drainage and there was then a reduction in the following years due to the progressive deterioration of the secondary system of subsoiling designed to improve the soil structure. Supporting evidence of increased peak flows following the drainage of more permeable soils also comes from studies at Cockle Park in northern Britain[20] and Ellingen in Central Germany.[21]

To identify factors influencing drainage response, the results of field drainage experiments under temperate northern European climates were analyzed in terms of their site characteristics.[22,23] This included topography, precipitation, drainage depth and spacing, natural (i.e., pre-drainage) soil water regime, and the soil properties. The only characteristic distinguishing sites, where drainage increased peak flows from those where they were reduced, were those relating to the soil water regime before drainage. The experimental sites all had similar land practices on the drained and the undrained land.

Drainage reduced peak flows on sites that had wetter soils and with poor natural drainage, and significant amounts of storm runoff were generated as overland flow and near-surface flow in the thin upper layers of the soil. These sites had higher topsoil clay contents and shallower depths to a poorly permeable subsoil horizon. When artificially drained, the surface saturation was largely eliminated, greatly increasing the soil water storage capacity.

In contrast, at sites with more permeable, loamy soils that were not routinely saturated before drainage, natural storm flow occurred predominantly by slower subsurface flow, and the artificial drainage pipes provided more rapid flow routes leading to increases in peak outflows.

The findings are summarized in Figure 1. This shows the topsoil texture, together with the effect of drainage on peak flows, and provides the engineer or conservationist with an initial guide to predict the effect on flows of the drainage of a site, based on knowledge of the predrainage site characteristics.

Further insights into the factors controlling the impact of drainage may be obtained by the application of modeling techniques to investigate the important interaction between soil properties and climate in determining soil water regimes. DRAINMOD[24] was applied to two of the field sites with similar climates: a heavy clay soil at Grendon and a more permeable loam at Withernwick. The model was applied to each site using actual field values of drain and soil parameters, and the simulated peak flows from drained and undrained land were compared for similar rainfall inputs. The results showed a 70% lower median peak flow after drainage of clay soil and an increase of 40% in the median peak flow from the more permeable land.[23]

The modeled fluxes and water stores confirmed that the reduction in peaks from the clay soil after drainage was achieved by a change in storm runoff generation from overland flow (caused by soil saturation) to subsurface drainflow. For the loamy soil, the model indicates that the increase in peak subsurface flow rates was due to the steeper hydraulic gradients created by the closer-spaced artificial drains.

The model also demonstrated the effect of different climatic conditions. If the loam soil site at Withernwick had double the normal rainfall (1200 mm yr⁻¹ instead of 600 mm yr⁻¹), the resulting increase in ground wetness would be sufficient to generate substantial amounts of overland flow on the undrained land. Artificial drainage in this case would then reduce peak flows—exactly as happens for a clay soil (where, in contrast, the ground wetness is caused by the low soil permeability). Using the model in this way enables these effects of site characteristics to be explored in an objective manner. The overall dominant criterion—the amount and frequency of surface runoff from undrained land—can be assessed in terms of both soil properties and climatic characteristics.

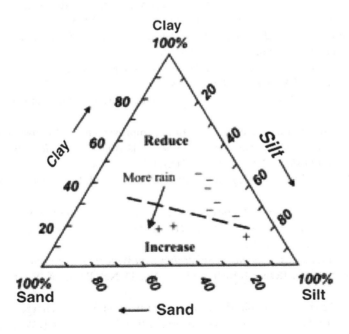

FIGURE 1   Observed impact of pipe drainage on downstream peak flows (increase/reduce), showing the importance of soil texture. Model simulations of climate changes indicate that higher rainfall and wetter ground conditions will shift the balance towards drainage schemes reducing peak flows. See text for details.

Conclusions

The effect of subsurface drainage on peak flows depends upon site wetness. If the water table is close to the surface (due to high rainfall or poor permeability), natural flows occur either over the surface or through the upper, more permeable layers of the soil. Drainage will increase soil water storage capacity and hence the amount of water that can infiltrate, thereby reducing surface runoff and peak storm flows. If the water table is deeper, due to a dry climate or due to more permeable soils, natural flows will occur through the body of the soil. In this case, artificial drainage will increase peak flows as a result of the shorter flow paths and steeper hydraulic gradients.

It must be noted that these conclusions depend upon the scale of the drainage considered. At the river catchment scale, main channel improvements will undoubtedly increase the speed of flow routing, and the timing of arrival of flows from different subcatchments will influence the peak discharge at the point of interest. The relative importance of field drainage and main drainage channels will vary with storm size: field drainage being dominant for small and medium storms, but main channel improvements becoming dominant for large events. In extreme situations where the rainfall intensity exceeds the infiltration capacity of the soil, the effects of the subsurface drains will be minimal, but the associated improved watercourses will rapidly carry away the surface runoff.

Overall, it seems likely that in large catchments, drainage schemes with substantial associated surface drainage and main channel improvements will lead to higher flow peaks downstream, even though locally, the effect of drainage may be to lower the peak flows.

References

1. Skaggs, R.W.; van Schilfgaarde, J., Eds. *Agricultural Drainage*; Agron Monograph 38; ASA, CSSA, and SSSA: Madison, WI, 1999.
2. Whiteley, H.R. Hydrologic implications of land drainage. Can. Water Res. J. **1979**, *4*, 12–19.
3. Serrano, S.E.; Whiteley, H.R.; Irwin, R.W. Effects of agricultural drainage on streamflow in the Middle Thames River, Ontario, 1949–1980. Can. J. Civil Eng. **1985**, *12*, 875–885.
4. Bailey, A.D.; Bree, T., Eds. Effect of improved land drainage on river flows. *Flood Studies Report—5 Years on*; Thomas Telford: London, 1981; 131–142.
5. Rycroft, D.W., Ed. The hydrological impact of land drainage. 4th International Drainage Workshop, Cairo; ICID- CHD, CEMAGREF; 1990; 189–197.
6. Oberlin, G. Influence du drainage et de l'assainissement rural sur l'hydrologie. CEMAGREF Bull. **1981**, *285*, 45–56.
7. Warmerdam, P.M.M., Ed. The effect of drainage improvement on the hydrological regime of a small representative catchment area in the Netherlands. *Application of Results from Representative and Experimental Basins*; UNESCO Press: Paris, 1982; 318–338.
8. Burke, W. Aspects of the hydrology of blanket peat in Ireland. Int. Assoc. Hydrol. Sci. **1975**, *105*, 171–181.
9. Wilcock, D.N. The hydrology of a peatland catchment in N Ireland following channel clearance and land drainage. In *Man's Impact on the Hydrological Cycle in the U.K.*; Hollis, G.E., Ed.; Geo Abstracts: Norwich, 1979; 93–107.
10. Seuna, P.; Kauppi, L., Eds. *Influence of Subdrainage on Water Quantity and Quality in a Cultivated Area in Finland*; Water Research Institute Publ. No. 43; Nat. Board of Waters: Helsinki, Finland 1981.
11. Harms, R.W. The effects of artificial subsurface drainage on flood discharge. In *Hydraulic Design in Water Resources Engineering: Land Drainage*; Smith, K.V.H., Rycroft, D.W., Eds.; Computational Mechanics Publication: Southampton, 1986; 189–198.
12. Denton, J. Bailey On the discharge from underdrainage and its effects on the arterial channels and outfalls of the country. Proc. Inst. Civil Eng. **1862**, *21*, 48–130.

13. Van der beken. The development of the theory and practice of land drainage in the 19th century. In *Water for the Future*; Wunderlich, W.O., Prins, J.E., Eds.; A.A. Balkema: Rotterdam 1987; 91–99.

14. Schwab, G.O.; Thiel, T.J.; Taylor, G.S.; Fouss, J.L. Tile and surface drainage of clay soils 1. Hydrologic performance with grass cover USDA. Agric. Res. Serv. Bull. **1963**, 935.

15. Robinson, M.; Beven, K.J. The effect of mole drainage on the hydrological response of a swelling clay soil. J. Hydrol. **1983**, *63*, 205–223.

16. Harris, G.L.; Goss, M.J.; Dowdell, R.J.; Howse, K.P.; Morgan, P. A study of mole drainage with simplified cultivation for autumn sown crops on a clay soil. II. Soil water regimes, water balances and nutrient loss in drain water, 1978–80. J. Agric. Sci. **1984**, *102*, 561–581.

17. Armstrong, A.C.; Garwood, E.A. Hydrological consequences of artificial drainage of grassland. Hydrol. Process. **1991**, *5*, 157–174.

18. Robinson, M.; Mulqueen, J.; Burke, W. On flows from a clay soil—Seasonal changes and the effect of mole drainage. J. Hydrol. **1987**, *91*, 339–350.

19. Robinson, M.; Ryder, E.L.; Ward, R.C. Influence on stream- flow of field drainage in a small agricultural catchment. J. Agric. Water Manage. **1985**, *10*, 145–148.

20. Armstrong, A.C., Ed. *The Hydrology and Water Quality of a Drained Clay Catchment—Cockle Park, Northumberland*; Report RD/FE/10; MAFF: London, 1983.

21. Schuch, M. Regulation of water regime of heavy soils by drainage, subsoiling and liming and water movement in this soil. In *International Institute for Land Reclamation and Improvement*, Proc. International Drainage Workshop; Wesseling, J., Ed.; Wageningen, Paper 1.14, 1978; 253–267.

22. Robinson, M., *Impact of Improved Land Drainage on River Flows*; Institute of Hydrology Report 113; Wallingford, U.K., 1990. ISBN 0-948540-24-9. http://nora.nerc.ac.uk/7349.

23. Robinson, M.; Rycroft, D.W. The impact of drainage on streamflow. *Agricultural Drainage*; Skaggs, R.W.; van Schilfgaarde, J., Eds.; Agron Monograph 38; ASA, CSSA, and SSSA: Madison, WI, 1999; Chap. 23, 767–800.

24. Skaggs, R.W. Drainage simulation models. *Agricultural Drainage*; Skaggs, R.W., van Schilfgaarde, J., Eds.; Agron Monograph 38; ASA, CSSA, and SSSA: Madison, WI, 1999; Chap. 13, 469–500.

44

Drainage: Soil Salinity Management

Introduction 439
Drainage Conditions 439
Drainage Requirement 440
    Saline Soils • Sodic Soils
Drainage System Design 441
    Relief Drains • Drain Depth • Drain Spacing • Drainage
    Wells • Saline Seeps
References 443

Glenn J. Hoffman

Introduction

Soil water must drain through the crop root zone when salinity is a hazard to prevent salts from increasing to levels detrimental to crop production. Drainage occurs whenever irrigation and rainfall provide soil water in excess of the soil's storage capacity. In humid regions, rainfall normally satisfies crop water requirements and precipitation infiltrating into the soil in excess of this requirement leaches (drains) salts present below the crop root zone. In subhumid areas, rainfall is often inadequate in amount or temporal distribution to satisfy crop needs and irrigation is implemented. For arid regions, rainfall is never abundant and the preponderance of the crop water requirement must be provided by irrigation. Regardless of the climate, if soluble salts are present, water in excess of that needed to satisfy the crop water requirement must be provided to leach excess salts. Leaching may be accomplished continuously or at intervals, depending on the degree of salinity control required. It may take decades or as little as one season, depending on the hydrogeology of the area, but without drainage, agricultural productivity cannot be sustained where salinity is a threat. For a more complete discussion on drainage design for salinity control, the reader is referred to Hoffman and Durnford.[1]

Drainage Conditions

All soils have an inherent ability to transmit soil water provided a hydraulic gradient exists. If the hydraulic gradient is positive downward, drainage occurs. Soils with compacted layers, fine texture, or layers of low hydraulic conductivity may be so restrictive to downward water movement that drainage is insufficient to remove excess salts. In some areas, the hydrogeology may be such that the hydraulic gradients are predominantly upward. This leads to water logging and salination.

Before designing a man-made drainage system, the natural drainage rate should be determined. If the natural hydraulic gradient causes soil water to drain out of the crop root zone, the capacity of the artificial system can be reduced, thereby decreasing the cost for drainage. In some situations, upward flow into the crop root zone from a shallow aquifer can significantly increase the drainage requirement.

The upward movement of groundwater leads to salination as the water evaporates at the soil surface, leaving salts behind. If upward flow is ignored, the drainage system may be inadequate. Regardless of the source, an artificial drainage system will not function unless it is below the surface of the water table.

Drainage Requirement

Saline Soils

The amount of drainage required to maintain a viable irrigated agriculture depends on the salt content of the irrigation water, soil, and groundwater; crop salt tolerance; climate; soil properties; and management. At present, the only economical means of controlling soil salinity is to ensure an adequate net downward flow of water through the crop root zone to a suitable disposal site. If drainage is inadequate, harmful amounts of salt can accumulate.

In irrigated agriculture, water is supplied to the crop from irrigation, rainfall, snow melt, and upward flow from groundwater. Water is lost through evaporation, transpiration, and drainage. The difference between water inflows and outflows is the change in soil water storage. A water balance, expressed in terms of equivalent depths (D) of water, can be written as

$$D_s = D_i + D_r + D_g - D_e - D_t - D_d \tag{1}$$

where the subscripts s, i, r, g, e, t, and d designate storage, irrigation, rainfall and snow melt, groundwater, evaporation, transpiration, and drainage, respectively. The corresponding salt balance, where S is the amount of salt and C is salt concentration, can be expressed as

$$S_s = D_i C_i + D_r C_r + D_g C_g + S_m + S_F - D_d C_d - S_p - S_c \tag{2}$$

with S_s being salt storage, S_m is the salt dissolved from minerals in the soil, S_f indicates salt added as fertilizer or amendment, S_p is precipitated salts, and S_c is the salt removed in the harvested crop.

Rarely do conditions prevail long enough for steady state to exist in the crop root zone. However, it is instructive to assume steady state to understand the relationship between drainage and salinity. If upward movement of salt, the term ($S_m + S_f - S_p - S_c$), and the change in salt storage are all essentially zero, then the salt balance Eq. (2) can be reduced to

$$D_d C_d = D_i C_i + D_r C_r \tag{3}$$

The leaching fraction, L, is the ratio of the amount of water draining below the crop root zone, D_d, and the amount applied, $D_i + D_r$. The ratio of the salt concentration entering and leaving the root zone can also be used to estimate L. Since C_r is essentially zero.

$$L = C_i/C_d = D_d/D_i + D_r \tag{4}$$

The concept in Eq. (4) is important because it illustrates the relationship between leaching fraction and salinity.

The minimum leaching fraction that a crop can endure without yield reduction is termed the leaching requirement, L_r. The leaching requirement is the minimum amount of drainage required to prevent excess accumulations of salt that result in loss of crop yield. Several models have been proposed to estimate the drainage (leaching) requirement. Of the four models tested,[2] the one presented in Figure 1 agrees well with measured values of the drainage requirement through the range of agricultural interest. The drainage requirement given in Figure 1 is the fraction of the volume of applied water that must pass through the crop root zone as a function of the salinity of the applied water and the salt tolerance of the crop.

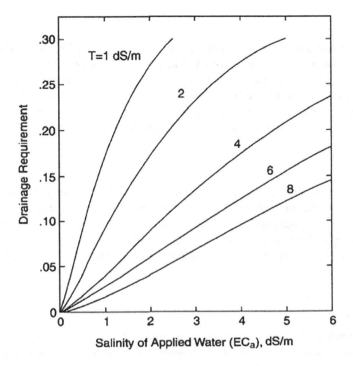

FIGURE 1 Drainage requirement as a function of the salinity of the applied water (reported as the volume weighted electrical conductivity) and the salt tolerance threshold value for the crop (T).
Source: Adapted from Hoffman and Van Genuchten.[15]

Sodic Soils

A soil is said to be sodic if an excessive concentration of sodium causes a deterioration of soil structure. The impact of excess sodium is a reduction in hydraulic conductivity and crust formation. Sodic conditions decrease the rate of drainage. Before a sodic soil can be restored to full productivity the excess sodium in the soil must be replaced with calcium or magnesium. This process frequently requires copious amounts of leaching to reclaim the soil. The design of an artificial drainage system that may be required, however, is based upon the long-term requirement for drainage as estimated in Figure 1 rather than the anticipated high drainage requirements for reclaiming a sodic soil.

Drainage System Design

There are three types of subsurface systems used to control soil salinity: relief drains, shallow wells, and interceptor drains. Relief drains, usually consisting of perforated corrugated plastic tubes buried in a regularly spaced pattern, is the most common subsurface system. Laterals for relief drains are typically placed 2.0–3.5 m deep and are spaced horizontally ten to hundreds of meters apart where salinity is a hazard. Shallow wells, called tube wells in some regions, can also be used to lower the water table by allowing pumping from shallow, unconfined aquifers. Tube wells are spaced at distances of a few hundred meters to several kilometers and may be a few meters to a hundred meters deep. Interceptor drains are used to remove excess soil water from saline seeps. Frequently, one subsurface drain, properly located at the upslope side of the seep, is sufficient. Regardless of the type of drainage system, the depth of the water table must be maintained low enough that (1), salts in the soil profile move to the water table (2), the rate of water movement by capillary flow to the soil surface because of evaporation is minimal, and (3), upflow of saline groundwater into the root zone is prevented.

Relief Drains

A relief drainage system consists of a main drain, collector drains, and field drains (laterals). The main drain is frequently a surface stream or an open drainage canal. Collectors and laterals are usually buried in a regular parallel pattern. Either open ditches or perforated pipes can serve as collectors and laterals. Open ditches are not normally installed now because they occupy land, are difficult to maintain, and are only capable of shallow drainage. Laterals are up to 300 m long and terminate in a collector drain. Both single- and double-sided entries by laterals into a collector are common.

Drain Depth

Subsurface drains are installed much deeper for salinity control in arid regions than drains for water table control in humid regions. The goal for salinity control is to place the drains deep to limit salination of the root zone by capillary upflow. Drains are placed at depths of 2.0–3.5 m in arid regions.[3] The appropriate drain depth depends upon the depth capacity of the installation machinery, the location of a shallow soil layer that impedes water movement, and anticipated benefits compared to additional costs of deeper installation.

Drain Spacing

The spacing between laterals is often estimated using simple drainage design equations. Drain spacing determinations can be based on criteria of steady-state, falling- water-table, or fluctuating-water-table conditions.[4] For large drainage projects or where more accurate values are desired, computerized drainage design models are available. An early computer model developed by Skaggs[5] has been altered by several for irrigated conditions.[6,7] Other models present drainage designs for irrigated areas based on optimization,[8] decision support systems,[9] or reuse of drainage water.[10]

Drainage Wells

Shallow or tube wells offer a viable alternative to relief drains when the aquifer has sufficient transmissivity to provide a significant yield of drain water and the vertical permeability between the crop root zone and the aquifer is adequate. Under these conditions, tube wells have the advantages of being able to lower the water table to greater depths than relief drains and also provide supplemental water for irrigation if the quality is appropriate.

Because drainage wells can be installed at convenient locations within the area to be drained and can be operated either continuously or intermittently, the management of a system of drainage wells is more versatile than relief drains. Relief drains are typically a passive drainage system relying on gravity and designed to operate continuously.

Economic comparisons between the costs of drainage wells and relief drains vary. It is generally found that relief drains have lower construction and operation costs.[11] However, Mohtadullah[12] showed tube wells were a better economic choice than relief drains for the Indus Basin.

Saline Seeps

The occurrence of saline water at the soil surface downslope from a recharge area is referred to as a saline seep. Saline seeps can occur because of the reduction of evapotranspiration that occurs when grasses or forests are converted to cropland in the upland (recharge) areas of a watershed. Dryland farming practices that include fallow periods tend to aggravate the seepage problem. Salination occurs as water infiltrating in the upper elevations of the watershed moves through salt-laden substrate on its path to a discharge site at a lower elevation. In the discharge area of the seep, crop growth is reduced or the plants killed by an intolerable level of salinity. Saline seeps can be distinguished from other saline soil conditions by their recent origin, relatively local extent, saturated soil profile, and sensitivity to precipitation and cropping systems.[13] Saline seeps occur throughout the Great Plains of North American and in Australia, India, Iran, Turkey, and Latin America.[14]

Planting crops in the recharge area that consume soil water before it percolates below the crop root zone will prevent saline seeps. Failing this, improved drainage may provide a solution. Installing an interceptor subsurface drain immediately upslope from the saline seep is frequently a successful solution. Interceptor drains to control seepage should be installed as deep as practical. If the layer restricting soil water flow is not too deep, placing the interceptor drain just above this layer is the most effective location.

References

1. Hoffman, G.J.; Durnford, D.S. Drainage design for salinity control. *Agricultural Drainage;* Agronomy Monograph No. 38; American Society of Agronomy: Madison, WI, 1999; 579–614; Chap. 17.
2. Hoffman, G.J. Drainage required to manage salinity. J. Irrig. Drain. Div., American Society of Civil Engineers, New York **1985**, 111, 199–206.
3. Ochs, W.J. *Project Drainage Issues,* Proceedings of the 3rd International Workshop on Land Drainage, Columbus, OH, Dec 7–11, 1987; E-83–E-88. 202.
4. Bouwer, H. Developing drainage design criteria. *Drainage for Agriculture;* Agronomy Monograph No. 17; American Society of Agronomy: Madison, WI, 1974; 67–79.
5. Skaggs, R.W. *A Water Management Model for Artificially Drained Soils,* Report 267; Water Resources Research Institute, North Carolina State University: Raleigh, NC, 1980.
6. Chang, A.C.; Hermsmeir, L.F.; Johnston, W.R. *Application of DRAINMOD on Irrigated Cropland,* Paper No. 81–2543; American Society of Agricultural Engineers: St. Joseph, MI, 1981.
7. Skaggs, R.W.; Chescheir, G.M. Application of drainage simulation models. *Agricultural Drainage;* Agronomy Monograph No. 38; American Society of Agronomy: Madison, WI, 1999; 529–556; Chap. 15.
8. Knapp, K.C.; Wichlens, D. Dynamic optimization models for salinity and drainage management. *Agricultural Salinity Assessment and Management;* American Society of Civil Engineers: New York, 1990; 530–548; Chap. 25.
9. Gates, T.K.; Wets, R.J.-B; Grisiner, M.E. Stochastic approximation applied to optimal irrigation and drainage planning. J. Irrig. Drain. Div., Am. Soc. Civil Eng. **1989**, 115, 488–510.
10. El-Din El-Quosy, D.; Rijtema, P.E.; Boels, D.; Abdel- Khalik, M.; Roest, C.W.J.; Adbel-Gawad, S. Prediction of the quantity and quality of drainage water by the use of mathematical modeling. *Land Drainage in Egypt;* Drainage Research Institute: Cairo, Egypt, 1989; 207–241.
11. Zhang, W. Drainage inputs and analysis in water master plans. Proceedings of the 4th International Drainage Workshop, Cairo, Egypt, Feb 1990; 181–187.
12. Mohtadullah, K. Interdisciplinary planning, data needs and evaluation for drainage projects. Proceedings of the 4th International Drainage Workshop, Cairo, Egypt, Feb 1990; 127–140.
13. Brown, P.L.; Halvorson, A.D.; Siddoway, F.H.; Mayland, H.F.; Miller, M.R. *Saline-Seep Diagnosis, Control, and Reclamation,* USDA Conservation Research Report No. 30; 1983.
14. Halvorson, A.D. Management of dryland saline seeps. *Agricultural Salinity Assessment and Management;* American Society of Civil Engineers: New York, 1990; 372–392; Chap. 15.
15. Hoffman, G.J.; Van Genuchten, M.Th. Soil properties and efficient water use: water management for salinity control. *Limitations to Efficient Water Use in Crop Production;* American Society of Agronomy: Madison, WI, 1983; 73–85; Chap. 2C.

45

Lakes: Restoration

Introduction 445
Recultivation Methods.. 448
Conclusions... 457
Acknowledgments.. 457
References.. 458

Anna Rabajczyk

Introduction

The 21st century is anticipated as the century of the environment. In this regard, it is crucial to remedy pollution and conserve resources, particular with respect to aqueous ecology, for fostering a healthy environment. Water reservoirs, whether natural or artificial, differ in methods of lake basin formation. The former exist as a result of physical and physicochemical processes occurring in nature; the latter appear due to human interference. Natural water reservoirs of relatively slow exchange of the liquid found therein are defined as lakes.[1,2]

Artificial reservoirs, in turn, also referred to as barrier lakes, are formed by closing river valleys with water dams or drops as damming structures. Several of these reservoirs may also be natural barrier lakes with regulated outflows.[3]

According to their functions, three main types of water reservoirs may be distinguished:[2]

- Dry reservoirs: periodical storage of water during the passage of flood peaks
- Flow-through reservoirs: maintenance of steady retention levels
- Retention reservoirs: storage of water at times of surplus in order to utilize it at other times (subtypes include flood control, navigation, power engineering, water equalization, municipal, industrial, agricultural, or rock waste control reservoirs)

Another classification, which takes into account land configuration, includes mountain, submountain, and lowland reservoirs. The first of these are usually the deepest, with short and high dams. The lowland ones occupy the largest areas and are less deep.[2]

Such considerable diversification of both formation methods and localities causes lakes to be characterized by varying parameters of lake basin morphometry, water transfer speed and routes, water quality therein, etc. Lake shape and dimensions condition lake susceptibility to degradation. Wind speed is higher over large lakes than over small ones, which results in stronger waving and better water oxygenation. Waters in lakes whose elongation follows the most frequent wind directions are also better mixed than in those with transverse orientation.[4]

Low values of the mean-to-maximum depth ratio suggest that large depths are found in a small part of the lake; hence, they may not be of considerable importance for lake functioning. The occurrence of thermal stratification and the size of the littoral zone that uses the biogenes found in the water are both dependent on lake depth.

Each lake, together with the surrounding land and watercourses, forms a catchment characterized by parameters such as topographic features, plant cover, basal complex, and climate conditions, which determine the volume of water resources in a given area. A catchment shaped by low atmospheric precipitation and lowland topographic features provides less water than one with high precipitation and mountainous topographic features, facilitating surface wash. The direction of precipitation waters, accompanied by a variety of compounds, is determined by the basal complex that conditions the river network density and, thus, inflow and outflow volumes. The vegetation found in a particular area stores the water and then gives it off to the atmosphere via the process of transpiration.[2]

At moderate latitudes, water exchange in reservoirs occurs twice yearly: in spring and in autumn. During spring circulation, winter-specific temperature distribution changes when bottom temperatures reach 4°C and upper layers of the reservoir become cooler. As a result of more intensive solar radiation and air warming, surface water temperatures rise. As they reach 4°C, the temperatures of the vertical section equalize because water temperature at the reservoir bottom is 4°C.[5]

As a consequence of wind-induced circulation currents, the entire water mass is mixed (spring homothermy). Further warming of surface waters causes near-surface waters to become lighter than the near-bottom waters whose temperature reaches 4°C (summer stagnation), and an upper warm layer called the epilimnion is formed. The lower layer of cold water is called the hypolimnion. In the metalimnion, i.e., the lake middle or transition water layer, a sharp temperature drop occurs at that time.[6]

In autumn, surface waters cool until they reach a state in which, upon reaching the temperature of 4°C, they attain density identical to that of near-bottom waters. In a way analogous to spring, waters are mixed by wind-induced motion (autumn homothermy). Further cooling of surface waters makes them obtain temperature that approximates 0°C, as well as lesser density than that of near-bottom waters. Subsequently, on the surface, the ice cover is formed (winter stagnation).[5]

Thermal stratification is related to oxygen content stratification in reservoir waters, thus affecting organic life growth. Because the epilimnion is a layer of intensive wind-induced water mixing as well as easy solar radiation access, it displays better oxygenation than deeper layers. As a result of intensive photosynthesis, in summer, water may become supersaturated with oxygen. Spring and autumn circulations cause oxygen to be distributed over the entire water mass.[5,7]

Temperature also conditions other processes that occur in the lake, including the rates of all chemical, biological, and physical reactions.[5] Higher temperature, for example, intensifies the processes of phosphorus release from bottom sediments (Figure 1).

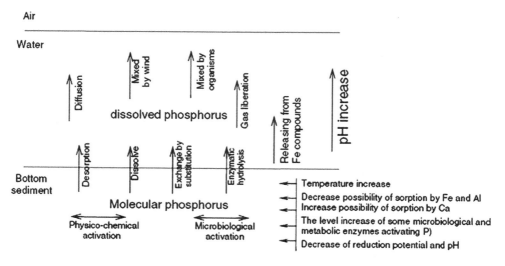

FIGURE 1 Phosphorus release from bottom sediments affected by temperature.
Source: Bajtkiewicz-Grabowska and Zdanowski.[8]

Water temperature increase in the reservoir decreases water viscosity and density. This is particularly significant for plankton organisms as it accelerates the sedimentation speed and changes locomotive conditions. Consequently, the rates of chemical and biochemical reactions increase, and solubility of most substances is enhanced as well, while gas solubility is reduced. The intensification of the processes occurring in the reservoir results in increased use of oxygen. Photosynthetic processes are accelerated, which may lead to cyclical supersaturation of upper water layers. A sudden water temperature rise by 10°C may lead to thermal shock or even death of organisms.[9]

Temperature increase may cause spontaneous discharge of bottom sediments and intensify gas liberation to the extent that the material collected at the bottom is loosened and discharged into water.

Access to solar radiation conditions the growth of autotrophs in the water, as well as that of the organisms situated behind them in the food chain, because the fluctuations in producer numbers affect level 1 (and further) consumer numbers. The propagation of solar radiation in the lake is affected by lake size and shape, i.e., morphometry. For instance, it determines thermal stratification of lake waters as well as chemical, physical, and biological processes that occur in reservoirs.[10]

Natural water reservoirs are efficiently operating ecosystems, capable of maintaining internal homeostasis despite the operation of adverse factors. Yet, human proximity and in particular the effects of human existence have far-reaching consequences (frequently negative) for the environment. Human activity takes forms against which a lake cannot defend naturally. Pollutants rich in biogenic elements from a variety of sources (including municipal waste, agricultural fertilizers, degraded woodland and communication routes) are discharged into lakes and rivers, causing considerable increase in the fertility of lake waters and rivers that serve as lake outflows. In raw municipal waste, for instance, nitrogen concentrations are contained in the 25–50 mgdm^{-3} range, while total phosphorus concentrations are in the range 4.0–12 mgdm^{-3}.[3] According to Vollenweider,[11] limiting concentrations of phosphorus and nitrogen compounds in waters, in excess of which mass algal growth may occur, amount to 0.01 mg Pdm^{-3} and 0.3 mg Ndm^{-3}, respectively.

For the assessment of external factors affecting a particular lake, an assessment of the catchment in terms of yearly delivery of biogenic (N and P) compounds into lake waters must be performed. The calculations are based on the formulas:[3,11]

$$L = I/P_j \left(g \cdot yr^{-1} \cdot m^{-2} \right) \tag{1}$$

$$I = I_{pr} + I_{pl} + I_{pz} + I_o + I_k + I_w + I_l \left(kg \cdot yr^{-1} \right) \tag{2}$$

where P_j is lake area (m^2), I is total load, I_{pr} is grassland load, I_{pl} is woodland load, I_{pz} is built-up area load, I_o is atmospheric precipitation load, I_k is bathers load, I_w is anglers load, and I_l is linear source load.

The values of external nitrogen and phosphorus loading in a particular lake are then compared with Vollenweider's criteria,[11] which define permissible and hazardous nitrogen and phosphorus loadings to a lake:

$$L_{d(N)} = 15 \left(25 \, z^{0.6} \right)$$

$$L_{n(N)} = 15 \left(50 \, z^{0.6} \right)$$

$$L_{d(P)} = 25 \, z^{0.6}$$

$$L_{n(P)} = 50 \, z^{0.6}$$

where z is mean lake depth, $L_{d(N)}$ is permissible reservoir N loading (mg·yr^{-1}·m^{-2}), $L_{d(P)}$ is permissible reservoir P loading (mg·yr^{-1}·m^{-2}), $L_{n(N)}$ is hazardous reservoir N loading (mg·yr^{-1}·m^{-2}), $L_{n(P)}$ is hazardous reservoir P loading (mg·yr^{-1}·m^{-2}), 25 is the limiting loading rate for oligo- and mesotrophic reservoirs, and 50 is the limiting loading rate for meso- and eutrophic reservoirs.

External loading of a reservoir with biogenic compounds per area unit is a major indicator specifying the reservoir's trophy. Trophy increase is based primarily on increased concentrations of biogenic compounds such as nitrogen and phosphorus, and this phenomenon is referred to as eutrophication. It is worth noting that the process of eutrophication is a most natural phenomenon. Organic substances are delivered to the lake starting from the moment of its formation and deposited at the bottom in the form of bottom sediments. Yet, due to intensified pollutant inflow, the phenomenon of eutrophication begins to intensify until it becomes a major threat to lakes.

Recultivation Methods

The growing eutrophication rate of water reservoirs and the increasing number of degraded reservoirs create the demand for effective prevention methods. In the 1960s, attempts at recultivation of degraded lakes were made. However, adequate know-how resulting from experience in this field was lacking, particularly because each water reservoir operates under different internal and external conditions. The current know-how and further developments in other fields of science, not only in natural sciences, have made it possible to explore theoretical issues related to water reservoir recultivation on the basis of the practical experience obtained.

As water ecosystems, lakes are located in land depressions; hence, they provide natural receiving water for the pollutants that come from the catchment area. Due to the functions that they perform in human environment and economy, it is crucial to take appropriate steps in order to improve the quality of degraded waters or protect those in better condition. To this end, scientific, organizational, and technological actions are undertaken jointly, known as recultivation. This process consists in elimination of possible chemical contamination of waters and sediments, improvement of oxygen and nutrient balance, maintenance of flora and fauna at a level appropriate for a given water ecosystem, and provision of engineering elements of flow regulation in the form of desired depths, erosion control, and speeds.[12–15]

The idea of lake recultivation is to restore the previous functions of lakes (e.g., water storage, recreation, household, and agriculture), as well as physical, chemical, and biological features approximating as closely as possible the natural ones. The selection of an appropriate method is determined by the diversity of individual lakes, differences in the ways and scopes of pollution, together with their location in a catchment. The type of the recultivation method used depends on the reservoir size, the nature of the fauna and flora resident therein, connection with watercourses, and proximity to clean water reservoirs in the vicinity of the reservoir under recultivation.[12,14]

Recultivation of water reservoirs is a four-phase process, including the phases of preparation, planning, implementation, and verification monitoring (Figure 2). The duration of actions aimed at restoration of utility values to a particular area, from the initial phase to the final stage of implementation, is typically measured in years.

The first stage is the collection of information as well as the development of a detailed inventory of the immediate catchment. The following may also be helpful: soil maps; maps of localities with erosion-susceptible land and soils, utility or arable lands adjacent to rivers and reservoirs, and critical terrain (with its specific problems); maps of areas where surface waters supply groundwaters; and urbanization plans for the building in catchment areas and industrial development trends.

In view of the above, an individualized approach to each reservoir and planning a complete ecosystem restoration process are necessary. While taking any recultivation actions, it should be remembered that physical, chemical, external, and internal environmental factors may cause profound changes, destabilize the water ecosystem, and push it into one of the two alternative stable states. It must also

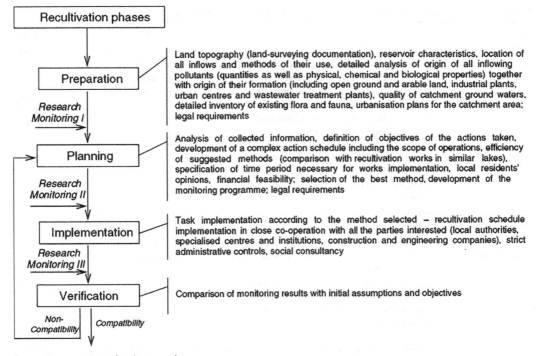

FIGURE 2 Recultivation phases.

be borne in mind that biological elements and processes conditioned thereby, which occur in a water ecosystem, stabilize and perpetuate its status quo.

A significant element of recultivation is the specification of a monitoring program, i.e., catchment control at critical points as well as in terms of registration of environmental quality indicators important for the recultivation process. This program assumes the establishment of observation sites, equipped with control and measurement apparatus. After a set monitoring time, all data from field observation sites are covered in a technical study that enables developing a representation of a catchment demonstrating the origin of possible pollutants and the problems to be tackled. Following from an extensive data base, including input from various sources, and complete with monitoring program results, it becomes possible to enter the definition phase for particular elements and actions necessary to achieve recultivation effect for a particular reservoir.

Nevertheless, it must be stressed that Research Monitoring I usually differs from the subsequent phases of monitoring, which results from task implementation at individual stages. During the inventorying process, maximum quantities of data concerning pollutant sources and migration routes, as well as the quantity and quality of substances delivered, ought to be collected. In contrast, the objective of Monitoring II is to obtain information on whether the selected recultivation method is well adjusted to the lake's individual characteristics.

To improve the situation in a degraded lake, it is not enough to restore its condition to that from before the disturbance: much feedback in the lake perpetuates the postchange status, and lakes tend to demonstrate resilience, i.e., immunity to recultivation actions. Very often, the condition for the application of any recultivation method is prior reduction of phosphorus concentration in the water to the 0.050–0.100 mg $PO_4 \cdot dm^{-3}$ level and phosphorus removal/ inactivation in sediments (internal import inhibition).[12–14]

TABLE 1 Most Commonly Used Lake Recultivation Methods and Techniques with Their Characteristics

Method/Technique	Characteristics	Model Application	Literature
Dredging	Bottom sediment removal from the entire lake, deepest waters and inflow (in flow-through lakes); complete sediment removal to reach mother rock guarantees radical improvement in water quality; necessity of thorough physico-chemical tests to determine the chemical composition of the sediments, their thickness and distribution in the reservoir (specification which sediment layer and in which lake parts ought to be removed); a sensitive issue is to find sites for storage and management of extracted sediments treated as hazardous	Mission Lake, Kansas; Lake Elkhorn, Columbia; Saluda Lake, South Carolina; Chain Lake, British Columbia	[16–19]
Macrophyte reintroduction	Bentonite is a good macrophyte substratum, which enables macrophyte layer reconstruction as well as restoration of destroyed trophy web structures	Alderfen, Barton, Belaugh and Cockshoot Broads, Norfolk;	[20]
Flushing, rewashing	Provision of clean, nutrient-deficient water, removal of strongly eutrophicated hypolimnion waters and replacing them with well-oxygenated waters from outside the lake; requires a source of clean water in the vicinity of the lake; lake water exchange ought to be performed a couple of times a year	Lake Veluwe, the Netherlands; Willow Lake, Arizona	[21,22]
Macrophyte removal	Comprises artificial lowering of lake water level, lake icing in winter, in-freezing of plant stalks into ice, raising the river level and plant uprooting, collection of floating plant remains and removal from the lake; the quantities of phosphorus thus removed are much lower than the loads introduced to the reservoir; effective in the case of lakes which are strongly polluted with biogenes, accumulated mostly in littoral vegetation	Lakes: Mary, Ida and Bass, Minnesota; Boreal Lake, Ontario	[23,24]
Hypolimnion water removal	Procedure possible in flow-through reservoirs; consists in liquid removal from the profundal layer via a hose; a disadvantage of this method is the pollution of the watercourse discharged from the lake; during near-bottom water removal, the sediments which remain in chemical balance with them become depleted; the process may cause reservoir pollution with H_2S, accumulated in the sediments and temperature increase near the bottom, which leads to hypolimnion oxygen depletion	Bled, Yugoslavia; Lake Kortowo, Olsztyn	[14,25]
Sparging, oxygenation (aeration)	Introduction of a duct into the water through which air is pumped, without disturbance (near-bottom liquid temperature is low: slow rate of organic matter decomposition) or with disturbance of natural lake stratification (entire water mass mixing and water oxygenation: near-bottom layer temperature increase, intensification of chemical and biological processes, including enhanced mineralization and internal supply); enhances hypolimnion oxygen conditions as well as enforcing lake water circulation	Canyon Lake, Teksas; Lake Commabbio, Lombardy; Kieleckie Bay, Kielce	[26–28]

(Continued)

TABLE 1 (*Continued*) Most Commonly Used Lake Recultivation Methods and Techniques with Their Characteristics

Method/Technique	Characteristics	Model Application	Literature
Artificial destratification	Destruction or prevention of the lake's thermal stratification, stimulating mass algal growth in the top (warmed and lighted) water layer; very often linked to sparging	Canyon Lake, California; Lake Catherine, Arkansas; Lake Starodworskie, Olsztyn	[29–31]
Nutrient deposition and deactivation in water and sediments	Possible in small reservoirs; does not guarantee permanent phosphorus removal from water; enduring results depend on coagulant properties ($FeCl_3$, $Al_2(SO_4)_3$, $FeSO_4$); optimum water pH 6–8 and high Redox potential level; phosphorus inactivation in sediments is possible due to application of a mixture of bentonite clay with lanthanum, which becomes a phosphorus bonding element upon adsorption on bentonite	Kielecki Bay, Kielce; Jessie Lake, Alberta; Lake Głęboczek, Tuchola; Lake Sønderby, Sønderby	[28,32–34]
Sediment isolation (capping)	Possible in small reservoirs where water transfer is low and does not cause mobility of the introduced layer; used in order to counter chemical exchange in the sediments-water system; physical isolation: sand and foil; chemical isolation: $+Al^{+3}$ diatomite, bentonite or other clayey minerals	Great Lakes, North America; Onondaga Lake, Central New York; Venice Lagoon, Porto Marghera	[35–37]
Precipitation (chemical methods)	Use of herbicides, including algaecides (e.g. barley straw, tree leaves) to fight algae; short-lasting procedure results due to appearance of other algae in place of the removed blooms; causes introduction of large amount of hazardous substances into water	Shoecraft Lake, Washington; Eight lakes located in the eastern and western portions of southern Michigan; Joliet Junior College Lake (JJC Lake), Illinois	[38–40]
Biomanipulation	Change in living conditions of organisms or quantitative ratios in a given ecosystem by means of several food chain dependencies: limiting the populations or complete elimination of individual groups of organisms (e.g. increase in the amount of zooplankton and introduction of selected fish species in order to reduce algal population, introduction of silver carp to the lake to limit the growth of phytoplankton; introduction of white amur to eliminate excess of macrophysical vegetation; introduction of predators such as pike, pike perch or perch in order to limit plankton-eating fish populations); use of biomanipulation requires thorough analysis of dependencies occurring in a given ecosystem	Lake Terra Nova, Loosdrecht Lakes ; Lake Eymir, Ankara	[41–43]
Fish catch	Regular catch of fast-growing fish (important is good cooperation with the lake management) enables reducing lake fertility (so-called trophy) and systemic enhancement of water quality; effective in the case of lakes which are strongly polluted with biogenes, accumulated mostly in the ichtiofauna	Lake Fure, Copenhagen; Lake Ringsjön, Skåne County	[44,45]
Seston removal/ catch	Pumping out water at sites richest in seston, its subsequent filtration in appropriate apparatus; in barrier lakes, zooplankton removal may be accomplished by water inflowing through bottom culverts when zooplankton density is highest at the bottom	Lake Jussi, Pikkjärv	[46]

Source: Cooke,[12] Klapper,[13] Sengupta and Dalwani,[14] and Gupta et al.[15]

As a rule, each catchment and its environment are distinct, but from a practical viewpoint, it is possible to distinguish several actions, techniques, and methods enabling renewal or protection of the status quo of the lake at issue (Table 1).

The most popular methods that provide significant effects in biogene removal are those based on sparging and dredging. Reservoir water sparging enables reduction of excessive algal and waterweed growth. Aeration is performed with the use of either mechanical apparatus or compressed air. Water sparging by means of mechanical apparatus may be performed by water spraying or direct penetration of air into the reservoir.[26,27]

Spraying occurs when water passes along a perforated hose through which it is catapulted into the air in the form of fine droplets. This effect can also be achieved by directing water to purpose-made diffusers. The most important element of this technique is the generation of adequately sized water droplets. The finer the water droplets that penetrate into the atmosphere, the larger the area of gas exchange between the droplet and the air, thanks to which better aeration effects are obtained—more oxygen is transferred into the water.[12–14]

The apparatus used for direct transfer of oxygen into a reservoir includes, for example, aerators with horizontal and vertical rotation axes, including surface, turbine- driven subsurface, and combination aerators. Aerators may also be classified according to paddle-wheel shapes: some may be straight, whereas others have single- or multisided curvatures.

The simplest is the surface aerator. Its operation relies on a turbine or rotor pumping the water upwards and then downwards, thus causing turbulence and splashing on water surface. Oxygen is delivered through water droplets in contact with atmospheric air. This type of aerator is used in ponds, water reservoirs, and rivers.

Combination aerators, which display features of both surface and turbine-driven aerators, are characterized by high capacity for oxygen transfer and provide good sparging results in deep reservoirs. In this type of devices, the surface rotor and the turbine are usually powered by the same aggregator, which causes high power demand, high noise levels, and possible icing during operation.[13,14]

The other types of aeration devices have vertical axes, which enables operation in both open and closed setups. In open aerators, the liquid is centrifugally catapulted, which causes the occurrence of the suction force that sucks the air into the so-called hopper and inter-paddle-wheel space. At the outlet of this space, a mixture of air and liquid appears and is then splashed on the surface of the aerating chamber. This construction provides for the mixing of the liquid and the air in the entire volume. In closed aerators, the liquid is propelled onto paddle-wheels by means of a pump. The air is thus sucked from the atmosphere by means of suction resulting from the flowing liquid at set sites in the inter-paddle-wheel channel.[47]

The compressed air aerator consists of a blower or a fan, sources of compressed air, and distribution ducts and sparging devices. In order to obtain the largest possible contact area between the two phases, a variety of porous elements are used for air dispersion. They are made of ceramic or artificial materials, and are pipe or plate shaped.[47]

In the process of selection of lake aeration systems and methods, the following factors ought to be taken into consideration:

- Oxygen concentration in the water and at the bottom of water reservoirs, which depends on water reservoir eutrophication degree, oxygen deficit volume in bottom sediments, with allowances being made for sediment resuspension, temperature, spring and autumn circulation intensity, and possible waving effects.
- Nominal result of aerator operation, which depends on total nominal hydraulic efficiency, reservoir depth at installation site, reservoir area, air volume fed into the aerator, and aerator construction type, such as the following:
  - Pump and surface dispersion
  - Pump and reversal-to-bottom with surface dispersion (with a thin water jacket)

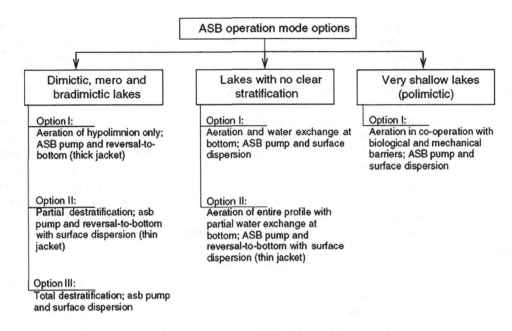

FIGURE 3   ASB operation mode options depending on lake type.

- Pump and reversal-to-bottom (with a thick water jacket)
- Pump with deep suction—in all the above types[48-50]

Overall change to oxygen and dynamic conditions at aerator installation site, i.e., genuine efficiency, depends on lake type and operation mode selected, i.e., aerator construction, as exemplified by the air-stream bottom aerator (ASB) (Figure 3).

During the method selection process, it is also necessary to answer the question of what immediate effect is expected from aeration since the procedure of aeration alone can bring positive results only in few cases.

An equally popular method of lake recultivation is dredging. During sediment removal, it is necessary to construct single or multiple thick-structure islands. Scarps, or island shores, should be equipped with bottom baffle piers that will constitute initial base for the benthos. At the early stage of island construction, the circumference wall should not be closed completely. Through the gap, waters overlying the sediment will be discharged. The gap ought to be secured with sedimentation barriers as well as a set of directional ASBs operating in the countercurrent to the outlet from the closed area. The applied cofferdam will only be temporary. At the end of island filling, the outlet gap ought to be closed with a cofferdam. Upon completion of sediment dredging, the whole ought to be covered with stable material.[48-50]

Upon completion of engineering work, red osier (for instance) needs to be planted. This type of islands can be used for recreation as well as natural purposes. Appropriate, separated parts of the islands can become nesting areas for many bird species. Sediments remaining at the bottom, whose thickness should not exceed 20 cm, may be immobilized by means of a baffle pier grid.

In the course of recultivation procedures, supply points ought to be perpetually monitored. If biogenes are found to be introduced via these routes, it is necessary to place biological barriers, possibly active (with aeration) at inflow inlets.[51,52]

The negative influence of surface wash can also be limited by barriers of appropriate plantings parallel to the shoreline. When forming the islands, it would be advisable to dredge a couple of local deepest waters (ca. 5–6 m). At those sites, an aerator is placed to disable the occurrence of oxygen deficits at windless periods. Artificial sparging enhancement ought to be sustained until balance is obtained between food and consumers, as well as between producers and reducers. Hence, in order to reduce power costs, in near-shore island zones and non-developed shore segments, installation of artificial reef as initial base for benthos is recommended. The ichtiopopulation status and composition also need to be considered.[48–50]

Another method aimed at restoration of good water quality is chemical bonding and inactivation of nitrogen and phosphorus. This system, used in Scandinavia, has been adapted to Polish conditions and tested at several sites, e.g., Lake Dlugie in Olsztyn. The phosphorus inactivation method is recommended for those lakes in which, despite cutoff of external sources of biogenic compounds, high fertility is sustained through deposition of these compounds from bottom sediments. Used primarily in shallow lakes, this method relies on removal of excess phosphorus from the pelagic zone and trapping it in bottom sediments. It also increases sorption capacity of degraded lake bottom sediments. Strongly eutrophicated lakes are characterized by high biogenic compound content, which causes blooms (increased phytoplankton population). At this degree of degradation, the lake no longer fulfills its recreational or economic functions. Chemical deposition of phosphorus by means of aluminum (PAX) and ferric (PIX) coagulants reduces the quantity of biogenic (nutrient) compounds, thus limiting algal growth intensity, which results in improvement of water quality and transparency.[53,54]

In the case of shallow reservoir recultivation, it is necessary to mitigate the deoxygenated layer of benthic water, the so-called oxicline. The oxicline generally occurs in the summer, appearing as a result of oxygen depletion in the course of chemical reactions. These reactions occur at the contact zone between bottom sediment and water, depending, among others, on ambient temperature: the warmer the water, the faster the reactions proceed. Under the circumstances, more and more frequently, actions are taken to place aerators equipped with a phosphorus inactivation system in reservoirs. This method has also been used in the Kielce Lake, where a pulverizing aerator was installed in May 2008. It is powered with wind energy and has a built- in installation dosage system for iron sulfate, which bonds phosphorus.[28]

Most efforts to alleviate the detrimental and undesirable effects of eutrophication on aquatic systems address the problem of P reduction in the inflows. Despite many cases of success in the reduction of nutrient loads of lakes in recent years, the expected corresponding reduction of phytoplankton abundances has often been delayed by many years. Sometimes, lakes have been reported to be resistant in their response to loading reductions. Recycling of P from the sediments becomes more important if the P in inflows is reduced.[55]

Phosphate inactivation in bottom sediments enables the elimination of internal transport in the reservoir (release of biogenes accumulated in bottom sediments into the pelagic zone), which is the cause of self-maintenance of high trophy levels—a feature of overfertile lakes, conditioning the occurrence of blue-green algae blooms. This means that as a result of the application of this method, the cause is removed, rather than the effects of excess of biogenes in the water. This testifies to greater efficiency of the above technique.[56]

The only alternative for application of the phosphorus inactivation method in bottom sediments is lake dredging. This is a technique that is equally effective, but very expensive, long-lasting, and accompanied by the problem of storage and further treatment of the removed sediments. It must be stressed that the storage site must not be located in the lake catchment area so that effluent waters together with their biogene load cannot find their way back to the reservoir. Similarly, treatment of the sediments is connected with the need for a recipient, transportation, etc. The phosphorus inactivation method in bottom sediments is far less costly, is incomparably faster at application, and provides enduring results of reduced phosphorus concentration in the water.[56,57]

The oldest method of phosphate inactivation in sediments, with simultaneous mineralization of organic matter, is the Riplox method, which yields good results. The required labor intensity and technical complexity make it extremely unpopular. Other techniques are its modifications, the Prote method being a case in point. The similarity lies in the fact that in both cases, air and the flocculant are simultaneously supplied to bottom sediments. A major difference is prior induction of intense sediment resuspension and flocculant application to a very carefully delineated sediment layer. The responsibility for the success of lake recultivation with this method lies with an effective device, providing precision dosage of flocculants to lake bottom sediments.[56,58]

To that end, several purpose-made mixes, containing a variety of substances as well as enzyme complexes, selected bacteria, biological activators, complete with extended-surface mineral carriers and stabilizers, have been developed. A preparation introduced into a water reservoir strongly activates the detritus food chain and increases the participation of autotrophic bacteria in matter circulation.[56,58]

The effects are so enduring that the following results are observed over 2 years:[56,58]

- Acceleration of organic matter decomposition in the sediments: quantitative reduction of accumulated bottom sediments and deceleration of natural reservoir shallowing
- Elimination of anaerobic zones in the hypolimnion and acceleration of sediment surface oxidation
- Enhanced elimination of biogenic compounds from matter circulation
- Blue-green algae bloom intensity control
- Boosting the growth of submerged water vegetation.
- Increased fish biomass[56,58]

Achieving enduring improvement of reservoir condition takes time and efficient implementation of the stages aimed at reduction of external supply, internal supply (elimination of anaerobic zones), and the quantity of biogenes in circulation (trophy reduction).

The experience in terms of external supply reduction has shown that the main condition that determines the efficiency of this technique is possibly the fastest reduction of external biogene supply. If this requires improvement of wastewater management in the reservoir catchment, the issue may seem problematic and distant in time due to the need for a wastewater treatment plant to be modernized or constructed. In contrast, the application of the Trigger-1 and Trigger-2 biopreparations in already-existing household, municipal, or industrial treatment plants enables limitation of pollutant inflow into the lake. Because the elimination of 1 kg of phosphorus and nitrogen in the process of wastewater treatment, even at plants with inadequate technical equipment, is far easier and cheaper than in the water environment, it is not advisable to wait for the construction to be completed, instead using all the methods to enable the reduction of external supply.[59]

In reservoirs where storage of large internal loads at stagnation periods has already occurred, the occurrence of anaerobic zones above the bottom is observed in the water. This triggers the mechanism of internal supply, i.e., releasing the previously accumulated biogenes from bottom sediments. This leads to accelerated degradation of the reservoir, although initially, only increased reservoir productivity is noted (primary and secondary productivity). Natural processes that occur in the reservoir are accelerated several times and result in rapid expansion of anaerobic zones. The methods applied to check this phenomenon, such as aeration, are expensive and carry a high risk of failure. A less costly method that does not require complex technical procedures is the application of a biopreparation, such as Trig-ger-3max, which, without changing water stratification, eliminates anaerobic zones in the hypolimnion, reducing the internal supply and eliminating anaerobic zones.

Recently, many biopreparations containing saprophytic microorganisms are commercially available and commonly offered for use in ponds, lakes, and reservoirs. These preparations usually consist of selected bacterial strains immobilized on a mineral carrier. Sometimes, the preparations are also enriched by bacterial enzymes such as biocatalyzators. Some preparations are even enriched by nutrients as growth stimulants at the beginning of growth after addition into the water. However, in most cases, parallel aeration or sediment oxidation by adding another electron acceptor such as nitrate will be necessary to support the growth and activity of added microbes. Otherwise, the method hardly brings any effects. With regard to many bacterial extracellular exudates that have been reported to inhibit the growth of cyanobacteria, a direct effect on cyanobacterial development may be possible as well.

However, no scientific proof of such an effect in the case of application of these commercially available biopreparation is available.[56,60]

All biogenes that occur in a water reservoir ecosystem are in circulation, the major stages of which are transformations of the dissolved phase into the solid phase and back into the dissolved phase. The main mechanisms of this change are the initial placement of elements in the biomass (solid and molecular phase), and their subsequent release from mortified organisms and fecal matter (dissolved phase). Phosphorus is the quickest to be released, and consequently, it circulates in the epilimnion, passing many times from the dissolved phase to the solid before falling onto the sediments. The application of a biopreparation will thus enable a reduction of the quantity of biogenes in circulation, thus limiting lake trophy.[48]

More and more frequently, attempts are made to develop and apply sustainable methods of water reservoir qualitative enhancement, based on "bio" structures. A case in point is the bio-hydro structure (biofilter filling), as well as active biological and mechanical filters, which may be used in open water reservoirs, thanks to which the processes of mineralization and biomass growth are induced.[48,61]

Components of biofilter filling are manufactured as standard panels. They are five-layer grids, made of wide strips, with single net-mesh dimensions. Each subsequent layer is shifted in relation to the previous layer in two directions, by half a net-mesh module.[48]

When this packet is placed vertically, the vertical strip plane is perpendicular to the panel plane. The horizontal strip plane is differently inclined to the panel plane. This geometric arrangement enables directing the stream upward or downward, depending on flow direction. Consequently, this structure may be used as, for instance, a sedimentation element, a lift component, or a resuspension barrier and initial base for the complex of plant (phytobenthos) and animal (zoobenthos) organisms, which occupy the water reservoir bottom.[61]

Strip surface is rough (knurled) and made of chemically neutral plastic, which enables occupancy by a variety of sedentary species. Despite the fact that the extended surface of such filters amounts to $122 \text{ m}^2 \cdot \text{m}^{-3}$, the filter displays a low degree of flow suppression and relatively large vacant internal space. Due to this, upon placement in the pelagic zone, it very soon becomes the habitat of a very wide range of periphyton. By being an ecological niche in its own right, it also acts as an attractant to many sedentary species as well as fish fry. Above the pelagic zone, autonomously, on the top elements of the panels, a wide variety of hydrophilous plants settle.[48]

In the 1980s, at Biotechnika Company, the occupancy extent of these filters was studied in Lakes Karczemne and Klasztorne. On average, one standard panel came to be occupied by biomass in the amount of ca. 70 kg. If such an occupied filter is removed from the reservoir at the right moment, then, together with the biomass, the built-in biogenic compound load will also be removed. The certainty of the effect achieved, in contrast to chemical methods for example, consists in the fact that the application of chemical deposition does not guarantee that, under certain conditions (pH, redox potential), the sorption capacity of metals in the coagulants applied will not be destroyed. As data demonstrate, this is a very frequent occurrence and then, apart from the still unsolved problem of phosphorus removal, the additional "foreign" load in the form of metal compounds is found in the water. Then, the phenomenon of system adjustment to new conditions, which can moreover take a completely unknown direction, arises.[48]

Sparging, in connection with bio-hydro barriers, also offers several options, including passive biological barriers as well as active biological and mechanical filters. If a lake receives a strongly polluted watercourse, the inlet of the watercourse may be separated from the remaining area with a barrier (consisting of several layers to make it most advantageous) made of bio-hydro structures placed jointly, streamwise—first downwards and then upwards. Mechanically, this setup will function as a sedimentation barrier, which will, due to autonomous occupancy, after a short period of time, also and perhaps primarily become a biological barrier.

On account of its structure, despite initial passivity, as occupancy progresses, it becomes an attractant for an increasing variety of species, in most cases beneficial to further development of reservoir biocenosis and further recultivation procedures. During the observation of this type of barrier placed in Lake Rybnickie, it was noted that it had become a "refuge," initiating changes in both animal and plant populations. A double-positive effect was thus achieved: the inflowing pollutant load that degraded the lake has been set to work towards beneficial renewal of the reservoir.

In Lake Rybnickie, the positive result of barrier operation was visible as soon as after 1 year of working. Positive results of water quality tests restored the recreational value of the reservoir, and after many years, the State Sanitary and Epidemiological Board ultimately permitted bathing.[48]

Conclusions

Fertility or abundance of a water reservoir with biogenic mineral substances gradually increases in time, and may be either a natural process or one resulting from human activity. Natural eutrophication proceeds slowly, whereas anthropogenic eutrophication is characterized by a rapid change rate. At the initial stage, an increase of primary and secondary productivity is observed, but as eutrophication progresses, the efficiency of secondary productivity decreases and algal biomass gradually accumulates. The reservoir basin gradually fills with sediments and the water volume decreases, which, even at no biogenic inflow, simultaneously accelerates eutrophication. Further succession leads to the reservoir being turned into land.

In order to preserve a lake, human intervention becomes necessary. All recultivation strategies require sensible planning, time, and, frequently, considerable financial outlay. Their results are not always certain, but the current know-how concerning the subject enables appropriate selection and implementation of an effective recultivation process in a particular lake. The type of the method and technique used depends on the size of the reservoir, the nature of fauna and flora resident therein, the character of the catchment, and the type of adjacent land development. Effectiveness and efficiency of the work performed depend on accurate assessment of the reservoir status and adoption of an appropriate action plan.

While analyzing the efficiency of various procedures carried out in the world's lakes, it must be stated that before embarking on recultivation, it is necessary to

- Define the current trophic status of the lake as well as causes and sources of its degradation basing on physicochemical and biological tests of lake waters, inflows and outflows, complete with an analysis of sediment chemical composition, considering exchange processes between sediments and water.
- Specify the feasibility of the water reservoir recultivation in view of its natural, recreational, and economic value.
- Design and perform protective procedures consisting of the elimination, or at least limitation, of individual pollution sources.
- Develop the most advantageous concept of a recultivation method that ought to make allowances for the trophic status as well as morphometric and hydrological conditions of the reservoir, together with possible uses of margin land, its development, ownership status, access to the shore and a power supply, duration of the process, and financial outlay required.
- Ensure that recultivation will be performed by an expert team under the supervision of an experienced limnologist at all times.
- Document the course of recultivation and its effects through the results of monitoring carried out at particular stages.
- Deem recultivation unfeasible in view of its failing to offer an opportunity for water quality enhancement, if it is found that excess loading of the lake is impossible to eliminate.
- Have a close connection with appropriate regulations, enabling the protection of a lake subject to recultivation procedures from further degradation processes.

Acknowledgments

This study has been financed in part by the State Committee for Scientific Research (KBN) (Grant No. N N305 306635).

References

1. USEPA. *The Lake and Reservoir Restoration and Guidance Manual,* 2nd Ed. (OWRS, EPA 440/4-90-006); United States Environmental Protection Agency, Office of Water: Washington, DC, 1990.
2. Bajkiewicz-Grabowska, E.; Mikulski Z. *Hydrology*; PWN: Warszawa, 2006.
3. Leonard, J.; Crouzet, P. *Lakes and Reservoirs in the EEA Area*; European Environment Agency: Copenhagen, 1998.
4. Viekman, B.E.; Flood, R.D.; Wimbush, M.; Faghri, M.; Asako, Y.; Van Leer, J.C. Sedimentary furrows and organized flow structure: A study in Lake Superior. Limnol. Oceanogr. **1992**, *37* (4), 797–812.
5. French, R.H.; McCutcheon, S.C.; Martin, J.L. Environmental hydraulics (Ch. 5). In *Hydraulic Design Handbook*; Mays, L., Ed.; McGraw-Hill Professional: New York, 1999;–5.33.
6. Wetzel, R.G. *Limnology*, 2nd Ed.; Saunders College Publishing: Fort Worth, 1983.
7. Martin, K.L.; McCuthceon, S.C. *Hydrodynamics and Transport for Water Quality Modeling*; CRC Press: Boca Raton, FL, 1999.
8. Grabowska, E.; Zdanowski, B. Phosphorus retention in lake section of Struga Siedmiu Jezior. Limnol. Rev. **2006**, *6,* 5–12.
9. Tipton, M.J.; Brooks, C.J. The dangers of sudden immersion in cold water (Ch. 3). In *Survival at Sea for Mariners, Aviators and Search and Rescue Personnel (RTO-AG- HFM-152)*; RTO/NATO, NATO Research and Technology Organization: Neuilly-sur-Seine Cedex, France, 2008,–3.10.
10. Lenntech, Water Treatment Solutions. Water ecology FAQ frequently asked questions, http://www.lenntech.com/wa-ter-ecology-faq.htm (accessed May 2011).
11. Vollenweider, R. Advances in defining critical levels for phosphorus in lake eutrophication. Mem. I[st]. Ital. Idrobiol. **1976**, *33,* 53–83.
12. Cooke, G.D. *Restoration and Management of Lakes and Reservoirs*, 3rd Ed.; CRC Press, Taylor and Francis Group: Boca Raton, 2005.
13. Klapper, H. Technologies for lake restoration. Papers from Bolsena Conference (2002). Residence time in lakes: Science, management, education. J. Limnol. **2003**, *62* (suppl. 1), 73–90.
14. Sengupta, M.; Dalwani, R., Eds. Hypolimnic Withdrawal for Lake Conservation, Proceedings of Taal2007: The 12th World Lake Conference, Jaipur, India, 2008; 812–818.
15. Gupta, A.K.; Shrivastva, N.G.; Shrama, A. Pollution Technologies for Conservation of Lakes; Proceedings of Taal2007: The 12th World Lake Conference, Jaipur, India; Sengupta, M.; Dalwani, R., Eds., 2008; 894–905.
16. Lake Dredging. Illinois EPA, Northeastern Illinois Planning Commission, Chicago, Illinois, 1998.
17. Helmke, D. Dredging: Kansas Takes First Steps to Reclaim Reservoirs, The Kansas Lifeline, Seneca, Kansas, 2010, 96–99.
18. Maxwell, A.O. Upstate's Saluda Lake Revived through Grassroots Conservation Effort. USDA-NRCS, http:// www.sc.nrcs.usda.gov/news/saluda_lake.html (accessed May 2011).
19. Murphy, T.P.; Macdonald, R.H.; Lawrence, G.A.; Mawhinney, M. Chain lake restoration by dredging and hypolimnetic withdrawal. In *Aquatic Restoration in Canada*; Backhuys Publishers: Leiden, the Netherlands, 1999; 195–211.
20. Kelly, A. Appendix 5. History of lake restoration. In Lake Restoration Strategy for The Broads, Broads Authority, The Broads—A Member of the National Part Family, Norwich Norfolk, 2008.
21. Hosper, H.; Meyer, A.-L. Control of phosphorus loading and flushing as restoration methods for Lake Veluwe, the Netherlands. Aquat. Ecol. **1986**, *20* (1–2), 183–194.
22. Mankiewicz, P.S.; Mankiewicz, J.A. *Ecological Engineering and Restoration Study. Flushing Meadows Lakes and Watershed*. The Gaia Institute: City Island Avenue, Bronx, NY, 2002.

23. Cross, T.K.; McInerny, M.C.; Davis, R.A. Macrophyte Removal to Enhance Bluegill, Largemouth Bass and Northern Pike Populations. Minnesota Department of Natural Resources, St. Paul, Investigation Report 415, 1992.

24. Rabasco, R. Trophic effects of macrophyte removal on fish populations in a boreal lake. A thesis submitted to the Faculty of Graduate Studies of the degree of Master of Natural Resources Management, Natural Resources Institute, The University of Manitoba, Canada, 2000.

25. Dunalska, J. Influence of limited water flow in a pipeline on the nutrients budget in a lake restored by hypolimnetic withdrawal method. Pol. J. Environ. Stud. **2002**, *11* (6), 631–637.

26. Anderson, M.A.; Paez, C.; Men, S. Sediment nutrient flux and oxygen demand study for Canyon Lake with assessment of in-lake alternatives. Final report: Canyon Lake Nutrient Flux and In-Lake Alternatives, Dept. of Environmental Sciences, Univ. of California, 2007.

27. Riccardi, N.; Mangoni, M. Chemical consequences of oxygenation in a shallow eutrophic lake studied with meso- cosms. J. Aquat. Ecosyst. Health **1996**, *5* (1), 63–71.

28. Rabajczyk, A.; Jóźwiak, M. Littoral zone flora versus quality of Kielecki Bay waters. Ecol. Chem. Eng. A, **2008**, *15* (12), 1359–1368.

29. Fast, A.W. Lake Aeration System for Canyon Lake, California, Lake Elsinore and San Jacinto Watersheds Authority (LESJWA), Riverside, CA, 2002.

30. Kothandaraman, V.; Roseboom, D.; Evans, R.L. Pilot Lake Restoration Investigations—Aeration and Destratification in Lake Catherine. Illinois State Water Survey, Urbana, Illinois, 1979.

31. Lossow, K.; Gawrońska, H.; Jaszczult, R. Attempts to use wind energy for artificial destratification of Lake Starod- worskie. Pol. J. Environ. Stud. **1998**, *7* (4), 221–227.

32. Itasca Soil and Water Conservation District, Jessie Lake Nutrient TMDL for Jessie Lake. Wenck Associates, Inc., Minnesota, 2011.

33. Łopata, M.; Gawrońska, H. Effectiveness of the polymictic Lake Głęboczek in Tuchola restoration by the phosphorus inactivation method. Pol. J. Nat. Sci. **2006**, *21* (2), 859–870.

34. Reitzel, K.; Hansen, J.; Andersen, F.Ø.; Hansen, K.S.; Jensen, H.S. Lake restoration by dosing aluminum relative to mobile phosphorus in the sediment. Environ. Sci. Technol. **2005**, *39* (22), 4234–4140.

35. Palermo, M.R.; Maynord, S.; Miller, J.; Reible, D.D. Guidance for In-Situ Subaqueous Capping of Contaminated Sediments, Assessment and Remediation of Contaminated Sediments (ARCS) Program, Great Lakes National Program Office, US EPA 905-B96-004, 1998.

36. Draft Onondaga Lake Capping and Dredge Area and Depth Initial Design Submittal. Parsons, Anchor QEA, 2009; P:\ Honeywell-SYR\444576 2008 Capping\09 Reports\9.3 December 2009_ Capping and Dredge Area and Depth IDS\ Appendices\Appendix A -RA Delineation\Attachment A- 1.doc (accessed May 2011).

37. Bona, F.; Cecconi, G.; Affiotti, A. An integrated approach to assess the benthic quality after sediment capping in Venice Lagoon, Aquat. Ecosyst. Health Manage **2000**, *3*, 379–386.

38. Wagner, K.J. The Practical Guide to Lake Management in Massachusetts. Executive Office of Environmental Affairs, Commonwealth of Massachusetts, Boston, 160 pp, 2004.

39. Madsen, J.D.; Getsinger, K.D.; Owens, Ch.S. Whole Lake Fluridone Treatments For Selective Control of Eurasian Wa- termilfoil: II. Impacts on Submersed Plant Communities, Lake and Reservoir Management, **2002**, *18* (3), 191–200.

40. Mitchell, J. STS CONSULTANTS, LTD. Joliet Junior College Lake. Diagnostic/Feasibility Study and Restoration Plan. Phase 1 B Report, 2007.

41. Van de Haterd, R.J.W.; Ter Heerdt, G.N.J. Potential for the development of submerged macrophytes in eutrophicated shallow peaty after restoration measures. Hydrobiologia **2007**, *584*, 277–290.

42. Bontes, B.M.; Per, R.; Ibelings, B.W.; Boschker, H.T.S.; Middelburg, J.J.; Van Donk, E. The effects of biomanipulation on the biogeochemistry, carbon isotopic composition and pelagic food web relations of a shallow lake. Biogeosciences **2006**, *3*, 69–83.

43. Beklioglu, M.; Ince, O.; Tuzun, I. Restoration of the eutrophic Lake Eymir, Turkey, by biomanipulation after a major external nutrient control I. Hydrobiologia **2003**, *489*, 93–105.

44. Frederiksborg County, Final report: Restoration of Lake Fure—A nutrient-rich lake near Copenhagen. LIFE02NAT/ DK/8589, Hillerod, 2006.

45. Hamrin, S.F. Planning and execution of the fish reduction in Lake Ringsjön. Hydrobiologia, **1999**, *404*, 59–63.

46. Terasmaa, J. Seston Fluxes and Sedimentation Dynamics in Small Estonian Lakes. Tallinn University, Dissertations on Natural Sciences 11, Tallinn, 2005.

47. Wilson, D.E.; Beutel, M. Final Report: Review of the Feasibility of Oxygen Addition or Accelerated Upwelling in Hood Canal. Washington, Brown and Caldwell, 2005.

48. Sadecka, Z.; Waś, J. Non-invasive methods of reservoir restoration—Perspective. In *Wastewater and Sludge Treatment Utilization;* Sadecka, Z.; Myszograj, S., Eds.; University of Zielona Góra Press: Poland, 2008; 247–260.

49. Company WBWW-BIOPAX Sp. z o.o, http://wbww- biopax.pl/ (accessed May 2011).

50. Company Biopax.pl, http://www.biopax.pl/cm.php?id=16 (accessed May 2011).

51. Allied Biological Inc. New York, available at http://www.alliedbiological.com/lake_managementandrestoration_ specialtyservices.htm (accessed May 2011).

52. Environmental Consulting and Technology, Inc., Carpenter Lake Restoration Project, Final Report, Corporate Of- fices—Gainesville, FL, 2008.

53. Brzozowska, R.; Gawrońska, H. Effect of the applied restoration techniques on the content of organic matter in the sediment of Lake Dlugie. Limnol. Rev. **2006**, *6*, 39–46.

54. Brzozowska, R.; Gawrońska, H. The influence of a longterm artificial aeration on the nitrogen compounds exchange between bottom sediments and water in Lake Długie. Oceanol. Hydrobiol. Stud., Int. J. Oceanogr. Hydrobiol. **2009**, 37 (1), 113–119.

55. Van Donk, E.; Hessen, D.O.; Verschoor, A.M.; Gulati, R.D. Re-oligotrophication by phosphorus reduction and effects on seston quality in lakes. Limnology **2008**, *38*, 189–202.

56. Drábková, M. Methods for control of the cyanobacterial blooms development in lakes. Dissertation thesis, RECE- TOX—Research Centre for Environmental Chemistry and Ecotoxicology, Brno, Czech Republic, 2007.

57. Salgot, M. Eutrophication in Lake Maryut: Diagnosis and actions proposal. ANNEX 1, Alexandria Lake Maryiut Integrated Management (ALAMIM), Alexandria, Egypt, 2009.

58. Quaak, M.; Does, J.; Boers, P.; Vlugt, J. A new technique to reduce internal phosphorus loading by in-lake phosphate fixation in shallow lakes. Hydrobiologia **1993**, *253* (1–3), 337–344.

59. VWP Individual Permit No. 08–0572, Part I—Special Conditions, DEQ, 2008, http://stauntonriverwatch.com/VWPP%20Final%20Part%20I%20special%20conditions.pdf (accessed May 2011).

60. Duval, R.J.; Anderson, L.J.W. Laboratory and greenhouse studies of microbial products used to biologically control algae. J. Aquat. Plant Manage. **2001**, *39*, 95–98.

61. Gulati, R.D.; Pires, L.M.D.; Van Donk, E. Lake restoration studies: Failures, bottlenecks and prospects of new ecotechnological measures. Limnology **2008**, *38*, 233–247.

46

Wastewater Use in Agriculture: Policy Issues

Introduction 461
Wastewater Is a Resource in Waterscarce Settings 462
    Wastewater Use in Developed Countries • Wastewater Use in Developing
    Countries
Policies Are Needed to Ensure Wise Use 465
Policy Issues in Developed Countries 466
Policy Issues in Developing Countries 466
Policy Options Include Treatment and Non-Treatment Alternatives 467
Policy Interventions Should Focus on Reducing Risk 468
    Farmers and Their Families • Agricultural Communities • Consumers of
    Farm Products
Motivating Safe Practices along the Wastewater Exposure Pathway 472
Examples of Public Policies 474
Conclusions 475
Acknowledgments 476
References 476

Dennis Wichelns

Introduction

Treated and untreated wastewaters have been used for many years to irrigate farms, landscapes, and golf courses in both developed and developing countries. Public officials in arid countries have promoted wastewater use primarily to extend their limited national water supplies. Officials in other countries have promoted the benefits of using treated wastewater in agriculture or landscape maintenance as an alternative to discharging the wastewater into a stream or ocean. Their rationale is based partly on cost and partly on environmental implications. Treating wastewater is expensive, and a portion of the cost can be recovered by using the wastewater in a productive endeavor. From an environmental perspective, applying the treated wastewater to a farm field provides another layer of quality improvement before the water is finally returned to a stream or ocean.

Many farmers in developing countries use untreated wastewater for irrigation, often because it is the only source of water available. Many small-scale farmers obtain irrigation water from streams or ditches that are polluted with effluent from a nearby city, industry, or housing development. The farmers likely would prefer higher-quality water, but in most cases, they have no alternative source for irrigation. There can be agronomic value in the nutrients in the untreated wastewater (the nitrogen, phosphorus, and potassium), but there are also pathogens and chemicals that threaten the health of farmers, food vendors, and consumers. Irrigating with untreated wastewater is risky business in developing countries, yet it generates household income for families with limited livelihood alternatives. Most farmers using

461

untreated wastewater for irrigation likely would vote to continue using the wastewater, even if they understood all the risks, in the absence of an alternative, higher-quality water supply.

In a sense, farmers using untreated wastewater provide a public service by removing effluent from polluted streams and applying it to soils, thus reducing the pollutant load in downstream locations. Soils filter some of the undesirable constituents, and plants consume some of the nutrients. However, wastewater irrigation also generates risk for farm communities and consumers of farm products. Polluted canals and ditches and wastewater-irrigated fields create hazards in which children and other residents are exposed to harmful pathogens and chemicals. Consumers of farm produce also are at risk of illness when they handle and ingest contaminated vegetables, particularly when the food is eaten raw or prepared with inadequate care toward reducing contamination risk.

The goal of this entry in the encyclopedia is to describe policy issues pertaining to wastewater use in agriculture. The issues in developed countries are somewhat straightforward and mature. Public agencies have largely determined appropriate water quality criteria and implemented treatment protocols to support wastewater use in irrigation. Future issues will include refining those standards and protocols and continually evaluating the costs and benefits of alternative levels of wastewater treatment and use in agriculture and other activities. There will also be discussions of who should pay for wastewater treatment and who should have priority in receiving limited supplies. These issues involve costs, returns, and the division of economic rents, but they generally do not involve decisions that can support or destroy livelihood opportunities, whether intentionally or as the unintended consequences of seemingly beneficial policy choices.

Policy issues are more challenging in developing countries, where most of the wastewater used for irrigation is untreated and much of the use is informal and unintentional. Farmers, communities, and consumers are at risk from harmful constituents in the untreated wastewater, yet each group also obtains critically important benefits. Farmers generate financial returns that enhance their livelihoods and improve the economic status of farm communities. Consumers gain nutritional value by having affordable access to locally grown fresh vegetables. The public, more generally, benefits also when farmers divert effluent for use in irrigation, rather than allowing it to continue flowing downstream.

There are no easy—and affordable—policy choices regarding the use of untreated wastewater in developing countries. Public funding for treating all wastewater will not be available in many countries within the foreseeable future. Lacking the treatment alternative, public agencies must identify measures that will reduce the risks of using untreated wastewater, while still maintaining a notable portion of the benefits that accrue to farmers, consumers, and the larger community. In this review of policy issues pertaining to wastewater use in irrigation, we focus largely on issues involving the informal use of untreated wastewater in developing countries. The issues in such countries are more challenging, and the potential rewards of identifying and implementing successful risk reduction measures are substantial.

Wastewater Is a Resource in Waterscarce Settings

With increasing demands on water resources in arid and humid regions, the role of wastewater as a resource in municipal, industrial, and agricultural applications will continue to gain importance in many countries. Large numbers of farmers irrigate with wastewater in developed and developing countries, yet we do not have good statistics on the areal extent or the value of output generated through wastewater irrigation. This is due partly to the difficulty of identifying and measuring farm areas irrigated fully or partially with treated or untreated wastewater. In most countries, wastewater is used to irrigate vegetables, fruits, cereals, cotton, and fodder crops (Table 1). Fruits and vegetables are often consumed without cooking and thus pose a higher risk of illness due to pathogens or chemicals in wastewater.

TABLE 1 Examples of Wastewater Reuse in Selected Countries

Region and Country	Examples of Use
Africa	
Tunisia	Citrus, fodder
Morocco	Vegetables, fodder
Americas	
Argentina	Vegetables, fodder
Chile	Vegetables, grapes
Mexico	Vegetables, cotton, fodder, parks, and greenbelts
Peru	Vegetables, cotton, fodder
United States	Vegetables, cereals, fodder, greenbelts, golf courses
Middle East	
Israel	Cotton
Kuwait	Vegetables, orchards, fodder
Saudi Arabia	Cereals, fodder, greenbelts
Asia	
India	Vegetables, cereals

Source: Gupta and Gangopadhyay.[1]

Wastewater Use in Developed Countries

Farmers and landscape managers have been using treated wastewater for many years in Australia, Israel, Spain, the United Arab Emirates, the United States, and other developed countries.[2–5] Treated wastewater is used directly in some applications and indirectly in others, often following storage in an aquifer or discharge to a drain or stream from which farmers obtain irrigation water at some distance from a water treatment facility.[6,7] Over the years, scientists and practitioners have improved their understanding of the impacts of using treated wastewater for irrigation, and public officials have implemented policies and regulations consistent with current knowledge and experience.[8–11]

Persistent water scarcity in the Middle East and North Africa has led many governments to advance the state of knowledge regarding irrigation with wastewater and desalinized water. Wastewater is a key irrigation resource in Israel, Jordan, Kuwait, Iran, Oman, and Saudi Arabia.[12,13] In Israel, where 73% of treated municipal sewage is recycled[10] the government has supported the construction of pipelines to deliver wastewater to agricultural areas where many farmers use drip irrigation systems to minimize potential harm from contact with the wastewater.[14] Treated effluent also is stored in aquifers before recovery and delivery to farmers.[10] In Jordan, 95% of the treated wastewater volume is used each year, primarily for irrigation in the Jordan Valley.[15] In Saudi Arabia and the United Arab Emirates, many farmers and municipal landscape agencies irrigate with secondary and tertiary treated wastewater.[16–18] Tunisia, Egypt, and Morocco also have notable programs in which farmers are provided with treated wastewater for use in irrigation.[13]

Treated wastewater is used extensively for irrigation also in Australia and in the arid southwest and the humid southeast of the United States. In Australia, as in Israel, treated municipal effluent is infiltrated into aquifers, and farmers later retrieve the water for irrigating horticultural crops.[19] Most of the irrigation with wastewater in the United States is found in California and Florida, where strict water quality standards are enforced to protect consumer health and to sustain positive market acceptance of crops produced in these heavily agricultural states.[13]

Wastewater Use in Developing Countries

Perhaps the greater challenge regarding wastewater use in agriculture in the future pertains to developing countries in which there is inadequate capacity to collect, treat, and distribute wastewater in ways that minimize potential harm to public health.[20] Millions of farmers in developing countries rely on untreated, partially treated, or haphazardly blended wastewater as a source of irrigation supply. They produce grains, vegetables, and fruits with the low-quality water, often for sale at attractive prices in urban markets.[21,22] The farmers and their family members who have contact with the untreated and partially treated wastewater are at risk of health damage due to biological and pathogenic constituents.[23] The consumers of leafy vegetables also are at risk if they do not take adequate precautions in preparing the vegetables prior to consumption.[24,25] In many cases, some consumers are unaware that vegetables have been produced using wastewater, and they have no experience of becoming ill after eating raw vegetables. Hence, they are not inclined to implement the necessary precautions.[26]

Worldwide, an estimated 800 million farmers are engaged in urban agriculture. One-fourth of these farmers produce crops for sale in local markets on small plots of land of which they might or might not have ownership, using low-quality irrigation water that often contains untreated wastewater.[27] Market demand for irrigated vegetables in tropical areas is strong, and those produced locally are particularly important in warm regions where modern transportation and storage facilities are largely unavailable. Farmers in urban and periurban areas can bring their highly perishable vegetables to market quickly, thus generating household income and providing an important source of vitamins for consumers.[24]

Many farmers in sub-Saharan Africa and south Asia rely on untreated wastewater for a portion of their irrigation supply. Some farmers utilize wastewater taken directly from tanker trucks or sewage lines, while others divert irrigation water from streams and ditches polluted with untreated municipal wastewater.[20,28] In many areas, farmers irrigate crops and sell them in local markets (Figure 1), with few precautions taken to minimize risks in production or consumption.[23]

More than 60% of the vegetables consumed in many West African cities are produced locally in urban and periurban areas.[29] In Kumasi, the second largest city in Ghana, urban farmers produce 90% of the

FIGURE 1 Selling wastewater-irrigated leafy vegetables in local markets generates revenue for small-scale farmers, while creating a risk of contamination for consumers.
Source: Sanjini de Silva, IWMI, Hyderabad.

lettuce and spring onions consumed in the city. With irrigation, urban farmers can harvest 9 to 11 crops per year on their small plots (0.1 hectare), thus generating annual incomes of $400 to $800 per year.[24] Farmers place substantial value on the availability and reliability of their irrigation supply, even though it contains wastewater from the city.

Farmers along the Musi River, downstream of Hyderabad, India, irrigate rice and fodder grass with untreated wastewater discharged into the largely dry riverbed from many locations in the city. Farmers rely on the wastewater as their only source of irrigation water.[30] While reporting negative health effects due to wastewater exposure, farmers also report that the availability of wastewater in the Musi River enhances their livelihoods.[31,32] Farmers irrigating with wastewater report greater occurrence of fever, headaches, and skin and stomach problems than farmers who do not use wastewater.[31] Although irrigation enhances incomes, substantial costs are imposed on households by illnesses that might be caused by exposure to wastewater. It is in such settings that effective policies are needed to enhance farm family health and to protect consumers from the potentially negative impacts of consuming wastewater-irrigated vegetables.

Farmers in periurban areas of Calcutta, India, irrigate about 5000 hectares of rice using untreated wastewater from the city's sewage collection system. Many farmers use wastewater for aquaculture as well, thus enabling them to generate a steady annual income and provide employment opportunities for local workers.[1] The sustainability of the system is uncertain, as much of the wastewater contains heavy metals that accumulate in soils and can be taken up by vegetables.

Untreated wastewater is used also by farmers in periurban areas of Addis Ababa, Ethiopia, where only 3% of the residents are served by a centralized sewer system.[33] The farmers appreciate the nutrient content of the wastewater in a region where the price of imported fertilizer has been increasing by as much as 20% per year. Controlled experiments on urine-irrigated plots in Ethiopia, and in trials on farm fields, have demonstrated the agronomic value of the nitrogen and phosphorus contained in untreated wastewater.[33]

Policies Are Needed to Ensure Wise Use

From a policy perspective, the use of wastewater in agriculture provides both opportunities and challenges that require public intervention. In one sense, wastewater is an effluent requiring treatment or disposal, subject to regulations that protect public health. In the absence of regulations, private generators of wastewater would have little incentive to reduce volume or to manage the flow of wastewater beyond their property line. Because wastewater generation is a negative externality in most settings, regulations and incentives are needed to minimize the potential harm from wastewater in the environment.

Wastewater management has public-good characteristics in that once it is provided, many members of society benefit. At the same time, it is difficult to exclude individuals from enjoying the benefits of a cleaner, healthier environment once the decision has been made to collect and treat all wastewater in a community. The non-rival nature of the benefits and the difficulty of exclusion provide the basis for managing wastewater treatment within the public sector.

The public-goods perspective is appropriate when viewing wastewater as an effluent requiring treatment or disposal. However, when viewing wastewater as a resource, there are notable private benefits for which individuals will be willing to invest time, effort, and funding to enhance their opportunities. The private-goods perspective pertains to both treated and untreated wastewater. Several water agencies in Australia, Israel, and the United States sell treated wastewater (directly or through an aquifer recharge program) to farmers and golf course owners who obtain private benefits through irrigation.[34,35] Often there is a price differential between treated wastewater and fresh water, thus providing a financial incentive for irrigators to select the treated wastewater.[36,37]

Farmers in developing countries also obtain private benefits, but the distribution of wastewater among them is much less formal, and the wastewater generally has not been treated. An estimated 80% of the sewage generated in developing countries is discharged untreated into the environment, and half the population is exposed to polluted water sources.[38,39] Many farmers acquire untreated wastewater when they divert irrigation water from a stream or ditch that carries effluent from a nearby city or from households in an urban, periurban, or rural area. Water diversions and the use of wastewater in such settings generate private benefits for the farmers. The public gains also as the farmers remove the low-quality water from streams and ditches. However, the primary motivation for farmers is to boost their productivity and increase their net returns in agriculture. By doing so, they risk the health of their families through exposure to untreated wastewater, and they create situations in which consumers also are at risk of eating harmful produce. Public policies are needed to reduce these risks and to optimize the management of wastewater from the public's perspective.

Policy Issues in Developed Countries

Current policy issues regarding wastewater use for irrigation are best described separately for developed and developing countries. In developed countries, the primary policy issues involve economics and finance. The public generally has already determined that protective water quality guidelines must be followed when using wastewater for irrigation. The technology for treating wastewater is well understood, and advances in technology that lower the costs and increase the benefits will be forthcoming in response to market demands and in accordance with government-sponsored research programs. The question will not be whether wastewater will be treated but who will pay for treatment and how much of the cost will be passed along to the users. In some areas, public officials also will seek to determine the economically optimal level of wastewater treatment, as a function of its intended uses.[40]

Key policy questions will also involve the appropriate levels of government involvement in wastewater treatment and reuse programs. Where the private, farm-level benefits of wastewater use are notable, farmers should be financially motivated to invest in production methods and develop market outlets that support irrigation with wastewater. Farmers should also be able and willing to pay for treated wastewater delivered by a public agency or water user association. Using contingent valuation methods in a survey of Greek farmers, Bakopoulou et al.[41] determined that 58% of the participants would be willing to pay half the fresh water price to purchase treated wastewater for irrigation. Small-scale farmers on the island of Crete expressed a greater willingness to pay for treated wastewater after attending a session in which they learned of the private and social benefits of wastewater irrigation.[42]

Farmers should also have incentive to invest in communal facilities for collecting, treating, and delivering wastewater to farms, as the price and availability of wastewater improve over time, relative to the price and availability of fresh water. Public agencies can hasten a farm-level switch from fresh water to wastewater with water pricing and investment incentives and also by informing farmers and consumers of the safety and benefits of irrigating with treated wastewater.

Policy Issues in Developing Countries

The primary policy questions in developing countries are more challenging, in part, because financial resources are limited and there are many competing demands on public funds. In addition, much of the wastewater irrigation takes place in decentralized, informal settings in which individual farmers gain access to wastewater very simply by diverting polluted water from a stream or ditch. Property rights to the water are not defined, and there is no communal agency or water user association that coordinates irrigation activities. Millions of individual farmers will be very reluctant to stop diverting polluted water for use in irrigation, given that their livelihoods currently depend on the sale of irrigated farm produce.

Public officials in developing countries must address the following question: How do we minimize the risks to farmers and consumers, while not destroying or severely diminishing the livelihoods of those farmers who currently irrigate with wastewater? This is not an easy question to answer. Public officials will be mindful of the benefits that farmers provide by diverting and using polluted water for irrigation. If not for that activity, larger volumes of wastewater would continue flowing downstream in

TABLE 2 Average Nutrient Availability in Human Excreta per Person, per Year

| Nutrient | Amount of Nutrient Available (kgs) | | | Amount of Nutrient Required to Produce 250 kg of Cereal (kgs) |
	In Urine (500 liters)	In Feces (50 liters)	Sum	
Nitrogen	4.0	0.5	4.5	5.6
Phosphorus	0.4	0.2	0.6	0.7
Potassium	0.9	0.3	1.5	1.2

many watercourses, creating greater risk for downstream residents and causing environmental harm over a larger area. Farmers who irrigate with wastewater generate one set of risks for their families and consumers while reducing another set of risks to residents downstream.

Public officials also will note that some of the pollutants in municipal wastewater (such as nitrogen, phosphorous, and potassium) are plant nutrients. In areas where inorganic fertilizer is costly and difficult for small-scale farmers to obtain, those nutrients have substantial agronomic value. The nutrients contained in the average amount of urine and feces generated by one person each year are sufficient to produce enough grain to sustain that person. Urine is a particularly good source of nitrogen, phosphorus, and potassium (Table 2). Efforts to collect and transport urine directly to farms, rather than discharging it to waterways, would allow farmers to manage their nutrient applications with greater care and accuracy.

In summary, farmers generate both private and public benefits when they divert polluted water from streams and ditches to irrigate crops in urban areas. The nutrient content of wastewater is sufficient to increase crop yields, provided that farmers do not apply excessive amounts of some nutrients, such as nitrogen.[1,43] Public officials in developing countries must determine how to sustain these beneficial aspects of wastewater irrigation and the livelihoods of farm families, while minimizing risks to those same families and the consumers of their produce.

Policy Options Include Treatment and Non-Treatment Alternatives

The policy options available to public officials for reducing the risks associated with wastewater irrigation in developing countries, while sustaining livelihood benefits, might be placed into four categories:

1. Improve and extend centralized wastewater treatment
2. Improve and extend decentralized wastewater treatment
3. Regulate (with enforcement) the use of untreated wastewater in agriculture
4. Complement existing wastewater use patterns with risk reduction interventions to protect farm families, communities, and consumers

The first category is likely the most costly and the least likely to be implemented along a reasonable timeline. There might be affordable opportunities in some settings within developing countries, in which new, large-scale wastewater treatment plants can be constructed to improve the quality of water available for agriculture. Yet it seems that if such opportunities were affordable, if they compared favorably with alternative public investments, and if an affordable source of finance were available, then such efforts would already be underway. It is difficult to imagine that the pace of investments in large, centralized wastewater treatment plants will be sufficient to improve water quality for many of the farmers who currently use wastewater for irrigation in developing countries.

The second category includes options that should be more affordable than building large, centralized wastewater treatment plants. The goal within this category is to identify opportunities for enhancing irrigation water quality at an appropriate scale and within a meaningful distance from the point of wastewater use. Small-scale wastewater treatment plants might be designed with the expressed purpose of making higher-quality water available for irrigation. The construction costs and operating criteria for such plants might be different—and less expensive—than those pertaining to centralized wastewater treatment plants that discharge water intended for uses outside agriculture.[44] For example, it is important to remove solids, salts, and pathogens from water intended for use in irrigation, but farmers can accommodate higher nutrient levels than wastewater users in municipal and industrial settings.

The third option likely will be challenging in many developing-country settings, given the decentralized, informal nature of wastewater use and the strong dependency of farm households on wastewater. Regulations will be politically unpopular, and enforcement will be difficult to achieve. In Syria, for example, the government disallows the irrigation of vegetables with wastewater, but compliance with the restriction is not complete. Syrian officials resort to destroying vegetable crops irrigated with wastewater when they find such situations. As a result, less than 7% of the area irrigated with wastewater near the city of Aleppo is in vegetable production.[27] The opportunity costs involved in planting and cultivating crops, only to have them destroyed by the government, can be substantial for farm households with limited sources of income.

The financial burden of treating wastewater in developing countries and the challenge of regulating wastewater use by farmers will remain substantial for the foreseeable future. Hence, many farmers will continue using wastewater, and their workers and families will remain at risk of infection while applying irrigation water. Consumers will remain susceptible to sickness caused by handling and consuming the irrigated produce. Given this near-term outlook, public agencies in developing countries should seek opportunities to reduce the risks of infection and sickness by intervening at selected stages of the process, which includes wastewater generation, capture, irrigation, crop production, harvest and handling, and food preparation and consumption. Thus, we focus on the fourth category of policy options—reducing risk to farm households, communities, and consumers.

Policy Interventions Should Focus on Reducing Risk

Conventional wastewater treatment might be viewed as the ultimate risk reduction measure when considering the use of wastewater in irrigation.[45] Establishing and enforcing water quality standards, in conjunction with a wastewater treatment program, can be effective in removing potentially harmful constituents. However, the cost of treating wastewater and enforcing water quality standards will exceed affordability in many developing countries. Recognizing this challenge, the World Health Organization (WHO) recommends shifting the policy focus from reliance on wastewater treatment and water quality standards to establishing health-based targets that might be achieved by implementing a range of risk-reducing interventions.[11,45]

The WHO[46] describes three sets of health protection measures pertaining to the three groups most susceptible to health impacts of wastewater irrigation: 1) farmers and their families; 2) consumers of farm products; and 3) agricultural communities. We consider each group in turn.

Farmers and Their Families

When delivering irrigation water or working in fields irrigated with wastewater, farmers, family members, and other farm workers can be exposed to microbial pathogens, including viruses, bacteria, helminths (nematodes and tapeworms), and protozoa.[47] Wastewater also can contain endocrine-disrupting chemicals, pharmaceutically active compounds, and residuals of personal care products.[48–51] Exposure to wastewater can result in skin irritation and diseases related to pathogens in human waste products. The WHO[46] recommends the following protective measures to be considered when designing public policies and intervention strategies:

1. Treating wastewater
2. Supporting the use of personal protective equipment.
3. Providing access to safe drinking water and sanitation on farms
4. Promoting good health and hygiene practices
5. Providing chemotherapy and immunization
6. Controlling disease vectors and intermediate hosts
7. Reducing contact with disease vectors

One or more of these measures would be helpful in breaking or disrupting the pathway of contamination from wastewater to farm family members and farm workers. However, success will be determined by how effectively the benefits of these measures are communicated to farmers and how aggressively farm workers adopt them. The farm-level cost of any measure also will be a key determinant of its successful adoption.

Agricultural Communities

In a sense, many residents of agricultural communities are susceptible to the same types of risks as farmers and their families, particularly if they utilize water in irrigation canals or ditches or if they have access to farm fields. In many irrigated areas, community residents use water from irrigation canals or ditches for cleaning clothes, washing livestock, and watering kitchen gardens (Figure 2).[52] Young children often swim or play in irrigation ditches, while some residents rely on irrigation canals as a source of household drinking water.[53] The lack of knowledge regarding the potential health risks in many rural and periurban settings and the scarcity of fresh water supplies create situations in which many residents are at substantial risk. The WHO[46] recommends the following measures to protect members of agricultural communities:

1. Treating wastewater
2. Restricting access to irrigated fields and canals and ditches

FIGURE 2 Community members are at risk of contamination from wastewater carried in local stream, as they withdraw water for cooking, cleaning, and other purposes.
Source: Ben Keraita, IWMI, Ghana.

3. Providing safe recreational water, particularly for adolescents
4. Providing safe drinking water and sanitation facilities to communities
5. Promoting good health and hygiene practices
6. Providing chemotherapy and immunization
7. Controlling disease vectors and intermediate hosts
8. Reducing contact with disease vectors

Several of these measures are similar to those recommended to protect farm families and farm workers, given the similarity in exposure opportunities on farms and in the larger community. Many of the challenges involved in implementing the measures and encouraging sustainable adoption also would be similar.

Consumers of Farm Products

In many settings, in the absence of policy intervention, consumers might be the least informed group regarding the potential health risks due to wastewater irrigation. They might be unaware that farmers using wastewater have produced some of the fruits and vegetables for sale in local markets. They might also be unaware that some of the farm produce carries harmful pathogens and chemicals or that cooking the produce might reduce the likelihood of damage from infectious pathogens. Given these considerations, the WHO[46] recommends the following measures to reduce the risk to consumers:

1. Treating wastewater
2. Restricting the crops that are irrigated with wastewater
3. Promoting irrigation practices that minimize contamination of plants
4. Implementing withholding periods that allow pathogens to die between the last irrigation and harvest
5. Promoting hygienic practices at food markets and during food preparation
6. Promoting good health and hygiene practices
7. Promoting produce washing, disinfection, and cooking
8. Providing chemotherapy and immunization

Although enforcement will be difficult, public agencies might consider disallowing wastewater irrigation of vegetables and other crops that consumers often eat without cooking. Leafy vegetables, such as lettuce and spinach, are particularly prone to accumulating pathogens on edible portions of the plant when wastewater is applied directly over the plants and when irrigators splash contaminated soil particles on the leaves (Figure 3).[54] Modifying the spouts of watering cans will reduce contamination by reducing the splashing of soil particles (Figure 4).[54] Drip irrigation on the soil surface or below ground will minimize contamination,[55] but many poor farmers will not have the funds to invest in such systems (Figure 5).

Withholding periods between the date of last irrigation and harvest are sensible approaches as well, but monitoring and enforcement might be problematic in areas where wastewater irrigation is prevalent. Some farmers report that irrigating lettuce on the morning of the day of harvest freshens the crop and enhances its appearance in local markets.[54] Encouraging farmers to change such practices will be challenging, particularly given the perishable nature of leafy vegetable crops. Farmers generally want to obtain the highest price possible and to sell their produce quickly, before its appearance and quality begin to fade.

Public efforts to improve hygienic practices and food preparation at homes and in the marketplace also will be challenging. In areas where small-scale farmers sell produce to small-scale vendors who resell the produce in a restaurant or fast-food outlet, individuals have little incentive to assume the extra cost of enhanced food treatment. This situation in which information is limited and asymmetric can be described also an externality involving producers and consumers. The benefits of a cleaner, safer food

FIGURE 3 Leafy vegetables are susceptible to contamination by pathogens in wastewater used for irrigation, particularly when watered with sprinklers or cans, as in this photograph from Ghana.
Source: Ben Keraita, IWMI.

FIGURE 4 Using a watering can spout with many small holes reduces the splashing of soil particles onto leaves of lettuce and other leafy vegetables.
Source: Ben Keraita, IWMI.

FIGURE 5 Drip irrigation minimizes contact of irrigation water with plant foliage, thus reducing the risk of contamination by pathogens in wastewater.
Source: Frank Rijsberman, IWMI.

supply accrue to consumers and communities, rather than to the farmers and food shop owners who will incur higher costs if they implement improved production, washing, and handling practices. Public policy is needed to ensure that farmers and vendors internalize the external costs of their activities.

Motivating Safe Practices along the Wastewater Exposure Pathway

From a policy perspective, it is helpful to view the use of wastewater in irrigation as an activity along a pathway on which many individuals and communities are exposed to potential contamination. The pathway begins with wastewater generation in households, companies, and industries. Farmers, fishers, and members of urban and rural communities who utilize water in streams or drains carrying wastewater are exposed to contamination as they conduct their activities. Farmers and laborers applying irrigation water or working in farm fields are exposed, as are community members who spend time in fields irrigated with wastewater. The exposure pathway continues from the farm to food processors, vendors, and consumers.

Along the wastewater exposure pathway, farmers and food vendors generate value as they produce and market crops that are desired by consumers. That value provides motivation for reducing risk, while maintaining irrigation opportunities. The value also can provide a partial source of finance for implementing measures to reduce risk. Both farmers and food vendors have a financial stake in sustaining their activities, and they should be able and willing to pay some portion of the cost of achieving acceptable risk levels. Given this perspective, cost-sharing programs in which farmers pay a portion of the cost of adopting safer irrigation methods might be helpful in promoting rapid adoption. Food vendors also might be offered cost-sharing programs for adopting measures that will improve hygiene and preparation of irrigated produce.

Local officials might also consider ways to motivate farmers and food vendors to participate actively in the study and enhancement of crop production and food preparation methods in areas where wastewater is used for irrigation. At present, it is likely that many farmers and vendors prefer to operate with minimal visibility, to reduce the probability that a public agency or consumer organization might endeavor to end the practice of wastewater irrigation. The public will gain value by encouraging farmers and vendors to explain their practices, maintain production records, gather information regarding water quality and food safety, and participate in training programs on irrigation and food preparation. Public officials might offer farmers greater security of land tenure or access to affordable credit, in exchange for active participation. Food vendors also might respond positively to the offer of affordable credit or the opportunity to obtain a license that enables them to sell food products in a desirable location for several years.

Consumers also can be expected to contribute some portion of the cost of reducing risk along the wastewater exposure pathway. Consumer willingness to pay market prices for fresh produce reflects the value generated by farmers and food vendors in making the produce available. Fresh fruits and vegetables enhance dietary intake in important ways, provided the produce is free of pathogens and chemicals. The net value consumers actually obtain from wastewater-irrigated produce will be diminished if they become ill due to interaction with pathogens or chemicals. Hence, they should be willing to pay a premium for produce that is certified free from potentially harmful constituents.

Local governments might consider implementing food safety assurance programs, in which trained professionals monitor and certify the production of farm produce in areas where wastewater is used for irrigation. The programs might be started on a voluntary basis, with the goal of eventually including all farmers using wastewater to produce vegetables for sale in local markets. Funds for the program could be generated through an assessment on the sale of farm products to food vendors, or at the point of sale from vendors to consumers. In either case, consumers would pay a portion of the cost of assuring food safety.

In summary, a key question for policy makers is how to enhance the values generated along the wastewater exposure pathway, while also encouraging farmers and food vendors to use a portion of those values to improve food safety and reduce farm-level and community risks. Poor farmers might initially be resistant to changing irrigation practices in ways that reduce contamination of their produce, but they should be encouraged to consider their financial situation with and without access to wastewater. To the extent that wastewater irrigation enhances farm-level net income, there should be some farm-level willingness and ability to invest in safe irrigation practices. Public officials might consider implementing cost-sharing programs to encourage farmers to invest in drip irrigation systems or other watering methods that reduce the risks associated with wastewater irrigation.

The same perspective applies with regard to food vendors. To the extent that their sales are enhanced through access to crops irrigated with wastewater, food vendors should be able and willing to invest some portion of their net returns in efforts to improve food safety. They might be encouraged to purchase vegetable washing devices and to clean all produce carefully before it is sold to consumers. A successful campaign of improving food safety might enhance food sales, as consumers learn of the campaign and as they suffer fewer health effects from consuming vegetables produced by local farmers.

Public officials also might consider implementing farm produce certification programs in which farmers who agree to improve their cultural practices in ways that reduce contamination are given special recognition. For example, one can envision a "consumer-safe" labeling program in which vegetables produced using risk-minimizing irrigation methods are given a label denoting that status. If implemented along with an innovative marketing campaign, consumers might be willing to pay a premium price for consumer-safe produce, thus rewarding farmers for investing in safe irrigation practices.

A similar program could be implemented for food vendors who purchase consumer-safe produce from farmers and then prepare the food for sale using safe handling and preparation practices. One can envision a consumer- safe label affixed to the storefronts of participating food vendors. Over time, if the program is successful, market forces might guide both farmers and food vendors to join the consumer-safe program, as the demand for non-labeled produce and food diminishes.

Examples of Public Policies

Helpful examples of public policies regarding wastewater use in irrigation are found in the Middle East and North Africa and other regions where farmers have been using treated and untreated wastewater for many years. In some countries, such as Egypt, the volume of municipal wastewater exceeds the treatment capacity, and large volumes of untreated wastewater enter agricultural drains.[56] The government attempts to manage the blending of treated and untreated wastewater with agricultural drainage water and the use of blended water by farmers, but success is limited by the scale of the problem and the strong demand for supplemental water supplies in the Nile Delta. Irrigation with treated wastewater will increase over time, with the expansion of wastewater treatment capacity. Egypt has developed a code of practice for using treated wastewater in agriculture (Table 3).

Several countries in the region, including Algeria, Cyprus, and Tunisia, do not allow the irrigation of vegetables with treated wastewater. Cyprus also disallows the irrigation of ornamental plants destined for sale in international markets.[57] Wastewater policies are well developed in Cyprus and Tunisia, where the governments actively support and regulate wastewater treatment and reuse. In Cyprus, the government pays for large portions of the cost of water treatment plants in cities and villages, while also paying for the distribution of wastewater to farmers (Table 4). Tunisia requires that industries comply with wastewater discharge standards designed to support reuse on farms, golf courses, and landscapes, and also for aquifer recharge (Table 4). Saudi Arabia plans to use all of its treated wastewater, primarily in agriculture. The city of Muscat in Oman has installed an extensive drip irrigation system for irrigating landscapes with treated municipal wastewater.[60]

In other regions, Italy has established water quality criteria regarding the use of treated wastewater on vegetables and grazing crops, while several autonomous provinces in Spain have developed legal prescriptions or recommendations regarding wastewater reuse.[61] Wastewater accounts for an estimated 41% of the irrigation water used on Spanish golf courses.[62]

The government of Botswana encourages greater reuse of wastewater in irrigation and mining, in part, by ending its policy of providing fresh water supplies at subsidized prices.[63] Botswana also is considering how to account for wastewater volumes within its national water accounting framework.[64]

Public officials in countries with little experience in regulating the use of wastewater in irrigation can gain value by reviewing the examples presented here and by considering ways to engage producers and consumers in active discussion of wastewater issues. As in many regulatory settings, the prospect of new rules and procedures regarding wastewater irrigation and food preparation will be viewed initially as a cost-increasing outcome that will harm the financial performance of individual farmers and food vendors. Hence, the rational strategy from an individual's perspective involves a combination of maintaining a low profile and quietly lobbying against the adoption of any new programs. Yet, in aggregate, net social welfare is decreased if the sum of damages from using wastewater in irrigation exceeds the sum of the benefits.

TABLE 3   The Cairo East Bank Effluent Re-use Project

Goal

Examine the potential for reusing in agriculture the treated wastewater from three sewage treatment plants that serve Cairo communities on the east bank of the Nile River.

Key Issues

1. What is the best way to distribute and allocate the treated wastewater?
2. What are the safest and most efficient ways to supply and use the wastewater?

Outcomes

1. A draft code of practice, which addresses irrigation requirements, legislative needs, water quality standards, health issues, and monitoring programs.
2. An institutional framework to ensure the design of safe and efficient water reuse projects.

Source: Angelakis et al.[58]

TABLE 4 Examples of Wastewater Policies in the Middle East

Cyprus

The government pays for 75% of the cost of water treatment plants in villages. In cities, the government pays the full costs of construction and operation of tertiary treatment plants.

In addition, the government pays the cost of distributing wastewater to farmers.

Egypt

The Ministry of Agriculture supports the restricted reuse of treated wastewater for irrigation of nonfood crops, including trees for timber and green belts designed to stabilize sand dunes. Large volumes of untreated wastewater flow into agricultural drains, where farmers withdraw water for irrigation.

Jordan

Wastewater treatment standards were introduced in 1982, and they have been modified in recent years. Vegetables may not be irrigated directly with treated wastewater.

Kuwait

Treated wastewater may not be used for landscape irrigation. It may be used only to irrigate "safe crops."

Saudi Arabia

The Kingdom plans to use all of its treated wastewater, primarily in agriculture. Other uses include landscape irrigation and aquifer recharge. Guidelines requiring secondary and tertiary treatment have been developed to support the unrestricted use of wastewater for irrigation.

Syria

Wastewater irrigation is restricted to fodder crops, industrial crops, and fruit trees, but enforcement is not complete.

Tunisia

Industries must comply with discharge standards to support intensive development of wastewater reuse on farms, golf courses, and landscapes, and also for aquifer recharge.

Source: Bakir.[59]

Perhaps the key to starting policy discussions is to demonstrate the potential gains in aggregate net benefits. Farmers, food vendors, and consumers can gain value together as they work with public officials to develop safe practices in crop production and food preparation. Individual farmers and food vendors will not be disadvantaged if everyone agrees to adopt safe practices and if consumers are willing to pay higher prices in return for safety assurances. Details regarding policy parameters and effective monitoring and enforcement programs can be developed over time, once all parties appreciate the potential gains in net benefits made possible through the safe and efficient use of wastewater in agriculture and the preparation of healthful food products.

Conclusions

Policy issues regarding the use of wastewater in irrigation are quite different in developed and developing countries. In developed countries, most municipal and industrial wastewater is treated, and thus, most of the wastewater used in agriculture is treated. Protective guidelines regarding the quality of wastewater used for irrigation have been in place for many years. Policy issues in developed countries pertain largely to financial and economic considerations regarding the improvement and expansion of wastewater treatment facilities. Public officials and water management agencies motivate greater use of wastewater by providing financial incentives and increasing public awareness of the safety and benefits of using treated wastewater on farms, golf courses, and urban landscapes.

Policy issues in developing countries also include financial and economic questions regarding investments in wastewater treatment. However, in many countries, the pace of such investments will not be sufficient to meet demand. Much of the wastewater generated in cities and rural areas will remain untreated for many years. As a result, farmers will continue to use untreated wastewater for irrigation, and its use will be largely unintentional and informal. Public officials must therefore implement risk reduction programs that protect farm families, communities, food vendors, and consumers from the potentially harmful effects of exposure to the pathogens and chemicals in untreated wastewater.

Policy options in developing countries will reflect a range of interventions along the pathway that includes wastewater generation, irrigation water capture and use, crop production and harvest, food preparation, and consumption. Public officials can implement risk-reducing guidelines and programs at each stage along the wastewater exposure pathway. For example, public officials can support improvements in wastewater treatment at the point of generation, when funds for such improvements are available. Officials also can call for changes in household and industrial production practices that would reduce the loads of harmful constituents in wastewater, thus reducing concentrations of those constituents in the irrigation water diverted from streams and ditches by farmers.

At the farm level, public agencies can provide technical assistance regarding water diversion and irrigation methods that would reduce potential exposure of farm workers to harmful pathogens and chemicals. Technical assistance regarding irrigation methods that reduce contamination of leafy vegetables and other produce consumed without cooking is essential for reducing risks to food vendors and consumers. Although difficult to enforce, regulations that establish a minimum time period between the dates of last irrigation and harvest would be helpful in reducing the risk of contamination from agricultural products.

Public officials in developing countries might also consider implementing certification programs for consumer-safe farm produce, particularly in markets where local farmers sell their irrigated vegetables. Public agencies can begin such programs, with support from farmers and food vendors, but eventually, market forces must arise to sustain them. Consumers must find value in certified produce, and they must be willing to pay a small premium that compensates farmers and vendors for their costs in providing the safer produce. Educational and marketing campaigns can be helpful in boosting demand for safe produce among consumers.

The policy issues we describe in this entry pertain largely to near-term strategies for minimizing the risk of negative health effects, while also enabling farmers to gain the potential benefits of using untreated and partially treated wastewater in agriculture. This approach is appropriate for countries that presently cannot afford to build, operate, and maintain a full complement of modern wastewater treatment facilities. Over time, as the demand for water in agriculture and other uses continues to increase, public officials in all countries should endeavor to provide wastewater treatment that matches end uses, including the irrigation of crops, landscapes, and golf courses. In developing countries, it will be necessary also to ensure that small-scale farmers retain access to a reliable source of irrigation water when the untreated and commingled wastewater they once relied on becomes unavailable, with the expansion of wastewater treatment programs.

Acknowledgments

I appreciate very much the helpful comments and suggestions of an anonymous reviewer.

References

1. Gupta, R.; Gangopadhyay, S.G. Peri-urban agriculture and aquaculture. Econ. Political Wkly. **2006**, *41* (18), 1757–1760.
2. Qian, Y.L.; Mecham, B. Long-term effects of recycled wastewater irrigation on soil chemical properties on golf course fairways. Agron. J. **2005**, *97* (3), 717–721.
3. Barker-Reid, F.; Harper, G.A.; Hamilton, A.J. Affluent effluent: Growing vegetables with wastewater in Melbourne, Australia—A wealthy but bone-dry city. Irrig. Drain. Syst. **2010**, *24* (1–2), 79–94.
4. Pedrero, F.; Kalavrouziotis, I.; Alarcón, J.J.; Koukoula- kis, P.; Asano, T. Use of treated municipal wastewater in irrigated agriculture—Review of practices in Spain and Greece. Agric. Water Manage. **2010**, *97* (9), 1233–1241.
5. Qadir, M.; Bahri, A.; Sato, T.; Al-Karadsheh, E. Wastewater production, treatment, and irrigation in Middle East and North Africa. Irrig. Drain. Syst. **2010**, *24* (1–2), 37–51.

6. Sheng, Z. An aquifer storage and recovery system with reclaimed wastewater to preserve native groundwater resources in El Paso, Texas. J. Environ. Manage. **2005**, *75* (4), 367–377.

7. Jamwal, P.; Mittal, A.K. Reuse of treated sewage in Delhi city: Microbial evaluation of STPs and reuse options. Resour., Conserv. Recycl. **2010**, *54* (4), 211–221.

8. Bixio, D.; Thoeye, C.; De Koning, J.; Joksimovic, D.; Savic, D.; Wintgens, T.; Melin, T. Wastewater reuse in Europe. Desalination **2006**, *187* (1–3), 89–101.

9. Salgot, M.; Huertas, E.; Weber, S.; Dott, W.; Hollender, J. Wastewater reuse and risk: Definition of key objectives. Desalination **2006**, *187* (1–3), 29–40.

10. Tal, A. Seeking sustainability: Israel's evolving water management strategy. Science **2006**, *213*, 1081–1084.

11. World Health Organization (WHO). *Guidelines for the Safe Use of Wastewater, Excreta and Greywater.* Volume 1: Policy and regulatory aspects; World Health Organization: Geneva, 2006a.

12. Akber, A.; Mukhopadhyay, A.; Al-Senafy, M.; Al-Haddad, A.; Al-Awadi, E.; Al-Qallaf, H. Feasibility of long-term irrigation as a treatment method for municipal wastewater using natural soil in Kuwait. Agric. Water Manage. **2008**, *95* (3), 233–242.

13. Hamilton, A.J.; Stagnitti, F.; Xiong, X.; Kreidl, S.L.; Benke, K.K.; Maher, P. Wastewater irrigation: The state of play. Vadose Zone J. **2007**, *6* (4), 823–840.

14. Oron, G. Water resources management and wastewater reuse for agriculture in Israel. In *Wastewater Reclamation and Reuse*; Asano, T., Ed.; CBC Press: Boca Raton, FL, 1998; 757–778.

15. van der Bruggen, B. The global water recycling situation. Sustainable Sci. Eng. **2010**, *2*, 41–62.

16. Hussain, G.; Al-Saati, A.J. Wastewater quality and its reuse in agriculture in Saudi Arabia. Desalination **1999**, *123* (2–3), 241–251.

17. Al-Rashed, M.F.; Sherif, M.M. Water resources in the GCC countries: an overview. Water Resour. Manage. **2000**, *14* (1), 59–75.

18. Al-Katheeri, E.S. Towards the establishment of water management in Abu Dhabi Emirate. Water Resour. Manage. **2008**, *22* (2), 205–215.

19. Dillon, P.; Pavelic, P.; Toze, S.; Rinck-Pfeiffer, S.; Martin, R.; Knapton, A.; Pidsley, D. Role of aquifer storage in water reuse. Desalination **2006**, *188* (1–3), 123–134.

20. Jiménez, B.; Drechsel, P.; Koné, D.; Bahri, A.; Raschid-Sally, L.; Qadir, M. Wastewater, sludge and excreta use in developing countries: An overview. In *Wastewater Irrigation and Health: Assessing and Mitigating Risk in Low-Income Countries*; Drechsel, P., Scott, C.A., Raschid-Sally, L., Redwood, M., Bahri, A., Eds.; Earthscan: London, 2010; 3–27.

21. Bradford, A.; Brook, R.; Hunshal, C.S. Wastewater irrigation in Hubli-Dharwad, India: Implications for health and livelihoods. Environ. Urbanization **2003**, *15* (2), 157–170.

22. Ensink, J.H.J.; Mahmood, T.; van der Hoek, W.; Raschid- Sally, L.; Amerasinghe, F.P. A nationwide assessment of wastewater use in Pakistan: An obscure activity or a vitally important one? Water Policy **2004**, *6* (3), 197–206.

23. Bos, R.; Carr, R.; Keraita, B. Assessing and mitigating wastewater-related health risks in low-income countries: An introduction. In *Wastewater Irrigation and Health: Assessing and Mitigating Risk in Low-Income Countries*; Drechsel, P. Scott, C.A., Raschid-Sally, L., Redwood, M., Bahri, A., Eds.; Earthscan: London, 2010; 29–47.

24. Keraita, B.; Drechsel, P.; Amoah, P. Influence of urban wastewater on stream water quality and agriculture in and around Kumasi, Ghana. Environ. Urbanization. **2003**, *15* (2), 171–178.

25. Amoah, P.; Drechsel, P.; Abaidoo, R.C.; Henseler, M. Irrigated urban vegetable production in Ghana: Microbiological contamination in farms and markets and associated consumer risk groups. J. Water Health **2007**, *5* (3), 455–466.

26. Tiongco, M.M.; Narrod, C.A.; Bidwell, K. Risk analysis integrating livelihood and economic impacts of wastewater irrigation on health. In *Wastewater Irrigation and Health: Assessing and Mitigating Risk in Low-Income Countries*; Drechsel, P., Scott, C.A., Raschid-Sally, L., Redwood, M., Bahri, A., Eds.; Earthscan: London, 2010; 127–145.

27. Qadir, M.; Wichelns, D.; Raschid-Sally, L.; McCornick, P.G.; Drechsel, P.; Bahri, A.; Minhas, P.S. The challenges of wastewater irrigation in developing countries. Agric. Water Manage. **2010,** *97* (4), 561–568.

28. Rutkowski, T.; Raschid-Sally, L.; Buechler, S. Wastewater irrigation in the developing world—Two case studies from the Kathmandu Valley in Nepal. Agric. Water Manage. **2007,** *88* (1–3), 83–91.

29. Drechsel, P.; Graefe, S.; Sonou, M.; Cofie, O.O. *Informal Irrigation in Urban West Africa: An Overview,* IWMI Research Report 102; International Water Management Institute: Colombo, Sri Lanka, 2006.

30. Van Rooijen, D.J.; Turral, H.; Biggs, T.W. Sponge city: Water balance of mega-city water use and wastewater use in Hyderabad, India. Irrig. Drain. **2005,** *54* (Suppl. 1), S81–S91.

31. Srinivasan, J.T.; Reddy, V.R. Impact of irrigation water quality on human health: A case study in India. Ecol. Econ. **2009,** *68* (11), 2800–2807.

32. Ensink, J.H.J.; Scott, C.A.; Brooker, S.; Cairncross, S. Sewage disposal in the Musi-River, India: Water quality remediation through irrigation infrastructure. Irrig. Drain. Syst. **2010,** *24* (1–3), 65–77.

33. Meinzinger, F.; Oldenburg, M.; Otterpohl, R. No waste, but a resource: Alternative approaches to urban sanitation in Ethiopia. Desalination **2009,** *248* (1–3), 322–329.

34. Mills, R.A.; Karajeh, F.; Hultquist, R.H. California's task force evaluation of issues confronting water reuse. Water Sci. Technol. **2004,** *50* (2), 301–308.

35. van Roon, M. Water localization and reclamation: Steps towards low impact urban design and development. J. Environ. Manage. **2007,** *83* (4), 437–447.

36. Hurlimann, A.; McKay, J. Urban Australians using recycled water for domestic non-potable use—An evaluation of the attributes price, saltiness, colour and odour using conjoint analysis. J. Environ. Manage. **2007,** *83* (1), 93–104.

37. Hurlimann, A.C. Water supply in regional Victoria Australia: A review of the water cartage industry and willingness to pay for recycled water. Resour. Conserv. Recycl. **2009,** *53* (5), 262–268.

38. United Nations Educational, Scientific, and Cultural Organization (UNESCO). Water for People, Water for Life - UN World Water Development Report. UNESCO Publishing: Paris, 2003; 36 pages.

39. Drechsel, P.; Evans, A.E.V. Wastewater use in irrigated agriculture. Irrig. Drain. Syst. **2010,** *24* (1–2), 1–3.

40. Fine, P.; Halperin, R.; Hadas E. Economic considerations for wastewater upgrading alternatives: An Israeli test case. J. Environ. Manage. **2006,** *78* (2), 163–169.

41. Bakopoulou, S.; Polyzos, S.; Kungolos, A. Investigation of farmers' willingness to pay for using recycled water for irrigation in Thessaly region, Greece. Desalination **2010,** *250* (1), 329–334.

42. Tsagarakis, K.P.; Georgantzis, N. The role of information on farmers' willingness to use recycled water for irrigation. Water Sci. Technol.: Water Supply **2003,** *3* (4), 105–113.

43. Toze, S. Reuse of effluent water—Benefits and risks. Agric. Water Manage. **2006,** *80* (1–3), 147–159.

44. van Lier, J.B.; Huibers, F.P. From unplanned to planned agricultural use: Making an asset out of wastewater. Irrig. Drain. Syst. **2010,** *24* (1–2), 143–152.

45. Keraita, B.; Drechsel, P.; Konradsen, F. Up and down the sanitation ladder: Harmonizing the treatment and multiple- barrier perspectives on risk reduction in wastewater irrigated agriculture. Irrig. Drain. Syst. **2010,** *24* (1–2), 24–35.

46. World Health Organization (WHO). *Guidelines for the Safe Use of Wastewater, Excreta and Greywater.* Volume 2: Wastewater use in agriculture; World Health Organization: Geneva, 2006.

47. Toze, S. Water reuse and health risks—Real vs. perceived. Desalination **2006,** *187* (1–3), 41–51.

48. Ternes, T.A.; Bonerz, M.; Herrmann, N.; Teiser, B.; Andersen, H.R. Irrigation of wastewater in Braunschweig, Germany: An option to remove pharmaceuticals and musk fragrances. Chemosphere **2007,** *66* (5), 894–904.

49. Lapen, D.R.; Topp, E.; Metcalfe, C.D.; Li, H.; Edwards, M.; Gottschall, N.; Bolton, P.; Curnoe, W.; Payne, M.; Beck, A. Pharmaceutical and personal care products in tile drainage following land application of municipal biosolids. Sci. Total Environ. **2008**, *399* (1–3), 50–65.

50. Siemens, J.; Huschek, G.; Siebe, C.; Kaupenjohann, M. Concentrations and mobility of human pharmaceuticals in the world's largest wastewater irrigation system, Mexico City- Mezquital Valley. Water Res. **2008**, *42* (8–9), 2124–2134.

51. Topp, E.; Monteiro, S.C.; Beck, A.; Coelho, B.B.; Boxall, A.B.A.; Duenk, P.W.; Kleywegt, S.; Lapen, D.R.; Payne, M.; Sabourin, L.; Li, H.; Metcalfe, C.D. Pharmaceutical and personal care products in tile drainage following land application of municipal biosolids. Sci. Total Environ. **2008**, *396* (1), 52–59.

52. Meinzen-Dick, R.; van der Hoek, W. Multiple uses of water in irrigated areas. Irrig. Drain. Syst. **2001**, *15* (2), 93–98.

53. Senzanje, A.; Boelee, E.; Rusere, S. Multiple use of water and water productivity of communal small dams in the Limpopo Basin, Zimbabwe. Irrig. Drain. Syst. **2008**, *22* (3–4), 225–237.

54. Keraita, B.; Konradsen, F.; Drechsel, P. Farm-based measures for reducing microbiological health risks for consumers from informal wastewater-irrigated agriculture. In *Wastewater Irrigation and Health: Assessing and Mitigating Risk in Low-Income Countries;* Drechsel, P., Scott, C.A., Raschid-Sally, L., Redwood, M., Bahri, A., Eds.; Earthscan: London, 2010; 188–207.

55. Capra, A.; Scicolone, B. Recycling of poor quality urban wastewater by drip irrigation systems. J. Cleaner Prod. **2007**, *15* (16), 1529–1534.

56. Abdel-Dayem, S.; Abdel-Gawad, S.; Fahmy, H. Drainage in Egypt: A story of determination, continuity, and success. Irrig. Drain. **2007**, *56* (Suppl. 1), S101-S111.

57. Lawrence, P.; Adham, S.; Barrott, L. Ensuring water reuse projects succeed—Institutional and technical issues for treated wastewater reuse. Desalination **2002**, *152* (1–3), 291–298.

58. Angelakis, A.N.; Marecos do Monte, M.H.F.; Bontoux, L.; Asano, T. The status of wastewater reuse practice in the Mediterranean Basin: Need for guidelines. Water Res. **1999**, *33* (10), 2201–2217.

59. Bakir, H.A. Sustainable wastewater management for small communities in the Middle East and North Africa. J. Environ. Manage. **2001**, *61* (4), 319–328.

60. Bazza, M. Wastewater recycling and reuse in the Near East Region: Experience and issues. Water Sci. Technol.: Water Supply **2003**, *3* (4), 33–50.

61. Angelakis, A.N.; Bontoux, L.; Lazarova, V. Challenges and prospectives for water recycling and reuse in EU countries. Water Sci. Technol.: Water Supply **2003**, *3* (4), 59–68.

62. Rodriguez Diaz, J.A.; Knox, J.W.; Weatherhead, E.K. Competing demands for irrigation water: Golf and agriculture in Spain. Irrig. Drain. **2007**, *56* (5), 541–549.

63. Swatuk, L.A.; Rahm, D. Integrating policy, disintegrating practice: Water resources management in Botswana. Phys. Chem. Earth **2004**, *29* (15–18), 1357–1364.

64. Arntzen, J.W.; Setlhogile, T. Mainstreaming wastewater through water accounting: The example of Botswana. Phys. Chem. Earth **2007**, *32* (15–18), 1221–1230.

47

Water: Total Maximum Daily Load

Introduction 481
Regulation of Point Sources 482
    Statutory Trigger: A "Discharge of a Pollutant" • NPDES Permit
    Program • Section 404 Dredge and Fill Permit Program • Role of Water
    Quality Standards in Point Source Permitting, before the TMDL Process
State Management of Nonpoint Sources 485
    State Nonpoint Source Management Plans • Role of Water Quality Standards
    in Nonpoint Source Management before the TMDL Process
Role of TMDLs in Achieving Water Quality Standards 486
    Introduction • Section 303 Impaired Waters List • Setting the TMDL and
    State Water Planning • Two Examples: Applying a TMDL
Conclusion 489
References 489

Robin Kundis Craig

Introduction

This entry explains what a "total maximum daily load" (TMDL) is and how TMDLs help regulators to achieve the ultimate goals of the Federal Water Pollution Control Act (FWPCA), more commonly known (since the 1977 amendments) as the Clean Water Act.[1] A TMDL is both a regulatory and an informational tool that helps regulators to "restore and maintain the chemical, physical, and biological integrity of the Nation's waters"[2] by ensuring that each water body in the United States meets its water quality standards. As such, the Act's TMDL provisions also force regulators to take a broader look at water body integrity, and hence, the TMDL requirements could prompt state water quality managers in particular to move water quality regulation beyond individual source requirements to a more systemic and watershed-based pollution control program.

Congress first enacted the FWPCA in 1948.[1] However, early versions of the FWPCA left water quality regulation almost entirely to the states, providing almost nothing that qualifies as a true regulatory program, although the 1965 Water Quality Act unsuccessfully tried to prompt states to adopt water quality standards programs.[1] In 1972, however, Congress significantly amended the Act in order to better protect water quality, significantly expanding the federal role in water quality regulation and creating mandatory regulatory requirements.[1] Specifically, Congress relied upon three main mechanisms to improve the quality and integrity of the Nation's waters: two federal permit programs that regulate the discharges of pollutants from point sources and state nonpoint source control programs that are supposed to manage more diffuse sources of water pollution, such as agricultural runoff.[1]

The Act's two permit programs and the state nonpoint source control programs are source-focused mechanisms for improving water quality—that is, they control water pollution by regulating or managing

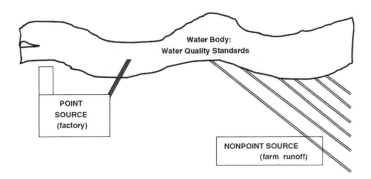

FIGURE 1   Clean Water Act overview.

the sources of that pollution.[3] Point sources—that is, "discernible, confined, and discrete conveyances," like pipes[4]—receive one of two kinds of permits. National Pollutant Discharge Elimination System (NPDES) permits dictate limits on how much of what kinds of pollutants the source can discharge into the water, predominantly based on technology-based effluent limitations.[1] Section 404 "dredge and fill" permits limit both how much dredged or fill material a source can discharge into waters and the broader impacts of construction projects in waters.[1] Similarly, when states manage nonpoint sources, they generally dictate how those nonpoint sources should manage water pollution, such as through best management practices (BMPs).[1]

However, neither the permit programs nor the state nonpoint source management programs dictate specific goals for water quality regulation—that is, nothing in the permit limitations or in BMPs specifies the ultimate water quality goals that their regulatory mechanisms are intended to achieve. Moreover, a single set of national water quality goals would not suffice, because water bodies naturally differ in their water quality characteristics. A headwater stream in the Rocky Mountains is not the Mississippi River, and Lake Erie is not Crater Lake.

Instead, the Clean Water Act uses water quality standards to determine the individual water quality goals for specific water bodies[1] (see Figure 1). The regulatory issue then becomes the following: how do regulators connect the Clean Water Act's two regulatory programs and the state nonpoint source management programs to these water quality standards? Although other mechanisms also exist,[1] the TMDL is the ultimate calculation that the Clean Water Act prescribes to connect water quality standards—the water quality goals for particular water bodies—and the Act's source-based regulatory mechanisms.

A TMDL is, literally, the "total maximum daily load"—that is, the maximum amount of a given pollutant that can be added to a particular water body on a daily basis while still having that water body meets its water quality standards. Individual states and the federal EPA use TMDLs when the national technology-based effluent limitations in NPDES permits, the water quality requirements in Section 404 dredge and fill permits, and the standard state nonpoint source control requirements are not stringent enough to allow a particular water body to achieve its water quality standards. Proper employment of a TMDL can require adjustments to the discharge permits, adjustments to the nonpoint source control requirements, or both.

Regulation of Point Sources

Statutory Trigger: A "Discharge of a Pollutant"

Total maximum daily loads cannot be understood without understanding the Clean Water Act's overall regulatory structure. As a starting point, the Act's regulatory provisions prohibit any "discharge of a pollutant" except as in compliance with the Act.[5] The Act defines "discharge of a pollutant" to be

"(A) any addition of any pollutant to navigable waters from any point source, (B) any addition of any pollutant to the waters of the contiguous zone or the ocean from any point source other than a vessel or other floating craft." [4] The Act does not define "addition," but case law has defined that term to mean any human-controlled contribution of pollutants to regulated waters.[1] The Act defines "pollutant" broadly to mean "dredged spoil, solid waste, incinerator residue, sewage, garbage, sewage sludge, munitions, chemical wastes, biological materials, radioactive materials, heat, wrecked or discarded equipment, rock, sand, cellar dirt and industrial, municipal, and agricultural waste discharged into water."[4] Case law has made clear that the Act does not apply to small, isolated waters.[6] By regulation, the "navigable waters," the "contiguous zone," and the "oceans" are defined so as to include almost everything else,[4] but in 2006, the U.S. Supreme Court issued an indecisive opinion regarding the test that the agencies and the courts should use to determine whether other smaller waters should qualify as "navigable waters" under the Act.[7] Finally, a "point source" is "any discernible, confined and discrete conveyance," like a pipe.[4]

NPDES Permit Program

In order to comply with the Clean Water Act, point sources that discharge pollutants must get a permit. The Act's most generally applicable permit program is the NPDES permit program, which applies to any "discharges of a pollutant" except discharges of dredged or fill material.[8,9] At the federal level, the EPA has the primary authority to implement the NPDES permit program,[8] although it has delegated much of its permitting authority to the individual states.[10] Even where states issue the NPDES permit, however, the EPA retains supervisory authority and can override state authority to issue individual permits.[8,11]

When a point source gets an NPDES permit, most of the requirements governing its discharge will be based on national effluent limitations[1] (see Figure 2). Effluent limitations are "end of the pipe," numerical limitations on the concentrations of pollutants that a discharger can discharge.[4] The EPA generally sets effluent limitations on an industry-wide basis.[5] Moreover, the effluent limitations are technology based. For example, the EPA currently sets most effluent limitations for most industries on the basis of the "best available technology economically achievable" for each category of industry.[5] In contrast, effluent limitations for publicly owned treatment works (or sewage treatment plants) are based on secondary treatment of the sewage. Where the EPA has not issued national technology-based effluent limitations, states write equivalent limits into individual permits.

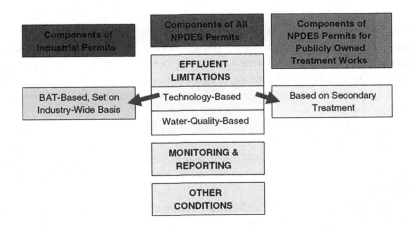

FIGURE 2 Components of an NPDES Permit.
Source: Adopted from U.S. EPA.[32]

Section 404 Dredge and Fill Permit Program

The Clean Water Act's second permit program, the Section 404 dredge and fill permit program, applies only to discharges of dredged or fill material.[9] At the federal level, the U.S. Army Corps of Engineers received the primary authority to implement this permit program, although the EPA retains oversight authority and can veto Section 404 permits.[9] States can also acquire authority to issue some Section 404 permits,[9] but to date, only two states—Michigan and New Jersey—have chosen to do so.[12] As a result, the Army Corps issues most Section 404 permits.

Under the Army Corps' regulations, "the term discharge of dredged material means any addition of dredged material into, including redeposit of dredged material other than incidental fallback within, the waters of the United States"[13]; "dredged material" is "material that is excavated or dredged from waters of the United States."[13] In turn, "[t]he term discharge of fill material means the addition of fill material into waters of the United States,"[13] where "fill material" is "material placed in waters of the United States where the material has the effect of: (1) Replacing any portion of a water of the United States with dry land; or (2) Changing the bottom elevation of any portion of a water of the United States."[13] Thus, Section 404 permits most often regulate construction in or filling of waters of the United States, especially wetlands.

In issuing an individual Section 404 permit (general permits are available for several kinds of smaller activities with limited impact on the environment), the Army Corps reviews the proposed activity against two sets of permitting criteria: the EPA's Section 404(b) Guidelines[9,14] and the Corps' public interest review requirements.[15] Both sets of criteria provide the Army Corps and, in its oversight capacity, the EPA with means to deny individual Section 404 permits to activities that are too damaging to the environment. However, the more common result of the agencies' review of projects in light of the Guidelines and the public interest criteria is that the permit applicant will have to modify the project's design, engage in mitigation, or both in order to receive the Section 404 permit.

Role of Water Quality Standards in Point Source Permitting, before the TMDL Process

The states received the primary authority to set water quality standards.[16] As noted, water quality standards establish the ultimate goals of water pollution regulation for individual water bodies. Specifically, water quality standards "consist of the *designated uses* of the navigable waters involved and the *water quality criteria* for such waters based upon such uses" (emphasis added).[16]

As far as designated uses are concerned, Congress directed states to consider waters' uses for public water supply, fish and wildlife support, recreation, agriculture, industry, and navigation.[16] However, states are free to designate other uses, as well (see Figure 3).

Water quality criteria, in turn, can be either numeric or narrative. State water quality criteria for toxic pollutants, for instance, have often been narrative in form—typically, "no toxic pollutants in toxic amounts." However, the EPA is strongly encouraging states to move to numeric water quality criteria for all pollutants of concern, and states setting numeric water quality criteria often borrow the EPA's water quality criteria guidelines.[17]

Whether numeric or narrative, however, the water quality criteria are supposed to specify the water quality, along multiple parameters (temperature, salinity, turbidity, toxicity, oxygen content, and so forth), that is necessary to achieve the waterway's designated uses.[16] For example, waters designated for salmon populations generally require cold, clear water, whereas waters designated for industrial use need not be so pristine.

In choosing the designated uses of particular water bodies and then establishing the associated water quality criteria necessary to achieve and support those uses, states particularize the meaning of the Clean Water Act's general goal of "chemical, physical, and biological integrity"[2] for specific water bodies. In theory, a state could create different water quality standards for every single water body in the

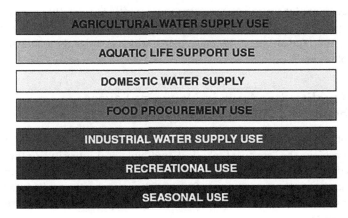

FIGURE 3 An example of Designated Uses: IOWA.

state. In practice, however, states tend to create categories of water bodies and then set water quality standards on that categorical basis.

Although the states received the primary authority to set water quality standards, the EPA retains the authority to set those standards for states that refuse to do so.[16] For example, in 2010, the EPA proposed numeric water quality criteria for nutrients (nitrogen and phosphorus) for the state of Florida.[18] The EPA also has the authority to disapprove state water quality standards that are inconsistent with the Act's requirements.[16]

Besides setting the water quality goals for individual water bodies, water quality standards also directly influence the two point source permitting programs. For example, in NPDES permitting, if the industry-wide technology-based effluent limitations are insufficient to achieve the water quality standards for a particular water body, the state or the EPA should instead use water quality–based effluent limitations in the permits of point sources discharging into that water body[19] (see Figure 2). The Army Corps, in turn, must ensure that projects requiring a Section 404 dredge and fill permit will not violate state water quality standards, and states can ensure that the project will comply with state water quality standards through their Section 401 certification authority.[20–22]

State Management of Nonpoint Sources

State Nonpoint Source Management Plans

The federal permit programs apply only to point sources—that is, "discernible, confined, and discrete conveyances"[4]—of water pollution. In contrast, a nonpoint source of water pollution is any source that is not a point source. Common nonpoint sources are uncontrolled and contaminated runoff or snowmelt or deposition onto water of air pollution.[23]

The Clean Water Act leaves the states in charge of managing nonpoint source water pollution. Initially, states were supposed to address nonpoint source pollution through their Section 208 area-wide waste treatment management programs.[24] These plans emphasized the construction of treatment works.[24] However, they were also supposed to do the following: 1) "identify, if appropriate, agriculturally and silviculturally related nonpoint sources of pollution" and establish controls for them; 2) identify and control "mine related sources of pollution"; 3) identify and control "construction activity related sources of pollution"; and 4) identify and control salt water intrusion problems.[24]

Dissatisfied with states' progress in controlling nonpoint source pollution under Section 208, Congress in 1987 added Section 319 to the Clean Water Act, which encouraged states to enact more specific and detailed nonpoint source management programs.[25] Section 319 nonpoint source management plans are supposed to include BMPs for all relevant nonpoint sources in the state, enforcement and assistance programs, and annual milestones against which to judge the state's progress in reducing nonpoint source pollution.[25]

Role of Water Quality Standards in Nonpoint Source Management before the TMDL Process

The ultimate goal of nonpoint source management, as in the regulatory provisions of the Clean Water Act generally, is to achieve the "chemical, physical, and biological integrity" of the Nation's waters,[2] which, as noted, is defined as a practical matter through the states' water quality standards. Thus, state nonpoint source management programs are often keyed to state water quality standards. Idaho, for example, explicitly requires that the designated BMPs for nonpoint sources ensure compliance with state water quality standards.[26]

As a more elaborate example, Washington explicitly ties its forest practice rules, which prescribe BMPs for forest activities that cause nonpoint source pollution, to its water quality standards, requiring the forest practice rules to be protective of water quality standards in waters affected by forestry-related nonpoint source pollution.[27] Once that connection is made, however, a forestry activity that complies with the forest practice rules is deemed to be in compliance with the state's water quality requirements.[27]

Role of TMDLs in Achieving Water Quality Standards

Introduction

As noted, Congress intended the regulatory mechanisms in the Clean Water Act "to restore and maintain the chemical, physical, and biological integrity of the Nation's waters."[2] The state-set water quality standards define such integrity for particular water bodies, and hence, water quality standards are the measure of whether the Act's goals have been achieved for an individual water body.

The Act's TMDL process provides a means for adjusting both permit requirements and the nonpoint source management requirements for the particular sources that contribute pollution to a water body that cannot achieve its water quality standards through the application of "normal" source-based requirements. The TMDL thus provides the information necessary to "ratchet down" the Clean Water Act's regulatory requirements to meet the water quality needs of specific water bodies.

Section 303 Impaired Waters List

The first step in the Clean Water Act's TMDL process is the state's creation of a Section 303(d) impaired waters list.[17] Specifically, the Act requires each state to identify water bodies or segments within its boundaries for which the effluent limitations "are not stringent enough to implement any water quality standard applicable to such waters."[17] States vary considerably in how many impaired waters they have.

Impaired waters are also known as "water-quality-limited water bodies." It is important to remember, however, that the designations of impairment depend on the following: 1) the exact water quality standards that each state has decided to apply to a particular water body; and 2) how the state decides to segment its water bodies in general. Given these particularities of state law, the number of impaired water bodies is actually a poor basis for attempting to compare actual water quality across state lines. For example, the fact that Pennsylvania has thousands of impaired water bodies may mean that most waters in Pennsylvania have poor water quality—but it could also mean that Pennsylvania has chosen very stringent water quality standards, applied water quality standards to very small segments of rivers and streams, and/or monitors water quality quite comprehensively. On the other hand, the fact that Alaska has identified less than 100 impaired waters might mean that water bodies in Alaska are generally pristine—but it could also mean that Alaska applies its water quality standards to entire rivers and lakes (creating fewer regulatory "water bodies"), has less stringent water quality standards than Pennsylvania, and/or monitors water quality less comprehensively than Pennsylvania.

Once a state identifies water bodies within its borders that do not meet their applicable water quality standards, it must rank those impaired water bodies for the purpose of establishing TMDLs for them.[16] The priority ranking must "tak[e] into account the severity of the pollution and the uses to be made of such waters." [16] States have considerable discretion in ranking their impaired water bodies. At the same time, the state must also "identify those waters or parts thereof within its boundaries for which controls on thermal discharges … are not stringent enough to assure protection and propagation of a balanced indigenous population of shellfish, fish, and wildlife."[16] States submit their lists of impaired waters to the EPA biennially,[28] and these lists—known as the Section 303(d) lists—are subject to the EPA's approval.[16]

Setting the TMDL and State Water Planning

Once a state has identified water-quality-impaired water bodies within its borders, establishing a TMDL is mandatory, and the EPA must establish the required TMDLs if the state fails to do so.[16] In establishing a TMDL, the state estimates, "in accordance with the priority ranking, the total maximum daily load" of all pollutants contributing to the violation of the relevant water quality standards that the EPA has determined can be subject to such a numeric calculation.[16] The state must establish the TMDL "at a level necessary to implement the applicable water quality standards with seasonal variations and a margin of safety which takes into account any lack of knowledge concerning the relationship between effluent limitations and water quality."[16] Thus, the TMDL should be somewhat conservative, allowing for both scientific uncertainty and natural variations in water quality.

Under the EPA's regulations, "TMDLs may be established using a pollutant-by-pollutant or biomonitoring approach. In many cases both techniques may be needed. Site-specific information should be used wherever possible."[28] Moreover, the state must establish TMDLs for "all pollutants preventing or expected to prevent attainment of water quality standards."[28]

Similarly, for thermally impaired waters, the state must estimate "the total maximum daily thermal load required to assure protection and propagation of a balanced, indigenous population of shellfish, fish, and wildlife."[16] Statutory factors for setting a thermal TMDL include "normal water temperatures, flow rates, seasonal variations, existing sources of heat input, and the dissipative capacity of the identified waters or parts thereof."[15] Moreover, like other TMDLs, thermal TMDLs are set conservatively, because they must include a margin of safety to account for scientific uncertainty.[16]

Once a state has established its TMDLs, it submits them to the EPA. The EPA must approve or disapprove the TMDLs within 30 days of submission.[16] If the EPA disapproves of the TMDLs, it has 30 days to establish its own TMDLs.[16]

For each TMDL established, the state must allocate the total daily load of the relevant pollutant among the sources that contribute that pollutant to the impaired water. The EPA recognizes that the TMDL's total pollutant load must be allocated among three sources (see Figure 4): background or "natural" sources of the pollutant; nonpoint sources of the pollutant (the "load allocation," or LA); and point sources of the pollutant (the "waste load allocation," or WLA).[29]

These allocations then provide the basis for strengthening the regulation of the relevant sources so that the water body can achieve its water quality standards. For example, the WLA provides the total amount of the relevant pollutant that the point sources can discharge. As a result, it becomes the basis for calculating water quality–based effluent limitations that should then be incorporated into the point sources' NPDES permits.[29] Moreover, the Clean Water Act limits the states' abilities to amend these

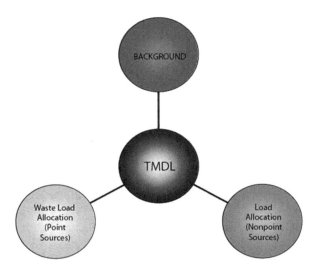

FIGURE 4   Allocation of the TMDL.

TMDL-based effluent limitations so long as the water body remains impaired: such standards "may be revised only if (i) the cumulative effect of all such revised effluent limitations based on such total maximum daily load or waste load allocation will assure the attainment of such water quality standard, or (ii) the designated use which is not being attained is removed. ..."[16] Removing designated uses is difficult, however, because the Act incorporates an antidegradation policy.[16,30]

The LA, in turn, allows for the possibility that the state will also amend its nonpoint source management requirements—especially for waters impaired entirely or predominantly as a result of nonpoint source pollution. As the EPA recognizes in its regulations, "[i]f Best Management Practices (BMPs) or other nonpoint source pollution controls make more stringent load allocations practicable, then wasteload allocations can be made less stringent. Thus, the TMDL process provides for nonpoint source control tradeoffs." [29]

If the EPA approves the state's TMDLs, or if the EPA establishes TMDLs for the state, the state also incorporates those TMDLs into its continuing planning process for water quality.[16] The state's continuing planning process includes the following: 1) effluent limitations and schedules of compliance, including effluent limitations stringent enough to meet water quality standards; 2) the applicable Section 208 area-wide waste management plans; 3) the TMDLs; 4) procedures for revision; 5) authority for intergovernmental cooperation; 6) mechanisms to implement new and revised water quality standards, including schedules of compliance; 7) controls for the disposal of residual wastes from wastewater treatment; and 8) an inventory and priority ranking of treatment works.[16]

Two Examples: Applying a TMDL

The potential impacts of a TMDL, and the potential tradeoffs between point source regulation and nonpoint source regulation, are best illustrated by example (see Table 1). Suppose that, in its Section 303 water quality standards, State A has designated both the Blue River and the Green River as cold-water rivers to support native trout populations. To support this designated use, State A establishes water quality criteria for sediment. Moreover, it includes standard effluent limitations for sediment in the NPDES permits that it issues to point sources. State A's nonpoint source management program includes extensive BMPs for farming, which has been a traditional source of water pollution in the state. However, forestry is a relatively new industry in State A, and the state has not amended its nonpoint source management program to account for water pollution from those activities.

TABLE 1 Sources of Sediment in the Blue River and Green River in State A

Source	Blue River	Green River
Background/natural	5 kg/day from cliff erosion	10 kg/day from upstream erosion
Point sources	500 kg/day from 4 confined animal feeding operations, each of which has an NPDES permit	None
Nonpoint sources	10 kg/day from runoff over a small farm	600 kg/day from 3 large timber operations along the river

During its most recent biennial Section 303(d) review of impaired waters, State A discovers that both the Blue River and the Green River are violating their water quality standards for sediment, causing harm to the native trout populations. Through the TMDL process, State A estimates that 100 kg of sediment can be added to each river each day without violating the sediment water quality standard for trout, leaving an ample margin of safety and allowing for seasonal variations, such as increased sediment runoff during the rainy season.

State A then identifies the sources contributing sediment to each river (Table 1).

Given these sources, State A should take a different approach to implementing the sediment TMDL on the Blue River than it does to implementing the sediment TMDL on the Green River. To ensure that the Blue River will meet its water quality standards, State A will need to adjust the point sources' *WLA*. As a result, to ensure compliance with the Blue River's water quality standards, State A will need to amend each of the point sources' NPDES permits to incorporate more stringent water quality–based effluent limitations for sediment.

In contrast, to ensure that the Green River will meet its water quality standards, State A will have to address the nonpoint sources' *LA*. Most obviously, State A should amend its nonpoint source management plan to impose nonpoint source control requirements, such as BMPs, on the timber companies. Such BMPs may, for example, require the use of buffer zones, require selective logging instead of clear-cutting, or limit the amount of road building and land clearing allowed.

Conclusion

Congress designed the TMDL process to ensure that every "water of the United States" would eventually enjoy chemical, physical, and biological integrity, as defined by the states in water quality standards. While TMDLs have been part of the Act since 1972, states ignored the TMDL process for many years.[31] Instead, citizen litigation drove the establishment and implementation of TMDLs, and many states are still working their way through court orders that require them to set TMDLs.[31]

Given that litigation context, TMDL implementation focused initially on adjusting the effluent limitations in point sources' Clean Water Act permits. However, as it becomes clear that nonpoint source pollution is the most significant remaining source of water quality impairment, TMDLs are increasingly becoming the mechanism that encourages states to adequately control those nonpoint sources of water pollution, as well.

References

1. Craig, R.K. *The Clean Water Act and the Constitution*, 2nd Ed.; Environmental Law Institute Press: Washington, DC, 2009.
2. Federal Water Pollution Control Act, § 101, 33 U.S.C. § 1251 (2006).
3. Craig, R.K. Climate Change, Regulatory Fragmentation, and Water Triage. University of Colorado Law Review **2008**, *79*(3), 825–927.
4. Federal Water Pollution Control Act, § 502, 33 U.S.C. § 1362; 2006.
5. Federal Water Pollution Control Act, § 301, 33 U.S.C. § 1311; 2006.

6. Solid Waste Agency of Northern Cook County v. U.S. Army Corps of Engineers, 531 U.S. 159; 2001.

7. Rapanos v. United States, 547 U.S. 715 (2006).

8. Federal Water Pollution Control Act, § 402, 33 U.S.C. § 1342; 2006.

9. Federal Water Pollution Control Act, § 404, 33 U.S.C. § 1344; 2006.

10. Office of Water, U.S. Environmental Protection Agency. *National Pollutant Discharge Elimination System (NPDES),* available at http://cfpub.epa.gov/npdes/statestats.cfm. (Accessed May 11, 2012).

11. Champion International Corp. v. U.S. Environmental Protection Agency, 850 F.2d 182 (4th Cir. 1988).

12. Office of Water, U.S. Environmental Protection Agency. *State or Tribal Assumption of the Section 404 Permit Program,* available at http://water.epa.gov/type/wetlands/outreach/fact23.cfm (accessed May 2012).

13. 33 C.F.R. § 323.2; 2009.

14. 40 C.F.R. §§ 230.1–230.80; 2009.

15. 33 C.F.R. §§ 320.1, 320.4; 2009.

16. Federal Water Pollution Control Act, § 303, 33 U.S.C. § 1313; 2006.

17. Federal Water Pollution Control Act, § 304, 33 U.S.C. § 1314; 2006.

18. U.S. Environmental Protection Agency. *Water Quality Standards for the State of Florida's Lakes and Flowing Rivers,* 75 Fed. Reg. 4174; (Jan. 26, 2010).

19. Federal Water Pollution Control Act, § 302, 33 U.S.C. § 1312; 2006.

20. Federal Water Pollution Control Act, § 401, 33 U.S.C. § 1341; 2006.

21. PUD. No. 1 of Jefferson County v. Washington Department of Ecology, 511 U.S. 700; 1994.

22. S.D. Warren Co. v. Maine Board of Environmental Protection, 547 U.S. 370; 2006.

23. Office of Water, U.S. Environmental Protection Agency. *What Is Nonpoint Source Pollution?,* available at http://water.epa.gov/polwaste/nps/whatis.cfm (accessed May 2012).

24. Federal Water Pollution Control Act, § 208, 33 U.S.C. § 1288; 2006.

25. Federal Water Pollution Control Act, § 319, 33 U.S.C. § 1329; 2006.

26. Idaho Code § 39–3620; 2009.

27. Revised Code of Washington § 90.48.420; 2009.

28. 40 C.F.R. § 130.7; 2009.

29. 40 C.F.R. § 130.2; 2009.

30. 40 C.F.R. § 131.12; 2009.

31. Houck, O.A. *The Clean Water Act TMDL Program: Law, Policy, and Implementation;* Environmental Law Institute Press: Washington, DC, 1999.

32. U.S. EPA, *NPDES Permit Document: Components,* available at http://www.epa.gov/waterscience/standards/academy/supp/permit/page6.htm.

48

Watershed Management: Remote Sensing and GIS

Introduction 491
Inventorying and Monitoring 492
Research and Development 492
Challenges and Opportunities Ahead 494
Conclusions 495
References 496

A.V. Shanwal
and S.P. Singh

Introduction

The need for natural resource conservation must be considered in any agricultural development plans involving conversion of new land use to increase production. An action plan to minimize natural resources degradation must be based on the principles of sustainability so that soil can be handed over to the future generation under better conditions than received from the previous generation. Thus, management practices must be ecologically sound, economically feasible, and socially and politically acceptable.

Natural resources degradation can be contained by adopting the watershed as a hydrological unit for development and management. This approach is multidisciplinary, broad based, and intensive vis-a-vis the simple "Seed and Fertilizer Approach." Despite the investment, it is an economic approach in the long run.[1]

The degradation of improperly used watersheds happens because of increase in soil erosion, decline in soil productivity, reduction in livestock-carrying capacity, decline in forest cover and perturbation in ecological equilibrium, and reduction in biodiversity. The problem requires a hightech solution with back-up policy decisions.

Remote sensing (RS) and geographic information system (GIS) are important technologies for addressing challenges in sustainable management of natural resources. Remotely sensed data (e.g., aerial photographs and satellite imagery) can be used to obtain information on soils, land use, vegetation, slope gradient, runoff, erosion, etc. However, recent developments in multispectral scanner, radar system, and a multitude of quantitative techniques for analyzing and processing such data provide opportunities for data acquisition through RS and array of techniques for data analyses.

GIS is an important tool for tracking spatial data. GIS draws the composite map by superimposing the data and image files obtained from traditional methods and satellite imageries, allows us to develop, analyze, and display spatially explicit information, and gives us the ability to deal with the larger spatial scales in process such as soil erosion and drainage (500 acres) and regional landscape (several million acres).

Inventorying and Monitoring

Despite efforts at inventorying and monitoring, conventional techniques provide only sketchy information on resources in a watershed, their location, and spatial and dynamic distribution. Of inventorying and monitoring, the latter is probably the major new thrust for resource managers and scientists.

Improved management practices are not thought of spontaneously prior to testing and implementation. Such practices have evolved over time through experimentation, experience, and trial and error. A case in point is the Bunga watershed in Ambala, Haryana, India.[10] As was expected, numerous components did not work properly. Neither the rate of runoff nor the outflow from the reservoir was in harmony in the initial stage. High rate of siltation of Sukhana Lake near Chandigarh, Haryana, India, due to severe soil erosion in the catchment area is another example of dire need for inventorying of data prior to starting any project.[11] Because of high siltation, the ponded area of Sukhana Lake decreased by about 30ha between 1980 and 2000. During this period, the courses of Sukhana and Kansal streams feeding the lake also changed considerably. The siltation of Sukhana Lake is the result of poor vegetation cover and heavy runoff in the catchment within the Shiwalik Hills.[11]

It is thus important that RS and GIS specialist and the resource managers must work together to determine the specific information needed (e.g., the nature of earth surface cover and their characteristics; the location, size, terrain, and other characteristics of the watershed area involved). Relevant information needed includes the format (e.g., maps, tables, scales, etc.); time frame for both collecting and processing the data; level of accuracy and reliability; costs of obtaining and interpreting/processing the data.

Research and Development

Using watershed, landscape or ecosystem approaches have broad support as a means in achieving sustainable use of natural resources and integrating objectives on a practical scale. Yet, addressing watershed management issues on large scales requires enhanced technical capabilities and modern set of tools and facilities.

Landsat-1 was launched in 1972 and the Landsat Thematic Mapper (TM) type of data has been in existence only since 1984. Thus, a tremendous progress has been made over short time in developing effective methods of processing and analyzing such data. The last decade of 20th century witnessed rapid advances and significant increase in the operational use of remotely sensed data.[3]

Landsat-7 launched in 1998 carries Enhanced Thematic Mappers (ETM) and pointable sensor with improved spatial resolution and signal-to-noise ratio (Table 1). The Earth Observing System consists of a morning (AM) and an afternoon (PM) components and carries five separate sensors including MODIS, a 36-narrow band imager with 1-km spatial resolution.[2] These sensors are significantly improved in their capability to map watershed information classes. These advanced sensors and computational capability have new requirements for satellite data processing algorithms. Added to the image analysis system, there is also a need for RS algorithms for estimating biophysical parameters necessary to drive and validate the watershed process models.[5]

ETM: enhanced thematic mapper; HRVIR: high resolution visible and middle infrared; LISS: linear imaging self scanner system; SAR: synthetic aperture radar; VIRS: visible infrared scanner; TMI: TRMM microwave imager; CERES: clouds earth's radiation energy system; AVHRR: advanced very high resolution radiometer; OCTS: ocean color and temperature scanner; AVNIR: advanced visible and near infrared radiometer; ASTER: advanced space borne thermal emission and reflectance radiometer; MODIS: moderate resolution imaging spectrometer-nadir; PAN: panchromatic; MS: multi spectral; IFOV: instantaneous field of view.[5]

During the 1980s, advances in computer hardware, particularly speed and data storage, catalyzed the development of software for handling spatial data. One of the most significant products of this period of rapid technological change was the development of GIS (Table 2). It has made a tremendous impact in identifying strategies of watershed development by manipulating and analyzing individual "layers"

TABLE 1 Recent RS Satellites with Advanced Sensors

Satellite	Country	Sensor Name	Sensor Spectrum	IFOV
Landsat-7	U.S.A.	ETM$^+$	0.45 μm ~ 12.50 μm (8 bands)	30 m
SPOT-4	France	HRVIR	0.50 μm ~ 1.75 μm (5 bands)	10 m
IRS-1D	India	PAN	0.50 ~ 0.75	5.8 m
		LISS-III	0.45 ~ 0.86 1.55 ~ 1.70 (5 bands)	24.0 m
		WIFS	0.67 ~ 0.86	188 m
RADARSAT	Canada	SAR	5.3 GHz	25 m × 28 m
TRMN	Japan	VIRS	0.63 ~ 10.7 (5 bands)	2km
		TMI	10.65 GHz ~ 85.5 GHz	5.6 km × 3.8 km
		CERES	0.3 μm ~ 50 μm (3 bands)	25 km
		LIS	0.7774 nm	5 km
NOAA-M	U.S.A.	AVHRR/3	0.58 μm ~ 12.40 μm (6 bands)	0.5 km
CRSS	U.S.A.	PAN	0.45 μm ~ 0.90 μm	0.82 m
		MSS	0.45 μm ~ 0.90 μm (4 bands)	3.20 m
ADEOS	Japan	OCTS	0.402 μm ~ 12.5 μm (12 bands)	700 m
		AVNIR	0.40 μm ~ 0.92 μm (5 bands)	8 m
EOS-AM	U.S.A.	ASTER	0.52 μm ~ 11.3 μm (3 bands)	15 m
EOS-PM	U.S.A.	MODIS-N	0.659 μm ~ 14.24 μm (5 bands)	250 m
EOSAT	U.S.A.	PAN	0.45 μm ~ 0.90 μm	1 m
		MS	0.45 μm ~ 0.90 μm (4 bands)	4 m

Note: ETM: enhanced thematic mapper; HRVIR: high resolution visible and middle infrared; LISS: linear imaging self scanner system; SAR: synthetic aperture radar; VIRS: visible infrared scanner; TMI: TRMM microwave imager; CERES: clouds earth's radiation energy system; AVHRR: advanced very high resolution radiometer; OCTS: ocean color and temperature scanner; AVNIR: advanced visible and near infrared radiometer; ASTER: advanced space borne thermal emission and reflectance radiometer; MODIS: moderate resolution imaging spectrometer-nadir; PAN: panchromatic; MS multi spectral; IFOV: instantaneous field of view.
Source: JSRS Remote Sensing Note.[5]

of spatial data and providing tools for analyzing and modeling the interrelationship among layers. The GIS also provides a means of displaying complex watershed information in a comprehensible manner.

The GIS also provides a means of predicting the outcomes of alternative courses of action, from both spatial and temporal perspective and in a timely and cost-effective manner. However, it does not preclude the need for monitoring the ground truth as a guide to development of future management practices.

The last decade of the 20th century witnessed rapid advances and a significant increase in the operational use of RS and GIS in watershed management.[8] Much of this increase in operational use is due to the continued integration of RS, GIS, GPS (global positioning system), and Crop Model (CM) techniques.[4] These four technologies form a powerful, interrelated combination (Figure 1). In watershed management, there has been a continuous increase in the use of RS data to provide input to new GIS database, upgrade existing database, and monitor land use changes.

TABLE 2 Different Types of GIS

GIS	Origin	CPU		Data Model			Applications	
		PC	WS	Vector	Raster	Analysis	DTM	Network
ARC=INFO	ESRI	˅	˅	˅	˅	˅	˅	˅
MGE	Intergraph		˅	˅	˅	˅	˅	˅
Geo=SQL	Generation 5 Technology	˅	˅	˅	˅	˅	˅	˅
GFIS	IBM	˅	˅	˅	˅	˅	˅	˅
IDRISI	Clark University, U.S.A.	˅	˅	˅	˅	˅	˅	˅
GRASS	GRASS Information Center	˅	˅	˅	˅	˅	˅	˅
ERDAS	ISRI	˅	˅	˅	˅	˅	˅	˅
GRAMM++	IIT Bombay, India	˅	˅	˅	˅	˅	˅	˅

Source: Maguire[6] and Morehouse.[7]

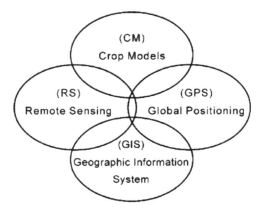

FIGURE 1 Relationship between RS, GIS, GPS, and CM.

In addition to providing input to GIS database, the RS data enables significant improvements in classification accuracies. Further, GPS capabilities provide effective cartographic control to the GIS database, and will enable field plots to be located efficiently and accurately in the data set.

The integration of GIS and CMs has proven very successful for alternative land use planning in the watershed. The important ones are DSSAT (Decision Support System for Agro-technology Transfer), AEGIS/WIN (Agricultural and Environmental Information System for Windows), IDSS (Intelligent Decision Support System), and DSSE (Decision Support System Engine). These simulation models are capable of predicting the potential yields of crops (rice, cassava, potato, sugarcane, sunflower, maize, wheat, barley, millet, sorghum, soybean, peanut, drybeans, tomato, and chickpea) in different physiographic units of a particular watershed. In addition, several GIS-based simulation models have been developed for natural resources management and are used in watershed management.[9] The important ones are SNAP (Scheduling and Network Analysis Program), LUCAS (Land Use Analysis System), and LANDSIM and LANDIS (Land Information System). All these models involve database (nonspatial), GIS, model base (simulation analysis), and a GUI (Graphic User Interface).

Challenges and Opportunities Ahead

Rapid advances in RS and GIS technologies can bring about quantum leap in watershed management. Yet the challenge lies in assuring that RS and GIS technologies continue to serve the practitioners and users, but not vice versa.

However, GIS and RS are both a panacea and a Pandora's box. These are panacea because of the promise to meet the challenges of resource inventory and monitoring, and planning and policy analysis. These are Pandora's box because of the numerous pitfalls of using the tools wrongly, capturing the data poorly, miscommunicating information, conveying incorrect results, and overselling the capabilities.

The dilemma can be addressed by a careful integration of RS, GIS, and other simulation models in watershed management and with due consideration of the following:

- Designing the classification system as totally exhaustive, mutually exclusive, and hierarchical
- Determining the temporal and spatial scale of the watershed by accommodating GIS and RS at multiple scales without compromising flexibility and quality of a project
- Identifying the appropriate data sources (video, aerial photography, satellite imagery, airborne scanner, etc.), for different land cover
- Assessing and reporting the accuracy of the data needs
- Limiting the scope of the project according to the budget and schedule
- Standardizing the formats needed for the exchange of information across projects and eliminating duplication

Future changes may include the following:

1. Developing effective techniques of integration and analysis of data from various sources such as AVHRR and Landsat TM or SPOT data or Landsat TM plus satellite radar data.
2. Converting research into operational applications in watershed management.
3. Developing effective expert system to assist the analyst.
4. Educating and training the user community in principles and theory of these technologies so that they can use these powerful tools wisely, appropriately, and effectively.

Anticipated developments and opportunities in these technologies comprise the following:

- Economic availability of satellite optical sensor data with improved spatial, spectral, radiometric, and temporal resolutions
- Availability of operational multifrequency, multipolarization synthetic aperture radar data from satellite altitudes
- Improvement in computer storage and processing capabilities
- Better understanding and use of combined data from optical, microwave, and other remote sensors
- Integration of RS, GIS, GPS, and CM technologies
- Increased use of expert systems for data analysis

These developments will improve the quality and characteristics of the data and analytical capabilities in the area of watershed management. A combination of knowledgeable resource managers and practical farmers with the technological tools and data available to them holds great promise for identifying effective watershed management technologies.

Conclusions

A variety of information about the characteristics and condition of the area is needed for judicious management of watershed. Aerial photography has been used since the 1950s for obtaining information on soils, land use, vegetation, slope, runoff, erosion, etc. The advent of space research through satellite, multispectral scanner, and radar data, and quantitative analytical techniques of processing such data has increased the array of data, analysis procedure, and results that can be obtained using RS capabilities.

During the 1980s, advances in computer hardware for processing and data storage catalyzed the development of software for handling spatial and image digital data. These technologies played an important role in the development of GIS for natural resource management, especially in preparing composite map superimposing the data and image files.

During the 1990s, significant development in GIS technology and integration of RS, GIS, GPS, and Crop Simulation Model techniques have created additional complexity and opportunities of using various data sources and analysis techniques for obtaining the information needed by the resource managers. Yet, there are some critical issues that need to be addressed for integration of RS, GIS, and other technologies to obtain information needed for sustainable natural resource management through watershed development.

References

1. Dhruva Narayana, V.V.; Sastry, G.; Patnaik, U.S. *Watershed Management*; Indian Council of Agricultural Research: New Delhi, 1990.
2. Hall, G.F. Adaptation of NASA remote sensing technology for regional-level analysis of forested ecosystem. In *Remote Sensing and GIS in Ecosystem and Management*; Sample, V.A., Ed.; Island Press: Washington, D.C., 1994; 287–314.
3. Hoffer, M.R. Challenges in development and applying remote sensing to ecosystem management. In *Remote Sensing and GIS in Ecosystem Management*; Sample, V.A., Ed.; Island Press: Washington, D.C., 1994; 25–40.
4. Hoogenboom, G.; Wilkins, P.W.; Tsuji, G.Y. *Discussion Support System for Agrotechnology Transfer*; University of Hawaii: Honolulu, Hawaii, 1999; Vol. 4.
5. JSRS Remote Sensing Note. Japan Association on Remote Sensing, University of Tokyo: Japan, 1998.
6. Maguire, D.J. The Raster GIS design model—a profile of ERDAS. Comput. Geosci. **1992**, *18* (4), 463–470.
7. Morehouse, S. The ARC/INFO geographic information system. Comput. Geosci. **1992**, *18* (4), 435–441.
8. Murai, S. *GIS Work Book*; Japan Association of Surveyors: Tokyo, Japan, 1996.
9. Sample, V.A. *Remote Sensing and GIS in Ecosystem and Management*; Island Press: Washington, D.C., 1994.
10. Singh, G.; Mittal, S.P.; Dhruva Narayana, V.V.; Agnihotri, R.C. *Watershed Management Plan (Vil. Bunga)*; Central Soil and Water Conservation Research and Training Institute: Dehra Dun, 1983.
11. Sohal, H.S. Soil and water conservation measures for sediment control in Sukhana Lake, Chandigarh. In *Fifty Years of Research and Sustainable Resource Management in Shiwalik*; Mittal, S.P., Aggarwal, R.K., Samra, J.S., Eds.; Central Soil and Water Conservation Research and Training Institute Research Center: Chandigarh, India, 2000; 259–266.

49

Wetlands: Conservation Policy

Introduction 497
Relationship between Policy and Wise Use 497
What Are Wetland Policies, and Are They Needed? 498
Guidelines 498
    Implementation Strategies • Global Review of the Status of Wetland Policies
    and Strategies • Policies
Conclusions 501
References 501

Clayton Rubec

Introduction

Meeting the challenge of conserving wetlands requires comprehensive national policies to provide a foundation for domestic action and a framework for international and national cooperation. Such policy is valuable as countries seek to address the management and habitat requirements for wildlife and other natural resources, such as soil and water, as well as human needs. Implementation of a national wetland policy is a key feature of the wise use principles of the Convention on Wetlands of International Importance (the Ramsar convention). However, such policy remains an elusive goal for many of the 145 nations that today (May 2005) are contracting parties to this global environmental treaty.

Responding to recommendations by the Convention, in 1998–1999 the author led a team of writers in preparing guidelines for developing and implementing national wetland policies.[1] These Policy Guidelines complement the convention's guidance on wetland legislation.[2] The following entry provides highlights of the guidelines and reports on the status of wetland policy development around the world, effective mid-1999.

Relationship between Policy and Wise Use

The wise use principles are a hallmark of the Ramsar convention. "Wise use" applies not only to sites listed as wetlands of international importance (as of May 2005 covering over 125 million hectares at 1429 sites), but also to all wetlands in the territory of contracting parties. These *principles* help contracting parties improve institutional and organizational arrangements, address legislative and policy needs, increase knowledge and awareness of wetland values, inventory and monitor the status of wetlands, identify program priorities, and develop action plans for specific sites as components of a national wetland policy.

The formulation of national wetland policies sometimes involves a lengthy and complex process. Political, jurisdictional, institutional legal, and financial constraints affect policy formulation in addition to social and economic factors that continue to contribute to wetland loss while the policy process is under way.

What Are Wetland Policies, and Are They Needed?

A policy can simply be a document. However, making is also a process involving consensus building, encapsulation of ideas and commitments, implementation, accountability, and review complemented by legislation, strategies, and operational programs. It is a mechanism for an administration to capture the public will or mandate on an issue, and refine it with its own vision.

A national wetland policy is nationwide in scope but it may be developed at several levels of government. In Australia and Canada, for example, both the federal government and state/provincial governments have developed wetland conservation policies. This reflects the federal nature of these two nations, wherein constitutional authority for natural resources management (including wetlands) is divided between the levels of government.

Wetlands are seldom explicitly covered in other natural resource management policies such as for water, soil, forest, land, or agriculture at a national level. Development of a "stand-alone" wetland policy and/or strategy can be an important step in recognizing and solving wetland problems. A wetland policy recognizes wetlands as ecosystems requiring different approaches not masked under other sectoral management objectives. Articulation of goals and objectives for these ecosystems identifies clear responsibilities of the government and a public expectation that the government will deliver these commitments.

Wetland policy objectives need to focus on a variety of themes as they become the image of the policy. However, practical implementation of the Policy may result in only one or two of these objectives receiving the greatest public attention. For example, Canada's announcement of its federal wetland policy in 1992 contained seven objectives but "no net loss of wetland functions" has proven to be its catch phrase.

Guidelines

The Guidelines review the key steps and issues that may arise in both developing and implementing a National Wetland Policy. These include over 20 detailed sections defining the purpose of such an initiative, organizing a suitable process, deciding how to present the content of the policy document, and developing strategies for implementation and monitoring. The text is complemented by seven wetland policy essays: 1) *defining stakeholders'*, 2) *consultations;* 3) *wetland policies within a federal state;* 4) *sectoral policies and legislation;* 5) *compliance strategies;* 6) *role of nongovernment organizations;* and the 7) *development and coordination process.* The guidelines provide a reference against which all nations can review their wetland action plans and strategies at the national level.

A team of contributors with governmental or nongovernmental work experience and expertise in wetland policy development prepared the guidelines. The team included writers from Ramsar national authorities in Australia, Canada, Trinidad and Tobago, Uganda, and the United States of America. Contributors from Bird Life International, University of Massachusetts, IUCN Environmental Law Centre, and Wetlands International were also involved.

Implementation Strategies

A National Wetland Policy includes specific implementation strategies that demonstrate the priorities of the government, but also fosters the cooperation and involvement of other interests. Linkages between these strategies and national water, soil, biodiversity, and sustainable development policy initiatives are explored in the guidelines.

An analysis of the strategies used in selected National Wetland Policies is summarized in Table 1. These include the policies/action plans of Australia, Cambodia, Canada, Columbia, Costa Rica, Finland, France, Jamaica, Malaysia, Peru, Trinidad and Tobago, and Uganda. These initiatives have many common strategic approaches including: a) ensuring public awareness and education; b) developing

TABLE 1 Implementation Strategies in Selected National Wetland Policies

Country	1	2	3	4	5	6	7	8	9	10	11	12
Australia	X	X	X	X	X	X						
Cambodia		X	X		X	X		X		X		X
Canada	X		X	X	X	X	X					
Colombia		X	X		X	X		X	X		X	X
Costa Rica	X		X		X			X	X	X		X
Finland	X	X		X	X		X				X	X
France		X	X		X						X	X
Jamaica		X					X	X	X			
Malaysia		X	X		X	X	X	X	X	X		X
Peru		X			X	X		X	X			X
Trinidad and Tobago	X		X	X	X		X	X				
Uganda		X	X		X	X		X	X		X	X

Policy strategies:
  1. Management of national wetland networks
  2. Integration with other policies such as water, soil and forests
  3. Public awareness and education
  4. Partnerships
  5. Science, monitoring, assessment, and research
  6. International commitments
  7. Managing special sites
  8. Administration and institutions, capacity building
  9. Enforcement, regulation, and legislation
  10. Financial mechanisms
  11. Restoration of degraded sites
  12. Sustainable use and conservation

cooperation and partnerships between levels of government from national to local; c) developing and supporting legislation and interrelated land and water use policies and programs; d) implementing wetland site management responsibilities; e) developing a sound basis for the policy through scientific research and expertise; f) developing institutional and financial capacity for policy implementation; and g) meeting international commitments. These strategies have been drafted to evoke a clear vision and acceptance across the nation.

Global Review of the Status of Wetland Policies and Strategies

Significant progress is evident globally in the development of National Wetland Policies since the Ramsar Convention focused attention on this issue in 1987.[1,3,4] Meetings of the contracting parties every three years allow regular review of the status of wetland policies. The Contracting Parties last met in November 2002 when there were 123 Contracting Parties while now (May 2005) there are 145 Contracting Parties.

Policies

As of November 2002, 60 (56%) of the 107 Ramsar Contracting Parties that submitted national reports to the Eighth Meeting of the Contracting Parties to the Convention on wetlands indicated that they were engaged in development or implementation of a National Wetland Policy. Between 1987 and 2002, the number of nations with a National Wetland Policy officially adopted grew from 0 to 41. An additional 19 nations indicated that a National Wetland Policy was in draft or under consideration. However, 47 (44%)

TABLE 2   Evolution of Ramsar Convention on National Wetland Policies/Strategies[a]

Status of National Wetland Policies or Strategies	1987 Regina COP3	1990 Montreux COP4	1993 Kushiro COP5	1996 Brisbane COP6	1999 San Jose COP7	2002 Valencia COP8
National wetland policies Policy/strategy adopted	0	0	3	6	12	41
Policy/strategy in draft or under consideration	2	5	12	21	30	19
No policy/strategy activity reported	15	40	36	65	72	47
Number of national reports tabled at this COP	17	45	51	92	98	107

[a] As of May 2005, the Ramsar Convention had 145 Contracting Parties and will next report on this issue in November 2005.

TABLE 3   Summary of National Wetland Policies and Strategies by Ramsar Regions (2002)

Ramsar Region	National Wetland Policy or Strategy Adopted by Government	National Wetland Policy or Strategy in Preparation or under Consideration
Africa	Congo Rep., Côte d'Ivoire, Senegal, South Africa, Uganda	Algeria, Botswana, Burkina Faso, Chad, Comoros, Dem. Rep. Congo, Egypt, The Gambia, Ghana, Guinea, Kenya, Malawi, Mali, Morocco, Namibia, Niger, Togo, Zambia
Asia	Indonesia, Japan, Thailand, Vietnam	Bangladesh, Cambodia, P.R. China, Georgia, India, Rep. Korea, Malaysia, Mongolia, Philippines, Russia,[a] Turkey
Europe	Belgium, Bulgaria, Denmark, Estonia, Finland, France, Greece, Iceland, Malta, Monaco, Netherlands, Norway, Romania, Sweden, Switzerland, United Kingdom	Austria, Belarus, Croatia, Czech Rep., Germany, Hungary, Ireland, Italy, Latvia, Lithuania, Poland, Portugal, Russia,[a] Slovak Rep., Slovenia, Ukraine, F.R. Yugoslavia
Neotropics (Central and South America and Caribbean)	Colombia, Costa Rica, Jamaica, Peru, Trinidad and Tobago	Argentina, Bahamas, Chile, Ecuador, Guatemala, Honduras, Nicaragua, Panama, Paraguay, Venezuela
North America	Canada, United States	Mexico
Oceania (Australia and Pacific)	Australia, New Zealand	—

[a] Russia straddles Asia and Europe.

of the Ramsar Contracting Parties did not yet report any actions being taken in support of National Wetland Policy development. A number of nations, particularly those with a commonwealth or federal make-up, reported wetland policies and strategies at the sub-national level also. National Wetland Policies have also been called a "National Wetland Strategy" or "Action Plan."

Table 2 summarizes the status of the development and adoption of National Wetland Policies from 1987 through November 2002. This table was developed by reviewing the reports and conference papers that summarize Convention activities every three years by each country. In 1987, only two nations indicated that they were involved in developing any sort of national wetland policy/strategy. By 2002, this number grew to at least 60 nations. Table 3 summarizes the countries by Ramsar's regions (Africa, Asia, Europe, Neotropics, North America, and Oceania) that have either a *national wetland policy* or *national wetland strategy* adopted, being drafted or considered up to May 2005.

Conclusions

Development and implementation of a national wetland policy in the 145 countries that have acceded to the Ramsar Convention is proceeding throughout the World. The Convention's Wise Use Principles and Guidelines on National Wetland Policy,[1] complemented by guidelines for wetland legislation,[2] are effective tools in fostering the completion and use of National Wetland Policies and Strategies as important cornerstones of this Convention. At least 56% of the Ramsar Convention's Contracting Parties now are implementing or developing national wetland policies/strategies.

One of the most interesting aspects of the Ramsar Convention is its capacity to foster sharing of experience. Interchanges by wetland policy experts are now occurring internationally that involve short-term invited visits or sabbaticals, informal exchange of documents, confidential advice, and review of draft policy. Consultation workshops, working with non-government groups, meeting with senior government officials, exploring funding mechanisms, and assistance in drafting of text have also been involved. Experience gained in one nation's development of these policies can be shared and local expertise enhanced, filling a need among the Ramsar family.

References

1. Rubec, C.D.A.; Pritchard, D.; Mafabi, P.; Nathai-Gyan, N.; Phillips, B.; Mahy, M.; Lynch-Stewart, P.; Chew, R.; Cintron, G.; Larson, J.; Ramakrishna, S. *Guidelines for Developing and Implementing National Wetland Policies;* Report to COP7 Ramsar Convention. Handbook No. 2; Ramsar Bureau: Gland, Switzerland, 2000.
2. Shine, C.; Glowka, L. *Reviewing Laws and Insti.tutions to Promote the Conservation and Wise Use of Wetlands*; IUCN Environmental Law Centre Report to COP7 Ramsar Convention. Handbook No. 3; Ramsar Bureau: Gland, Switzerland, 2000.
3. Rubec, C.D.A. Status of national wetland policy development in ramsar nations. Proceedings, Sixth Meeting of the Conference of the Contracting Parties. Convention on Wetlands, Brisbane, Australia, 1996; Vol. 10/12A: 22–29.
4. Ramsar convention secretariat, National Wetland Policies in Ramsar Contracting Parties. Information Report to Eighth Meeting of the Contracting Parties to the Convention on Wetlands, Gland, Switzerland, 2002.

VI

ENT: Environmental Management Using Environmental Technologies

50

Irrigation Systems: Subsurface Drip Design

Introduction 505
System Design 506
    Site, Water Supply, and Crop • Lateral Type, Spacing, and Depth • Special
    Requirements
System Components 508
    Pumps, Filtration, and Pressure Regulation • Laterals and
    Emitters • Chemical Injection • Air Entry and Flushing
Operation and Maintenance 509
    Operation • Maintenance
Conclusion 510
References... 510

Carl R. Camp, Jr.
and Freddie L.
Lamm

Introduction

Subsurface drip irrigation (SDI) is generally defined as the application of water below the soil surface through emitters, with discharge rates in the same range as drip irrigation.[1] While this definition is not specific regarding depth below the soil surface, most SDI laterals are installed at a depth sufficient to prevent interference with surface traffic or tillage implements and to provide a useful life of several years as opposed to annual replacement of surface or near-surface drip laterals.

Development of drip irrigation accelerated with the availability of plastics following World War II, primarily in Great Britain, Israel, and the United States. SDI was part of drip irrigation development in the United States beginning about 1959, especially in Hawaii and California. While early drip irrigation products were relatively crude by modern standards, SDI devices were being installed in both experimental and commercial farms by the 1970s. As drip irrigation products improved during the 1970s and early 1980s, surface drip irrigation grew at a faster rate than SDI, probably because of emitter plugging problems and root intrusion. However, interest in SDI increased during the early 1980s, increased rapidly during the last half of the 1980s, and continues today, especially in areas with declining water supplies, with environmental issues related to irrigation, and where wastewater is used for irrigation. Initially, SDI was used primarily for sugarcane, vegetables, tree crops, and pineapple in Hawaii and California. Later, SDI use was expanded to other geographic areas and to agronomic and vines crops, including cotton, corn, and grapes. Currently in the United States, the major uses of SDI are for cotton, processing tomato, corn, and onions using various installation depths and other design aspects.[2]

SDI has the advantage of multiple-year life, reduced interference with cultural practices, dry plant foliage, and a dry soil surface. Multiple-year life allows amortization of the entire system cost over several years, often more than ten. There are a few systems in the United States that have achieved 20 years are more in longevity.[3] If all system components are installed below tillage depth, surface cultural practices can be accomplished with minimal concern for system damage. Dry soil surfaces can reduce weed growth in arid climates and may reduce evaporation losses of applied water. Because the plant canopy is not irrigated, the foliage remains dry, which may reduce the incidence of disease. SDI is also very adaptable to irregularly shaped fields and low-capacity water supplies that may provide design limitations with other irrigation systems.

The major disadvantages of SDI include system cost, difficulty in locating and repairing system leaks and plugged emitters, and poor soil water manipulation near the soil surface. Most system components are installed below the soil surface and are neither easy to locate nor directly observable. In a properly designed and managed SDI system, the soil surface should seldom be wet. Consequently, seed germination, especially for small seeds, can be very difficult.

SDI systems offer considerable flexibility, both in design and operation. For example, SDI systems can apply small, frequent water applications, often multiple times each day, to very specific sites within the soil profile and plant root zone. Fertilizers, pesticides, and other chemical amendments can be applied via the irrigation system directly into the active root zone, often at a modest increase in equipment cost. In many cases, the operational cost may be less than that for applying these chemicals via conventional surface equipment.

System Design

Site, Water Supply, and Crop

Design of subsurface drip systems is similar to that of surface drip systems, especially with regard to hydraulic characteristics.[3,4] Specific crop and soil characteristics are used in the design process to select emitter spacing and flow rate, lateral depth and spacing, and the required system capacity. Emitter properties and lateral location are influenced by soil properties such as texture, soil compaction, and soil layering because these affect the rate of water movement through the soil profile and the subsequent wetting pattern for each emitter.

The water supply capacity directly affects the design of an SDI system. The size of the irrigated field or zone is often determined by the water supply capacity. For example, in some humid areas, high-capacity wells are not available but multiple low-capacity wells can be distributed throughout a farm. Fortunately, the design of SDI systems can be economically adjusted to correspond to the field size and shape, to the available water supply capacity, and to other factors. Water supply quality should be tested by an approved laboratory before proceeding with system design. This information is needed for the proper design and management of the water filtration and treatment system. Some water supplies require frequent or intermittent injection of acids and/or chlorine. Other saline and/or sodic water supplies may require treatment or special management. As water supplies become more limited, treated wastewater is becoming an increasingly important alternative water supply that can be applied through SDI systems. Camp[5] listed several reports that emphasized water supplies (saline, deficit, and wastewater) for SDI systems.

The SDI system is usually designed to satisfy peak crop water requirements, which vary with specific site, soil, and crop conditions. When properly designed and managed, SDI is one of the most efficient irrigation methods, providing typical application efficiencies exceeding 90%. In comparison with other methods of irrigation, reported yields with SDI were equal to or greater than those with other irrigation methods. Generally, water requirements with SDI are similar or slightly lower than those with other irrigation methods. In some cases, water savings of up to 40% have been reported.[5] However, unless more specific information is available, it is usually best to use standard net water requirements for the location when designing SDI systems.

Lateral Type, Spacing, and Depth

SDI lateral depth for various cropping systems is normally optimized for prevailing site conditions and soil characteristics.[3,5] Where systems are used for multiple years and tillage is a consideration, lateral depths vary from 0.20 to 0.70 m. Where tillage is not a consideration (e.g., turfgrass, alfalfa), depth is sometimes less (0.10–0.40 m). Lateral spacing also varies considerably (0.25–5.0 m), with narrow spacing used primarily for turfgrass and wide spacing used for vegetable, tree, or vine crops. In uniformly spaced row crops, the lateral is usually located under either alternate or every third midrow area (furrow). For crops with alternating row spacing patterns, the lateral is located about 0.8 m from each row, usually in the narrow spacing of the pattern.

The lateral should be installed deep enough to prevent damage by tillage or injection equipment but shallow enough to supply water to the crop root zone without wetting the soil surface. Generally, laterals in SDI systems are placed at depths of 0.1–0.5 m, at shallower depths in coarse-textured soils, and at slightly deeper depths on finer-textured soils. The selection of emitter spacing and flow rate is influenced by crop rooting patterns, lateral depth, and soil characteristics. It is also desirable to select an emitter spacing that provides overlapping subsurface wetted zones along the lateral for most row crops. For wider spaced crops such as trees and vines, emitters are normally located near each plant and may have wider spacings that do not provide overlapping patterns. Lateral spacing is determined primarily by the soil, crop, and cultural practice and should be narrow enough to provide a uniform supply of water to all plants.

Special Requirements

Site topography must be considered in system design and selection of components as with any irrigation system, but SDI is suitable for most sites, ranging from flat to hilly. For sites with considerable elevation change, especially along the lateral, pressure-compensating emitters should be used.

Two special design requirements for SDI systems, which are significantly different from those for surface drip systems, are the needs for flushing manifolds and air entry valves. Flushing manifolds are needed to allow frequent flushing of particulate matter that may accumulate in laterals. Air relief valves are needed to prevent aspiration of soil particles into emitter openings when the system is depressurized. These valves must be located in sufficient number and at the higher elevations for each lateral or zone to prevent negative pressures within the laterals.

Emitter plugging caused by root intrusion is a major problem with some SDI systems but can be minimized by chemicals, emitter design, and irrigation management. Chemical controls include the use of herbicides, either slow-release compounds embedded into emitters and filters or periodic injection of other chemical solutions (concentrated and/or diluted) into the irrigation supply. Periodic injection of acid and chlorine for general system maintenance can also modify the soil solution immediately adjacent to emitters and reduce root intrusion. In some cases, emitters plugged by roots may be cleared via injection of higher concentrations of chemicals, such as acids and chlorine.

Emitter design may also affect root intrusion. Smaller orifices tend to have less root intrusion but are more susceptible to plugging by particulate matter. Some emitters are constructed with physical barriers to root intrusion. Root intrusion appears to be more severe when emitters are located along dripline seams, which can be an area of preferential root growth. However, root intrusion problems appear to be greater for emitters, driplines, and porous tubes that are not chemically treated.

Irrigation management can also be used to influence root intrusion by controlling the environment immediately adjacent to the emitter. High-frequency pulsing that frequently saturates the soil immediately surrounding the emitter can discourage root growth in that area for some crops but not others. Conversely, deficit irrigation sometimes practiced to increase quality or maturity, or to control vegetative growth, can increase root intrusion in lower rainfall areas because of high root concentrations in the soil zone near emitters.

System Components

Pumps, Filtration, and Pressure Regulation

Pump requirements for SDI are similar to those for other drip irrigation systems, meaning water must be supplied at a relatively low pressure (170–275 kPa) and flow rate in comparison to other irrigation methods. Because of the flushing requirement for SDI systems, a flow velocity of about 0.3 m/s must be achievable, either by reducing the zone size while using the same pumping rate or by increasing the pumping rate without changing the zone size.

Water filtration is more critical for SDI systems than for surface drip systems because the consequences of emitter clogging are more severe and more costly. Generally, the better the water quality, the less complex the filtration system required. Surface and recycled or wastewater supplies require the most elaborate filtration systems. However, good filtration is the key to good system performance and long life and should be a major emphasis in system design. Filtration systems range from simple screen filters for relatively clean water to more elaborate and complex disc and sand media filters for poorer quality water.

The pressure regulation requirement in SDI systems is similar to that in surface drip systems. When non-pressure-compensating emitters are used on relatively flat areas, pressure is typically regulated within the system supply lines (main and/or submain) using pressure-regulating valves. When pressure-compensating emitters are used, typically on more hilly terrain, the pressure within the system supply lines is controlled at a higher, but more variable, pressure that is within the recommended input pressure range for the emitters used. Water pressure should be monitored on a regular basis at the pump or supply port and at various locations throughout the SDI system, especially at the both ends of laterals.

Laterals and Emitters

Many types of driplines have been used successfully for SDI and most have emitters installed as an integral part of the dripline. This is accomplished by one of three methods: (1) molded indentions created during the fusing of dripline seams, (2) prefabricated emitters welded inside the dripline, or (3) circular prefabricated in-line emitters installed during extrusion. Regardless of the emitter used, dripline wall thickness and expected longevity must be considered along with other design factors in selecting the lateral depth. Flexible, thin-walled driplines typically are installed at shallow depths and normally have a shorter expected life. Thicker-walled, flexible driplines have been used successfully for several years provided they are installed deep enough to avoid tillage, cultivating, and harvesting machinery but shallow enough to prevent excessive deformation or permanent collapse of the dripline by machinery or soil weight. Rigid tubing with thicker walls can be installed at deeper depths without deformation and is often used on perennial crops or annual crops for longer time periods (>10 years). Some driplines are impregnated with bactericides or other chemicals to reduce the formation of sludge or other material that could plug emitters.

Chemical Injection

Subsurface drip systems offer the potential for precise management of water, nutrients, and pesticides if the system is properly designed and managed. The marginal cost to add chemical injection equipment is generally competitive with other, more conventional application methods. Water and fertilizers can be applied in a variety of modes, varying from multiple continuous or pulsed applications each day to one application in several days. Choice of application frequency depends upon several factors, including soil characteristics, crop requirements, water supply, system design, and management strategies. If labeled for the purpose, some systemic pesticides and soil fumigants can be safely injected via SDI systems. Use of the SDI system for chemical applications has the potential to minimize exposure to workers and the environment, reduce the cost of pesticide rinse water disposal, and improve precision of application to the desired target (root pests). Injection of other chemicals, such as acids and chlorine, is often required to clean and maintain emitters in optimum condition. However, a high level of management with system automation and feedback control is required to minimize chemical movement to the groundwater when chemicals are used.

Air Entry and Flushing

Air entry valves must be installed at higher elevations in SDI systems to prevent the emitter from ingesting soil particles that could plug emitters when the system is depressurized. Typically, air entry valves are located in water supply lines near the head works or control station and in both the supply and flushing manifolds. In some cases, such as turf or pasture, air entry valves may be installed below the soil surface and enclosed within a protective box. Flushing valves installed on the flushing manifold are required to control periodic system flushing.

Operation and Maintenance

Operation

SDI systems can be operated in several modes, varying from manual to fully automated. Overall, SDI systems are probably more easily automated than many other types of irrigation. One reason is that most are controlled from a central point using electrical or pneumatic valves and controllers that vary from a simple clock system to microprocessor systems, which are capable of receiving external inputs to initiate and/or terminate irrigation events.

Irrigation scheduling is as important for SDI systems as for any other type of irrigation. Choosing to initiate an irrigation event and how much water to apply during each event depends on crop, soil, and irrigation system type and design. Factors that affect those decisions include soil water storage volume, sensitivity of the crop to water stress, irrigation application rate, weather conditions, and water supply capacity. Camp[5] discussed several irrigation scheduling methods that have been used successfully with SDI. However, the important point is that a science-based scheduling method can conserve the water supply and increase profit.

If seed germination and seedling establishment and growth are critical, especially in arid climates when initial soil water content is not adequate, either sprinkler or surface irrigation is often used for germination. However, the need for two systems increases cost and decreases economic return. If subsurface drip is used for germination, an excessive amount of irrigation is often required to wet the seed zone for germination, which could result in excessive leaching and off-site environmental effects as well as increased cost. Surface wetting can also occur when the emitter flow rate exceeds the hydraulic conductivity of the soil surrounding the emitter, but wetted areas are often not uniform.

Because salts tend to accumulate above the lateral, high salt concentrations may occur between the lateral and soil surface in arid areas where rainfall is not available to leach the salts downward. Salts may also be moved under the row when laterals are placed under the furrow.[6] Supplemental sprinkler irrigation may be required in some areas to control salinity if precipitation is inadequate for leaching during several consecutive years.

Maintenance

Often, SDI systems must have a long life (>10 years) to be economical for lower value crops. Thus, appropriate management strategies are required to prevent emitter clogging and protect other system components to ensure proper system operation. Locating and repairing/replacing failed components is much more difficult and more expensive with SDI systems than with surface systems because most system components are buried, difficult to locate, and cannot be directly observed by managers. Consequently, operational parameters such as flow rate and pressure must be measured frequently and used as indicators of system performance. Good system performance requires constant attention to maintain good water quality, proper filtration, and periodic system flushing to remove particulate matter that could plug emitters. Periodic evaluation of SDI system performance in relation to design performance can identify problems before they become serious and significantly affect crop yield and quality.

Conclusion

Although there is a general consensus that use of SDI is increasing on a worldwide basis, this growth is difficult to document. In the 10-year period (2003–2013), SDI in the United States increased by 89% from 164,017 to 310,361 ha according to USDA-NASS irrigation surveys.[2] Use of SDI should continue to increase in the future, depending primarily upon the economic and water conservation benefits in comparison to other irrigation methods. As water supplies become more limited, the high application efficiency and water-conserving features of SDI should increase its application. Also, SDI offers potential advantages such as reduced odors and exposure to pathogens when using recycled domestic and animal wastewater. The SDI technology offers the capability to precisely place water, nutrients, and other chemicals in the plant root zone at the time and frequency needed for optimum crop production. With proper design, installation, and management, SDI systems can provide excellent irrigation efficiency and reliable performance with a system life of 10–20 years.

References

1. ASABE. *Soil and Water Terminology*. 49th Ed.; ASAE: St. Joseph, MI, 2001; 970–990.
2. Lamm, F. R. 2016. Cotton, tomato, corn, and onion production with subsurface drip irrigation – A review. *Trans. ASABE 59* (1), 263–278.
3. Lamm, F.R. and C.R. Camp. 2007. Subsurface drip irrigation. In *Microirrigation for Crop Production - Design, Operation and Management*. F.R. Lamm, J.E. Ayars, and F.S. Nakayama (Eds.), Elsevier Publications: Amsterdam; 473–551.
4. ASAE. *Design and Installation of Microirrigation Systems*. 49th Ed.; ASAE: St. Joseph, MI, 2001; 903–907.
5. Camp, C.R. Subsurface drip irrigation: A review. *Trans. ASAE* **1998**, *41* (5), 1353–1367.
6. Ayars, J.E.; Phene, C.J.; Schoneman, R.A.; Meso, B.; Dale, F.; Penland, J. Impact of bed location on the operation of subsurface drip irrigation systems. In Microirrigation for a Changing World, *Proceedings of the Fifth International Microirrigation Congress*, Orlando, FL, April 2-6, 1995; Lamm, F.R., Ed.; ASAE: St. Joseph, MI, 1995; 141–146.

51

Recent Approaches to Robust Water Resources Management under Hydroclimatic Uncertainty

J. Pablo Ortiz-
Partida, Mahesh
L. Makey,
and Alejandra
Virgen-Urcelay

Introduction 511
Robust Reservoirs Management 512
Integrated Framework 513
Moving Forward 513
Conclusions and Future Needs 514
References 515

Introduction

Managing water resources under hydroclimatic uncertainty is a primary focus with respect to water resources management (Brown et al. 2015; Poff et al. 2015). There is a scientific consensus that climate change is going to impact current water systems in a variety of ways. For example, expected impacts on California's water resources include increase in evaporation rates and water demands (Hayhoe et al. 2004); reduction in crop yield as a consequence of higher temperatures (Pathak et al. 2018); less precipitation falling as snow and more as rain translating into wetter wet seasons and drier dry seasons (Mallakpour et al. 2018), which may force operators to lower reservoir levels to allow for space and reduce the risk of flood events; increase in water temperatures affecting aquatic and riparian ecosystem (Poff et al. 2012); and sea-level rise, which will increase seawater intrusion into coastal aquifers (Ferguson and Gleeson 2012). Meanwhile, there is, however, uncertainty on their magnitude and extent, which raise the question of how we can develop management strategies that account for the range of possible future alternatives.

Water systems modeling has been crucial to answer such questions because of the advancement of computational power. Water systems modeling was often deterministically developed, considering some variables such as temperatures, precipitation, streamflow, or water demands as known parameters. Despite recognizing their hydrologic variability, the only available data are often treated as a statistical description of hydrologic variables resulting into unreliable forecasts. Deterministic models treat stochastic parameters as known quantities reducing – to some extent – the complexity of the model (Chen et al. 2018). Depending on the application of the model, such assumptions can be sustained. Nevertheless, the simplifications cannot retain all the essential characteristics of the original data and may lead to unsatisfactory results given the complex behavior of a system (Puente et al. 2018). Climate change is bringing extreme hydroclimatic events of low probabilities but high impacts that – as noted – deterministic approaches do not capture well (Farmer and Vogel 2016; Philbrick and Kitanidis 1999). Thus, there is a need for integrating the seemingly random behavior of precipitation, temperature, streamflow, and other sometimes-unknown variables (e.g., soil properties, water quality) for developing water management strategies suitable for a variety of future socioeconomic and climatic alternatives.

This chapter describes recent approaches from the scientific literature that incorporate hydroclimatic uncertainty to develop robust water management strategies. This chapter is divided into three sections. The first section describes the development of robust reservoir operations, a main technical area in water resources. The second section describes an approach to achieve water systems sustainability that goes beyond common performance criteria (e.g., economic, reliability) by integrating stakeholder-defined performance metrics within the modeling framework. The last section outlines how to move forward and incorporate uncertainty into the decision-making process, a crucial component pertaining to real-life activities.

Robust Reservoirs Management

Reservoirs around the world provide storage to supply water for urban, agriculture, and industry sectors; flood protection; hydropower; and support recreational activities. However, in some cases, these benefits came at the cost of environmental and social degradation (WCD 2000). Leaving aside the controversy of building dams, it is important to recognize that most of the current reservoirs will persist despite their contribution to the degradation of river ecosystems. Given the uncertainty and variability of hydroclimatic conditions under climate change, droughts and floods may worsen, turning reservoir management vital to prevent or reduce expected drought and flood events (Cristina and Tullos 2017).

Modifying reservoir operations offers an opportunity for mitigating hydrologic responses to climate change as current operation rules tend to be static and based on historical inflows and outflows observations. Given the new recurrence and magnitude of hydroclimatic events, some reservoir operation rules are no longer suitable for managing drought periods and floods or for reaching the full potential of a reservoir (Howard 1999; Moy et al. 1986). New reservoir operations that incorporate projected hydroclimatic variability may be an effective strategy for reducing the impacts of changes in water supplies and demands (Vonk et al. 2014). Reservoir operations developed with new available tools and information contribute to increased resilience of water management systems and ecosystem restoration.

Alternative approaches to improve long-term reservoir operations have utilized multi-objective genetic programming (Ashofteh et al. 2015), machine learning (Herman and Steinschneider 2018), and stochastic modeling (Ermoliev et al. 2019) to incorporate climate change projections and a broader set of observations and forecasts. Such approaches link hydroclimatic observations and predictions with water resources management decisions to improve tradeoffs among human and environmental water supply and flood management objectives.

Another recently proposed framework presents a two-stage stochastic optimization model that maximizes regional economic benefits as a function of reservoir deliveries to water users (Ortiz-Partida et al. 2019). The first-stage decisions allocate water based on an expected deficit for the whole system, while in the second stage, the now known deficit is allocated in a way that maximizes economic benefits. This framework was applied to the single reservoir system of Luis L. Leon dam in the Rio Conchos, the main tributary to the Rio Grande of North America. The model result is a set of robust monthly reservoir releases that can cover most of the water demands – including the environment – and reduce the frequency and magnitude of flood events under a wide range of hydroclimatic conditions.

Results from these different approaches suggest that robust operations could improve long-term planning by making the water system more reliable and resilient, and less vulnerable to extreme hydroclimatic events. However, these technically complex approaches tend to have bigger scale implications and usually lack stakeholder involvement, making them difficult to implement.

Integrated Framework

Attaining water sustainability needs to go beyond historically used criteria (e.g., economic performance) and integrate socioeconomic, environmental, and climate components. As such, decision makers are seeking for robust decisions with satisfactory performance across a large range of plausible futures. To make plans that are flexible and enhance long-term decision-making, various model-based frameworks have been proposed (Kwakkel et al. 2016). One of the frameworks that distinguishes from others is decision scaling (DS). DS integrates stakeholder-defined performance metrics that allow for inclusion and empowerment to foster collaboration across historically conflicting perspectives (Brown et al. 2011, 2012). In addition, DS attempts to find alternative solutions for multi-objective water resources systems under a variety of nonstationary hydroclimatic conditions and modeling uncertainties. Steinschneider et al. (2015) advanced previous applications of DS to identify long-term planning alternatives with the use of Monte Carlo simulations and by quantifying uncertainty in each stage of the modeling chain. Such a novel approach efficiently selects future climate realizations to assess the effects of hydrologic modeling uncertainty. Its limitation lies on a computational burden and required high-performance computing facilities to handle large multi-reservoir systems.

DS has been further developed by Poff et al. (2015) into the eco-engineering decision-scaling (EEDS) framework that explicitly and quantitatively explores tradeoffs across management actions under hydroclimatic uncertainty and incorporates ecological and engineering performance metrics. EEDS has five main steps that include (1) developing a clear definition of performance criteria, unacceptable thresholds, and management options; (2) building a systems model that relates hydroclimatic variables with the defined performance criteria; (3) performing a vulnerability analysis under a variety of hydroclimatic and social alternatives to discard unacceptable options and identify robust management strategies; (4) evaluating management options on the basis of their ability to satisfy both engineering and ecosystem performance under uncertain future conditions; and lastly, (5) selecting an alternative or developing new performance criteria based on political and institutional feasibility through converting the framework into an iterative process. As a case study around Coralville Dam, Poff et al. (2015) identified the most robust management strategy that would satisfy economic and ecological goals (Figure 1).

In truly achieving robustness and avoiding undesirable consequences, Herman et al. (2015) argue that (1) alternative decisions should be searched and not prespecified, (2) uncertainty should be discovered with sensitivity analysis and not assumed, and (3) stakeholders should carefully select their satisfactory measure of robustness based on their problem-specific performance requirements.

Moving Forward

Robust water management strategies need to go beyond water data and incorporate uncertainty in socioeconomic, environmental, and climate variables to be capable of assessing a broader range of future alternatives. Moreover, other information related to global climate change, population growth, land-use change, globalization, or even political stability makes long-term water resources planning difficult. Projected information of these variables is pertinent to long-term sustainable water resources strategies. A lack of information regarding interactions among these variables often misleads outputs from models entailing uncertainties to be adopted by water managers, planners, and decision makers. A first step towards creating robust water management strategies is to improve data collection and monitoring, which can be costly, suggesting that benefits and tradeoffs between investments in monitoring and development should be considered properly (Wada et al. 2016).

Adequate performance of long-term water resources management highly depends on a modeling chain that links technical, socioeconomic, environmental, and climate components. This is because each stage of the modeling chain is vulnerable to uncertainty. Such uncertainties include nonstationarity in future climate and potential shifts in large-scale synoptic circulation (Sheridan and Lee 2010) as

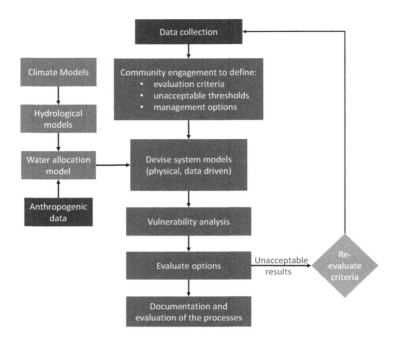

FIGURE 1   Eco-engineering decision scaling (EEDS). (Adapted from Poff et al. (2015).)

well as nonstationarity in water demands (Kang and Lansey 2012). As observed and experienced, it is difficult to propose long-term robust approaches without better prediction of climatic shifts and water demand variability (Korteling et al. 2013; Matrosov et al. 2013; Prudhomme et al. 2010; Turner et al. 2014). In fact, these frameworks attempt to bring the pieces together for decision-making rather than accounting each component individually.

To meet diverse water demands, Groves et al. (2019) recently applied "robust decision making (RDM)" to (a) identify water management strategies to reduce vulnerabilities in the Colorado River Basin and (b) develop robust investment strategies for the Green Climate Fund that reduce greenhouse gas emissions. This work suggests that RDM addresses long-term policy challenges associated with multifaceted, nonlinear, natural, and socioeconomic systems for defining policy pathways in a beneficial way. Existing water system models still need to address issues related to current human impact modeling and associated indicators and their limitations in representing regional water management. In this way, one can integrate land use and climate interaction to integrate human behavior into a large-scale modeling and to fill the gap among human–water management information. Advancing water system models coupled with human–water systems over a wide range of temporal and spatial scales will be beneficial for water science community and for understanding the climate and Earth system science communities.

Conclusions and Future Needs

Current water management strategies and infrastructure were not designed for some of the hydroclimatic changes that we have observed in the last decades. Robust water resources management strategies provide an opportunity to cope and adapt to the changing climate. This chapter presents different approaches to incorporate hydroclimatic uncertainty for developing robust water resources management strategies. Technically complex approaches tend to have bigger scale implications and usually lack stakeholder involvement, which leads to simplified social and environmental components. Long-term planning with stakeholder involvement tends to simplify climate processes. Ultimately, the social, economic, environmental, and climate components must be considered for developing sustainable and resilient water management systems. Despite potential model and data limitations, one important step is to account more realistically for the nature of socioeconomic and environmental components to identify associated uncertainties and data gaps. Such considerations may lead to accurately assess future sustainability of water use under the current envisaged population growth, ecosystems degradation, and climate change.

References

Ashofteh, Parisa-Sadat, Omid Bozorg Haddad, and Hugo A. Loáiciga. 2015. "Evaluation of Climatic-Change Impacts on Multiobjective Reservoir Operation with Multiobjective Genetic Programming." *Journal of Water Resources Planning and Management* 141 (11): 04015030. doi:10.1061/(ASCE)WR.1943-5452.0000540.

Brown, Casey, Yonas Ghile, Mikaela Laverty, and Ke Li. 2012. "Decision Scaling: Linking Bottom-up Vulnerability Analysis with Climate Projections in the Water Sector." *Water Resources Research* 48 (9). doi:10.1029/2011WR011212.

Brown, Casey M. Jay R. Lund, Ximing Cai, Patrick M. Reed, Edith A. Zagona, Avi Ostfeld, Jim Hall, Gregory W. Characklis, Winston Yu, and Levi Brekke. 2015. "The Future of Water Resources Systems Analysis: Toward a Scientific Framework for Sustainable Water Management." *Water Resources Research* 51: 6110–24. doi:10.1002/2015WR017114.

Brown, Casey, William Werick, Wendy Leger, and David Fay. 2011. "A Decision-Analytic Approach to Managing Climate Risks: Application to the Upper Great Lakes1." *Journal of the American Water Resources Association (JAWRA)* 47 (3): 524–34. doi:10.1111/j.1752-1688.2011.00552.x.

Chen, Cheng, Chuanxiong Kang, and Jinwen Wang. 2018. "Stochastic Linear Programming for Reservoir Operation with Constraints on Reliability and Vulnerability." *Water* 10 (2): 175.

Cristina, M. Mateus, and Desiree Tullos. 2017. "Reliability, Sensitivity, and Vulnerability of Reservoir Operations under Climate Change." *Journal of Water Resources Planning and Management* 143 (4): 04016085. doi:10.1061/(ASCE)WR.1943-5452.0000742.

Ermoliev, Yu., T. Ermolieva, T. Kahil, M. Obersteiner, V. Gorbachuk, and P. Knopov. 2019. "Stochastic Optimization Models for Risk-Based Reservoir Management." *Cybernetics and Systems Analysis* 55 (1): 55–64. doi:10.1007/s10559-019-00112-z.

Farmer, William H., and Richard M. Vogel. 2016. "On the Deterministic and Stochastic Use of Hydrologic Models." *Water Resources Research* 52 (7): 5619–5633. doi:10.1002/2016WR019129.

Ferguson, Grant, and Tom Gleeson. 2012. "Vulnerability of Coastal Aquifers to Groundwater Use and Climate Change." *Nature Climate Change* 2 (5): 342.

Groves, David G., Edmundo Molina-Perez, Evan Bloom, and Jordan R. Fischbach. 2019. "Robust Decision Making (RDM): Application to Water Planning and Climate Policy." In Decision Making under Deep Uncertainty: From Theory to Practice, edited by Vincent A. W. J. Marchau, Warren E. Walker, Pieter J. T. M. Bloemen, and Steven W. Popper, 135–63. Springer: Cham.

Hayhoe, Katharine, Daniel Cayan, Christopher B. Field, Peter C. Frumhoff, Edwin P. Maurer, Norman L. Miller, Susanne C. Moser, et al. 2004. "Emissions Pathways, Climate Change, and Impacts on California." *Proceedings of the National Academy of Sciences of the United States of America* 101 (34): 12422. doi:10.1073/pnas.0404500101.

Herman, Jonathan D., Patrick M. Reed, Zeff B. Harrison, and Gregory W. Characklis. 2015. "How Should Robustness Be Defined for Water Systems Planning under Change?" *Journal of Water Resources Planning and Management* 141 (10): 04015012. doi:10.1061/(ASCE)WR.1943-5452.0000509.

Herman, Jonathan D., and S. Steinschneider. 2018. "Atmospheric Rivers and Reservoir Control: Combining Short-Term Forecasts with Clustering of Weather Regimes to Inform Conjunctive Management in California." In, 2018:H21G-02. https://ui.adsabs.harvard.edu/abs/2018AGUFM.H21G..02H.

Howard, Charles D.D. 1999. "Death to Rule Curves." *29th Annual Water Resources Planning and Management Conference, Proceedings*, 1–5. doi:10.1061/40430(1999)232.

Kang, Doosun, and Kevin Lansey. 2012. "Multiperiod Planning of Water Supply Infrastructure Based on Scenario Analysis." *Journal of Water Resources Planning and Management* 140 (1): 40–54.

Korteling, Brett, Suraje Dessai, and Zoran Kapelan. 2013. "Using Information-Gap Decision Theory for Water Resources Planning under Severe Uncertainty." *Water Resources Management* 27 (4): 1149–72.

Kwakkel, Jan H., Marjolijn Haasnoot, and Warren E. Walker. 2016. "Comparing Robust Decision-Making and Dynamic Adaptive Policy Pathways for Model-Based Decision Support under Deep Uncertainty." *Environmental Modelling & Software* 86: 168–83.

Mallakpour, Iman, Mojtaba Sadegh, and Amir AghaKouchak. 2018. "A New Normal for Streamflow in California in a Warming Climate: Wetter Wet Seasons and Drier Dry Seasons." *Journal of Hydrology* 567 (December): 203–11. doi:10.1016/j.jhydrol.2018.10.023.

Matrosov, Evgenii S., Ashley M. Woods, and Julien J. Harou. 2013. "Robust Decision Making and Info-Gap Decision Theory for Water Resource System Planning." *Journal of Hydrology* 494: 43–58.

Moy, Wai-See, Jared L. Cohon, and Charles S. ReVelle. 1986. "A Programming Model for Analysis of the Reliability, Resilience, and Vulnerability of a Water Supply Reservoir." *Water Resources Research* 22 (4): 489–98.

Ortiz-Partida, J. Pablo, Taher Kahil, Tatiana Ermolieva, Yuri Ermoliev, Belize Lane, Samuel Sandoval-Solis, and Yoshihide Wada. 2019. "A Two-Stage Stochastic Optimization for Robust Operation of Multipurpose Reservoirs." *Water Resources Management* 33 (11): 3815–3830. doi:10.1007/s11269-019-02337-1.

Pathak, B. Tapan, L. Mahesh Maskey, A. Jeffery Dahlberg, Faith Kearns, M. Khaled Bali, and Daniele Zaccaria. 2018. "Climate Change Trends and Impacts on California Agriculture: A Detailed Review." *Agronomy* 8 (3). doi:10.3390/agronomy8030025.

Poff, Boris, Karen A. Koestner, Daniel G. Neary, and David Merritt. 2012. "Threats to Western United States Riparian Ecosystems: A Bibliography." *Gen. Tech. Rep.* RMRS-GTR-269. Fort Collins, CO: US Department of Agriculture, Forest Service, Rocky Mountain Research Station. 78 p. 269.

Philbrick, C. Russ, and Peter K. Kitanidis. 1999. "Limitations of Deterministic Optimization Applied to Reservoir Operations." *Journal of Water Resources Planning and Management* 125 (3): 135–142. doi:10.1061/(ASCE)0733-9496(1999)125:3(135).

Poff, N. LeRoy, Casey M. Brown, Theodore E. Grantham, John H. Matthews, Margaret A. Palmer, Caitlin M. Spence, Robert L. Wilby, et al. 2015. "Sustainable Water Management under Future Uncertainty with Eco-Engineering Decision Scaling." *Nature Climate Change* 6: 25. doi:10.1038/nclimate2765; https://www.nature.com/articles/nclimate2765#supplementary-information.

Prudhomme, Christel, Robert L. Wilby, S. Crooks, Alison L. Kay, and Nick S. Reynard. 2010. "Scenario-Neutral Approach to Climate Change Impact Studies: Application to Flood Risk." *Journal of Hydrology* 390 (3–4): 198–209.

Puente, Carlos E., Mahesh L. Maskey, and Bellie Sivakumar. 2018. "Studying the Complexity of Rainfall Within California Via a Fractal Geometric Method." In *Advances in Nonlinear Geosciences*, Anastasios A. Tsonis (Ed.) 519–42. Springer: Cham.

Sheridan, Scott C., and Cameron C. Lee. 2010. "Synoptic Climatology and the General Circulation Model." *Progress in Physical Geography* 34(1): 101–9.

Steinschneider, Scott, Rachel McCrary, Sungwook Wi, Kevin Mulligan, Linda O. Mearns, and Casey Brown. 2015. "Expanded Decision-Scaling Framework to Select Robust Long-Term Water-System Plans under Hydroclimatic Uncertainties." *Journal of Water Resources Planning and Management* 141(11): 04015023.

Turner, Sean W.D., David Marlow, Marie Ekström, Bruce G. Rhodes, Udaya Kularathna, and Paul J. Jeffrey. 2014. "Linking Climate Projections to Performance: A Yield-based Decision Scaling Assessment of a Large Urban Water Resources System." *Water Resources Research* 50(4): 3553–67.

Vonk, E., YuePing Xu, Martijn J. Booij, Xujie Zhang, and Dionysius C.M. Augustijn. 2014. "Adapting Multireservoir Operation to Shifting Patterns of Water Supply and Demand." *Water Resources Management* 28(3): 625–43.

Wada, Y., M. Flörke, N. Hanasaki, S. Eisner, G. Fischer, S. Tramberend, Y. Satoh, et al. 2016. "Modeling Global Water Use for the 21st Century: The Water Futures and Solutions (WFaS) Initiative and Its Approaches." *Geoscientific Model Development* 9 (1): 175–222. doi:10.5194/gmd-9-175-2016.

WCD. 2000. *Dams and Development: A New Framework for Decision-Making: The Report of the World Commission on Dams*. World Commission on Dams: London.

52

Rivers: Restoration

Introduction 519
Methods Used in River Recultivation 521
Conclusion 530
Acknowledgments 531
References 531

Anna Rabajczyk

Introduction

River ecosystems play a significant role in the human environment and economy. They are characterized by a remarkable variety of plant and animal organisms, while simultaneously acting as ecological corridors and exerting considerable landscaping and recreational impact. The riverbed, riparian zones, and the valley support vegetation, whereas the mass of the flowing water, the bottom soil substratum, riverbanks, and riparian land support a rich generic and onthogenic variety of fauna. The occurrence of these organisms is connected with microhabitats formed owing to the river's morphological diversity, resulting from natural processes occurring over the centuries.[1,2] The naturalness of the river course, including its various curvatures, their radiuses and central angles; the diversification of flow volume, directions, and velocity; the fluctuations of the river water level; and the riverbed bottom type all affect the appearance and maintenance of the morphological diversity. All these factors are subject to change in time and along the river course under the influence of human activity, but also of natural phenomena and processes. If natural processes predominate, living organisms adapt to them and the entire natural system remains stable, due to which rivers can create favorable conditions for the development of various organisms and river ecosystems display very high biological diversity.[2,3]

However, over the centuries, humans have been using rivers, regulating and straightening their courses, building water drops, clearing riverside carrs, and draining floodplains in order to enhance river navigability, and expand farming acreage and area allocated to industrial and housing infrastructure. In many cases, this has led to changing natural watercourses into homogeneous channels deprived of natural value, whose function has become limited to water removal. Simultaneously, the rapid progress of land development, together with farming and industrial advances, has become the source of many pollutants, considerably diversified in terms of quantity and quality, and frequently hazardous to the functioning and life of plants, animals, and humans.[1–5]

Widespread environmental degradation has forced the authorities to take appropriate steps aimed at the introduction of principles of sustainable development to provide social welfare, understood as a possibility to satisfy fundamental needs not only by the present-day but also by future generations, with a focus on eco-safety. Appropriate policies and legislative regulations are being introduced to enable the determination of water quality, and guidelines are being developed for actions to be taken to improve it.[5,6]

One of the objectives in water resources management is to restore and maintain good water quality. First and foremost, it is necessary to specify the subject of river maintenance. To specify the types and

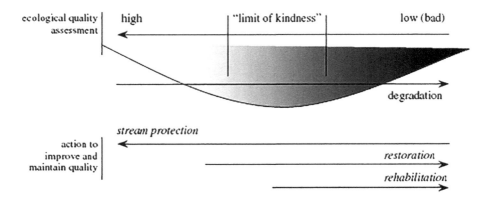

FIGURE 1   Types and scopes of actions geared at river protection and maintenance in terms of quality.
Source: Based on Phillips et al.[7]

scopes of actions geared at river protection and maintenance, it may prove helpful to use a diagram that shows river and stream quality in terms of the so-called limit of kindness (Figure 1).[7]

In the case of very good, good, and moderate quality of water ecosystems, with focus on very good and good quality, only stream protection is justified since actions need to be taken to enable both maintenance and protection of high-quality natural river ecosystems. Another type of action is the recovery of conditions, or restoration, used in the case of rivers characterized by moderate quality, whose "limit of kindness" suggests the hazard of rapid deterioration of the ecosystem's conditions. It is vital to take actions geared at the restoration of natural (primary) characteristics and values from a period taken as a reference point, including, but not limited to

- Natural water quality range
- Natural sedimentation and flow regime (including seasonal fluctuations in a year and multiyear seasons, according to flood patterns)
- Natural riverbed geometry and its stability, determined according to stability ratios, from that period
- Natural riparian vegetation
- Restoration of indigenous aquatic flora and fauna[2,5–9]

Rehabilitation, in turn, pertains to cases of more serious degradation of natural ecosystems and/or when complete restoration of natural features is impossible (e.g., due to technical or economic reasons). Rehabilitation comprises enhancement and improvement of major environmental features of the watercourse so that valuable characteristics thereof are created to define ecological quality.

Another method is to implement remediation, which does not lead to the reconstruction of the primary ecosystem but to the construction of a completely new, different one (Figure 2).[7] This pertains primarily to strongly urbanized (urban) areas, in particular to situations when

- Natural watercourses became degraded in the past and are currently being "brought back to life."
- Ditches were used as storm drains and, partly, as sewage sludges in the past, and are now, upon the introduction of new, closed systems of municipal sewers, often treated as natural.[8–11]

It ought to be stressed that, as a result of adequate actions, it is possible to create new ecosystems, of high biological functionality, yet depleted in terms of biodiversity

River recultivation can only be effective when approached holistically—as an element of comprehensive protection and maintenance of water ecosystems. Consequently, as part of the recultivation process, the planned and implemented actions ought to

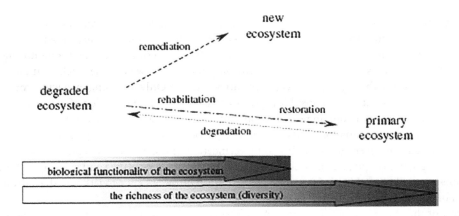

FIGURE 2 Actions aimed at restoration of relevant features of a water ecosystem.
Source: Based on Phillips et al.

- Be integrated, i.e., a particular watercourse (a river segment or the entire river subject to recultivation), together with its appurtenant ecosystem, ought to form one whole, immune to possible environmental impact (seasonal high water, depleted oxygen content, etc.), clearly, up to a limit.
- Reflect the catchment's capacity for natural self-renewal.
- Take into account the physical structure of a given watercourse that fosters the development of desirable flora and fauna (morphological changes, riverbed straightening, or cutoffs from ecosystems may promote degradation).
- Take into account the influence of catchment area users on a particular water body (this is related to the impact of industry, urban areas, or commercial centers).
- Take into account all the reasons for the existing situation, in particular the immediate causes of degradation (including toxic waste discharge, changes to land use, expanded hard road surface in proportion to green areas, and water-logged area elimination).
- Be based on strict guidelines and their methodological implementation, and on adequate monitoring of actions taken to improve the situation.[8–17]

Methods Used in River Recultivation

Adequate implementation of individual actions requires clearly specified methods, techniques, and technical procedures. However, first of all, a conceptual framework ought to be established to specify the principles of operation in terms of

- An assessment of the current status, consistent with legal requirements as well as local/domestic conditions and limitations resulting from both data availability and historical and cultural contexts.
- Principles of operation for the restoration of those features of the ecosystem that determine its capacity for the maintenance of continuity, as well as the major characteristics that constitute the value of ecological quality.
- Following the establishment of the conceptual framework and set of principles, detailed standards are to be developed that will enable the methodological, technical, and technological implementation of appropriate solutions. That is why specialists in a variety of scientific disciplines (such as ecology, water environment biology, hydrology, hydraulics, geomorphology, construction engineering, spatial planning, communications, and sociology) ought to participate in the works.[11,18–26]

Each catchment and its environment are, as a rule, unique, and each case requires an individualized approach, hence the first step is to collect information on the catchment structure and development from a number of sources. This should be complemented with results of research monitoring over a specified period (such as several months or a hydrological year) as well as the catchment area located before the final part of the segment subject to recultivation.[27,28] Only by basing on a wide range of data collected from several sources is it possible to pass on to the phase of specification of individual elements and procedures necessary to obtain recultivation results in a particular watercourse.

The development of criteria for an assessment of a particular watercourse's quality enables standard assessment of current river quality, followed by verification of the actions implemented as part of the recultivation process (Figure 3). These should take into account not only the valid worldwide, European, and domestic regulations, but also the uniqueness resulting from locality, together with environmental and economic functions.[29]

In view of the above, preparatory recultivation works cover watercourse assessment in accordance with adopted criteria, on the basis of which it is possible to describe the status quo and pinpoint short-comings in relation to an upgraded status. A partial function assessment is based on the lowest rating

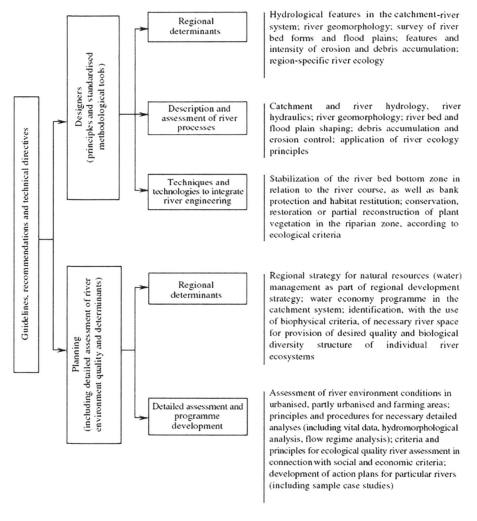

FIGURE 3 Guidelines and recommendations for planning and development of river recultivation programs.

for the criteria appurtenant with that function; the segment rating covers the function's lowest partial rating, with allowances being made for the weakest-link principle. The content of this kind of a document enables fast identification of hotspots whose adjustment will allow for value enhancement of a particular watercourse segment. On this basis, appropriate procedures and diverse actions are selected.

As part of complex works toward river ecosystem renewal, actions are taken to influence various elements of the environment (Figure 4). This course of recultivation allows the objective of river renewal to be achieved. In most cases, single-tasking actions do not yield expected results, or the results take much longer to become visible than they would in the case of a wider spectrum of actions.[29]

Multitasking actions ensure high diversity in the methods and techniques that may be, and are, used in the process of river recultivation. Yet, according to recultivation assumptions, the performed operations ought to enable the intensification of watercourses' natural capacity for cleansing. Consequently, possible works include the following[30–36]

- Construction of aeration-enhancing drops and step falls
- River course regulation
- Elimination of isolated still water pools where oxygen depletion is easy to occur

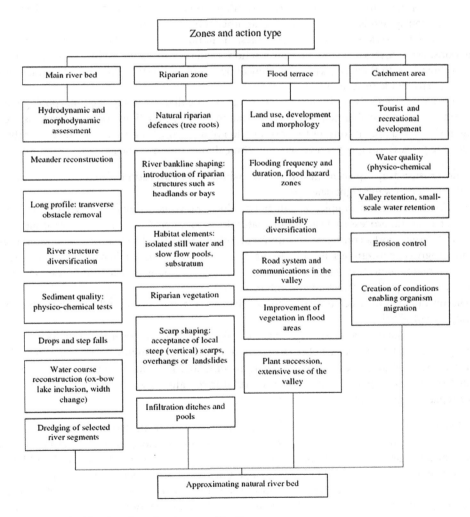

FIGURE 4 Example of multitasking for river recultivation.
Source: Bojarski et al.[30]

- Dredging sediments in selected river segments
- Formation of riparian vegetation strips
- Removal of excess vegetation from the riparian zone
- Formation of infiltration pools and ditches along river- banks to enable both surface runoff limitation and precipitation water pretreatment
- Establishment of small-scale water retention facilities
- Precipitation water management in urban areas
- Groundwater remediation

Not all cases require a full range of actions; it is not always and not everywhere that implementation of all assumptions will be possible. That is why the recultivation process ought to be implemented in stages. First and foremost, the most vital actions need to be taken, i.e., those that will yield optimum natural results, and whose implementation is acceptable to other water users. Of particular importance is, however, the development of strategic action plans that ought to outline both further actions aimed at bringing waters close to nature and result monitoring of the operations performed, complete with reactions adequate to the results obtained.

A complex approach to the problem of river recultivation is renaturization, understood as restoring the river, previously regulated or reshaped in another way, into a state close to the natural (existing before the regulation or found in nature). At times, the notion of river revitalization is used to describe these actions. Yet, it covers a narrower scope of actions than renaturization and is conceived as restoring ecological functions to the river, without technical operations to restore the natural characteristics of the riverbed, riverbank, or valley.[33,34]

One of the major objectives of renaturization is to restore diversified riverbed morphology to regulated rivers, which may be obtained, e.g., through shaping a more curvilinear river course, cross-sections varying in shape and dimensions, or the introduction of diversified vegetation. Most frequently, this is a long-lasting process, comprising a variety of technical undertakings, such as

- Anti-erosion drop replacement with slides enabling communication of water organisms as well as elimination of water-tight defenses.
- Works, which—upon completion—do not produce ready-made elements of renaturized water but foster a natural process that can restore naturalness to waters (e.g., vegetation planting).
- Maintenance works consisting in minor corrections to natural transformations when these are not progressing toward a state close to the natural; replacement of fallen trees and bushes.
- Water maintenance during the renaturization process (maintenance and conservation procedures, water quality protection and enhancement).
- Discontinuation of selected actions concerning water management and leaving them to natural impact only.[33–38]
- A river functionality analysis suggests that a significant role in river naturalness restoration is played by vegetation. As part of renaturization operations, the following are implemented: riparian plant structure, tree planting in riparian zones and flood areas, introduction of vegetation into the existing riparian technical structures (the so-called revitalization of existing structures or replacement with plant defenses).[36]

Conducted recultivation procedures, both technical and biological, constitute a long-lasting and difficult-to-design process whose course requires expert supervision, including result monitoring. Hence, riparian vegetation that improves riverbank stability and carries high natural and landscape importance ought to be distributed in stories comprising trees, bushes, perennial grasses, and gramineous plants already found on the site, which meet the necessary requirements. New replacements and seedings of species from nearby habitats are also needed. Tall vegetation ought to be rarefied so that gramineous plants receive enough light. The interface of the tree-covered area and the river ought to be as long as possible, taking tree thinning into account.[27–29]

It must be stated that, when selecting flora and fauna elements, local species are recommended, as their functioning is conditioned by local hydraulic and hydrological determinants. In the spatial distribution of vegetation, they may fulfill natural functions (e.g., those related to ornithology). The introduction of elements from beyond a given ecological system may cause unwanted disturbance in the environment. Moreover, the plants selected ought not to display antagonisms.[2,5,36]

The elongation of river bankline fosters the dovetailing of the riverbed with the adjacent land strip. Over newly formed water habitats, vegetation plantings accelerate the watercourse's integration with the landscape. This is of particular significance in urban areas. In addition, slides situated in the bottom to break up the watercourse gradient, together with islands and boulders to disturb the flow, serve to diversify the current. Hence, they enable the enrichment of the streambed structure and the formation of various minor biotopes for the animal world.[36–38]

The objective of technical actions is to eliminate obstacles that prevent or limit the course of natural fluvial processes. These works are meant to initiate a return to conditions approximating the natural, and affect both abiotic and biotic conditions.[36]

Upon removal of obstacles to shaping morphological conditions in accordance with natural laws, the second stage of renaturization follows. This is a spontaneous transformation implemented by nature as, due to succession and natural transformation, the environment comes to approximate the natural. Most frequently, this spontaneous process comprises riverbed transformations and shaping hydromorphological conditions, as well as vegetation and animal organism development. It is long lasting and may occasionally take from 10 to 15 years, or even up to 30 years.

Recultivation through renaturization ought to make allowances for the multifunctionality of rivers and valleys.[37,39] The chances for implementation of renaturization procedures in rivers increase when these actions, apart from enhancement of the natural state of the rivers, bring other benefits, including economic, municipal, recreational, etc. A case in point may be the improvement of flood control by restoration of the valley retention function. The municipal aims of renaturization may be, e.g., the enhanced self-cleansing capacity of the waters, as well as upgrading the protection status of particular infrastructure. The objective of renaturization in urban areas may be, in turn, the enhancement of a city's (and its surroundings') landscape value, thanks to the restored naturalness of existing water bodies and through the establishment of new ones. These actions are desirable in many cities; however, at the same time, they encounter much difficulty posed by urbanization.[36,40–43]

The most effective, from the natural viewpoint, are renaturization actions conducted in riverbeds and riparian zones in order to increase the diversity of riverbed morphology. This diversity is among the most significant natural features of rivers. It results from water and debris movement as well as the closely related phenomena of erosion and sedimentation. Increased morphodynamic activity may be achieved through elimination or limitation of factors that restrict the process: changes to the river course, riparian defenses and the long profile, or inducing the removal of obstacles disturbing the river debris movement regime.[36,37]

The recreation of the natural shape of the river course (altered as a result of river regulation) is rarely possible as ox-bow lakes are usually buried or permanently cut off from the regulated river. Due to these reasons, as part of renaturization procedures, reclamation usually comprises only fragments of ox-bow lakes, or provision of organisms with opportunities for mobility between ox-bow lakes and the main riverbed. Another option is that a regulated (straightened) river receives a curvilinear shape, resembling those found in natural rivers.[36,39–45]

However, even this kind of local recovery of the river course is very often impossible and the restoration of naturalness is limited to a reconstruction of structures characteristic of natural rivers, i.e., bays, headlands, islands, and local barriers that affect velocity distribution, leading to enhancement of riverbed morphology. This pertains, in particular, to large regulated rivers in which renaturization operations may only cover segments of the riverbank, elimination or replacement of riparian defenses, clearing lateral river arms, renewal of intergroyne space, or partial reclamation of floodplains.[10,14,30]

A significant element in restoration of good ecological quality to rivers is to increase the diversity of crosssection forms. Under natural conditions, this diversification results from the complex hydraulic structure of a water stream as it runs along a curving route. If the restoration of the curvilinear river system is deemed impossible, in order to diversify bottom topography, facilities that steer the water stream (deflectors, digger logs, or appropriately shaped drops) are introduced. Diversification of cross-section width and placement of barriers in the riverbed to enable changing the flow velocity, thus enhancing the diversity of the riverbed morphology, are also practicable.[35–40]

The diversified river banklines thus obtained enable the expansion of the interface between the substratum and water, causing the appearance of a variety of biotopes: a system of fissures in the watercourse gravel substratum, the water area, the interface between water and land, the water level fluctuation zone, the riparian area, and other humid habitats.

The research conducted suggests that, as a result of riverbed renaturization, in the cases of minor lowland rivers with sandy bottoms, where riparian impact significantly affects flow resistance, flow resistance increases. This changes several parameters characterizing water flow and debris drag transport conditions. If riverbed dimensions and flow regime do not change, the increase in scarp coarseness will cause increased falls of the water plane, riverbed filling, and stream impact on the bottom. At constant flow intensity, the drag rate will increase and cause deep riverbed erosion. In view of the above, in river renaturization projects, it is necessary to check not only the riverbed capacity but also the debris drag conditions.[33–40]

To maintain and enhance the ecological quality of rivers, regardless of renaturization procedures, other actions also ought to be taken. A case in point may be adequately conducted river maintenance and conservation works.

These works ought to take into account major requirements of natural protection. Special attention must be paid to the methods used in technical solutions for river development and use. The principles of environmentally friendly river regulation (natural river regulation) ought to be implemented, basing on the preservation of riverbed morphological diversity, through restricted riverbed transformation. This results in the limitation of adverse effects of planned actions on the natural environment. Natural regulation solutions are based on the following principles[36,38,40–49]:

- The river route ought to remain unchanged, and crosssections ought to preserve their diverse forms and dimensions.
- River rerouting ought to be seen as an exception, which may be justified only by very important arguments and lack of alternative solutions.
- River bed structures, i.e., islands, slower flow areas, river outwash, silts, and other riverbed diversifications, should not be eliminated.
- Riverbanks ought to be reinforced only in those segments where the occurring erosion may cause major hazards.

The application of environmentally friendly river regulation principles enables obtaining a compromise solution that will ensure the protection of the most valuable natural resources as well as the fulfillment of major economic goals.

Yet, if riverbed regulation is needed, the solutions should be based on the following principles[36,38,40–49]:

- Natural riverbed protection: rechanneling is done through initializing the meander pool by imposing a triangular section.
- Smoothing natural riverbed meanders by means of linear structures with cross-beams: dams are replaced with river walls that only reinforce concave banks and are pitched against the existing banks (apart from cases when this solution cannot be used).
- Meandering, including riverbank walls or short structures resembling groyne heads (only for riparian protection, low impact on riverbed flows and the riverbed itself).
- Transverse structures that stabilize the riverbed bottom: made as gentle-sloped erratic riffles (1:10 is the optimum gradient, tested in practice), thus preserving the biological passability of the

watercourse; if possible, the location of the riffle is selected so that it can also act as a cataract and not only as bottom gradient reducer and stabilizer.

- Stone backfill with appropriate scarp inclination from 1:2.5, the recommended ratio being 1:4 and lower (wherever possible): more capacious watercourse bed, accessible to people and animals; regulative structures are more durable.
- Preservation of the riverbed's and the river bottom's morphological form: construction of stone backfill from the meander pool bottom (elimination of the necessity to disturb the natural gravel bottom shield during the formation of partitions from these alluviums; the structures are made without partitions).
- Stone backfill durability and visual appeal enhancement: boulder placement and turfing initialization by soil filling of interboulder space and grass seeding according to pertinent formulas.
- Application of defenses made of stone netting baskets: only for protection of building structures situated in the vicinity of the watercourse bank.
- River bar or island leveling or correction—riverbank reinforcement.

As part of the recultivation process, it is necessary to take action in the field of precipitation water management, which is important in view of the current change in climate, and increasing freshwater use and degradation. The construction of retention and infiltration basins enables establishment of the so-called small-scale water retention facilities at the sites where precipitation occurs.

Enhancement of individual forms of retention may be obtained through a variety of methods. Briefly speaking, technical, planning, and agrotechnical procedures may be distinguished. The group of technical methods includes the majority of hydrotechnology and melioration works, aimed at limitation of surface water runoff, e.g.,[5,37,39]

- Surface water retention through the construction of small water reservoirs; lake damming; and construction of damming structures in watercourses, ditches, and channels.
- Regulation of water runoff from drainage systems and ditch networks.
- Increase in underground water reservoir supply through the construction of infiltration ponds and wells.
- Application of adequate methods for rainwater removal from sealed surfaces (roofs, squares, and streets), enabling water infiltration in adjacent nonsealed areas.
- Limitation of rapid surface water runoff through renaturization of small watercourses and reconstruction of floodplains where possible due to economic (farming) reasons.

A significant role in water management, including water quality protection and enhancement, is played by relevant spatial arrangement in rural areas, particularly in terms of formation of such systems in which precipitation and meltwater runoff does not occur rapidly. The methods used in this respect are defined as planning. These include, but are not limited to the following[26–29,30–39]:

- Development of an appropriate arrangement of arable fields, green areas, and woodland in the catchment.
- Peat land, swamp, and wetland protection and reconstruction.
- Formation of protective vegetation strips (bushes and trees: increased catchment woodiness), reconstruction of as many as possible ecological sites, including aquatic gardens.
- Reconstruction of floodplains in river valleys.
- Establishment of protective areas with appropriate facilities.
- The application of appropriate agrotechnical methods, including observance of the Code of Good Agricultural Practice, may contribute to improvement and reconstruction of natural conditions in rivers. The basic actions in this respect, which form the so-called group of agrotechnical methods, include but are not limited to the following[45,48–50]:
- Soil retention increase through soil structure enhancement and soil humus content increase (correct tillage, agromelioration procedures, fertilizing, and liming).

- Surface runoff limitation through erosion control and aftercrops planting.
- Evapotranspiration decrease through appropriate plant selection and limitation of soil surface evaporation.

The suggested methods for catchment retention capacity enhancement do not introduce any significant change to the natural water regime, but only necessary corrections aimed at improving the water balance structure without disturbing the ecosystem's biological balance. Also in terms of precipitation water management, complex solutions are recommended, which comprise urban area drainage, including roofing, downspouts, and underground pipes, thus providing for rapid and effective rainwater removal to a retention and infiltration box system.[27–29,32,50–54] There are several solutions available in the field of rational rainwater removal:

1. Infiltration: used in permeable and poorly permeable soils, with the maintenance of the minimum 1-m distance from the bottom of the box, which is wrapped with non-woven geotextile to the groundwater level; the water infiltrates to the soil; the most popular solution in rainwater management.
2. Retention: used in firm grounds and at high groundwater levels; the water is stored for subsequent runoff to storm drains or waterways; the box system is wrapped with a geomembrane, which prevents water infiltration into the soil.
3. Storage: an alternative solution for open surface reservoirs; precipitation water is stored in boxes wrapped with geomembrane, and may be used later, e.g., as fire water supply.

The variety of methods and techniques that may be used in the recultivation process enables their adaptation to individual needs and objectives (Table 1).

The works in the field of watercourse restoration are subject to numerous technical, legal, administrative, economic, and natural constraints. Under the circumstances, a compromise is needed between approximating nature and achieving economic results. If a river's naturalness has been lost, for instance, owing to the building of hydrotechnical facilities, it cannot be fully restored without the demolition of these facilities. Consequently, that would mean the community's return to conditions that, in the past, made living difficult (floods or swamping) to the extent that led to decisions being taken to improve the situation. While undertaking recultivation procedures, it is necessary to bear in mind that the scope will be limited in proportion to what might be implemented if there is no need to meet the economic requirements. That means that the recovery of natural values in the scope in which they are found in rivers untampered with by humans will not be achieved.

Serious limitations for river recultivation may also result from lack of space for appropriate actions. "Human nature" conflicts most frequently arise in situations when morphological diversification of the riverbed is intended, including addition of a curve to a straightened river; riverbed diversification through the introduction of changeable bottom widths or scarp gradients; or the introduction of islands, headlands, bays, breaches, precipices, and beach segments. The reason is that these require the taking of land adjacent to the river and disturbing the riverbanks, whose shaping is perceived differently by land owners and by ecologists. There is little chance of reaching a compromise on these issues, so the best solution would be to purchase the strip of riparian land where appropriate actions, such as river bankline diversification or allowing for riverbank erosion, may be taken.

Recultivation actions may also be constrained by technical reasons. The restoration of natural elements to rivers, including cutoff meanders, ox-bow lakes, or connections between the river, land depressions, and marshes, will only become possible if, after the regulation, these structures have not been completely demolished and the land has not been developed. Relatively frequent are cases when the existing, necessary riverbed facilities as well as the ways in which they are used (e.g., the river running across a built-up area, a walled riverbed, a link between the river and the water economics system) render transformation of rivers into more natural ones completely unfeasible or feasible only to a small extent.

TABLE 1 Sample Procedures Geared at River Restoration with Their Characteristics

Method/ Technique	Model Application	Characteristics	Literature
Dredging	Buffalo River	Primarily relies on natural sedimentation processes after dredging to achieve dredge residual performance standards established during detailed design; remediation area was approximately 164 acres, including 138 acres in the Buffalo River and 26 acres in the City Ship Canal	[55]
	Eel River	Excess sand and all peat excavated during site restoration work were placed in the former sand borrow pits around the perimeter of the bogs; insecticide-affected peat was buried beneath a variable layer of sand; once filled—sand borrow pits into the surrounding upland forest; a layer of top soil was imported and placed as a final cap over the former borrow pits, revegetated with native plantings, and monitored over time to ensure growth and survival	[56]
Capping	Buffalo River	Introduced for remediation of the end of the City Ship Canal, beyond the limits of the authorized navigation channel, to isolate underlying sediment contaminants and provide a clean sediment surface; capping depths and cap materials were designed to optimize and enhance habitat restoration plans while providing adequate protection against damage from root penetration; cap placement was performed by either of the following methods: 1) extending a navigational channel to the downstream end; 2) through hydraulic means; or 3) in dry conditions using earth moving equipment, by temporary sheeting and dewatering the proposed cap area and using the adjoining upland areas for material handling	[55]
Renaturization	Izara River	Along a 14-km-long segment, the riverbank structures were removed; there appeared gravel river bars, shoals, main riverbed forks, and stagnant waters; in selected segments, modern fish ladders were built and riparian communities originated	[57]
	Mistelbach Stream	Concrete gutters were removed from stream bottom and banks; so was the irregular stream bank shaping; islands and widenings were installed; stones were introduced into the streambed bottom; slides (ramps) were built and planted with vegetation	[58]
	Enz River	Irregular widenings were formed (at two sites, forking the river by means of islands); the bottom shape was diversified, the scarp inclination of 1:4–1:5 was obtained; the watercourse gradient was increased to 2.7%; dead willow fascine was used; structures were erected with the use of reed beds; plantings of tree shoots were inserted into the substratum, and seedings were made (vegetation was mostly obtained from floodplains and nearby areas)	[59,60]
	Kwacza River	Riparian alders were felled; six bypasses were made for the existing riverbed; in the riverbed, fallen trees were retained, providing valuable natural microhabitats; in the current, artificial structures in the forms of stone piles were placed; wooden and stone deflectors were made, as well as wooden log drops and in-stream digger logs made of tree trunks (deflectors); riverbank defenses were made of tree trunks; the connection between the major riverbed and the ox-bow lake was reconstructed; bays and stagnant water pools were formed; several hundreds of tons of gravel and stones were used to increase bottom material graining so that its composition would approximate the natural salmonid spawning sites	[61]
	Skjern River	Elimination of hydrotechnical structures, restoration of the river's meandering course, and purchase of farming land adjacent to the river	[62]

(Continued)

TABLE 1 (*Continued*)   Sample Procedures Geared at River Restoration with Their Characteristics

Method/ Technique	Model Application	Characteristics	Literature
Rehabilitation	River Frome	Dig a borrow pit to create a supply of clay; use clay to raise the riverbed level (specific depth to be agreed on site); win gravels from embankments (old river dredgings or borrow pit); return gravel to the river creating riffles and gravel features; spoil created from bank reprofiling used to infill the borrow pit (less material available—in the original pit volume, a small pond or wetland was created), giving varied habitats	[63]
	River Rother	Restore the loop to these dimensions tempering them with uncovered in-channel features (basic river archaeology using remnant channel features as a guide in favor of theoretical section design); fix the channel features at the upstream and downstream end of the system to prevent channel adjustment between the canalized section and the restored naturalized section (headward recession, etc.); restore the effect of the lock gates and system to prevent any passage of flow down the canal and divert all flows through the loop	[64]
Restoration	River Avon	Physical features and geomorphological processes of the watercourse habitat were re-established at six river restoration demonstration sites; approximately 7 km of river and banks was enhanced and woody debris introduced; an estimated 0.36 ha of new spawning area for Atlantic salmon, bullhead, and lamprey was created (Atlantic salmon are already spawning on the new gravel)	[65]
Restoration of native terrestrial plants	Niagara River	Restore extirpated species to gorge; close unauthorized trails; upgrade existing trails with water bars, fencing, signs; initiate trail steward program; selectively remove aggressive nonnative species from the vicinity of gorge and gorge rim; use flow restriction or diffusion devices on outfalls; eliminate perched outfalls; restore water flow to Devil's Hole; reduce acidification; restore original character of cove; exclude birds (removal of nests, egg oiling); reduce soil acidification; reduce direct trampling or pulling of plants by gulls; reduce input of alien seeds in material brought in by gulls	[66]
Remediation	Sumida River	Waterfront amenity improvement coexisted with flood prevention; gently sloping levees and high standard levees were constructed as substitutes for linear levees to increase durability; river edge terraces (promenades) were improved and levees were forested to improve water amenities; improvement of the sewage system and activities of citizens' groups	[67]
Revitalization	River Brent	Restored two sections of this river by remeandering the straightened channel along its original route, creating a backwater channel, and naturalizing the river's banks; linked up with the earlier restoration work to enhance the entire park	[68]
	River Thames	Activities covered by the tributaries were carrying pollutants, e.g., creation of a natural channel in place of a buried culvert; creation of sustainable instream, marginal, wetland, and floodplain habitats; potential for improved biodiversity in the area and increased flood storage capacity; reconnected the two stretches of seminatural channel up- and downstream of Brookmill Park	[69,70]

Conclusion

The recultivation of flowing surface waters may be based on tested case studies from other regions of the country or abroad. Yet, it should be remembered that each environment, while similar in terms of the problems encountered, is unique. The adaptation of solutions must then be carried out with due care.

More and more often, an ideal model of a watercourse is taken as a starting point during the recultivation planning process. The model corresponds to natural conditions of the watercourse, unchanged by humans. It is only on that basis that its current state, and deficits, or degradation rating, are assessed. Subsequently, taking into account socioeconomic boundary conditions, such as the degree of valley occupancy, navigation, fire safety, and environmental protection requirements, development goals are set, i.e., the character of the watercourse after recultivation is determined. The actions taken are based on a variety of methods and techniques, individually adapted to the local situation and the specified objectives. They include technical, engineering, biological, and hydrological procedures. Nonetheless, the scope of the works conducted must demonstrate multitasking and be capable of complex inclusion of selected environmental elements, at the same time stimulating their reconstruction process.

It must be stressed that recultivation procedures do not usually yield direct economic benefits, which constitutes another obstacle. The improvement of a river's natural condition or enhancement of landscape values in themselves are not sufficient arguments in favor of beginning the recultivation.

Such actions, understandable and convincing from the point of view of the natural environment and its needs, tend to breed substantial controversy if local economic expectations are considered. It is then vital to recognize hydrological and hydraulic conditions, and describe the phenomenon of water supply to the land subject to recultivation procedures, together with recognition of farming water needs in the area. Social and sociological surveys that would answer the questions related to the intentions of the farmers themselves, the prospects for local agriculture, possibility of land sale, etc., are also needed.

The chief obstacle is indeed found in the mentality of local communities who expect the proprietor of flowing waters to fulfill the protective obligations. These include deepening and narrowing the riverbed and the streambed so that they no longer flood; reinforcing riverbanks and securing private facilities built nearby; and creating or establishing conditions for recreational river use. Yet, the responsibility to protect the environment as our shared asset remains a shared obligation.

Acknowledgments

The present study has been financed in part by the State Committee for Scientific Research (KBN), grant no. N N305 306635.

References

1. Petts, G.; Calow, P. *River Restoration*; Blackwell Sciences: Oxford, U.K., 1996.
2. Rao, R.J. Biological resources of the Ganga River, India. Hydrobiologia **2001**, *458*, 159–168.
3. U.S. Army Corps of Engineers. *Environmental Design Handbook*; Upper Mississippi River System Environmental Management Program: Rock Island, IL, 2006.
4. Theiling, C.H. Defining ecosystem restoration potential using a multiple reference condition approach; Upper Mississippi River System, USA, Theses and Dissertations, University of Iowa, 2010.
5. Zalewski, M.; Harper, D.M.; Robarts, R.D., Eds. Proceedings of the International UNESCO Workshop "Aquatic Habitats in Integrated Urban Water Management" within the framework of the International Hydrological Programme (IHP) and the Man and the Biosphere (MAB) Programme, Ecohydrology & Hydrolobiology. 2005, 5(4), pp. 93.
6. European Parliament and the Council. Directive 2000/60/ CE of the European Parliament and of the Council of 23 October 2000 establishing a Framework for Community Action in the Field of Water Policy; Official Journal of the European Communities, L 327: Luxembourg, 2000.
7. Phillips, N.; Bennett, J.; Moulton, D. Queensland Environmental Protection Agency, *Principles and Tools for Protecting Australian Rivers,* Land & Water Australia, Canberra 2001.
8. Jasperse, P. Policy networks and the success of lowland stream rehabilitation projects in the Netherlands. In *Rehabilitation of Rivers. Principles and Implementation*; Waal, L.C., Large, A.R.G., Wade, P.M., Eds.; John Wiley & Sons; Chichester, U.K., 1998, 14–29.

9. Redondo, M.D. Social impact assessment for river restoration: A more sustainable perspective, Thesis presented in part-fulfillment of the degree of Master of Science in accordance with the regulations of the University of East Anglia, Norwich, 2003.

10. Schanze, J.; Olfert, A.; Tourbier, J.T.; Gersdorf, I.; Schwager, T. Existing Urban River Rehabilitation Schemes (Work package 2), Final Report, Urban River Basin Enhancement Methods, 2004.

11. Gore, J.A.; Shields, F.D. Can large rivers be restored? BioScience **1995**, *45* (3), 142–152.

12. Brierley, G.; Fryirs, K.; Outhet, D.; Massey, C. Application of the River Styles framework as a basis for river management in New South Wales, Australia. Appl. Geogr. **2002**, 22, 91–122.

13. Wagner, I.; Marsalek, J.; Breil, P. *Aquatic Habitats in Sustainable Urban Water Management, Science, Policy and Practice;* Urban Water Series—UNESCO-IHP, Paris, France; Taylor & Francis: Leiden, the Netherlands, 2008

14. Stalzer, W. Institutional Challenges in the Danube River Basin: Inputs from the International Commission for the Protection of the Danube River, MRC International Conference, 2–3 April 2010; Hua Hin, Thailand.

15. Reeve, D.E. Predicting morphological changes in rivers, estuaries and coasts, Executive Summary, Report number T05–07-03, Members of the FLOOD*site* Consortium, University of Plymouth, 2009.

16. Nakamura, K.; Tockner, K.; Amano, K. River and wetland restoration. Lessons from Japan. BioScience **2006**, *56* (5), 419–429.

17. River Restoration Project, Texas State University, San Marcos, http://www.aquarena.txstate.edu/Wetlands/Restoration-grants.html (accessed January 2011).

18. ARRN's Roundtable Meeting, How to Develop Technology and Guideline for River Restoration through Networks, ARRN Asian River Restoration Network; Foundation for Riverfront Improvement and Restoration and CTI Engineering Co., Ltd.: Seoul, Korea, 2011.

19. Groundwork Thames Gateway London South, Feasibility Study, Chinbrook Meadows River Naturalisation Scheme; London Borough of Lewisham: London, 2001.

20. Wheaton, J.M. *Review of River Restoration Motives and Objectives;* unpublished review, Southampton, U.K., 2005.

21. Herricks, E.E. Water quality issues in river channel restoration. In *River Channel Restoration: Guiding Principles for Sustainable Projects;* Brookes, A., Shields, F.D., Eds.; John Wiley & Sons: Chichester, U.K., 1996; 179–201.

22. River Restoration Centre. *River Restoration Manual of Techniques,* 1999, http://www.therrc.co.uk/rrc_manual.php (accessed May 2012).

23. U.S. Army Corps of Engineers. *Upper Mississippi River System Environmental Design Handbook;* U.S. Army Corps of Engineers: Rock Island, IL, 2006.

24. Bolton, S.; Shellberg, J. *Executive Summary: Ecological Issues in Floodplains and Riparian Corridors,* White paper prepared for Washington Department of Fish and Wildlife, Washington Department of Ecology and Washington Department of Transportation, Seattle, Washington; University of Washington, Center for Streamside Studies, Seattle, Washington, 2001.

25. Rhoads, B.L.; Herricks, E.E. Naturalization of headwater streams in Illinois: Challenges and possibilities. In *River Channel Restoration: Guiding Principles for Sustainable Projects;* Brookes, A., Shields, F.D., Eds.; John Wiley & Sons: Chichester, U.K., 1996; 331–369.

26. Sear, D.A. The sediment system and channel stability. In *River Channel Restoration: Guiding Principles for Sustainable Projects;* Brookes, A., Shields, F.D., Eds.; John Wiley & Sons: Chichester, U.K., 1996; 149–179.

27. De Waal, L.; Wade, P.M.; Large, A., Eds.; *Rehabilitation of Rivers. Principles and Implementation;* John Wiley & Sons: London, 1998.

28. Jormola, J., Ed. *Urban Rivers, Flood Risk Management and River Restoration,* Gumiero, B.; Rinaldi, M.; Fokkens, B., (Eds), 4th ECRR Conference on River Restoration, ECRR - European Centre for River Restoration, CIRF - Centro Italiano per la Riqualificazione Fluviale, Venice, San Servolo Island, 2008; Chapter 12, pp. 889–948.

29. Wisconsin DNR. *Wisconsin Water Quality Report to Congress*, WDNR Pub WT-924–2010, 2010.

30. Bojarski, A.; Jeleński, J.; Jelonek, M.; Litewka, T.; Wyżga, B.; Zalewski, J. *Good Practice in Maintaining Mountain Rivers and Streams*; Ministry of the Environment, Department of Water Resources: Warsaw, Poland, 2005 (in Polish).

31. Hulse, D.; Gregory, S. Integrating resilience into floodplain restoration. Urban Ecosyst. **2004**, *7*, 295–314.

32. Jones, M.; Bradley, W.; Dunsford, D. An integrated approach to flood defence design. River Restor. News Newsl. River Restor. Centre **2002**, *12*, 5.

33. Kondolf, G.M.; Downs, P.W. Catchment approach to planning channel restoration. In *River Channel Restoration: Guiding Principles for Sustainable Projects*; Brookes, A., Shields, F.D., Eds.; John Wiley & Sons: Chichester, U.K., 1996; 129–149.

34. Pedroli, B.; de Blust, G.; van Looy, K.; van Rooij, S. Setting targets in strategies for river restoration. Landsc. Ecol. **2002**, *17* (Suppl. 1), 5–18.

35. Darby, S.; Sear, D. *River Restoration, Managing the Uncertainty in Restoring Physical Habitat*; John Wiley & Sons, Ltd., 2008.

36. Doratli, N.; Hoskara, S.O.; Fasli, M. An analytical methodology for revitalization strategies in historic urban quarters: A case study of the Walled City of Nicosia, North Cyprus. Cities **2004**, *21* (4), 329–348. doi: 10.1016/ j.cities.2004.04.009.

37. Żelazko, J. Rivers and valleys restoration, infrastructure and ecology of the countryside, Polish Academy of Sciences, Cracow Branch, Technical Committee on Rural Infrastructure, 2006, No. 01/04/2006, 11–31.

38. Davis, J.R. Revitalization of a northcentral Texas river, as indicated by benthic macroinvertebrate communities. Hydrobiologia **1997**, *345*, 95–117.

39. Osugi, T.; Tate, S-I.; Takemura, K.; Watanabe, W.; Ogura, N.; Kikkawa, J. Ecological research for the restoration and management of rivers and reservoirs in Japan. Landsc. Ecol. Eng. **2007**, 3, 159–170.

40. Ryder, D.; Boulton, A.; De Deckker, P. Setting goals and measuring success: Linking patterns and processes in stream restoration. Hydrobiologia **2005**, 552, 147–158.

41. Williams, C.A.; Cooper, D.J. Mechanisms of riparian cottonwood decline along regulated rivers. Ecosystems **2005**, *8*, 382–395.

42. Wang, W.; Tang, X.Q.; Huang, S.L.; Zhang, S.H.; Lin, C.; Liu, D.W.; Che, H.J.; Yang, Q.; Scholz, M. Ecological restoration of polluted plain rivers within the Haihe river basin in China. Water Air Soil Pollut. **2010**, *211*, 341–357.

43. Zhang, L. Ecological water requirement based on ecological protection and restoration targets in the lower reaches of the Heihe River, northern China. Front. For. China **2009**, *4* (3), 263–270. doi: 10.1007/s11461-009-0052-0.

44. Buijs, A.E. Public support for river restoration. A mixed- method study into local residents' support for and framing of river management and ecological restoration in the Dutch floodplains. J. Environ. Manage. **2009**, *90*, 2680–2689.

45. Bar-Or, Y. Restoration of the rivers in Israel's coastal plain. Water Air Soil Pollut. **2000**, *123*, 311–321.

46. Willis, K.G.; Garrot, G.D. Angling and recreation values of low-flow alleviation in rivers. J. Environ. Manage. **1999**, *57* (2), 71–83.

47. Lüderitz, V.; Jüpner, R.; Müller, S.; Feld, C.K. Renaturalization of streams and rivers—The special importance of integrated ecological methods in measurement of success. An example from Saxony-Anhalt (Germany). Limnology **2004**, *34*, 249–263.

48. Nowak, M.C. *Burnt River Subbasin Plan, DRAFT*. Cat Tracks Wildlife Consulting. Prepared for Northwest Power and Conservation Council, 2004.

49. Arlettaz, R.; Lugon, A.; Sierro, A.; Werner, P.; Kéry, M.; Oggier, P.-A. River bed restoration boosts habitat mosaics and the demography of two rare non-aquatic vertebrates. Biol. Conserv. **2011**, *144*, 2126–2132.

50. Junker, B.; Buchecker, M. Aesthetic preferences versus ecological objectives in river restorations, Landsc. Urban Plan. **2008**, *85*, 141–154.

51. Ivannikov, F.A.; Prokofieva, T.V. Artificial soil-like bodies of a river valley and their evolution in an urban environment (based on the example of the Moscow river valley). Moscow Univ. Soil Sci. Bull. **2010**, *65* (4), 145–150.

52. Pollard, P.; Devlin, M.; Holloway, D. Managing a complex river catchment: A case study on the River Almond. Sci. Total. Environ. **2001**, *265*, 343–357.

53. Rohde, S.; Kienast, F.; Bürgi, M. Assessing the restoration success of river widenings: A landscape approach. Environ. Manage. **2004**, *34* (4), 574–589.

54. Connelly, N.A.; Knuth, B.A.; Kay, D.L. PROFILE, public support for ecosystem restoration in the Hudson River Valley, USA. Environ. Manage. **2002**, *29* (4), 467–476. doi: 10.1007/s00267-001-0033-Z.

55. Jun, K.S.; Kim, J.S. The Four Major Rivers Restoration Project: Impacts on river flows. KSCE J. Civil Eng. **2011**, *15* (2), 217–224. doi: 10.1007/s12205-011-0002-x.

56. Feasibility Study by Buffalo River, New York, Draft Final. ENVIRON International Corporation, MACTEC Engineering and Consulting, Inc., LimnoTech, 2010.

57. Eer River Headwaters Restoration Project, Phase 1—Cranberry Bog and River Restoration, Massachusetts Department of Environmental Protection, Bureau of Resource Protection—Wetlands and Waterways, Plymouth, 2008.

58. How to reconcile the interests of man and nature? The Regional Water Management Board in Krakow, available at http://www.krakow.rzgw.gov.pl/ (accessed May 19, 2011) (in Polish).

59. Chardon, W.J.; Koopmans, G.F., Eds. Critical evaluation of options for reducing phosphorus loss from agriculture. IPW 4—Proceedings of the 4th Int. Phosphorus Workshop, Wageningen, the Netherlands, 2004.

60. Larson, P. Restoration of river corridors: German experiences. In: *River Restoration;* Petts, G.E., Calow, P., Ed.; Blackwell: Cambridge, U.K., 1996; 124–143.

61. Meixner, H.; Schnauder, I.; Bölscher, J.; Iordache, V., Eds. Hydraulic, sedimentological and ecological problems of multifunctional riparian forest management RIPFOR, guidelines for end-users. Berl. Geogr. Abh. **2006**, *66*, 1–168.

62. Obolewski, K.; Glińska-Lewczuk, K.; Osadowski, Z.; Kobus, S. Water quality monitoring of agricultural River catchments In the example of the Kwacza River Valley (Middle Slupia River basin). Polish Journal of Environmental Studies, **2008**, *17* (2c), 39–46.

63. Alwan, A.A.; Appiah-Kubi, G.; Majland-Kristensen, P. *The Possible Impact of the Restoration of River Skjern;* Environmental Studies; University of Aarhus: Aarhus, Denmark, 2001.

64. River Frome Rehabilitation Plan, Environment Agency, Newsletter No. 2, 2010.

65. Bramley, M. Development of Engineering in River Restoration. River Restor. News Newsl. River Restor. Centre **2002**, *12*, 2–3.

66. STREAM—River Avon cSAC demonstrating strategic restoration and management LIFE05 NAT/UK/000143, STREAM Newsletter no. 7, 2007.

67. Niagara Power Project (Ferc No. 2216), Feasibility Study for the Restoration of Native Terrestrial Plants in the Vicinity of the Niagara Gorge, New York Power Authority, 2008.

68. Wada, A. Sumida River, Japan. Japan River Restoration Network JRNN, Panel on Urban Water/River Front Redevelopment, World Bank Tokyo Office, May 30th, 2011.

69. Environment Agency. Bringing Your Rivers Back to Life. *A Strategy for Restoring Rivers in North London;* Environment Agency, Rio House, Waterside Drive, Aztec West, Almondsbury, Bristol, 2006.

70. Environment Agency. The London Rivers Action Plan, A tool to help restore rivers for people and nature, available at http://www.therrc.co.uk/lrap.php (accessed May 2011).

71. Environment Agency. *River Restoration. A Stepping Stone to Urban Regeneration Highlighting the Opportunities in London;* Environment Agency: Bristol, U.K., 2002

53

Waste: Stabilization Ponds

Application of Waste Stabilization Pond Systems 535
Types of WSPS and Their Specific Uses 535
Processes in WSPS 536
    Anaerobic Ponds • Facultative Ponds • Kinetics of the Processes in
    Facultative Ponds • Maturation Ponds • Nutrient Removal in WSPs
Design of WSPs 539
    Design Parameters • Estimation of Water Flows and BOD
    Concentrations • Design of Anaerobic Ponds • Design of
    Facultative Ponds • Design of Maturation Ponds for Fecal Coliform
    Removal • Helminth Egg Removal • Water Quality and WSPs

References 541

Sven Erik Jørgensen

Application of Waste Stabilization Pond Systems

A more detailed review of waste stabilization ponds (WSPs), their kinetics, and design can be found in Kayombo et al.[1]

WSPs are large, shallow basins in which raw sewage is treated entirely by natural processes involving both algae and bacteria.[2] They are used for sewage treatment in temperate and tropical climates and represent one of the most cost-effective, reliable, and easily operated methods for treating domestic and industrial wastewater. WSPs are very effective in the removal of fecal coliform bacteria. Sunlight energy is the only requirement for its operation. Further, it requires minimum supervision for daily operation, by simply cleaning the outlets and inlet works. The temperature and duration of sunlight in tropical countries offer an excellent opportunity for high efficiency and satisfactory performance for this type of water-cleaning system. They are well suited for low-income tropical countries where conventional wastewater treatment cannot be achieved due to the lack of a reliable energy source. Further, the advantage of these systems, in terms of removal of pathogens, is one of the most important reasons for its use.

Types of WSPS and Their Specific Uses

WSP systems comprise a single string of anaerobic, facultative, and maturation ponds in series, or several such series in parallel. In essence, anaerobic and facultative ponds are designed for removal of biochemical oxygen demand (BOD), and maturation ponds are designed for pathogen removal, although some BOD removal also occurs in maturation ponds and some pathogen removal occurs in anaerobic and facultative ponds.[3,4] In most cases, only anaerobic and facultative ponds will be needed for BOD removal when the effluent is to be used for restricted crop irrigation and fishpond fertilization, as well as when weak sewage is to be treated prior to its discharge to surface waters. Maturation ponds are only required when the effluent is to be used for unrestricted irrigation, thereby having to comply with the World Health Organization guideline of >1000 fecal coliform bacteria/100 mL. The WSP does not require mechanical mixing, needing only sunlight to supply most of its oxygenation. Its performance may be measured in terms of its removal of BOD and fecal coliform bacteria.

Processes in WSPS

Anaerobic Ponds

Anaerobic ponds are commonly 2–5 m deep and receive wastewater with high organic loads (i.e., usually greater than 100 g/m² 24h, equivalent to more than 3000 kg/ha/ day for a depth of 3 m). They normally do not contain dissolved oxygen or algae. In anaerobic ponds, BOD removal is achieved by sedimentation of solids and subsequent anaerobic digestion in the resulting sludge. The process of anaerobic digestion is more intense at temperatures above 15°C. The anaerobic bacteria are usually sensitive to pH <6.2. Thus, acidic wastewater must be neutralized prior to its treatment in anaerobic ponds. A properly designed anaerobic pond will achieve about a 40% removal of BOD at 10°C and more than 60% at 20°C. A retention time of 1.0–1.5 days is commonly used.

Facultative Ponds

Facultative ponds (1–2 m deep) are of two types: Primary facultative ponds that receive raw wastewater, and secondary facultative ponds that receive particle-free wastewater (usually from anaerobic ponds, septic tanks, primary facultative ponds, and shallow sewerage systems). The process of oxidation of organic matter by aerobic bacteria is usually dominant in primary facultative ponds or secondary facultative ponds. The processes in anaerobic and secondary facultative ponds occur simultaneously in primary facultative ponds, as shown in Figure 1. It is estimated that about 30% of the influent BOD leaves the primary facultative pond in the form of methane.[2] A high proportion of the BOD that does not leave the pond as methane ends up in algae. This process requires more time, more land area, and possibly 2–3 weeks water retention time, rather than 2–3 days in the anaerobic pond. In the secondary facultative

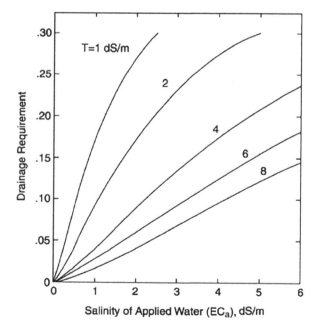

FIGURE 1 Pathways of BOD removal in primary facultative ponds.
Source: After Marais.[2]

pond (and the upper layers of primary facultative ponds), sewage BOD is converted into "algal BOD" and has implications for effluent quality requirements. About 70%–90% of the BOD of the final effluent from a series of well-designed WSPs is related to the algae they contain.

In secondary facultative ponds that receive particle-free sewage (anaerobic effluent), the remaining non-settleable BOD is oxidized by heterotrophic bacteria (*Pseudomonas, Flavobacterium, Achromobacter,* and *Alcaligenes* spp). The oxygen required for oxidation of BOD is obtained from photosynthetic activity of the micro-algae that grow naturally and profusely in facultative ponds. Facultative ponds are designed for BOD removal on the basis of a relatively low surface loading (100–400 kg BOD/ha/day), in order to allow for the development of a healthy algal population, since the oxygen for BOD removal by the pond bacteria is generated primarily via algal photosynthesis. The facultative pond relies on naturally growing algae. The facultative ponds are usually dark green in color because of the algae they contain. Motile algae (*Chlamydomonas* and *Euglena*) tend to predominate the turbid water in facultative ponds, compared to non-motile algae (*Chlorella*).

The algal concentration in the pond depends on nutrient loading, temperature, and sunlight, but is usually in the range of 500–2000 μg chlorophyll-*a* per liter.[3,4] Because of the photosynthetic activities of pond algae, there is a diurnal variation in the dissolved oxygen concentration. The dissolved oxygen concentration in the water gradually rises after sunrise, in response to photosynthetic activity, to a maximum level in the mid-afternoon, after which it falls to a minimum during the night, when photosynthesis ceases and respiratory activities consume oxygen. At peak algal activity, carbonate and bicarbonate ions react to provide more carbon dioxide for the algae, leaving an excess of hydroxyl ions. As a result, the pH of the water can rise to above 9, which will kill fecal coliform. Good water mixing, which is usually facilitated within the upper water layer by strong wind, ensures a uniform distribution of BOD, dissolved oxygen, bacteria, and algae, thereby leading to a better degree of waste stabilization.

Kinetics of the Processes in Facultative Ponds

The growth of the mixed culture was studied at concentrations ranging between 200 and 800 mg COD/L, in a series of batch static reactors. From laboratory data, the specific growth rate (μ) was determined, using the modified Gompertz model.[1,5] There are several growth models in the literature used to evaluate growth parameters, first-order reaction, logistic models, etc. The maximum specific growth rate (μ_{max}) and half saturation coefficients (K_s) were calculated using the Monod kinetic equation, which is the generally applied expression:

$$\mu = \left(\mu max\right)^{*} S / \left(S + \left(K_s\right)\right)$$

S in this equation is the substrate concentration usually expressed as mg/L BOD5 or COD.

The maximum observed growth rate (μ_{max}) for heterotrophic bacteria was 3.8 L/24 hr, with a K_s of 200 mg/L when COD was applied as unit for S.

The μ_{max} for algal biomass, based on volatile suspended solids, was 2.7 L/24 hr, with a K_s of 110 mg COD/L. The μ_{max} of algae, based on chlorophyll-*a*, was 3.5 L/24 hr with a K_s of 50 mg COD/L. The indicated constants were all found on the basis of literature values that were found from data. The regression coefficients for the constants were all 0.96–0.99 according to the literature.

The observed specific substrate removal by heterotrophic bacteria varied between the concentrations of substrate used, with an average value of 0.82 (mg COD/mg biomass). Thus, the determined Monod kinetic parameters are useful for defining the operation of secondary facultative ponds. The specific substrate utilization rate in the bioreactors was directly proportional to the specific growth rate.

Although the Monod equation was developed for the pure culture of bacteria growing in a single organic substrate, the results may be used correctly also for a mixed culture of algae and heterotrophic bacteria growing in a heterogeneous-mixed organic carbon. The K_s values obtained are the characteristic of the wastewaters containing a complex mixture of organic carbon. The Monod constants for algae depend on the method used in determining the biomass as indicated above.

Knowledge of the effects of the pH on the growth rate of algae and heterotrophic bacteria, and its subsequent impacts on the degradation of organic matter, is useful for better operation and design of secondary facultative ponds. The kinetics of microbial growth and product formations are influenced by diurnal variation in the pH in the pond resulting from diurnal variations in the carbon dioxide.[6] Algae require large quantities of dissolved carbon dioxide during photosynthesis, causing a depletion of carbon dioxide (CO_2) and leading to a shift in the carbonate-bicarbonate ($CO3^{2-}$–$HCO3^-$) equilibrium, resulting in an increase in pH due to the formation of hydroxyl (OH^-) ions.[7] The effects of pH on the growth rate of heterotrophic bacteria and algae in secondary facultative ponds were investigated, using batch growth at a pH value between 5 and 11. The optimum pH was found to be between 7.0 and 8.0. The results indicate that, for a pH value higher than 8, the chlorophyll-*a* content decreases. However, the specific growth rate of heterotrophic bacteria and algae was high with a pH value between pH 6.5 and 8. At a pH value above 9, the specific growth rates of both biomasses decrease.

Oxygen tension in WSPs is an operational parameter with a great deal of daily and hourly variation. The rate of oxygen production is a function of the algal concentration. Because algal growth is both light and temperature dependent, the rate of oxygen production (photosynthetic) follows the same pattern. Temperature also is a parameter exhibiting marked seasonal and daily variation in WSPs. It influences photosynthesis, the growth of microorganisms, and the bio-decomposition of organic carbon in the system. The fluctuation in pH influences the kinetics of microbial growth, species competition, and product fomations in the pond.[6] Microbial species can grow within a specific pH range, which typically extends over 3 to 4 pH units, with an optimum growth rate near the midpoint of the range. Values of pH up to 11 are not uncommon in WSPs, with the highest levels being reached in the late afternoon. These results lead to the conclusion that the most useful rate that may be used to govern the processes from the WSP should be those determined from the combined influence of the various forcing functions.

Maturation Ponds

The maturation ponds, usually 1–1.5 m deep, receive the effluent from the facultative ponds. Their primary function is to remove excreted pathogens. Although maturation ponds achieve only a small degree of BOD removal, their contribution to nutrient removal can be significant. Maturation ponds usually show less vertical biological and physicochemical stratification, and are well oxygenated throughout the day. The algal population in maturation ponds is much more diverse than in the facultative ponds. The algal diversity generally increases from pond to pond along the series.[3] Although fecal bacteria are partially removed in the facultative ponds, the size and numbers of the maturation ponds especially determine the numbers of fecal bacteria in the final effluent. There is some removal of solids-associated bacteria in anaerobic ponds, principally by sedimentation. The principal mechanisms for fecal bacterial removal in facultative and maturation ponds are now known to be

1. Time and temperature
2. High pH (>9)
3. High light intensity, combined with high dissolved oxygen concentration

Time and temperature are the two principal parameters used in designing maturation ponds. Fecal bacterial die-off in ponds increases with both time and temperature. High pH values (above 9) occur in ponds, due to rapid photosynthesis by pond algae, which consumes CO2 faster than it can be replaced by bacterial respiration. As a result, carbonate and bicarbonate ions dissociate, as follows:

$$2\,HCO_3^- \rightarrow CO_3^{2-} + H_2O + CO_2$$
$$CO_3^{2-} + H_2O \rightarrow 2\,OH^- + CO_2$$

The resulting CO_2 is fixed by the algae, and the hydroxyl ions accumulate, often raising the pH to values even above 10. Fecal bacteria (with the notable exception of *Vibrio cholerae*) die very quickly at pH values higher than 9.[8] The role of high light intensity and high dissolved oxygen concentration has recently been elucidated. Light of wavelengths between 425 and 700 nm can damage fecal bacteria by being absorbed by the humic substances ubiquitous in wastewater. They remain in an excited state sufficiently long to damage the cell. Light-mediated die-off is completely dependent on the presence of oxygen and is enhanced at high pH values. Thus, the sun plays a threefold role in directly promoting fecal bacterial removal in WSP and in increasing the pond temperature, and more indirectly by providing the energy for rapid algal photosynthesis. This not only raises the pond pH value above 9, but also results in high dissolved oxygen concentrations, which are necessary for its third role, namely, promoting photo-oxidative damage.

Nutrient Removal in WSPs

In anaerobic ponds, organic nitrogen is decomposed to ammonia. Thus, the effluent from anaerobic ponds usually has higher concentrations of ammonia than what is found in raw sewage. In facultative and maturation ponds, ammonia is incorporated into algal biomass. At high pH values, ammonia leaves the pond through volatilization. There is little evidence for nitrification (and hence denitrification, unless the wastewater has a high nitrate content).[3] This is due to the fact that the population of the nitrifying bacteria is low because of the lack of physical attachment sites in the aerobic zone. Total nitrogen and ammonia removal from WSP can reach 80% and 95%, respectively.[3] Phosphorus removal in WSP is associated with its uptake by algal biomass, precipitation, and sedimentation. Mara[3] suggested that the best way to remove much of the phosphorus in the wastewater by WSP is to increase the number of maturation ponds. However, both nitrogen and phosphorus must be removed in order to prevent eutrophication in receiving water bodies. The common practice in the design of the WSP is not based on nutrient removal; rather, it is based on BOD and fecal coliform removal.

Design of WSPs

Design Parameters

There are four important design parameters for WSP, including temperature, net evaporation, flow, and BOD. The climate also is important in as much as the processes responsible for BOD5 and fecal bacterial removal are temperature dependent. Further, algal photosynthesis depends on solar insulation, which is a function of latitude and cloud cover. Cloud cover periods are seldom a problem because the solar insulation during the day in tropical and subtropical regions generally greatly exceeds the saturation light intensity of the algae in the ponds. The design temperature usually is the mean air temperature in the coolest month (or quarter). The pond water is usually 2–3°C warmer than the air temperature in the cool season, with the reverse also being true in the warm season.

Because the bacteria responsible for treatment are mesophilic, high temperatures are not a problem. However, low temperatures can be, since they slow down the treatment process. In the case of the methanogenic bacteria (crucial to anaerobic digestion), methane production virtually ceases below temperatures of 15°C. Thus, in areas where the pond temperature remains below 15°C for more than a couple of months of the year, careful consideration should be given to deciding whether or not anaerobic units are needed. Net evaporation (evaporation minus rainfall) must be taken into account during the design of facultative and maturation ponds, but not for anaerobic ponds.[9] Anaerobic ponds generally have a scum layer, which effectively prevents significant evaporation of the water.

Total nitrogen and free ammonia (NH_3, rather than $NH_4^+ + NH_3$) are important in the design of wastewater-fed fish ponds. Typical concentrations of total nitrogen in raw domestic wastewater are 20–70 mg N/L, and total ammonia ($NH_4^+ + NH3$) concentrations are 15–40 mg N/L. Fecal coliform numbers are important if the pond effluent is to be used for unrestricted crop irrigation or for fishpond fertilization. Grab samples of the wastewater may be used to measure the fecal coliform concentration if wastewater exists.

Estimation of Water Flows and BOD Concentrations

The mean water flows should be carefully estimated, since they have direct effects on the size of the ponds and the construction costs. A suitable design is 85% of the in-house water consumption. The BOD may be measured if wastewater exists, based on 24 hr flow weighted data. Alternatively, the BOD may be estimated from the value 30 to 70 g/(capita/24 hr), with rich communities producing more BOD than poor communities. In medium-sized towns, a value of 50 g/(capita/24 hr) is more suit-able.[10,11] A typical design figure for an urban area in a developing country would be 40–50 g BOD_5/(capita/24 hr).[9] Although it is dangerous to generalize, in view of the wide variations that can be expected with differing social customs, religion, etc., a BOD_5 per capita contribution of 40 g/day with a wastewater contribution of about 100 L/(capita/24 hr) is probably a reasonable initial estimate where there is a household water supply; however, flows also could be considerably less. The usual range of fecal coliform in domestic wastewater is 10^7–10^8 fecal coliform/100 mL, with a suitable design value being 5×10^7/100 mL.

Design of Anaerobic Ponds

The anaerobic ponds are designed on the basis of a recommended volumetric loading between 100 and 400 g BOD_5/m^3/24 hr, in order to maintain anaerobic conditions. Once the organic loading is selected, the volume of the pond is then determined and the hydraulic retention time is calculated. A retention time of less than 1 day should not be used for anaerobic ponds. If it occurs, however, a retention time of 1 day should be used, and the volume of the pond should be recalculated accordingly.

Design of Facultative Ponds

Facultative ponds can be designed on the basis of kinetic or empirical models. The rate at which the organic matter is oxidized by bacteria is a fundamental parameter in the rational design of biological wastewater treatment systems. It has been found that BOD removal often approximates first-order kinetics; that is, the rate of BOD removal (rate of oxidation of organic matter) at any time is proportional to the quantity of BOD (organic matter) present in the system at that time. The simple approach to the rational design of facultative ponds assumes that they are completely mixed reactors in which BOD_5 removal follows first-order kinetics[12] with the kinetic constant, k_1 at 20°C to be equal to 0.3/day.[13] For the temperature dependence, the Arrhenius expression is used, with the constant between 1.01 and 1.09—an average value of 1.05 is recommended to be applied generally. A minimum retention time value of 5 days should be adopted for temperatures below 20°C, and 4 days for temperature above 20°C. This is to minimize hydraulic short-circuiting and to give algae sufficient time to multiply (i.e., to prevent algal washout).

Design of Maturation Ponds for Fecal Coliform Removal

The method of Marais[2] is generally used to design a pond series for fecal coliform removal. This assumes that fecal coliform removal can be reasonably well represented by a first-order kinetic model in a completely mixed reactor. The numbers of fecal coliform/100 mL in the effluent and influent are used as unit in the applied first-order equation. k_T, the first-order rate constant for fecal coliform removal, is 2.6 L/24 hr, and it is highly dependent on the temperature. If an Arrhenius expression is used to calculate the influence of the temperature, an Arrhenius constant as high as 1.19 should be applied. Thus, k_T changes by 19% for every change in temperature of 1°C. The first-order equation represents the removal of fecal coliform in a series of ponds as a whole reasonably well. Maturation ponds require careful design to ensure that the fecal coliform removal is sufficient. Marais,[2] and Mara,[13] recommend a value of 3 days as the minimum retention time of the maturation ponds, although at temperatures below 20°C, values of 4–5 days are preferable. The BOD loading on the first maturation pond must be checked and must not be higher than that on the preceding facultative pond; in fact, it is preferable that it be significantly lower. The maximum BOD loading in the first maturation pond should normally be 75% of that on the preceding facultative pond. It is not necessary to check the BOD loadings on subsequent maturation ponds, as the non-algal BOD contribution to the load is very low. The loading on the first maturation pond is calculated on the assumption that 70% of unfiltered BOD has been removed in the preceding anaerobic and facultative ponds (or 80% for temperatures above 20°C). Mara and Pearson[11] and Mare[14] also suggested 90% cumulative removal in anaerobic and facultative ponds, and then 25% in each maturation pond, for temperatures above 25°C (80% and 20%, respectively, for temperatures below 20°C), when the BOD is based on filtered BOD values.

Helminth Egg Removal

Helminth eggs are normally removed by sedimentation, with the process occurring in the anaerobic or primary facultative ponds. If the final effluent is to be used for restricted irrigation, it is necessary to ensure that it contains no more than one egg per liter.

Water Quality and WSPs

By the right design of WSPs, it is possible to obtain close to 90% removal of BOD_5 (i.e., a BOD_5 of about 10–20 mg/L), a phosphorus removal of about 20%–35%, and a total nitrogen removal of at least 25%–50%. Usually, nitrate is removed very effectively. The organic nitrogen is easily oxidized to ammonia but ammonia may be removed by a relatively low efficiency due to insufficient oxygen concentrations in the ponds. By aeration (even rather moderate aeration), for instance, in the last facultative pond, it is possible to increase the nitrogen removal to almost 85% by oxidation of ammonium to nitrate, which, as indicated above, is removed usually with high efficiency. By careful design of the maturation ponds, the removal of coliform bacteria will be satisfactory.

Kayombo et al.[1] present several models that are able to give a fully acceptable design of WSPs. The same reference gives several equations for an empirical design, but as the model can consider interactions of several factors of importance for the design, it is recommendable to apply models for the design of WSPs whenever it is possible.

WSPs are, with very few exceptions, only applied in the tropical region, because the treatment results would not be sufficient during the winter months in the temperate zone, where the industrialized countries are using mostly high-technological treatment methods.

References

1. Kayombo, S.; Mbwette, T.S.A.; Katima, J.H.Y.; Ladegaard, N.; Jørgensen, S.E. Waste Stabilization Ponds and Constructed Wetlands. Design Manual Published by Danida and UNEP, 2004.
2. Marais, G.V.R. Dynamic behavior of oxidation ponds. Proceedings, 2nd International Symposium for Wastewater Lagoon, Missouri Basin Engineering Health Council and Federal Water Quality Administration, University of Kansas, Lawrence, 1970; 15–46.
3. Mara, D.D. Waste stabilization ponds. Problems and controversies. Water Qual. Int. **1987**, *1*, 20–22.
4. Mara, D.D. *Sewage Treatment in Hot Climates*; John Willey & Sons: New York, 1976; Vol. 120, 149.
5. Kayombo, S.; Mbwette, T.S.A.; Mayo, A.W.; Katima, J.H.Y.; Jorgensen, S.E. Modelling diurnal variation of dissolved oxygen in waste stabilization ponds. J. Ecol. Model. **1999**, *127*, 21–31.

6. Fritz, J.J.; Middleton, A.C.; Meredith, D.D. Dynamic process modelling of wastewater stabilization ponds. J. Water Pollut. Control Fed. **1979**, *51* (11), 2724–2742.

7. Neel, J.K.; McDermott, J.H.; Monday, C.A. Experimental lagoon of raw sewage at Fayette, Missouri. J. Water Pollut. Control Fed. **1961**, *33* (6), 603–641.

8. Pearson, H.W.; Mara, D.D.; Mills, S.W.; Smallman, D.J. Factors determining algal populations in waste stabilization ponds and the influence of algae on the performance. Water Sci. Technol. **1987**, *19* (12), 131–140.

9. Arthur, J.P. Notes on the Design and Operation of Waste Stabilization Ponds in Warm Climates of Developing Countries. Technical Paper No. 7, World Bank, Washington, DC, 1983.

10. Mara, D.D.; Pearson, H. Artificial freshwater environment: Waste stabilization ponds. Biotechnology **1986**, 8–16.

11. Mara, D.D.; Pearson, H. *Waste Stabilization Pond. Design Manual for Mediterranean Europe*; WHO Regional Office for Europe: Copenhagen, 1987.

12. Marais, G.v.R.; Shaw, V.A. A rational theory for the design of sewage stabilization ponds in tropical and sub-tropical areas. Trans. S. Afr. Inst. Civil Eng. **1961**, 3, 205.

13. Mara, D.D. *Design Manual for Waste Stabilization Ponds in India*. Ministry of Environment and Forests, National River Conservation Directorate, Lagoon Technology International: Leeds, U.K., 1997.

54

Wastewater Treatment Wetlands: Use in Arctic Regions 5-Year Update

Introduction 543
State of Current Knowledge and Practice 544
A Discussion of Performance 545
  Case Study of Paulatuk Treatment Wetland • Background
  Information • Methods • Results • Discussion
Potential for CWs 553
Modeling Treatment Wetlands 555
Conclusion 556
References..556

Colin N. Yates,
Brent Wootton, and
Stephen D. Murphy

Introduction

In the last 50 years, conventional and alternative wastewater treatment systems (e.g., lagoons and tundra wetlands) have been used in remote Arctic communities in Canada for wastewater treatment. Current knowledge of the performance of these systems is limited, as little research has been conducted and regulatory monitoring has been poorly documented or not observed at all. Also, in the past, the rational design process of treatment systems in Arctic communities has not acknowledged cultural and socioeconomic aspects, which are important for the long-term management and performance of the treatment facilities in Canadian Arctic communities.

Wastewater treatment in extreme cold climate regions of world, such as the Canadian Arctic, is a comparatively recent exercise in the history of wastewater treatment. The term extreme cold climate wastewater treatment is used in this entry as cold climate wastewater has been coined for treatment in more temperate locations where mean annual ambient air temperatures are above 0°C. In comparison, the regions discussed in this entry all have mean annual ambient air temperatures well below –5°C, and often below –10°C.

Most communities in the far north regions of Canada, as well as other localities globally such as Greenland, have primarily relied on dilution of small volumes of wastewater by the receiving environment. However, with rapid population growth occurring in Arctic regions because of resource extraction and exploration, risk to the receiving environment has become a concern. In communities where wastewater treatment technologies have been employed, performance evaluations of the systems are now required to determine whether they can cope with the growing demand on them. This entry includes an overview of wastewater treatment in the Canadian Arctic with a specific emphasis on the use of tundra wetlands to treat municipal wastewater from the region's remote communities.

The corresponding objectives of this entry are as follows: (1) to provide a detailed description of current wastewater treatment practice in the Canadian Arctic with special consideration on the use of wetland systems; (2) to convey the current knowledge of wastewater treatment performance using natural wetlands in the Canadian Arctic; (3) to explore the potential for the application of ecotechnologies, namely, constructed or engineered wetlands in remote Arctic communities; and (4) to discuss a case study on a natural treatment wetland (Paulatuk, Northwest Territories, Canada) as an example to illustrate performance as well as describe other challenges associated with remote extreme cold climate wastewater treatment.

State of Current Knowledge and Practice

Current knowledge of performance of Canadian Arctic wastewater treatment systems is largely restricted to governmental reports, consultant reports, and other sources of gray literature. Only a few peer-reviewed documents exist to contribute to our current understanding of performance and are confined to the performance of lagoons.

In the Canadian Arctic, wastewater treatment facilities such as lagoons and wetlands are largely designed and managed using southern engineering standards, adopting design models to reflect Arctic temperature.[1–3] Since the 1970s, our knowledge of wastewater treatment in remote Canadian Arctic communities has grown very little despite a half-century of operation. Much of our understanding has been developed from site-specific consultant and government reports[4,5] and only a few peer-reviewed entries, as well as conference proceedings.[6,7]

The entries that have been listed above primarily address the performance of lagoon treatment systems. Only Doku and Heinke discuss the potential for greater use of natural and constructed wetlands (CWs) to treat wastewater in Northern Canada in detail.[8] The work of Dubuc et al. is one of very few papers to investigate long-term performance of treatment wetlands in Northern Canada, with a study on a hydro-construction camp along the 55th parallel in the province of Quebec.[9] To date, no long-term monitoring of treatment wetlands has occurred in the Canadian Arctic, except for research conducted by the authors in several wetland systems between the years 2008 and 2010. Unfortunately, in the past 10 years, additional long-term studies have not continued. Nor has there been any extensive discussion or study of mechanistic functions of tundra wetlands to treat wastewater in peer-reviewed literature except for some sporadic example works conducted since 2013.[10,11] Still, the closest approximation for mechanistic functionality in Arctic treatment wetlands is drawn from cold temperate climate regions of southern Canada, Scandinavia, and northern United States; examples from extensively studied locations being from Minot Wetland in North Dakota and Houghton Lake wetland in Michigan.[12,13] Only an entry by Kadlec and Johnson addresses some mechanistic function in a Canadian Arctic treatment wetland but does not provide significant background data.[1] Furthermore, much of the current knowledge on plant and microbial influence on wastewater treatment in the Arctic derives from smaller-scale fertilizations and carbon cycling studies in different Arctic environments.[14,15]

The authors would also like to note the lack of attention that has been given to the planning practice of wastewater treatment in the Canadian Arctic. In both peer-reviewed and gray literature, evidence of planning practice for wastewater treatment has been minimal. Only Ritter and Johnson directly touch upon the issue of planning and wastewater management in remote Arctic communities in North America.[16,17] More recently, more discussion has been made on public health risks in the Canadian Arctic with respect to wastewater.[18,19] The remainder of current thought on the subject relies on contributions from indirect sources on waste management and contamination in the Arctic.[20,21] Unpublished work by author Yates discusses these issues in great detail and identifies several primary factors that have compounded the problem of sanitation in Arctic communities, including the climate, their remote localities, and in some cases physiographic features. A few experts in sanitation and wastewater treatment have proposed and implemented conventional techniques in the Canadian Arctic with varying degrees of success in terms of (1) performance of the technology and (2) acceptance or understanding by the community.[6,23,24] Despite this, there remains a great deal of uncertainty with regard to which approaches are most suitable for Arctic communities. This uncertainty is because there is limited knowledge of how Arctic environments respond to increased loads of nutrients, other pollutants, and water, and how conventional systems respond to Arctic conditions.[7] Therefore, appropriate loading rates and predictions of expected performance are still largely speculative. Socioeconomic issues are also an ever-present concern, especially in the Canadian Arctic. This is related to a lack of resources and trained personnel as well as other factors.[16] Furthermore, there is distrust caused by a lack of communication and discussion between aboriginal groups in the Canadian Arctic and government agencies over treatment approaches and, more recently, concerns over compliancy to existing and proposed new regulatory standards.[25]

As previously described, there is a paucity of seasonal and long-term performance data of tundra treatment wetlands. Therefore, unsurprisingly, we know even less with respect to treatment mechanisms in tundra treatment wetlands. Attention still needs to be given to testing alternative technologies for wastewater treatment, such as CWs or engineered wetlands in these remote communities as well as emerging technologies that may provide low cost, effective, and socioculturally acceptable. The following discussion presents some evidence on the capability of tundra wetlands' ability to treat municipal wastewater.

A Discussion of Performance

Cold climate treatment wetlands have been identified as a significant area of interest for those studying treatment wetlands in the past three decades.[26] Vymazal also identified the important role that natural wetlands historically have played in our understanding of wetland function for wastewater treatment.[26] However, because of the growing knowledge of the importance of wetland function and values early in the adoption of wetlands to treat wastewater, their use has largely ceased except in controlled conditions[29,30] and in a few other locations around the world.[22] The tundra wetlands in the Canadian Arctic are among those still used to treat wastewater.

During a period over about 5 years (2008–2013), significant effort and funding were made available to characterize wastewater treatment wetlands in the Canadian Arctic for performance and to make recommendations on how to best manage these extreme cold climate systems. This work was largely conducted by CAWT (Fleming College) and Dalhousie. In the following discussion, a case study is used to illustrate the performance of Arctic tundra wetland wastewater treatment the authors conducted during that time. The authors also draw on some preliminary results collected in other tundra treatment wetlands that they have studied throughout the Canadian Arctic (Figure 1) and provide discussion on the potential for CWs.

Case Study of Paulatuk Treatment Wetland

This case study documents unpublished data collected by the authors in 2009 in a remote community of the Canadian Arctic (see summary of water quality data from Paulatuk and other selected wetland sites in Table 1). The Hamlet of Paulatuk is located in the Northwest Territories, Canada (69°N 124°W). The system is composed of a facultative lake and wetland serving approximately 294 residents.[31] Since this time, the population has decreased to 265 residents.[32] Wastewater from households and a small number of businesses is trucked to the facultative lake daily. The community is only accessible by aircraft, as few physical transportation corridors exist between the majority of communities in the Canadian Arctic (Figure 2).

Background Information

In 2007, it was estimated that approximately 11,200 m^3 of wastewater was being discharged into the lake (~31 m^3/day). The lake is estimated to have a volume of 103,000 m^3.[33] Basic estimates of effluent flow rate from a preferential flow channel as measured by Yates and Wootton showed a rate of 1.2 m^3/day.[34]

TREATMENT WETLAND PERFORMANCE AND PHYSICAL CHARACTERIZATION

FIGURE 1   Map of Arctic Canada and its regions. Locations of each of the communities where treatment wetlands were studied are located in Nunavut and Northwest Territories. (Created by Noreen Goodliff.)

TABLE 1   Influent and Effluent Concentrations from Selected Tundra Treatment Wetlands in Arctic Canada

	Taloyoak			Paulatuk			Gjoa Haven		
	Influent	Effluent	% Removal	Influent	Effluent %	% Removal	Influent	Effluent	% Removal
Ammonia (NH_3N)	4.58	0.127	97	3.19	0.01	99	102	5.75	94
Total phosphorus	3.86	0.324	91	2.42	0.04	98	12.3	2.08	94
$cBOD_5$	12	3	75	40	2	95	138	3	83
Total suspended solids	–	–	–	35	3	91	22	3	97
E. coli	1.30E+03	2.40E+01	98	2.85E+03	1.00E+00	99	2.42E+03	3.30E+01	86

Note: All values are expressed as milligrams per liter (mg/L) except for *E. coli,* which was recorded as colony-forming units (cfu)/100 mL.

The wetland ranged from 40 to 80 m in width, extending approximately 350 m from the facultative lake to the Arctic Ocean (Figures 3–8). The wetland was characterized as wet-sedge tundra, dominated by *Carex* and *Poa* spp. In drier upland areas along the wetland boundaries, *Salix* spp. were observed to be dominant. The highest daily maximum temperature for the area is 15°C for July; mean annual temperature is approximately –9.2°C.[35] Paulatuk's wetland hydraulic loading rate into the facultative lake was estimated at 31 m³/day, assuming that the amount of water continuously discharged from the facultative lake is equal to the amount as it receives a Carbonaceous Biochemical Oxygen Demand ($cBOD_5$) areal loading rate of 0.9 kg/ha/day. However, the flow rate into the wetland is much less because of evaporation and some loss into groundwater. Precipitation and runoff from neighboring hillsides may add to the flow through the wetland but only a minimal amount as this region only receives 84 mm of precipitation

FIGURE 2 Aerial view of Paulatuk facultative lake and treatment wetland. The wetland is located on the upper right of the photo. Effluents flow to the right and a distance of approximately 350 m to the Arctic Ocean.
Source: Photo courtesy of Aboriginal Affairs and Northern Development Canada North Mackenzie District.

FIGURE 3 View of facultative lake flowing into the beginning of the wetland system.

FIGURE 4   View of productive vegetation in wetland.

FIGURE 5   View of wetland showing discrete boundaries.

FIGURE 6 Upslope view of wetland.

FIGURE 7 Downslope view of wetland.

FIGURE 8   View of wetland discharging into the Arctic Ocean.

from June to October.[35] Background water quality concentrations for $cBOD_5$, NH_3–N, and *Escherichia coli* from a nearby reference wetland were measured to be 2 mg/L, 0.01 mg/L, and 9 colony-forming units (cfu)/100 mL, respectively.

Methods

The authors conducted an extensive characterization analysis on the treatment wetland, collecting surface and subsurface water samples throughout the wetland. A onetime sampling of the site was conducted over a week in early September 2009 with a sample size of $n = 41$. Spatial interpolative analysis was conducted to illustrate concentration changes throughout the wetland. Measurements for $cBOD_5$, NH_3–N, and *E. coli* were taken. Samples were shipped within 24 hours to an ISO 17025 accredited laboratory and processed using *Standard Methods for the Examination of Water and Wastewater*. Tension spline was used to conduct interpolation analyses mapping of water quality parameters.

Results

The interpolation analysis showed that treatment for all parameters was found to be occurring in the first 50–100 m of the wetland. After 150 m, flows of wastewater were difficult to detect, as wastewater appeared to be flowing evenly at low velocities across much of the wetland. With a basic understanding of the rate of flow and loading of the wetlands and interpolation of concentration of specific wastewater parameters, it is possible to discuss the performance of the system. Treatment of wastewater was observed to occur primarily in the upper portions of the wetlands, with concentrations quickly dissipating to background levels (Figure 9).

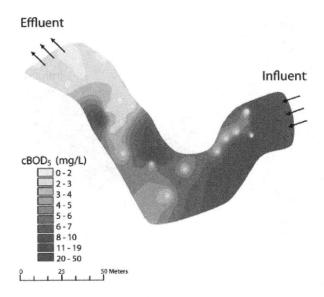

FIGURE 9 Concentration gradients of cBOD$_5$ in the Paulatuk treatment wetland.

Expected effluent concentrations for the identified active portions of the wetlands were calculated using a first-order kinetic model at 10°C. We calculated expected effluent concentrations using P-k-C* in order to determine expected effluent values for cBOD$_5$. The van't Hoff–Arrhenius equation as described in Crites and Tchobanoglous was adopted[31]:

$$\frac{d(\ln k)}{dT} = \frac{E}{RT^2} \tag{1}$$

The P-k-C* model is described in Campbell and Ogden as follows[32]:

$$As = \frac{Q(\ln Co - \ln Ce)}{k_t * d * n} \tag{2}$$

The k_t value for the P-k-C* model was determined by using a k_{10} value of 1.0; the θ-factor used was 1.14. A high θ-factor was deemed appropriate for extreme temperature cases as determined for a Minnesota horizontal subsurface flow wetland with a temperature range from 1°C to 17°C, as outlined in Kadlec and Wallace.[30]

Expected cBOD$_5$ effluent values calculated for the Paulatuk wetland were 3 mg/L. In the Paulatuk wetland, cBOD$_5$ levels were actually observed to decrease rapidly (Table 1) within the first 50–100 m. This was likely due to the low influent levels from pretreatment facilities, allowing the top end of the wetland to assimilate or treat remaining organic matter. The influent demand was observed to be 40 mg/L. Effluent demands were observed to be 2 mg/L. However, a level of 2 mg/L was observed to consistently appear less than halfway down the wetland during the study. Effluent oxygen demands were observed to be the same as background conditions measured in the nearby reference sites.

Nutrient parameters, specifically ammonia (NH$_3$–N), also decreased rapidly down each of the wetlands; concentrations quickly dissipated within the first 50 m of the wetland, again likely due to the low influent concentrations (Figure 10).

Observed NH$_3$–N concentration in the influent was 3.19 mg/L. Effluent concentration was observed to be 0.01 mg/L. Like cBOD$_5$, NH$_3$–N concentrations dissipated rapidly, achieving background concentrations in approximately 150 m. Pathogen removal by the wetland was also observed to occur quickly in

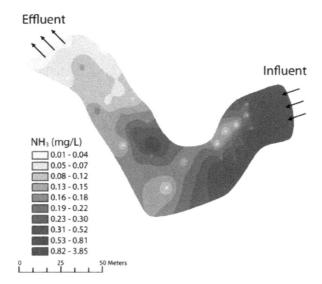

FIGURE 10   Concentration gradients of NH₃–N in the Paulatuk treatment wetland.

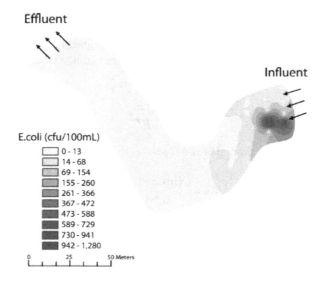

FIGURE 11   Concentration gradients of *E. coli* in the Paulatuk treatment wetland.

the pretreated wastewaters entering the wetlands in Paulatuk. Concentrations of *E. coli* were observed to be low at the influence of the wetland: 2.85E3 cfu/100 mL. *E. coli* was observed to be quickly removed, only observed in trace amounts within a third of the wetland distance. Variation throughout the wetland is likely explained by natural background concentrations from local wildlife (Figure 11).

Discussion

Microbial activity and plant growth influenced by soil temperature are likely the most important mechanisms for the treatment of wastewater in Arctic wetlands.[36] Arctic soil is known to be an excellent sink of organic matter and nutrients, immobilizing nutrients within the frozen matrix and within the microbial community.[37] Phosphorus has been shown to be bound to soils in the Arctic, rendering it unavailable for plant uptake.[38] However, it is unknown how much the soil matrix is responsible for "treatment" by locking nutrients. Fertilization studies in various Arctic habitats, including wet-sedge tundra, have shown that in nutrient-limiting conditions, plant communities respond to increased nutrient input based on small nutrient additions,[39–41] especially when nutrients were added simultaneously,[27] as would be the case with wastewater. Some Arctic plants have even demonstrated the ability to uptake organic forms of N because mineralization of organic material is slow due to low soil temperatures.[39] Plants in tundra treatment wetlands, such as those presented here, may be utilizing the readily available nutrients in such a manner, which may explain low values of inorganic N in a similar study conducted on Arctic treatment wetlands.[28] A study by Yates et al. showed that plants may play a significant role in treatment.[11] However, actual nutrient uptake rates in these systems have not been studied to determine rate or percentage of nutrients discharged into the system or taken up by the plant community. The microbial community may also play an equal role in the uptake of available nutrients in wastewater in these Arctic wetlands. Similar to plant communities, microbial activity is generally limited by temperature and available nutrients. Arctic microbial species are more efficient at lower temperatures than temperate microbial species; Arctic species continue to transform nutrients throughout the winter.[42,43] Hobbie and Chapin also suggested that microbial activity is able to utilize nutrients in soils at temperatures as low as –5°C.[36] Nutrient uptake at low temperatures was recently validated by Edwards and Jefferies.[16] These observations likely contribute to the rapid increase in wetland performance from late June to early July due to increases in microbial populations as a result of additional nutrient availability in still semi-frozen soils observed by Yates et al.[28] Whether winter microbial activity is sufficient to continue to mineralize organic matter and nutrients is unknown. It is likely that the microbial community would not be able to significantly consume the excess nutrient and organic loads at the top of the wetland, resulting in the gradual infilling of organics at the influence.

Natural ultraviolet radiation plays an important role in the disinfection of wastewater in surface wetlands and lagoon systems in more temperate systems. The long exposure of sunlight in the Arctic during the summer months in theory should promote increased disinfection. However, if water temperatures are not optimal, lysis of bacteria may not occur, as cold temperatures appear to stabilize populations, at least in lagoon environments.[2] In wetlands that do not maintain large areas of open water, solar radiation cannot penetrate the water column because of the plant canopy,[44] which is often the case in tundra wetlands that maintain dense stands of *Carex*.

Finally, sedimentation of solids occurs on the wetland surface, in various preferential flow channels throughout the wetlands, and also through entrapment in vegetation. Personal observations in field notes show accumulations of organic matter in many of the wetlands surveyed throughout the Arctic. As discussed earlier, decomposition rates by the microbial community are not as high as deposition rates. Although it was observed that water quality was low in organic load throughout much of the wetland, it would be expected that deposition would occur further down the wetland in the future.

Potential for CWs

The authors have also studied numerous other tundra wetland wastewater treatment systems throughout the Canadian Arctic observing similarly positive results. Examples of preliminary results can be found on Chesterfield Inlet's tundra treatment wetland,[28] as well as a subsurface flow CW in Baker Lake,[45] the latter being examined in some detail in the following discussion. Data from other communities that the authors have studied are included below and exemplify in/out performance of tundra treatment wetlands (Table 1).

Current findings in Arctic wetlands correspond with the results presented by Andersson et al. and Kadlec, demonstrating that natural wetlands can effectively polish pretreated wastewater, often in a much small area than was originally calculated by mass balance equations.[13,46] The performance results observed further prove the ability of wetlands, specifically natural systems, to treat wastewater in a cold climate, but more significantly, these early results demonstrate the resilience of wetlands to produce low-concentration effluents following approximately 9 months of frozen conditions. Although early studies by the authors did not aim to determine which mechanisms are largely responsible for tundra wetlands' high efficacy, specialized bacteria and macrophytes that have evolved in the low-temperature conditions of the Canadian Arctic are likely candidates. More importantly, these results indicate that ecotechnologies such as CWs could potentially be an appropriate technology for extreme cold climate wastewater treatment.

CWs or engineered wetlands have been applied around the world in numerous climates.[26,47–49] Most definitions of CWs simply acknowledge a CW as a man-made structure that emphasizes the natural characteristics of wetlands to transform and absorb contaminants.[30] Vymazal provides a similar definition: "CWs are engineered systems that have been designed and constructed to utilize natural processes involving wetland vegetation, soils, and the associated microbial assemblages to assist in treating wastewaters."[47] Throughout this entry, CWs are defined in the following manner: CWs are engineered systems that are lined to prevent significant exfiltration of wastewater into the underlying ground prior to passing through the system and maintain mechanisms to control influent and effluent flow. The preceding definition is used because various measures have been used to manipulate flow regimes in tundra wetlands in the Canadian Arctic. For example, some tundra systems have made use of some engineered structures, such as berms and inflow/outflow pipes, or make use of natural liners, such as bedrock, to contain or control flow. These systems are described as augmented natural wetlands.

Early belief was that cold climate conditions would not allow wetlands to optimally treat wastewater, and therefore, treatment wetlands would not find a place in cold climate wastewater treatment.[48] Studies from both North America and Scandinavia have largely shown that this has not been the case, and in most instances, only minor impediments to treatment have been observed.[50–52] Despite early convictions, CWs have shown great promise as alternative low-cost technologies to treat wastewater in remote, economically challenged regions and small communities even in cold climatic conditions.[53–55] However, despite their extensive successful use in cold temperate climates around the world,[48,51,52] they have yet to be tested in extreme cold climate conditions, like the Canadian Arctic. Communities in the Canadian Arctic in theory make excellent candidates for alternative wastewater treatment technologies, because of limited economic resources, physiographic characteristics, and trained personnel to operate and maintain more conventional mechanical treatment facilities.[18] For a number of decades, communities in the Canadian Arctic have been using tundra wetlands to treat their wastewater.[1,56] Although our knowledge is growing and understanding of treatment performance and mechanisms of the tundra wetlands is limited, some evidence has shown excellent (sometimes orders of magnitude below regulatory standards) removal for regulated wastewater effluent parameters during the summer months.[1,28] Because of the socioeconomic conditions and the extensive use of tundra wetlands in the Canadian Arctic, CWs warrant experimentation in this region. However, testing of these systems in the Canadian Arctic will require research trials on different design specifications to account for low soil/water temperatures, short frost-free period, slow rate of decomposition of organic matter, and therefore slow mineralization rates of various nutrients. Engineered designs to optimize existing tundra wetlands, augmented natural wetlands to increase hydraulic residency time (HRT), and increase active treatment zones (decrease areal loading rates) have been adopted in a few instances in the Arctic; Cambridge Bay, Nunavut, is an example of one such system.[1] However, the Arctic does provide an environment that in theory provides optimal treatment conditions, for example, 24 hours of sunlight in summer, and plants and bacteria that have evolved in a cold and nutrient-limited environment, giving them characteristics that allow them to utilize excess nutrients efficiently.

As tundra wetlands are extensively used in the Canadian Arctic, CWs have excellent potential to act as low-cost technologies for Arctic communities. The authors studied the performance of the first known experimental engineered HSSF system in the Canadian Arctic. The system demonstrated very promising results in its first year (2009) of operation despite high loading rates; observed reductions of wastewater concentrations were 25%, 31%, 52%, 99.3%, 99.3%, and 5% for $cBOD_5$, COD, TSS, *E. coli*, total coliforms, and TP, respectively.[45] In 2010, the system was operated with lower loading rates. It was expected that the system would achieve greater reductions, but this was not the case. Concentrations in the wetland effluent were observed to be greater than the effluent. Based on these observations, it was concluded that high organic loading prior to biofilm and plant establishment and high organic loading during the first year of study saturated the system with organics, resulting in the release of solids and unmineralized nutrients into a less concentrated influent. Overall, the HSSF system did not perform as expected but did demonstrate the potential for use in remote Arctic communities. Further investigations of other CW designs should be undertaken in the future.

The issues discussed to this point have largely pertained to wastewater management in the Canadian Arctic. However, they do reflect similar conditions in remote communities elsewhere in the world, cold and warm climates alike. Similar scenarios have been recently described by Jenssen in Greenland, which have similar demographics to the Canadian Arctic.[56] In the mid-1990s, rural communities in Estonia were also facing similar challenges with insufficient treatment facilities because of shortcomings in economic resources.[57] Denny and Kivaisi both describe how CWs could have potential for developing countries, specifically those in Africa.[55,58] As suggested by the previous examples, the discussion of wastewater management for developing countries is prevalent in the literature, yet little attention until this point has been directed towards remote underdeveloped communities in the Canadian Arctic. It is important to note that many common themes run throughout all of these regions that are not dissimilar to those described in the Canadian Arctic; these include but are not limited to low-economic capacity, absence of skilled labor, and complex sociocultural environments. The recommendations made for wastewater management in the Canadian Arctic contribute to knowledge development in all remote regions globally. Also, a similar set of approaches that include the use of an adaptable management framework, and accounting for differences in understanding from the experts in the field (e.g., engineers and planners) designing systems to those adopting the technology (e.g., wastewater operators) in the communities, could be easily adopted or tested outside the Canadian Arctic. The continued optimized use of wetlands, particularly CWs, is one approach that could be more extensively explored in all cases.

Modeling Treatment Wetlands

To support the United Nations Millennium Development Goals of reversing the loss of environmental resources as well as halving by 2015 the proportion of people without access to safe drinking water and sanitation, the United Nations Environment Programme promotes and facilitates environmentally sound technologies (ESTs). Natural wetlands and CWs are considered to be ESTs for wastewater treatment. Design and use of treatment wetlands are facilitated with the use of models to replicate processes and provide practical tools for sanitation.

SubWet, software for the design of subsurface horizontal flow artificial wetlands for water quality improvement and treatment, was originally developed by United Nations Environment Programme–Division of Technology, Industry and Economics–International Environmental Technology Centre (UNEP–DTIE–IETC). After being successfully used as a design tool in 15 cases in Tanzania, it was felt that the model should be upgraded for cold climate application. The Centre for Alternative Wastewater Treatment of Fleming College further developed a new version in collaboration with UNEP–DTIE–IETC, creating SubWet 2.0 to accommodate temperate and cold climatic conditions including summer Arctic and temperate winter conditions.[58] The model simulates removal of nitrogen (including nitrogen in ammonia, nitrate, and organic matter), phosphorus, and BOD_5 in milligrams per liter and the corresponding removal efficiencies in percentage. Design inputs in the model for the wetland are as follows: width, length, depth, slope, % particulate matter, precipitation factor (total flow rate included precipitation/flow rate of treated water), hydraulic conductivity, and selected flow rate (in cubic meters per day).

The forcing functions applied for the model are the following: volume of wetland; flow of water; porosity; input of BOD_5, ammonium, nitrate, total phosphorus, and organic nitrogen; fraction of BOD_5, phosphorus, and organic N as suspended matter; average oxygen concentration; and average temperature. The length of model simulations must be indicated as number of days. The model uses

16 parameters, for example, the temperature coefficient of nitrification, and has a version for warm climate and a version for cold climate. The difference between the two versions is only the set of parameter values. The model uses 25 differential process equations. The model calculates the values of BOD_5-out, nitrate-out (NIT-out), ammonium-out (AMM-out), total phosphorus-out (TPO-out), and organic nitrogen-out (ORN-out) and can be shown in the form of both tables and graphs.

SubWet 2.0 has been developed to support decision-making processes by assisting experts and water managers in the design and evaluation of CWs to improve water quality and treat domestic wastewater. Furthermore, SubWet 2.0 can also be used as a tool to improve the efficiency of low- or non-performing systems. Due to its characteristics, this software is very useful for training technicians and students who are interested in modeling natural and artificial wetlands used for wastewater treatment and can also be used as a tool by engineers and regulators.

Finally, some other technologies have emerged to try solving the challenge effectively and economically that are worth noting including chemical coagulation, UV, and peracetic acid disinfection in Greenland[59] and Bioelectrochemical Anaerobic Sewage Treatment, or BEAST. However, little data have been published on the success of these technologies to date in small Arctic communities.

Conclusion

Wastewater treatment in extreme cold climates will continue to be a challenge into the future until more knowledge is gained in the field and new technological advancements are tested in these regions. Socioeconomic barriers will likely persist even longer and will only become exacerbated in the short term as regions such as Canada's northern territories cope with rapid growth. However, as the authors have described, tundra wetlands can provide preferred levels of treatment of domestic wastewater in remote Canadian Arctic communities. The successful use of natural systems indicates that further experimentation of the use of CWs and engineered wetlands in extreme cold climates is warranted. Finally, much study is still required in tundra wetlands to identify the primary mechanistic functions responsible for the removal of nutrients and organic matter from wastewater in these systems.

References

1. Kadlec, R.H; Johnson, K. Cambridge Bay, Nunavut, Wetland planning study. *J. North. Territories Water Waste Assoc.* **2008**, *9* (Fall), 30–34.
2. Prince, D.S.; Smith, D.W.; Stanley, S.J. Intermittent-discharge lagoons for use in cold regions. *J. Cold Reg. Eng.* **1995**, *9* (4), 183–194.
3. Heinke, G.W.; Smith, D.W.; Finch, G.R. Guidelines for the planning and design of waste-water lagoon systems in cold climates. *Can. J. Civ. Eng.* **1991**, *18* (4), 556–567.
4. Dillon Consulting Ltd. *Hamlet of Paulatuk: Operation and Maintenance Manual, Sewage and Solid Waste Disposal.* Report 04-3332, 2004, available at http://www.docstoc.com/docs/45286151/Appendix-H---Operations-and-Main-tenance-Manual-Sewage-and-Solid-Waste (accessed March 2012).
5. Environment Canada. *Sewage lagoons in cold climates.* Report EPS 3/NR/1. Government of Canada. 1985. 98–&.
6. Miyamoto, H.K.; Heinke, G.W. Performance evaluation of an arctic sewage lagoon. *Can. J. Civ. Eng.* **1979**, *6*, 324–328.
7. Johnson, K; Wilson, A. *Sewage treatment systems in communities and camps of the Northwest Territories and Nunavut Territory.* 1st Cold Regions Specialty Conference of the Canadian Society for Civil Engineering, Regina, Canada, June 2–5, 1999.
8. Doku, I.A; Heinke, G.W. Potential for greater use of wetlands for waste treatment in Northern Canada. *J. Cold Reg. Eng.* **1995**, *9* (2), 75–88.

9. Dubuc, Y.; Janneteau, P.; Labonte, R.; Roy, C.; Briere, F. Domestic waste-water treatment by peatlands in a northern climate—A water-quality study. *Water Resour. Bull.* **1986**, *22* (2), 297–303.

10. Hayward, J.; Jamieson, R.; Boutilier, L.; Goulden, T.; Lam, B. Treatment performance assessment and hydrological characterization of an arctic tundra wetland receiving municipal wastewater. *Ecol. Eng.* **2014**, *73*, 786–797.

11. Yates, C.; Wootton, B.; Varickanickal, J.; Cousins, S. Testing the ability to enhance nitrogen removal at cold temperatures with *C. aquatilis* in a horizontal subsurface flow wetland system. *Ecol. Eng.* **2016**, *94*, 344–351.

12. Hammer, D.A.; Burckhard, D.L. Low temperature effects on pollutant removals at Minot's wetland. In *Natural Wetlands for Wastewater Treatment in Cold Climates*; Mander, U., Jenssen, P., Eds.; WIT Press: Boston, MA, 2002.

13. Kadlec, R.H. The Houghton Lake wetland treatment project. *Ecol. Eng.* **2009**, *35* (9), 1285–1286.

14. Shaver, G.R.; Chapin, F.S. Long-term responses to factorial, NPK fertilizer treatment by Alaskan wet and moist tundra sedge species. *Ecography* **1995**, *18* (3), 259–275.

15. Arens, S.J.T.; Sullivan, P.F.; Welker, J.M. Nonlinear responses to nitrogen and strong interactions with nitrogen and phosphorus additions drastically alter the structure and function of a high arctic ecosystem. *J. Geophys. Res., [Biogeosci.]* **2008**, *113* (G3), G03S09.

16. Edwards, K.A.; Jefferies R.L. Nitrogen uptake by *Carex aquatilis* during the winter–spring transition in a low Arctic wet meadow. *J. Ecol.* **2010**, *98* (4), 737–744.

17. Ritter, T.L. Sharing environmental health practice in the North American Arctic: A focus on water and wastewater service. *J. Environ. Health* **2007**, *69* (8), 50–55.

18. Johnson, K. The social context of wastewater management in remote communities. *Environ. Sci. Eng. Mag.* **2010**, (2), 28–30.

19. Daley, K.; Jamieson, R.; Rainham, D.; Hansen, L. Wastewater treatment and public health in Nunavut: A microbial risk assessment framework for the Canadian Arctic. *Environ. Sci. Pollut. Res.* **2018**, *25* (33), 32860–32872.

20. Daley, K.; Castleden, H.; Jamieson, R.; Furgal, C.; Ell, L.; Water systems, sanitation, and public health risks in remote communities: Inuit resident perspectives from the Canadian Arctic. *Soc. Sci. Med.* **2015**, *135*, 124–132.

21. Berkes, F.; Berkes, M.K.; Fast, H. Collaborative integrated management in Canada's north: The role of local and traditional knowledge and community-based monitoring. *Coastal Manage.* **2007**, *35* (1), 143–162.

22. Environment Canada. *National Consultation Report: Wastewater Consultations.* Government of Canada, 2009, available at http://www.ec.gc.ca/eu-ww/default.asp?lang=En&n=4B94EB4A-1 (accessed March 2012).

23. Dawson, R.N.; Grainge, J.W. Proposed design criteria for waste-water lagoons in arctic and subarctic regions. *J. Water Pollut. Control Fed.* **1969**, *41* (2P1), 237–246.

24. Grainge, J.W. Arctic heated pipe water and waste water systems. *Water Res.* **1969**, *3* (1), 47–71.

25. Johnson, K. *Inuit Position Paper Regarding the CCME Canada-wide Strategy for the Management of Municipal Wastewater Effluent and Environment Canada's Proposed Regulatory Framework for Wastewater.* Inuit Tapiriit Kanatami: Ottawa, 2008.

26. Vymazal, J. Constructed wetlands for wastewater treatment: Five decades of experience. *Environ. Sci. Technol.* **2011**, *45* (1), 61–69.

27. Gough, L.; Wookey, P.A.; Shaver, G.R. Dry heath arctic tundra responses to long-term nutrient and light manipulation. *Arctic Antarctic Alpine Res.* **2002**, *34* (2), 211–218.

28. Yates, C.N.; Wootton, B.; Jprgensen, S.E.; Santiago, V.; Murphy, S.D. Natural Wetlands for Treating Municipal Wastewater in the Canadian Arctic: A Case Study of the Chesterfield Inlet, Nunavut Wetland. *Proceedings of the 12th International Conference on Wetland Systems for Water Pollution Control*, Venice, Italy, October 12–17, 2010.

29. Mander, Ü.; Jenssen, P.D. *Constructed Wetlands for Wastewater Treatment in Cold Climates*; WIT Press: Boston, MA, 2003.

30. Kadlec, R.H.; Wallace, S. *Treatment Wetlands,* 2nd Ed.; CRC Press: Baton Rouge, LA, 2009.

31. Statistics Canada. 2012. *Paulatuk, Northwest Territories (Code 6101014) and Region 1, Northwest Territories (Code 6101) (Table).* Census Profile. 2011 Census. Statistics Canada Catalogue no. 98-316-XWE. Ottawa. Released February 8, 2012, available at http://www12.statcan.ca/cen-sus-recensement/2011/dp-pd/prof/index.cfm?Lang=E (accessed March 2012).

32. Statistics Canada. 2017. *Paulatuk, HAM [Census subdivision], Northwest Territories and Northwest Territories [Territory] (table).* Census Profile. 2016 Census. Statistics Canada Catalogue no. 98-316-X2016001. Ottawa. Released November 29, 2017.

33. Wootton, B.; Durkalec, A.; Ashley, S. *Canadian Council of Ministers of the Environment Draft Canada-wide Strategy for the Management of Municipal Wastewater Effluent: Inuvialuit Settlement Region Impact Analysis*; Inuit Tapiritt Kanatami: Ottawa, 2008.

34. Yates, C.N.; Wootton, B. *Treatment Wetland Performance and Physical Characterization: Paulatuk, NWT and Pond Inlet, NT;* Environment Canada—Wastewater Technology Centre: Burlington, 2010; 1–103.

35. Environment Canada. Canadian Climate Normals. 2010, available at http://www.climate. weatheroffice.gc.ca/climate_normals/index_e.html (accessed July 2010).

36. Hobbie, S.E.; Chapin, F.S. Winter regulation of tundra litter carbon and nitrogen dynamics. *Biogeochemistry* **1996**, *35* (2), 327–338.

37. Schmidt, I.K.; Jonasson, S.; Michelsen, A. Mineralization and microbial immobilization of N and P in arctic soils in relation to season, temperature and nutrient amendment. *Appl. Soil Ecol.* **1999**, *11* (2–3), 147–160.

38. Mack, M.C.; Schuur, E.A.G.; Bret-Harte, M.S.; Shaver, G.R.; Chapin, F.S. Ecosystem carbon storage in arctic tundra reduced by long-term nutrient fertilization. *Nature* **2004**, *431* (7007), 440–443.

39. Chapin, F.S.; Moilanen, L.; Kielland, K. Preferential use of organic nitrogen for growth by a nonmycorrhizal arctic sedge. *Nature* **1993**, *361* (6408), 150–153.

40. Hobbie, S.E.; Gough, L.; Shaver, G.R. Species compositional differences on different-aged glacial landscapes drive contrasting responses of tundra to nutrient addition. *J. Ecol.* **2005**, *93* (4), 770–782.

41. Edwards, K.A.; McCulloch, J.; Kershaw, G.P.; Jefferies, R.L. Soil microbial and nutrient dynamics in a wet Arctic sedge meadow in late winter and early spring. *Soil Biol. Biochem.* **2006**, *38* (9), 2843–2851.

42. Larsen, K.S.; Grogan, P.; Jonasson, S.; Michelsen, A. Dynamics and microbial dynamics in two subarctic ecosystems during winter and spring thaw: Effects of increased snow depth. *Arctic Antarctic Alpine Res.* **2007**, *39* (2), 268–276.

43. MacIntyre, M.E.; Warner, B.G.; Slawson, R.M. *Escherichia coli* control in a surface flow treatment wetland. *J. Water Health* **2006**, *4* (2), 211–214.

44. Yates, C.N.; Wootton, B.; Jørgensen, S.E.; Santiago, V.; Murphy, S.D. Annak, Ijji, Imaqsuk: Early Findings from The First Experimental Constructed HSSF Wetland in the Canadian Arctic. *Proceedings of the 12th International Conference on Wetland Systems for Water Pollution Control,* Venice, Italy, October 12–17, 2010.

45. Andersson, J.L.; Wittgren, H.B.; Kallner, S.; Ridderstolpe, P.; Hagermark, I. Wetland Oxelosund, Sweden—The first five years of operation. In *Natural Wetlands for Wastewater Treatment in Cold Climates;* Mander, U., Jenssen, P., Eds.; WIT Press: Boston, 2002.

46. Vymazal, J. Constructed wetlands for wastewater treatment. *Ecol. Eng.* **2005**, *25* (5), 475–477.

47. Wittgren, H.B.; Maehlum, T. Wastewater treatment wetlands in cold climates. *Water Sci. Technol.* **1997**, *35* (5), 45–53.

48. Maehlum, T.; Stalnacke, P. Removal efficiency of three cold-climate constructed wetlands treating domestic wastewater: Effects of temperature, seasons, loading rates and input concentrations. *Water Sci. Technol.* **1999**, *40* (3), 273–281.

49. Giaever, H. Treatment performance of a multistage constructed wetland at 69 degrees north, Norway. *J. Environ. Sci. Heal. A.* **2000**, *35* (8), 1377–1388.

50. Jenssen, P.D.; Maehlum, T.; Krogstad, T.; Vrale, L. High performance constructed wetlands for cold climates. *J. Environ. Sci. Heal. A.* **2005**, *40* (6–7), 1343–1353.

51. Wallace, S; Parkin, G.; Cross, C. Cold climate wetlands: Design and performance. *Water Sci. Technol.* **2001**, *44* (11–12), 259–265.

52. O'Hogain, S. Reed bed sewage treatment and community development/participation. In *Wastewater treatment, Plant Dynamics and Management in Constructed and Natural Wetlands*; Springer: New York, 2008.

53. Merlin, G.; Pajean, J.L.; Lissolo, T. Performances of constructed wetlands for municipal wastewater treatment in rural mountainous area. *Hydrobiologia* **2002**, *469* (1–3), 87–98.

54. Kivaisi, A.K. The potential for constructed wetlands for wastewater treatment and reuse in developing countries: A review. *Ecol. Eng.* **2001**, *16* (4), 545–560.

55. Wootton, B.; Yates, C.N. Wetlands: Simple and effective wastewater treatment for the north. *Meridian Newsl.* **2010**, (Fall/Winter), 12–16.

56. Jenssen, P. *Wastewater treatment in cold/arctic climate with a focus on small scale and onsite systems.* 28th Alaska Health Summit, Anchorage, Alaska, January 10–13, 2011.

57. Tenson, J. Wastewater treatment in Estonia: Problems and opportunities. *Recycl. Res.: Proc. Sec. Int. Conf. Ecol. Eng. Wastewater Treat.* **1996**, *5* (6), 65–71.

58. United Nations Environment Program. 2009. Download software here: http://www.unep.or.jp/Ietc/Publications/Water_Sanitation/SubWet2/index.asp (accessed March 2012).

59. Chhetri, R. K.; Klupsch, E.; Andersen, H. R.; Jensen, Pernille, E. Treatment of Arctic wastewater by chemical coagulation, UV and peracetic acid disinfection. *Environ. Sci. Pollut. Res.* **2018**, *25* (33), 32851–32859.

55

Wastewater Treatment: Biological

Introduction 561
Wastewater Treatment Options 562
    Pretreatment/Preliminary Treatment • Primary Treatment • Secondary
    Treatment • Tertiary/Advanced Treatment
Biological Treatment Options 563
    Aerobic Processes • Anoxic Processes • Anaerobic
    Processes • Suspended Growth Processes • Attached Growth Process
Aerobic Biological Waste Treatment Processes 565
    Activated Sludge Process • Aeration Tanks • Secondary Clarifiers
Important Operating Parameters in Activated Sludge Systems 566
    Solid Retention Time • Sludge Volume Index • Dissolved Oxygen
    Concentration • Food-to-Microorganism Ratio • Organic Loading
    Rate • Common Microorganisms in Activated Sludge Systems
Attached Growth Processes 568
Anaerobic Wastewater Treatment Processes 568
    UASB Reactors • Important Operating Parameters in Anaerobic
    Reactors • pH • Waste Composition • Temperature • Loading
    Rate • Retention Time • Toxicity • Granule Deterioration
Biological Removal of Nitrogen 572
    Sharon Process • Anaerobic Ammonium Oxidation • Combined Nitrogen
    Removal • Canon Process • Biological Phosphorus Removal
Conclusion 574
References 574

Shaikh Ziauddin
Ahammad, David
W. Graham, and
Jan Dolfing

Introduction

Human activities and population growth have placed the environment under increasing stress. Furthermore, indiscriminate use of natural resources is accompanied by increased local and global pollution levels, which are reflected in imbalances in our ecosystems. The generation of large quantities of wastewater with a high organic content and toxicants is one obvious product of excessive consumption. It has been known for many years that environmental discharges of high loads of organic matter can result in oxygen depletion in receiving waters due to stimulated microbial activity. This oxygen depletion and the presence of trace toxicants found in wastes also negatively influence ecosystems, including reduced biodiversity and environmental health. Therefore, negative environmental impacts have driven our need to understand the effect of pollution on water bodies and develop proper measures to reduce discharges, including treatment processes.

Different technologies are available to treat wastes. However, biological wastewater treatment methods are most valuable because their economic benefits are high, especially when coupled with waste stabilization and resource recovery. The optimal treatment processes depend on the waste type and treatment goals. Wastewater generally originates from two sources: 1) domestic wastewater from gray water, toilets, and other domestic activities; and 2) industrial wastewater, generated by industries during the normal course of activity, which often rely on the local sewerage systems for waste processing. Therefore, the composition of wastewater, including quantity and constituents, varies considerably from place to place, depending on suite of sources, social behavior, the type and number of industries within a catchment, climatic conditions, water consumption, and the nature of the wastewater collection system. Given this variety, wastewater treatment processes must be innately versatile, but also sometimes must be tailored to the specific waste and conditions. The purpose of this entry is to describe different biological treatment methods and then discuss their relative capacities to treat different wastes on the basis of waste characteristics and the desire for resource recovery.

Wastewater Treatment Options

Special handling and treatment of wastes have been performed for thousands of years in response to their perceived importance, although approaches have changed as perceptions have changed over history. In 4th century B.C. in Greece, the *Athenian Constitution* written by Aristotle[1] proscribed provisions for the appropriate handling of sewage. Concern was based on aesthetics, probably odors, because relationships between domestic wastes and health were not yet known. It was not until the mid-1800s that links between wastes and human health became more apparent, which led to a progression of waste management approaches and technologies to address health concerns.

Treatment technologies evolved slowly over time, including physical, chemical, and biological approaches, many of which are still used in different sectors. Physical methods are based on the application of physical forces, such as screening, mixing, flocculation, sedimentation, flotation, filtration, and gas transfer. Alternately, chemical processes treat contaminants by adding chemicals or by stimulating specific chemical reactions. Precipitation, adsorption, and disinfection are common examples of chemical treatment methods. Physical and chemical methods are often combined, especially in industrial treatment scenarios. In contrast to physiochemical processes, biological processes remove organic contaminants (e.g., biodegradable organic material) largely through microbiological activity. Commonly used biological treatment methods include aerobic treatment in ponds, lagoons, trickling filters, and activated sludge plants,[2] and anaerobic treatment[3,4] in similar reactor systems. Processes that combine anaerobic and aerobic unit operations are also common.[5]

The best overall treatment approach depends on the source and nature of waste, such as production rates, constituents, and relative concentrations. As such, optimal process trains and designs should be as simple as possible in design and operation, while being efficient in removing key pollutants and minimizing energy consumption and negative by-products. More complex operations are only used when absolutely necessary.

Within a typical treatment plant, each type of treatment has a different purpose. For example, the main objective of biological treatment is to treat soluble organic matter in the wastes, which often requires physical pretreatment to remove solids before biological treatment.[2] For domestic wastewater, the main objective is to reduce the organic content and, in growing numbers of cases, secondary nutrients (nitrogen, N; phosphorus, P). For industrial wastewaters, the objective is usually to remove or reduce the concentration of organic compounds, especially specific toxicants that can be present in some wastes, which is why chemical processes are also included in industrial treatment systems. However, biological processes are almost always used when possible.

Biological degradation of organics is accomplished through the combined activity of microorganisms, including bacteria, fungi, algae, protozoa, and rotifers. To maintain the ecological balance in the receiving water, regulatory authorities have set standards for the maximum amount of the undesirable compounds present in the discharge water. In a typical wastewater treatment plant, the following steps are carried out to achieve the desired quality of the effluent before it can be safely discharged into the receiving water.

Pretreatment/Preliminary Treatment

Pretreatment is primarily used to protect pumping equipment and promote the success of subsequent treatment steps. Pretreatment devices such as screen and/or grit removal systems are designed and implemented to remove the larger suspended or floating solids, or heavy matter that can damage pumps. Sometimes, froth flotation is also used to remove excessive oils or grease in the wastes.

Primary Treatment

Most of the settleable solids are removed from the wastewater by simple sedimentation, a purely physical process. In this process, the horizontal velocity of the water through the settle is maintained at a level that provides solids adequate time to settle and floatable material be removed from the surface. Therefore, primary treatment steps consist of settling tanks, clarifiers, or flotation tanks, which send separated solids to digestion units and supernatant to subsequent, typically microbiological, treatment units.

Secondary Treatment

Secondary treatment uses microbial communities, under varying growth conditions, to biochemically decompose organic compounds in the waste that have passed from primary treatment units. An array of reactors are employed for biological treatment, which include suspended biomass, biofilm, fixed-film reactors, and pond or lagoon systems.

Secondary Clarification

Most biological treatment processes produce excess biomass through the conversion of waste carbon to new cells. As such, before the final treatment steps, such as disinfection or nutrient removal, solids must be separated from the secondary treatment effluents. This is usually by settling, but membranes are also employed. The separated solids are either recycled back to the head of the process train or sent to digesters for solids reduction and processing, depending on the type of the digester system.

Tertiary/Advanced Treatment

Advanced or tertiary treatment consists of processes that are designed to achieve higher effluent quality than attainable by conventional secondary treatment methods. These include polishing steps such as activated carbon adsorption, ion exchange, reverse osmosis, electrodialysis, chemical oxidation, and nutrient removal. Although not technically a tertiary process, final effluent disinfection is often performed after secondary or tertiary treatment using chlorination, ultraviolet methods, ozonation, and other methods designed specifically to kill residual organisms in the wastewater after all previous treatment steps.

Biological Treatment Options

Biological processes are classified according to the primary metabolic pathways present in the dominant different microorganisms active in the treatment system. As per the availability and utilization of oxygen, the biological processes are classified as aerobic, anoxic, and anaerobic.

Aerobic Processes

Treatment processes that occur in the presence of molecular oxygen (O_2) and use aerobic respiration to generate cellular energy are called aerobic processes. They are most metabolically active, but also generate more residual solids as cell mass.

Anoxic Processes

These are processes that occur in the absence of free molecular oxygen (O_2) and generate energy through anaerobic respiration. Microorganisms use combined oxygen from inorganic material in the waste (e.g., nitrate) as their terminal electron acceptor. Anoxic processes are common biological nitrogen removal systems through denitrification.[2]

Anaerobic Processes

These are the processes that occur in the absence of free or combined oxygen, and result in sulfate reduction and methanogenesis. They usually produce biogas (i.e., methane) as a useful by-product and tend to generate lower amounts of biosolids through treatment.

Apart from a classification based on microbial metabolism and/or oxygen utilization, biological wastewater treatment processes also can be classified based on the growth conditions in the reactor (see Figure 1). In this case, the two main categories are suspended growth and attached growth processes.

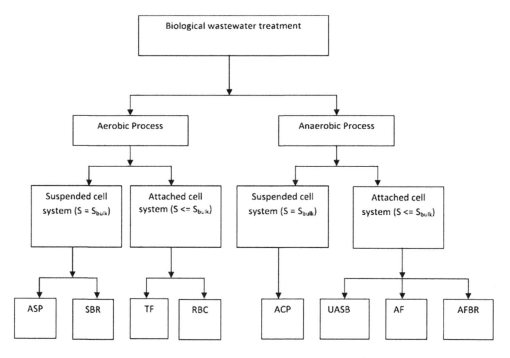

FIGURE 1 Different biological treatment processes. S, substrate concentration available to microorganisms; S_{bulk}, substrate concentration in the bulk of the liquid; ASP, activated sludge process; SBR, sequencing batch reactor; TF, trickling filter; RBC, rotating biological contactor; ACP, anaerobic contact process; AF, anaerobic filter; UASB, upflow anaerobic sludge blanket; AFBR, anaerobic fluidized bed reactor.

Suspended Growth Processes

In these processes, the microorganisms, which are responsible for the conversion of waste organic matter to simpler compounds and biomass, are maintained in suspension within the liquid phase. However, there are different types of aerobic and anaerobic suspended growth processes. Aerobic processes include activated sludge, aerated lagoons, and sequencing batch reactors, whereas anaerobic processes include bag digesters, plug-flow digesters, stirred-tank reactors, and baffled reactors with organisms primarily in the liquid phase.

Attached Growth Process

In these processes, the microorganisms responsible for degrading the waste are attached to surfaces (e.g., stones, inert packing materials), or are self-immobilized on flocs or granules in the system. Attached growth processes can be aerobic or anaerobic. Aerobic attached growth processes include trickling tilters, roughing filters, rotating biological contactors, and packed-bed reactors. Anaerobic systems include upflow packed-bed reactors, down-flow packedbed reactors, anaerobic rotating biological contactors, anaerobic fluidized bed reactors, upflow anaerobic sludge blanket (UASB) reactors, and various hybrid anaerobic reactors (HAR). UASBs are widely used reactors for the anaerobic treatment of industrial and domestic wastewater.

Aerobic Biological Waste Treatment Processes

Typical aerobic waste treatment systems provide a location where microbes are exposed to molecular oxygen (O_2) to oxidize complex organics present in the waste, producing carbon dioxide, simple organics, and new cell biomass. The activated sludge process (ASP) is very well known and the most widely used biological treatment process in developed countries.

Activated Sludge Process

Classic ASPs are aerobic suspended cell systems. Mineralization of waste organic compounds is accompanied by the formation of new microbial biomass and sometimes the removal of inorganic compounds, such as ammonia and phosphorus, depending on the particular process design. Activated sludge processes were first conceived in the early 1900s with the word "activated" referring to solids that catalyze the degradation of the waste. It was subsequently discovered that the "activation" part of the sludge was a complex mixture of microorganisms. The liquid in activated sludge systems is called the "mixed liquor," which includes both wastewater and the resident organisms.

There have been several incarnations of the ASP. The most common designs use conventional, step aeration, and continuous-flow stirred-tank reactors.[2] A conventional ASP consists of standard pretreatment steps, an aeration tank, and a secondary clarifier, an example of which is shown in Figure 2. The aeration tank can be aerated by subsurface or surface aerators designed to supply adequate dissolved oxygen to the water for the microorganisms to thrive. The wastewater flows through the tank and resident microorganisms consume organic matter in the wastewater. The aeration tank effluent flows to the clarifier where the microorganisms are removed. The clarifier supernatant is then transferred to disinfection or treatment units, and then ultimately discharged to the receiving water. Biosolids from the settler are recycled back to the head of the treatment system or sent to digesters for further processing.

Aeration Tanks

Aeration tanks are usually designed uncovered, open to the atmosphere. Air is supplied to the microorganisms by two primary methods: mechanical aerators or diffusers. Mechanical aerators, such as

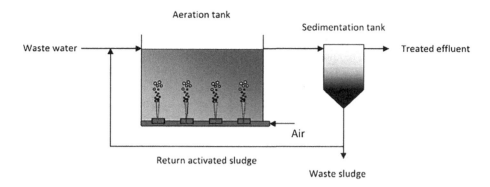

FIGURE 2   Activated sludge process.

surface aerators and brush aerators, aerate the surface of the water mechanically and promote diffusion of oxygen to water from the atmosphere. The concentration of dissolved oxygen in the liquid can be controlled by adjusting the speed of the rotors. Both mechanical aerators and diffusers are the largest energy consumers in aerobic biological wastewater treatment processes. Diffusers bubble air directly into the tank at depth and are usually preferred because of higher oxygen transfer efficiencies.

As previously indicated, aeration provides O_2 to the microorganisms and also serves to mix the liquor in the tank. Although complete mixing is desired, there are usually "dead zones" in the tank where anaerobic/anoxic conditions develop in poorly mixed areas. It is desirable to keep these zones to a minimum to minimize undesired odors and also problems with sludge bulking, which can reduce settling efficiency in secondary clarifiers.

Secondary Clarifiers

Clarifiers are used to separate the biomass and other solids coming out of the aeration tank by means of gravity settling. The flow rate of the liquid is maintained in such a way that the upflow velocity of the liquid is less than the settling velocity of the biosolids present in the liquid. As noted, some of the settled biosolids are returned back to the aeration tank to increase the solids' contact time with the wastes and also maintain the desired biomass levels in the aeration tank.

Important Operating Parameters in Activated Sludge Systems

Key operating parameters and typical values for activated sludge systems are provided in Table 1. All parameters ultimately are used to guide and pseudo-control biosolids levels, and they profoundly affect process performance. The total suspended solids in the aeration tank are known as mixed-liquor suspended solids (MLSS). This term refers to the amount of solids in a certain volume of the water (usually milligram of solids per liter). The actual biomass fraction of the solids is estimated as the solids that can be volatilized at 550°C. The volatile fraction is known as mixed-liquor volatile suspended solids (MLVSS). Therefore, MLVSS is frequently used as a proxy for the active biomass treating the waste. MLVSS ranges from about 70% to 90% of the MLSS concentration in most activated sludge systems.[6]

Solid Retention Time

The most important design parameter in activated sludge systems is the mean cell residence time of cells in the reactor, also known as the sludge age or solid retention time (SRT). The SRT can be controlled by manipulating the rate at which excess sludge is wasted and is influenced by hydraulic flow conditions through the reactor. It is the ratio of the total solids in the system and the total solids leaving the system.

TABLE 1 Typical Design Parameters for ASP

Process Components or Variables	Typical Values	Reference
Aeration tank		[2]
Depth (m)	5–8	
Width (m)	7–12	
SRT (day)	5–15	[2]
MLSS (kg/m3)	1500–4000	[9]
SVI (kg/m3)	40–150	[7]
F/M	0.2–0.4	[6]
Organic loading rate (kg COD/m3day)	20–60	[6]
Oxygen requirement (kg/kg COD removed)	1.4–1.6	[9]

$$SRT = VX/(QX_e + Q_w X_w)$$

where SRT is the mean cell residence time (day); V is the volume of aeration basin (e.g., L); X is the mixed liquor suspended solids concentration (mg/L); Q is the volumetric flow rate (e.g., L/day); X_e is the effluent suspended solids concentration (mg/L); Q_w is the waste sludge flow rate (e.g., L/day); and X_w is the waste sludge suspended solids concentration (mg/L).

Sludge Volume Index

The sludge volume index (SVI) is another key parameter and used to describe the settling characteristics of the sludge. The SVI is expressed as the volume occupied by 1 g of sludge (mL/g) after 30 min of settling time. Well-settled sludge normally yields a clear separation between the water and the sludge. However, if the sludge has any problems, such as bulking, pinpoint floc formation of tiny, poorly settling floc, or ashing, the interface between the sludge and the water may not be seen clearly. Such conditions usually result from problems in the aeration tank and cause reduced effluent quality because of poor settling in the clarifier.

Dissolved Oxygen Concentration

Microorganisms in an activated sludge system require adequate oxygen to oxidize organics in the waste. The basic oxidation reaction for organics degradation can be approximated as (stoichiometry not provided)

$$CHON + O_2 + microorganisms \rightarrow CO_2 + H_2O + NH_3 + more\ microorganisms$$

Organics are consumed by microorganisms, and new microbial cells are synthesized with ratio of organisms produced relative to the organics consumed being the sludge yield. As noted, oxygen is supplied by mechanical aerators or diffusers in the aeration tank. Required oxygen levels in the system depend on the process, but the design goal is to minimize oxygen addition due to energy costs. The dissolved oxygen concentration can be controlled by either adjusting the speed of the air pump or throttling the air pipes. Air pumps are more widely used to aerate the wastewater because of their lower operational and maintenance costs.

Food-to-Microorganism Ratio

The food-to-microorganism ratio (F/M) is a good indicator for designing and regulating the operation of the aeration tank.[7] The F/M ratio is expressed as the amount of organic biodegradable material [milligrams of 5-day biological oxygen demand (BOD_5)] available for the amount of microorganisms present (mg MLVSS) per day.

$$F/M = (QS_o)/X$$

where F/M is the food-to-microorganism ratio (day^{-1}); S_o is the influent BOD_5 concentration (mg/L); X is the MLVSS concentration (mg/L); and Q is the volumetric flow rate (L/day).

The targeted F/M ratio for any treatment system varies depending on the design of the system, and values can range widely. However, since influent BOD cannot be controlled, MLVSS is typically modulated by varying the return activated sludge rate from the secondary clarifier, the goal being to maintain an optimum F/M ratio for specific activated sludge design.

Organic Loading Rate

The amount of organic matter in wastewater is commonly measured by BOD5, chemical oxygen demand (COD), or the total organic carbon content.[8,9] If there are excess organics in the influent or inadequate organisms in the aeration tank, incomplete treatment will result.

Common Microorganisms in Activated Sludge Systems

Activated sludge is a complex mixture of broadly differing microorganisms.[10] Major categories are as follows: bacteria, fungi, algae, protozoa (e.g., flagellates, ciliates, and rotifers), and viruses. Viruses and pathogenic bacteria are often present in wastewater, which is the primary reason for having post-biological disinfection steps in treatment plants.

Attached Growth Processes

Attached growth processes, such as trickling filters (Figure 3), can achieve similar treatment objectives as activated sludge systems. Conversion processes in these systems are typically mass transport limited: microorganisms in the outer layers of the biofilm contribute most to the overall substrate removal. The support material in trickling filters is chosen to provide sufficiently large pore spaces to allow air through the trickling filter regardless of biofilm growth and water trickling down the filter. Wastewater is distributed using rotary arms at the top and then trickles down the filter. Trickling filters are mainly used for the oxidation of carbon and ammonia, but can also achieve denitrification when convection of air through the system is optimized.[11]

Anaerobic Wastewater Treatment Processes

Anaerobic treatment technologies are widely practiced in different industries on the basis of their requirement and suitability. The processes have some advantages and disadvantages in treating different wastes, and few of them are summarized in Table 2. Under anaerobic conditions, organic matter is degraded through the sequential and syntrophic metabolic interactions of various trophic groups of prokaryotes, including fermenters, acetogens, methanogens, and sulfate-reducing bacteria (SRB).[12,13] Metabolic interactions between these microbial groups lead to the transformation of complex organic compounds to simple compounds such as methane, carbon dioxide, hydrogen sulfide, and ammonia.[14] The digestion process is essentially accomplished in four major reaction stages involving different microorganisms in each stage.[15,16]

Stage 1: Hydrolysis—The organic waste material mainly consists of carbohydrates, proteins, and lipids. Complex and large substances are broken down into simpler compounds by the activity of the microbes

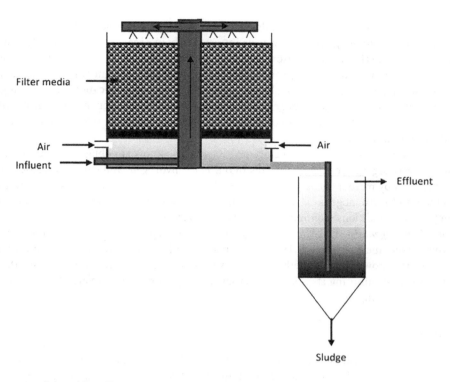

FIGURE 3   Aerobic trickling filter.

TABLE 2   Advantages and Disadvantages of Anaerobic Wastewater Treatment

Advantages

High efficiency: Good removal efficiency can be achieved in the system, even at high loading rates and low temperature.

Simplicity: The construction and operation of these reactors are relatively simple.

Flexibility: Anaerobic treatment can easily be applied on either a very large or a very small scale.

Low energy consumption: As far as no heating of the influent is needed to reach the working temperature and all plant operations can be done by gravity, the energy consumption of the reactor is almost negligible.

Energy recovery: Energy is produced during the process in the form of methane.

Low sludge production: Sludge production is low, well stabilized, and has good dewatering property.

Low nutrient and chemical requirement: Especially in the case of sewage, an adequate and stable pH can be maintained without addition of chemicals.

Disadvantages

Low pathogen and nutrient removal: Pathogens and nutrients are partially removed and hence post-treatment is needed.

Long start-up: Due to low growth rate of methanogenic organisms, the start-up takes longer time.

Possible bad odor: Hydrogen sulfide is produced. Proper handling of biogas is required to avoid bad smell.

Necessity of post-treatment: Post-treatment of the anaerobic effluent is generally required to reach the discharge standards for organic matter and pathogen.

Source: Data from Seghezzo et al.[22]

and the extracellular enzymes released by these microbes. The hydrolysis or solubilization is mainly done by hydrolytic microbes such as *Bacteroides, Bifidobacterium, Clostridium,* and *Lactobacillus.* These organisms hydrolyze complex organic molecules (cellulose, lignin, proteins, lipids) into soluble monomers such as amino acids, glucose, fatty acids, and glycerol. These hydrolysis products are used by the fermentative acidogenic bacteria in the next stage.[14,17]

Stage 2: Acidogenesis—Fermentative acidogenic bacteria convert simple organic materials such as sugars, amino acids, and long-chain fatty acids into short-chain organic acids such as formic, acetic, propionic, butyric, valeric, isobutyric, isovaleric, lactic, and succinic acids; alcohols and ketones (ethanol, methanol, glycerol, and acetone); carbon dioxide; and hydrogen. Generally, acidogenic bacteria have high growth rates and are the most abundant bacteria in any anaerobic digester.[18] The high activity of these organisms implies that acidogenesis is never the rate-limiting step in the anaerobic digestion process.[19] The volatile acids produced in this stage are further processed by microorganisms characteristic for the acetogenesis stage.

Stage 3: Acetogenesis—In this stage, acetogenic bacteria, also known as obligate hydrogen-producing acetogens, convert organic acids and alcohols into acetate, hydrogen, and carbon dioxide, which are subsequently used by methanogens and SRB. There is a strong symbiotic relationship between acetogenic bacteria and methanogens. Methanogens and SRB use hydrogen, which helps achieve the low hydrogen pressure conditions required for acetogenic conversions.[20]

Stage 4: Methanogenesis—It is the final stage of anaerobic digestion where methanogenic archaea convert the acetate, methanol, methylamines, formate, and hydrogen produced in the earlier stages into methane. The growth rate of methanogens is very low, and therefore, in most cases, this step is considered as the rate-limiting step of the anaerobic process, although there are also examples where hydrolysis is rate limiting.[21]

UASB Reactors

The most common and widely used anaerobic reactor is the UASB reactor.[22] It is an attached, self-immobilized cell system, which consists of a bottom layer of packed sludge bed (sludge blanket) and an upper liquid layer, as shown in Figure 4.[23]

Wastewater flows upward through a sludge bed consisting of bacterial aggregates floating blanket, and the microbes present in the sludge bed convert the complex organic materials to methane, carbon dioxide, and hydrogen.[24] The granular sludge (1–5 mm in diameter) has high biomass content (MLVSS) and specific activity, and good settling properties. The upward flow of the liquid inside the reactor is obtained by means of effluent recirculation. Because of the high density of biomass present in the self-immobilized granular sludge, the reactor is able to support a high SRT, which is diverse from the hydraulic retention time (HRT) and require no support material. The major drawback of the UASB is the requirement of high HRT to achieve desired biodegradation. Maintenance of high HRT demands huge reactor volume. These problems are overcome by using HAR where the advantages of AFBR are coupled with UASB operation by maintaining a high upflow velocity (4–8 m/hr) inside the reactor.[25] With higher upflow velocity, better mass transfer is obtained in the reactor, which reflects on the higher degradation with less HRT operation. The main purpose of these reactors is to achieve better degradation of waste and increase the production of biogas (methane) in a substantially reduced-size anaerobic reactor.

Important Operating Parameters in Anaerobic Reactors

Different operating parameters such as pH, temperature, HRT, and nutrients, among others, and their disturbances can manifest in case of industrial wastewaters treatment in anaerobic reactors, even under normal operational conditions.[26,27] Some of these factors are discussed below.

pH

The optimum degradation is achieved when the pH value of wastewater in the digester is maintained between 6.5 and 7.5. In the initial period of fermentation, as large quantities of organic acids are produced by acidogens and acetogens, a drop in pH occurs inside the digester. This low pH condition inhibits methanogens and subsequently reduces methane production. As the digestion proceed, the pH

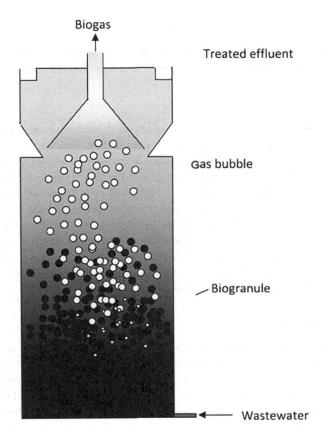

FIGURE 4 Upflow anaerobic sludge blanket reactor.

increases owing to the conversion of organic nitrogen to NH_4. When the methane production level is stabilized, the pH range remains buffered between 7.2 and 7.8.[28,29]

Waste Composition

To attain optimum degradation, wastewaters have to be nutritionally balanced in terms of carbon (C), nitrogen (N), phosphorous (P), and sulfur (S). The C/N/P ratio of 700:5:1 is recommended for efficient anaerobic digestion.[30] A fairly high concentration of acetate is required to prevent SRB outcompeting methanogens for acetate and hydrogen.[31]

Temperature

Methanogens are inactive at extremely high and low temperatures.[32] Few psychrophilic methanogens have been discovered, which can grow at a temperature range of 4–6°C.[33] Most of the methanogens can grow well from 25°C to 65°C temperatures.[34] The optimum temperature for the growth of the mesophilic methanogens is 35–37°C.[34] When the ambient temperature goes down to 10°C, gas production virtually stops. Satisfactory gas production takes place in the mesophilic range, from 30°C to 40°C.

Loading Rate

High organic loading rate may lead to acid accumulation and reduction of methane production. Similarly, if the plant is underfed, the gas production will also be low.[15]

Retention Time

The retention time depends on the growth rate of the microbial population and reactor configuration (attached cell or suspended cell system), waste strength, and waste composition.

Toxicity

The presence of toxicants in the wastewater, such as oxygen (lethal to obligate anaerobes), ammonia, chlorinated hydrocarbons, aromatic hydrocarbons, heavy metals, and long-chain fatty acids, among several others, may also result in occasional failures of anaerobic digesters.[15] The presence of trace amount of metals (e.g., nickel, cobalt, molybdenum) also stimulates the growth of microbes. Excess volatile fatty acid (VFA) concentrations are reported to inhibit the growth of several microbial species.[35] The undissociated forms of VFA can diffuse across the cell membrane and dissociate intracellularly, which results in reduction in growth rate.[35,36] The 50% inhibition of acetoclastic-methanogenesis in granular sludge was observed at a concentration of 13,000, 3,500, and 15,000 mg/L of acetate, propionate, and butyrate, respectively.[37] Small amounts of sulfide, a vital sulfur source, are beneficial for methanogens.[38] Acetoclastic methanogens are the most sensitive in terms of sulfide inhibition. Fifty percent inhibition was observed at total sulfide concentrations of 220–980 mg/L over the pH range 6.5–8.0.[39]

Granule Deterioration

Lipids present in the wastewater creates problem by forming long-chain fatty acids during hydrolysis in the anaerobic reactor. Long-chain fatty acid imparts toxic effect to acetogenic and methanogenic microbes. It also becomes adsorbed onto the sludge, inducing sludge flotation and resulting in washout.[40] Some long-chain fatty acids also act as surfactant at neutral pH and obstruct the floc formation by lowering the surface tension between water and the hydrophobic bacteria and promote their washout.[41] Addition of polyelectrolytes (calcium salts) may prevent inhibition to some extent, but it does not prevent flotation.[42]

Biological Removal of Nitrogen

The conventional biological nitrogen removal is a two- step process, nitrification followed by denitrification. The process is slow due to low microbial activity and yield. Nitrification involves a chemolithoautotrophic oxidation of ammonia to nitrate under strict aerobic conditions. This oxidation is a result of two sequential oxidative stages: ammonia to nitrite (ammonia oxidation) and nitrite to nitrate (nitrite oxidation). Different microorganisms involved in these stages use molecular oxygen as an electron acceptor and carbon dioxide as carbon source. The oxidation of ammonia to nitrite is performed by nitrifier microorganisms such as *Nitrosomonas, Nitrosococcus, Nitrosospira, Nitrosovibrio,* and *Nitrosolobus.* In the nitrite oxidation stage, *Nitrobacter, Nitrospira, Nitrospina, Nitrococcus,* and *Nitrocystis* are known to be involved in the production of nitrate.[10,43] Ammonia uptake rate varies according to reactor configuration, substrate type, and influent ammonium concentration. Denitrification is the second stage of the nitrogen removal process. It is a heterotrophic bioconversion process carried out by the heterotrophic denitrifiers under anoxic conditions. The oxidized nitrogen compounds (NO_2^- and NO_3^-) are reduced to nitrogen gas by the denitrifiers that use nitrite and/or nitrate as terminal electron acceptors and organic matter as carbon and energy source. *Pseudomonas, Alcaligenes, Paracoccus, Thiobacillus,* and *Halobacterium* are commonly found in dentrification systems.[44]

Few advanced processes, including partial nitrification, anaerobic ammonium oxidation (Anammox) and autotrophic nitrogen removal (Canon) are also being practiced in different treatment plants according to the characteristics of the wastewater. A combined system of partial nitrification and Anammox is advantageous as no extra carbon addition is needed, a negligible amount of sludge is produced, and less energy and oxygen are required compared with the conventional two-stage process.[45]

Sharon Process

The Sharon (single-reactor high-activity ammonium removal over nitrite) process is used for removal of ammonia through nitrite formation.[45,46] In this process, both autotrophic nitrification and heterotrophic denitrification take place in a single reactor with intermittent aeration. The denitrification in the Sharon process is achieved by adding methanol as a carbon source. Although the process is not suitable for all wastewaters due to a high temperature dependency, the Sharon process is suitable for removing nitrogen from waste streams having high ammonia concentrations (>0.5 g/L).

Anaerobic Ammonium Oxidation

Anaerobic ammonium oxidation (Anammox) is a highly exergonic, lithoautotrophic biological conversion process where ammonia becomes converted to nitrogen by the activity of a group of planctomycete bacteria.[47] These microorganisms use CO_2 as the sole carbon source and have a capability to oxidize ammonia to gaseous nitrogen by using nitrite as the electron acceptor in an anoxic condition.

Combined Nitrogen Removal

Ammonia-rich wastewater can be treated by Anammox, which requires nitrite as precursor. Thus, before feeding into the Anammox process, ammonia has to be preoxidized to nitrite. Thus, a partial Sharon process can be used before the Anammox process to improve the nitrogen removal efficiency. Partial nitritation (conversion of 55%–60% of ammonium to nitrite) is achieved in the Sharon process without heterotrophic denitrification. Nitrite-rich waste is then treated in an Anammox reactor. In the partial Sharon-Anammox digester, overall 83% ammoniacal nitrogen removal can be obtained from the waste stream has a total nitrogen load of 0.8 kg N/m³/day.[48]

Canon Process

The Canon (completely autotrophic nitrogen removal over nitrite) process is also the combination of partial nitritation and Anammox processes. In this process, two groups of aerobic and anaerobic microorganisms (e.g., *Nitrosomonas* and planctomycetes) perform two sequential reactions in a single and aerated reactor. The nitrifiers consume oxygen and oxidize ammonia to nitrite. Consumption of oxygen creates an anoxic condition the Anammox process needed. The performance of the Canon process is very much dependent on operational parameters such as dissolved oxygen, biofilm thickness, nitrogen-surface load, and temperature.[49]

Biological Phosphorus Removal

The removal of phosphorus from the wastewater by the biological means is known as biological removal of phosphorus. The groups of microorganisms that are largely responsible for phosphorus removal are known as the polyphosphate-accumulating organisms (PAOs). These organisms are able to store phosphate as intracellular polyphosphate, leading to phosphorus removal from the bulk liquid phase through PAO cell removal in the waste activated sludge. Enhanced biological phosphorus removal can be achieved through the ASP by recirculating sludge through anaerobic and aerobic conditions.[50] Unlike most other microorganisms, PAOs can take up carbon sources such as VFAs under anaerobic conditions, and store them intracellularly as carbon polymers, namely poly-β-hydroxy-alkanoates (PHAs). The energy for this biotransformation is mainly generated by the cleavage of polyphosphate and release of phosphate from the cell. Reducing power is also required for PHA formation, which is produced largely through the glycolysis of internally stored glycogen.[51] The principal advantages of biological phosphorous removal are reduced chemical costs and less sludge production as compared with chemical precipitation.

The different types of bacteria used in biological phosphorus removal are *Acinetobacter, Pseudomonas, Microlunatus phosphovorus, Aeromonas,* and *Lampropedia. Acinetobacter calcoaceticus* has a very high capacity to intracellularly accumulate polyphosphate from various activated sludges. It can accumulate phosphate of an amount of 0.9%–1.9% of dry cell weight.

Conclusion

Biological treatment processes have a proven track record of dealing adequately with various kinds of wastes generated by human activities. They mimic natural processes occurring in streams and rivers. Waste treatment processes are increasingly engineered in such a way that they perform this task efficiently with a minimal input of energy. Traditionally, treatment has relied on technological approaches designed to mimic aerobic processes occurring in the water column of streams and rivers. To become truly sustainable, however, we must move away from energy-consuming aerobic processes and switch to anaerobic treatment processes, again mimicking natural processes, but now those occurring in the anaerobic sediments of the aforementioned streams and rivers. For example, there is a new focus in the water industry to integrate these two processes into systems where the waste is initially digested in an anaerobic step followed by an aerobic polishing step. Only by integrating these two processes, and variants thereof such as partial nitrification and Anammox wastewater treatment, will waste treatment become truly energy efficient and sustainable. Finally, it should be noted that anaerobic digestion to methane is not the only sustainable option. Great strides are now being made in microbial fuel cell technology within waste treatment with chemical energy from wastes being captured as electricity. All told, we are finally beginning to see again that wastes are not problems to be solved but are valuable resources, and new technologies continue to be developed to capture this capacity.

References

1. Van de Kraats, J. Editorial. Eur. Water Pollut. Control **1997**, *7*, 3–4.
2. Tchobanoglous, G.; Burton, F.L.; Stensel, H.D.; Metcalf and Eddy, Inc. Fundamentals of biological treatment. In *Wastewater Engineering: Treatment, Disposal, Reuse,* 6th Ed.; McGraw-Hill: New York, 2003.
3. Lettinga, G. Anaerobic digestion and wastewater treatment systems. Antonie van Leeuwenhoek **1995**, *67*, 3–28.
4. Lettinga, G. Sustainable integrated biological wastewater treatment. Water Sci. Technol. **1996**, *33*, 85–98.
5. Jewell, W.J. Resource-recovery wastewater treatments with biological systems. Proceedings of the Workshop on Sustainable Municipal Wastewater Treatment Systems, ETCWASTE, Leusden, the Netherlands, 1996; 67–101.
6. Qasim, S.R. *Wastewater Treatment Plants: Planning, Design and Operation*; Holt, Rinehart and Winston: New York, 1985.
7. WPCF, Natural Systems for Wastewater Treatment. *Manual for Practice,* prepared by Task Force on Natural Systems, Sherwood C. Reed, chairman, Water Pollution Control Federation, Alexandria VA, 1990.
8. American Public Health Association. *Standard Methods for the Examination of Water and Wastewater,* 17 th Ed.; American Public Health Association: Washington, DC, 1989.
9. Water Environment Federation. *Design of Municipal Wastewater Treatment Plants,* 4th Ed.; Water Environment Federation: Alexandria, VA, 1998; Manual of Practice No. 8.
10. Rittmann, B.E.; McCarty, P.L. *Environmental Biotechnology: Principles and Applications*; McGraw-Hill: New York, 2001.
11. Henze, M.; van Loosdrecht, M.C.M.; Ekama, G.A.; Brdjanovic, D., Eds. *Biological Wastewater Treatment. Principles, Modelling and Design*; IAW Publishing: London, 2008.

12. Zehnder, A.J.B., Ed. *Biology of Anaerobic Microorganisms*. John Wiley and Sons: New York, 1988.

13. Colleran, E.; Finnegan, S.; Lens, P. Anaerobic treatment of sulphate-containing waste streams. Antonie van Leeuwenhoek **1995**, *67* (1), 29–46.

14. Polprasert, C. *Organic Wastes Recycling*; Wiley: Chichester, 1989.

15. Stafford, D.A.; Wheatley, B.I. *Anaerobic Digestion*; Applied Science Pub. Ltd.: London, 1979.

16. Bitton, G. *Wastewater Microbiology*, 4th Ed.; Wiley- Blackwell: New York, 1994.

17. Speece, R.E. Anaerobic biotechnology for industrial wastewater treatment. Environ. Sci. Technol. **1983**, *17*, 416A–427A.

18. Zeikus, J.G. Microbial populations in digesters. In *Anaerobic Digestion*; Stafford, D.A., Wheatley, B.I., Hughes, D.E., Eds.; Applied Science Pub. Ltd.: London, U.K., 1980; 61–89.

19. Gujer, W.; Zehnder, A.J.B. Conversion processes in anaerobic digestion. Water Sci. Technol. **1983**, *15*, 127–167.

20. Dolfing J. Acetogenesis. In *Biology of Anaerobic Microorganisms*; Zehnder A.J.B., Ed.; John Wiley and Sons: New York, 1988; 417–468.

21. Gavala, H.N.; Angelidaki, I.; Ahring, B.K. Kinetics and modelling of anaerobic digestion process. Adv. Biochem. Eng. Biotechnol. **2003**, *81*, 57–93.

22. Seghezzo, L.; Zeeman, G.; van Lier, J.B.; Hamelers, H.V.M.; Lettinga, G. A review: The anaerobic treatment of sewage in UASB and EGSB reactors. Bioresour. Technol. **1998**, *65*, 175–190.

23. Lettinga, G.; van Velsen, A.F.M.; Hobma, S.W.; de Zeeuw, W.; Klapwijk, A. Use of up-flow sludge blanket (USB) reactor concept for biological wastewater treatment, especially for anaerobic treatment. Biotechnol. Bioeng. **1980**, *22*, 699–734.

24. Schink, B. Principles and limits of anaerobic degradation: Environmental and technological aspects. In *Biology of Anaerobic Microorganisms*; Zehnder, A.J.B., Ed.; John Wiley and Sons: New York, 1988; 771–846.

25. Ahammad, Sk. Z.; Gomes, J.; Sreekrishnan, T.R. A comparative study of two high cell density methanogenic bioreactors. Asia Pac. J. Chem. Eng. **2011**, *6*, 95–100.

26. Dolfing J. Granulation in UASB reactors. Water Sci. Technol. **1986**, *18* (12), 15–25.

27. Punal, A.; Lema, J.M. Anaerobic treatment of wastewater from a fish-canning factory in a full-scale upflow anaerobic sludge blanket (UASB) reactor. Water Sci. Technol. **1999**, *40* (8), 57–62.

28. Gerardi, M.H. *The Microbiology of Anaerobic Digesters*; John Wiley and Sons, Inc., Hoboken, New Jersey, USA, 2001.

29. Kim, In S.; Hwang, Moon H.; Jang, Nam J.; Hyun, Seong H.; Lee, S.T. Effect of low pH on the activity of hydrogen utilizing methanogen in bio-hydrogen process. Int. J. Hydrogen Energy **2004**, *29*, 1133–1140.

30. Sahm, H. Anaerobic wastewater treatment. Adv. Biochem. Eng. Biotechnol. **1984**, *29*, 84–115.

31. Lawrence, A.W.; McCarty, P.L.; Guerin, F.J.A. The effects of sulfides on anaerobic treatment. Air Water Int. J. **1966**, *110*, 2207–2210.

32. McHugh, S.; Carton, M.; Collins, G.; O'Flaherty, V. Reactor performance and microbial community dynamics during anaerobic biological treatment of waste waters at 16–37°C. FEMS Microbiol. Ecol. **2004**, *48*, 369–378.

33. Nozhevnikova, A.N.; Zepp, K.; Vazquez, F.; Zehnder, A.J.B.; Holliger, C. Evidence for the existence of psychrophilic methanogenic communities in anoxic sediments of deep lakes. Appl. Environ. Microbiol. **2003**, *69* (3), 1832–1835.

34. Bergey D.H.; Holt, J.G.; Krieg, N.R.; Sneath, P.H.A. *Bergey's Manual of Determinative Bacteriology*, 9th Ed.; Lippincott Williams and Wilkins: Philadelphia, PA, 1994.

35. van den Heuvel, J.C.; Beeftink, H.H.; Verschuren, P.G. Inhibition of the acidogenic dissimilation of glucose in anaerobic continuous cultures by free butyric-acid. Appl. Microbiol. Biotechnol. **1988**, *29* (1), 89–94.

36. Gyure, R.A.; Konopka, A.; Brooks, A.; Doemel, W. Microbial sulfate reduction in acidic (pH-3) strip-mine lakes. FEMS Microbiol. Ecol. **1990**, *73* (3), 193–201.

37. Dogan, T.; Ince, O.; Oz, N.A.; Ince, B.K. Inhibition of volatile fatty acid production in granular sludge from a UASB reactor. J. Environ. Sci. Health A **2005**, *40* (3), 633–644.

38. Daniels, L.; Belay, N.; Rajagopal, B.S. Assimilatory reduction of sulfate and sulfite by methanogenic bacteria. Appl. Environ. Microbiol. **1986**, *51* (4), 703–709.

39. O'Flaherty, V.; Colohan, S.; Mulkerrins, D.; Colleran, E. Effect of sulphate addition on volatile fatty acid and ethanol degradation in an anaerobic hybrid reactor. II: Microbial interactions and toxic effects. Bioresour. Technol. **1999**, *68* (2), 109–120.

40. Hwu, C.S.; van Beek, B.; van Lier, J.B.; Lettinga, G. Thermophilic high-rate anaerobic treatment of wastewater containing long-chain fatty acids: Effect of washed out biomass recirculation. Biotechnol. Lett. **1997**, *19* (5), 453–456.

41. Daffonchio, D.; Thaveesri, J.; Verstraete, W. Contact angle measurement and cell hydrophobicity of granular sludge from upflow anaerobic sludge bed reactors. Appl. Environ. Microbiol. **1995**, *61*, 3676–3680.

42. Hanaki, K.; Matsuo, T.; Nagase, M. Mechanisms of inhibition caused by long chain fatty acids in anaerobic digestion process. Biotechnol. Bioeng. **1981**, *23*, 1591–1560.

43. Teske, A.; Alm, E.; Regan, J.M.; Toze, S.; Rittmann, B.E.; Stahl, D.A. Evolutionary relationship among ammonia- and nitrite-oxidizing bacteria. J. Bacteriol. **1994**, *176*, 6623–6630.

44. *Zumft W.G. The denitrifying prokaryotes. In* The Prokaryotes. A Handbook on the Biology of Bacteria: Ecophysiology *Isolation Identification Applications;* Balows, A, Truper, H.G., Dworkin, M., Harder, W., Schleifer, K.H., Eds.; 2nd Ed.; Springer-Verlag: New York, 1992; Vol. 1, 554–582.

45. Jetten, M.S.M.; Schmid, M.; Schmidt, I.; Wubben, M.; Van Dongen, U.; Abma, W. Improved nitrogen removal by application of new nitrogen-cycle bacteria. Rev. Environ. Sci. Biotechnol. **2002**, *1*, 51–63.

46. Hellinga C.; Schellen A.A.J.C; Mulder J.W.; van Loosdrecht, M.C.M. The Sharon process: An innovative method for nitrogen removal from ammonium-rich wastewater. Water Sci. Technol. **1998**, *37*, 135–142.

47. Jetten, M.S.M.; Wagner, M.; Fuerst, J.; van Loosdrecht, M.C.M.; Kuenen, J.G.; Strous, M. Microbiology and application of the anaerobic ammonium oxidation (anammox) process. Curr. Opin. Biotechnol. **2001**, *12*, 283–288.

48. Jetten, M.S.M.; Horn, S.J.; Van Loosdrecht, M.C.M. Towards a more sustainable municipal wastewater treatment system. Water Sci. Technol. **1997**, *35*, 171–180.

49. van Loosdrecht M.C.M. *Recent Development on Biological Wastewater Nitrogen Removal Technologies,* In Proceedings of the International Conference on Wastewater Treatment for Nutrient Removal and Reuse (ICWNR'04); Bangkok, Thailand, 2004.

50. Barnard J.L. Biological nutrient removal without addition of chemicals. Water Res. **1975**, *9* (5–6), 485–490.

51. Mino, T.; van Loosdrecht, M.C.M.; Heijnen, J.J. Microbiology and biochemistry of the enhanced biological phosphate removal process. Water Res. **1998**, *32* (11), 3193–3207.

56

Wastewater Treatment: Conventional Methods

Pollution Problems Associated with Wastewater.....................................577
Wastewater Treatment 577
Upgrading of Existing Wastewater Treatment Plants............................578
Selection of Treatment Method...579
References.. 581

Sven Erik Jørgensen

Pollution Problems Associated with Wastewater

The water pollution problems associated with municipal and industrial wastewaters include their content of

- Nutrients (nitrogen and phosphorus) causing eutrophication
- Biodegradable organic matter causing oxygen depletion
- Bacteria and viruses affecting the sanitary quality of water, which is of particular importance, when the water is used for bathing, swimming, and drinking purposes
- Heavy metals, mainly lead, zinc, and cadmium from gutters, heavy metals from fungicides and other agricultural chemicals, and a wide range of other heavy metals in minor concentrations
- Refractory organic matter, originating from industries, hospitals, the use of pesticides, and the use of a wide spectrum of household entries

Wastewater Treatment

Tables 1 and 2 provide an overview of the wide range of conventional wastewater treatment methods, their efficiencies, and costs. Clearly, there is, with good approximations, a method available to virtually any of the aforementioned problems. This overview presents the conventional methods. To give a complete overview of all available methods, it is necessary to supplement this overview with a summary of low-cost methods and recently developed methods of wastewater treatment. Moreover, ecological engineering methods are not included. Particularly, the application of constructed wetland offers an attractive alternative to some of the methods presented here.

Industrial wastewaters can cause the same water pollution problems as municipal wastewater. In addition, they may contain higher concentrations of toxic organic and/ or inorganic compounds (particularly heavy metals and persistent organic pollutants). However, it is necessary in most cases to solve the problems associated with industrial wastewater at the source. They can hardly be solved by municipal wastewater treatment methods. It is also the general legislation all over the world today that industries are obliged to treat wastewater before it is discharged to the public sewage system. In many countries, the practice of the polluter-has-to-pay system has forced industries to solve their pollution problems to keep their production costs low. The major portion of toxic substances is therefore today removed by the industries themselves, at least in most industrialized countries. They are only partially removed, if at all, at municipal wastewater treatment plants, as this could contaminate the sludge produced at municipal wastewater treatment plants, thereby eliminating the possibility of the use of the sludge as a soil conditioner.

Application of the methods identified in Tables 1 and 2 gives only approximate results, and the indications should therefore be used with caution. However, first estimates such as those shown in the tables are useful for evaluations of various alternative solutions to wastewater pollution problems. The mentioned biological treatment may either be an activated sludge plant or a trickling filter of different design.

The cost of treating $100\,m^3$ of wastewater is based on approximate estimations, as the included cost (labor, electricity, and so on) varies from place to place, and is furthermore highly dependent on the size of the waste-water treatment plant. The costs are calculated as the running costs (electricity, labor, chemicals, and maintenance) plus 10% of the investment to cover interest and annual appreciation. The annual water consumption of one person in an industrialized country corresponds to approximately $100\,m^3$. For comparison, the treatment of municipal wastewater on waste stabilization ponds (WSPs) amounts to $3–$12/100 m^3$, and that on constructed wetlands amounts to $6–$18/100 m^3$. In most cases, WSPs cannot achieve a BOD_5 reduction above 85%–88%. In some cases, more than 90% efficiency is required. Constructed wetlands can, in most cases, with the right design, offer a fully acceptable water quality, particularly when they are used after WSPs.

The removal of high concentrations of biodegradable organic matter at the source is most often strongly recommended, since it is usually much more cost-effective to remove these pollutants, at least partially, when they are present in high concentrations. High concentrations of biodegradable organic matter are found in wastewater from slaughterhouses, starch factories, fish industries, dairies, and canned food industries.

The listed methods are often used in combinations of two or more steps to obtain the overall removal efficiency required by the most cost-moderate solution. The methods can of course also be applied in combination with cleaner technology or ecotechnology. Because wastewater treatment is often costly, it is recommended in the planning phase to examine *all* possible combinations of treatment options in order to identify the most feasible and appropriate one. For a more comprehensive presentation of all the features of the methods listed in Tables 1 and 2.[1]

A combination of treatment methods is needed in most cases, and the most applied combinations for treatment of municipal wastewater are presented in the entry "Municipal Waste Water."

Upgrading of Existing Wastewater Treatment Plants

Many existing municipal wastewater treatment plants were constructed years or decades ago, and may not meet today's higher standards. Nevertheless, upgrading existing wastewater treatment plants is possible and may be more cost-moderate than building new ones.[2,3] Because the funding allocated to pollution abatement is often limited, the overall effect of upgrading wastewater treatment plants that can be upgraded with sufficient efficiency will be to the benefit of the environment. An attractive solution is often to introduce *tertiary treatment* by chemical precipitation and flocculation in an existing mechanical–biological treatment plant, with the addition of chemicals and flocculants before the primary sedimentation phase. The installation costs for this solution are minor, and the additional running costs are limited to the costs of chemicals. The result is an 85%–95% removal of phosphorus at low cost. Similarly, nitrification and denitrification, ensuring an 80%–85% removal of nitrogen, can be realized with the installation of additional capacity for biological treatment (the overall water retention time in the plant is increased by 4–18 hr, depending on the standards and composition of the wastewater), which is considerably less costly than the installation of a completely new treatment plant.[4,5]

TABLE 1 Survey of Generally Applied Wastewater Treatment Methods

Method	Pollution Problem	Efficiency	Costs ($/100 m3)
Mechanical treatment	Suspended matter removal	0.75–0.90	3–5
	BOD5 reduction	0.20–0.35	
Biological treatment	BOD5 reduction	0.70–0.95	25–40
Flocculation	Phosphorus removal	0.3–0.6	6–9
	BOD5 reduction	0.4–0.6	
Chemical precipitation	Phosphorus removal	0.65–0.95	10–15
$Al_2(SO_4)_3$ or $FeCl_3$	Reduction of heavy metal concentrations	0.40–0.80	
	BOD5 reduction	0.50–0.65	
Chemical precipitation	Phosphorus removal	0.85–0.95	12–18
$Ca(OH)_2$	Reduction of heavy metal concentrations	0.80–0.95	
	BOD5 reduction	0.50–0.70	
Chemical precipitation	Phosphorus removal	0.9–0.98	12–18
and flocculation	BOD5 reduction	0.6–0.75	
Ammonia stripping	Ammonia removal	0.70–0.95	25–40
Nitrification	Ammonium is oxidized to nitrate	0.80–0.95	20–30
Active carbon	COD removal (toxic substances)	0.40–0.95	60–90
adsorption	BOD5 reduction	0.40–0.70	
Denitrification after	Nitrogen removal	0.70–0.90	15–25
nitrification			
Ion exchange	BOD5 reduction (e.g., proteins)	0.20–0.40	40–60
	Phosphorus removal	0.80–0.95	70–100
	Nitrogen removal	0.80–0.95	45–60
	Reduction of concentrations		10–25
Chemical oxidation	Oxidation of toxic compounds	0.90–0.98	60–100
(e.g., with Cl_2)			
Extraction	Heavy metals and other toxic compounds	0.50–0.95	80–120
Reverse osmosis	Removes pollutants with high efficiency, but is expensive		100–200
Disinfection methods	Reduction of microorganisms	High, can hardly be indicated	6–10
Ozonation + active carbon adsorption	Removal of refractory compounds	0.5–0.95	100–120

Selection of Treatment Method

Any removal efficiency of pertinent parameters (BOD_5, nutrients, bacteria, viruses, toxic organic compounds, color, taste, heavy metals) can be obtained with a suitable combination of the available treatment methods, as is evident from Tables 1 and 2. However, which removal efficiencies are needed in a considered case? Because wastewater treatment is costly, the maximum allowable concentrations should not be set significantly lower than the lake or reservoir receiving the effluents can tolerate. The ban of phosphate detergents to decrease phosphorus concentrations in municipal wastewater treatment plant effluents is a point to consider in this context, as the treatment costs can be reduced considerably by the introduction of phosphorus- free detergents. On the other hand, it might be even more expensive to install an insufficient treatment plant. Thus, the potential effects of a wide range of possible pollutant inputs on water quality and on the receiving aquatic ecosystem should be assessed as the basis for selecting an acceptable option. This requires a quantification of the impacts of various possible pollutant inputs and the consideration of a wide range of solutions. All processes and components affected significantly by the impacts should be included in the quantification. It is usually very helpful to develop a water quality/ecosystem model to assist in the selection of specific

TABLE 2   Efficiency Matrix Relating Pollution Parameters and Wastewater Treatment

	Suspended Matter	BODs	COD	Total Phosphorus	Ammonium Nitrogen	Total Nitrogen	Heavy Metals	E. coli	Color	Turbidity
Mechanical treatment	0.75–0.90	0.20–0.35	0.20–0.35	0.05–0.10	~0	0.10–0.25	0.20–0.40	–	0.80–0.98	–
Biological treatment[a]	0.75–0.95	0.65–0.90	0.10–0.20	0.05–0.10	~0	0.10–0.25	0.30–0.65	Fair	~0	–
Chemical precipitation	0.80–0.95	0.50–0.75	0.50–0.75	0.80–0.95	~0	0.10–0.60	0.80–0.98	Good	0.30–0.70	0.80–0.98
Ammonia stripping	~0	~0	~0	~0	0.70–0.96	0.60–0.90	~0	~0	~0	~0
Nitrification	~0	~0	~0	~0	0.80–0.95	0.80–0.95	~0	Fair	~0	~0
Active carbon adsorption®	–	0.40–0.70	0.40–0.70	~0.1	High[b]	High[b]	0.10–0.70	Good	0.70–0.90	0.60–0.90
Denitrification after nitrification	~0	–	–	~0	–	0.70–0.90	~0	Good	~0	–
Ion exchange	–	0.20–0.50	0.20–0.50	0.80–0.95	0.80–0.95	0.80–0.95	0.80–0.95	Very good	0.60–0.90	0.70–0.90
Chemical oxidation	–	Corresponding to oxidation	~0	~0	~0	~0	~0	~0	0.60–0.90	0.50–0.80
Extraction	–	Corresponding to extraction of toxic compounds	~0	~0	~0	~0	0.50–0.95	~0	~0	~0
Reverse osmosis[a]	–	See Table 1								
Disinfection methods	–	Much corresp. to appl. of chlorine, ozone, etc.						Very high	0.50–0.90	0.30–60

[a] Depends on the composition.

[b] As chloramines.

environmental treatment methods. It is important to emphasize that a model has an uncertainty in all its predictions that must be considered in making a final decision. Thus, it is essential to use safety factors to the benefit of the environment, in order to ensure that the selected treatment methods will have the anticipated effects. If the uncertainty is not taken into account for the sake of economy, as is unfortunately often done, the investment may be wasted because the foreseen recovery of the ecosystem will not be realized.

A problem in many developing countries is the relatively high cost of wastewater treatment. This cost justifies the application of "soft technology" or "ecotechnology," but proper planning at an early phase and the consideration of all predictable problems will always offer a wider range of cost-effective possibilities and may allow the prevention of pollution problems before they occur.

Corrections at a later stage, when pollution has already degraded the water quality and associated ecosystems, are possible, but will always be more expensive than the costs of proper wastewater treatment at an early stage. This is due in part to the fact that the accumulation of pollutants in an ecosystem over time will always cause additional problems and therefore result in additional costs. Thus, preventing pollution at an early stage is better than curing pollution at a later stage. Removal of phosphorus from wastewater at an early stage, for example, is always beneficial since the surplus phosphorus will accumulate in the sediments to a large extent, allowing its remobilization back into the water column under certain chemical conditions in the water body.

Completely new approaches have emerged in regard to sustainable development. For instance, serious consideration is being given to separation toilets in some locations, which collect urine separately from feces, thereby allowing utilization of the septic urine as fertilizer.

The selection of proper wastewater treatment methods for point sources of pollution is summarized in the following points:

- Develop models for the impacts of the wastewater on freshwater ecosystems, considering the impacts on the water quality and the entire lake ecosystem.
- Apply the model to identify the maximum allowable pollutant concentration in the treated wastewater. Any uncertainty associated with the model predictions should be reflected in identifying the lower maximum allowable concentrations.
- Select the combination of available treatment methods able to meet the standards at the lowest costs without impacting the proper operation of the plant.
- If the investment needed for a proper solution to a problem cannot be provided, the application of cost-moderate technology that will reduce the accumulation of pollutants in the aquatic ecosystem should be considered. Any measures taken at an early stage will inevitably reduce the costs at a later stage.

References

1. Jørgensen, S.E. *Principles of Pollution Abatement*; Elsevier: Amsterdam, 2000; 520 pp.
2. *Novotny, V.; Somlyódy, L., Eds.* Remediation and Management of Degraded River Basins with Emphasis on Central and Eastern Europe; *Springer Verlag: Berlin, Germany, 1995.*
3. Loosdrecht M. van. Upgrading of waste water treatment process for integrated nutrient removalxThe BCFS process. Water Sci. Technol. **1998**, *37* (9), 37–46.
4. Hahn, H.H.; Muller, N. Factors affecting water quality of (large) rivers past experiences and future outlook. In *Remediation and Management of Degraded River Basins with Emphasis on Central and Eastern Europe*; Novotny V., Somlyódy L., Eds.; Springer Verlag: Berlin, Germany, 1995; 385–426.
5. Henze, M.; Ødegaard, H. Wastewater treatment process development in Central and Eastern Europe—Strategies for a stepwise development involving chemical and biological treatment. In *Remediation and Management of Degraded River Basins with Emphasis on Central and Eastern Europe*; Novotny, V., Somlyódy, L., Eds.; Springer Verlag: Berlin, Germany, 1995; 357–384.

57

Water and Wastewater: Filters

Introduction 583
Treatment of Water and Wastewater 584
Filtration Process 585
Types of Granular Filters 587
  Slow Sand Filter • Rapid Gravity Filter with Coagulant
  Aid • Pressure Sand Filter
Cartridge Filter 589
Biofilters 589
Biological Aerated Filters 589
Trickling Filters 589
Membrane Filtration 590
  Future Developments of MBR • Applicability
Ultrafiltration 593
  Performance of UF • Ultrafiltration System Design • Ultrafiltration in
  Water Treatment
Reverse Osmosis System 594
Integrated Membrane System 596
Water and Wastewater Purification 596
Eco-Filtration Systems 596
Constructed Wetlands 597
Soil Biotechnology 597
Potential and Scope of Eco-Technology 597
Soil Scape Filter 598
Green Bridge Technology 599
Other Emerging Filtration Techniques 599
  Disc Filters • Pleated Filtration Panel innovation
Conclusion 600
References.. 601

Sandeep Joshi

Introduction

Windhoek, Namibia, a city in an arid region, suffered chronic freshwater scarcity. In the late 1960s, due to prolonged droughts, the only alternative for survival was to use treated wastewater for drinking purposes. In 1968, it became the first city in the world to directly supplement its drinking-water supplies with treated wastewater.[1] Windhoek has approximately 2,00,000 residents, with yearly increase of the population of about 5%. Since 1973, an epidemiological study of Windhoek residents has shown no adverse health impacts associated with drinking reclaimed wastewater. It provides 30% of the demand.

In 2002, this treatment plant was rebuilt, and the wastewater effluent is now the primary source for drinking water. This plant treats 24,000 m³/day and has preozonation, flocculation, and filtration units comprising a rapid sand filter, a biological activated carbon filter, and two granular activated carbon (GACs) filters in a series, followed by ultrafiltration (UF).[2] This is a classic example of significance of filtration systems in water and wastewater treatment.

Water treatment commonly refers to the actions taken to purify water for domestic and industrial purposes. Generally, processes used for water treatment are physicochemical. Wastewater or sewage treatment processes are mainly biological. Most of the treatment plants purifying surface waters might have two stage processes—filtration and disinfection. For the wastewaters, normally, two more processes are added—pre- and postclarification. Water quality as far as suspended solids are concerned is achieved by filtration. Most of the filters are composed of granular media such as sand, anthracite, soil, etc. These filters have the ability to produce high-quality water when perfectly operated. Clarification and filtration processes are complimentary to each other.

Traditionally, settling of solids was the main process of clarifying water. Filtration was also practiced in ancient times. The earliest filters seem to be very simple—infiltration wells excavated near the banks or lakeshores. Wells on the seashore usually were dug tapping sweet water aquifers flowing from the land to the sea. The technology is modernized with addition of sophisticated UF or reverse osmosis for better-quality water. Filters can be classified based on the following: 1) filtration rate—slow, rapid, and pressured; 2) direction of flow—downflow, upflow, and biflow; 3) filtration medium—sand, soil, and synthetic media; 4) layers of media—single, dual, and multimedia; 5) input water quality—water, sewage, industrial effluents, etc.

Water or wastewater treatment is the process of removing undesirable chemical or biological contaminants. These can be gaseous or solid constituents. Concentration of gaseous contaminants is dependent on factors like solubility, temperature, partial pressure, etc. They can be removed physically by diffusion or temperature increment Size of solid contaminants may range from millimeters to angstroms—less than a nanometer. These contaminants are different types of solids that can be categorized based on their sizes or physicochemical properties. Origin of these solids can be nonbiological or biological. They can be divided into two—depending on their chemical composition—organic and inorganic, as shown in Figure 1. Suspended or dissolved solids having different sizes are shown in Figure 2. Solids larger than suspended solids are termed as floating solids like plastic, paper, cloth, etc. They can be easily removed by putting a screen across the path of water flow, but smaller particles need finer sieves.

Treatment of Water and Wastewater

Treatment includes physical, chemical, and biological processes to remove physical, chemical, and biological contaminants. Its objective is to produce an environmentally safe fluid waste stream (or treated effluent) and solid waste (or treated sludge, http://en.wikipedia.org/wiki/Sludge) suitable for disposal or

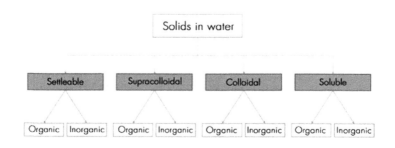

FIGURE 1   Classification of solids in water.

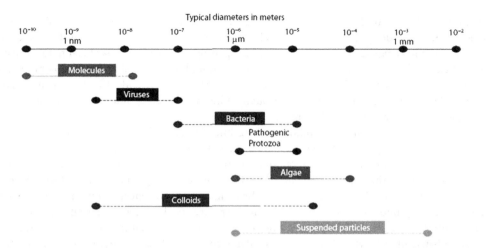

FIGURE 2 Different sizes of solid contaminants in water.

reuse (usually as farm fertilizer). Using advanced technology, it is now possible to reuse sewage effluent for drinking water. Singapore has advanced to convert sewage into drinking water on mass scale.

In general, a conventional water treatment plant usually consists of physical treatment (screening, sedimentation, flotation, and/or filtration) and chemical treatment (pH adjustment, coagulation-flocculation process, oxidation- reduction process, adsorption process).[3] The degree of the complexity of the treatment plant also depends on the quality of raw water and treated water requirement. In industrial processing, water is used in numerous applications requiring likewise different qualities of water. Examples of different uses are cooling water, water for rinsing and chemical production, boiler feed water, purified water, and water for injection. The growth in population, the increasing costs of treatment and distribution, contamination of fresh water sources, and the sophistication of end users somehow force the development of better water treatment technology.[4,5]

Filtration Process

Filtration means a process of solid-liquid separation using some form of physical barrier. In sedimentation or floatation, no barrier is used. Depending on other supportive mechanisms, the filters can be divided into mechanical or biological filters keeping the built-in solid–liquid stage. In 1685, Porzio, an Italian physician, used a filter to protect the health of soldiers in the Austro–Turkish war of that year. In 1790, James Peacock was granted the first patent for an upflow filter with graded support gravel and reverse- flow wash. James Simpson built a 1-acre slow sand filter, and it was commissioned satisfactorily in 1829. In 1858, the use of a rake was proposed to break the schmutzdecke (surface clogging layer). Holly, in the United States in 1871, patented the reverse-flow wash downflow filter, and in 1877, Cook developed the concept of a battery filter for keeping standby options during the washing of used filters.[6] In India, similarly, an underground filtration flow system was developed in Pune in 18th century to cater the people's needs for fresh and clean water.[7]

It has been noticed that very small particles and large particles are predominantly removed by diffusion and straining mechanisms, respectively. Five different mechanisms other than simple straining[8]

Generally, in conventional water or wastewater treatment plants, filtration is considered as the core of the process in which the solids are substantially removed. Incoming flow to the filters is clearly non-potable or turbid, but after filtration, it becomes clean and transparent. If the water is disinfected, then it can be used for human consumption.

The process of filtration is composed of passing water through a granular bed of sand or any other suitable medium at a designed flow rate. To achieve the desired quality of filtered water, the filtration system has to remove particles, and the sizes of the openings are crucial. It has been noticed that very small particles and large particles are predominantly removed by diffusion and straining mechanisms, respectively. Five different mechanisms other than simple straining[9] are identified in the filtration process when particles in a flow of water pass through a filter, viz., interception, sedimentation, diffusion, hydrodynamic action, and inertia (Figure 3).

Particles moving uniformly may collide with a grain of filter media; their transport is intercepted. Efficient interception is possible by decreasing the size of the media or increasing the size of the particle to be removed. Larger particles are removed more efficiently. Sedimentation is a result of gravitational forces acting on the particles depending on the ratio of settling velocity and velocity of the flow approaching the media. Diffusion as a result of Brownian motion is effective only for small particles. It becomes more pronounced with increasing temperature. Hydrodynamic action arising from the velocity gradient separates small particles from the flow. This is comparatively nonsignificant. Inertia (impaction) is useful only when hydrodynamic diversion is less. At less velocity, inertia is not significant. In addition to these mechanisms, coagulation, flocculation, adsorption, and absorption play major roles in the effectiveness of a filtration system.

Filtration with a granular bed is composed of indivisible grains that rest on each other, leaving voids in between. Dirt is removed when water flows through these voids. The real filter media are composed of irregular grains-often unpolished rounded edges. Sands are the most readily available material from natural sources such as the sea coast and riverbeds. Granular media can be obtained by crushing materials such as basalt, slag, and anthracite. The voidage increases with angularity of the material, e.g., crushed anthracite has more voidage (45%) than sand (38%) after gentle tapping. Filtration media retain solids while permitting water having turbidity less than 0.2 NTU (nephelometric turbidity unit) to pass.

Solids are collected on the surface of the filter bed, forming a surface mat or "mud cake." This needs to be regularly cleaned to avoid resistance to flow. A granular bed filter having a mass of leading edges and space for accumulation of deposits helps in retaining the grains that are no longerin the line of flow. They are like silt on the sand bed of a river. These materials are comparatively cheaper than fibrous materials that are used for large-scale water or wastewater applications. If the grain does not dissolve in

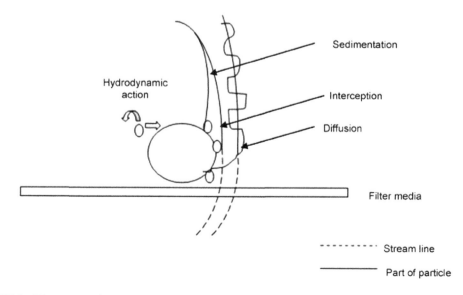

FIGURE 3   Filtration mechanisms.

water, then its chemical nature is not important. Therefore, structurally weak materials such as activated carbon (granular or powdered) can be used for filtration purposes.

Filters operate best normally at a low concentration of solids as a result of limited volume of pores (about 40% of the bed volume). A granular bed filter having a mass of leading edges and space for accumulation of deposits helps in retaining the grains that are no longer in the line of flow. Therefore, such filters operate at 0.2 bar (2 m head), but media such as a cloth, paper, porous sinter, or membrane tend to accumulate deposits, and then it requires high pressure (up to 7 bars) to maintain its permeability. The depth in simple granular filters may be 600–1000 mm for its desired effectiveness. Failure of the filters is associated with cleaning the mechanism either manually or mechanically with water or air backwash.

Types of Granular Filters

There are three basic types of granular filters, viz., sand filters, rapid gravity (RG) filters, and pressure filters. Slow sand filters—operating at low loading rates—are the oldest variety of filters. These filters made up of fine sand have two processes—physical straining and biological action. Rapid filters function at higher loading rates using coarse media with a higher permeability. In addition to sand, GAC is used to adsorb chemicals dissolved in water. Coagulation helps in effective removal of solids by RG filters. Simple RG filters use single-medium sand, but a multimedia version uses two or more types of media. These are cleaned by reversing the flow of water to wash out the dirt. Pressure filters are the variety operating under pressure in large closed vessels. These are normally used for treating groundwater sources, which are pumped from boreholes. Various types of filters are shown in Figure 4.

Slow Sand Filter

Slow sand bed filters are very-low-flow-rate filters. They are used in rural areas or where the water flow rate required is low. A slow sand filter can even remove turbidity and suspended solids. Pathogens are also removed, additionally producing water with very reasonable taste and no color. An operational advantage of slow sand filters is that no backwashing is done. They require few operator skills. Sometimes, slow sand filters are used in the treatment of sewage as a final polishing stage for the treated effluent and as a bed to dewater the sludge produced during treatment.

Rapid Gravity Filter with Coagulant Aid

Purification of water for drinking purposes is done using chemicals to flocculate as much particulate matter to be removed as quickly as possible in the filter. Coagulation of the particulate matter is achieved by adding highly charged cations, such as aluminum sulfate (alum) and small amounts of charged polymer chains. Coagulation and flocculation processes require contact time and agitation in tanks before the water is filtered so that a reasonably sized floc can be formed. This process is pH dependent. Thus, it is necessary to adjust the pH of the water to ensure efficient removal of water. A rapid gravity sand filter strains out the floc and the particles, including bacteria trapped within them. The medium of the filter has grain sizes to allow the water to pass through the filter rapidly. This filter is backwashed with clean water on a regular basis as the pores are clogged by particulate matter (flocs), indicated by a rapid drop-off in the flow rate of water through the filter. This backwash water is run into settling tanks so that the floc can settle out, and it is then disposed of as residue. Supernatant water is then run back into the treatment process or disposed of as a wastewater stream.

Pressure Sand Filter

Applying pressure to the water passing through a sand filter will give a greater flow rate, while using smaller grains of sand in a filter allows a greater surface area of filtration medium to remove

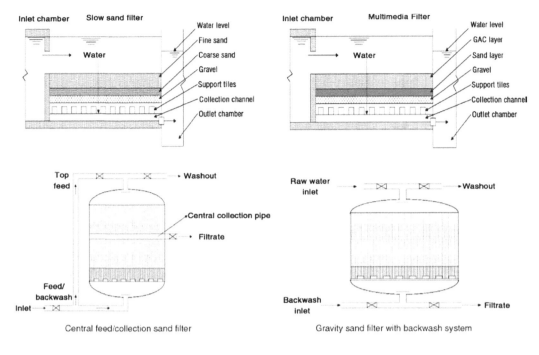

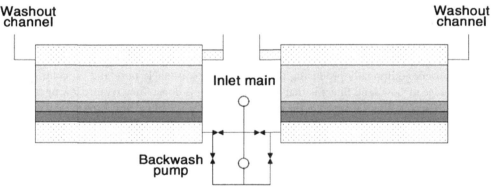

FIGURE 4   Different types of filters.

particulate matter. Smaller grain size requires more pressure to drive the water through the sand bed. Sand grain size in pressure filters is typically 0.6–1.2 mm, and large particulates (>100 pm) are removed. This causes rapid blinding of the filter with floc and the need for too-frequent backwashing. Pressure sand filters are typically 0.6–1.8 m in depth and operate under a maximum flow rate of about 9m^3/m^2/hr under a feed pressure of between 2 and 5 bars. Buildup of particulate solids causing an increase in pressure across the bed for a given flow rate will need to be backwashed when the pressure drop is around 0.5 bar. Particulate solids are then washed away with the backwash water. It is diverted to a settling tank to separate the solids from water. A small amount of sand can be lost in the backwashing process. Then, the sand bed may need to be topped up from time to time. Most pressure filters in industry or large water works employ an automated multiport valve together with sophisticated pressure and flow sensors.

Cartridge Filter

This type of filter has a removable housing, into which different types of filtration elements can be placed. A domestic cartridge filter element will often be rated at 30–50 μm or larger, whereas specialist industrial filter elements may be rated at 5 μm or less. Some cartridge filters are made effective by using some resins that are designed to remove specific contaminants from the water, such as nitrates, fluoride, or lead. Granulated activated carbon is used in filters to remove color, odor, volatile organics, and chlorine from water.

Biofilters

Biological treatment or biofiltration has become the mainstay of wastewater treatment systems over a period of time because of simplicity. Odor control using biofiltration has been patented from the 1950s for huge trickling filter plants and soil filters. Biofilters have primary applications for odor control at wastewater treatment plants and composting operations. They are being used also in the treatment of volatile organic compounds (VOCs), as an innovative method to treat toxic air emissions from industrial processes. The chronological development of biofilters is as shown below:

- 1923: To treat odorous compounds, Bach thought of using a biologically active biofilter to control emissions of H_2S from a wastewater treatment plant.
- 1955: A biofilter was applied to treat odorous emissions in low concentrations in Germany.
- 1959: A soil bed was installed at a sewage treatment plant in Nuremberg for the control of odors from an incoming sewer main.
- 1960s: Biofiltration was first used for the treatment of gaseous pollutants both in Germany and the United States.
- 1970s: Biofiltration became widespread in Germany.
- 1980s: Biofiltration is used for the treatment of toxic emissions and VOCs from industry.

Biological Aerated Filters

Biological aerated filters or biofilters combine filtration with biological carbon removal and nitrification or denitrification processes. A biologically aerated filter usually includes a reactor filled with a filter medium, which is either in suspension or supported by a gravel layer at the foot of the filter. The dual purpose of this medium is to support highly active biomass that is attached to it and to filter suspended solids. Carbon reduction and ammonia conversion occur in aerobic mode and are sometimes achieved in a single reactor, while nitrate conversion occurs in anoxic mode. Biological aerated filters are operated either in up- flow or downflow configuration.

Trickling Filters

The first trickling filter became operational in 1893. The concept of trickling filters was developed from contact filters—which were watertight basins filled with stones. Clogging, long resting period, and relatively low loading rate were some of the limitations of contact filters. The modern trickling filter is composed of a permeable medium (rock or plastic media) for the attachment of bacteria. Through this biofilter, wastewater percolates—trickles down. The filter medium is either rock (size varies from 25 to 100 mm) or slag or plastic packing materials. The depth of the rock medium is about 1–2.5 m. Wastewater is distributed on the top of the rock bed by rotary distributor. Plastic media are either round or square or irregularly shaped. Depth of plastic media varies from 4 to 12 m. Depending on packing, the trickling filter can be categorized into three groups, viz., vertical flow, cross-flow, and random.

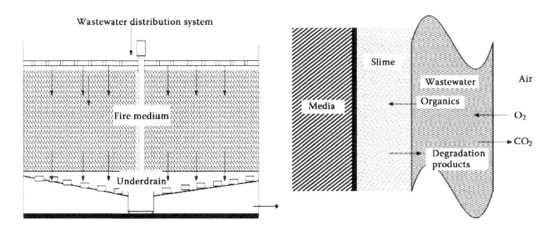

FIGURE 5   Trickling filter and slime layer.

The components of a trickling filter are the following:

- A bed of filter medium on which a layer of microbial slime is developed
- A container that houses the bed of filter medium
- Distribution of wastewater over the filter medium
- Removal and disposal of any sludge from the treated effluent

Filters are provided with underdrains for collection of treated wastewater and biosolids detached from the media. Underdrain systems are important as far as air circulation is concerned. Organic matter is degraded by the microorganisms present in the biological film or slime layer attached to the filter media (Figure 5). The outer layer of biofilm, about 0.1–0.2 mm, is aerobic, and the layer near the surface of filter media becomes anaerobic due to diffused oxygen and thus cannot penetrate the full depth of the slime layer. The inner layer is starved of nutrition of pollutants as the slime layer thickens. Endogenous degradation of bacteria in the inner layer leads to sloughing due to shear velocity of trickling water. Facultative microorganisms dominate the biological communities in the slime layer of the trickling filter. With these organisms, fungi, algae, protozoans, worms, insect larvae, and sometimes snails also dwell in the trickling filter.[9,10] Their presence in the filter is useful to maintain the bacterial population in a state of high growth or high rate of food consumption.

An excellent material for trickling filter is that having more surface area per unit volume, low cost, durability, and no tendency for clogging. Until the 1960s, a commonly used material was high-quality granite or blast furnace slag. River gravel or crushed stones were also used. All underdrains are developed to maintain and facilitate air ventilation. Sufficient airflow in the system is a key to successful operation of the trickling filter. In case of natural ventilation, the temperature difference between the ambient air and air in the pores of media is the key reason for air circulation. Good design may provide airflow at least 0.3 m³/minm³ of the filter area in either direction. Trickling filters can be classified on the following bases: 1) frequency of hydraulic and organic loading—slow rate, high rate, batch, intermittent, continuous; 2) nature of filtering media—natural, synthetic; 3) direction of inflow—downflow, up-flow, sectionalized; 4) internal environment of the filter— aerobic, anaerobic, anoxic; 5) depth of media—shallow, deep; 6) method of ventilation—natural, forced; and 7) extent of treatment expected—complete, roughing, polishing.

Membrane Filtration

The major breakthrough in the development of membrane technology was recorded in the late 1950s. The membrane bioreactor (MBR) has developed quite rapidly to become this important. The idea of combining sludge digestion with a very fine filter was first developed in the mid-1960s by Dorr Oliver, as a system with flat-plate membranes in a sidestream loop. This development was made possible by the appearance on the market of commercial-scale microfiltration (MF) and UF membranes. Membrane bioreactors have served industries that can afford an expensive system. The submerged membrane was innovated in Japan in 1989, which led to a Japanese government initiative to find better ways of wastewater treatment. Submerged or immersed MBRs used less energy than the sidestream version. This version entered the European market in the mid-1990s. Since 1990, the numbers of MBR installations have grown exponentially.

Technologies of secondary biological treatment systems for municipal wastewater generally involve microorganisms suspended in the wastewater to treat it. They have some deficiencies, including the difficulty of growing the exact types and numbers of microorganisms and the physical requirement of a large site. Application of MFMBRs surmounts many limitations of conventional systems. These membrane filtration systems combine the suspended biological growth with solids removal via filtration (Figure 6).

The basic MBR consists of two processing steps—a bioreactor, in which aerobic bacteria digest organic material in the presence of dissolved oxygen, and a membrane module, through which relatively pure water separates from the suspension of organic matter and bacteria. These two units may be set up to run in succession (i.e., the liquid flows first through the bioreactor, where it is held for as long as necessary for the reaction to be completed, and then through a loop containing the membrane separation stage), with a recycle of some of the separated sludge to the bioreactor. This is often called sidestream (or external) operation. Membranes are suspended in the slurry in the bioreactor, which is appropriately partitioned to achieve the correct airflows, with the surplus sludge withdrawn from the base of the bioreactor at a rate to give the required sludge retention time. This is termed as a submerged (or immersed) MBR.

Membranes can be designed for and operated in small spaces and with high removal efficiency of contaminants such as total suspended solids, nitrogen, phosphorus, biochemical oxygen demand (BOD), and bacteria. The membrane filtration system in effect can replace the secondary clarifier and sand filters in a typical activated sludge treatment system for large sewage treatment plants. Membrane filtration allows a higher biomass concentration to be maintained, thereby allowing smaller bioreactors to be used.

Membrane filtration involves the cross-flow of water containing pollutants across a membrane. Waste matters left behind do not accumulate at the membrane surface. They are removed from the system by gravity for disposal. Water passing through the membrane is called the permeate or filtrate, while water with more concentrated pollutants is called the concentrate, retentate, or reject. Membranes are composed of cellulose or other polymers having a number of pores. The requirement is that the membranes

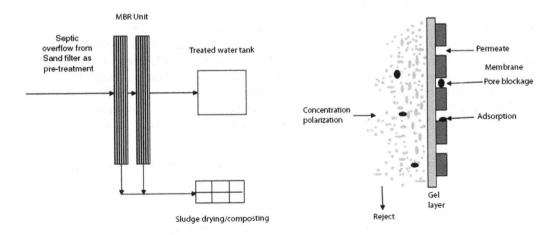

FIGURE 6 Typical flow chart of membrane filtration for sewage treatment and filtration through membrane.

filter out particles and microorganisms more than 1 μm (0.001 mm) in size. Two configurations most often used are hollow fibers grouped in bundles or flat plates for MBR designs.

Membrane filtration is an alternative to remove waterborne pathogens in water treatment, replacing the disinfection step in conventional water treatment plants to avoid risks and side effects of chemicals. Membrane processes can be classified in many ways, i.e., based on their nature, structure, or driving force. Hydrostatic pressure differences are used in MF and nanofiltration (NF), as well as in reverse osmosis (RO) and gas separation as a driving force for mass transport through the membrane. The MBR takes the place of the whole secondary stage of conventional activated sludge processing—and does it better and in a much smaller space. The excess biodegradation products can be removed for subsequent treatment.

The sludge settlement stage of the conventional secondary process is a slow one, so the removal of the clear liquid from the slurry by membrane filtration is a better option. A major advantage of the MBR system is that it can operate at a much higher concentration of solids in the bioreactor than that of a conventional activated sludge plant. An MBR plant can work effectively at concentrations of mixed liquor suspended solids typically in the range of 8000–12,000 mg/L (or 0.8%–1.2%) and can be extended up to 3%, whereas conventional activated sludge plants work at about 2000–3000 mg/L, because of the limitations on settling. This high sludge concentration capability enables an MBR system to deal effectively with strong industrial wastes, especially in places where water is short and factories are seeking to close their water cycles. It also results in a much smaller digestion tank, and thus a much smaller footprint for the whole plant. The smaller space footprint of an MBR plant will make it much more attractive for construction in developed urban areas.

The longer sludge retention times in the MBR permit the destruction of molecules difficult to biodegrade, such as detergents. With proper system design, nitrogen and phosphorus content can also be significantly reduced. The quality of water produced as effluent by an MBR is excellent and usually well under the local consent limits. Figures of less than 2 mg/L for BOD and less than 0.5 mg/L for total suspended solids are readily achievable, along with ammoniacal nitrogen of less than 0.5 mg/L and turbidity of less than 0.2 NTU. Particularly relevant are the capabilities of UF membranes (the most common form used) for removing pathogens such as protozoa, bacteria, and most viruses.

Most membrane filtration processes require quite a high transmembrane pressure in order to produce an acceptable permeate flow rate. However, an MBR operates with a low differential, about 0.5 bar. This can be provided by a vacuum pump, sucking on the permeate discharge line, through a receiver, or by the hydrostatic head of a deep bioreactor tank, or by a low level of pressurization of the tank.

Future Developments of MBR

The MBR (including its most common form, submerged MBR—not to be confused with the moving bed bioreactor) is a fascinating combination of two steps of waste treatment: the bacterial digestion of waste organic matter and the separation of the treated effluent from residual suspended solids. Its prime advantage is that it can constitute the whole of the secondary phase of the traditional three- phase municipal sewage treatment process and can eliminate some parts of the tertiary phase.

In the long history of wastewater treatment (over 110 years), the MBR is quite a recent invention, and so, not surprisingly, it is still in a period of intense development. Key areas of system investigation include the following:

- Aeration system design, to ensure minimal energy consumption and maximum mixing and scouring
- Use of enriched air or pure oxygen for aeration
- The possible use of non-membrane MF media
- The design and manufacture of MBR membranes truly defying fouling, enabling a realistic "fit-and-forget" situation

- Development of an MBR capable of continuous use
- Use of MBR "in reverse" to add nutrients in a controlled fashion to remove gases from bioreactor
- Extraction of specific pollutants from bioreactor zone
- Development of an MBR for use in gray water recycling
- Conversion of MBR from an aerobic process to anaerobic operation, thus producing more energy (in the form of methane) than the plant consumes, the anaerobic digestion UF system

A major disadvantage of MBR systems is the higher capital and operating costs than conventional systems for the same quantum of wastewater. Operation and maintenance costs include membrane cleaning, fouling control, and eventual replacement of membrane Energy costs are also higher because of the need for air scouring to control bacterial growth on the membranes. Additionally, the waste sludge from such a system may have a low settling property.

Applicability

Membrane bioreactor systems are also well suited for some industrial and commercial applications. The high- quality treated water produced by MBRs makes them particularly applicable to reuse applications and for surface water discharge applications requiring extensive nutrient (nitrogen and phosphorus) removal. Designers of MBR systems require only basic information about the wastewater characteristics (e.g., influent characteristics, effluent requirements, flow data) to design an MBR system. Depending on effluent requirements, certain supplementary treatment steps can be clubbed with the MBR system, e.g., chemical coagulation before the primary settling tank or before the secondary settling tank, before the MBR or final filters for phosphorus removal.

Ultrafiltration

Ultrafiltration is a fractionation technique that can simultaneously concentrate macromolecules or colloidal substances in a process stream.[11] It can be described as simultaneous purification, concentration, and fractionation of macromolecules or fine colloidal suspensions.

Ultrafiltration membranes are developed from both organic (polymer) and inorganic materials. The preference of a given polymer as a membrane material is dependent on very specific properties such as molecular weight, chain flexibility, and chain interaction. Some of these materials are polysulfone, polyethersulfone, polyvinylidene fluoride, polyacrylonitrile, cellulosics, polyimide, polyetherimide, and aliphatic polyamides. Inorganic materials have also been used such as alumina and zirconia.[9] The structure of a UF membrane can be symmetric or asymmetric. Thickness of symmetric membrane ranges from 10 to 200 µm. Resistance to mass transfer is determined by the thin top layer. Membrane material should have chemical, thermal, and mechanical stability. Specification of UF is determined based on molecular weight cut-off—a measure of membrane pore dimensions—to describe the retention capabilities of UF membrane. It is the molecular mass of a macrosolute (typically, polyethylene glycol, dextran, or protein) for which the membrane has a retention capability greater than 90%.

Performance of UF

One of the critical factors determining the overall performance of aUF system is the rate of solute or particle transport in the feed side. Steady state conditions are achieved when the convective transport of solute to the membrane is equal to the sum of the permeate flow plus the diffusive back-transport of the solute.[12] The accumulation of solutes/particles at the membrane surface can affect the permeate flux in two ways.[13] First, the accumulated solute can generate an osmotically driven fluid flow back across the membrane from the permeate side toward the feed side, thereby reducing the net rate of solvent transport. Second, the solutes/particles can irreversibly foul the membrane due to specific physical and/or chemical interactions, thereby generating an additional hydraulic resistance. These interactions can be attributed to one or more of the following mechanisms: 1) adsorption; 2) gel layer formation; and 3) plugging of the membrane pores. The severity depends on the membrane material, the nature of solutes, and other variables such as pH, ionic strength, solution temperature, and operating pressure.[14]

Ultrafiltration System Design

Ultrafiltration is a low-pressure system operating at transmembrane pressures of 0.5–5 bars. A UF system comprises series/parallel modules operating according to various modes, ranging from an intermittent single-stage system to a continuous multistage system.[15] Ultrafiltration membranes can be fabricated essentially in tubular or flat sheet forms. Two major types of UF modules can be used, i.e., hollow fibers (capillary) and spiral wound. Other modules are plate and frame, tubular, rotary modules, vibrating modules, and Dean vortices. Operation of a UF membrane can be performed in two different service modes, viz., dead-end flow and cross-flow. The dead-end flow approach allows optimal recovery of feed water in about a 95%–98% range, but is generally limited to feed streams of low suspended solids (<1 NTU). The cross-flow mode differs from the dead-end mode in that there is an additional flow—the concentrate. The cross-flow mode of operation typically results in lower recovery of feed water, about 90%–95%.[16]

Ultrafiltration in Water Treatment

Ultrafiltration can be used to replace clarification steps, i.e., coagulation, sedimentation, and filtration, in a conventional water treatment plant. Thus, it can be defined as a clarification and disinfection membrane operation. All particulate biological contaminants, viz., viruses and bacteria, are rejected. The main advantages of low- pressure UF membrane processes compared with conventional clarification processes are no need for chemicals and size-exclusion filtration for constant quality of treated water in terms of particle and microbial removal.

Source water quality directly impacts UF membrane performance. Depending on the quality of raw water, UF can be operated as single operation or in combination with another process (coagulation, adsorption, etc.), or hybrid membrane system (UF/MF). In water application, UF can be the main process or used as pretreatment, for example, in an RO system.[17] It is a preferred alternative to conventional technology to remove water-borne pathogens in the preparation of drinking water.[16]

Ultrafiltration alone is not very effective for removing dissolved substances in general. It has limited capability in removing organic matter. Thus, the use of powdered activated carbon (PAC) in combination with a UF membrane is attracting increasing interest for the removal of organic compounds in drinking-water treatment.[17] It is found that the combination of UF with PAC/GAC could improve the removal of organics and other micropollutants such as agrochemicals. Another potential application of UF is to produce ultrapure water, acting as a pretreatment of RO unit. Microfiltration/UF can be considered as an alternative water treatment in aquaculture.[18] A sufficient supply of good-quality water is essential to any aquaculture operation. Ultrafiltration generates highly pathogen-free water.

Reverse Osmosis System

Reverse osmosis is a hi-tech filtration method that removes many types of large molecules and ions from solutions by applying pressure to the solution. Dissolved solids are not removed by any other filtration technique. The result is that solute is retained on the pressurized side of the membrane, and the pure solvent is allowed to pass to the other side. Selectively, this membrane does not allow large molecules or ions through the pores (holes) but allows smaller components of the solution (such as the solvent) to pass freely. The process of osmosis through semipermeable membranes was first observed in 1748 by Jean Antoine Nollet. For the following 200 years, osmosis was only a phenomenon observed in the laboratory. In 1949, the University of California at Los Angeles first investigated desalination of seawater using semipermeable membranes.

In the normal osmosis process, the solvent naturally moves from an area of low solute concentration, through a membrane, to an area of high solute concentration. The movement of a pure solvent to equalize solute concentrations on each side of a membrane generates pressure, i.e., osmotic pressure. Applying an external pressure to reverse the natural flow of pure solvent is the RO. The process is very similar to membrane filtration, but the only difference is between flow of solvents and filtration. Reverse osmosis involves a diffusive mechanism so that separation efficiency is dependent on solute concentration, pressure, and water flux rate.[19] The membranes used for RO have a dense barrier layer in the polymer matrix where most separation occurs. In most cases, the membrane is designed to allow only water to pass through the dense layer, while preventing the passage of solutes such as salt ions. This process requires high pressure, usually 2–17 bars (30–250 psi) for fresh and brackish water and 40–70 bars (600–1000 psi) for seawater (http://en.wikipedia.org/wiki/Reverse_osmosis-cite_note-2http://en.wikipedia.org/wiki/Reverse_osmosis-cite_note-water-0).

One of the first membrane applications for the utilization of membrane technology was the conversion of seawater into drinking water by RO. AnRO system separates dissolved solutes (includes single charged ions, such as Na^+, Cl^-) from water. Reverse osmosis can be described as a diffusion-controlled process. Physical holes may not exist in an RO membrane, which differentiates it from other filtration systems. An RO membrane is very hydrophilic; therefore, water will be able to readily diffuse into and out of the polymer structure of the membrane. An RO membrane is capable of rejecting contaminants as small as 0.001 µm.[20]

Four types of modules are used for RO membrane, i.e., plate and frame, tubular, hollow fiber, and spiral wound. However, the spiral-wound element is the most common by far for the production of drinking water. Reverse osmosis configurations include single-stage, two-stage, and two-pass systems. A two-stage system is common for brackish water use, where it is necessary to increase the overall recovery ratio.[21] Nowadays, the RO system has become a popular water treatment technology in industry requiring desalination and in residential units to improve the taste of water and to remove unhealthy contaminants. Reverse osmosis has increased the water supply by making possible the use of brackish waters. Recent advances in membrane materials and pretreatment have made RO desalination economically attractive even for seawater. The scale of membrane applications is now very large; plants with capacity in excess of 19 million liters per day (MLD) are common.[22] An RO desalination plant with a capacity of 100 MLD is being developed in Chennai, India.[23]

Reverse osmosis membranes are comparatively more sensitive than thermal desalination processes to scaling, fouling, and chemobiological attacks. The susceptibility to fouling is one of the major shortcomings of the RO membrane. Hence, though RO is an energy-efficient alternative to thermal processes, it still continues to face competition due to the requirements of pretreatment. Pretreatment is important when working with RO and NF membranes due to the nature of their spiral-wound design (Figure 7). Material is engineered in such a fashion as to allow only one-way flow through the system. Since accumulated material can-not be removed from the membrane surface systems, they are highly susceptible to fouling. Therefore, pretreatment is a necessity for any RO or NF system.

Pretreatment has four major components:

- *Screening of solids.* Suspended solids to be removed to prevent fouling of the membranes by fine particles or biological growth to reduce the risk of damage to high- pressure pump components.
- *Cartridge filtration.* String-wound polypropylene filters used to filter particles between 1 and 5 mm.
- *Dosing.* Oxidizing biocides, such as chlorine added to kill bacteria, followed by bisulfite dosing to deactivate the chlorine, which otherwise can destroy a thin-film composite membrane.
- *Prefiltration pH adjustment.* If the pH, hardness, and alkalinity in the feedwater result in a scaling tendency (estimated using the Langelier saturation index), acid is dosed to maintain carbonates in their soluble carbonic acid form. Use of antiscalant is recommended[24] (http://en.wikipedia.org/wiki/Reverse_osmosis - cite_note-8).

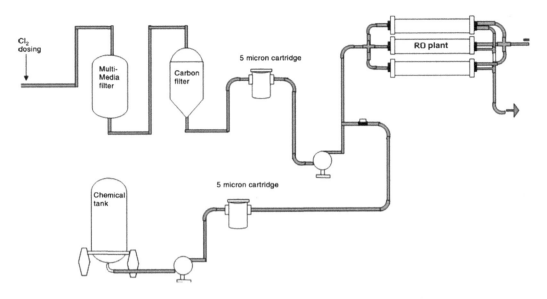

FIGURE 7   Pretreatment for RO system.

Integrated Membrane System

The integrated membrane system (IMS) design approach to water treatment systems has some significant advantages over RO systems designed with conventional pretreatment. The pretreatment of feed water prior to RO is intended to lower the silt density index, remove excessive turbidity or suspended solids, and adjust and control the pH.[25] The choice of IMS system depends on the fouling properties of the feed water, which may necessitate additional (pre)treatment based on the local circumstances. The combination of UF with a precoagulation at low dose helps in controlling the UF membrane fouling and providing filtered water in a steady-state condition.[26] Operation of an RO seawater unit at a higher flux and recovery rate enables optimization of the RO process and reduction of water cost.[27]

Water and Wastewater Purification

Rainwater collected from storm drains can be purified using reverse osmosis. Treated water can be used for landscape irrigation and industrial cooling. It works out to be an acceptable solution to the problem of water shortages. It is also used to purify the effluent and brackish groundwater. The effluent in larger volumes should be treated in an effluent treatment plant first, and then the clear effluent is subjected to reverse osmosis system. The process of reverse osmosis can be used for the production of deionized water. The RO process for water purification does not require thermal energy. Flow through an RO system can be regulated by a high-pressure pump. In 2002, the Singapore government developed NEWater using reverse osmosis to treat domestic wastewater before discharging the NEWater back into the treated water storages.

Eco-Filtration Systems

Zero-electricity systems with comparatively negligible maintenance can be developed using the ecosystem approach. Constructed wetlands, soil scape filters (vertical eco-filtration), green bridge (horizontal eco-filtration), etc., are useful in absorbing the pollutants and rendering them nontoxic through various degradative pathways to such an extent that they become nutrients for various groups of organisms.

Various applications like in situ treatment are evolving to cope with the pollution from nonpoint sources for which there is no economically viable solution in conventional systems.

Constructed Wetlands

Constructed wetlands can either be surface-flow or subsurface-flow and horizontal- or vertical- flow systems.[28] They may include engineered reedbeds and plant beds and belong to the family of phytorestoration and eco-technologies. A high degree of biological improvement takes place due to bioassimilation of pollution. These systems can act as a primary, secondary, and sometimes tertiary treatment. They are known to be highly productive systems as they copy natural wetlands, called the "kidneys of the earth" for their fundamental recycling capacity of the hydrological cycle in the biosphere. Robust and reliable, their treatment capacities improve as time goes by, the opposite of conventional treatment plants whose machinery ages with time. They are being increasingly used, although adequate and experienced designs are more fundamental than for other systems, and space limitation may impede their use (Figure 8).

Soil Biotechnology

Some advancement is being developed by researchers in erstwhile soil-based technologies for odor control in the 20th century. A process called soil biotechnology (SBT), developed in the Indian Institute of Technology in Mumbai, has encouraging results in process efficiency enabling total water reuse, due to extremely low operating power requirements of less than 50 J per kg of treated water.[32] Typically, SBT systems can achieve chemical oxygen demand (COD) levels less than 10 mg/L from sewage input of 400 mg/L COD. Soil biotechnology plants exhibit high reductions in COD values and bacterial counts as a result of the very high microbial densities available in the media. Unlike conventional treatment plants, SBT plants produce insignificant amounts of sludge, precluding the need for sludge disposal areas that are required by other technologies.[29]

Potential and Scope of Eco-Technology

Engineering applications of ecological principles and succession of biological communities are very useful inconsuming organic and inorganic pollutants from the water and bioconverting them into nontoxic form.[30,31] The consortia of organisms at different trophic levels utilize pollutants as nutrients.

FIGURE 8 Flourishing artificial wetland wastewater treatment system (courtesy of SERI, Pune, India).

Biodynamics of aqua-environ-equilibrium

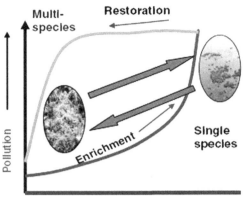

FIGURE 9 Aquatic biochemical equilibrium.

These eco-transformations, eco-conversions, and degradation or bioutilization of pollutants to nutrients are part of ecological and biogeochemical cycles. An attempt has been made to apply natural flora and fauna in a well-designed manner to develop technologies like Green Bridge, Green Lake Eco-Systems, Green Channel, BIOX (Biological Oxidation) and Stream Eco-Systems.

Many growing cities need cost-effective and less energy-intensive treatment methodology to control their pollution emanating from point and nonpoint sources. The eco-technological treatment systems, vertical filtration systems for point sources, and horizontal filtration systems promise results with minimal electricity as compared with conventional aerobic or anaerobic treatment systems.

Eco-filtration (vertical and horizontal) systems are based on the principle of living systems in action for pollution treatment. Efficiency of treatment is dependent on expression of multispecies intelligence—adaptability of organisms to changing environmental conditions while continuing the degradation or conversion of pollutants into utilizable form. When changes occur in the natural systems due to external inputs, biogeochemical cycles and food chains are reorganized and balanced. A new dynamic order emerges in the eco-filtration systems suitable to the environmental changes superimposing on it (Figure 9).

Detritus-feeding organisms in the eco-filtration systems consume the pollutants and wastes (biodegradation products) generated from this process and are useful for green plants. Secondly, the green plants absorb carbon dioxide from the atmosphere and produce oxygen, which is transferred to the ecoreactor in the rootzone area. Thus, the pollutants are transferred to natural cycles, i.e., biogeochemical cycles, and carbon gets stored in vegetation and subsequently in the soil. Plants store carbon in the form of live biomass. Once plants die, the biomass becomes a part of the food chain again and eventually enters the soil as soil carbon. This is a natural process that does not need electricity at all or any chemical for coagulation or activated carbon for adsorption. This makes the eco-filtration system very economical and routine operations very simple.

Soil Scape Filter

This is the simulation of natural filtration of wastewater through the ecofert—mixed cultures and fragmented rock materials. A soil scape filter contains layers of bioactive (i.e., biologically activated) organic material—bacterial mixtures derived from nontoxic and nonhazardous materials. This vertical

FIGURE 10   Sewage treatment using soil scape filter (courtesy of SERI, Pune, India).

eco-filtration process harnesses the ecological principles of interactions and interrelationships of biota with their environment and eco-transformations of substrates into assimilable form by treating, transforming, and detoxifying the pollutants using solar energy[33] (Figure 10).

Green Bridge Technology

Green Bridge technology uses filtration power of biologically originated cellulosic/fibrous material in combination with sand, gravels, root systems of green plants, and microorganisms.[34] It is an innovative approach to minimize the cost of pollution treatment when the cellulosic/ fibrous materials like coconut coir, dried water hyacinth, or aquatic grasses are compacted and woven to form a bridge/porous wall-like structure strengthened by stones and sand (Figure 11). All the floatable and suspended solids are trapped in this biological bridge, and the turbidity of flowing water is reduced substantially. The green plants growing there help in absorption of soluble substances, including heavy metals.

Other Emerging Filtration Techniques

Disc Filters

With disc filters, the divergence in woven filtration media and flow pattern makes each application unique. Both inside-out and outside-in configurations perform to expectations, considering that these devices have been marketed for the relatively clean waters of municipal tertiary applications. Inside-out technique provides better sustainability and life cycle cost. One major drawback of these systems is the biological growth that occurs on the inside face (the filtrate side) of the filter cloth. It is common with unchlorinated filter feed/secondary effluents. Another problem of these systems is the need to shut down the process in order to accomplish the washing of cloth/filtering medium. These negative aspects result in lengthy rinsing schedule, sizeable washwater volumes, and redundancy requirements to ensure continuous wastewater treatment.

FIGURE 11   Green Bridge installation for ecological restoration of Ahar River of Udaipur, India (courtesy of SERI, Pune, India).

Pleated Filtration Panel innovation

Inside-out and outside-in techniques rely on filtration media configured in a flat panel design. The effective filtration area is therefore the available area that is perpendicular to the direction of flow and should not account for any impervious surfaces of supporting structures such as frames, panel boxes, and metal works.

The disc filter with inside-out configuration brings a pleated design to the filtration panels, increasing the effective filtration area by 40% over flat panel designs. This also augments the overall strength of the filtration panel, allowing higher operating headloss. The pleated configuration makes the filtration panel less prone to deformities of the cloth media. Combination of higher filtration surface area and depth of the disc filter leads to high throughput capacity. The internal assembly of the disc/drum has a sliding cover of stainless steel for ease of inspection and maintenance. This is beneficial during installations in any weather conditions.

Conclusion

Technology can be related on the basis of—type of solids—colloidal or dissolved—chemical or biological species—capital and operational costs. Various types of sand filters are useful in treating surface or subsurface waters for drinking purposes. They can be supplemented with GAC or membrane filtrations to yield high-quality water required for industrial processes and production. Trickling filters, biofilters, and eco-technological filters have applications in treating domestic wastewaters before being discharged into the receiving water bodies or for land irrigation. These wastewater filters can be used as polishing systems also to improve the quality of secondarily treated water by reducing the expenditures on tertiary treatment. Eco-technological filters are found to be useful in treating industrial wastewaters by reducing the ecological toxicity of polluted waters. Disc filters or pleated filters are in the development stage. All these filters can be selected to treat waters or wastewaters depending on their source, the desired quality of treated water, land availability, electricity, skilled manpower requirements, etc.

References

1. WHO. Water Sanitation and Health. Available at http://www.who.int/water_sanitation_health/sanitproblems/en/index5.html (accessed on August 16, 2011).
2. IEES. International Ecological Engineering Society. Reclaiming wastewater in Windhoek, Namibia. Available at http://www.iees.ch/cms/index.php?option=com_content&task=view&id=24&Itemid=64 (accessed on August 16, 2011).
3. Kurita Water Industries, Ltd. *Kurita Handbook of Water Treatment*; Kurita Water Industries, Ltd, Tokyo, 1985. pp. 4.1–4.25.
4. Anselme, C.; Jacobs, E.P. Ultrafiltration. In *Water Treatment Membrane Processes*; Mallevialle, J., Odendaal, P.E., Wiesner, Eds.; American Water Works Association Research Foundation. Lyonnaise des Eaux, Water Research Commission of South Africa: McGraw-Hill, New York, 1996.
5. Clever, M.; Jordt, F.; Knauf, R.; Rabiger, N.; Rudebusch, M.; Hilter-Scheibel, R. Process water production from river water by ultrafiltration and reverse osmosis. In Proceedings of the Conference on Membranes in Drinking and Industrial Water Production; Desalination Publications: L'Aquila, Italy, 2000; Vol. 1.
6. Stevenson, D.G. *Water Treatment Unit Processes*. Imperial College Press. London. 1998.
7. Joshi, S. Unpublished study. 2000.
8. James, M. Montgomery Consulting Engineers Inc. *Water Treatment Principles and Design*; John Wiley and Sons, New York, 1985.
9. Hawkes, H.A. *The Ecology of Wastewater Treatment*; MacMillan: New York, 1985.
10. Higgins, I.J.; Burns, R.G. *The Chemistry and Microbiology of Pollution*; Academic: London, 1975.
11. Cheryan, M. *Ultrafiltration Handbook*; Technomic Publishing Company, Inc., Lancaster, Basel, 1986.
12. Mulder, M. *Basic Principles of Membrane Technology*. Kluwer Academic Publishers, Dordrecht, 1996.
13. Aimar, P.; Meireles, M.; Bacchin, P.; Sanchez, V. Fouling and concentration polarization in ultrafiltration and microfiltration. Paper presented at the ASI NATO Meeting, Curia, Portugal, March 1993.
14. Jonsson, G. Boundary layer phenomena during ultrafiltration of dextran and whey protein solutions. Desalination **1984**, *51*, 61–77.
15. Jönsson, A.S.; Trägardh, G. Fundamental principles of ultrafiltration. Chem. Eng. Process **1990**, *27*, 67–81.
16. Aptel, P.; Buckley, C.A. Categories of membrane operations. *Water Treatment Membrane Processes*; Mallevialle, J., Odendaal, P.E., Wiesner, M.R., Eds.; American Water Works Association Research Foundation, Lyonnaise des Eaux, Water Research Commission of South Africa: McGraw-Hill, New York, 1996.
17. Bates, W.T. Capillary UF as RO pretreatment. Hydranau- tics; 1999, available at http://www.membranes.com.
18. Lainé, J.M.; Vial, D.; Moulart, P. Status after 10 years of operation—Overview of UF technology today. In Proceedings of the Conference on Membranes in Drinking and Industrial Water Production; Desalination Publications, 2000; Vol. 1, 17–25.
19. Côté, P.; Cadera, J.; Adams, N.; Best, G. Monitoring and maintaining the integrity of immersed ultrafiltration membranes used for pathogen reduction. Water Supply **2002**, *2* (5–6), 307–311.
20. Campos, C., I. Baudin, J.M. Lainé. 2001. Adsorption performance of powdered activated carbon—Ultrafiltration systems. Abstract. Water Supply **2001**, *1* (5–6), 13–19.

21. Wenten, I.G. 2004. Industrial membrane application in Indonesia. Case study: Clean production in cassava starch industry. In Regional Symposium on Membrane Science and Technology Johor, Malaysia, 2004.

22. Crittenden, J.; Trussell, R.; Hand, D.; Howe, K.; Tchobano- glous, G. *Water Treatment Principles and Design,* 2nd Ed; John Wiley and Sons: New Jersey, 2005.

23. Taylor, J.S.; Jacobs, E.P. Reverse osmosis and nanofiltration. *Water Treatment Membrane Processes;* Mallevialle, J., Odendaal, P.E., Wiesner, M.R., Eds.; American Water Works Association Research Foundation, Lyonnaise des Eaux, Water Research Commission of South Africa: McGraw-Hill, New York, 1996, pp. 9.1–9.70.

24. Fawzi, N.; Al-Enezi, G. Design consideration of RO units: Case studies. Desalination **2002**, *153,* 281–286.

25. Buckley, C.A.; Hurt, Q.E. Membrane applications: a contaminant-based perspective. *Water Treatment Membrane Processes;* Mallevialle, J., Odendaal, P.E., Wiesner, M.R., Eds.; American Water Works Association Research Foundation, Lyonnaise des Eaux, Water Research Commission of South Africa: McGraw-Hill, New York, 1996.

26. Anon. Chennai desal project will open new opportunities: IVRCL in The Hindu. Business Line. Available at http://www.thehindubusinessline.com/todays-paper/tp-corporate/article1000069.ece (accessed on August 2011).

27. Malki, M. Optimizing scale inhibition costs in reverse osmosis desalination plants, Int. Desalin. Water Reuse Q. **2008**, *17* (4), 28–29, available at http://www.membranechemicals.com/english/Optimizing%20Operational%20Costs%20in%20Rerse%200smosis%20Desalination%20Plants%20%202008.pdf.

28. Ebrahim, S.; Abdel-Jawad, M.; Bou-Hamad, S.; Safar, M. Fifteen years of R&D program inseawater desalination at KISR. Part I. Pretreatment technologies for RO systems. Desalination **2001**, *135,* 141–153.

29. Brehant, A.; Bonnelye, V.; Perez, M. Comparison of MF/ UF pretreatment with conventional filtration prior to RO membranes for surface seawater desalination. Desalination **2002**, *144,* 353–360.

30. Glueckstern, P.; Priel, M.; Wilf, M. Field evaluation of capillary UF technology as a pretreatment for large seawater RO systems. Desalinatination **2002**, *147,* 55–62.

31. USEPA. *Constructed Wetlands: Treatment of Municipal Wastewater,* EPA/625/R-99/010; Office of Research and Development: Ohio, 2000.

32. Kadam, A.; Ozaa, G.; Nemadea, P.; Duttaa, S.; Shankar, H. Municipal wastewater treatment using novel constructed soil filter system. Chemosphere **2008**, *71* (5), 975–981.

33. Joshi, S. Ecotechnological treatment for industrial wastewater containing heavy metals. J. Indian Assoc. Environ. Manage. **2000**, *27* (2), 98–102.

34. Joshi, S.; Joshi, S. Ecotechnological applications for the control of lake pollution. In Proceedings of TAAL 2007: the 12th World Lake Conference held at Jaipur organized by Ministry of Environment and Forests of India and ILEC, New Delhi. Ed. Sengupta, B. and Dalwani, R.; 2008; pp. 864–867.

58

Wetlands: Constructed Subsurface

Introduction 603
Technology Background 604
    HF Systems • VF Systems
Hybrid Systems 609
Use for Various Types of Wastewater 609
Conclusions 610
Acknowledgments 610
References 611

Jan Vymazal

Introduction

Constructed wetlands (CWs) with subsurface flow have been used to treat various types of wastewater for more than four decades.[1] In general, there are two types of CWs with subsurface flow: horizontal flow (HF) and vertical flow (VF).[2–4]

In HF CWs, the wastewater is continuously fed in at the inlet and flows slowly through the porous medium under the surface of the porous filtration material planted with emergent vegetation to the outlet where it is collected before leaving via a water level control structure.[4,5] As a consequence of the continuous saturation, the majority of the processes responsible for pollutant removal are either anoxic or anaerobic.[3] VF CWs are fed intermittently on the wetland surface. Wastewater then percolates down through the bed and is collected by a drainage network at the bottom. The bed drains completely and that allows air to refill the bed. Thus, VF CWs provide greater oxygen transfer into the bed, thus allowing for aerobic processes, such as nitrification, to occur.[4]

Horizontal and vertical CWs could be combined in several stages in so-called hybrid systems. These systems use the advantages of both types of systems to enhance removal of pollution, especially nitrogen. Hybrid systems may comprise various types of CWs but the combination of VF and HF CWs is the most common.[6]

During the 1980s and the early 1990s, the CWs with subsurface flow were used primarily in Europe for treatment of municipal or domestic wastewater.[7] During the late 1990s and 2000s, subsurface flow CWs spread throughout the world and were used to treat also various industrial and agricultural effluents, landfill leachate, and various stormwater runoff waters.[2,3]

Technology Background

HF Systems

Horizontal subsurface flow CWs are lined trenches or beds about 0.8 m deep, filled with porous materials such as pea gravel (Figure 1) and planted with wetland emergent macrophytes. The filtration material should be coarse enough to sustain subsurface flow. In the early systems designed by Seidel, coarse sand was proposed as filtration material.[8] The later concept proposed by Kickuth[9] used heavy cohesive soils that provided high level of treatment but failed to sustain subsurface flow due to clogging problems.[10] Based on the experience in the United Kingdom in the late 1980s,[4] the HF CWs have used gravel since then, usually with size fraction between 5 and 20 mm. During passage through the filtration material, the wastewater comes into contact with a network of aerobic, anoxic, and anaerobic zones. Most of the bed is anoxic/anaerobic due to permanent saturation of the beds. The aerobic zones occur around roots and rhizomes that leak oxygen into the substrate.[4,5]

The macrophytes growing in CWs have several properties in relation to the treatment process that make them an essential component of the design.[11] The macrophytes, among other functions, provide oxygen to the rhizosphere through the radial oxygen loss,[12] insulate filtration bed during winter periods,[13] provide substrate for the growth of attached bacteria,[14] can contribute to nutrient removal when harvested,[3,15] and exudate various antimicrobial compounds.[16] Therefore, they should 1) be tolerant of high organic and nutrient loadings; 2) have rich belowground organs (i.e., roots and rhizomes) in order to provide substrate for attached bacteria and oxygenation (even very limited) of areas adjacent

FIGURE 1 Constructed wetlands with horizontal subsurface flow. Left: filtration bed sealed with plastic liner, right: filtration bed filled with crushed rock with inflow and outflow zones filled with stones.
Source: Photos courtesy of Jan Vymazal.

to roots and rhizomes; and 3) have high aboveground biomass for winter insulation in cold and temperate regions and nutrient removal via harvesting.[17,18] The most frequently used plant for HF CWs is Common reed (***Phragmites australis***) (Figure 2).[3]

A large number of wetland systems have shown an exponential decrease of pollutant concentration level with the distance through the wetland from inlet to outlet. The observation is consistent with a first-order removal model, with the removal rate being proportional to the pollutant concentration. The removal can thus empirically be described with first-order plug-flow kinetics:[2,4,19]

$$(\ln C_i - \ln C_o) = A\, k_A / Q \tag{1}$$

$$C_o = C_i \, \exp^{(-A\, kA/Q)} \tag{2}$$

or for the *k-C** model:[20]

$$\ln\left[\left(C_i - C^*\right)/C_o - C^*\right)\right] = A\, k_A / Q \tag{3}$$

where
A = area of the bed (m²)
Q = average flow rate (m³day⁻¹)
C_i = inflow concentration of the pollutant (mg L⁻¹)
C_o = outflow concentration of the pollutant (mg L⁻¹)
k_A = first-order areal rate constant (m day⁻¹)
C^* = background concentration of a pollutant (mg L⁻1),

which is a concentration caused by an internal release of particulate and dissolved biomass to the water.[20]

As BOD (biochemical oxygen demand) was the primary target, the value of first-order areal rate constant k_A used for the design was k_{BOD}. The formerly proposed value of 0.19 m day⁻¹ by Kickuth[9,19] resulted in too small area of the bed and consequently lower treatment effect. The extensive field measurements in Denmark in the mid-1980s found the average k_{BOD} value of 0.083±0.017 m day⁻¹.[21] Cooper et al.[4] reported values between 0.067 and 0.1 m day⁻¹ in the United Kingdom and Vymazal and Kröpfelová[3] reported an average k_{BOD} value of 0.087 m day⁻¹ for 31 systems in the Czech Republic. This generally

FIGURE 2 Common reed (*Phragmites australis*) growing in a HF constructed wetland.
Source: Photo courtesy of Jan Vymazal.

TABLE 1   First-Order Areal Rate Constant kA (m day–1) for Total
Suspended Solids (TSS), Total Phosphorus (TP), Total Nitrogen (TN) and
Ammonia-Nitrogen (NH4-N) Reported by Various Authors in the
Literature

Reference	TSS	TP	TN	NH4-N
[3]	0.085	0.026	0.025	0.024
[10]		0.025	0.033	
[22]	2.74	0.033	0.074	0.093

means that the value of A is about $5\,m^2$ per one population equivalent (PE) for mechanically pretreated municipal or domestic wastewater. However, HF CWs may be designed with other target parameters than BOD_5 and therefore appropriate constants were developed for other pollutants as well (Table 1). When a CW is designed with more target parameters at the same time, e.g., BOD5, TSS, nitrogen, and phosphorus, then it is necessary to select the largest area according to all four calculations.

At present, the state-of-the-art k-C^* model seems to be best available design tool if the designer makes sure that all the assumptions are fulfilled and if he or she is aware of many pitfalls in the model.[22] One of the most important uncertainties comes from the fact that background concentrations inevitably change over the operation period as the system matures. In addition, Stein et al.[23] suggested that C^* is temperature dependent. Kadlec and Knight[20] suggested C^* values (in mg L^{-1}) for BOD_5 ($3.5 + 0.053C_i$), TSS ($7.8 + 0.063C_i$), TP (0.02), and TN (1.5).

Recently, the so-called P-k-C^* model has been developed:[2]

$$\left(C_o - C^*\right) / \left(C_i - C^*\right) = 1 / \left(1 + k_A / Pq\right)^P \tag{4}$$

where

k_A = modified first-order areal constant (m day^{-1})
P = apparent number of TIS (tanks in series)
q = hydraulic loading rate (HLR) (cm day–1)

In HF CWS, pollution is removed through a complex of aerobic, anoxic, and anaerobic processes that occur simultaneously in the filtration bed with aerobic processes being restricted only to the thin area adjacent to macrophyte roots.[3,24,25] The treatment efficiency of HF CWs has recently been thoroughly evaluated by Vymazal and Kröpfelová.[3] Organic compounds are removed primarily by microbial processes; particulate and colloid organics may be filtered out through the wastewater passage through the bed. Suspended solids are removed in HF CWs via filtration and sedimentation. Systems that have been designed with a specific area of about $5\,m^2\,PE^{-1}$ have usually very high treatment efficiency for organics and suspended solids. The removal efficiency in most cases equals the efficiency of conventional treatment systems. However, contrary to conventional treatment systems, e.g., activated sludge systems, HF CWs can handle wastewaters with very low concentrations of organics such as wastewaters from combined sewer systems.[26,27]

Removal of nitrogen is usually low due to lack of oxygen in filtration bed, which limits nitrification of ammonia-nitrogen.[28] On the other hand, HF CWs provide suitable conditions for denitrification due to anoxic/anaerobic conditions in the filtration beds. Plant uptake is responsible for removal of only small part of the inflow nitrogen loading and usually does not exceed 10% of the inflow.[29,30] For municipal sewage, the removal of ammonia-N usually varies between 20% and 40%.[3]

HF CWs are seldom built with phosphorus being the primary target of the treatment, and therefore, materials with relatively low sorption capacity but high hydraulic conductivity such as river gravel or crushed rock are commonly used.[3] In this case, removal efficiency for municipal sewage is usually between 20% and 50%. In order to achieve high phosphorus removal, it is necessary to select materials with high P adsorption capacity. Such materials may include minerals with reactive Fe or Al hydroxide or oxide groups on their surfaces, or calcareous materials that can promote precipitation of Ca-phosphate,[31-33] lightweight aggregates,[34] industrial by-products and waste products such as electric arc furnace steel slags[35,36] or fly ash.[31] By using materials with high sorption capacity, the phosphorus removal efficiency may exceed 90%, but it is necessary to keep in mind that the sorption capacity is always limited and the material will have to be replaced in the future in order to keep a high rate of P removal.

Microbial pollution is removed through a complex of physical, chemical, and biological factors, which all participate in the reduction of the number of bacteria of anthropogenic origin. Physical factors include mechanical filtration, straining, adsorption, and sedimentation. Chemical factors include oxidation, exposure to biocides excreted by some plants, and adsorption to organic matter; biological removal factors include antibiosis, predation by nematodes, protozoa and zooplankton, attack by lytic bacteria and viruses, and natural die-off.[3,37-40]

One of the major threats for the long-run treatment performance of HF CWs is clogging and subsequent surface flow[41,42] and therefore efficient mechanical pretreatment should always be installed before the wastewater enters the filtration beds. For small systems, septic tank is usually used; for larger systems, Imhoff tank is suitable. In case of combined sewer systems, i.e., systems where wastewater is mixed with stormwater runoff, grit chamber should be included (Figure 3).

VF Systems

The earliest form of VF CW is that of Seidel in Germany in the 1960s where vertical filters were used to oxygenate anaerobic septic tank effluents.[43] Interest in the particular process seemed to wane, but it has been revived in the 1990s because of the need to produce beds that nitrify.[4]

FIGURE 3 Mechanical pretreatment: screens and horizontal grit chamber. (**Source:** Photo courtesy of Jan Vymazal).

VF CWs comprise a flat bed of sand or graded gravel planted with macrophytes (Figure 4). Contrary to HF CWs, vertical systems are fed intermittently with large batches on the bed surface. The distribution system should allow for even wastewater distribution across the bed (Figure 5). The intermittent feeding creates conditions where the filter completely drains, thus enabling oxygen to diffuse into the bed. On the other hand, VF CWs do not provide suitable conditions for denitrification to complete conversion to gaseous nitrogen forms that escape to the atmosphere. The major purpose of macrophyte presence in VF CWs is to help maintain the hydraulic conductivity of the bed.[4]

The early VF systems have usually been composed of several stages with two to four beds in the first stage that were fed with wastewater in rotation. Such VF systems are now called first-generation VF systems.[44] The early VF systems were frequently the first stage of the hybrid systems.[45,46] Recently, VF systems with only one bed have been used. These systems are called second-generation VF CWs or compact VF beds.[47,48]

FIGURE 4 Distribution of wastewater on the surface of a vertical flow constructed wetlands. Experimental station, Carrión de los Céspedes, Spain. (**Source:** Photo courtesy of Jan Vymazal).

FIGURE 5 Vertical flow CWs at Saint Thomé, France. (**Source:** Photo courtesy of Jan Vymazal).

VF CWs require pretreatment in order to prevent clogging, but in France, hundreds of systems are fed directly with raw sewage onto the first stage, allowing for easier sludge management in comparison to dealing with primary sludge from an Imhoff tank.[49,50] While compact VF systems are usually small, VF systems treating raw sewage in France are often much larger. The calculation of VF bed area has evolved in years, but at present, the area varies between $2\,m^2\,PE^{-1}$ (split into $1.2\,m^2$ and $0.8\,m^2$ for the first and second stage, respectively) in France[49] and $4\,m^2\,PE^{-1}$ in Austria and Germany.[51–53] Danish design recommendations are set for $4\,m^2\,PE^{-1}$.[54] One of the major threats of good performance of VF systems is clogging of the filtration substrate,[55,56] and therefore, it is important to properly select the filtration material and the HLR and distribute the water evenly across the bed surface. HLRs reported in the literature vary greatly between about $3\,cm\,day^{-1}$ up to $150\,cm\,day^{-1}$,[47,57] but to avoid flooding due to clogging, HLRs should probably be kept below $80\,cm\,day^{-1}$.[58]

VF CWs provide high removal of organics and suspended solids and are also effective in removing ammonia- N. However, they have very limited capacity to denitrify; therefore, the removal of total nitrogen is about the same as for HF CWs. Removal of phosphorus is also within the same range as for HF CWs unless special media with high sorption capacity are used.[3]

Hybrid Systems

Various types of CWs may be combined in order to achieve higher removal efficiency, especially for nitrogen. Many of these systems are derived from original hybrid systems developed by Seidel at the Max Planck Institute in Krefeld, Germany.[43] The design consists of two stages, several parallel VF beds followed by two or three HF beds in series. The VF wetland is intended to remove organics and suspended solids and to provide nitrification, while in the HF wetland, denitrification and further removal of organics and suspended solids occur.

In the early 1980s, several hybrid systems of Seidel's type were built in France with the systems at Saint Bohaire and Frolois being the best described.[59] A similar system was built in 1987 in the United Kingdom at Oaklands Park,[60] and in the 1990s and the early 2000s, VF–HF systems were built in many countries such as Norway,[61] Ireland,[62] Estonia,[63] or Thailand.[64]

In the mid-1990s, Brix and Johansen[65] introduced a HF–VF hybrid system with a large HF bed placed first and a small VF bed as the second stage. In the first bed, removal of organics and suspended solids takes place while nitrification takes place in the second VF wetland. In order to remove nitrate produced during nitrification, the water have to be recirculated to the front end of the system where denitrification can take place in the less aerobic HF bed using the raw feed as a source of carbon needed for denitrification.[65] The hybrid systems of this type have been reported from other European countries,[66,67] North America,[68] or Asia.[69]

Use for Various Types of Wastewater

Subsurface flow CWs have predominantly been used to treat municipal (secondary as well as tertiary treatment) or domestic wastewater.[1–4] However, recently, they have been used to treat a variety of wastewaters including industrial and agricultural effluents, landfill leachate, or stormwater runoff (Table 2).

When organics and suspended solids are the primary target of the treatment, the HF CWs are usually used. These systems perform effectively and the operation and maintenance costs are very low as compared to conventional treatment systems. When removal of ammonia is required, VF CWs are more effective. These systems usually require electricity, but despite that, the operation and maintenance costs are still relatively low. However, to remove both ammonia- and nitrateN, hybrid CWs provide a suitable and effective option. Hybrid CWs are also often used to treat more complex wastewaters (Table 2).

TABLE 2 Examples of the Use of Subsurface CWs for Various Types of Wastewater

Type of Wastewater	Location	Type of CWs	Reference
Petrochemical	United States, China	HF	[70,71]
	Pakistan	VF	[72]
Chemical industry	United Kingdom	HF	[73]
	Portugal, United States, Germany	VF	[74–76]
Paper and pulp wastewaters	United States	HF	[77]
Abattoir	Mexico, Ecuador	HF	[78,79]
	Poland	VF–HF	[80]
Textile industry	Australia	HF	[81]
Tannery industry	Portugal	HF	[82]
Mixed industrial	Slovenia	HF–VF–HF	[83]
Food industry	Slovenia, Italy	HF	[84,85]
Distillery and winery	India, Italy	HF	[86,87]
Pig farm	Australia, Lithuania	HF	[88,89]
	Thailand	VF–HF	[90]
Fish farm	Canada, Germany	HF	[91,92]
Dairy	United States, Germany	HF	[93,94]
	The Netherlands	VF	[95]
	France, Japan	VF–HF	[96,97]
Highway runoff	United Kingdom	HF	[98]
Airport runoff	United States	HF	[99]
	Canada	VF	[100]
Nursery runoff	Australia	HF	[101]
Greenhouse	Korea	HF–VF–HF	[102]
Landfill leachate	Poland	HF	[103]
	Australia	VF	[104]
	Slovenia	VF–HF	[105]
Compost leachate	France	VF–HF	[94]

Conclusions

Constructed treatment wetlands have evolved during the last five decades into a reliable treatment technology applicable to all types of wastewater including sewage, industrial and agricultural wastewaters, landfill leachate, and stormwater runoff. The subsurface CW technology started in Germany but spread quickly to all continents. At present, subsurface constructed treatment wetlands provide a viable alternative to conventional treatment systems.

The subsurface technology has substantially improved over the years, but there are still gaps that need to be solved. Among others, the low-cost filtration medium that provides high removal of phosphorus is still to be determined, and the problem of system longevity and filtration material replacement or cleaning are under investigation. Recently, also more attention has been paid to dual-purpose or multipurpose services of constructed treatment wetlands such as flood control, carbon sequestration, or wildlife habitat. Another gap where a lot of work has been done recently is the modeling of HF CWs.[106]

Acknowledgments

The study was supported by grant No. QH81170 "Complex Evaluation of the Effect of Area Protection on Important Water Management Localities" from the Ministry of Agriculture of the Czech Republic.

References

1. Vymazal, J. Constructed wetlands for wastewater treatment: Five decades of experience. Environ. Sci. Technol. **2011,** *45* (1), 61–69.
2. Kadlec, R.H.; Wallace, S.D. *Treatment Wetlands,* 2nd Ed.; CRC Press: Boca Raton, Florida, 2008.
3. Vymazal, J.; Kröpfelová, L. *Wastewater Treatment in Constructed Wetlands with Horizontal Sub-Surface Flow;* Springer: Dordrecht, the Netherlands, 2008.
4. Cooper, P.F.; Job, G.D.; Green, M.B.; Shutes, R.B.E. *Reed Beds and Constructed Wetlands for Wastewater Treatment;* WRc Publications: Medmenham, Marlow, U.K., 1996.
5. Brix, H. Treatment of wastewater in the rhizosphere of wetland plants—The root zone method. Water Sci. Technol. **1987,** *19* (10), 107–118.
6. Vymazal, J. Horizontal sub-surface flow and hybrid constructed wetlands systems for wastewater treatment. Ecol. Eng. **2005,** *25,* 478–490.
7. Vymazal, J.; Brix, H.; Cooper, P.F.; Green, M.B.; Haberl, R., Eds. *Constructed Wetlands for Wastewater Treatment in Europe;* Backhuys Publishers: Leiden, the Netherlands, 1998.
8. Seidel, K. Phenol-Abbau in Wasser durch *Scirpus lacustris* L. wehrend einer versuchsdauer von 31 Monaten. Naturwissenschaften **1965,** *52* (13), 398–406.
9. Kickuth, K. Abwasserreinigung im Wurzelraumverfahren. Wasser, Luft Betr. **1980,** *11,* 21–24.
10. Brix, H. Denmark. In *Constructed Wetlands for Wastewater Treatment;* Vymazal, J., Brix, H., Cooper, P.F., Green, M. B., Haberl, R., Eds.; Backhuys Publishers: Leiden, the Netherlands, 1998; 123–152.
11. Brix, H. Do macrophytes play a role in constructed treatment wetlands? Water Sci. Technol. **1997,** *35* (5), 11–17.
12. Brix, H. Gas exchange through the soil-atmosphere interface and through dead culms of *Phragmites australis* in a constructed wetland receiving domestic sewage. Water Res. **1990,** *24,* 377–389.
13. Mander, Ü.; Jenssen, P., Eds. *Constructed Wetlands for Wastewater Treatment in Cold Climates;* WIT Press: Southampton, U.K., 2003.
14. Vymazal, J.; Ottová, V.; Balcarová, J.; Doušová, H. Seasonal variation in fecal indicators removal in constructed wetlands with horizontal subsurface flow. In *Constructed Wetlands for Wastewater Treatment in Cold Climates;* Mander, Ü., Jenssen, P., Eds.; WIT Press: Southampton, U.K., 2003; 239–258.
15. Adcock, P.; Ganf, G.G. Growth characteristics of three macrophyte species growing in natural and constructed wetland system. Water Sci. Technol. **1994,** *29,* 95–102.
16. Vincent, G.; Dallaire, S.; Lauzer, D. Antimicrobial properties of roots exudate of three macrophytes: *Mentha aquatica* L., *Phragmites australis* (Cav.) Trin. and *Scirpus lacustris* L. In Proceedings of the 4th International Conference Wetland Systems for Water Pollution Control, Guangzhou, China, November 6–10, 1994; ICWS '94 Secretariat, 1994; 290–296.
17. Čížková-Končalová, H.; Květ, J.; Lukavská, J. Response of *Phragmites australis, Glyceria maxima* and *Typha latifo-lia* to addition of piggery sewage in a flooded sand culture. Wetlands Ecol. Manage. **1996,** *4,* 43–50.
18. Květ, J.; Dušek, J.; Husák, Š. Vascular plants suitable for wastewater treatment in temperate zones. In *Nutrient Cycling and Retention in Natural and Constructed Wetlands;* Vymazal, J., Ed.; Backhuys Publishers: Leiden, the Netherlands, 1999; 101–110.
19. Kickuth, R. 1981. Abwasserreinigung in Mosaikmatritzen aus aeroben und anaeroben Teilbezirken. In *Grundlagen der Abwasserreinigung;* Moser, F., Ed.; Verlag Oldenburg: Munchen, Wien, 1981; 639–665.
20. Kadlec, R.H.; Knight, R.L. *Treatment Wetlands;* CRC Press: Boca Raton, Florida, 1996.
21. Brix, H. Constructed wetlands for municipal wastewater treatment. In *Global Wetlands: Old World and New;* Mitsch, W.J., Ed.; Elsevier: Amsterdam, 1994; 325–333.

22. Rousseau, D.P.L.; Vanrolleghem, P.A.; De Pauw, N. Model-based design of horizontal subsurface flow constructed wetlands: a review. Water Res. **2004**, *38*, 1484–1493.

23. Stein, O.R.; Biederman, J.A.; Hook, P.B.; Allen, W.C. Plant species and temperature effects on the k-C^* firstorder model for COD removal in batch-loaded SSF wetlands. Ecol. Eng. **2006**, *26*, 100–112.

24. Imfeld, G.; Braeckevelt, M.; Kuschk, P.; Richnow, H.H. Monitoring and assessing processes of organic chemicals removal in constructed wetlands. Chemosphere **2009**, *74*, 349–362.

25. Garcia, J.; Rousseau, D.; Morato, J.; Lesage, E.; Matamoros, V.; Bayona, J.M. Contaminant removal processes in subsurface-flow constructed wetlands: A review. Crit. Rev. Environ. Sci. Technol. **2010**, *40*, 561–661.

26. Vymazal, J. Horizontal sub-surface flow constructed wetlands Ondřejov and Spálené Poříčí in the Czech Republic—15 years of operation. Desalination **2009**, *246*, 226–237.

27. Green, M.B.; Martin, J.R. Constructed reed beds clean up stormwater overflows on small wastewater treatment works. Water Environ. Res. **1996**, *68*, 1054–1060.

28. Vymazal, J. Nitrogen removal in constructed wetlands with horizontal sub-surface flow-can we determine the key process? In *Nutrient Cycling and Retention in Natural and Constructed Wetlands*; Vymazal, J., Ed.; Backhuys Publishers: Leiden, the Netherlands, 1999; 1–17.

29. Obarska-Pempkowiak, H.; Gajewska, M. The dynamics of processes responsible for transformation of nitrogen compounds in hybrid wetlands systems in a temperate climate. In *Wetlands—Nutrients, Metals and Mass Cycling*; Vymazal, J., Ed.; Backhuys Publishers: Leiden, the Netherlands, 2003; 129–142.

30. Browning, K.; Greenway, M. Nutrient removal and plant biomass in a subsurface flow constructed wetland in Brisbane, Australia. Water Sci. Technol. **2003**, *48* (5), 183–189.

31. Drizo, A.; Frost, C.A.; Grace, J.; Smith, K.A. Physico-chemical screening of phosphate removing substrates for use in constructed wetlands. Water Res. **1999**, *33*, 3595–3602.

32. Molle, P.; Liénard, A.; Grasmick, A.; Iwema, A. Phosphorus retention in subsurface constructed wetlands: Investigations focused on calcareous materials and their chemical reactions. Water Sci. Technol. **2003**, *48* (5), 75–83.

33. Ádám, K.; Krogstad, T.; Vråle, L.; Søvik, A.K.; Jenssen, P. Phosphorus retention in the filter materials shellsand and Filtralite®—Batch and column experiments with synthetic P solution and secondary wastewater. Ecol. Eng. **2007**, *29*, 200–208.

34. Jenssen, P.D.; Krogstad, T. Design of constructed wetlands using phosphorus sorbing lightweight aggregate (LWA). In *Constructed Wetlands for Wastewater Treatment in Cold Climates*; Mander, Ü., Jenssen, P., Eds.; WIT Press: Southampton, U.K., 2003; 260–271.

35. Mann, R.A.; Bavor, H.J. Phosphorus removal in constructed wetlands using gravel and industrial waste substrate. Water Sci. Technol. **1993**, *27*, 107–113.

36. Xu, D.; Xu, J.; Wu, J.; Muhammad, A. Studies on the phosphorus sorption capacity of substrates used in constructed wetland systems. Chemosphere **2006**, *63*, 344–352.

37. Decamp, O.; Warren, A.; Sanchez, R. The role of ciliated protozoa in subsurface flow wetlands and their potential as bioindicators. Water Sci. Technol. **1999**, *40* (3), 91–98.

38. Neori, A.; Reddy, K.R.; Číšková-Končalová, H.; Agami, M. Bioactive chemicals and biological–biochemical activities and their functions in rhizospheres of wetland plants. Bot. Rev. **2000**, *66*, 350–378.

39. Stevik, T.K.; Geir Ausland, K.A.; Hanssen, J.F. Retention and removal of pathogenic bacteria in wastewater percolating through porous media: A review. Water Res. **2004**, *38*, 1355–1367.

40. Seidel, K. Macrophytes and water purification. In *Biological Control of Water Pollution*; Tourbier, J., Pierson, R.W., Eds.; Pennsylvania University Press: Philadelphia, 1976; 109–122.

41. Knowles, P.; Dotro, G.; Nivala, J.; Garcia, J. Clogging in sub-surface flow treatment wetlands: Occurrence and contributing factors. Ecol. Eng. **2011**, *37*, 99–112.

42. Cooper, D.; Griffin, P.; Cooper, P. Factors affecting the longevity of sub-surface horizontal flow systems operating as tertiary treatment for sewage effluent. Water Sci. Technol. **2005**, *51* (9), 127–135.

43. Seidel, K. Neue Wege zur Grundwasseranreicherung in Krefeld.Vol. II. Hydrobotanische Reinigungsmethode. GWF Wasser/Abwasser **1965**, *30*, 831–833.

44. Cooper, P.F. The performance of vertical flow constructed wetland systems with special reference to the significance of oxygen transfer and hydraulic loading rates. Water Sci. Technol. **2005**, *51* (9), 81–90.

45. Burka, U.; Lawrence, P. A new community approach to wastewater treatment with higher water plants. In *Constructed Wetlands in Water Pollution Control*; Cooper, P.F., Findlater, B.C., Eds.; Pergamon Press: Oxford, U.K., 1990; 359–371.

46. Lienard, A.; Boutin, C.; Esser, D. Domestic wastewater treatment with emergent helophyte beds in France. In *Constructed Wetlands in Water Pollution*; Cooper, P.F., Findlater, B.C., Eds.; Pergamon Press: Oxford, U.K., 1990; 183–192.

47. Weedon, C.N. Compact vertical flow reed bed system—First two years performance. Water Sci. Technol. **2003**, *48* (5), 15–23.

48. Arias, C.A.; Brix, H. Initial experience from a compact vertical flow constructed wetland treating single household wastewater. In *Natural and Constructed Wetlands: Nutrients, Metals and Management*; Vymazal, J., Ed.; Backhuys Publishers: Leiden, the Netherlands, 2005; 52–64.

49. Molle, P.; Liénard, A.; Boutin, C.; Merlin, G.; Iwema, A. How to treat raw with constructed wetlands: An overview of French systems. Water Sci. Technol. **2005**, *51* (9), 11–21.

50. Paing, J.; Voisin, J. Vertical flow constructed wetlands for municipal septage treatment in French rural areas. Water Sci. Technol. **2005**, *51* (9), 145–155.

51. Langergraber, G.; Prandstetten, C.; Pressl, A.; Rohrhofer, R.; Haberl, R. Removal efficiency of subsurface vertical flow constructed wetlands for different organic loads. In Proceedings of the 10th International Conference Wetland Systems for Water Pollution Control; MAOTDR: Lisbon, Portugal, 2006; 587–597.

52. ÖNORM B 2505. *Bepflanzte Bodenfilter (Pflanzenkläran-lagen)—Anwendung, Bemessung, Bau und Betrieb* (Subsurface-flow constructed wetlands—Application, dimensioning, installation and operation); Österreichisches Normungsinstitut: Vienna, Austria, 2005 (in German).

53. DWA. *Grundsätze für Bemessung, Bau und Betrieb von Pflanzenkläranlagen mit bepflanzten Bodenfiltern zur biologischen Reinigung kommunalen Abwassers*; Arbeitsblatt DWA-A 262; DWA—Deutsche Vereinigung für Wasserwirtschaft, Abwasser und Abfall e.V., Hennef, Germany, 2006 (in German).

54. Brix, H.; Johansen, N.H. *Retningslinier for etablering af beplantede filteranlæg op til 30 PE* (Guidelines for vertical flow constructed wetland systems up to 30 PE); Økologisk Byfornyelse og Spildevandsrensning N. 52, Miljøstyrelsen, Miljøministeriet (in Danish), 2004.

55. Chazarenc, F.; Merlin, G. Influence of surface layer on hydrology and biology of gravel bed vertical flow constructed wetlands. Water Sci. Technol. **2005**, *51* (9), 91–97.

56. Platzer, C.; Mauch, K. Soil clogging in vertical-flow reed beds—Mechanisms, parameters, consequences, and solutions? Water Sci. Technol. **1997**, *35* (5), 175–181.

57. Cooper, P.F.; Smith, M.; Maynard, H. The design and performance of a nitrifying vertical-flow reed bed system. Water Sci. Technol. **1997**, *35* (5), 215–222.

58. Johansen, N.H.; Brix, H.; Arias, C.A. Design and characterization of a compact constructed wetland system removing BOD, nitrogen and phosphorus from single household sewage. In Proceedings of the 8th International Conference Wetland Systems for Water Pollution Control; University of Dar es Salaam: Tanzania, 2002; 47–61.

59. Boutin, C. Domestic wastewater treatment in tanks planted with rooted macrophytes: Case study, description of the system, design criteria, and efficiency. Water Sci. Technol. **1987**, *19* (10), 29–40.

60. Burka, U.; Lawrence, P. A new community approach to wastewater treatment with higher water plants. In *Constructed Wetlands in Water Pollution*; Cooper, P.F., Findlater, B.C., Eds.; Pergamon Press: Oxford, U.K., 1990, 359–371.

61. Mæhlum, T.; Stålnacke, P. Removal efficiency of three cold-climate constructed wetlands treating domestic wastewater: Effects of temperature, seasons, loading rates and input concentrations. Water Sci. Technol. **1999,** *40* (3), 273–281.

62. O'Hogain, S. The design, operation and performance of a municipal hybrid reed bed treatment system. Water Sci. Technol. **2003,** *48* (5), 119–126.

63. Öövel, M.; Tooming, A.; Mauring, T.; Mander, Ü. School-house wastewater purification in a LWA-filled hybrid constructed wetland in Estonia. Ecol. Eng. **2007,** *29,* 17–26.

64. Kantawanichkul, S.; Neamkam, P. Optimum recirculation ratio for nitrogen removal in a combined system: Vertical flow vegetated bed over horizontal flow sand bed. In *Wetlands: Nutrients, Metals and Mass Cycling*; Vymazal, J., Ed.; Backhuys Publishers: Leiden, the Netherlands, 2003; 75–86.

65. Brix, H.; Johansen, N.H. Treatment of domestic sewage in a two-stage constructed wetland—Design principles. In *Nutrient Cycling and Retention in Natural and Constructed Wetlands*; Vymazal, J., Ed.; Backhuys Publishers: Leiden, the Netherlands, 1999; 155–163.

66. Masi, F.; Martinuzzi, N. (2009). Constructed wetlands for the Mediterranean countries: Hybrid systems for water reuse and sustainable sanitation. Desalination **2009,** *215,* 44–55.

67. Obarska-Pempkowiak, H.; Gajewska, H.; Wojciechowska, E. Application, design and operation of constructed wetland systems. In Proc. Workshop Wastewater Treatment in Wetlands. Theoretical and Practical Aspects; I. Toczylowska, I., Guzowska, G., Eds.; Gdańsk University of Technology Printing Office: Gdansk, Poland, 2005; 119–125.

68. Belmont, M.A.; Cantellano, E.; Thompson, S.; Williamson, M.; Sánchez, A.; Metcalfe, C.D. Treatment of domestic wastewater in a pilot-scale natural treatment system in central Mexico. Ecol. Eng. **2004,** *23,* 299–311.

69. Singh, S.; Haberl, R.; Moog, O.; Shrestha, R.R.; Shrestha, P.; Shrestha, R. Performance of an anaerobic baffled reactor and hybrid constructed wetland treating high-strength wastewater in Nepal—A model for DEWATS. Ecol. Eng. **2009,** *35,* 654–660.

70. Wallace, S.D. On-site remediation of petroleum contact wastes using subsurface-flow wetlands. In *Wetlands and Remediation II*; Nehring, K.W., Brauning, S.E., Eds.; Battelle Press: Columbus, OH, 2002; 125–132.

71. Ji, G.; Sun, T.; Zhou, Q.; Sui, X.; Chang, S.; Li, P. Constructed subsurface slow wetland for treating heavy oil-produced water of the Liaohe Oilfield in China. Ecol. Eng. **2002,** *18,* 459–465.

72. Aslam, M.M.; Malik, M.; Baig, M.A.; Qazi, I.A.; Iqbal, J. Treatment performance of compost-based and gravel-based vertical flow wetlands operated identically for refinery wastewater treatment in Pakistan. Ecol. Eng. **2007,** *30,* 34–42.

73. Sands, Z.; Gill, L.S.; Rust, R. Effluent treatment reed beds: Results after ten years of operation. In *Wetlands and Remediation*; Means, J.F., Hinchee, R.E., Eds.; Battelle Press: Columbus, OH, 2000; 273–279.

74. Machate, T.; Noll, H.; Behrens, H.; Kettrup, A. Degradation of phenanthrene and hydraulic characteristics in a constructed wetland. Water Res. **1997,** *31,* 554–560.

75. Tan, K.; Jackson, W.A.; Anderson, T.A.; Pardue, J.H. Fate of perchlorate-contaminated water in upflow wetlands. Water Res. **2004,** *38,* 4173–4185.

76. Novais, J.M.; Martins-Dias, S. Constructed wetlands for industrial wastewater treatment contaminated with nitro-aromatic organic compounds and nitrate at very high concentrations. In Proc. Conf. The Use of Aquatic Macrophytes for Wastewater Treatment in Constructed Wetlands; Dias, V., Vymazal, L., Eds.; ICN and INAG: Lisbon, Portugal, 2003; 277–288.

77. Thut, R.N. Feasibility of treating pulp mill effluent with a constructed wetland. In *Constructed Wetlands for Water Quality Improvement*; Moshiri, G.A., Ed.; Lewis Publishers: Boca Raton, FL, 1993; 441–447.

78. Poggi-Varaldo, H.M.; Gutiérez-Saravia, A.; Fernández-Villagómez, G.; Martínez-Pereda, P.; Rinderknecht-Seijas, N. A full-scale system with wetlands for slaughterhouse wastewater treatment. In *Wetlands and Remediation II*; Nehring, K.W., Brauning, S.E., Eds.; Battelle Press: Columbus, OH, 2002; 213–223.

79. Lavigne, R.L.; Jankiewicz, J. Artificial wetland treatment technology and its use in the Amazon River forests of Ecuador. In Proc. 7th Internat. Conf. Wetland Systems for Water Pollution Control; University of Florida: Gainesville, 2000, 813–820.

80. Soroko, M. (2005). Treatment of wastewater from small slaughterhouse in hybrid constructed wetlands system. In Proceedings of the Workshop Wastewater Treatment in Wetlands. Theoretical and Practical Aspects; Toczylowska, I., Guzowska, G., Eds.; Gdańsk University of Technology Printing Office: Gdansk, Poland, 2005; 171–176.

81. Davies, T.H.; Cottingham, P.D. The use of constructed wetlands for treating industrial effluent. In Proceedings of the 3rd International Conference Wetland Systems in Water Pollution Control; IAWQ and Australian Water and Wastewater Association: Sydney, NSW, Australia, 1992; 53.1–53.5.

82. Calheiros, C.S.C.; Rangel, A.O.S.S.; Castro, P.K.L. Constructed wetland systems vegetated with different plants applied to the treatment of tannery wastewater. Water Res. **2007**, *41*, 1790–1798.

83. Zupančič Justin, M.; Vrhovšek, D.; Stuhlbacher, A.; Griessler Bulc, T. Treatment of wastewater in hybrid constructed wetland from the production of vinegar and packaging of detergents. Desalination **2009**, *246*, 100–109.

84. Vrhovšek, D.; Kukanja, V.; Bulc, T. Constructed wetland (CW) for industrial waste water treatment. Water Res. **1996**, *30*, 2287–2292.

85. Mantovi, P.; Marmiroli, M.; Maestri, E.; Tagliavini, S.; Piccinini, S.; Marmiroli, N. Application of a horizontal subsurface flow constructed wetland on treatment of dairy parlor wastewater. Bioresour. Technol. **2003**, *88*, 85–94.

86. Billore, S.K.; Singh, N.; Ram, H.K.; Sharma, J.K.; Singh, V.P.; Nelson, R.M.; Das, P. Treatment of a molasses based distillery effluent in a constructed wetland in central India. Water Sci. Technol. **2001**, *44* (11/12), 441–448.

87. Masi, F.; Conte, G.; Martinuzzi, N.; Pucci, B. Winery high organic content wastewaters treated by constructed wetlands in Mediterranean climate. In Proc. 8th Internat. Conf. Wetland Systems for Water Pollution Control; University of Dar-es-Salaam, Tanzania and IWA, 2002; 274–282.

88. Finlayson, M.; Chick, A.; von Oertzen, I.; Mitchell, D. Treatment of piggery effluent by an aquatic plant filter. Biol. Wastes **1987**, *19*, 179–196.

89. Strusevičius, Z.; Strusevičiene, S.M. Investigations of wastewater produced on cattle-breeding farms and its treatment in constructed wetlands. In Proceedings of the International Conference Constructed and Riverine Wetlands for Optimal Control of Wastewater at Catchment Scale; Mander, Ü., Vohla, C., Poom, A., Eds.; University of Tartu, Institute of Geography: Tartu, Estonia, 2003; 317–324.

90. Kantawanichkul, S.; Neamkam, P. Optimum recirculation ratio for nitrogen removal in a combined system: Vertical flow vegetated bed over horizontal flow sand bed. In *Wetlands: Nutrients, Metals and Mass Cycling*; Vymazal, J., Ed.; Backhuys Publishers: Leiden, the Netherlands, 2003; 75–86.

91. Comeau, Y.; Brisson, J.; Réville, J.P.; Forget, C.; Drizo, A. Phosphorus removal from trout farm effluents by constructed wetlands. Water Sci. Technol. **2001**, *44* (11/12), 55–60.

92. Schulz, C.; Gelbrecht, J.; Rennert, B. Treatment of rainbow trout farm effluents in constructed wetland with emergent plants and subsurface horizontal water flow. Aquaculture **2003**, *217*, 207–221.

93. Drizo, A.; Twohig, E.; Weber, D.; Bird, S.; Ross, D. Constructed wetlands for dairy effluent treatment in Vermont: two years of operation. In Proceedings of the 10th International Conference Wetland Systems for Water Pollution Control; MAOTDR 2006; Lisbon, Portugal, 2006; 1611–1621.

94. Kern, J.; Brettar, I. Nitrogen turnover in a subsurface constructed wetland receiving dairy farm wastewater. In *Treatment Wetlands for Water Quality Improvement*; Pries, J., Ed.; CH2M Hill Canada Limited: Waterloo, Ontario, 2002; 15–21.

95. Veenstra, S. the Netherlands. In *Constructed Wetlands for Wastewater Treatment in Europe*; Vymazal, J., Brix, H., Cooper, P.F., Green, M.B., Haberl, R., Eds.; Backhuys Publishers: Leiden, the Netherlands, 1998; 289–314.

96. Kato, K.; Koba, T.; Ietsugu, H.; Saigusa, T.; Nozoe, T.; Ko-bayashi, S.; Kitagawa, K.; Yanagiya, S. Early performance of hybrid reed bed system to treat milking parlour wastewater in cold climate in Japan. In Proceedings of the 10th International Conference Wetland Systems for Water Pollution Control; MAOTDR 2006; Lisbon, Portugal, 2006; 1111–1118.

97. Reeb, G.; Werckmann, M. First performance data on the use of two pilot-constructed wetlands for highly loaded non-domestic sewage. In *Natural and Constructed Wetlands: Nutrients, Metals and Management*; Vymazal, J., Ed.; Backhuys Publishers: Leiden, the Netherlands, 2005; 43–51.

98. Revitt, D.M.; Shutes, R.B.E.; Jones, R.H.; Forshaw, M.; Winter, B. The performance of vegetative treatment systems for highway runoff during dry and wet conditions. Sci. Total Environ. **2004,** *334–335,* 261–270.

99. Karrh, J.D.; Moriarty, J.; Kornue, J.J.; Knight, R.L. Sustainable management of aircraft anti/deicing process effluents using a subsurface-flow treatment wetland. In *Wetlands and Remediation II*; Nehring, W., Brauning, S.E., Eds.; Battelle Press: Columbus, OH, 2002; 187–195.

100. McGill, R.; Basran, D.; Flindall, R.; Pries, J. Vertical-flow constructed wetland for the treatment of glycol-laden stormwater runoff at Lester B. Pearson International Airport. In Proceedings of the 7th International Conference Wetland Systems for Water Pollution Control; University of Florida and IWA: Lake Buena Vista, FL, 2000; 1080–1081.

101. Headley, T.R.; Huett, D.O.; Davison, L. The removal of nutrients from plant nursery irrigation runoff in subsurface horizontal-flow wetlands. Water Sci. Technol. **2001,** *44* (11–12), 77–84.

102. Seo, D.C.; Hwang, S.H.; Kim, H.J.; Cho, J.S.; Lee, H.J.; DeLaune, R.D.; Jagsujinda, A.; Lee, S.T.; Seo, J.Y.; Heo, J.S. Evaluation of 2- and 3- stage combinations of vertical and horizontal flow constructed wetland for treating greenhouse wastewater. Ecol. Eng. **2008,** *32,* 121–132.

103. Wojciechowska, E.; Obarska-Pempkowiak, H. Performance of reed beds supplied with municipal landfill leachate. In *Wastewater Treatment, Plant Dynamics and Management in Constructed and Natural Wetlands*; Vymazal, J., Ed.; Springer: Dordrecht, the Netherlands, 2008; 251–265.

104. Headley, T.R.; Davison, L.; Yeomans, A. Removal of am-monia-N from landfill leachate by vertical flow wetland: A pilot study. In Proceedings of the 9th International Conference Wetland Systems for Water Pollution Control; ASTEE: Lyon, France, 2004; 143–150.

105. Bulc, T.G. Long term performance of a constructed wetland for landfill leachate treatment. Ecol. Eng. **2006,** *26,* 365–374.

106. Langergraber, G.; Giraldi, D.; Mena, J.; Meyer, D.; Peňa, M.; Toscano, A.; Brovelli, A.; Korkusuz, E.A. Recent developments in numerical modeling of subsurface flow constructed wetlands. Sci. Total Environ. **2009,** *407,* 3931–3943.

Wetlands: Sedimentation and Ecological Engineering

Problems of Sedimentation in Wetlands 617
    Sediment Accretion • Reduced Conductivity of the
    Bed • Phosphorus Accumulation
Particle Types and Sedimentation Processes in Wetlands 618
Ecological Engineering Solutions to Sedimentation 619
Conclusions 620
References...620

Timothy C. Granata
and J.F. Martin

Problems of Sedimentation in Wetlands

Sediment Accretion

The accretion of particulate material in wetlands limits the lifetime of these systems, advancing their succession to a terrestrial ecosystem as a function of sedimentation rates. Three sources of particulate material can be identified: suspended particles derived from the inflow, production of microbial particles in the wetland, and the production of detritus from macrovegetation. The latter two increase with increased primary production in wetlands. The former is a function of hydraulic loading, defined as the flow, Q, per unit area of wetland, A, i.e., QA^{-1}. Natural wetlands are prominent components of riparian ecosystems that exist either as floodplains, receiving periodic submersion, or as part of the riparian corridor, receiving continuous, though varying, inflow. In both cases, flow transports suspended sediments to wetlands. Suspended material is often deposited as current velocities decrease in broad, shallow expanses of wetlands. Sediment transport is a function of total flow (and thus rainfall intensity) in a watershed but also depends on the slope of the landscape and the susceptibility of soils to erosion. Steep slopes with highly unconsolidated soils will contribute to large suspended sediment loads during high runoff events. For low-flow events suspended sediment concentrations do not exceed 20 mg/L while for high flows they can exceed 80 mg/L.[1] In extreme cases of accretion, sedimentation rates of 26.67 cm/yr can bury emergent grasses because sediment accumulation outpaces plant growth.[2] In contrast, constructed wetlands are designed to treat wastewater having steady and high suspended solid concentrations, ranging from 20 to 75 mg/L, and volumetric ratios of settable solids from 5 to 20 mL/L.[3]

Reduced Conductivity of the Bed

Wetlands can experience reduced infiltration rates as particulate matter clogs bottom substrates. Generally, a reduction of hydraulic conductivity occurs with increasing hydraulic loading to the wetland. As a consequence of lower current velocities and higher sedimentation rates with increasing distance from the inlet, the hydraulic conductivity is lowest near the inlet and highest near the outlet. A general design criterion for wetlands is a hydraulic conductivity 1500 m/ day based on bed of 1.25–2.5 cm diameter gravel.[4]

Phosphorus Accumulation

The accretion of organic material on the bed results in wetlands with a high ion exchange capacity. The beds are partially responsible for removal of phosphates by sorption until they become saturated. The movement of phosphorus through the bed is slower than the hydraulic conductivity as a result of the storage (exchange) capacity of the bed.[5] Plant uptake of phosphates is limited because a reduced hydraulic conductivity decreases infiltration and thus transport of phosphates to plant roots.[6] A more consistent process for phosphate removal is by adsorption onto suspended sediments and their subsequent deposition.[7] Phosphorus (P) loading is directly proportional to suspended sediment concentrations and hydraulic loading. It is estimated that 40 mg P accompanies each gram of suspended solids entering wetlands.[1] Higher loading rates occur for flows over landscapes rich in P, such as agricultural lands. During periods of inundation, wetlands are characterized by aerobic and anaerobic soil zones. In anaerobic zones the decreased redox potential leads to an increase in the solubility of particulate phosphorus, which may be discharged from the wetlands. This dynamic feedback between removal of P by sedimentation and release of P after particulate accretion often leads to variable retention of P by wetlands. Despite mineralization of particulate P, wetlands are generally sinks for P, with removal as high as 97%.[8] The retention of P, however, decreases as P loading increases[9] and often decreases through the life of the wetland.[10]

Particle Types and Sedimentation Processes in Wetlands

The size distribution of particles in a wetland varies from nanometer-sized colloidal material through micrometer phytoplankton and heterotrophic organisms to millimeter- and centimeter-sized leaf detritus and sediments. For discrete particles, settling results from the force balance between particle weight (F_w) and drag on the particle such that

$$F_w = F_D = \rho C_D A_{pr} \frac{w^2}{2} = \left(\rho_P - \rho\right)V_g \tag{1}$$

where C_D is the drag coefficient on a 2D surface, A_{pr} is the projected area of the particle, w is the settling speed, ρ_P and ρ are the particle and water densities, respectively, V is the volume of the particle, and g is the gravitational acceleration (9.8 m/sec^2). Solving for settling rate, w, gives

$$w = \sqrt{2\frac{\left(\rho_P - \rho\right)V_g}{\rho A_{pr} C_D}} \tag{2}$$

Thus, the settling rate of a particle is proportional to the excess density of the particles over that of water and the length scale of the particle defined by VA_{pr}^{-1}. Providing flow is not turbulent, specified as a Reynolds number Re <0.1, and particle shape is unimportant to the drag; C_D can be approximated as

$$C_D = \frac{24}{Re} = \frac{24v}{2Rw} \tag{3}$$

where R is the radius of the particle. Assuming a spherical particle, the above equation reduces to

$$w = \frac{2\rho' g R^2}{9v} \tag{4}$$

the celebrated Stokes' Law for sinking particles. Here the excess density is written as

$$\rho' = \frac{\rho_P - \rho}{\rho} \tag{5}$$

For biological material, excess density is less than 0.1%, and size is the determining factor. For this reason small microbial organisms will sink slower than larger detrital material. The excess density of suspended sediments can reach 260%, indicating that the density is responsible for the high settling rates of suspended sediments.

Sedimentation in a wetland will occur if the settling speed is greater than the surface loading rate per area of wetland, or $w > QA^{-1}$. Given the large spectrum of particle types in natural flows, w is usually calculated based on the size and density of the particle type to be removed. Dividing the depth by the sinking rate gives the retention time in the wetland. Particles that are vertically well mixed as they enter a wetland will have a uniform concentration distribution comprised of the spectrum of particle sizes. If the settling rate of the particle, w_p, is greater than w, the particle will be deposited in the wetland cell. For particles with $w_p < w$, the fraction of particles that will be removed is X_p. Thus, the total particle removal by a wetland can be calculated as

$$FR = (1 - X_c) + QA^{-1} \sum X_r \Delta x \qquad (6)$$

where $1 - X_c$ is the fraction of particles removed with sinking speeds $w_p > w$ and $QA^{-1} \Sigma X_r \Delta x$ is the fraction with a settling speed $w_p < w$ that are removed over a distance Δx, along the path of the flow.[3] The crucial step in determining the sedimentation rate in wetlands is first determining the settling rate of each concentration fraction, either by settling columns or size-concentration measurements.

Often aggregations of particles form in wetlands,[11] a process called flocculation. Sedimentation rates of aggregates are dominated by organic particles of fractal dimension.[12] Aggregates tend to settle faster than their discrete component particles, and the only way to determine removal rate is by carefully transferring flocs to a settling column and determining the percent removal as a function of the height of the column and time.[3]

In wetlands with a high concentration of suspended solids, settling is affected by the contact between particles. This would occur in the bottom sediments of most wetlands where two settling processes can be identified: hindered and compression settling.[3] The hindered zone is marked by a large gradation in particle concentrations, which is less than the total concentrated in the compression zone. By plotting the concentration of particles over the height of a settling column as a function of time, a break point can be found in the time from hindered settling to compression settling.[3]

Because large, heavy particles settle out first, the highest sedimentation rates in wetlands are near the inlets. As mentioned above, the hydraulic conductivity through the bottom sediment decreases as clogging occurs. Because clogging is also a function of distance from the inlet, the most accurate estimate of the hydraulic conductivity is to measure the change in water surface elevation with increasing distance from the inlet. However, as Kadlec and Watson[11] show, this is not a simple function of Darcy's Law but must include evapotranspiration in the wetland. For newly constructed wetlands, the hydraulic conductivity of the bottom substrate (gravel, sand, and mud) cannot be used because clogging will result from sediment accretion.[4]

Ecological Engineering Solutions to Sedimentation

To keep suspended particles entrained by flows from accruing in wetlands and clogging bottom sediments, one or more settling basins can be included between the inlet channel and the wetland cell. This would have two effects: first, to collect all but the finest and least dense suspended particles and, second, to remove phosphorus from the inflow.

To overcome the problem of detrital accumulation in wetlands, woody plants could be substituted for grasses and periodically harvested. This would not only reduce the amount of biomass accrued in the wetland but would also increase the efficiency of nutrient uptake in the unclogged root zone and provide a potentially marketable resource. Because phytoplankton are the most abundant nutrient filters in a wetland and have intrinsically low sinking rates, a wetland could be designed with low retention of suspended plankton to further reduce sedimentation while improving nutrient removal. A settling basin could then be sized to accumulate these nutrient-rich particles for harvesting before nutrients are remineralized. To keep flocs of particles from forming in wetlands, cells could be mixed with aerators or inlet pumps. This would break up flocs, which would be exported as discrete particles.

Conclusions

Because wetlands are always shallow to promote macrophtye growth, they essentially act as flat plate collectors of sinking particles. The dominant particles in wetlands are suspended particles in the inflow and biomass of vegetation resulting from growth in the wetlands. The processes of sedimentation can reduce a wetland's storage capacity, its efficiency to retain nutrients, and its lifetime. Future efforts should be aimed at enhancing nutrient removal while reducing sedimentation. This can be done with engineered structures, such as settling basins and grit chambers, or by less costly technologies such as stilling wells. More innovative green solutions are on the horizon.

References

1. Porter, K.S. Nitrogen and phosphorous. In *Food Production, Waste, and the Environment*; Ann Arbor Science Publishers Inc.: Ann Arbor MI, 1975; 372.
2. Chung, C.-H. Ecological engineering of coastlines with salt-marsh plantations. In *Ecological Engineering: An Introduction to Ecotechnology*; Mitsch, W.J., Jørgenson, S.E., Eds.; John Wiley and Sons: New York, 1989; 472.
3. Metcalf, Eddy. *Wastewater Engineering: Treatment, Disposal, Reuse*; McGraw Hill: New York, 1991; 1334.
4. Watson, T.J.; Choate, K.D. Hydraulic conductivity of onsite constructed wetlands. 9th National Symposium of Individual and Small Community Sewage Systems, Fort Worth, Texas, Mar 11–14, 2001.
5. Kadlec, R.H.; Knight, R.L. *Treatment Wetlands*; CRC Press/Lewis Publishers: Boca Raton, FL, 1996; 839.
6. Davies, T.H.; Cottingham, P.D. Phosphorus removal from wastewater in a constructed wetland. In *Constructed Wetlands for Water Quality Improvement*; Moshiri, G.A., Ed.; Lewis Publishers: Boca Raton, FL, 1993; 632 pp.
7. Richardson, C.J.; Craft, C.B. Efficient phosphorus retention in wetlands: fact or fiction? In *Constructed Wetlands for Water Quality Improvement*; Moshiri, G.A., Ed.; Lewis Publishers: Boca Raton, FL, 1993; 632.
8. Mitsch, W.J.; Gosselink, J.G. *Wetlands*; Van Nostrand Reinhold Co.: New York, 2000; 920 pp.
9. Mitsch, W.J.; Reeder, B.C.; Klarer, D.M. Wetlands for nutrient control: Western Lake Erie. In *Ecological Engineering: An Introduction to Ecotechnology*; Mitsch, W.J., Jørgenson, S.E., Eds.; John and Wiley Sons: New York, 1989; 472 pp.
10. White, J.S.; Bayley, S.E.; Curtis, P.J. Sediment storage of phosphorus in a northern prairie wetland receiving municipal and agro-industrial wastewater. Ecol. Eng. **2000**, *14*, 127–138.
11. Kadlec, R.H.; Watson, J.T. Hydraulics and solids accumulation in a gravel bed treatment wetland. In *Constructed Wetlands for Water Quality Improvement*; Moshiri, G.A., Ed.; Lewis Publishers: Boca Raton, FL, 1993; 632 pp.
12. Kilps, J.R.; Logan, B.E.; Alldregde, A.L. Fractal dimensions of marine snow determined from image analysis of in situ photographs. Deep-Sea Res. **1994**, *41* (8), 1159–1169.

60

Wetlands: Treatment System Use

How Do Wetlands Work?...621
Treatment Wetland Types 622
    Constructed vs. Natural Wetlands
Free-Water vs. Submerged-Bed Wetlands..622
Treatment Processes...623
Design Considerations...625
VSB Wetlands...625
FWS vs. VSB Systems...626
Operation and Maintenance 626
Conclusion 626
Acknowledgments..626
References..626

Kyle R. Mankin

How Do Wetlands Work?

Wetlands can be used to treat wastewater because they process contaminants. However, they treat wastewater more slowly than traditional treatment plants. Oxygen, and the manipulation of oxygen levels, is a primary concern for wastewater treatment because many of the necessary biological and chemical treatment processes require oxygen. Traditional treatment plants can easily manipulate oxygen levels by pumping air into the wastewater. Oxygen enters wetlands by slower, natural processes. Increasing oxygen concentration, by increasing wastewater contact with air, plant roots, or photosynthetic algae, often can enhance the processing ability of wetlands.

When considering wastewater treatment by constructed wetlands, five contaminant groups are of primary importance: sediments, organic matter, nutrients, pathogenic microbes, and metals. Wetlands slow down water movement, allowing sediments to settle out of the water. Organic matter can be processed, or decomposed, by highly competitive microbes. Less competitive microbes called nitrifiers process nitrogen. Both microbe types require oxygen. Because the nitrifiers are less competitive, oxygen levels become very important to insure that both organic matter and nitrogen are fully processed. The other two, pathogenic microbes and metals, are more situational, related to the specific waste being treated. Wetlands treat pathogenic microbes by detaining them until they naturally die off, are eaten by other predatory organisms in the wetland, or are exposed to UV radiation near the water surface. Metals are processed by being adsorbed to other particles and settling out of the water.

The remainder of this entry further explains wetland processes and design considerations. References are provided for more in-depth information.

Treatment Wetland Types

Constructed vs. Natural Wetlands

Wetlands constructed as treatment systems differ from natural wetlands in several important ways. Constructed wetlands usually are built with uniform depths and shapes designed to provide consistent detention times and maximize contaminant removal. In contrast, natural wetlands are irregular in depth and shape, which causes irregular flow, allows water to by-pass the shallow treatment zones by moving through the deeper channels, and leads to less effective treatment. In addition, water-quality regulations in the United States dictate that if a natural wetland is associated with an existing water body of the United States, as most are, wastewater discharges into the wetland must meet specific quality standards, similar to other water bodies. Wetlands constructed as wastewater treatment systems typically are located in uplands where wetlands did not exist before and are not subject to inflow water-quality regulations. Natural wetlands are *not* recommended for use as treatment wetlands.

Constructed wetlands increasingly are being used for wastewater treatment in a variety of applications (Table 1). Examples can be found of wetlands being used to treat municipal sewage, urban runoff, onsite residential wastewater, animal feedlot and barnyard runoff, cropland runoff, industrial wastewater, mine drainage, and landfill leachate. Each application takes advantage of a combination of physical, chemical, and biological processes characteristic of natural wetlands to reduce the concentration of contaminants in water. Such contaminants include sediments, organic materials, nutrients (particularly nitrogen and phosphorus), metals, microbial pathogens, and pesticides.

Free-Water vs. Submerged-Bed Wetlands

Constructed wetlands have two common types. Free-water surface (FWS) wetlands (also called surface-flow wetlands) have plants that grow in a shallow layer of water over a soil substrate (Figures 1 and 2). The location of the plants in the system can vary: the plants can float on the water surface with their roots suspended in the water (free-floating macrophyte systems); they can be rooted in the soil with the

TABLE 1   North American Wetlands as of 1994

Wastewater Type	Quantity	Size (ha)		
		Minimum	Median	Maximum
Agricultural	58	0.0004	0.1	47
Industrial	13	0.03	10	1093
Municipal	159	0.004	2	500
Stormwater	6	0.2	8	42
Other	7	3	376	1406

Source: Kadlec et al.[1]

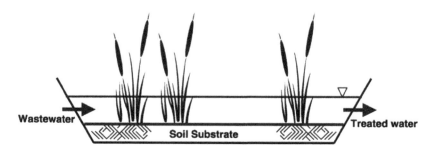

FIGURE 1   FWS wetland with emergent macrophytes.

FIGURE 2 A three-cell, FWS wetland for treating dairy wastewater. This system is in its first year of operation; plants were recently established.
Source: Photo by Peter Clark.

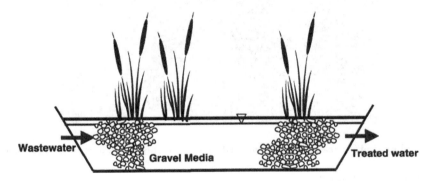

FIGURE 3 VSB wetland with emergent macrophytes.

entire leaves and stems below the water surface (submerged-macrophyte systems); they can be rooted in the soil having leaves and stems that rise above the water surface (emergent macrophyte systems); or the wetland may use a combination of planted and open-water zones. About two-thirds of existing wetlands as of 1994 were FWS.[1] In vegetated submergedbed (VSB) wetlands (also called subsurface flow wetlands or rock-plant filters), plants are rooted in a porous media, such as sand or gravel, and water flows through the media in either horizontal or vertical direction (Figures. 3 and 4). About one-quarter of treatment wetlands were VSB systems.[1] However, these systems are currently used in thousands of smaller-scale, onsite residential applications in the United States that do not appear in this database.

Treatment Processes

Many wastewaters entering constructed wetlands must be pretreated to avoid excessive contaminant loading, particularly of mineral and organic solids. Pretreatment technologies include septic tanks for onsite systems or anaerobic lagoons for animal waste, municipal, or mine-drainage treatment systems. In each case, the anaerobic condition in the pretreatment process reduces production of additional algae solids. Typical contaminant levels entering treatment wetlands are summarized in Table 2.

The wetland type impacts the processes used to retain or remove contaminants. In a VSB system, wastewater flows through pore spaces of the media and comes into direct contact with the roots of plants. In a FWS system, water flows across the media surface and contacts plant stems and leaves. In either system, solid particles, including sediments (clay and silt particles and colloids) and organic matter (manure particles, organic residues, and algae or other phytoplankton), settle out of the water column or are trapped or filtered as water passes through a wetland. Contaminants that are adsorbed

FIGURE 4   VSB wetland for treating onsite residential wastewater. This system uses gravel media and variety of wetland plants.
Source: Photo by Barbara Dallemand.

TABLE 2   Wetland Influent Concentrations

Wastewater Type	bod$_5$ (mg/L)	TSS (mg/L)	TN (mg/L)	nh$_4$-n (mg/L)	no$_3$-n (mg/L)	TP (mg/L)	FC (per 100 mL)
Residential-septic tank[2]	129–147	44–54	41–49	28–34	0–0.9	12–14	105.4–106.0
Municipal-primary[2]	40–200	55–230	20–84	15–40	0	4–15	105.0–107.0
Municipal-pond[2]	11–35	20–80	8–22	0.6–16	0.1–0.8	3–4	1008–1056
Livestock13 [avg.]	263	585	254	122	3.6	24	1.6×105
Livestock13 [median]	81	118	274	60	1.1	20	1.7×103
Landfill leachate[4]	312–729	241–7840	287–670	254–2074	0–3	0.9	

Note: BOD5 = 5-day biochemical oxygen demand, TSS = total suspended solids, TN = total nitrogen, NH4-N = ammonium nitrogen, NO3-N = nitrate nitrogen, TP = total phosphorus, FC = fecal coliform bacteria. FWS wetlands

to sediments (e.g., P, NH4, fecal bacteria) or absorbed within organic solids (e.g., nutrients) are also removed. However, these constituents can be re-suspended or desorbed back into the wetland water. This natural cycling of materials is an important function of wetlands, although it makes system design and interpretation of treatment complex.

Once entrapped, organic materials and associated contaminants are decomposed in wetlands by microbial and chemical transformations. In the degradation process, microbes use oxygen. The amount of oxygen used is related to the amount of organic material in the water. The controlled measurement of biochemical oxygen demand (BOD) is a common way to illustrate the amount of organic matter in water. When wastewater lacks oxygen, or is anaerobic, it requires the addition of oxygen to degrade organic matter. Oxygen is also required for transformation of ammonium to nitrite and nitrate (nitrification), whereas anaerobic conditions are required for transformation of nitrate to nitrogen gas (denitrification). Aerobic wetland conditions often remove metals by aerobic oxidation of iron; subsequently iron hydroxides and other metals precipitate in the wetland.[5] Although some oxygen diffuses into a wetland from the air, a common assumption is that oxygen also is transported through wetland plants and made available to microbes in close proximity to leaky roots.[6] This mechanism may be less important than once thought, though.[2] Treatment wetlands are thought to function effectively because they combine anaerobic zones in the water column with aerobic zones near the water interfaces with air and roots. However, because the microbes that break down organic carbon can out compete nitrifiers for oxygen, nitrogen removal in higher strength wastewaters is often low.

Design Considerations

Design and resulting effectiveness of constructed wetlands (Table 3) depend upon many factors: climate (precipitation, temperature, growing season, evapotranspiration), wastewater characteristics (constituents, loading, flow rate, and volume), topography, and wildlife activity. Wetland designs must specify total area; the number, depth, and size of wetland cells; hydraulic retention times; vegetation types and coverage; inlet and outlet configuration and location; and internal flow patterns.[2] Details for design can be found in numerous references[2,7-11] and some elements are discussed here.

VSB Wetlands

Properly designed VSB systems can achieve high removal rates. Treatment in a VSB wetland is governed by system residence time and wastewater contact with media and plant-root surfaces. Because of this, depth is a critical dimension and is often chosen according to the rooting depth of the selected plant (e.g., cattails: 30 cm; reeds: 40 cm; bulrush: 60 cm). Once depth is chosen, crosssectional area (and thus wetland width) is selected to assure adequate flow rates. Then, volume (and thus wetland length) is determined from the retention time needed to treat the wastewater to the desired quality. Proper design of inlet and outlet control structures helps maintain uniform flow patterns and depth, avoids problems with clogging and freezing, and minimizes system operation and maintenance (O and M) problems. High loading from influent solids and clogging can lead to surface flows and poor treatment. VSB systems must receive influents that are pretreated to remove solids (e.g., septic tank and effluent filter or anaerobic lagoon).

Properly designed FWS systems also can achieve high removal rates. Design typically follows one of two methods. The areal loading approach allows a designer to select the wetland surface area according to the influent load and the desired effluent quality.[13] Another approach allows a designer to select the wetland area by knowing the biological reaction rate, wastewater concentration, and flow rate along with selected water depth and target outflow water quality.[7,8] Again, depth is a critical dimension and is governed by plant tolerance to standing water and treatment objectives.

TABLE 3 Wetland Treatment (%)

Wastewater Type	BOD5	TSS	TN	nh₄-n	TP	FC	Metals
Municipal[11] [avg.]	74	70	53	54	57	—	—
Livestock[13] [avg.]	65	53	42	48	42	92	—
Landfill leachate[12] [range]	11–90	45–97	7–45	13–88	—	—	8–95+

Note: BOD5 = 5-day biochemical oxygen demand, TSS = total suspended solids, TN = total nitrogen, NH_4–N = ammonium nitrogen, TP = total phosphorus, FC = fecal coliform bacteria, metals = Fe, Cu, Pb, Ni, or Zn.

FWS vs. VSB Systems

Selection of the most appropriate wetland system depends on wastewater characteristics, treatment requirements, and site constraints. VSB systems generally require less land area, are less susceptible to freezing and mosquito problems, and have no exposed wastewater at the surface (avoiding contact-related health problems). FWS systems are less expensive to construct (without the cost of media), have greater potential for wildlife habitat, and are easier to maintain if solids accumulate.

Operation and Maintenance

O and M of treatment wetlands are relatively simple. The goal of an O and M plan is to assure that the wetland system continues to operate as planned, designed, and constructed. Several sources provide specific O and M guidance,[14] and most design manuals also contain such guidelines. Operation should be consistent with treatment objectives while maintaining structural integrity of the system, uniform flow conditions, and healthy vegetation as well as minimizing odors, nuisance pests and insects. Most maintenance plans require such items as checking water levels, checking for evidence of leaks or wildlife damage, and maintaining plant health on a weekly or monthly basis.

Conclusion

Constructed wetlands are complex natural-treatment systems that are well suited for many applications. They are low in cost and maintenance, provide significant reductions of many contaminants, and offer an aesthetic appearance. More work is needed to characterize treatment processes in constructed wetlands and improve design procedures to account for variability in wastewater and climate.

Acknowledgments

The author extends his gratitude to Kristina M. Boone, Associate Professor of Agricultural and Environmental Communications, Kansas State University for her significant review contribution to this entry, and to those at Kansas State University who have supported wetlands research and education.

References

1. Kadlec, R.H., Knight, R.L., Reed, S.C., Ruble, R.W., Eds. *Wetlands Treatment Database (North American Wetlands for Water Quality Treatment Database)*; EPA/600/C-94/002; U.S. Environmental Protection Agency: Cincinnati, OH, 1994.
2. US EPA. *Constructed Wetlands Treatment of Municipal Wastewaters*, EPA/625/R-99/010; U.S. Environmental Protection Agency: Cincinnati, OH, 2000.
3. *CH2M Hill and Payne Engineering. Constructed Wetlands for Livestock Wastewater Management: Literature Review, Database, and Research Synthesis;* Gulf of Mexico Program— Nutrient Enrichment Committee: Montgomery, AL, 1997.
4. Kadlec, R.H. Constructed wetlands for treating landfill leachate (Ch. 2). In *Constructed Wetlands for the Treatment of Landfill Leachates;* Mulamoottil, G., McBean, E.A., Rovers, F., Eds.; Lewis Publishers: Boca Raton, FL, 1999.
5. US EPA. *Engineering Bulletin—Constructed Wetlands Treatment;* EPA/540/S-96/501; U.S. Environmental Protection Agency: Cincinnati, OH, 1996.
6. Brix, H. Do macrophytes play a role in constructed treatment wetlands? WaterSci. Technol. **1997**, *35* (5), 11–17.
7. Kadlec, R.H.; Knight, R.L. *Treatment Wetlands;* Lewis Publishers: Boca Raton, FL, 1996.

8. Reed, S.C.; Crites, R.W.; Middlebrooks, E.J. *Natural Systems for Waste Management and Treatment*, 2nd Ed.; McGraw-Hill: New York, 1995.

9. Steiner, G.R., Watson, J.T., Eds. *General Design, Construction, and Operation Guidelines: Constructed Wetlands Wastewater Treatment System for Small Users Including Individual Residences, Technical Report TVA/ MW-93/10*; Tennessee Valley Authority: Chattanooga, TN, 1993.

10. Powell, G.M.; Dallemand, B.L.; Mankin, K.R. *Rock-Plant Filter Design and Construction for Home Wastewater Systems*; MF-2340, http://www.oznet.ksu.edu/library/h20ql2/mf2340.pdf Kansas State University Cooperative Extension Service: Manhattan, KS, 1998.

11. Sievers, D.M. *Design of Submerged Flow Wetlands for Individual Homes and Small Wastewater Flows*; Special Report 457; Missouri Small Wastewater Flows Education and Research Center, 1993.

12. Mulamoottil, G., McBean, E.A., Rovers, F., Eds. *Constructed Wetlands for the Treatment of Landfill Leachates*; Lewis Publishers: Boca Raton, FL, 1999; 69, 86, 157, 212.

13. USDA-NRCS. *Constructed Wetlands for Agricultural Wastewater Treatment: Technical Requirements*; U.S. Department of Agriculture-Natural Resource Conservation Service: Washington, DC, 1991.

14. Powell, G.M.; Dallemand, B.L.; Mankin, K.R. *Rock-Plant Filter Operation, Maintenance, and Repair*, MF-2337; Available at http://www.oznet.ksu.edu/library/h20ql2/mf2337.pdf Kansas State University Cooperative Extension Service: Manhattan, KS, 1998.

VII

NEC: Natural Elements and Chemicals Found in Nature

61

Cyanobacteria: Eutrophic Freshwater Systems

Anja Gassner and
Martin V. Frey

Factors Leading to Cyanobacterial Dominance 631
Consequences of Cyanobacterial Blooms 632
Monitoring and Management of Algal Blooms 632
Conclusions 633
References 633

Factors Leading to Cyanobacterial Dominance

The taxonomic composition of phytoplankton communities, the abundance and the relative dominance of the different species and groups present, undergo seasonal changes. This process of continuous community change is termed succession. Under undisturbed conditions, most phytoplankton populations are of relatively short duration. Typically, the growth and decline cycle of one specific population lasts, on average, 4 to 8 weeks. The "seasonal paradigm" of phytoplankton succession[1] describes the typical pattern of phytoplankton succession corresponding to the prevailing nutrient cycle in temperate, undisturbed lakes: a spring maximum of diatoms, sometimes followed by a second maximum in the autumn, an early summer maximum of Chlorophyceae (green algae) and a late summer maximum of Cyanophyta (blue-green algae).

It is generally accepted that with excess nutrients in the water column, in particular phosphorus, the phytoplankton flora deviates from the traditional seasonal community pattern with a shift toward cyanobacterial dominance. However, it must be stressed that nutrient limitation does not, in itself, provide cyanobacteria with the ability to become dominant; it is the combination of a multitude of abiotic and biotic factors. Enrichment experiments demonstrated that the maximum biomass of temperate lakes is ultimately limited by the phosphorus supply.[2] Increasing supplies of phosphorus lead to an increase of phytoplankton growth until other essential nutrients become limited. The first nutrient to become limited after phosphorus is usually nitrogen. Cyanobacteria are the only species that are able to fix atmospheric nitrogen. Whereas other algae become nitrogen limited, the ascendancy of nitrogen-fixing cyanobacteria is favored.

Apart from their ability of fixing atmospheric nitrogen, cyanobacteria feature some adaptations that enable them to outcompete other species. Eutrophic conditions result in large suspended stocks of phytoplankton, which reduce light penetration. Cyanobacteria possess gas vacuoles to control buoyancy. When subjected to suboptimal light conditions, they respond by increasing their buoyancy (regulated by the rate of photosynthesis) and move nearer to the surface and hence to the light.[2] Additionally, the possession of chlorophyll *a* together with phycobiliproteins allows them to harvest light efficiently and to grow in the shade of other species. Cyanobacteria are supposed to be more tolerant of high pH conditions and have an additional selective advantage at times of high photosynthesis because of their ability to use CO_2 as carbon source.[3] Some genera are able to offset the effects of photoinhibiting UV radiation encountered by near-surface populations. The resistance to photoinhibition is achieved by producing increased amounts of carotenoid pigments, which act as "sunscreens."[4] Once established, cyanobacteria are able to inhibit the growth of other algae by producing secondary metabolites that are toxic to species of other genera.[5]

Consequences of Cyanobacterial Blooms

Like any phytoplankton, bloom proliferation of blue-green algae reduces water quality in terms of human water use but also results in a reduction in diversity of the aquatic species assemblage at all trophic levels. The presence of "pea soup green" water, the accumulation of malodorous decaying algal cells, and the buildup of sediments rich in organic matter lead to user avoidance with the associated problems and implications for water quality management. The most obvious sign of an advanced blue-green algae bloom is the formation of green "scum," which leads to deoxygenation of underlying waters, subsequent fish kills, foul odors, and lowered aesthetic values of affected waters.[6] In addition, certain genera and species produce taste and odor compounds, typically geosmin and 2-methyl iso-borneol, which cause non-hazardous but unpleasant problems for suppliers and users of potable water.[4]

The most serious public health concerns associated with cyanobacteria arise from their ability to produce toxins. Since the first published reported incidence of mammal deaths related to a toxic cyanobacterial bloom in 1978, more than 12 species belonging to nine genera of blue-green algae have been implicated in animal poisoning.[7] For human exposure, routes are the oral route via drinking water, the dermal route during recreational use of lakes and rivers, or consumption of algal health food tablets. Toxins produced in a random and unpredictable fashion by cyanobacteria are called cyanotoxins and classified functionally into hepatotoxins, neurotoxins, and cytotoxins. Additionally, some cyanobacteria produce the lesser toxic lipopolysaccharides (LPS) and other secondary metabolites that may be of potential pharamacological use.[8] One of the most tragic encounters of humans with cyanobacterial toxins led to the deaths of 60 dialysis patients due to contaminated water supply used in a hemodialysis unit.[9] Presently, a drinking water guideline of 1 ng L^{-1} of toxin has been developed and implemented only for microcystin-LR.[4] Haider et al.[8] stress that the biggest challenge for water treatment procedures for the removal of cyanobacterial toxins is that one is faced with soluble and suspended substances. Thus, the most common treatment, chlorination, in general has been found not to be an effective process in destroying cyanotoxins.

Monitoring and Management of Algal Blooms

Drinking water treatment strategies are not always successful in removing algal toxins. Thus, detection of early-stage (emergent) blooms of cyanobacteria, especially if the bloom has not started to produce toxins, is important to allow municipalities and recreation facilities to implement a response plan. It has been shown that remote sensing technology can be used to estimate the concentration and distribution of cyanobacteria through measurement of the concentration of the pigment phycocyanin.[10]

Once detected, the growth of nuisance algae is prevented by the use of chemicals; the commonest is copper sulfate. Other algicides include phenolic compounds, amide derivatives, quaternary ammonium compounds, and quinone derivatives. Dichloronaphthoquinone is selectively toxic to blue-greens. The inherent problem of algicides is that on cell lysis, toxins contained in the algae cell are released into the surrounding water. In 1979, almost 150 people had to be hospitalized for treatment of liver damage after a reservoir contaminated with *Cylindrospermopsis* was treated with copper sulfate.[4] Biological control by zooplankton is, in principle, possible, although not always practical or effective because of the low nutrient adequacy, toxicity, and inconvenient size and shape of most blue-green algae.

The only zooplankton reported to successfully graze on blue greens is *Daphnia* sp., but it tends to decrease with increasing nutrient content of the water[11] More effective is the use of microorganisms, as certain chytrids (fungal pathogens) and cyanophages (viral pathogens) specifically infest akinetes and other heterocysts, whereas Myxobacteriales (bacterial pathogens) can affect rapid lysis of a wide range of unicellular and filamentous blue-greens, although heterocysts and akinetes remain generally unaffected.[12]

The consensus regarding the management of blue-green algal blooms is the management of excess nutrient loads into receiving water bodies.[13,14] Management options can be divided into two broad categories: catchment management (decrease of nutrient export) or lake management (decrease of internal nutrient supply). Catchment options are, e.g., management of urban and agricultural runoff, biological and chemical treatment of wastewater, nutrient diversion, and implementation of legislation. Lake management options are dredging, chemical sediment treatment, and biomanipulation.[13]

Conclusions

Cyanobacteria pose a serious threat to ecosystem health and human livelihood. From a human perspective, the most serious threat associated with blue-greens are their toxins. Routes for human exposure are the oral route via drinking water, the dermal route during recreational use of lakes and rivers, or consumption of algal health food tablets. Removal of these algae and their toxins from water bodies poses a great logistical problem. However, it is important to understand that the proliferation of blue-greens and thus the presence of their toxins is a response to human-induced "cultural" eutrophication. Increasing awareness of the need of proper watershed management is urgently needed among municipalities and stakeholders, especially because chlorination has been shown not to be very effective in removing toxins from the water.

References

1. Reynolds, C.S. *The Ecology of Freshwater Plankton*; Cambridge University Press: Cambridge, 1983.
2. Havens, K.E.; James, R.T.; East, T.L.; Smith, V.H. N:P ratios, light limitation, and cyanobacterial dominance in a subtropical lake impacted by non-point source nutrient pollution. Environ. Pollut. **2003**, *122* (3), 379–390.
3. Shapiro, J. Current beliefs regarding dominance by blue-greens: the case for the importance of CO_2 and pH. Proceedings of the International Association of Theoretical and Applied Limnology, Verh. Int. Verein. Theor. Angew. Lim-nol. **1990**, *24*, 38–54.
4. *Chorus, I.; Bartram, J., Eds.; WHO (World Health Organization) Toxic Cyanobacteria in Water: A Guide to Their Public Health Consequences, Monitoring and Management, 1st Ed. E and F Spon: London, 1999.*
5. Carr, N.G.; Whitton, B.A. The biology of blue-green algae. Bot. Monogr. Vol. 9. University of California Press, Berkeley.
6. Codd, G.A. Cyanobacterial toxins, the perception of water quality, and the prioritisation of eutrophication control. Ecol. Eng. **2000**, *16* (1), 51–60.
7. Carmichael, W.W. Health effects of toxin-producing cyanobacteria: "The cyanoHABs." Hum. Ecol. Risk Assess. **2001**, *7*, 1393–1407.
8. Haider, S.; Naithani, V.; Viswanathan, P.N.; Kakkar, P. Cyanobacterial toxins: a growing environmental concern. Chemosphere **2003**, *52* (1), 1–21.
9. Pouria, S.; de Andrade, A.; Barbosa, J.; Cavalcanti, R.; Barreto, V.; Ward, C. Fatal microcystin intoxication in haemo-dialysis unit in Caruaru, Brazil. Lancet **1998**, *352*, 21–26.
10. Richardson, L.L. Remote sensing of algal bloom dynamics. BioScience **1996**, *44*, 492–501.
11. Epp, G.T. Grazing on filamentous cyanobacteria by *Daphnia pulicaria*. Limnol. Oceanogr. **1996**, *41* (3), 560–567.

12. Philips, E.J.; Monegue, R.L.; Aldridge, F.J. Cyanophages which impact bloom-forming cyanobacteria. J. Aquat. Plant Manage. **1990**, *28,* 92–97.

13. Harding, W.R.; Quick, A.J.R. Management options for shallow hypertrophic lakes, with particular reference to Zeekoevlei, Cape Town. S. Afr. J. Aquat. Sci. **1992**, *18* (1/2), 3–19.

14. Herath, G. Freshwater algal blooms and their control: comparison of the European and Australian experience. J. Environ. Manage. **1997**, *51,* 217–227.

62

Estuaries

Introduction 635
Origins of the Problems of Estuaries 636
Environmental Assessment of Estuarine Ecosystems 637
    Chemical Analyses in Environmental Matrices • Bioassay Methods for
    the Evaluation of Toxic Effects of Contaminants • Estuarine Community
    Structure
New Tools 642
    Ecosystem Functioning or Ecosystem Structure? • Biomarkers
How Can We Solve the Problem? 643
    Environmental Quality Standards and Guidelines • How the Biota Will
    Respond to Restoration • Predictive Risk Assessment of New and Existing
    Chemicals
Conclusions 644
References 645

Claude
Amiard-Triquet

Introduction

In the whole world, estuaries are the seat of the same paradox: they are among the most productive ecosystems while being strongly impacted by anthropogenic activities. Considering the value of the world's ecosystem service and natural capital, estuaries are among the most productive systems.[1] More than 76% of all commercially and recreationally important fish and shellfish species are estuarine dependent.[2] By using the specific search term "estuary" in the Marine Pollution Bulletin database, 553 titles emerged to date (July 2011), highlighting the importance of pollution threats to these ecosystems. This estuarine paradox is well exemplified in Figure 1, showing the mouth of the Loire estuary, one of the big estuaries along the northwestern Atlantic coast. On the north bank, the town of Saint Nazaire (65,000 inhabitants) is highly industrialized, being particularly famous for its shipyards. Between Saint Nazaire and the urban community of Nantes (580,000 inhabitants) 60 km upstream, industrial areas receive large plants, such as a company producing fertilizers, an oil refinery, and an electrothermal plant. Agricultural areas destined to meadows and livestock farming, open fields, and specialized cultures (market gardening, horticulture, wine yards, etc.) are inserted between industrial areas. The south bank is less urbanized, less industrialized, and consequently less artificialized. As shown in the picture, at low tide, large mudflats emerge. They are home to rich invertebrate fauna, including small crustaceans, bivalves, and worms, which are at the bases of food chains leading to different fish species (flounder, eel, sole, sea bass, etc.) of economic interest and to vast populations of wading birds.

Worldwide, there are some 1200 major estuaries (discharges of 10 m³/s), with a total area of approximately 50 million ha.[1] Tidal flats and estuaries have been widely degraded and lost. Sea traffic has required the fitting of dikes, the dredging of channels, and land reclamation to support harbors and industry with drastic ecological consequences such as the reduction of mudflats and also wetlands, which are important for the protection of water quality and are ecosystems of floristic and faunistic interest.

FIGURE 1 The Loire estuary near its mouth showing the contrast between the north bank, urbanized and industrialized, and the south bank, with intertidal mudflats that are crucial for feeding of fish and wading birds.

Origins of the Problems of Estuaries

Drivers of loss and change to wetland ecosystems have been listed in the Millennium Ecosystem Assessment.[1] Influx of nutrients due to river transport is at the basis of the biological wealth of estuarine areas, but on the other hand, the fluxes of reactive (biologically available) nitrogen and high phosphorus loading from anthropogenic sources result in eutrophication.[3] The degradation of primary producers responsible for eutrophication is oxygen consuming as is also the degradation of anthropogenic influx of NH_4 (present in wastewater treatment effluents) into nitrates. Thus, episodes of hypoxia may be observed, and in many estuaries they are also favored by the presence of a natural entrapment zone [estuarine turbidity maximum (ETM)] where the accumulation of suspended organic and mineral matter enhances bacterial activity.[4] In estuarine environments, bacteria are generally attached to particles, and their behavior is related to the dynamics of suspended matter.[5] Bianchi[3] reports that the abundance and production of virus-like particles (viriobenthos) in sediments are considerably higher than those found for water column virus particles (virioplankton).

The processes controlling movement of sediments in estuaries have been described by Bianchi.[3] The location of the ETM is generally considered to be controlled by tidal amplitude, volume of river flow, and channel bathymetry. Due to rapid and high sedimentation rates in the ETM, the accumulation of particles in the benthic boundary layer can result in the formation of mobile and fluid muds, [3] but resuspension episodes frequently occur, for instance, during spring tide and storm events.

Urban and industrial activities are responsible for the input of contaminated effluents and microbes in the aquatic environment, whereas agriculture results in diffuse inputs of pesticides and fertilizers. Contaminants are a source of concern for the growth and reproduction of cultivated species as well as a risk for the health quality of seafood products, particularly for pollutants able to biomagnify in food chains (e.g., methylmercury, polychlorinated biphenyls [PCBs]). In industrial areas, the role of estuaries as waterways leads to dredging activities in order to keep a convenient bathymetry. Because dredged sediments are loaded with a cocktail of many classes of contaminants, dredging activities are a serious source of concern.

The *Eisler's Encyclopedia of Environmentally Hazardous Priority Chemicals*[6] provides 142 instances when using the search terms "estuary" or "estuarine." Bianchi[3] also provides a review of the literature dealing with nutrients and contaminants entering estuaries: trace metals, either essential at low doses (e.g., Cu, Zn) or nonessential and toxic even at low doses (e.g., Hg, Pb); PCBs, very stable so they are still present despite their ban; and polycyclic aromatic hydrocarbons (PAHs). Bianchi[3] also reports that inputs of these multiple stressors can interact to reduce, enhance, and/or mask the individual effects of each. Emerging contaminants of concern in coastal and estuarine environments have been reviewed by Hale and La Guardia.[7] They include brominated flame retardants, natural and synthetic estrogens, alkylphenol ethoxylates, and associated degradation products, pharmaceuticals, and personal care products. The recent advances in nanotechnology and the increasing use of nanomaterials in every sector of society have resulted in uncertainties regarding environmental impacts. Deposit-feeding benthic estuarine organisms may be particularly at risk given the likelihood of nanoparticles to agglomerate, aggregate, and settle.[8]

The public and scientists realized environmental problems of estuaries first because of the disastrous depletion of populations of migratory fish, especially salmon and eels. This blindness is partly explained by the erroneous confidence in the following: 1) the power of dilution of pollutant fluxes in huge volumes of seawater at river mouths; and 2) the ability of estuarine species to cope with chemical stress as they are able to cope with dramatic changes in ecological conditions (salinity, temperature, turbidity, oxygen). On the contrary, organisms living at the limits of their tolerance to natural variations are generally more sensitive to any additional stress.[9,10]

Environmental Assessment of Estuarine Ecosystems

Environmental assessment is classically based on a triad of analysis including the following: 1) chemical analyses in environmental matrices; 2) bioassay methods for the evaluation of toxic effects of contaminants; and 3) biological responses at the level of community structure. This approach was particularly well developed as the Sediment Quality Triad[12] since in the aquatic environment, sediment is the main store for most contaminants entering ecosystems. In estuaries, all assessment techniques need to consider the unique and complex dynamic processes at work, in particular, salinity effects, but also key estuarine parameters such as temperature, pH, dissolved O_2, redox potential, hydrodynamics and sedimentary processes, and their spatiotemporal changes.[13,14] These parameters, indeed, strongly influence estuarine biogeochemical processes, controlling contaminant exchanges between sediment and the water column as well as contaminant bioavailability and toxicity.[13] They also govern the biological features of estuarine communities according to the concept of the estuarine quality paradox.[15,16]

Legislation has been adopted on a worldwide scale to determine the ecological integrity of surface waters including estuaries (United States' Clean Water Act [CWA], 1972; European Community Water Framework Directive, [WFD]).[17] The so-called ecological quality status (EcoQS)[17] must be assessed by comparison with undisturbed conditions, but at the beginning of the third millennium, the formulation of reference values is highly questionable. Anyway, effects at the community level become significant only after severe environmental degradation has already occurred, thus leading to expensive remediation processes. In this context, the methodology of biomarkers provides predictive tools applicable much earlier in any environmental degradation process.[18,19]

Chemical Analyses in Environmental Matrices

Knoery and Claisse[20] have described the experience of the French marine chemical monitoring network over the last 33 years. They highlight the rationale for using different environmental matrices, namely, water, sediment, and biota (suspension feeders such as *Mytilus* sp. and *Crassostrea gigas*, as recommended in NAS[21]), to describe spatiotemporal variations of chemical pollution (PAHs, PCBs, organochlorinated pesticides, metals). Despite being recognized in the WFD, water is not a suitable matrix to answer this aim due to analytical problems at environmentally very low levels of contaminants combined with the low spatial and temporal representativeness of water samples collected at insufficient frequencies (even if recent progress in the field of passive samplers allows the measurement of time-weighted average concentrations; see, for instance, Mills et al.[22]). Using analyses in organisms, which reflect past exposure in the environment, with higher contaminant levels due to bioaccumulation, tremendously increases the reliability of data. This strategy was complemented by monitoring chemical contaminants in sediments for integration over several years. The public can access the data as time series plots through Ifremer's Web site at http://wwz.ifremer.fr/envlit/resultats/surval__1 and as maps at http://wwz.ifremer.fr/var/envlit/storage/documents/parammaps/contaminants-chimiques/index.html. The main estuaries are clearly among the most contaminated sites.

However, the gross concentrations of contaminants in any of the compartments of the environment have poor ecotoxicological significance. A greater knowledge of the biogeochemical cycling in estuaries, which involves the transformation, fate, and transport of chemical substances, is critical in understanding the effects of contaminants.[3] The role of bacteria in biogeochemical cycles is well established, and in the field of pollution studies, a focus has been made on the role of bacteria in the methylation of mercury, a process that is responsible for a higher bioavailability and a higher toxicity of this element for aquatic organisms.[6] More generally, there is clearly a need to put more "bio" into biogeochemical cycles (Ouddane et al. in Ref. [14]).

Bioassay Methods for the Evaluation of Toxic Effects of Contaminants

In their book devoted to *Coastal and Estuarine Risk Assessment,* Newman, Morris, and Hale[7] underlined the predominant freshwater focus for risk assessment and discussed the use of more abundant data for freshwater species to predict consequences to saltwater species. They concluded that improving the current practice effectively will require substantially more data for marine species. In an inventory of marine biotest methods, estuarine species, belonging to different taxa, are listed among relevant biological models.[23] On the copepod *Acartia tonsa*—known as highly tolerant toward salinity and temperature fluctuations—combined stresses from suboptimal temperature and salinity and toxic stress were tested.[24] However, toxicological stress altered the established response pattern of *A. tonsa* toward temperature and salinity, confirming that organisms living in stressful or extreme environments may experience profound interaction effects of anthropogenic and natural stress factors. On the other hand, a review performed by the European Centre for Ecotoxicology and Toxicology of Chemicals[25] suggests a reasonable correlation between the ecotoxicological responses of freshwater and saltwater biota.

Bioassays are used for the determination of predicted no-effects concentrations (PNECs) as described in detail in the Technical Guidance Document on Risk Assessment (TGD) in support of European Commission regulations.[26] The determination of the PNECs takes into account the number of bioassay results available for different species representative of different trophic levels and the availability of long-term toxicity data. However, the notion of long term in the framework of the methodology of bioassays[23] differs from the notion of long term when considering ecological consequences at higher levels of biological organization.

The relevance of the TGD[26] for in situ assessment in estuarine environments was tested by determining local PNECs for atrazine in a highly polluted estuary, the Seine estuary, France. For downstream, middle, and upstream sections of this estuary, key species important for local food chains were selected. Because the number of ecotoxicological data was restricted at the local scale compared with the global scale, a higher security factor needed to be used. In fact, differences between European Union PNECs[27] and local PNECs were limited.[28]

Estuarine Community Structure

Estuarine Quality Paradox

In agreement with Dauvin[15] and Elliott and Quintino[16] describe this paradox as follows: "the dominant estuarine faunal and floral community is adapted to and reflects the high spatial and temporal variability of highly naturally-stressed areas. However, this community has features very similar to those found in anthropogenically-stressed areas, thus making it difficult to detect anthropogenically-induced stress in estuaries. Furthermore, as estuaries are naturally organically rich, the biota thus is similar to anthropogenically-organic rich areas. Because of this, there is a danger that any indices based on these features and used to plan environmental improvements will be flawed." Although numerous bioindicators and indices are used to define the EcoQS of coastal waters, very few of them were developed specifically for environments with a mosaic of conditions and salinity levels. Natural and anthropogenic stressors make estuaries highly heterogeneous environments, which may generate thousands of potential combinations (Figure 2). According to the WFD, the elements of biological quality must be determined by comparison with reference conditions. Because of habitat heterogeneity in estuaries, it has been proposed that reference conditions should be habitat specific in order to reflect natural gradients.[29]

Tolerance and the Monitoring of Communities

This topic has been recently reviewed by Berthet et al.[11] A biological community is composed of different species, the inherent sensitivity of which toward a given toxicant is highly variable. Thus, in a contaminated environment, the most sensitive organisms are lost, whereas tolerant organisms are maintained. The variability of interspecific responses is used in the concept of the pollution-induced community tolerance developed by Blanck, Wängberg, and Molander[31] and revisited by Tlili and Montuelle.[11] Depending on the functional role of different species in the community, different indirect effects of tolerance may be expected. If a sensitive species is a prey or a host species, its extinction will lead to a depletion of its predator or symbiont populations. On the other hand, the extinction of a sensitive species that is a competitor or a predator of a tolerant species will favor the latter.

Determining ecological integrity to answer the needs of legislations (CWA, WFD) typically emphasizes analyses of phytoplankton, macroalgae, angiosperms, benthic macrofauna, and ichthyofauna. In this aim, existing data were revisited, whereas new data were obtained, for instance, in the framework of the WISER Project (Water bodies in Europe: Integrative Systems to assess Ecological status and

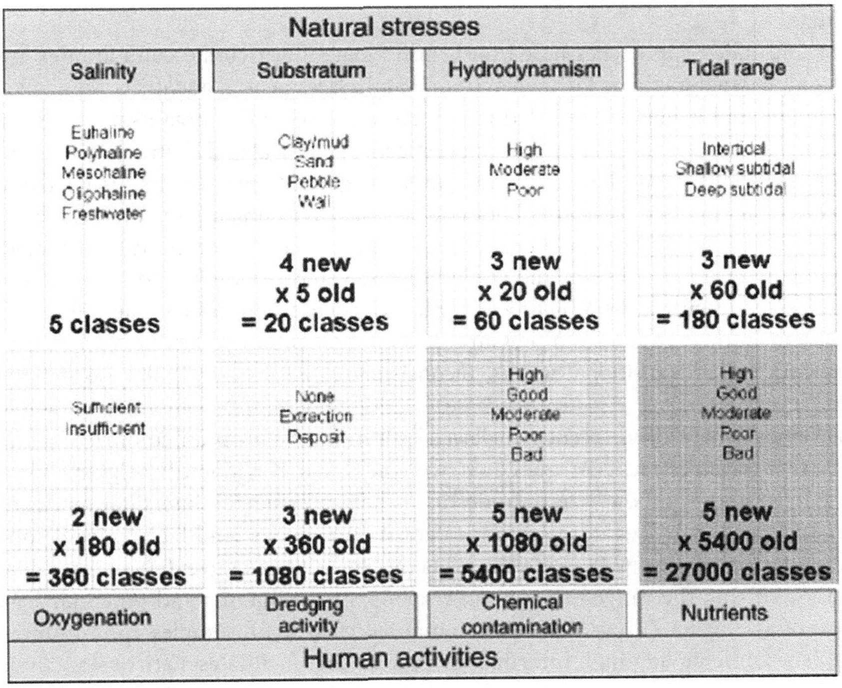

Natural stresses			
Salinity	Substratum	Hydrodynamism	Tidal range
Euhaline Polyhaline Mesohaline Oligohaline Freshwater	Clay/mud Sand Pebble Wall	High Moderate Poor	Intertidal Shallow subtidal Deep subtidal
5 classes	**4 new x 5 old = 20 classes**	**3 new x 20 old = 60 classes**	**3 new x 60 old = 180 classes**
Sufficient Insufficient	None Extraction Deposit	High Good Moderate Poor Bad	High Good Moderate Poor Bad
2 new x 180 old = 360 classes	**3 new x 360 old = 1080 classes**	**5 new x 1080 old = 5400 classes**	**5 new x 5400 old = 27000 classes**
Oxygenation	Dredging activity	Chemical contamination	Nutrients
Human activities			

FIGURE 2 Heterogeneity of estuarine conditions that can potentially result from interactions between natural stressors and anthropogenic activities.
Source: Adapted from Dauvin et al.,[30] with permission.

Recovery, http://www.wiser.eu/programme/). Only a few results had already been published in international journals, but many reports have been prepared, and an inventory was made by Courrat, Foussard, and Lepage[32] for the different categories of biotic communities in the framework of BEEST (for good ecological status of large estuaries), a research project funded by the French ministry of environment, http://seine-aval.crihan.fr/web/pages.jsp?currentNodeId=7).

Concerning phytoplankton, the WISER project aims at developing assemblage phytoplankton metrics, including the potential use of pigment data, taking into account sources of uncertainty on the determination of biomass and community composition due to spatiotemporal heterogeneity. In the ETM region of macrotidal estuaries, high light attenuation results in low primary production, at least as phytoplankton in the water column.[33,34] Thus, it has been proposed to explore the feasibility of using microphytobenthos, an important component in primary estuarine production.[35] The spatiotemporal variability of microphytobenthos on intertidal mudflats is largely unknown because traditional techniques based upon spot sampling techniques do not allow a sound assessment of large-scale distribution. It is only recently that remote sensing techniques have been applied to obtain synoptic information on microphytobenthos distribution in estuarine intertidal zones.[36,37]

Due to phytoplankton paucity in the ETM region, it is interesting to propose another indicator for the water column. In estuaries, the omnivorous copepod *Eurytemora affinis* is especially prevalent in ETM regions worldwide(Forget-Leray et al. in Ref. [14]).[33] Many references quoted in Souissi and Devreker[38] show that the life cycle of certain species is well described and responds to the presence of pollutants or toxins. These authors also highlight that estuarine copepods, restricted to water masses with a well-defined salinity, are not affected by spatiotemporal natural fluctuations as much as benthic species. This limitation of confounding factors favors clear responses to stress. An international inquiry about the potential use of zooplankton as an indicator of estuarine water quality (22 participants from 13 countries in Europe, North America, and Asia) was made in the framework of BEEST.[38] Because of its abundance and key role in the estuarine trophic web, zooplankton was recognized as a good candidate.

Indicators and index approaches based on benthic macroinvertebrate communities are the most consistently emphasized methods for the environmental assessment of aquatic ecosystems. According to Dauvin, Bellan, and Bellan-Santini,[39] this is because macrobenthic organisms are "relatively non-mobile and therefore useful for studying the local effects of physical and chemical perturbations; some of these species are long-lived; their taxonomy and their quantitative sampling is relatively easy; and there is extensive literature on their distribution in specific environments and on the effect of the various stresses that these organisms could encounter." In the framework of the WISER project, 13 single metrics (abundance, number of taxa, and several diversity and sensitivity indices) and 8 of the most common indices used within the WFD for benthic assessment were tested. The different indices are largely consistent in their response to pressure gradient (preliminary classification based on professional judgment), but in transitional waters, inconsistencies between indicator responses were most pronounced.[40]

Fish-based indices have been recently reviewed.[32,41] In most European countries, sound scientific bases exist for regional assessment (Basque country, [42] the United Kingdom, [43] Belgium, [44] France[45,46]).

Amiard-Triquet and Rainbow[14] have underlined that on estuarine intertidal mudflats, bacteria, microalgae, and meiofauna (including foraminiferans, nematodes, and harpacticoid copepods) are relatively abundant compared with macrofaunal taxa. They also tend to exhibit higher diversity than macrofauna, increasing the range of potential responses to pollutants and potential sensitivity for the detection of an impact. Thus, data can be obtained with small samples and minimized impact on the study site while having high information content that facilitates statistical analysis. In addition, because of short generation times and low dispersal characteristics of most of these taxa, rapid population responses to environmental perturbations may be expected. On the other hand, such microscopic taxa often require specialist taxonomic skills, expertise that is of decreasing availability throughout the world.

Question of the Reference Site

All index developers invest a large amount of effort on the formulation of reference values, that is, the quality or conservation value given to pristine, undisturbed, condition, or reference status.[41] At the best, sites as clean as possible may be used as references, but of course, they will not be found among large estuaries, strongly occupied by human populations, but only in small estuaries, less affected by human impacts, but which also differ from large ones by many natural factors.[14] These confounding factors, which cannot be controlled (salinity, granulometry, organic content, hydrodynamics, food availability, etc.), must at least be measured in parallel with ecological and ecotoxicological determinations to mitigate the interpretation of indicators and indices. In this context, the WFD provides a list of hydromorphological, chemical, and physicochemical parameters that must be monitored as sustaining biological parameters.

The notions of biological or ecological integrity and the good chemical and ecological status, which are, respectively, at the basis of the U.S. CWA and the WFD, are still topics of discussion. They are based on an ecocentric point of view including the conservation of biodiversity with reference to a nearly undisturbed situation, in agreement with the concepts of climax and resilience, which have been developed in the 1960s but today are reexamined by scientific ecologists.[47,48] For nonscientists involved in ecology, reaching the good status may be more or less perceived as the recovery of a kind of Garden of Eden, the so-called pristine areas that in fact no longer exist at the present time. Sociological inquiries carried out in France in the framework of the BEEST project showed that for users and inhabitants living in estuarine areas, points of view differ depending on their relation to estuaries: fishing, social and recreational activities, aesthetic relationship, etc.

Concurrently to the ecocentric approach, an anthropocentric approach (Figure 3), based on the conservation of goods and services, may be proposed. It leads to the concept of good ecological potential

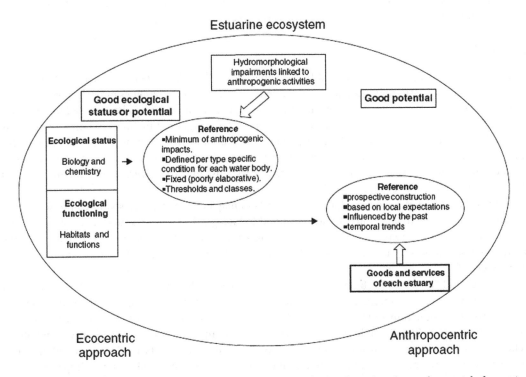

FIGURE 3 From the ecological integrity (CWA) or good ecological status (WFD) to the good potential of estuaries. **Source:** Adapted from Synthèse du projet BEEST Vers une approche multicritère du Bon Etat des grands ESTuaires atlantiques.[49]

rather than good ecological status, allowing a prospective construction including, for instance, expectations and demands of the public or uncertainties like global change. It is also in agreement with the WFD, which mentions limitations that must be taken into account (environmental and social needs, excessive costs, maintaining people's security, higher general interest) and explicitly imposes the information, consultation, and involvement of the public, including users.

Assessing the good potential of an estuary needs tools other than those useful to describe only the situation at a given time (Figure 3). It is indispensable to characterize the key elements of ecological functioning through the following: 1) by considering the concept of functional habitat; 2) by using indicators allowing a better integration of functional aspects of estuarine ecosystems; and 3) by searching out evolutionary trends rather than quality thresholds, which are not really relevant in this highly variable context.[49]

New Tools

Ecosystem Functioning or Ecosystem Structure?

According to Elliott and Quintino,[16] "functional characteristics either as well as or rather than structural ones should be used in detecting environmental perturbations in estuaries." Indicators described above may be used for a better characterization of ecosystem functioning. Phytoplankton and microphytobenthos may serve as proxies for primary production, and in addition, Underwood and Kromkamp[35] underscore that microphytobenthic biofilms may play an important role in biogeochemical cycles (exchanges between sediment deposits and water phase, effects on bacterial processes). Because of its central place in the trophic web, zooplankton is a useful indicator of good ecological functioning rather than an indicator of structure.[38] Fish-based indices are generally constructed by using ecological and trophic guilds, thus allowing the description and explanation of transitional waters' community structure and estuarine functionalities.[46] For instance, taking into account juvenile marine fish can allow the characterization of the role of a given estuary as nursery.

Biomarkers

Despite biomarkers having been defined by Depledge in 1994[50]: "A biochemical, cellular, physiological or behavioral variation that can be measured in tissue or body fluid samples or at the level of whole organisms that provides evidence of exposure to and/or effects of, one or more chemical pollutants (and/ or radiations)," they are still "new" since their use is limited or ignored in environmental regulations (CWA, WFD) even if they are well incorporated in some monitoring procedures, for instance, under the OSPAR Convention. They suffer mainly from three kinds of criticisms: 1) their responsiveness to confounding factors; 2) their insufficient specificity of response toward a given class of chemical; and 3) their lack of ecological relevance.

As for other parameters reviewed above, biomarkers may be responsive to spatiotemporal changes of natural conditions. In addition, they also respond to biological parameters such as the size, sex, and reproductive status of organisms. Thus, before using a sentinel species, it is needed to determine the sources of fluctuations[51,52] to limit them and, for those that cannot be controlled, to determine correction factors.[53]

"Specific" and "nonspecific" biomarkers may complement each other to assess the health status of estuarine ecosystems.[14] Among nonspecific biomarkers, those that are linked directly or indirectly to the success of reproduction and the sustainability of populations have been termed "biomarkers of ecological relevance."[54] They include changes in lysosomal stability, immunotoxicity, energy metabolism, behavior, endocrine disruption, and genotoxicity. When such impairments are demonstrated, it is necessary to identify the main classes of contaminants which may be responsible for such impairments by using a multibiomarker approach based on "more specific biomarkers" (e.g., metallothioneins induced by different metals, delta aminolevulinic acid dehydratase, specific to lead toxicity; imposex of gastropods in the presence of tributyltin). Then, the quantification and characterization (specific fractions, metabolites) of chemicals in water, sediments, and biota may allow the validation of hypotheses based on biomarker studies.

How Can We Solve the Problem?

Preventing and controlling pollution must be based on a combined approach aiming at reducing pollution at source by determining emission limit values and environmental quality standards (EQSs, e.g., WFD). Environmental management aiming at the improvement of chemical and ecological quality in estuaries must be based on robust risk assessments. Retrospective risk assessments are performed when sites have potentially been impacted in the past. When they show a degradation of environmental quality, the restoration of degraded habitats and ecosystems must be addressed. Prospective or predictive risk assessments aim at assessing the future risks of anthropogenic pressure such as climate change or chemical releases into the environment.

Environmental Quality Standards and Guidelines

Establishing EQSs (concentration in water, sediment, or biota that must not be exceeded) is a major tool to protect the aquatic environment and human health. The procedures recommended for their determination[17] are similar to those described for PNECs in the TGD,[26] and very often, EQSs are identical to PNECs. In Europe, EQSs for priority substances and certain other pollutants in surface waters have been recently published.[27] However, for persistent, bioaccumulative, and toxic pollutants, it is not possible to ensure protection against indirect effects and secondary poisoning by EQS for surface water alone. It is therefore appropriate to establish EQS for sediment and biota. The PNECoral is the quality standard for biota tissue with respect to secondary poisoning of top predators as an objective of protection. According to the OSPAR/International Council for the Exploration of the Sea (ICES) experts,[55] the risk to marine predators is calculated as the ratio between the concentration in their food (marine fish) and the no-effect concentration for oral intake (PNECoral-predator) with the help of bioaccumulation models (Abar-nou in Ref. [14]). However, in estuaries, the presence of top predators is limited, and the major risk is for human consumers, which is a reason why oyster culture or fishing is forbidden in certain sites as a consequence of the overrun of quality standards for human health.

At this stage, no EQSs were set in sediments under the WFD partly because the significance of gross concentrations of potentially toxic chemicals in sediments is not easily established, and this is particularly difficult in estuaries where the mobility and bioavailability of contaminants is deeply affected by salinity changes.[13] In the United States, the Washington State Department of Ecology has published sediment management standards.[56] Bioavailability may be examined by using the methods described by Hansen et al.[57] for PAH mixtures and U.S. Environmental Protection Agency (EPA)[58] for metals. When EQSs are not available, sediment quality guidelines (SQGs) are commonly used by official organisms. such as the effects-range low (ERL) and the effects-range median (ERM) values that were derived for the National Oceanic and Atmospheric Administration (NOAA) with numerous modeling, laboratory, and field studies performed in marine and estuarine sediments.[59] Australian sediment quality guidelines were adapted from ERL/ERM.[60] The threshold effect level (TEL) and the probable effect level (PEL) initially described by MacDonald et al.[61] are also used by the NOAA.[62]

Mean SQG quotients can be calculated by dividing chemical concentrations in sediments by their respective SQGs and calculating the mean of quotients for individual chemicals. The resulting index provides a method of accounting for the presence and concentrations of multiple chemicals in sediments, relative to their effects-based guidelines.[56]

Sediment quality guidelines are also of great use in the regulation of dredging activities. This activity is particularly well regulated at the international level, and all these regulations share the same philosophy. For details, the reader can refer to the revised OSPAR guidelines for the management of dredged material.[63]

Environmental monitoring is indispensable to verify if EQSs are honored or if it is necessary to adopt decisions in order that the environment and human health may be adequately protected. Once again, the procedures are comparable in different areas in the world, at least in economically developed countries, and for instance, the Coordinated Environment Monitoring Programme (CEMP) provides a monitoring manual Web page that is regularly updated to take into account scientific and technical advances provided by the OSPAR commission in collaboration with the ICES (http://www.ospar.org/content/content.asp?menu=00900301400135_000000_000000).

How the Biota Will Respond to Restoration

From a large number of examples examined by Borja et al.,[64] including the recovery from a list of different stressors, it appears that only in a few cases, recovery can take less than 5 years, but more frequently, a minimum of 15–25 years is needed to attain the original biotic composition, and diversity may lag far beyond that period. Similarly, Hering et al.[65] conclude that it cannot be expected that European aquatic ecosystems will fully recover within 15 or even 30 years from over a century of degradation. Considering such situations, the common saying "an ounce of prevention is worth a pound of cure" makes sense.

Predictive Risk Assessment of New and Existing Chemicals

Predictive risk assessment aims at assessing the future risks from chemical releases into the environment. In the United States, the Federal Toxic Substances Control Act gives the EPA authority to regulate, and even ban, the manufacture, use, and distribution of both new and existing chemicals. In Europe, a significant improvement has occurred recently with a new chemical policy, REACH (Registration, Evaluation, and Authorization of Chemicals).[66] The pros and cons of REACH have been analyzed, with some details in Hansen[24] and Verdonck et al.[67]

Risk assessment under REACH will be carried out mainly by applying existing methodologies.[26] The risk quotient approach needs the determination of predicted environmental concentrations (PECs) and PNECs. Very simplistically, if the PEC/PNEC ratio is greater than 1, the substance is considered to be of concern, and risk reduction measures must be envisaged. According to Verdonck et al.,[67] uncertainty is hidden and concealed in risk quotient numbers that appear to be certain and, therefore, create a false sense of certainty and protectiveness. These authors propose strategies to improve uncertainty analysis in risk assessment of chemicals. Management obstacles have been summarized by Hansen,[24] who considers that "there are few areas of applied science where the connection between science and policy is as difficult and complicated as in the area of risk management of chemicals." She underlines the lack of communication of uncertainties between assessors and decision makers. Even in the cases where there is an important set of data on a given chemical, there may be little consensus regarding what the actual risk is. Quoting Chapman,[68] Hansen[24] bewails that the effect of all this uncertainty is generally the continued production and use of possible harmful chemicals. Even with "old" contaminants very well studied, the complexity of field situations often leads to a lack of decision. This is well illustrated by Hameedi et al.:[62] despite substantive scientific evidence that points to copper as a contaminant of concern in the St. Lucie Estuary, it is debatable as to how much the observed levels of copper in the estuary are adversely affecting the flow of products (such as fish) or services (such as recreation) of the estuary. This approach poses the problem of what is environmental risk for different actors, as previously mentioned in the section "The Question of the Reference Site."

Conclusions

In aquatic environments, an important set of methods and techniques is available to provide decision makers with tools allowing the prevention, control, and remediation of pollution inflows and impacts. However, because estuaries are highly dynamic and complex systems, the application of these methods is particularly complicated, generating a high degree of uncertainty, whatever the category of assessment tools considered among the classical triad of analysis (chemistry, bioassays, community structure) as well as other assessment strategies such as biomarkers. Many authors have underlined that only a few toxicity data are available for estuarine species compared with marine and, moreover, freshwater species. Thus, toxic impacts on estuarine biota are often derived from toxicological parameters determined in non-estuarine species.

Among tolerant organisms, some are keystone species, with important roles in ecosystem functioning, on which numerous species will depend. Depending on the harshness of natural conditions in different ecosystems, the number of species able to fulfill the same functional role is strikingly variable. As estuarine species are much less numerous than freshwater or marine species, it is evident that the extinction of even a small number of species is sufficient to affect ecosystem functioning.

The state of knowledge would allow considerable improvement of risk assessment and monitoring. The integration of scientific advances in regulations and management needs decade(s). Some of the reasons for that are the higher technicity and the higher cost of certain new tools. Is the society ready to pay for better conservation of the environment? Furthermore, for what kind of conservation: recovery of the Garden of Eden or sustainability of estuaries for the benefits of nature to human beings?

References

1. MEA (Millennium Ecosystem Assessment). *Ecosystems and Human Well-Being: Synthesis.* Island Press, World Resources Institute: Washington, DC, 2005.
2. Scott, G.; Klaine, S.J. Marine and estuarine toxicology and chemistry. Environ. Toxicol. Chem. **2001**, *20* (1), 1–2.
3. Bianchi, T.S. *Biogeochemistry of Estuaries.* Oxford University Press: Oxford, U.K., 2007.
4. Uncles, R.J.; Joint, I.; Stephens, J.A. Transport and retention of suspended particulate matter and bacteria in the Humber–Ouse estuary, United Kingdom, and their relationship to hypoxia and anoxia. Estuaries **1998**, *21* (4A), 597–612.
5. Berthe, T.; Touron, A.; Leloup, J.; Deloffre, J.; Petit, F. Faecal-indicator bacteria and sedimentary processes in estuarine mudflats (Seine-France). Mar. Pollut. Bull. **2008**, *57* (1–5), 59–67.
6. Eisler, R. Eisler's Encyclopedia of Environmentally Hazardous Priority Chemicals; Elsevier: Amsterdam, The Netherlands, 2007.
7. Newman, M.C.; Morris, H.R., Jr.; Hale R.C. *Coastal and Estuarine Risk Assessment*; Lewis Publishers: Boca Raton, FL, USA, 2002.
8. Klaine, S. J.; Alvarez, P. J. J.; Batley, G. E.; Fernandes, T. F.; Handy, R. D.; Lyon, D. Y.; Mahendra, S.; McLaughlin, M. J.; Lead, J. R. Nanomaterials in the environment: Behavior, fate, bioavailability, and effects. Environ. Toxicol. Chem. **2008**, *27* (9), 1825–1851.
9. Hummel, H.; Bogaards, R.; Bek, T.; Polishchuk, L.; Amiard-Triquet, C.; Bachelet, G.; Desprez, M.; Strelkov, P.; Sukhotin, A.; Naumov, A.; Dahle, S.; Denisenko, S.; Gantsevich, M.; Sokolov, K.; De Wolf, L. Sensitivity to stress in the bivalve *Macoma balthica* from the most northern (Arctic) to the most southern (French) populations: Low sensitivity in Arctic populations because of genetic adaptations? Hy0drobiologia **1997**, *355* (1–3), 127–138.
10. Heugens, E.H.W.; Hendriks, A.J.; Dekker, T.; van Straalen, N.; Admiraal, W. A review of the effects of multiple stressors on aquatic organisms and analysis of uncertainty factors for use in risk assessment. Crit. Rev. Toxicol. **2001**, *31* (3), 247–284.
11. Amiard-Triquet, C.; Rainbow, P.S.; Romeo, M., Eds. 2011. *Tolerance to Environmental Contaminants*; CRC Press: Boca Raton, FL, 2011.
12. Chapman, P.M. The Sediment Quality Triad approach to determining pollution-induced degradation. Sci. Total Environ. **1990**, *97–98*, 815–825.
13. Chapman, P.M.; Wang, F. Assessing sediment contamination in estuaries. Annual review. Environ. Toxicol. Chem. **2001**, *20* (1), 3–22.

14. Amiard-Triquet, C.; Rainbow, P.S., Eds. *Environmental Assessment of Estuarine Ecosystems. A Case Study;* CRC Press: Boca Raton, FL, 2009.

15. Dauvin, J.C. Paradox of estuarine quality: Benthic indicators and indices in estuarine environments, consensus or debate for the future. Mar. Pollut. Bull. **2007**, *55* (1–6), 271–281.

16. Elliott, M.; Quintino, V. The estuarine quality paradox, environmental homeostasis and the difficulty of detecting anthropogenic stress in naturally stressed areas. Mar. Pollut. Bull. **2007**, *54* (6), 640–645.

17. WFD (2000). Directive 2000/60/EC of the European Parliament and Council of 23 October 2000 establishing a framework for policy in the field of water (JO L 327 of 22 December 2000); European Commission: Brussels.

18. Allan, I.J.; Vrana, B.; Greenwood, R.; Mills, G.A., Roig, B.; Gonzalez, C. A "toolbox" for biological and chemical monitoring requirements for the European Union's Water Framework Directive. Talanta **2006**, *69* (2), 302–322.

19. Hagger, J.A.; Jones, M.B.; Lowe, D.; Leonard, P.; Owen, R.; Galloway, T.S. Application of biomarkers for improving risk assessments of chemicals under the Water Framework Directive: A case study. Mar. Pollut. Bull. **2008**, *56* (6), 1111–1118.

20. Knoery, J.; Claisse, D. Insights from 30 years of experience in running the French marine chemical monitoring network. ICES-CM 2010 / F08.

21. NAS. *The International Mussel Watch;* National Academy of Sciences: Washington, DC, 1980.

22. Mills, G.A.; Greenwood, R.; Allan, I.J.; Lopuchin, E.; Brümmer, J.; Knutsson, J.; Vrana B. Application of passive sampling techniques for monitoring the aquatic environment. In *Analytical Measurements in Aquatic Environments;* Namiesnik, J., Szefer, P., Eds.; CRC Press: Boca Raton, FL, USA, 2010; 41–68.

23. Nendza, M. Inventory of marine biotest methods for the evaluation of dredged material and sediments. Chemo-sphere **2002**, *48* (8), 865–883.

24. Hansen, M.E. Risk Assessment—The Role of Combined Environmental and Toxic Stress. MS Thesis, Roskilde University, 2006.

25. ECETOC. *Risk Assessments in Marine Environments,* Technical report no. 82; European Centre for Ecotoxicology and Toxicology of Chemicals (ECETOC): Brussels, 2001.

26. TGD. Technical Guidance Document on Risk Assessment in support of Commission Directive 93/67/EEC on Risk Assessment for new notified substances, Commission Regulation (EC) n° 1488/94 on Risk Assessment for existing substances and Directive 98/8/EC of the European Parliament and of the Council concerning the placing of biocidal products on the market; European Commission. Joint Research Centre. EUR 20418 EN/2, 2003.

27. EC 2008. Directive 2008/105/EC of the European Parliament and Council of 16 December 2008 on environmental quality standards in the field of water policy, amending and subsequently repealing Council Directives 82/176/EEC, 83/513/EEC, 84/156/EEC, 84/491/EEC, 86/280/EEC and amending Directive 2000/60/EC of the European Parliament and of the Council (JO L 348 of 24 December 2008); European Commission: Brussels.

28. Bocquené, G. Risque environnemental d'origine chimique dans l'estuaire de Seine, 2011. Available at http://seine-aval.crihan.fr/web/attached_file/componentId/kmelia324/attachmentId/26836/lang/fr/name/_Contamination%20de%20l%27estuaire_n5_reduit.pdf

29. De Paz, L.; Patricio, J.; Marques, J.C.; Borja, A.; Laborda, A.J. Ecological status assessment in the lower Ebro estuary (Spain). The challenge of habitat heterogeneity integration: A benthic perspective. Mar. Pollut. Bull. **2008**, *56* (7), 1275–1283.

30. Dauvin, J.C.; Bachelet, G.; Barillé, A.L.; Blanchet, H.; de Montaudouin, X.; Lavesque, N.; Ruellet, T. Benthic indicators and index approaches in the three main estuaries along the French Atlantic coast (Seine, Loire and Gironde). Mar. Ecol. **2009**, *30*, 228–240.

31. Blanck, H.; Wängberg, S.A.; Molander, S. 1988. Pollution-induced community tolerance—A new tool. In *Function Testing of Aquatic Biota for Estimating Hazards of Chemicals;* Cairns, J.J., Pratt, J.R., Eds.; American Society for Testing and Materials: Philadelphia, PA, USA, 1988; 219–230.

32. Courrat, A.; Foussard, V.; Lepage, M. Les indicateurs DCE estuariens. Etat des lieux à l'échelle européenne en décembre 2010, http://seine-aval.crihan.fr/web/attached_file/componentId/kmelia324/attachmentId/24968/lang/fr/name/axel_rapport_synthese_indicateur_mario.pdf.

33. Lloyd, S.S. Zooplankton Ecology in the Chesapeake Bay Estuarine Turbidity Maximum, with Emphasis on the Calanoid Copepod Eurytemora affinis. PhD thesis, University of Maryland, 2006.

34. Scholle, J.; Dau, K. Reference Conditions of Biological Quality Components in Accordance with the EU Water Framework Directive in Coastal and Transitional Waters in Nl, D and DK. Status of Development and Comparison with the Targets of the Trilateral Wadden Sea Plan, Harbasins report; 2007; 65 pp., http://www.harbasins.org/fileadmin/inhoud/pdf/Final_Products/WP1/1.2/WFD_Implementation_Status_in_NL-D-DK_Nov.2007.pdf.

35. Underwood, G.J.C.; Kromkamp, J. Primary production by phytoplankton and microphytobenthos in estuaries. Adv. Ecol. Res. **1999**, *29*, 93–153.

36. Lerouxel, A.; Blandin, E.; Rosa, P.; Launeau, P.; Rincé, Y.; Barillé, L. Cartographie du microphytobenthos de l'estuaire de la Loire par télédétection infra-rouge; Project BEEST, Rapport; 2010, http://seine-aval.crihan.fr/web/attached_file/componentId/kmelia324/attachmentId/24963/lang/fr/name/axe%203_rapport_microphytobenthos.pdf.

37. van der Wal, D.; Wielemaker-van den Dool, A.; Herman P.M.J. Spatial synchrony in intertidal benthic algal biomass in temperate coastal and estuarine ecosystems. Ecosystems **2010**, *13* (2), 338–351.

38. Souissi, S.; Devreker, D. Le zooplankton peut-il être utilisé comme indicateur de la qualité des eaux estuariennes?; Rapport Projet BEEST, 2010, http://seine-aval.crihan.fr/web/attached_file/componentId/kmelia324/attachmentId/24978/lang/fr/name/axe3_rapport_zooplancton.pdf.

39. Dauvin, J.C.; Bellan, G.; Bellan-Santini, D. Benthic indicators: From subjectivity to objectivity—Where is the line? Mar. Pollut. Bull. **2010**, *60* (7), 947–953.

40. Borja, A.; Barbone, E.; Basset, A.; Borgensen, G.; Brkljacic, M.; Elliott, M.; Garmendia, J.M.; Marques, J.C.; Mazik, K.; Muxika, I.; Neto, J.M.; Norling, K.; Rodriguez, J.G.; Rosati, I.; Rygg, B.; Teixeira, H.; Trayanova, A. Response of single benthic metrics and multimetric methods to anthropogenic pressure gradients, in five distinct European coastal and transitional ecosystems. Mar. Pollut. Bull. **2011**, *62* (3), 499–513.

41. Perez-Domínguez, R.; Maci, S.; Courrat, A.; Borja, A.; Neto, J.; Elliott, M. *Review of Fish-Based Indices to Assess Ecological Quality Condition in Transitional Waters*, WISER report; 2010; 31 pp., http://www.wiser.eu/down-load/D4.4-1.pdf.

42. Uriarte, A.; Borja, A. Assessing fish quality status in transitional waters, within the European Water Framework Directive: Setting boundary classes and responding to anthropogenic pressures. Estuar. Coast. Shelf Sci. **2009**, *82* (2), 214–224.

43. Coates, S.; Waugh, A.; Anwar, A.; Robson, M. Efficacy of a multi-metric fish index as an analysis tool for the transitional fish component of the Water Framework Directive. Mar. Pollut. Bull. **2007**, *55* (1–6), 225–240.

44. Breine, J.; Quataert, P.; Stevens, M.; Ollevier, F.; Volckaert, F.A.M.; Van den Bergh, E.; Maes, J. A zone-specific fish-based biotic index as a management tool for the Zee-schelde estuary (Belgium). Mar. Pollut. Bull. **2010**, *60* (7), 1099–1112.

45. Courrat, A.; Lobry, J.; Nicolas, D.; Laffargue, P.; Amara, R.; Lepage, M.; Girardin, M.; Le Pape, O. Anthropogenic disturbance on nursery function of estuarine areas for marine species. Estuar. Coast. Shelf Sci. **2009**, *81* (2), 179–190.

46. Delpech, C.; Courrat, A.; Pasquaud, S.; Lobry, J.; Le Pape, O.; Nicolas, D.; Boët, P.; Girardin, M.; Lepage, M. Development of a fish-based index to assess the ecological quality of transitional waters: The case of French estuaries. Mar. Pollut. Bull. **2010**, *60* (6), 908–918.

47. Lévêque, C.; Mounolou, J.C. *Biodiversity.* Wiley: Chichester, 2004.

48. Hillebrand, H.; Matthiessen, B. Biodiversity in a complex world: Consolidation and progress in functional biodiversity research. Ecol. Lett. **2009**, *12* (12), 1405–1419.

49. Lévêque, C. Synthèse du projet BEEST Vers une approche multicritère du Bon Etat des grands ESTuaires atlantiques; http://seine-aval.crihan.fr/web/pages.jsp?currentNodeId=157, 2011.

50. Depledge, M.H. The rational basis for the use of biomarkers as ecotoxicological tools. In *Non Destructive Biomarkers in Invertebrates;* Fossi, M.C., Leonzio, C., Eds.; Lewis Publishers: Boca Raton, 1994; 261–285.

51. Kalman, J.; Buffet, P.E.; Amiard, J.C.; Denis, F.; Mouneyrac, C.; Amiard-Triquet, C. Assessment of the influence of confounding factors (weight, salinity) on the response of biomarkers in the estuarine polychaete *Nereis diversicolor.* Biomarkers **2010**, *15* (5), 461–469.

52. Fossi Tankoua, O.; Buffet, P.E.; Amiard, J.C.; Amiard-Triquet, C.; Mouneyrac, C.; Berthet; B. Potential influence of confounding factors (size, salinity) on biomarkers in the sentinel species *Scrobicularia plana* used in programmes monitoring estuarine quality. Environ. Sci. Pollut. Res. **2011**, *18*, 1253–1263, doi: 10.1007/s11356-011-0479-3.

53. Mouneyrac, C.; Linot, S.; Amiard, J.C.; Amiard-Triquet, C.; Métais, I.; Durou, C.; Minier, C.; Pellerin, J. Biological indices, energy reserves, steroid hormones and sexual maturity in the infaunal bivalve *Scrobicularia plana* from three sites differing by their level of contamination. Gen. Comp. Endocrinol. **2008**, *157* (2), 133–141.

54. Mouneyrac, C., Amiard-Triquet, C. Biomarkers of ecological relevance. In *Encyclopedia of Aquatic Ecotoxicology*; Blaise, C., Férard, J.F., Eds.; Springer: Berlin, 2012 (in press).

55. OSPAR Commission. Revised OSPAR guidelines for the management of dredged material. Reference 2004-08. Convention for the Protection of the Marine Environment of the North-East Atlantic; 2004; 30 pp., http://www.dredging.org/documents/ceda/downloads/environ-ospar-revised-dredged-material-guidelines.pdf.

56. Long., E.R.; Ingersoll, C.G.; MacDonald, D.D. Calculation and uses of mean sediment quality guideline quotients: A critical review. Environ. Sci. Technol. **2006**, *40* (6), 1726–1736.

57. Hansen, D.J.; Di Toro, D.M.; McGrath, J.A.; Swartz, R.C.; Mount, D.R.; Spehar, R.L.; Burgess, R.M.; Ozretich, R.J.; Bell, H.E.; Reiley, M.C.; Linton, T.K. *Procedures for the Derivation of Equilibrium Partitioning Sediment Benchmarks (ESBs) for the Protection of Benthic Organisms: PAH Mixtures,* EPA/600/R-02/013; Office of Research and Development, U.S. Environmental Protection Agency: Nar-ragansett, RI, 2003.

58. USEPA. Procedures for the Derivation of Equilibrium Partitioning Sediment Benchmarks (ESBS) for the Protection of Benthic Organisms: Metal Mixtures (Cadmium, Copper, Lead, Nickel, Silver and Zinc), EPA/600/R-0211. US Environmental Protection Agency: Washington, DC, 2005.

59. Long, E.R.; MacDonald, D.D.; Smith S.L.; Calder, F.D. Incidence of adverse biological effects within ranges of chemical concentrations in marine and estuarine sediments. Environ. Manage. **1995**, *19* (1), 81–97.

60. McCready, S.; Birch, G.F.; Long, E.R.; Spyrakis, G.; Greely, C.R. An evaluation of Australian sediment quality guidelines. Arch. Environ. Contam. Toxicol. **2006**, *50* (3), 306–315.

61. MacDonald, D.D.; Carr, R.S.; Calder, F.D.; Long, E.R.; Ingersoll, C.G. Development and evaluation of sediment quality guidelines for Florida coastal waters. Ecotoxicology **1996**, *5* (4), 253–278.

62. Hameedi, M.J.; Johnson, W.E.; Kimbrough, K.L.; Browder, J.A. Sediment contamination, toxicity and infaunal community composition in St. Lucie estuary, Florida based upon measures on the Sediment Quality Triad, 2006. Available at http://www.ccma.nos.noaa.gov/publications/sle_report.pdf.

63. OSPAR/ICES 2004. Workshop on the evaluation and update of background reference concentrations (B/RCs) and ecotoxicological assessment criteria (EACs) and how these assessment tools should be used in assessing contaminants in water, sediment and biota, The Hague, Feb 9–13 2004; Final rep. Moffat, C., Pijnenburg, J., Traas, T., Eds. http://www.ospar.org/documents/dbase/publications/p00214_BRC%20EAC%20Workshop.pdf.

64. Borja, A.; Dauer, D.M.; Elliott, M.; Simenstad, A. Medium-and long-term recovery of estuarine and coastal ecosystems: Patterns, rates and restoration effectiveness. Estuar. Coasts **2010**, *33* (6), 1250–1260.

65. Hering, D.; Borja, A.; Carstensen, J.; Carvalho, L.; Elliott, M.; Feld, C.K.; Heiskanen, A. S.; Johnson, R.K.; Moe, J.; Pont, D.; Solheim, A.L.; van de Bund, W. The European Water Framework Directive at the age of 10: A critical review of the achievements with recommendations for the future. Sci. Total Environ. **2010**, *408* (19), 4007–4019.

66. CEC Commission of the European Communities. Proposal for a regulation of the European Parliament and of the Council concerning the registration, evaluation, authorisation and restrictions of chemicals (REACH) establishing a European Chemical Agency and amending directive 1999/45/EC and regulation (EC) (on persistent organic pollutants). COM 644; 2003.

67. Verdonck, F.A.M.; Souren, A.; Van Asselt, M.B.A.; Van Sprang, P.A.; Vanrolleghem, P.A. Improving uncertainty analysis in European Union risk assessment of chemicals. Integr. Environ. Assess. Manage. **2007**, *3* (3), 333–343.

68. Chapman, A. Regulating chemicals—From risks to riskiness. Risk Anal. **2006**, *26* (3), 603–616.

63

Everglades

Kenneth L.
Campbell, Rafael
Munoz-Carpena,
and Gregory Kiker

Introduction 651
Everglades Water Management—Past, Present, and Future 651
    History • Environmental Issues • Restoration
Conclusion 653
References 653

Introduction

The Everglades of south Florida was originally a broad, shallow "River of Grass"[1] that extended from the south shore of Lake Okeechobee to Florida Bay at the southern tip of the state, east to the Coastal Ridge, and west to the Immokalee Ridge. Historically, the area was a vast sawgrass marsh, dotted with tree islands and interspersed with wet prairies and sloughs covering an area about 40 mi wide by 100 mi long. One of the unique regions of the world, it has steadily decreased in size and declined in health during the past century. Half its wetland area has been lost to agriculture and urban development and the remaining segments are impacted by lack of a clean, dependable water supply. Natural water flows have been diverted for irrigation, drinking water, and flood protection. The conveyance system of canals, levees, structures, and pumps developed for flood control has altered natural patterns of water flow and storage, adversely affecting food webs that supported a diverse ecosystem. Nutrient runoff from urban and agricultural sources is transported by the conveyance system to the remaining natural wetland areas, causing undesirable changes in flora and fauna. Hydroperiod changes have altered natural fire patterns and stimulated invasion of exotic species. A multi-agency state and federal task force has developed a Comprehensive Everglades Restoration Plan (CERP)[2] to address and reverse these major changes to this unique wetland ecosystem. The major hydrologic modifications to be addressed in the Everglades restoration include: 1) regain lost storage capacity; 2) restore more natural hydropatterns; 3) improve timing and quantities of fresh water deliveries to estuaries; and 4) restore water quality conditions. The Comprehensive Plan, considered the world's largest such project, includes more than 60 components proposed for implementation over a period of four decades with an estimated investment approaching $8 billion. State and federal legislation provides for a 50/50 cost share between the federal and state governments to implement the plan.

Everglades Water Management—Past, Present, and Future

History

Primitive canals were dug in portions of the Everglades as early as the late 1800s in attempts to reclaim fertile swampland for agriculture.[3] Early promoters and developers led people to believe that a productive subtropical agriculture was possible in the entire Everglades region. These early attempts at land reclamation were largely unsuccessful until the 1920s when a period of less than normal rainfall helped dry the region around Lake Okeechobee for farming. Following severe hurricane damage in the region in the late 1920s and again in 1947, the focus was shifted from land reclamation to flood protection and the Central and Southern Florida Flood Control Project was authorized and implemented beginning in 1948. Over the next 15 years, this project resulted in a perimeter dike around Lake Okeechobee and the extensive conveyance system of canals, levees, structures, and pumps currently in place. It also allowed development of the Everglades Agricultural Area (EAA), a highly productive, 700,000-acre region of organic soils in the northern Everglades used primarily for sugar cane and winter vegetable production.[4]

Environmental Issues

By the mid-1960s, concerns were already growing about conservation issues and adverse environmental impacts. Additional areas along the eastern border of the Everglades have since experienced urban encroachment. A total of about 1 million acres, roughly 50% of the Everglades wetlands, have been transformed for human uses during the past half-century. The 1700 mi of canals and levees in the region have interrupted connections between the central Everglades and the adjacent wetlands, resulting in over-drainage in some areas and excessive flooding in others. This system provides water supply, flood protection, water management, and other benefits to south Florida, but it must be modified to reduce the negative impacts on the environment. The current canal system works very effectively, discharging an average of 1.7 billion gal of water per day to the ocean and gulf. This discharge must be reduced if future urban, agricultural, and environmental demands for water are to be met.

Today's remaining Everglades have been significantly affected by the current water management system. Wading birds and other wildlife populations are greatly decreased. Tree islands, with their unique combination of wetland and terrestrial vegetation and wildlife, are considered to be an excellent indicator of the overall health of the Everglades. Many of these tree islands have disappeared from the northern Everglades over the past 50 years, and many others have been taken over by exotic vegetation. These effects are mainly due to changes in the quantity, quality, timing, and distribution of water that have occurred over the years as a result of changed water management. Water depth, duration, and timing are important to both wildlife and vegetation. The sawgrass wetlands of the Everglades developed under very low nutrient conditions with rainfall as the main source of phosphorus. Nutrient inflows, especially phosphorus, as a result of development and modified water management have influenced changes in vegetation type.[5] Where phosphorus concentrations have increased, sawgrass and spike rush have been replaced by cattail causing undesirable changes in the ecosystem. Native vegetation remains healthy where phosphorus concentrations are low.

Restoration

Restoration of the remaining Everglades depends upon a knowledge and understanding of the original conditions. Efforts are focusing on improving upstream water quality and the distribution, timing, depth, and flow of surface water into and through the Everglades. Early historical information sources, combined with further interpretation and analysis, are being used to estimate original drainage patterns and soil, topographic and vegetation conditions before canal drainage began in the late 1800s. Results of these studies indicate that the predrainage landscape of the Everglades probably was configured in subtle ridges and sloughs with two major flow pathways: a flow path southeastward to the Atlantic Ocean, and a southwestward flow path along Shark Slough to the Gulf of Mexico.[6] These flow patterns may have influenced the ridge and slough landscape configuration that is important to the health of the ecosystem. Redevelopment of these flow patterns and landscape configuration will be important to the restoration process. About 70% less water flows through the Everglades today compared to the historic Everglades system.

The main goal of Everglades restoration is to deliver the correct amount of water, with the correct quality, to the correct locations, and at the correct time.[7] Most of the water currently lost to the ocean or gulf will be stored in surface and subsurface storage areas until needed, when 80% of it will be allocated to the environment and 20% to increase urban and agricultural water supplies. Water to be stored for future use will be routed through surface storage reservoirs and wetland-based stormwater treatment areas to improve its quality. Additional water quality improvements can be expected from comprehensive integrated water quality planning efforts currently in progress. To restore water flow paths, more than 240 mi of canals and levees will be removed in the Everglades. This will allow more natural overland water flow in the remaining natural areas of the Everglades. Water held and released will be managed to match natural discharge patterns more closely. Operational plans will be developed in some areas to simulate natural rainfall patterns with water releases to improve the timing of water flowing through the Everglades ecosystem. These strategies are all being designed to enhance not only ecosystem restoration, but also urban and agricultural water supply and flood protection as part of the process of moving toward a more sustainable south Florida.

Conclusion

The Everglades landscape is a unique combination of subtropical wetlands and uplands, including sawgrass marshes, sloughs, wet prairies, tree islands, tropical hardwood hammocks, pinelands, and mangroves. It provides important habitat for many threatened and endangered species. Water management for flood control and water supply purposes has caused some areas to become drier and others to become wetter than normal. More than half of the original wetland area has been lost to agricultural and urban development. The introduction of increased nutrients resulting from this development has caused undesirable shifts in vegetation communities. Hydrologic changes have altered the extent of naturally occurring fires and promoted the growth of exotic species. While the current water management system performs well for flood protection it must be modified to reduce adverse environmental impacts and conserve more fresh water to meet a variety of needs. A Comprehensive Everglades Restoration Plan received initial authorization in 2000 to begin the restoration of the south Florida ecosystem and provide for water-related needs of the region. This plan addresses the quantity, quality, distribution, and timing of water to the Everglades. A large amount of additional information regarding the Everglades is available on the web at http://www.sfwmd.gov/koe_section/2_everglades.html and http://www.evergladesplan.org/.

The following quote from the Comprehensive Everglades Restoration Plan web site[7] conveys the importance of the Everglades and the current restoration program. The significance of the remaining Everglades to the nation and the world has been affirmed time and again. Congress established Everglades National Park. The Everglades have also been designated an International Biosphere Reserve, a World Heritage Site, and a Wetland of International Significance. Identified as one of the world's major ecosystem types, the Everglades are home to 68 threatened or endangered plant and animal species. The benefits and functions of these plants and animals may never be known if we do not restore and protect their habitat. Saving the Everglades requires us to save the entire south Florida ecosystem. The ecological and cultural significance of the Everglades is equal to the Grand Canyon, the Rocky Mountains, or the Mississippi River. As responsible stewards of our natural and cultural resources, we cannot sit idly by and watch any of these disappear. The Everglades deserves the same recognition and support.

References

1. Douglas, M.S. *The Everglades: River of Grass;* Hurricane House, Coconut Grove, FL, 1947.
2. U.S. Army Corps of Engineers, Jacksonville District. *Central and Southern Florida Project Comprehensive Review Study Final Integrated Feasibility Report and Programmatic Environmental Impact Statement,* April 1999; The Corps: Jacksonville, FL, South Florida Water Management District: West Palm Beach, FL, 1999; 10 Vols.; http://www.evergladesplan.org/pub/restudy_eis.shtml (accessed March 2002).

3. Blake, N.M. *Land into Water—Water into Land: A History of Water Management in Florida*; University Presses of Florida: Tallahassee, FL, 1980.

4. Bottcher, A.B.; Izuno, F.T. *Everglades Agricultural Area (EAA): Water, Soil, Crop, and Environmental Management*; University Press of Florida: Gainesville, FL, 1994.

5. Reddy, K.R., O'Connor, G.A., Schelske, C.L., Eds.; *Phosphorus Biogeochemistry in Subtropical Ecosystems*; Lewis Publishers: Boca Raton, FL, 1999.

6. South Florida Water Management District. *Watershed Management: Everglades/Florida Bay*; http://www.sfwmd.gov/org/wrp/wrp_evg/ (accessed March 2002).

7. U.S. Army Corps of Engineers, Jacksonville District, South Florida Water Management District. *Rescuing an Endangered Ecosystem—the Journey to Restore America's Everglades*; http://www.evergladesplan.org/ (accessed March 2002).

64

Water Quality: Range and Pasture Land

Introduction 655
Physical Characteristics 655
    Suspended Sediment • Livestock Impacts on Soil • Livestock Impacts on
    Vegetation
Chemical Characteristics 657
    Dissolved Chemicals • Dissolved Oxygen
Biological Characteristics 657
Conclusions 658
    Limit Runoff and Erosion • Limit Direct Livestock Use of Waterways and
    Sensitive Riparian Areas
References 658

Thomas L. Thurow

Introduction

Livestock and clean water are two products that can be simultaneously obtained from range and pasture lands. This requires that ecological and hydrological principles be applied when crafting a grazing management strategy that is compatible with *predetermined* water quality goals. Making protection of water quality the starting point of land use planning is a philosophical foundation of 1972 U.S. Clean Water Act and subsequent amendments. This goal is operationalized by management agencies establishing total maximum daily load (TMDL) standards for waterways. A TMDL is a calculation of the maximum amount of a pollutant from all contributing point sources (a specific location such as a confined animal feedlot operation [CAFO]) and non-point sources (pollution that occurs over a wide area such as may originate from grazing).

Physical Characteristics

Suspended Sediment

Suspended sediment is the most pervasive non-point source pollutant from grazing lands. All waterways naturally contain some suspended sediment attributable to the geologic (natural) erosion influenced by stream type (primarily determined by the geology, topography, and location within the watershed) and ecological factors (e.g., climate, vegetation, soil). Therefore, formulation of TMDL suspended sediment standards must be catchment specific so that geologic erosion can be differentiated from accelerated erosion associated with human activities such as grazing management.

Grazing management can effect the erosion rate of a site primarily by influencing the degree to which livestock impact the soil and vegetation.

Livestock Impacts on Soil

Soil structure is the arrangement of soil particles and intervening pore spaces. The size of soil particles (aggregation) and their stability when wetted determines the porosity of the soil, which governs the rate at which water will enter the soil (infiltration). If the rainfall rate is greater than the infiltration rate, water will run off the site, carrying sediment with it.

Livestock trampling compacts the soil, increasing the bulk density (i.e., the pore volume is reduced resulting in decreased infiltration rate). The degree of damage associated with trampling at a particular site depends on soil type, soil water content, seasonal climatic conditions, and the intensity of livestock use.[1] Compacted trails form on sites where livestock traffic is concentrated. The density of trails tends to increase as the number of pastures is increased within an intensive rotation grazing system. Another common reason for trail formation is repeated movement to and from limited sources of water, mineral supplements, or shelter. The low infiltration rate of trails results in concentrated runoff, which may eventually create gullies. Roads across hilly range and pasture lands are also a serious erosion source, especially since they are often poorly designed and maintained.[2]

Another way livestock trampling causes surficial problems is by churning dry soil to dust. This is very detrimental to infiltration because the disaggregated soil particles are carried by water and lodge in the remaining soil pores making them smaller or sealing them completely. This "washed in" layer where clay particles clog soil pores is a common way that soil crusts are formed. Soil crusts can reduce infiltration by 90%, thereby dramatically increasing runoff and sediment transport.[3] Trampling a crusted soil does break the crust and incorporates mulch and seeds into the soil. However, this benefit is short lived because the subsequent impact of falling raindrops re-seals the soil surface after several minutes. To effectively address a soil-crusting problem, livestock grazing systems must concentrate on addressing poor aggregate stability, which is the cause of crusting. This requires protecting the soil surface from direct raindrop impact through maintaining vegetation cover and facilitating organic matter buildup in the soil via litter deposition.

Livestock Impacts on Vegetation

Direct raindrop impact on soil represents the greatest potential erosive force on grazing land; therefore it is very important that raindrop energy be dissipated by striking some form of cover before reaching the soil.[4] The amount of cover is positively associated with vegetation litter deposition. Litter slows overland flow, resulting in reduced ability to transport sediment. Litter also aids formation of stable aggregates (associated with high infiltration and low erosion rates) by binding soil particles together with adhesive byproducts produced by decaying litter and microbial synthesis.[5]

Grazing impacts on the vegetation community may be manifest by physical removal of standing vegetation through herbivory or through a gradual change in the composition of vegetation. As grazing pressure increases, the amount of cover and the amount of organic matter returned to the soil is reduced, resulting in an increased likelihood of runoff and erosion. Cover and infiltration rate tends to be greatest under trees and shrubs, followed in decreasing order by bunchgrass, shortgrass, and bare ground.[6] There is little impact on species composition with moderate or light grazing but composition change is great in response to heavy grazing, regardless of grazing strategy.[7] Often the change in species composition associated with heavy grazing is toward dominance by annuals or shortgrass species that have more runoff and erosion associated with them.[8] By the time erosion becomes obvious it may be too late to implement economically viable conservation options. Early recognition of a developing degradation pattern requires knowledge of range ecology, for the first signs of an impending erosion problem almost invariably are manifest by changes in plant density, composition, and vigor.[9]

Chemical Characteristics

Dissolved Chemicals

Nutrient loss from grazing lands via leaching or runoff is normally negligible, i.e., less than the input of nutrients from rainfall.[10] Most of the dissolved chemical constituents in runoff are contributed from the soil. Nutrients and organic matter adsorbed to the soil particles are also lost via erosion. Therefore, the most important role of a grazing system in nutrient loss is manifest through land use activities that alter the volumes or timing of runoff and erosion.[11]

Most of the nitrogen in urine is lost via volatilization, and most of the nitrogen in feces is sequestered by microorganisms or eventually transferred to soil organic matter. Nitrate is very mobile during heavy rain periods but loss by leaching is probably insignificant on most grasslands.[12] Feces contain almost all of the phosphorus excreted by livestock. Phosphorus is very resistant to leaching as it is rapidly precipitated or absorbed by other soil minerals. Nitrogen or phosphorus contamination of waterways is only of imminent concern when livestock are allowed to congregate near waterways.[13] Because of this concern, the U.S. Environmental Protection Agency interpretation of the Clean Water Act has deemed location of feedlots near waterways an unacceptable practice.

Dissolved Oxygen

Dissolved oxygen decreases when organic matter, such as animal manure, is added to water. This decrease occurs because biological decomposition processes consume available oxygen, as does oxidation of other reduced compounds such as ammonium. Excessive additions to surface water of nutrients such as nitrogen or phosphorus lead to eutrophication, often expressed by enhanced growth of aquatic plants and reduced water transparency (especially due to increases in algae). As the aquatic plants decay the microbes consume oxygen, lowering the concentration of oxygen available needed to support higher forms of aquatic life such as macroinvertrebrates and fish.

Biological Characteristics

The primary types of pathogens associated with livestock and wildlife feces are bacteria (e.g., *Campylobacter jejuni, Escherichia coli, Leptospira interrogans, Salmonella* spp.) and water-borne protozoa (e.g., *Cryptosporidia parvum, Giardia duodenalis*). These infectious pathogens can pose potential health risks to human drinking water supplies. Environmental fluctuation in temperature and soil moisture of grazing land creates a harsh environment for bacteria and the oocycsts of protozoa. Fecal coliforms can survive for several months in soil but can survive for up to a year within feces.[14] There is a rapid mortality of most oocysts when feces are deposited on land,[15] however, viable oocysts can be transported overland, especially when fresh feces are washed by an intense storm.[16] Once pathogens reach a water body, the threat of contamination may last from days to months,[17] with freshwater sediments being the site of greatest concentration and survival.[18]

Few detailed studies have explicitly studied the link between livestock grazing and water-borne pathogens. Much of the research has relied upon indicator coliforms that are more easily cultured but have been shown to be poorly correlated with some types of pathogenic bacteria.[14] Furthermore, many wildlife species harbor the same pathogens that livestock do, thus the natural occurrence of pathogens must be considered when analyzing water quality and making the relationship to livestock use of an area. The greatest threat of pathogen contamination of waterbodies occurs when livestock are allowed to concentrate along streams.[19] In situations where risk of bacteriological contamination is unacceptable, it is necessary to restrict livestock access to streams or riparian areas. Livestock use of these sensitive sites can be significantly reduced through development of water supply away from streams.[20]

Conclusions

Two broad objectives must be achieved to protect water quality associated with range and pasture grazing.

Limit Runoff and Erosion

Suspended sediment is the most common pollutant associated with grazing. Best management practices (BMPs) to limit runoff and erosion rely on maintenance of soil structure. Vegetation provides the organic matter necessary to enhance formation of stable aggregates and provides the cover to dissipate the erosive force of direct raindrop impact. Appropriate range and pasture grazing systems are designed to maintain vegetation cover and composition by adjusting intensity, frequency, and season of use. Flexibility needs to be built into grazing systems to adjust for unexpected fluctuation in the climate or market prices. The underdevelopment of climate and market risk management planning and policy regarding grazing plans is perhaps the most formidable threat to progress in improving water quality since these variables continue to be used as an excuse for water quality deterioration and/or the lack of progress in improving it.[21]

Limit Direct Livestock Use of Waterways and Sensitive Riparian Areas

Contamination of waterways by nutrients and pathogens is a predominant concern only on sites that allow livestock to congregate near water. On sites with limited water distribution, livestock tend to stay in the vicinity of water so long as forage is available. This increases the likelihood of excrement being deposited directly into the waterway. It also causes deterioration of the soil structure and plant community near the waterway, resulting in accelerated runoff and erosion. Streambanks and moist soil around springs and streamside meadows are particularly susceptible to erosion damage and compaction. Livestock impacts to streams and riparian sites can be limited by providing water, mineral supplements, and shelter at locations away from natural water sources. Special fencing or livestock herding may also be needed to protect sensitive areas from excessive use at critical times. Another reason for protecting wetland or riparian sites is that they serve as vegetation buffer strips that slow runoff and trap sediment before it reaches a waterway.

References

1. Warren, S.D.; Blackburn, W.H.; Taylor, C.A., Jr. Effects of season and stage of rotation cycle on hydrologic condition of rangeland under intensive rotation grazing. J. Range Manag. **1986**, *39*, 486–491.
2. Sutterlund, D.R. *Wildand Watershed Management*; Ronald Press: New York, 1972.
3. Boyle, M.; Frankenberger, W.T., Jr.; Stolzy, L.H. The influence of organic matter on soil aggregation and water infiltration. J. Prod. Agric. **1989**, *2*, 290–299.
4. Hudson, N. *Soil Conservation*; Cornell University Press: Ithaca, New York, 1981.
5. Thurow, T.L.; Blackburn, W.H.; Taylor, C.A., Jr. Infiltration and interrill erosion responses to selected livestock grazing strategies, Edwards plateau, Texas. J. Range Manag. **1988**, *41*, 296–302.
6. Thurow, T.L.; Blackburn, W.H.; Taylor, C.A., Jr. Hydrologic characteristics of vegetation types as affected by livestock grazing systems, Edwards plateau, Texas. J. Range Manag. **1986**, *39*, 505–509.
7. Ellison, L. Influence of grazing on plant succession of rangelands. Bot. Rev. **1960**, *26*, 1–78.
8. Thurow, T.L.; Blackburn, W.H.; Taylor, C.A., Jr. Some vegetation responses to selected livestock grazing strategies, Edwards plateau, Texas. J. Range Manag. **1988**, *41*, 108–114.
9. Thurow, T.L. Hydrology and erosion. In *Grazing Management: An Ecological Perspective*; Heitschmidt, R.K., Stuth, J.W., Eds.; Timber Press: Portland, Oregon, 1991; 141–159.

10. Menzel, R.G.; Rhoades, E.D.; Olness, A.E.; Smith, S.J. Variability of annual nutrient and sediment discharges in runoff from Oklahoma cropland and rangeland nutrient and bacterial pollution. J. Environ. Qual. **1978**, *7,* 401–406.

11. Robbins, W.D. Impact of unconfined livestock activities on water quality. Trans. Am. Soc. Agric. Eng. **1979**, *22,* 1317–1323.

12. Woodmansee, R.G.; Allis, I.; Mott, J.J. Grassland nitrogen. Ecol. Bull. **1981**, *33,* 443–462.

13. Khaleel, R.; Keddy, D.R.; Overcash, M.R. Transport of potential pollutants in runoff water from land areas receiving animal wastes: a review. Water Resour Res. **1979**, *14,* 421–436.

14. Bohn, C.C.; Buckhouse, J.C. Coliforms as an indicator of water quality in wildland streams. J. Soil Water Conserv. **1985**, *40,* 95–97.

15. Walker, J.J.; Montemagno, C.D.; Jenkins, M.B. Source water assessment and nonpoint sources of acutely toxic contaminants: a review of research related to survival and transport of *Cryptosporidium parvum.* Water Resour. Res. **1998**, *34,* 3383–3392.

16. Tate, K.W.; Atwill, E.R.; George, M.R.; McDougald, N.K.; Larsen, R.E. *Cryptosporidium parvum* transport from cattle fecal deposits on California rangelands. J. Range Manag. **2000**, *53,* 295–299.

17. Stephenson, G.R.; Street, L.V. Bacterial variations from a southwest Idaho rangeland watershed. J. Environ. Qual. **1978**, *7,* 150–157.

18. Burton, G.A.; Gunnison, D.; Lanza, G.R. Survival of pathogenic bacteria in various freshwater sediments. Appl. Environ. Microbiol. **1987**, *53,* 633–638.

19. Larsen, R.E.; Miner, J.R.; Buckhouse, J.C.; Moore, J.A. Water quality benefits of having cattle manure deposited away from streams. Biores. Technol. **1994**, *48,* 113–118.

20. Miner, J.R.; Buckhouse, J.C.; Moore, J.A. Will a water trough reduce the amount of time spent in the stream? Rangelands **1992**, *14,* 35–38.

21. Thurow, T.L.; Taylor, C.A., Jr. Viewpoint: the role of drought in range management. J. Range Manag. **1999**, *52,* 413–419.

65

Water: Drinking

Introduction 661
Inorganic Components and Pollutants of Drinking Water 662
Organic Pollutants of Drinking Water 665
Characteristics of Some Water Bacteria and Pathogens in
    Drinking Water 671
Microbiological Research Methods for Analysis of Drinking Water 672
References 673

Marek Biziuk
and Matgorzata
Michalska

Introduction

Water is a source and a basis of life on earth, and is a substance vital and necessary for humans and their well-being. It is a main part of our daily diet, transporting many necessary macro- and microelements in the human body. Excluding fat, water composes approximately 70% of the human body by mass. We drink 2 to 3 L of water every day; thus, the quality of drinking water is important. Humans can survive for several weeks without food, but for only a few days without water. Additionally, clean water is becoming a much desirable and expensive product. Desertification of large areas of the world, increasing industrialization, and intensification of agriculture result in a large input of toxic organic and inorganic compounds, significantly reducing the quality of surface waters, which are a source of drinking water for a large part of the world population. Other dangers include biological pollution by bacteria and viruses, human fecal pathogens, and parasites, causing waterborne diseases. Over large parts of the world, humans have inadequate access to potable water and use sources contaminated with unacceptable levels of dissolved chemicals or bacteria and viruses. Thus, water, especially drinking water, can create a serious hazard to human health and life, as well as to the life and health of flora and fauna. Sources of drinking water include spring water and groundwater; surface water from rivers, streams, lakes, and glaciers; water from precipitation (rain, snow); and seawater after desalination. All industrialized countries have 100% access to safe drinking water (data from the United Nations Children's Fund from 2000); however, some countries, especially those from Africa (Kenya, Uganda), have limited access.[1] The population percentages from chosen countries with limited access to safe drinking water are listed in Table 1. It was estimated that in 2006, about 1.1 billion people lacked proper drinking water. Waterborne diseases were estimated to cause 1.8 million deaths each year.[2]

Unfortunately, quality water from groundwater supply, which is practically non-renewable, is diminishing. In addition, anthropogenic groundwater pollution has become a fact of life. The majority of anthropogenic water pollutants are toxic not only to humans but also to animals and plants. These pollutants should be removed or destroyed during water treatment, and determined and monitored in surface and tap water. Water pollutants can be divided into physical, biological, radioactive, inorganic, and organic.

TABLE 1   Percentage of Population with Access to Safe Drinking Water (2000)

Country	%	Country	%	Country	%
Algeria	89	Indonesia	78	South Africa	86
Azerbaijan	78	Iran	92	South Korea	92
Brazil	87	Iraq	85	Sudan	67
Chile	93	Kenya	57	Syria	80
China	75	Mexico	88	Turkey	82
Cuba	91	Morocco	80	Uganda	52
Egypt	97	Peru	80	Venezuela	83
India	84	Philippines	86	Zimbabwe	83

Source: United Nations Children's Fund (UNICEF).[1]

For many millennia, the basic criteria of water suitability for drinkable purposes was organoleptic analysis—checking the taste, smell, and appearance of water. These criteria are included, even now, in guidelines for drinking water quality; however, they had to be dramatically verified when in 1854 it was established in London that water from the well at Brad Street, very popular because of its taste, smell, and appearance, was the source of a cholera epidemic.[3] Threshold limit values for bacteriological contamination of water (the so-called coliform count) were the first standard introduced for controlling drinking water quality (in 1914 in the United States). The presence of coliform bacteria is the most convenient for a class of harmful fecal pathogens. This standard is still valid today. Microbial pathogenic parameters are very important because of their immediate health risk. The following were included in guidelines for drinking water quality: metals and anions since 1925, organic compounds since 1942, and radioactive substances since 1962. The Public Health Service and later the Environmental Protection Agency (EPA) have prepared such guidelines for the United States.[4,5] The World Health Organization (WHO)[6] worldwide and the European Community (EC)[7,8] in Europe also elaborate and publish such guidelines. On these guidelines, each country defines its maximum admissible concentrations (MACs) of contaminants in drinking water. Each member state is responsible for establishing the required policing measures to ensure that the legislation is implemented.[9]

Drinking water should fulfill the following requirements:

- It should be clear, without smell, colorless, and refreshing in taste
- It should not contain pathogenic bacteria; animal parasites and their larvae or eggs; toxic compounds; and excessive quantities of calcium, magnesium, iron, or manganese compounds
- It should be easily accessible, be always of good quality, and occur in appropriate amounts
- It should be permanently protected from contamination.
- It should contain substances that are necessary for human life in the proper amounts

Inorganic Components and Pollutants of Drinking Water

The problem of the presence of mineral components and anions in drinking water is very complicated because some elements are necessary for human life.[10–12] They play a significant role in physiological processes and metabolic functions. We can divide these into two groups:

- Macroelements, such as carbon, hydrogen, oxygen, nitrogen (the main components of organic matter), as well as sulfur, potassium, magnesium, calcium, sodium, phosphorus, and chlorine
- Microelements, such as boron, cobalt, chromium, copper, fluorine, iron, lithium, molybdenum, manganese, nickel, rubidium, selenium, silicon, vanadium, and zinc (which also participate in the metabolic function and are components of enzymes), as well as hormones, vitamins, etc

Table 2 shows some macro- and microelements, their recommended daily intake, and the effects of deficiency in these elements. Some of these elements are toxic at higher concentrations, and very often

TABLE 2 Macro- and Microelements, Their Recommended Daily Intake, and the Effects of Deficiency

Element	Recommended Concentration in Drinking Water in Poland (mg/L)	Recommended Daily Intake (mg)	Effects of Deficiency
Ca	—	900–1200	Osteoporosis, rickets, neurological disorders
Fe	0.2	14–18	Anemia
P	—	700–900	Muscular weakness, bone aches, lack of appetite
Mg	30–125	300–370	Gastrointestinal problems, muscular cramps
Na	200	1000	Gastrointestinal problems, diarrhea, weakness, headache
K	—	2–3.5	Muscular weakness, neurological disorders
Zn	—	13–16	Hair loss, skin lesions, reduced resistance to infection
Cu	2	2–2.5	Anemia, heart arrhythmia, blood vessel rupture
F	1.5	1.5–4	Tooth decay
I	—	0.15–0.16	Underactive thyroid, goiter
Se	0.01	0.06–0.075	Cancer, heart diseases, reduced resistance to microbiological and viral infection, reduced antibody production

the difference between an indispensable and a toxic dose is very small. Insufficient or excessive daily intake of these elements can cause harmful effects (see Tables 2 and 3).[9,11,12] In drinking water, we can find all elements and inorganic substances present in nature and introduced by industry, some not needed by organisms and toxic to humans. The most dangerous are lead, mercury, cadmium, and arsenic (the so-called Big Four), but the following are hazardous as well: aluminum, antimony, asbestos, barium, boron, chromium, cobalt, copper, cyanides, manganese, nickel, silver, tin, vanadium, and zinc.[12–49] The harmfulness of metals is mainly due to their persistence in the environment and biomagnification in the food chain, as well as their tendency to accumulate in selected tissues. Metals can be transported with water and air, and thus can endanger humans also through the water they drink. In Table 3, the toxicity of the most dangerous elements, their daily intake, and their typical and recommended concentrations in drinking water are listed.[4–9,23–34]

Underground and even mineral water, used as a source of drinking water, are not as clean as we would expect. Very often, overly high natural concentrations of manganium, iron, fluorides, ammonia, and hydrogen sulfide occur, and these waters should be treated before consumption. Volatile compounds such as ammonium and hydrogen sulfide can be removed by aeration, whereas manganium and iron can be removed by precipitation in the form of hydroxides and filtration or biotechnological methods. Fluorides present the biggest problem.[50] The best solution is to mix high-fluoride water with water containing no or very small concentrations of fluorides. An additional problem concerns pipes made from asbestos-cement materials or lead, used in the past in water distribution systems. Lead can be diluted in water, particularly when water is treated by chlorination or ozonization. Asbestos can contaminate drinking water as a result of corrosion, and thereby accumulate in lung tissues, resulting in symptoms that can develop many years after exposure. Asbestos fibers in water are usually shorter than 10 pm, and can cause cancer of the gastrointestinal and urinary systems. When they are transferred from water to air, after water spillage and drying, asbestos fibers can cause lung and throat cancer upon inhalation. Such an episode was described by Webber,[51] when in Woodstock, New York, in water transported by eroded asbestos-cement pipes in the distribution system; 10,000 asbestos fibers per liter was detected in some samples and up to 0.12 fiber per milliliter was detected in household air.

All these elements and substances and their recommended MACs are listed in the WHO and EC Guidelines for Drinking Water Quality, or in the EPA and national MACs,[4–9] and have to be determined

TABLE 3 Toxicity of the Most Dangerous Elements, Their Daily Intake, and Typical and Recommended Concentrations in Drinking Water

Element	Concentration in Drinking Water (mg/L)				Daily Intake (mg)	Toxicity
	Typical	Recommended MAC				
		EC	WHO	Poland		
Al	0.02–0.1		0.1–0.2	0.2	20–45	Senile dementia of the Alzheimer's type
As	<0.01	0.01	0.01	0.01	0.01–0.02	Respiratory and gastrointestinal tract, skin, and central nervous system effects leading to coma and death, peripheral artery disease, cardiovascular disease, diabetes mellitus, various cancers, neurological effects; the inorganic form of arsenic (As^{3+}) is the most toxic, whereas elemental as well as organic arsenic compounds are considered to be virtually non-toxic
B	0.1	1	0.5	1	1.3–4.5	Dry skin and gastric disorders
Cd	0.002	0.005	0.003	0.003	0.03–0.04	Proteinuria, renal dysfunction, damage to the blood vessel system, lung cancer, low reproduction functions, emphysema, hypertension
Cr	0.002	0.05	0.05	0.05	0.04–0.08	Dermatitis, skin ulcers, respiratory and gastrointestinal problems, liver and kidney damage
Co	0.005	—	—	—	0.04–0.05	Gastrointestinal problems, heart dilation, secondary thrombosis, neurotoxicological symptoms
Cu	<0.1	2		2.0	2–4	Congenital disorder causing accumulation of copper in the liver, brain, and kidney resulting in hemolytic anemia and neurological abnormalities
Fe	0.001–0.5	—	—	0.2	12–16	Gastrointestinal problems, fibrosis, heart disease, abnormal glucose metabolism
Pb	0.002–0.01	0.01	0.01	0.025	0.1–0.5	Carcinogenic, anemia, hypertension, renal dysfunction, insomnia, weakness
Mn	0.02–0.05		0.4	0.05	3.7	Apathy, anorexia, insomnia, speech disturbance, central nervous system effects
Hg	0.0002–0.0003	0.001	0.006	0.001	0.6	Organic forms of mercury (methyl- and dimethylmercury)— the most toxic of the discussed metals; central nervous damage; DNA deformation; immunological impairment; mutagenic, carcinogenic, and teratogenic effects; coronary disease; brain damage
F	0.5	1.5	1.5	1.5	0.5–2	Osteoporosis, fluorosis, central nervous system effects, carcinogenic (osteosarcoma)
Ni	0.001–0.002	0.02	0.07	0.02	0.1–0.9	Dermatoses, potential carcinogenity
Se	<0.01	0.001	0.01	0.01	0.15	Atrophy and decompensation of the heart, renal glomerulonephritis, liver cirrhosis, anemia
V	4–222	—	—	—	2	Irritation of skin and eyes, discoloration of the tongue and oral mucosa

in drinking water. In Tables 2 and 3, recommended MACs for some metals from the WHO Guidelines for Drinking Water Quality and Polish standards are listed as examples. Excessive concentrations of pollutants, above maximum contaminant levels, not only warn consumers but also signal the existence of uncontrolled discharge of wastes, improperly operating treatment plant, lack of enforcement of legislation dealing with water management, or other violations of environmental laws. To be able to monitor environmental pollutants properly and effectively, analysts need a variety of methods for the isolation and determination of these contaminants, taking into account both physical-chemical properties of individual compounds and groups of compounds, as well as the characteristics of the matrices in which these compounds occur.

Consequently, there is a need for continuous monitoring of the degree of pollution of potable and surface waters by all these elements and substances. The following have been used mainly for final analysis: flame atomic emission spectrometry; flame atomic absorption spectrometry; electrothermal atomic absorption spectrometry; inductively coupled plasma (ICP) coupled with atomic emission, atomic absorption, or mass spectrometry (MS); and electrochemical methods (anodic stripping voltammetry).[12,24,52–63]

Organic Pollutants of Drinking Water

The majority of organic compounds in water are of natural origin. They represent a large and diversified group of mostly unidentified compounds such as humic and fulvic acids, tannins, peptides, amino acids, etc. Their total concentration varies from 0 to 1 mg/L in groundwater, from 1 to 5 mg/L in most surface waters, and from 20 to 25 mg/L in some waters with the highest concentration of organic matter.[20,52,53,63–67] These compounds are predominantly non-toxic, but they can be precursors of toxic compounds in the process of water treatment. The main hazard to life and health of humans and flora and fauna comes from anthropogenic organic compounds. The number of known organic compounds is now estimated to be about 16 million,[3] 2 million of which are produced by synthesis alone. Every year, approximately 250,000 new compounds are synthesized, about 1000 of which are manufactured on an industrial scale.[68] Presently, ca. 70,000 organic compounds are commercially available with an annual global production of 100–200 million tons. Approximately one- third of all organic compounds produced end up in the environment, including water. More than 700 chemical compounds, including more than 600 organic compounds,[64,69] many of which are biologically active, have been detected in some water supply samples. Using gas chromatography (GC)-MS, Coleman et al.[69] identified approximately 460 organic compounds in Cincinnati drinking water.

Particularly important pollutants among organic compounds are volatile organohalogen compounds and pesticides as a result of their common use, persistence in the environment, and toxicity. They are primarily anthropogenic. Volatile organohalogen compounds are used mainly as solvents, cleaning and degreasing agents, blowing agents, polymerization modifiers, and heat-exchange fluids. It is estimated that the annual global production of organohalogen solvents alone amounts to several million tons.[64] The most important source of organohalogen compounds in drinking water, particularly volatile ones, is water disinfection by chlorination[14,63–65,70–75] for killing pathogenic bacteria. It is a paradox that during this process, harmless and naturally occurring humic and fulvic compounds in water (the so-called precursors) are converted into organohalogen compounds, which are dangerous to human life and health (see Figure 1). The largest group of compounds formed during chlorination comprises trihalomethanes (THMs), i.e., trichloromethane (chloroform, the most abundant compound), bromodichloromethane, dibromochloromethane, and tribromomethane. Organobromine compounds are formed when the water being chlorinated contains a large amount of bromides or when the chlorine used for disinfection is contaminated with bromine. Hypobromous acid formed in the reaction of bromide ions with hypochlorous acid reacts with an organic matrix about 200 times faster than hypochlorous acid does.[76] The amount and kind of organohalogen compounds formed depend on the water pH, the amount of chlorine used, and the content of organic matrix (total organic carbon, TOC) in chlorinated water.

Apart from THMs, other organohalogen compounds are formed in the course of chlorination, in smaller quantities. Volatile organohalogen compounds such as tetrachloromethane, chloroethylene, 1,1-dichloroethylene, 1,1,2-tri-chloroethylene, tetrachloroethylene, 1,1,1-trichloroethane, and 1,2-dichloroethane are also commonly found in chlorinated water. In addition, more than 100 other derivatives after chlorination of humic substances have already been identified, including chlorinated

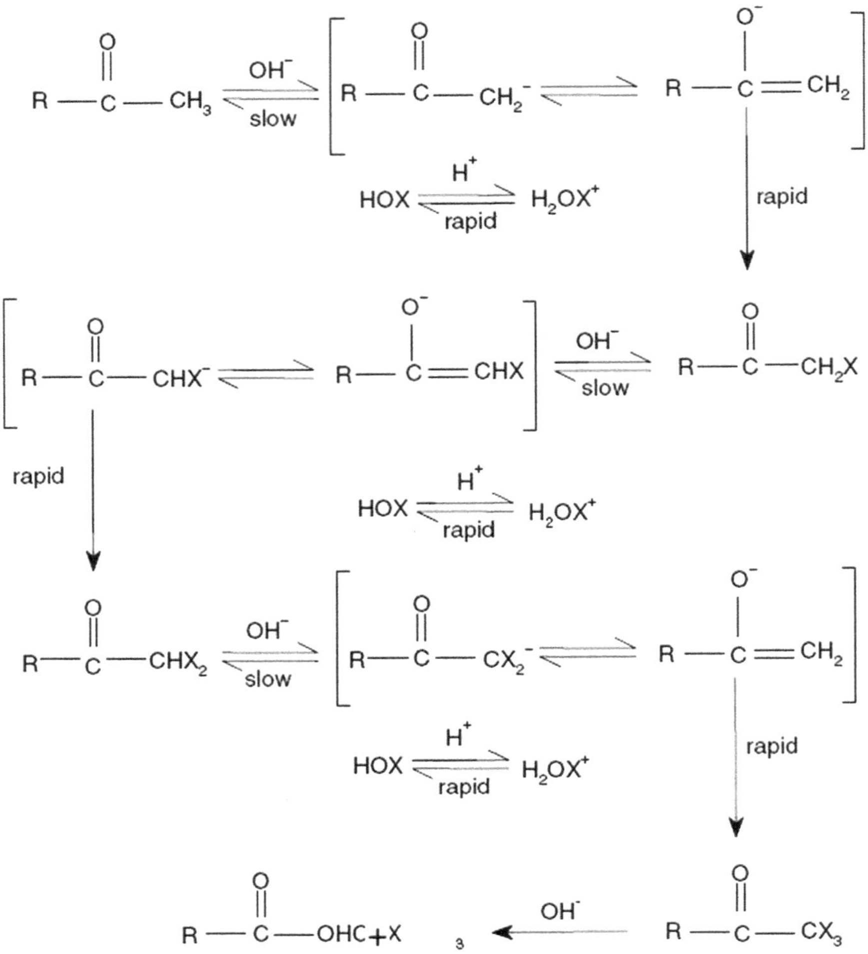

$$R-\overset{O}{\overset{\|}{C}}-CH_3 \underset{slow}{\overset{OH^-}{\rightleftharpoons}} \left[R-\overset{O}{\overset{\|}{C}}-CH_2^- \rightleftharpoons R-\overset{O^-}{\overset{|}{C}}=CH_2 \right] \xrightarrow{rapid}$$

$$HOX \underset{rapid}{\overset{H^+}{\rightleftharpoons}} H_2OX^+$$

$$\left[R-\overset{O}{\overset{\|}{C}}-CHX^- \rightleftharpoons R-\overset{O^-}{\overset{|}{C}}=CHX \right] \underset{slow}{\overset{OH^-}{\rightarrow}} R-\overset{O}{\overset{\|}{C}}-CH_2X$$

$$\xrightarrow{rapid}$$

$$HOX \underset{rapid}{\overset{H^+}{\rightleftharpoons}} H_2OX^+$$

$$R-\overset{O}{\overset{\|}{C}}-CHX_2 \underset{slow}{\overset{OH^-}{\rightleftharpoons}} \left[R-\overset{O}{\overset{\|}{C}}-CX_2^- \rightleftharpoons R-\overset{O^-}{\overset{|}{C}}=CH_2 \right] \xrightarrow{rapid}$$

$$HOX \underset{rapid}{\overset{H^+}{\rightleftharpoons}} H_2OX^+$$

$$R-\overset{O}{\overset{\|}{C}}-OHC+X_3 \xleftarrow{OH^-} R-\overset{O}{\overset{\|}{C}}-CX_3$$

FIGURE 1 The haloform reaction—THMs formation.

acetone, chlorinated acetonitrile, chloropicrin, chloral, chloroacetic acids, chlorinated ethers, chlorophenols, chlorinated ketones, etc.[71] Koch and Krasner[74] estimated that among organohalogen compounds formed during chlorination of water, 77% are THMs, 15% are haloacetic acids, 3% are halonitriles, 4% is trichloroacetaldehyde hydrate, and 1% are the remaining compounds. Typical concentrations of trichlorometh-ane (chloroform), the most frequently present compound in chlorinated water, varied from 1 to 30 µg/L. Volatile organohalogen compounds have been determined primarily in tap water;[3,63–65,68–76] however, they also occur in surface water, groundwater, rainwater,[65,70,77,78] or even in the water and ice of polar regions.[65] Several volatile organohalogen compounds fall into the category of known or suspected carcinogens. In Table 4, the carcinogenic potential of organohalogen compounds occurring in drinking water and recognized by the EPA[2,4,5] as toxic or carcinogenic is presented. The MAC for drinking water varied from 0.3 µg/L for 1,1-dichloroethene (WHO, Norway) to 100 µg/L (EU, WHO, EPA, United Kingdom) and 350 µg/ L (Canada) for the total THMs. The MAC in the directives of WHO, Poland, and Great Britain for chloroform in tap water is 30 µg/L. Besides the groups of compounds discussed above, large amounts of other toxic pollutants, including pesticides, polychlorinated biphenyls (PCBs), polycyclic aromatic hydrocarbons (PAHs), crude oil derivatives, phenols, endocrine disruptors, etc., can be present in drinking water. Pesticides enter water either directly

TABLE 4 Carcinogenic Potential of Organohalogen Compounds Occurring in Drinking Water and Recognized by the U.S. EPA as Toxic or Carcinogenic

Compound	Carcinogenic Potential (μg/kg body/day)	Certainty of Hazard
Chloromethane	2.5	S
Dichloromethane	2.2	3
Trichloromethane (causes hepatocellular carcinomas and liver and kidney tumor in mice, and renal tumor in rats after chronic exposure)	11	CA
Tetrachloromethane (induces preneoplastic changes in the liver of rats, and liver tumors in male rats and mice)	0.4	CA
Bromodichloromethane	55	S
Dibromochloromethane	250	S
Tribromomethane	1100	S
Dichloroiodomethane	2.2	S
1,2-Dichloroethane	9.4	CA
1,1,1-Trichloroethane	20	S
Vinyl chloride	0.017	CU
Trichloroethylene (causes increases in the incidence of hepatocellular carcinomas, pulmonary carcinoma, and malignant lymphomas in mice)	1.6	CA
Tetrachloroethene (induces high incidence of nephropathy in mice and rats, and hepatocellular carcinomas in mice)	0.73	CA
1,4-Dichlorobenzene	1.4	S
1,3-Dichlorobenzene	88	S
1,2-Dichlorobenzene	6.9	S
1,2,4-Trichlorobenzene	4.4	S
2,4-Dichlorophenol	1.7	S
Pentachlorophenol	40	S
Polychlorinated biphenyls	53	S

Note: CA, animal carcinogen; CU, human carcinogen; S, suspected carcinogen.

by their application for mosquito control, or indirectly from drainage of agricultural lands, permeation through soil, erosion, wastewater from pesticide production, municipal wastes (fungicides and bactericides), etc. Consequently, a number of papers have been published confirming the presence of pesticides and PCBs not only in surface waters but also in potable water, particularly in water from wells existing in agricultural areas, in rainwater, as well as in water and ice from polar regions.[66,79–81] For humans, the major route of exposure to these pollutants is the gastrointestinal system, mainly by way of food (because of bioaccumulation and biomagnification in the food chain), but also through drinking water. Pesticides, PCBs, phenols, and some PAHs are carcinogenic, mutagenic, and teratogenic, and cause cardiovascular, neurological, and other diseases;[14,20,64,66,69,82] thus, there is a need for continuous monitoring of the degree of pollution by these compounds of potable and surface waters. Owing to the possibility of bioaccumulation of organic compounds, even low concentrations can result in poisoning of an organism. WHO, the EC, and most countries, including Poland, introduced a MAC for some toxic organic compounds present in tap water.[4–9,83] In Table 5, the MACs for toxic organic compounds recommended by WHO and those established by Poland are listed.[6,9]

Consequently, there is a need for continuous monitoring of the degree of pollution of potable and surface waters by toxic anthropogenic organic compounds. To be able to monitor environmental pollutants properly and effectively, analysts need a variety of methods for the isolation and determination of these contaminants, taking into account both physical-chemical properties of individual compounds

TABLE 5 WHO Guideline Values for Organic Compounds That Are of Health Significance in Drinking Water (In Milligrams per Liter) and the Established Macs for Drinking Water in Poland

Compound	Concentration in Drinking Water (mg/L) Recommended MAC		Compound	Concentration in Drinking Water (mg/L) Recommended MAC	
	WHO	Poland		WHO	Poland
Acrylamide	0.0005	0.0001	Edetic acid (EDTA)	0.6	
Alachlor	0.02	0.0001	Endrin	0.0006	0.0001
Aldicarb	0.01	0.0001	Epichlorohydrin	0.0004	0.0001
Aldrin and dialdrin	0.00003	0.0001	Ethylbenzene	0.3	
Atrazine	0.002	0.0001	Fenoprop	0.009	0.0001
Benzene	0.01	0.001	Hexachlorobutadiene	0.0006	0.0001
Benzoja jpyrene	0.0007	0.00001	Isoproturon	0.009	0.0001
Bromodichloromethane	0.06	0.015	Lindane	0.002	0.0001
Bromoform	0.1		2-Methyl-4-chlorophenoxyacetic acid	0.002	0.0001
Carbofuran	0.007	0.0001	Mecoprop	0.01	0.0001
Carbon tetrachloride	0.004	0.002	Metoxychlor	0.02	0.0001
Chlordane	0.0002	0.0001	Metolachlor		0.0001
Chloroform	0.3	0.03	Microcystin LR	0.001	0.001
Chlorotoluron	0.03	0.0001	Molinate	0.006	0.0001
Chlorpyrifos	0.03	0.0001	Monochloramine	3	
Cyanazine	0.0006	0.0001	Monochloracetate	0.02	
2,4-Dichlorophenoxyacetic acid	0.03	0.0001	Nitrilotriacetic acid	0.2	
Dichlorodiphenylotrichloroethane and metabolites	0.001	0.0001	N-Nitrosodiethylamine	0.1	
Di(2-ethylhexyl)phthalate	0.008	0.02	Pendimethalin	0.02	0.0001
Dibromoacetonitrile	0.07		Pentachlorophenol	0.009	
Dibromochloromethane	0.1		Permethrin	0.3	0.0001
1,2-Dibromo-3-chloropropane	0.001		Perlproxylen	0.3	
1,2-Dibromomethane	0.0004		Simazine	0.002	0.0001
Dichloroacetate	0.05		Styrene	0.02	
Dichloroacetonitrile	0.02		2,4,5-Trichlorophenoxyacetic acid	0.009	
1,2-Dichlorobenzene	1		Terbuthylazine	0.007	0.0001
1,4-Dichlorobenzene	0.3		Toluene	0.7	
1,2-Dichloroethane	0.03	0.003	Trichloroacetate	0.2	
1,2-Dichloroethene	0.05		Tetrachloroethene	0.04	0.01
Dichloromethane	0.02		Trichloroethene	0.02	
1,2-Dichloropropane	0.04		2,4,6-Trichlorophenol	0.2	0.2
1,3-Dichloropropene	0.02		Trichloromethanes, total	1	0.15
Dichloroprop	0.1	0.0001	Vinyl chloride	0.0003	0.0005
Dimethoate	0.006	0.0001	Xylenes	0.5	
1,4-Dioxane	0.05		Pesticides, total		0.0005

and groups of compounds and the characteristics of the matrices in which these compounds occur. Volatile organohalogen compounds, pesticides, and other organic pollutants occur in water at relatively low concentration levels; thus, there is a need for the isolation of organic compounds from a complex aqueous matrix and their preconcentration.

Methods of isolation and preconcentration of organic compounds from water are closely associated with the kind of analytes, their volatility, polarity, stability, water solubility, solubility in organic solvents, etc. Numerous techniques for the isolation and enrichment of organic analytes have been developed, the most common ones being solvent extraction, solid sorbent extraction, techniques utilizing the distribution of solute among the liquid and the gaseous phase (headspace, purging), as well as less commonly used techniques such as freezing out, lyophilization, vacuum distillation, steam distillation, and membrane processes (reverse osmosis, ultrafiltration, dialysis).[53–58,61–70,84–91] A schematic diagram of utilization of various isolation techniques for the determination of organic compounds in water is shown in Figure 2. The only method of determination of organohalogen compounds in water that avoids the isolation and preconcentration step uses direct injection of an aqueous sample onto a GC column and an electron capture detector (direct aqueous injection-electron capture detection, DAI-ECD).[63,65,70,92–97] This method requires a special injector allowing cold on-column injection and special capillary columns (see Figure 3). Cooling of the injector prevents sample evaporation from the needle of a syringe before its withdrawal. The DAI-ECD technique has been successfully used for the determination of volatile organohalogen compounds containing one or two carbon atoms (the products

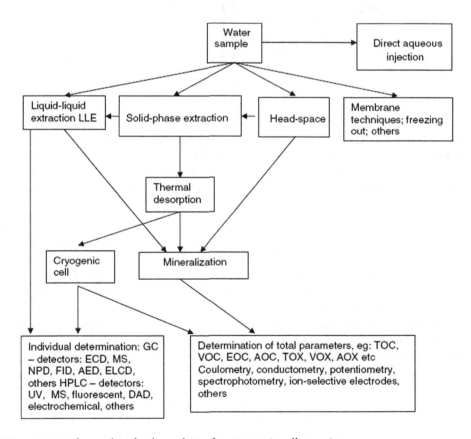

FIGURE 2 A general procedure for the analysis of trace organic pollutants in water.

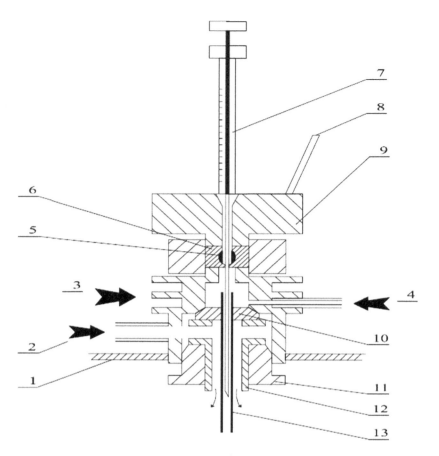

FIGURE 3 Schematic diagram of non-vaporizing septum-less, cold on-column injector: 1—the oven wall; 2—the secondary air cooling; 3—the principal air cooling system; 4—the carrier gas inlet; 5—the stainless steel rotating valve; 6—the valve seal; 7—the syringe; 8—the valve lever; 9—the upper part of injector; 10—the column seal; 11—the bottom part of injector; 12—the cooling jacket; 13—the capillary column.

of chlorination of humic substances; THMs) in tap water and surface water.[63,65,90,92–96] The method avoids the problems associated with incomplete recovery of the analytes during their isolation from the aqueous phase, the effect of potential contaminants when using solvent or solid-phase extraction, and losses of the analytes during the enrichment step. The main advantages of DAI-ECD include its simplicity (no isolation and preconcentration methods are necessary), repeatability, reduction of the possibility of sample contamination, and low detection limits (0.015–0.6 µg/L depending on the percentage of halogen in the compound). The detection limit of the method is related to the amount of the analyte in the sample injected onto the column and the volatility of the analytes.

For final determination, GC with specific detectors such as ECD, nitrogen-phosphorus detector (NPD), flame photometric detector, atomic emission detector (AED), and electrolytic conductivity detector (ELCD), or GC with universal detectors such as flame ionization detector (FID) or mass spectrometry (MS) is used. Liquid chromatography methods with ultraviolet (UV), MS, diode-array (DAD), electrochemical, or fluorescent detectors are also often used.[53–58,62,63,65–68,92–96] In Figure 2, two main ways of determining organic compounds in water are shown: determination of individual compounds or of total parameters. Total parameters such as total organic carbon or halogen (TOC, TOX), dissolved organic carbon or halogen (DOC, DOX), and suspended organic carbon or halogen (SOC, SOX) are used to characterize the content of organic compounds in water. Other parameters are defined in terms of the method of isolation of an organic fraction from water: volatile (VOC, VOX) or purgeable (POC, POX), extractable (EOC, EOX), and adsorbable (AOC, AOX) organic carbon or halogen. However, total parameters measuring the carbon content in an organic fraction are not particularly suitable as an estimate of anthropogenic water pollutants and their hazard to human health since a decisive majority of organic compounds in water are biogenic.

Characteristics of Some Water Bacteria and Pathogens in Drinking Water

Natural non-contaminated groundwater and surface water contain relatively low numbers of microorganisms. In general, groundwater has fewer bacteria than surface water. Microbial abundance and species composition depend on the depth of water-bearing stratum, soil type, and the geological structure and sanitary conditions in the given area. Aquatic bacteria are both autochthonic and allochtonic organisms. Autochthonic organisms, the so-called local microbes, reside permanently in the waters of a given water body, while allochtonic (alien) organisms are transferred into the water basin from the atmosphere and soil and through municipal and industrial wastewaters. Autochthonic microflora consists of autotrophic bacteria that are capable of photo- and chemosynthesis. It also includes heterotrophic organisms that cannot synthesize their own food. Purple as well as green bacteria are among the photosynthesizing autotrophs. Chemosynthetic bacteria are represented by nitrifying bacteria, iron- and sulfur-oxidizing bacteria, and hydrogen bacteria. Many species of autochthonic bacteria are heterotrophic organisms that feed exclusively on non-organic matter of allochthonous origin. Typical heterotrophic bacteria belong to the genera *Pseudomonas, Spirocheta, Vibrio,* and *Aeromonas.* Allochthonic organisms are mostly heterotrophic. They consist of saprophytes (feeding on organic compounds that are the products of decomposing remnants of plants and animals) as well as parasites (feeding at the expense of other live organisms). The most abundant saprophytic group is represented by a rod-shaped species such as *Escherichia coli, Aerobacter aerogenes,* and *Serratia marcescens.* Moreover, the rods from the genera *Bacillus* and *Clostridium* that become transferred from soil into the natural waters during rainfall can be classified as allochtonic aquatic bacteria. Water, which is not the natural environment of disease-causing microorganisms, plays a very important role in the process of their transfer. In cases when infectious diseases occur among the general public or when people become disease carriers, it is possible that potable water can become contaminated with the excrements of infected humans. The use of such water for drinking or food preparation and the exposure to it during washing may result in infection. The survival times of disease-causing organisms in the aquatic environment depend on many factors, inter alia, the pH of water, UV radiation, the presence of chemical compounds, and the level of water contamination. The main source of disease-causing microorganisms in natural waters is wastewater of human and animal origin. Pathogenic bacteria that become transferred into the water bodies through excrements, wastewater, and runoff from the fields are mainly related to digestive tract disorders in humans and animals.[98–100] The most typical bacteria of this kind come from the rod-shaped genus *Salmonella* and cause a variety of digestive system infections. The main genera of disease-causing bacteria that are transferred by water and the diseases caused by these bacteria are presented in Table 6. In water delivered by the water supply system, there can be microorganisms that have no influence on human health but form a film or membrane on the inside of the pipeline walls. The growth of such bacteria has a detrimental effect on water quality. A typical manifestation of biological fouling inside the water supply system is the altered color, turbidity, and undesirable taste and smell of water. In addition, the parameters of water exploitation may worsen owing to microbiological corrosion, plugging of the water-bearing stratum, and the lowered efficiency of the well.

TABLE 6   Main Kinds of Pathogenic Bacteria Found in Water and the Diseases They Cause

No.	Pathogenic Bacteria	Diseases
1	Rod-shaped Salmonella	Abdominal typhoid fever, Paratyphoid fever, salmonellosis
2	Rod-shaped Shigella	Bacterial dysentery
3	Curved rods Vibrio cholerae	Cholera
4	Francisella tularensis	Tularemia
5	Spirochaeta from Leptospira genus	Spirochetal jaundice, leptospirosis
6	Rod-shaped Escherichia coli	Acute dysentery, infection sticks of colon
7	Legionella pneumophila	Acute pneumonia, Legionnaires' disease
8	Aerobic, spore-forming bacilli Bacillus	Corneal abrasion, ankylosing spondylitis, feed poisoning, anthrax
9	Rod-shaped Mycobacterium	Tuberculosis
10	Yersinia enterocolitica	Yersiniosis

Microbiological Research Methods for Analysis of Drinking Water

The danger of humans contracting infections from water necessitates the permanent sanitary control of drinking water by bacteriologists and hygienists. Simple methods that can indicate the presence of pathogenic organisms are used because indirect methods are laborious and time consuming. Currently, in routine studies, intermediate methods based on indicative bacteria are used. The organism indicative of contamination has to comply with the following conditions:

- It has to be a permanent inhabitant of the intestine that occurs in large numbers
- It cannot be a pathogenic bacterium
- It cannot form spores
- It cannot multiply in water
- should be easy to detect
- It has to exist longer than pathogens in the natural environment (water, soil)
- It should be removed during the water conditioning in a similar way as the pathogen organisms

Water samples for microbiological analysis must always be collected in a sterile container. The samples should be refrigerated and transported to the testing laboratory as soon as possible. Fecal contamination of water can be determined by the presence of fecal coliforms or enterococci in a water sample by the multiple-tube technique. The bacteria can also be detected by the membrane filtration technique. The membrane filtration technique is based on the entrapment of the bacterial cells by a membrane filter (pore size, 0.22 or 0.45 μm). After filtration, the membrane with the microorganisms is placed on a petri dish containing the appropriate medium and incubated. After incubation at the correct temperature and time, characteristic colonies on the membrane surface are counted. Confirmatory tests are carried out where necessary. The membrane filtration method is especially useful for testing drinking water because large volumes of sample can be analyzed in a short time. The main indicating organisms of water fecal contaminations are *E. coli,* lacto-positive bacteria, coliform bacteria, thermotolerant *E. coli* types, fecal enterococcus, and *Clostridium* species. Several of the above-mentioned indicator conditions are fulfilled by *E. coli* and, to a lesser extent, by coliform bacteria and thermotolerant *E. coli* types. *E. coli* belongs to the family Enterobacteriaceae. It is a small, non-spore-forming, gram-negative, rod-shaped bacterium. It lives in the large intestine of humans and animals, where its density reaches a maximum of 10^9 cells per gram of fecal matter. These bacteria are grown at temperatures of 44–45°C on complex culture media. Usually, they are able to ferment lactose and mannitol with formation of acid and aldehyde, and they are able to produce indole from tryptophan. *E. coli* possesses two enzymes: β-galactosidase and β-glucoronidase; however, it does not form oxidase and does not hydrolyze urea. Lacto-positive bacteria are able to form colonies in aerobic conditions at 36±3°C on culture medium with lactose, with acid formation. Coliform bacteria are non-spore-forming, oxidase-negative, gram-negative, rod-shaped bacteria that are able to ferment, at 35–37°C, lactose to acids and aldehydes within 24–48 hr. Thermotolerant *E. coli* types are able to ferment lactose at 44–45°C. *Enterobacter, Citrobacter,* and *Klebsiella* also belong to this group. Coliform bacteria do not have to be directly connected with fecal contamination or with the occurrence of pathogens in drinking water. They could be present in drippings, in nutrient-rich water, in soil, and in decayed vegetable residues. These bacteria cannot occur in conditioned drinking water. Their presence in water suggests improper conditioning of water, secondary contamination, and excessive content of nutrient substances in conditioned water. These bacteria could be exploited as an indicator of the effectiveness of water conditioning. The frequency of indicative bacteria is determined to set the coliform titer—the lowest volume of water (milliliters) in which these bacteria are detected. According to ISO standards, the final Result is given as an index of coliform bacteria of *E. coli* in 100 mL of water tested. Fecal enterococci are spherical or oval, catalase-negative, gram-positive bacteria, occurring as a short chain. They can reduce 2,3,5-triphenyltetrazolic chloride to formazine and hydrolyze esculin at 44°C. They express D Lancefield antigen. The term *fecal Enterococcus* refers to those bacteria that occur in droppings of humans. Fecal enterococci that are not found in droppings of humans but occur in droppings of animals do not belong to the fecal enterococcus group. They die quickly in the external environment, faster than many pathogenic bacteria. The presence of fecal enterococcus in water can indicate fresh contamination of water; thus, it also proves the potential menace of pathogenic bacteria. Because these bacteria are resistant to drying, they could be helpful in routine control of water quality carried out after new water supply systems are built or after their repair. After *E. coli,* they could be the second fecal enterococcus that can be used as an indicative organism of contamination. Clostridial bacteria are sulfite-reducing, gram-positive, spore-forming bacilli that can survive in soil and water for a long time (even for many years). They are usually found in droppings, but they can originate from other sources. These bacteria are grown at 37°C. Breeding of these species can indicate a very "old" water contamination, after the death of all pathogenic bacteria. The spores of *Clostridium* are resistant to disinfection; thus, their presence in disinfected water may indicate a shortcoming in water conditioning. These bacteria are not recommended for the routine monitoring of water quality or detecting fresh contamination because of their length of survival. *Clostridium* might be present for a long time after the appearance of the contamination and far away from the place of this contamination. In sanitary analysis of drinking water, besides the presence of indicative bacteria, the general number of psychrophilic and mesophilic bacteria in 1 mL of water is determined. Culturing of psychrophilic bacteria is carried out on a solid agar culture medium for 72 hr at 22°C—the optimum conditions for these bacteria. A large quantity of these bacteria can indicate the inflow of organic substances to water, which creates favorable conditions for the growth of saprophytic bacteria. Culturing of mesophilic bacteria is performed on a solid agar culture medium for 24 h or 48 h at 37°C. The thermal optimum for these bacteria is the temperature of the human body, 37°C. The overall quantity of mesophilic bacteria consists of effluent bacteria and some soil bacteria. Their presence in water is a proof that contamination with domestic sewage and industrial waters has occurred.

References

1. United Nations Children's Fund (UNICEF). *Excerpt from Progress since the World Summit for Children: A Statistical Review*; UNICEF: New York, 2001.
2. U.S. Centers for Disease Control and Prevention. *Safe Water System: A Low-Cost Technology for Safe Drinking Water Fact Sheet*; World Water Forum 4 Update: Atlanta, GA, 2006.
3. *Water Treatment. Principles and Design*; Montgomery, J.M., Ed.; John Wiley and Sons: New York, 1985.
4. *Drinking Water Contaminants*; EPA 816-F-09–0004; EPA; May 2009.
5. EPA. *National Primary Drinking Water Regulations*; Code of Federal Regulations, 40 CFR Part 141; EPA; Electronic Code for Federal Regulations: June 3, 2010.

6. *Guidelines for Drinking-Water Quality,* 3rd Ed., incorporating first and second addenda; WHO: Geneva, Switzerland, 2006.

7. Kaika, M. The water framework directive: A new directive for a changing social, political and economic European Framework. Eur. Plann. Stud. **2003**, *11,* 299–316.

8. Directive 2000/60/EC of the European Parliament and of the Council of 23 October 2000 establishing a framework for community action in the field of water policy, Official Journal of the European Communities; 1.327/1: 22.12.2000.

9. Rozporzadzenie Ministra Zdrowia z dnia 29 marca 2007 r. w sprawie jakożci wody przeznaczonej do spozycia przez ludzi; Dziennik Ustaw Nr 61; Poz 417.

10. Synowiecki, J. Skladniki mineralne (Mineral components). In *Chemia Żywności (Food Chemistry)*; Sikorski, Z., Ed.; WN-T: Warsaw, Poland, 2000 (in Polish).

11. *Kompendium wiedzy o żywności, żywieniu i zdrowiu;* Gawecki, J., Mossor-Pietraszewska, T., Eds.; PWN: Warsaw, Poland, 2004 (in Polish).

12. Biziuk, M.; Kuczyńska, J. Mineral components in food— Analytical implications (Ch. 1). In *Mineral Components in Foods;* Szefer, P., Nriagu, J., Eds.; CRC Press: Boca Raton, FL, 2007; 1–31.

13. Eichler, W. *Gift in unserer Nahrung*; Kilda: Greven, Germany, 1982.

14. Bull, R.J. Carcinogenic and mutagenic properties of chemicals in drinking water. Sci. Tot. Environ. **1998**, *47,* 385–413.

15. Hartwig, A. Carcinogenicity of metal compounds: Possible role of DNA repair inhibition. Toxicol. Lett. **1998**, *102,* 235–239.

16. Kasprzak, K.S. Oxidative DNA and protein damage in metal-induced toxicity and carcinogenesis. Free Radic. Biol. Med. **2002**, *32,* 958–967.

17. Pourahmad, J.; O'Brien, P.J.; Jokar, F.; Daraei, B. Carcinogenic metal induced sites of reactive oxygen species formation in hepatocytes. Toxicol. In Vitro **2003**, *17,* 803–810.

18. Schwerdtle, T.; Walter, I.; Hartwig, A. Arsenite and its biomethylated metabolites interfere with the formation and repair of stable BPDE-induced DNA adducts in human cells and impair XPAzf and Fpg. DNA Repair **2003**, 1449–1463.

19. Tully, D.B.; Collins, B.J.; Overstreet, J.D.; Smith, C.S.; Dinse, G.E.; Mumtaz, M.M.; Chapin, R.E. Effects of arsenic, cadmium, chromium, and lead on gene expression regulated by a battery of 13 different promoters in recombinant HepG2 cells. Toxicol. Appl. Pharmacol. **2000**, *168,* 79–90.

20. Connell, D.W. *Bioaccumulation of Xenobiotic Compounds;* CRC: Boca Raton, FL, 1990.

21. Zukowska J.; Biziuk M. Methodological evaluation of method for dietary metal intake. J. Food Sci. **2008**, 73, 21–29.

22. Villaescusa, I.; Bollinger, J.C. Arsenic in drinking water: Sources, occurrence and health effects (a review). Rev. Environ. Sci. Biotechnol. **2008**, *7,* 307–323.

23. Ferguson, L.R. Natural, man-made mutagens and carcinogens in the human diet. Mutat. Res. **1999**, *443,* 1–10.

24. Moore, J.W. *Inorganic Contaminants of Surface Water. Research and Monitoring Priorities*; Springer: New York, 1991.

25. Nriagu, J.O.; Pacyna, J. Quantitative assessment of worldwide contamination of air, water and soils by trace metals. Nature **1988**, *333,* 134–139.

26. Rojas, E.; Herrera, L.A.; Poirier, L.A.; Ostrosky-Wegman, P. Are metals dietary carcinogens. Mutat. Res. **1999**, *443,* 157–181.

27. Kabata-Pendias, A.; Pendias, H. *Trace Elements in Soil and Plants,* 3rd Ed.; CRC Press: Boca Raton, FL, 2001.

28. Kabata-Pendias, A.; Mukherjee, A.B. *Trace Elements from Soil to Human;* Springer: Berlin, 2007.

29. Waalkes, M.P.; Misra, R.R. Cadmium carcinogenicity and genotoxicity. In *Toxicology of Metals*; Chang, L.W., Ed.; CRC Press: Boca Raton, FL, 1996; 231–243.

30. Cohen, M.D.; Kargacin, B.; Klein, C.B.; Costa, M. Mechanisms of chromium carcinogenicity and toxicity. Crit. Rev. Toxicol. **1993**, *23,* 255–281.

31. McLaughlin, M.J.; Parker, D.R.; Clarke, J.M. Metals and micronutrients—Food safety issues. Field Crops Res. **1999**, *60*, 143–163.

32. Pershagen, G. The carcinogenity of arsenic. Environ. Health Perspect. **1981**, *40*, 93–100.

33. Pocock, S.J.; Delves, H.T.; Ashby, D.; Shaper, A.G.; Clayton, B.E. Blood cadmium concentration in the general population of British middle-aged men. Hum. Toxicol. **1988**, *7*, 95–103.

34. Verougstraete, V.; Lison, D.; Hotz, P. Cadmium, lung a prostate cancer: A systematic review of recent epidemiological data. J. Toxicol. Environ. Health, Part B **2003**, *6*, 227–255.

35. Coogan, T.P.; Latta, D.M.; Snow, E.T.; Costa, M. Toxicity and carcinogenicity of nickel compounds. CRC Crit. Rev. Toxicol. **1989**, *19*, 341–384.

36. Soo, Y.O.; Chow, K.M.; Lam, C.W.; Lai, F.M.; Szeto, C.C.; Chan, M.H.; Li, P.K. A whitened face woman with nephrotic syndrome. Am. J. Kidney Dis. **2003**, *41*, 250–253.

37. Strenzke, N.; Grabbe, J.; Plath, K.E.; Rohwer, J.; Wolff, H.H.; Gibbs, B.F. Mercuric chloride enhances immunoglobulin E-dependent mediator release from human basophils. Toxicol. Appl. Pharmacol. **2001**, *174*, 257–263.

38. Ellingsen, D.G.; Bast-Pettersen, R.; Efskind, J.; Thomassen, Y. Neuropsychological effects of low mercury vapor exposure in chloralkali workers. Neurotoxicology **2001**, *22*, 249–258.

39. Godfrey, M.; Wojeik, D.; Cheryl, K. Apolipoprotein-E as a potential biomarker for mercury neurotoxicity. J. Alzheimer's Dis. **2003**, *5*, 189–195.

40. Mutter, J.; Naumann, J.; Sadaghiani, C.; Schneider, R.; Walach, H. Alzheimer disease: Mercury as a pathogenetic factor and apolipoprotein-E as a moderator. Neuroendocrinol. Lett. **2004**, *25*, 275–283.

41. Leonard, A.; Jacquet, P. Mutagenicity and teratogenicity of mercury compounds. Mutat. Res. **1983**, *114*, 1–18.

42. Zahir, F.; Rizwi, S.J.; Haq, S.K.; Khan, R.H. Low dose mercury toxicity and human health. Environ. Toxicol. Pharmacol. **2005**, *20*, 351–360.

43. Duker, A.A.; Carranza, E.J.M.; Hale, M. Arsenic geochemistry and health. Environ. Int. **2005**, *31*, 631–641.

44. Robson, M. Methodologies for assessing exposures to metals: Human host factors. Ecotoxicol. Environ. Saf. **2003**, *56*, 104–109.

45. Gochfeld, M. Cases of mercury exposure, bioavailability, and absorption. Ecotoxicol. Environ. Saf. **2003**, *56*, 174179.

46. Apostoli, P. Elements in environmental and occupational medicine. J. Chromatogr. B **2002**, *778*, 63–97.

47. Gebel, T. Genotoxicity of arsenical compounds. Int. J. Hyg. Environ. Health **2001**, *203*, 249–262.

48. Egan, S.K.; Tao, S.S-H.; Pennington, J.A.T.; Bolger, P.M. U.S. Food and Drug Administration's Total Diet Study: Intake of nutritional and toxic elements, 1991–1996. Food Addit. Contam. **2002**, *19*, 103–125.

49. Lee, H-S.; Cho, Y-H.; Park, S-O.; Kye, S-H.; Kim, B-H.; Hahm, T-S.; Kim, M.; Lee, J.O.; Kim, C-I. Dietary exposure of the Korean population to arsenic, cadmium, lead and mercury. J. Food Compos. Anal. **2006**, *19*, S31–S37.

50. Robberecht, H.; Van Cauwenbergh, R.; Bosscher, D.; Cornelis, R.; Deelstra, H. Daily dietary total arsenic intake in Belgium using duplicate portion sampling and elemental content of various foodstuffs. Eur. Food Res. Technol. **2002**, *214*, 27–32.

51. Ozsvath, D.I. Fluoride and environmental health: A review. Rev. Environ. Sci. Biotechnol. **2009**, *8*, 59–79.

52. Webber, J.S.; Syrotynski, S.; King, M.V. Asbestos-contaminated drinking water: Its impact on household air. Environ. Res. **1988**, *46*, 153–167.

53. Nemerow, N.L. *Stream, Lake, Estuary and Ocean Pollution;* Van Nostrand Reinhold Co.: New York, 1985.

54. *Handbook of Water Analysis;* Nollett, L.M.L., Ed.; Marcel Dekker: New York, 2000.

55. Fresenius, W.; Quentin, K.E.; Schneider, W., Eds. *Water Analysis;* Springer-Verlag: Berlin, Germany, 1988.
56. Leenheer, J.A. *Water Analysis;* Academic Press: Oxford, U.K., 1984.
57. Hewitt, C.N., Ed. *Instrumental Analysis of Pollutants;* Elsevier: London, 1991.
58. Hunt, D.T.E.; Wilson, A.L., Eds. *The Chemical Analysis of Water;* Royal Society of Chemistry: London, 1988.
59. Willard, H.H.; Merritt, L.L.; Dean, J.A.; Settle, F.A., Eds. *Instrumental Methods of Analysis;* Wadsworth: Belmont, CA, 1988.
60. Taylor, L.R.; Pap, R.B.; Pollard, B.D. *Instrumental Methods for Determining of Elements;* VCH: New York, 1994.
61. Bersier, P.M.; Howell, J.; Bruntlett, C. Advanced electroanalytical techniques versus atomic absorption spectrometry, inductively coupled plasma atomic emission spectrometry and inductively coupled plasma mass spectrometry in environmental analysis. Analyst **1994**, *119*, 219–233.
62. Wang, J. *Stripping Analysis. Principles, Instrumentation and Applications;* VCH: Deerfield Beach, FL, 1985.
63. Hellmann, H. *Analysis of Surface Waters;* Ellis Horwood: New York, 1987.
64. Biziuk, M.; Michalska, M. Analysis of drinking water. In *Methods of Analysis of Food Components and Additives;* Otles, S., Ed.; CRC Taylor & Francis: Boca Raton, FL 2005; 31–57.
65. de Kruijf, H.A.M.; Kool, J.H., Eds. *Organic Micropollutants in Drinking Water and Health;* Elsevier: Amsterdam, the Netherlands, 1985.
66. Biziuk, M.; Przyjazny, A. Methods of isolation and determination of volatile organohalogen compounds in natural and treated waters. J. Chromatogr. A **1996**, *733*, 417–448.
67. Biziuk, M.; Przyjazny, A.; Czerwiriski, J.; Wiergowski, M. Occurrence and determination of pesticides in natural and treated waters. J. Chromatogr. A **1996**, *754*, 103–123.
68. Moore, J.W.; Ramamoorthy, S. *Organic Chemicals in Natural Waters. Applied Monitoring and Impact Assessment;* Springer Verlag: New York, 1984.
69. Bruchet, A.; Legrand, M.F.; Anselme, C.; Mallevialle, J. Perspectives actuelles pour l'analyse des composés organiques volatils et non volatils dans les eaux naturelles ou en cours de traitement. Water Supply **1989**, *7*, 77–84.
70. Coleman, W.E.; Melton, R.G.; Kopfler, F.C.; Barone, K.A.; Aurand, T.A.; Jellison, M.G. Identification of organic compounds in a mutagenic extract of a surface drinking water by computerized gas chromatography-mass spectrometry system. Environ. Sci. Technol. **1980**, *14*, 576–588.
71. Biziuk, M. Gas chromatography by direct aqueous injection in environmental analysis. In *Encyclopedia of Analytical Chemistry;* Mayers, R.A., Ed.; Wiley: Chichester, U.K., 2000; Vol. 3, 2549–2587.
72. de Leer, W.B. *Aqueous Chlorination Products. The Origin of Organochlorine Compounds in Drinking and Surface Waters;* Delft University Press: Delft, the Netherlands, 1987.
73. Jolley, R.L.; Cotruvo, J.A.; Cummings, R.B.; Matice, J.S.; Jacobs, V.A., Eds. *Water Chlorination. Environmental Impact and Health Effects;* Ann Arbor Sc. Pub. Inc.: Ann Arbor, MI, 1980.
74. Rebhun, M.; Manka, J.; Zilberman, A. Trihalomethane formation in high-bromide Lake Galilee water. J. AWWA **1988**, *June*, 84–89.
75. Koch, B.; Krasner, S.W. Occurrence of disinfection byproducts in a distribution system. Proc. Water Qual. Technol. Conf. AWWA, Los Angeles, June 18–22, 1989; J. AWWA: Denver, 1989; 1203–1230.
76. Singer, P.C. DBPs in drinking water: Additional scientific and policy consideration for public health protection. J. AWWA **2006**, *October*, 73–80.
77. Peleg, M.; Yungster, H.; Shuval, H. *Developments in Arid Zone Ecology and Environmental Quality;* Balaban ISS: Philadelphia, PA, 1981; 271 pp.
78. Hirata, T.; Nakasugi, O.; Yoshioka, M.; Sumi, K. Groundwater pollution by volatile organochlorines in Japan and related phenomena in subsurface environment. Water Sci. Technol. **1992**, *25*, 9–16.

79. Stevens, A.A.; Dressman, R.C.; Sorrell, R.K.; Brass, H.J. Organic halogen measurements: Current uses and future prospects. J. AWWA **1985**, *April,* 146–152.

80. Beyer, A.; Biziuk, M. Environmental fate and global distribution of polychlorinated biphenyls Rev. Environ. Contam. Toxicol. **2009**, *201,* 137–158.

81. Dias-Cruz, M.S.; Llorca, M.; Barcelo, D. Organic UV filters and their photodegradates, metabolites and disinfection byproducts in the aquatic environment Trends Anal. Chem. **2008**, *27,* 873–887.

82. Eichhorn, P.; Knepper, T.P.; Ventura, F.; Diaz, A. The behavior of polar aromatic sulfonates during drinking water production: A case study on sulfophenyl carboxylates in two European waterworks. Water Res. **2002**, *36,* 2179–2186.

83. Biziuk, M.; Bartoszek, A. Environmental contamination of food. In *Carcinogenic and Anticarcinogenic Food Components;* Baer-Dubowska, W., Bartoszek, A., Malejka- Giganti, D.; CRC: Boca Raton, FL, 2006; 113–136.

84. Quevauviller, P.; Borchers, U.; Gawlik, B.M. Coordinating links among research, standardisation and policy in support of water directive chemical monitoring requirements. J. Environ. Monit. **2007**, *9,* 915–923.

85. Jakubowska, N.; Zygmunt, B.; Polkowska, Z.; Zabiegala, B.; Namiessnik, J. Sample preparation for gas chromatographic determination of halogenated volatile organic compounds in environmental and biological samples. J. Chromatogr. A **2009**, *1216,* 422–441.

86. Demeestere, K.; Dewulf, J.; De Witte, B.; Van Langenhove, H.Sample preparation for the analysis of volatile organic compounds in air and water matrices. J. Chromatogr. A **2007**, *1153,* 130–144.

87. Ras, M.R.; Borrull, F.; Marce, R.M. Sampling and preconcentration techniques for determination of volatile organic compounds in air samples. Trends Anal. Chem. **2009**, *28,* 347–361.

88. Tobiszewski, M.; Mechliriska, A.; Zygmunt, B.; Namiesnik, J. Green analytical chemistry in sample preparation for determination of trace organic pollutants. Trends Anal. Chem. **2009**, *28,* 943–951.

89. Namiesnik, J.; Górecki, T.; Biziuk, M. Isolation and preconcentration of volatile organic compounds from water. Anal. Chem. Acta **1990**, 237, 1–60.

90. Pawliszyn, J. *Solid Phase Microextraction—Theory and Practice*; Wiley: New York, 1997.

91. Pawliszyn, J. *Application of Solide Phase Microextraction*; The Royal Society of Chemistry: Cambridge, 1999.

92. Biziuk, M.; Czerwiriski, J. *Determination of dichloromethane, trichloromethane, tetrachloromethane, bromodichloromethane, dibromochloromethane, tribromomethane, trichloroethylene, tetrachloroethylene, 1,1,1-trichloroethane, I, 1,2,2-tetrachloroethane in water by gas chromatography using direct injection of the sample;* Polish Standard Method-PN-C-04549–1; Polish Committee for Standardization: Warsaw, Poland, 1998.

93. Astel, A.; Biziuk, M.; Przyjazny, A.; Namiesnik, J. Chemometrics in monitoring spatial and temporal variations in drinking water. Water Res. **2006**, *40,* 1706–1716.

94. Astel, A.; Astel, K.; Biziuk, M.; Namiesnik, J. Classification of drinking water samples using the Chernoff's faces visualization approach. Pol. J. Environ. Stud. **2006**, *15,* 691–697.

95. Biziuk, M.; Polkowska, Z.; Gorlo, D.; Janicki, W.; Namiessnik, J. Determination of volatile organic compounds in water intakes and tap water by purge and trap and direct aqueous injection-electron capture detection techniques. Chem. Anal. (Warsaw) **1995**, *40,* 299–307.

96. Biziuk, M.; Namiessnik, J.; Czerwinsski, J.; Gorlo, D.; Makuch, B.; Janicki, W. Occurrence and determination of organic pollutants in tap and surface waters of the Gdanssk district. J. Chromatogr. A **1996**, *733,* 171–184.

97. Starostin, L.; Witkiewicz, Z. Environmental water samples preparation for chemical analysis. Chem. Anal. (Warsaw) **1994**, *39,* 263–279.

98. Hench, K.R.; Bissonnette, G.K.; Sexstone, A.J.; Coleman, J.G.; Garbutt, K.; Skousen, J.G. Fate of physical, chemical, and microbial contaminants in domestic wastewater following treatment by small constructed wetlands. Water Res. **2003**, *37,* 921–927.

99. Golas, I.; Filipkowska, Z.; Lewandowska, D. Zmyslowska, I. Potentially pathogenic bacteria from the family *Enterobacteriaceae, Pseudomonas* sp. and *Aeromonas* sp. in waters designated for drinking and household purposes. Pol. J. Environ. Stud. **2002**, *11,* 325–330.
100. Nawrocki, J.; Bilozor, S. Uzdatnianie wody. *Procesy Chemiczne i Biologiczne;* PWN: Warsaw, Poland, 2000 (in Polish).

66

Water: Surface

Introduction 679
Pesticide Use..679
Pesticide Transport and Patterns of Occurrence......................................679
Effects of Pesticides 681
Gaps in Knowledge 682
References..682

Victor de Vlaming

Introduction

In numerous aquatic ecosystems across the United States losses of biodiversity and significant population declines of multiple species have been documented over the past 40 years. These aquatic ecosystems are imperiled due to human activities such as habitat destruction, damming and diversion of waters, introduction of exotic species, and release of toxic concentrations of chemicals, including pesticides, into surface waters.

Pesticide Use

Quantities of pesticides used in the United States have increased approximately 50-fold since the 1960s such that more than a billion pounds of pesticide active ingredients are used each year.[1] Hundreds of different pesticides have been developed for application in agricultural and other settings to control weeds, insects, fungus, and other pests. The characteristics of pesticides have evolved over the past 30 years from chemicals such as the organochlorines (OC) that are very persistent in the environment but tend to have relatively low toxicity, to chemicals such as organophosphorus (OP) and carbamate insecticides that are toxic at very low concentrations (e.g., parts per trillion) but generally are not very persistent. The majority of pesticide use is agricultural (70%–80% of total), but there is also use in urban areas for gardens, lawns, homes, and buildings, as well as in forestry, along roads and railways, and in various industrial and commercial situations.

Pesticide Transport and Patterns of Occurrence

Unfortunately pesticides do not always remain where they are applied. Off-site movement occurs irrespective of applications made according to label instructions. Off-site movement can occur by aerial drift of sprays, by evaporation, in storm and irrigation water runoff, and by seepage. Surface waters are vulnerable to pesticide contamination because most agricultural and urban areas drain into streams and rivers. Since pesticides are designed to be lethal to organisms, they pose a significant risk to biota when they enter aquatic ecosystems. Many of the currently used pesticides are so lethal that runoff of less than 1% of the quantities applied in a watershed into surface waters can have profound effects on aquatic biota.

The most extensive program for monitoring surface waters for chemical contaminants in the United States is the National Water-Quality Assessment (NAWQA) Program of the U.S. Geological Survey. Initiated in 1991, the focus in NAWQA has been on major watersheds distributed throughout the United States, encompassing 60%–70% of national water use. Pesticides have been one of NAWQA's top priorities. Much of the information summarized below was collected in the NAWQA program.[1–3]

Pesticides have been detected in every region of the United States where surface waters were analyzed for these chemicals. Although no individual study analyzed for every pesticide, a wide variety of pesticides including insecticides, herbicides, and fungicides have been identified in surface waters throughout the United States. The distribution of pesticides in surface waters generally follows geographic and seasonal agricultural patterns and also the influence of urban areas. That is, frequency of detections and highest concentrations are recorded in streams and rivers where agriculture is a major land use and pesticide use is intense. In most agricultural areas, the highest concentrations of pesticides occur as seasonal pulses, with duration of a few weeks to several months. Frequency of occurrence and highest concentrations also are associated with seasonal application patterns, being greatest coincident with or after applications. In urban-dominated streams, seasonal patterns are less obvious, pesticide concentrations being elevated for longer periods. In these urban streams, insecticides are detected at higher frequencies and concentrations than in streams draining agricultural areas. The largest areas where high quantities of pesticides are applied to crops occur in California, Florida, the Midwest, the lower Mississippi River Valley, and the coastal areas of the Southeast.

Herbicides are detected more frequently, and at the highest concentrations, than other pesticides in surface waters. Considering data collected from across the United States, herbicides occurring most frequently and at the highest concentrations are atrazine, simazine, metolachlor, prometon, DEA, alachlor, and cyanozine. The insecticides measured most frequently and at the highest concentrations are diazinon, chlorpyrifos, carbaryl, carbofuran, and malathion.

Pesticides are encountered most often and at elevated concentrations in streams and rivers draining agricultural areas just prior to and during the growing season. Also, a greater number of pesticides occur in surface waters coincident with these periods. These mixtures of pesticides have a potential for additive and synergistic adverse impacts on aquatic ecosystem health. An exposure pattern that is developing is one of long-term exposure to relatively low concentrations of pesticide mixtures punctuated with seasonal pulses of high concentrations, the effects of which are not currently known.

Off-site movement is a critical issue with regard to pesticide contamination of aquatic ecosystems. The relationship between quantities of pesticides applied (amount per unit area and total area of application) and detection frequency, as well as concentrations in surface waters was stated above. While this is a general principle, it does not apply to all pesticides and situations. Some pesticides, which are used rather extensively in agriculture, are seldom detected in surface waters. This low detection rate in surface waters relates to physical/chemical and/or degradation properties. Some such properties can result in pesticides adsorbing to particles (organic or soil) that may reduce off-site movement while different properties of other pesticides favor rapid degradation.

One might conclude that use of pesticides with a lower potential for off-site movement would reduce risk of impacting aquatic ecosystem health. While there is some truth in this idea, such physical/chemical properties often render pesticides more persistent. Furthermore, organic and soil particles to which pesticides are adsorbed can be transported by erosion (associated with rainfall, irrigation, or wind) into streams and rivers. Pesticides adsorbed to particles can settle into surface water sediments and have deleterious effects on bottom-dwelling organisms. Such physical/chemical properties also tend to result in bioaccumulation by aquatic organisms. These pesticides can bioaccumulate in aquatic species to levels that are detrimental and/or biomagnified to adverse levels in the food chain as contaminated organisms are eaten.

Examination of fish and bivalve tissues reveal that they are being exposed to a variety of bioaccumulable pesticides in both agricultural- and urban-dominated waterways. Residues of some pesticides in fish tissues can be such that they are harmful to human health. Organochlorine (e.g., DDT, chlordane), pyrethroid, and other hydrophobic insecticides are examples of pesticides that adsorb to organic materials, are persistent, bioaccumulate, and biomagnify. Unfortunately, there is a paucity of information on what pesticide tissue residue levels are deleterious to organisms or to other species that eat them.

Effects of Pesticides

The occurrence of pesticides in surface waters of the United States is widespread. What is the significance of this phenomenon? Adverse effects of chemicals, including pesticides, are determined by concentration, as well as by duration and frequency of exposure. The federal Clean Water Act (CWA) was enacted to protect human and aquatic organism health, requiring that no chemical can occur in surface waters at toxic concentrations. CWA requirements are implemented through enforceable water quality standards for specific chemicals and toxicity. Water quality standards and criteria have been established by various agencies for only a few pesticides. These standards and criteria are an estimate of a chemical concentration in water below which detrimental effects are not expected to occur. Comparing concentrations of a chemical in surface water to such standards and criteria provides an indication of potential antagonistic impacts on aquatic biota.

Standards for human health apply to treated drinking water supplied by community agencies. Therefore, the standards do not apply directly to most surface waters. While these standards do not pertain to concentrations of pesticides in surface waters, they do afford a benchmark to which measured pesticide concentrations can be compared. Chemical analyses of surface water samples do not include all pesticides; however, few pesticides included in analyses were detected at concentrations exceeding any drinking water standards. The pesticides most often exceeding standards are the triazine and acetanilide herbicides, atrazine, alachlor, cyanazine, and simizine.[1] Some, but not all, treatments of water to be used for drinking destroy or remove these herbicides.

For the more than 120 pesticides detected in surface waters there exist only 13 U.S. Environmental Protection Agency criteria developed for the protection of aquatic ecosystem health. For most currently used insecticides and for all herbicides there are no criteria for the protection of aquatic life. Canada has a larger number of aquatic life criteria, which are more stringent than U.S. criteria. U.S. pesticide aquatic life criteria are commonly exceeded in streams and rivers collecting from agricultural lands and/ or urban areas. Aquatic life criteria of four OP insecticides, azinphos-methyl, chlorpyrifos, diazinon, and malathion are the most frequently exceeded by concentrations in surface waters. Data collected from across the United States indicate that azinphos-methyl and chlorpyrifos exceed aquatic life criteria for more days per year than other insecticides monitored.[1,3] Major concerns regarding OP insecticides are that they are toxic to aquatic species at very low concentrations, different OPs repeatedly co-occur in surface waters, and their toxicity is additive.

In some regions of the country, where their use is high, carbamate insecticides are threats to aquatic biota. More than 20 years after being banned, OC insecticides continue to be detected in surface waters and sediments at concentrations that exceed aquatic life criteria. Especially in streams and rivers draining agricultural areas, but also in urbandominated streams, two or more pesticides often cooccur at concentrations that exceed their respective aquatic life criteria. Several studies from across the country reported high occurrences of diseased, deformed, and highly parasitized fish, as well as fish with a high incidence of tumors in surface waters where pesticide concentrations exceed aquatic life criteria and/or are elevated.

Measuring pesticide concentrations in surface waters does not furnish direct information of bioavailability (the percentage of the analytically measured amount of a chemical that produces toxic effects) or toxicity to aquatic biota. Toxicity testing of surface waters provides a direct measure of capacity of these waters to support healthy aquatic organisms. Standardized toxicity tests with aquatic species are available that measure lethal and sublethal (e.g., inhibition of reproduction, growth, etc.), effects. As a diagnostic tool for assessing water quality, toxicity testing has several merits. Toxicity tests afford an integrative measure of adverse effects of chemicals on organism health and viability, as well as the bioavailabity of chemicals.[4] Also, results of these toxicity tests have been reliable predictors of impacts on aquatic ecosystem biota. Chemical analyses of water cannot provide such information.

Statewide monitoring programs in California have disclosed that, on a seasonal basis, agricultural- and urban-influenced streams and rivers are lethal to test species.[4,5] Complex toxicological, chemical, and physical procedures (toxicity identification evaluation-TIE) that specifically identify the chemical(s) causing toxicity demonstrated that the pesticides most commonly responsible for the surface water toxicity are diazinon and chlorpyrifos. Carbofuran, malathion, carbaryl, methyl-parathion, thiobencarb, diuron, and molinate also have been shown to be causes of toxicity in California's surface waters. These surface water toxicity testing programs are not common in other regions of the United States. Such surface water toxicity is likely to occur in most urban streams and in surface waters collecting runoff from areas where agriculture is the predominant land use and where pesticides are used intensively.

Gaps in Knowledge

Pesticide adverse impacts on aquatic ecosystem health throughout the United States are most likely underestimated for several reasons. 1) The number of U.S. streams and rivers thoroughly monitored is very limited so that the distribution and extent of pesticide contamination is unknown. 2) Most investigations have been incomplete for one or more reasons, including too few sampling sites, sample collection was in frequent, and study duration was abbreviated. 3) None of these monitoring investigations included the complete range of pesticides that could impact aquatic biota. 4) Analytical detection limits for pesticides have been a problem in assessing detrimental effects on aquatic ecosystem health. In most monitoring projects analytical detection limits for many pesticides were higher than known toxic effects on aquatic species, as well as above aquatic life criteria. 5) Sublethal and delayed effects of pesticides, including bioaccumulation and biomagnification responses, generally are not evaluated. For example, several pesticides, including alachlor, atrazine, 2,4-D, metribuzin, trifluralin, aldicarb, carbaryl, parathion, some pyrethroids, benomyl, mancozeb, maneb, zineb, and ziram, commonly used in U.S. agriculture, have been shown to disrupt endocrine systems in some aquatic species.[6,7] Existing aquatic species toxicity screening procedures are inadequate for assessing endocrine disruption. 6) Seldom are indirect effects considered. For example, direct adverse effects on zooplankton during a period when they are critical food for larval fish could indirectly impact fish populations. 7) As stated above, aquatic organisms are exposed to pesticide mixtures in many watersheds across the United States. Assessments of impacts infrequently involve analysis of exposures to multiple chemicals.

Pesticides, especially insecticides, are having widespread impacts on surface water quality and aquatic ecosystem health throughout the United States. To reduce risks of pesticide impacts on aquatic ecosystem health, measures should be identified, developed, and implemented to eliminate or reduce off-site movement of these chemicals. More extensive and thorough monitoring of pesticides and of toxicity caused by these chemicals is advisable to assess the extent of pesticide-caused water quality degradation, as well as the effectiveness of remediation projects.

References

1. Larson, S.J.; Capel, P.D.; Majewski, M.S. *Pesticides in Surface Waters: Distribution, Trends, and Governing Factors;* Ann Arbor Press, Inc.: Chelsea, MI, 1997; 373.
2. Larson, S.J.; Gilliom, R.J.; Capel, P.D. Pesticides in Streams of the United States—Initial Results from National Water-Quality Assessment Program. *Water Resources Investigations Report 98-4222;* U.S. Geological Survey: Sacramento, CA, 1999; 92.
3. Gilliom, R.J.; Barbash, J.E.; Kolpin, D.W.; Larson, S.J. Testing water quality for pesticide pollution. Environ. Sci. Technol. **1999**, *33*, 164A–169A.

4. de Vlaming, V.; Connor, V.; DiGiorgio, C.; Bailey, H.C.; Deanovic, L.A.; Hinton, D.E. Application of whole effluent toxicity test procedures to ambient water quality assessment. Environ. Toxicol. Chem. **2000**, *19*, 42–62.

5. Kegley, S.; Neumeister, L. *Disrupting the Balance: Ecological Impacts of Pesticides in California*, Californians for Pesticide Reform, San Francisco, 1999; 99.

6. *Environmental Endocrine Disruptors: A Handbook of Property Data*; Lawrence, H.K., Ed.; John Wiley and Sons, Inc.: New York, 1997; 1232.

7. Moore, A.; Waring, C.P. Sublethal effects of the pesticide diazinon on olfactory function in mature male Atlantic salmon Parr. J. Fish Biol. **1996**, *48*, 758–775.

67

Wetlands

Introduction 685
Wetland Definitions 685
Wetland Types.. 686
Extent of Wetlands 689
References...689

Ralph W. Tiner

Introduction

Wetland is a universal term used to describe the collection of flooded or saturated environments that have been referred to as marshes, swamps, bogs, fens, salinas, pocosins, mangroves, wet meadows, sumplands, salt flats, varzea forests, igapo forests, bottomlands, sedgelands, moors, mires, potholes, sloughs, mangals, palm oases, playas, muskegs, and other regional and local names. It has been defined as a basis for inventorying these natural resources, for conducting scientific studies, and, in some countries, for regulating uses of these areas. Given that wetlands include a diverse assemblage of ecosystems, classification schemes have been developed to separate and describe these different systems and to group similar habitats. Wetlands provide a number of functions that are considered valuable to society (e.g., surface water storage to minimize flood damages, sediment retention and nutrient transformation to improve water quality, shoreline stabilization, streamflow maintenance, and provision of vital habitat for fish, shellfish, wildlife, and plants that yield food and fiber for people). Because of these values and the widespread recognition of wetlands as important natural resources, numerous wetland definitions and classification systems have been developed to inventory these resources around the globe. The purpose of this entry is to provide readers with an understanding of what wetlands are (wetland definition), how they vary globally (wetland types), and their extent as determined by various inventories. This entry should serve as a starting point for learning about wetlands, with the listed references being sources of more detailed information.

Wetland Definitions

Wetlands are aquatic to semiaquatic ecosystems where permanent or periodic inundation or prolonged waterlogging creates conditions favoring the establishment of aquatic life. Wetlands are often located between land and water and have, therefore, been referred to as ecotones (i.e., transitional communities). However, many wetlands are not ecotones between land and water, since they are not associated with a river, lake, estuary, or stream.[1] Wetlands may derive water from many sources, including groundwater, river overflow, surface water runoff, precipitation, snowmelt, tides, melting permafrost, and seepage from impoundments or irrigation projects.

While the term "wetland" has many definitions, all definitions have common elements (see Table 1; some definitions even include deepwater habitats.). The presence of water in wetlands may be permanent or temporary. Their water may be salty or fresh. Wetlands may be natural habitats or artificially created.

TABLE 1 Examples of Wetland Definitions Used for Inventories

Country/Organization	Wetland Definition [Source]
International/Ramsar	"areas of marsh, fen, peatland, or water, whether natural or artificial, permanent or temporary, with water that is static or flowing, fresh, brackish, or salt, including areas of marine water the depth of which at low tide does not exceed 6 m .... may incorporate riparian and coastal zone adjacent to wetlands, and islands or bodies of marine water deeper than 6 m at low tide lying within the wetlands."[2]
Australia	"areas of seasonally, intermittently, or permanently waterlogged soils or inundated land, whether natural or artificial, fresh or saline, e.g., waterlogged soils, ponds, billabongs, lakes, swamps, tidal flats, estuaries, rivers, and their tributaries.[12]
Canada	"land that is saturated with water long enough to promote wetland or aquatic processes as indicated by poorly drained soils, hydrophytic vegetation, and various kinds of biological activity which are adapted to a wet environment."[3]
U.S.	"lands transitional between terrestrial and aquatic systems where the water table is usually at or near the surface or the land is covered by shallow water." Wetland attributes include hydrophytic vegetation, undrained hydric soil, or saturated or flooded substrates.[4]

Source: Photos courtesy of U.S. Fish and Wildlife Service.

They range from shallow water environments to temporarily wet (i.e., flooded or saturated) areas. All are wet long enough and often enough to, at least, periodically support hydrophytic vegetation and other aquatic life (including anaerobic microbes), to create hydric soils or substrates, and to activate biogeochemical processes associated with wet environments.

Wetland Types

Differences in climate, soils, vegetation, hydrology, water chemistry, nutrient availability, and other factors have led to the formation of a multitude of wetland types around the globe. In general, wetlands are characterized by their hydrology (e.g., tidal vs. nontidal, inundation vs. soil saturation, frequency and duration of wetness), the presence or absence of vegetation (vegetated vs. nonvegetated), the type of vegetation (forested or treed, shrub, emergent, or aquatic bed), and soil type (e.g., organic vs. inorganic, peatland vs. nonpeatland). Table 2 presents brief descriptions of some North American types and Figure 1 shows examples of vegetated wetlands.

Various countries have devised classification systems for describing differences among their wetlands and for categorizing wetlands for natural resources inventories. Scientists have created systems to organize certain wetlands into meaningful groups for analysis and management (e.g., peatland classifications; see Tiner[1] for details). In 1998, the Ramsar Convention Bureau published a multinational classification system to provide consistency for inventorying wetlands and designating wetlands of international importance.[2] This system includes 11 types of marine or coastal wetlands (i.e., shallow water and intertidal habitats: permanent shallow marine waters; marine subtidal aquatic beds; coral reefs; rocky marine shores; sand; shingle or pebble shores; estuarine waters; intertidal mud, sand, or salt flats; intertidal marshes; intertidal forests; coastal brackish/saline lagoons; coastal freshwater lagoons). This system also includes 19 inland wetland types (i.e., permanently flooded aquatic habitats to intermittently flooded sites are represented: permanent inland deltas; permanent rivers/streams/creeks; seasonal, intermittent, or irregular rivers/streams/creeks; permanent freshwater lakes; seasonal or intermittent freshwater lakes; seasonal or intermittent saline/brackish/alkaline lakes and flats; permanent saline/brackish/alkaline marshes and pools; seasonal or intermittent saline/brackish/alkaline marshes and pools; permanent freshwater marshes and pools; seasonal or intermittent freshwater marshes and pools; nonforested peat-lands; alpine wetlands including meadows and temporary snowmelt waters; tundra wetlands; shrubby-dominated wetlands; freshwater tree-dominated wetlands on inorganic soils; forested peatlands; freshwater springs and oases; geothermal wetlands; and subterranean karst and cave hydrological systems). Lastly, nine man-made wetland types (aquaculture ponds; ponds; irrigated land

TABLE 2   Brief Nontechnical Descriptions of Some Wetland Types in North America

Wetland Type	General Description
Marsh	Herb-dominated wetland with standing water through all or most of the year, often with organic (muck) soils
Tidal marsh	Herb-dominated wetland subject to periodic tidal flooding
Salt marsh	Herb-dominated wetland occurring on saline soils, typically in estuaries and interior arid regions
Swamp	Wetland dominated by woody vegetation and usually wet for extended periods during the growing season
Mangrove swamp (Mangal)	Tidal swamp dominated by mangrove species
Peatland, mire, moor, or muskeg	Peat-dominated wetland
Bog	Nutrient-poor peatland, typically characterized by ericaceous shrubs, other woody species, and peat mosses
Fen	More or less nutrient-rich peatland, often represented by sedges and/or calciphilous herbs and woody species
Wet meadow	Herb-dominated wetland that may be seasonally flooded or saturated for extended periods, often with mineral hydric soils
Bottomland	Riverside or streamside wetland, usually on floodplain
Flatwood	Forested wetland with poorly drained mineral hydric soils located on broad flat terrain of interstream divides, common on coastal plains and glaciolacustrine plains
Farmed wetland	Wetland cultivated for rice, cranberries, sugar cane, mints, or other crops

Note: These types may be defined differently in other regions.

including rice paddies; seasonally flooded agricultural land; salt exploitation sites; water storage areas including impoundments generally more than 8ha; excavations; wastewater treatment areas; and canals and drainage channels) are also included in the system.

In North America, the Canadian and United States wetland classification systems were developed by government agencies interested in wetland conservation and management. The Canadian system emphasizes wetland origin (class), form, and vegetation in describing the wetland types.[3] Five wetland classes are recognized: bog, fen, marsh, swamp, and shallow water. Within each class, different forms and types are characterized. Eight general vegetation types are defined by the presence or absence of vegetation (treed, shrub, forb, graminoid, moss, lichen, aquatic bed, and nonvegetated). These types may be subdivided into other types (e.g., treed into coniferous or deciduous, shrub into tall, low, and mixed, graminoids into grass, reed, tall rush, low rush, and sedge, aquatic bed into floating and submerged). The U.S. Fish and Wildlife Service's wetland and deepwater habitat classification[4] is the official federal system used for mapping wetlands and for reporting the status and trends of wetlands in the U.S. The features separating wetlands include general ecological and physical factors and specific features such as vegetation, soil/substrate composition, hydrology, water chemistry, and human alterations. Classification follows a hierarchical approach with five main levels designated: ecological system (marine, estuarine, lacustrine, riverine, and palustrine), subsystem, class (vegetated: forested, scrub-shrub, emergent, and aquatic bed; nonvegetated: unconsolidated shore, rocky shore, streambed, and reef), subclass, and modifiers. The modifiers are used to describe a wetland's hydrology (water regime), pH and salinity (water chemistry), soils, and the influence of humans and beaver (special modifiers). Common types include estuarine intertidal emergent wetlands (e.g., salt and brackish marshes), estuarine intertidal unconsolidated shore (e.g., tidal flats and beaches), palustrine emergent wetlands (e.g., marshes, fens, and wet meadows), palustrine forested wetlands, and palustrine scrub-shrub wetlands (e.g., shrub bogs and shrub swamps).

A hydrogeomorphic approach (HGM) to wetland classification has also been developing in the U.S.[5] The HGM system emphasizes abiotic features important for assessing wetland functions. Seven

FIGURE 1 Some examples of North American wetlands: (a) tidal salt marsh, (b) inland marsh, (c) pothole marsh, (d) wet meadow/shrub swamp, (e) northern peatland, (f) bottomland swamp, (g) hardwood swamp, and (h) flat-wood wetland.
Source: Photos courtesy of U.S. Fish and Wildlife Service.

hydromorphic classes are identified: riverine, depressional, slope, mineral soil flats, organic soil flats, lacustrine fringe, and estuarine fringe. The U.S. Fish and Wildlife Service has adapted the HGM approach to provide additional modifiers to its classification system on a pilot basis. These HGM-type descriptors include landscape position (i.e., lotic, lentic, terrene, estuarine, and marine), landform (i.e., slope, basin,

TABLE 3 Estimates of the Current Extent of Wetlands in Different Regions of the World

Region/Country	Wetland Extent (ha)	Source
Africa	121,321,683–124,686,189	[13]
Asia	211,501,790–224,117,790	[14]
Central America		
Mexico	3,318,500 (very incomplete)	[15]
Europe		
Eastern	225,849,930	[16]
Western	28,821,979	[17]
Middle East	7,434,790	[18]
Neotropics	414,996,613	[19]
North America		
Canada	127,199,000–150,000,000	[20]
U.S.	114,544,800	[1]
Oceania	35,748,853	[21]
South America		
Tropical region	200,000,000	[22]

interfluve, floodplain, flat, island, and fringe), and water flow path (i.e., inflow, outflow, throughflow, bidirectional flow, isolated, and paludified).[6] These descriptors provide the required information to aid the evaluation of functions of wetlands across watersheds and large geographic areas.

Extent of Wetlands

Comprehensive wetland inventories do not exist in most countries. There are many inconsistencies among the inventories (e.g., different levels of effort, focus on particular types, and artificial wetlands such as rice paddies are often not included in wetland inventories).[7] Consequently, comparative analysis is fraught with problems. Nonetheless, Table 3 provides some perspective on the extent of wetlands in many regions. Most of the data came from a series of reports produced for the Bureau of the Ramsar Wetlands Convention.[8] Globally, estimates for wetlands range from about 750 million ha[7] to about 1.5 billion ha. Ten countries have over 2 million ha of peat-lands alone, with Canada leading at nearly 130 million ha (represents about 18% of the country) followed by the former U.S.S.R. at 83 million ha.[9,10] About a third of Finland is covered by peatlands (10 million ha). The Pantanal of South America, perhaps the largest wetland in the world, reportedly covers about 200,000 km^2 (or 2 million ha) during the wet season.[11]

References

1. Tiner, R.W. *Wetland Indicators: A Guide to Wetland Identification, Delineation, Classification, and Mapping;* Lewis Publishers, CRC Press: Boca Raton, FL, 1999; 392 pp.
2. *Ramsar Convention Bureau.* Information Sheet on Ramsar Wetlands: Gland, Switzerland, 1998.
3. National Wetlands Working Group. *Wetlands of Canada;* Ecological Land Classification Series No. 21, Land Conservation Branch, Canadian Wildlife Service; Environment Canada: Ottawa, Ont., 1988; 452 pp.
4. Cowardin, L.M.; Carter, V.; Golet, F.C.; LaRoe, E.T. *Classification of Wetlands and Deepwater Habitats of the United States;* FWS/OBS-79/31; U.S. Department of the Interior, Fish and Wildlife Service: Washington, DC, 1979; 131 pp.
5. Brinson, M.M. *A Hydrogeomorphic Classification for Wetlands;* Wetlands Research Program Tech. Rep. WRP-DE-4; U.S. Army Waterways Expt. Station: Vicksburg, MS, 1993; 103 pp.

6. Tiner, R.W. *Keys to Waterbody Type and Hydrogeomorphic-Type Wetland Descriptors for U.S. Waters and Wetlands, Operational Draft;* U.S. Department of the Interior, Fish and Wildlife Service; Northeast Region: Hadley, MA, 2000; 20 pp.

7. Finlayson, C.M.; Davidson, N.C. Global review of wetland resources and priorities for wetland inventory: summary report. In *Global Review of Wetland Resources and Priorities for Wetland Inventory;* Finlayson, C.M.; Spiers, A.G., Eds.; Supervising Scientist Report 144; Canberra, Australia, 1999; 9 pp.

8. Finlayson, C.M., Spiers, A.G., Eds.; *Global Review of Wetland Resources and Priorities for Wetland Inventory;* Supervising Scientist Report 144; Canberra, Australia, 1999.

9. Taylor, J.A. Peatlands of the British Isles. In *Mires: Swamp, Bog, Fen, and Moor; Regional Studies;* Gore, A.J.P., Ed.; Elsevier: Amsterdam, the Netherlands, 1–46.

10. Botch, M.S.; Massing, V.V. Mire ecosystems in the U.S.S.R. In *Mires: Swamp, Bog, Fen, and Moor; Regional Studies;* Gore, A.J.P., Ed.; Elsevier: Amsterdam, the Netherlands; 95–152.

11. Swarts, F.A., Ed. *The Pantanal of Brazil, Bolivia, and Paraguay;* Waterland Research Institute; Hudson MacArthur Publishers: Gouldsboro, PA, 2000; 287 pp.

12. Wetland Advisory Committee. *The Status of Wetlands Reserves in System Six;* Report of the Wetland Advisory Committee to the Environmental Protection Authority: Australia, 1977.

13. Stevenson, N.; Frazier, S. Review of wetland inventory information in Africa. In *Global Review of Wetland Resources and Priorities for Wetland Inventory;* Finlayson, C.M., Spiers, A.G., Eds.; Supervising Scientist Report 144; Canberra, Australia, 1999; 94 pp.

14. Watkins, D.; Parish, F. Review of wetland inventory information in Asia. In *Global Review of Wetland Resources and Priorities for Wetland Inventory;* Finlayson, C.M., Spiers, A.G., Eds.; Supervising Scientist Report 144; Canberra, Australia, 1999; 26 pp.

15. Olmstead, I. Wetlands of Mexico. In *Wetlands of the World: Inventory, Ecology, and Management Volume I;* Whigham, D.F., Dykyjova, D., Heiny, S., Eds.; Kluwer Academic Publishers: Dordrecht, the Netherlands, 1993; 637–677.

16. Stevenson, N.; Frazier, S. Review of wetland inventory information in Eastern Europe. In *Global Review of Wetland Resources and Priorities for Wetland Inventory;* Finlayson, C.M., Spiers, A.G., Eds.; Supervising Scientist Report 144; Canberra, Australia, 1999; 53 pp.

17. Stevenson, N.; Frazier, S. Review of wetland inventory information in Western Europe. In *Global Review of Wetland Resources and Priorities for Wetland Inventory;* Finlayson, C.M., Spiers, A.G., Eds.; Supervising Scientist Report 144; Canberra, Australia, 1999; 57 pp.

18. Frazier, S.; Stevenson, N. Review of wetland inventory information in the Middle East. In *Global Review of Wetland Resources and Priorities for Wetland Inventory;* Finlayson, C.M., Spiers, A.G., Eds.; Supervising Scientist Report 144; Canberra, Australia, 1999; 19 pp.

19. Davidson, I.; Vanderkam, R.; Padilla, M. Review of wetland inventory information in the Neotropics. In *Global Review of Wetland Resources and Priorities for Wetland Inventory;* Finlayson, C.M., Spiers, A.G., Eds.; Supervising Scientist Report 144; Canberra, Australia, 1999; 35 pp.

20. Davidson, I.; Vanderkam, R.; Padilla, M. Review of wetland inventory information in the North America. In *Global Review of Wetland Resources and Priorities for Wetland Inventory;* Finlayson, C.M., Spiers, A.G., Eds.; Supervising Scientist Report 144; Canberra, Australia, 1999; 35 pp.

21. Watkins, D. Review of wetland inventory information in Oceania. In *Global Review of Wetland Resources and Priorities for Wetland Inventory;* Finlayson, C.M., Spiers, A.G., Eds.; Supervising Scientist Report 144; Canberra, Australia, 1999; 26 pp.

22. Junk, W.J. Wetlands of tropical South America. In *Wetlands of the World: Inventory, Ecology, and Management Volume* I; Whigham, D.F., Dykyjova, D., Heiny, S., Eds.; Kluwer Academic Publishers: Dordrecht, the Netherlands, 1993; 679–739.

VIII

PRO: Basic Environmental Processes

68

Eutrophication

Eutrophication Problem 693
Growth of Phytoplankton 694
Solutions to the Eutrophication Problem 698
References 699

Sven Erik Jørgensen

Eutrophication Problem

Many aquatic ecosystems suffer from eutrophication: lakes, reservoirs, estuaries, lagoons, fjords, and bays. It is a worldwide problem and, together with the oxygen depletion problem, is probably the most serious pollution problem of aquatic ecosystems.

The word eutrophy is generally taken to mean "nutrient rich." In 1919, Nauman introduced the concepts of oligotrophy and eutrophy, distinguishing between oligotrophic lakes containing little planktonic algae and eutrophic lakes containing much phytoplankton. The eutrophication of aquatic ecosystems all over the world has increased rapidly during the last decade due to increased urbanization and the consequently increased discharge of nutrients. The discharges in industrialized countries are about 1.8 kg N/yr and 0.5 kg P/yr. The production of fertilizers has grown exponentially in this century, and the concentration of phosphorus in many lakes reflects this.

The word eutrophication is used increasingly in the sense of the artificial addition of nutrients, mainly nitrogen and phosphorus, to waters. Eutrophication is generally considered to be undesirable, but this is not always true. The green color of eutrophied lakes makes swimming and boating less safe due to the increased turbidity, and from an aesthetic point of view, the chlorophyll concentration should not exceed 100 mg/m^3. However, the most critical effect from an ecological point of view is the reduced oxygen content of the hypolimnion, caused by the decomposition of dead algae, particularly in the fall. Eutrophic aquatic ecosystems sometimes show a high oxygen concentration (at the surface during the summer time) but a low concentration of oxygen in the hypolimnion, which may be lethal to fish. The oxygen depletion in the hypolimnion will often imply that the eutrophication is more difficult to abate, because anaerobic sediment will more easily release its content of phosphorus. Iron (III) is reduced to iron (II) by anaerobic conditions. As iron (III) has a very insoluble phosphorus salt, while iron (II) phosphate is readily soluble, the phosphorus release by anaerobic conditions is dependent on the composition of the sediment, particularly of course the iron content. One of the most applied lake restoration methods is the pumping of air or oxygen to the hypolimnion, which is used to significantly reduce the release of phosphorus from the sediment.

Water makes up 75%–90% of the total wet weight of plant tissue. It means that except for oxygen and hydrogen, the relative composition on dry weight basis would be 4–10 times higher. For phytoplankton, the carbon, nitrogen, and phosphorus content on dry weight basis are, respectively, approximately 40%–60%, 6%–8%, and 0.75%–1.0%. Phosphorus is considered the major cause of eutrophication in lakes, as it was formerly the growth-limiting factor for algae in the majority of lakes. Its use has increased tremendously during the last decade. Nitrogen is limiting in a number of East African lakes as a result of the nitrogen depletion of soils by intensive erosion in the past. Nitrogen is furthermore often limiting in the coastal ecosystems, at least for part of the year. However, today, nitrogen may become limiting in lakes as a result of the tremendous increase in the phosphorus concentration caused by the discharge of wastewater, which contains relatively more phosphorus than nitrogen. While algae use 5–10 times more nitrogen than phosphorus (see the content of these two elements in phytoplankton above), wastewater generally contains only 3 times as much nitrogen as phosphorus. In lakes, a considerable amount of nitrogen is furthermore lost by denitrification (nitrate $\rightarrow N_2$). Lakes that have received wastewater for a longer period may therefore be limited by nitrogen. In environmental management, the key question is, however, not which element is the limiting factor but which element can be most easily controlled as a limiting factor. As phosphorus generally can be removed more easily and cheaply, it is often the most effective abatement of eutrophication to remove phosphorus effectively from wastewater discharge to the lake. Further details are given below.

It is a good management strategy for an abatement of eutrophication in aquatic ecosystems to find quantitatively all the sources of nitrogen and phosphorus. In most cases, based on this information, it is easy to find possible solutions and the corresponding costs. In this context, it is beneficial to apply an ecological model. Eutrophication models have been developed and applied for all these ecosystems. Many eutrophication models have been applied for environmental management, particularly for lakes and reservoirs. The *Handbook of Ecological Models Used in Ecosystem and Environmental Management*[1] gives a good overview of the available ecological models, including eutrophication models.

Growth of Phytoplankton

The growth of phytoplankton is the key process of eutrophication, and it is therefore important to understand the interacting processes that regulate growth. Primary production has been measured in great detail in a number of aquatic ecosystems and presents the synthesis of organic matter. The overall process can be summarized as follows:

$$Light + 6CO_2 + 6H_2O \rightarrow C_6H_{12}O_6 + 6O_2 \tag{1}$$

The composition of phytoplankton is not constant. The composition of phytoplankton and plants in general reflects to a certain extent the concentration of the water. If, e.g., the phosphorus concentration is high, the phytoplankton will take up relatively more phosphorus—this is called luxury uptake. Phytoplankton consists mainly of carbon, oxygen, hydrogen, nitrogen, and phosphorus: without these elements, no algal growth will take place. This leads to the concept of the limiting nutrient, which has been developed by Liebig as the law of the minimum. However, the concept has been considerably misused due to oversimplification. First of all, growth might be limited by more than one nutrient. The composition as mentioned above is not constant but varies with the composition of the environment. Furthermore, growth is not at its maximum rate until the nutrients are used, at which point growth stops, but the growth rate slows down as the nutrients become depleted.

The sequence of events leading to eutrophication has often been described as follows: Oligotrophic waters will have a ratio of N to P greater than or equal to 10, which means that phosphorus is less abundant than nitrogen for the needs of phytoplankton. If sewage is discharged into the aquatic ecosystem, the ratio will decrease, since the N-to-P ratio for municipal wastewater is 3:1, and consequently, nitrogen will be less abundant than phosphorus relative to the needs of phytoplankton, as indicated above. In this situation, however, the best remedy for the excessive growth of algae is not necessarily the removal of nitrogen from the sewage, because the mass balance might then show that nitrogen-fixing algae will give an uncontrollable input of nitrogen into the system. It is particularly the case if the aquatic ecosystem is a lake or reservoir. It is necessary to set up mass balances for each of the nutrients as already pointed out, and these will often reveal that the input of nitrogen from nitrogen-fixing blue-green algae, precipitation, and tributaries is contributing too much to the mass balance for the removal of nitrogen from the sewage to have any effect. On the other hand, the mass balance may reveal that the phosphorus input (often more than 95%) comes mainly from sewage, which means that it is better management to remove phosphorus from the sewage than nitrogen. Thus, what is important in environmental management is not which nutrient is most limiting, but rather, which nutrient can most easily be made to limit the algal growth.

The conceptual diagram in Figure 1 shows the nitrogen cycle, and Figure 2 shows the phosphorus cycle of aquatic ecosystems and illustrates the processes behind the cycling of these nutrients. As clearly pointed out above, it is always beneficial to use a mass balance to choose the most important components, forcing functions, and state variables to be considered in the management context. Let us illustrate the needed mass balance considerations by use of an example. Let us anticipate that it is an open question whether birds should be included in the selected management strategy (and in the eutrophication model). Birds may contribute considerably to the inputs of nutrients by their droppings. If the nutrients—nitrogen and phosphorus—coming from the birds' droppings are insignificant compared with the amounts of nutrients coming from drainage water, precipitation, and wastewater, inclusion of birds in the management strategy is only an unnecessary complication that only would contribute to the uncertainty. There are, however, a few cases where birds may contribute as much as 25% or at least more than 5% of the total inputs of nutrients. In such cases, it is of course important to include birds in the management and as a model component or at least as an important forcing function. A mass balance is always needed to uncover the main sources of a pollution problem—in the example of the inputs of nutrients. It is, however, rare (a good guess based on the author's experience is about 1% of aquatic ecosystems) that it is necessary to consider the droppings of birds as a significant source of eutrophication.

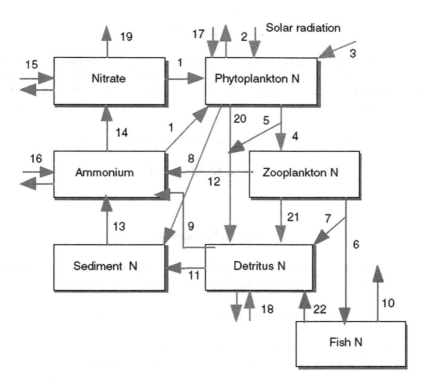

FIGURE 1 The conceptual diagram of a nitrogen cycle in an aquatic ecosystem. The processes that connect the state variables and forcing functions are the following: 1) uptake of nitrate and ammonium by algae; 2) photosynthesis; 3) nitrogen fixation; 4) grazing with loss of undigested matter; 5–7), predation and loss of undigested matter; 8) settling of algae; 9) mineralization; 10) fishery; 11) settling of detritus; 12) excretion of ammonium from zooplankton; 13) release of nitrogen from the sediment; 14) nitrification; 15–18) inputs/outputs; 19) denitrification; and 20–22) mortality of phytoplankton, zooplankton, and fish.

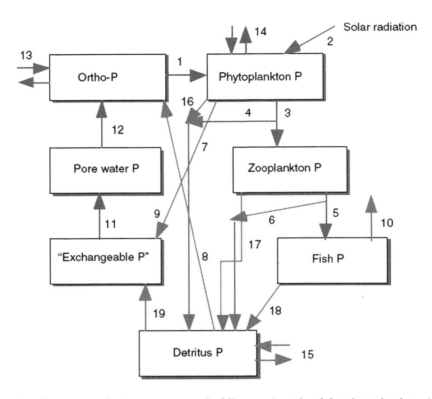

FIGURE 2 The phosphorus cycle. The processes are the following: 1) uptake of phosphorus by algae; 2) photosynthesis; 3) grazing with loss of undigested matter; 4–5) predation with loss of undigested material; 6, 7, 9) settling of phytoplankton; 8) mineralization; 10) fishery; 11) mineralization of phosphorous organic compounds in the sediment; 12) diffusion of pore water P; 13–15) inputs/outputs; 16–18) mortalities; and 19) settling of detritus.

The so-called Michaelis–Menten equation can be applied to describe the growth of phytoplankton:

$$gr = grmax \, NS / (kn + NS) \qquad (2)$$

Or

$$gr = grmax \, PS / (kp + PS) \qquad (3)$$

dependent on which nutrient is limiting, N or P. grmax, kn, and kp are parameters. If both nutrients are limiting in different periods of the year, the formulation is

$$gr = grmax \, \min \left(NS / (kn + NS), \ PS / (kp + PS) \right) \qquad (4)$$

The product or average of several limiting factors has also been proposed and applied.
    For the influence of the temperature, there are two possible formulations:

$$K_t^{(TEMP-20)} \left(\text{the so–called Arrhenius equation} \right) \qquad (5)$$

Or

$$\exp \left(A^* \left(TEMP - OPT \right) / \left(TEMP - MAXTEMP \right) \right) \qquad (6)$$

K_t is a parameter, which in most cases is between 1.04 and 1.06 and is on average 1.05. OPT is the optimum temperature for phytoplankton growth, and MAXTEMP is the maximum temperature. OPT, MAXTEMP, and A are all parameters that are different for different phytoplankton species.

The description of phytoplankton growth by equations using the constant stoichiometry approach is a simplification, because the phytoplankton growth is in reality a two-step process, as illustrated in Figure 3 and applied in the non-constant approach. The first step is uptake of nutrients, and the second step is growth of phytoplankton (increase of the biomass). The more correct description can be formulated mathematically by the following equations:

$$\text{Uptake rate } P = dPA/dt = PA^*maxupp^*\left(PS/\left(kp+PS\right)\right)$$

$$* \left(\left(PAMAX-PA\right)/\left(PAMAX-PAMIN\right)\right)$$

$$\text{Parallel for uptake rate } N \tag{7}$$

$$\text{Uptake rate } C = dCA/dt = CA^*maxupc^*\left(CS/kc+CS\right)^*$$

$$\left(\left(CAMAX-CA\right)/\left(CAMAX-CAMIN\right)\right)^* \left(\left(L/KL+L\right)-RESP\right.$$

$$\text{IF } L\langle L1, L \text{ is used; if } L2\rangle L>L1, L1 \text{ is used;}$$

$$\text{if } L>L2, L1+L2-L \text{ is used for } L \tag{8}$$

PA, NA, and CA are state variables that cover the amount of phosphorus, nitrogen, and carbon, in the form of phytoplankton, expressed as mg P, N, or C per liter of water. Notice that the unit is mg in 1 L of water. maxupp, maxupn, macups, kp, kn, kc, PAMAX, PAMIN, NAMAX, NAMIN, CAMAX, CAMIN, KL, L1, and L2 are all parameters. PAMAX, PAMIN, NAMAX, NAMIN, CAMAX, and CAMIN are,

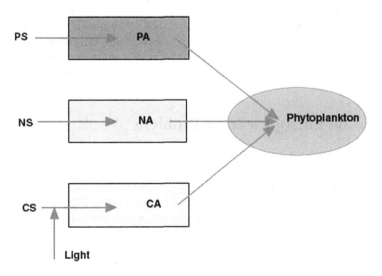

FIGURE 3 The two-step model of phytoplankton growth. The first step is uptake of nutrients PS, NS, and CS, followed by a growth of phytoplankton dependent on the nutrient concentrations in the phytoplankton cells, PA, NA, and CA. The uptake of carbon is dependent on light, while the uptake of phosphorus and nitrogen can take place even in darkness. It is a more physiologically correct description of phytoplankton growth than the (constant stoichiometry) approach.

however, known fairly well. They are the phytoplankton concentration times, respectively, 0.025, 0.005, 0.12, 0.05, 0.6, and 0.4 with good approximations. It is of course more difficult to calibrate the two-step growth equations than the constant stoichiometry approach due to the higher number of parameters in the NC equations, although the approximate knowledge that is available to the six parameters, PAMAX, PAMIN, NAMAX, NAMIN, CAMAX, and CAMIN, facilitates the calibration slightly. Notice that the uptakes of phosphorus, nitrogen, and carbon are, according to the equations, dependent both on the concentrations of the nutrients in the water and on the concentrations of nutrients in the cells.

The closer the nutrient concentrations in phytoplankton are to the minimum, the faster is the uptake. When a nutrient concentration, on the other hand, has reached the maximum value, the uptake stops. The carbon uptake opposite the uptake of phosphorus and nitrogen is dependent also on light, according to a Michaelis–Menten expression that includes the light prohibition. Finally, RESP covers the respiration. Of course, only carbon is involved in the respiration.

The growth process is quantified by the following equation:

$$\text{Growth} = \text{grmax}^* \, \text{phytoplankton}^* \, (min((PA - PAMIN)/$$

$$(PAMAX - PAMIN)), ((NA - NAMIN)/$$

$$(NAMAX - NAMIN)), ((CA - CAMIN)/$$

$$(CAMAX - CAMIN))) \qquad (9)$$

grmax is a parameter in line with the corresponding parameter in Eq. 1. Eq. 9 indicates that the higher the nutrient concentrations are compared with the minimum levels, the faster the growth is.

The phytoplankton growth model based on this approach has four state variables: PA, NA, CA, and phytoplankton. They are all assumed to be expressed in mg/L. As the minimum and maximum values are presumed to be a parameter times the phytoplankton concentration, they are also expressed in mg/L.

The two-step description is of course more difficult to calibrate, validate, and use generally, which of course raises the following question: when should the two-step description be applied instead of the more easily applicable constant stoichiometric approach? Di Toro and Conolly[2] have revealed that the need for the two-step description increases with the shallowness and eutrophication of the aquatic ecosystem; see Figure 4. It is definitely recommended for shallow, very eutrophied aquatic ecosystems to apply the two-step description, while it is hardly needed for deep mesotrophic or oligotrophic aquatic ecosystems.

Solutions to the Eutrophication Problem

It is usually necessary to apply a wide spectrum of methods to solve eutrophication problems, because the nutrients have many sources. The amount of nutrients discharged with the wastewater can be reduced by several wastewater treatment methods; see for instance the overview of conventional wastewater treatment methods and municipal wastewater and its treatment. Nutrients can be discharged by drainage water, including drainage water from agriculture, which may have high nutrient concentration; this can be removed or reduced by a number of ecological engineering methods, for instance, wetlands; see these entries.

It is possible to reduce the nutrient concentrations in the aquatic ecosystems by use of environmental technological methods, ecotechnological methods, cleaner technology, and environmental legislation. There are a number of cases where a consequent environmental strategy has solved the problems, but there are also many examples of insufficient environmental management, which has led to only a partial solution or no solution at all of the eutrophication problem. From experience, it can be concluded that quantitative nutrient balances, including all sources of nutrients, are the best starting point for a good

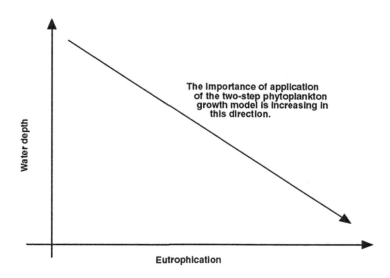

FIGURE 4 The need for the two-step description of phytoplankton growth increases with nutrient concentration and decreases with depth.

environmental management strategy, because the nutrient balances show clearly which sources are important to eliminate or reduce, and they facilitate a comparison of the costs for various management strategies. For further details about selection of an environmental management strategy, see Jørgensen[3] and Jørgensen et al. [4]

References

1. Jørgensen, S.E. *Handbook of Ecological Models Used in Ecosystem and Environmental Management*; CRC; Boca Raton, FL, 2011; 620 pp.
2. Di Toro, D.M.; Conolly, J.F. Jr. Mathematical models of water quality in large lakes, Part II: Lake Erie. U.S. Environmental Protection Agency, Ecological Research Series, EPA-600/3–80-065, 1980.
3. Jørgensen, S.E. *Principles of Pollution Abatement;* Elsevier: Amsterdam and Oxford, 2000; 520 pp.
4. Jørgensen, S.E.; Löffler, H.; Rast, W.; Straskraba, M. *Lake and Reservoir Management*; Elsevier: Amsterdam and Oxford, 2004; 503 pp.

69

Wastewater Use in Agriculture

Manzoor Qadir,
Pay Drechsel, and
Liqa Raschid-Sally

Introduction 701
Crop Selection and Diversification 702
Irrigation Management.. 704
Soil- and Health-based Interventions... 705
Conclusions... 706
References.. 707

Introduction

The volume of wastewater generated by domestic, industrial, and commercial water use has increased with population, urbanization, improved living conditions, and economic development. With the increase in wastewater generation, its productive use in agriculture has increased, particularly downstream of cities where often farmers have no or limited alternative sources of reliable irrigation water.[1–3] In developing countries, treated, partly treated, diluted, or untreated wastewater is used in urban and peri-urban areas to produce vegetables as a market-ready product. In addition, rice, wheat, fodder, and industrial crops are produced. In developed countries, wastewater is treated and mostly used for landscape irrigation, particularly in cities though extensive and well-regulated use in agriculture occurs as well.[4]

Despite official restrictions and potential health implications, farmers in many freshwater-scarce developing countries tend to opt for wastewater irrigation where available because (1) wastewater is a reliable or often the only water source available for irrigation throughout the year; (2) wastewater irrigation often reduces the need for fertilizer application as it is a source of nutrients; (3) wastewater use involves less energy cost even when pumping, if the alternative clean water source is from deep groundwater; and (4) wastewater generates additional benefits such as greater income generation from cultivation and marketing of high-value crops such as vegetables, which create year-round employment opportunities.[5,6]

As wastewater irrigation is in most instances part of the informal irrigation sector, authorities face challenges controlling or regulating the practice. The protection of consumer and farmer health and environment is especially the main concern.[7] Thus, sustainable use of wastewater must address three major aspects: pertinent policies, regulations, and institutional arrangements; wastewater treatment per intended reuse option; and risk management practices that eliminate or minimize the health and environmental impacts, particularly when wastewater treatment is limited. This entry addresses on-farm management practices with a focus on crop, water, and soil management interventions.

Crop Selection and Diversification

Where wastewater is used for irrigation, crop selection is determined by more factors than usual freshwater irrigation conditions. While the market potential of the crop is always an important criterion, the suitability of the crop to its biophysical environment requires more attention than is usually given. Even more important is the potential risk to consumers, which becomes a decisive factor influencing enforcement of regulations and crop restrictions. Especially if irrigated crops are eaten raw, as with salad greens, microbial contamination can be high; hence, different crops should be chosen. In guidelines addressing safe use of wastewater, it is advised to avoid cultivating crops eaten uncooked and prefer cereal and fodder crops, and trees.[8]

A particular challenge for farmers is the salinity of the wastewater as considerable variation exists among crops in their ability to tolerate saline conditions.[9] An appropriate selection of plant species capable of producing adequate biomass is vital while using saline water for irrigation.[10] Such selection is generally based on the ability of the species to withstand elevated levels of salinity in irrigation water and soil[9] while also providing a salable product or one that can be used on-farm.[11,12] The salt tolerance of a crop is not an exact value because it depends on several soil, crop, and climatic factors. This diversity can be exploited to identify local crops that are better adapted to saline and/or sodic soil conditions.[13,14]

Based on the linear response equation proposed by Maas and Hoffman,[9] crops can be characterized for their salt tolerances. The following are two parameters obtained from the Maas–Hoffman equation: (1) the maximum allowable soil salinity for a crop without yield reduction (the threshold soil salinity) and (2) the percent decrease in crop yield per unit increase in salinity beyond the threshold salinity level for the crop (the slope). Both values can be used to calculate relative yield for any given soil salinity exceeding the threshold level (Table 1).

Research efforts have led to the identification of several field crops, forage grasses and shrubs, biofuel crops, and fruit-tree and agroforestry systems, which can suit a variety of salt-affected environments.[15] Such systems linked to secure markets should support farmers in finding the most suitable

TABLE 1 Yield Potentials of Some Grain, Forage, Vegetable, and Fiber Crops as a Function of Average Root Zone Salinity

Common Name	Tolerance Based On	Yield Potential (%) at Specified Salinity (dS/m)		
		50%	80%	100%
Durum wheat	Grain yield	19	11	6
Barley	Grain yield	18	12	8
Cotton	Seed cotton yield	17	12	8
Rye	Grain yield	16	13	11
Sugar beet	Storage root	16	10	7
Wheat	Grain yield	13	9	6
Sorghum	Grain yield	10	8	7
Alfalfa	Shoot dry weight	9	5	2
Spinach	Top fresh weight	9	5	2
Broccoli	Shoot fresh weight	8	5	3
Eggplant	Fruit yield	8	4	1
Rice paddy	Grain yield	7	5	3
Potato	Tuber yield	7	4	2
Maize	Ear fresh weight	6	3	2

Source: Based on the salt tolerance data of different crops and percentage decrease in yield per unit increase in root zone salinity in terms of dS/m as reported by Maas and Grattan.[14]
Note: These data serve only as a guideline to relative tolerances among crops. Absolute tolerances can vary between varieties also depending on climate, soil conditions, and cultural practices.

and sustainable crop diversifying systems to mitigate any perceived production risks, while ideally also enhancing the productivity per unit of saline wastewater and protecting the environment.

Aside from excess salts in the water, there is an increasing possibility of heavy metal contamination where industrial effluent contributes to the water that farmers use. In terms of potential toxicity of the metals and metalloids, Hamilton et al.[16] classified them into four groups based on their retention in soil, translocation in plants, phytotoxicity, and potential risk to the food chain. They categorized cadmium, cobalt, selenium, and molybdenum as posing the greatest risk to human and animal health because they may appear in wastewater-irrigated crops at concentrations that are not generally phytotoxic; that is, farmers cannot rely on their plants dying before concentrations reach levels not recommended for humans.

Guidelines for maximum allowable levels of metals and metalloids in irrigation water are summarized in Table 2. Where these limits are exceeded, Simmons et al.[17] describe available methods for damage control.

TABLE 2 Recommended Maximum Concentrations (RMCs) of Selected Metals and Metalloids (mg/L) in Irrigation Water

Element	RMC[a]	Remarks
Aluminum	5.00	Can cause non-productivity in acid soils (pH < 5.5), but more alkaline soils at pH > 7.0 will precipitate the ion and eliminate any toxicity
Arsenic	0.10	Toxicity to plants varies widely, ranging from 12 mg/L for Sudan grass to less than 0.05 mg/L for rice
Beryllium	0.10	Toxicity to plants varies widely, ranging from 5 mg/L for kale to 0.5 mg/L for bush beans
Cadmium	0.01	Toxic at concentrations as low as 0.1 mg/L in nutrient solution for beans, beets, and turnips. Conservative limits recommended
Chromium	0.10	Not generally recognized as an essential plant growth element. Conservative limits recommended
Cobalt	0.05	Toxic to tomato plants at 0.1 mg/L in nutrient solution. It tends to be inactivated by neutral and alkaline soils
Copper[b]	0.20	Toxic to a number of plants at 0.1–1.0 mg/L in nutrient solution
Iron[b]	5.00	Non-toxic to plants in aerated soils, but can contribute to soil acidification and loss of availability of phosphorus and molybdenum
Lithium	2.50	Tolerated by most crops up to 5 mg/L. Mobile in soil. Toxic to citrus at low concentrations with recommended limit of <0.075 mg/L
Manganese[b]	0.20	Toxic to a number of crops at a few tenths to a few milligrams per liter in acidic soils
Molybdenum	0.01	Non-toxic to plants at normal concentrations in soil and water. Can be toxic to livestock if forage is grown in soils with high concentrations of available molybdenum
Nickel	0.20	Toxic to a number of plants at 0.5–1.0 mg/L; reduced toxicity at neutral or alkaline pH
Lead	5.00	Can inhibit plant cell growth at very high concentrations
Selenium	0.02	Toxic to plants at low concentrations and toxic to livestock if forage is grown in soils with relatively high levels of selenium
Zinc[b]	2.00	Toxic to many plants at widely varying concentrations; reduced toxicity at pH ≥ 6.0 and in fine textured or organic soils

Source: Modified from Ayers and Westcot.[22]

[a] The maximum concentration is based on a water application rate that is consistent with good irrigation practices (10,000 m³/ha/yr). If the water application rate greatly exceeds this, the maximum concentrations should be adjusted downward accordingly. No adjustment should be made for application rates less than 10,000 m³/ha/yr. The values given are for water used on a long-term basis at one site.

[b] Synergetic action of copper and zinc, and antagonistic action of iron and manganese have been reported in certain plant species' absorption and tolerance of metals after wastewater irrigation. If irrigation water contains high concentrations of copper and zinc, copper concentrations in the tissue may increase greatly. In plants irrigated with water containing a high concentration of manganese, manganese uptake in plants may increase, and consequently, the concentration of iron in the plant tissue may be reduced considerably. Generally, metal ion concentrations in plant tissue increase with concentrations in irrigation water. Metal ion concentrations in roots are higher than those in leaves[26] and metal ion concentrations such as cadmium concentrations in leafy vegetables are higher than those in non-leafy species.[27]

The examples show that the selection of crops for irrigation with wastewater has to consider a range of factors uncommon in freshwater irrigation. Finding the best compromise between the various criteria while targeting profits (i.e., high market demand) limits the choice of crop significantly in most cases.

Irrigation Management

In addition to the choice of the most appropriate crop(s), irrigation water management (ranging from water access to the type of irrigation, application rates, and scheduling) offers a variety of important management practices to address the particularities of wastewater irrigation.[15,17,18]

There are different ways in which crops are irrigated with wastewater, such as surface or flood irrigation, manual irrigation with watering cans, furrow irrigation, sprinkler irrigation, and micro-irrigation such as drip or trickle irrigation. There are also different ways of water access, from manual water fetching to pumping and gravity flow. Each method has a different level of risk for farmers (through farmers' contact with the irrigation water) and consumers [through the contact of the harvested (and eventually consumed) crop part and the water]. Several parameters for the evaluation of commonly used irrigation methods in relation to risk reduction are given in Table 3.[7,19] Key criteria are health risks, costs, and water use efficiency.

Flood irrigation is usually a low-cost method with, however, also low water use efficiency. Health protection is limited for the farmers and—if the crops are low growing—for the consumers as well. With medium level of health protection, furrow irrigation needs more soil preparation and is suitable when

TABLE 3 Parameters for Evaluation of Selected Irrigation Methods in Relation to Risk Reduction for Crops and Humans

Evaluation Parameter	Irrigation Method			
	Furrow Irrigation	Flood Irrigation	Sprinkler Irrigation	Drip Irrigation
Farmer exposure to pathogens	Low to medium	Medium to high	Low when sprinkler is off	Very low
Crop exposure to pathogens	Low if planted on ridge	High only for low-growing crops	High	Very low
Possibility of leaf damage from salts resulting in poor yield	No foliar injury as the crop is planted on the ridge	Some bottom leaves may be affected, but the damage is not so serious as to reduce yield	Severe leaf damage can occur resulting in significant yield loss	No foliar injury likely
Root zone salt accumulation with repeated applications	Salts tend to accumulate in the ridge, which could harm the crop	Depending on soil texture, salts might move vertically downwards or can accumulate in the root zone	Salt drainage is limited as water amounts are low. Surface crusting possible	Salt movement is radial along the direction of water movement. A salt wedge is formed between drip points; clogging of pipes can occur
Ability to maintain high soil water potential (risk of soil moisture stress)	Plants may be subject to stress between irrigations	Plants may be subject to water stress between irrigations	Not possible to maintain high soil water potential throughout the growing season	Possible to maintain a high and well-targeted soil moisture content throughout the growing season
Suitability to handle brackish wastewater without significant yield loss	Fair to medium. With good management and drainage, acceptable yields are possible	Fair to medium. Good irrigation and drainage practices and conditions can produce acceptable levels of yield	Poor to fair. Crops might suffer from leaf damage resulting in low yields	Excellent to good. Almost all crops can be grown with very little reduction in yield, unless the pipes clog

Source: Modified from Pescod.[19]

TABLE 4 On-Farm Options for Pathogen Reductions

Control Measure	Pathogen Reduction (Log Units)	Notes
Alternative safe water source	>6	Depends on availability of safe groundwater and/or alternative farm land
Crop restrictions	>6	Acceptance depending on controls and profit margin of the alternative crop
Drip irrigation	2–4	2-log unit reduction for low-growing crops, and 4-log unit reduction for high-growing crops
Pathogen die-off	0.5–2 per day	Die-off after last irrigation before harvest (rate depends on climate, crop type, etc.)
Slow sand filter	1–3	Depends on appropriate particle size
Furrow irrigation	1–2	Might reduce cropping density and yield
Reduced splashing	1–2	Splashing adds contaminated soil particles onto the crop, which can be avoided
Allow sedimentation in ponds and dugouts	1–2	During dry season via natural die-off. Reduction of helminths to less than 1 egg over 2–3 days possible

Source: Modified from WHO.[8]

there is a greater leaching need to remove high levels of salts. Irrigation with sprinklers is medium to high cost, does not require soil preparation, but has medium-level water use efficiency. Sprinkler irrigation systems have the advantage of reducing the amounts of water and salts applied to soil and crop.[20,21] The disadvantage compared to flood and furrow irrigation is that sprinklers distribute any water contaminants straight on the top of the plants. The same applies to the use of watering cans unless they are directed towards the roots. Overhead irrigation may also cause leaf burn under direct sunlight, from salts absorbed directly through wetted leaf surfaces,[22] which can be avoided by irrigating at night.[14,23,24]

Drip irrigation systems are costly, but highly efficient in water use along with the highest levels of health protection for farmers and consumers. The clogging of drippers on the other hand may limit the use of drip irrigation systems for many types of wastewater. Therefore, prior filtration is needed to prevent clogging of emitters.[25]

Irrigation frequency is an ambivalent issue in wastewater irrigation. Because soluble salts reduce the availability of water in almost direct proportion to their concentration, irrigation frequency should generally be high. This will help in maintaining the moisture content and salinity of irrigated soils at acceptable levels, which is important especially during seedling establishment.[23] On the other hand, a low frequency of irrigation—if possible even the cessation of irrigation for several days before harvest—supports the natural die-off of pathogens and is an important low-cost measure for health risk reduction.[7]

Other options for on-farm interventions addressing risks from pathogens are summarized in Table 4. Options to cope with salinity or heavy metals were recently summarized by Qadir and Drechsel[10] and Simmons et al.[17]

Soil- and Health-based Interventions

Good management practices play a crucial role in the preservation of key soil properties while irrigating with wastewater. Soil-based interventions are important, particularly in case of inorganic contaminants, which usually accumulate in the upper part of the soil because of strong adsorption and precipitation phenomena. For moderate levels of metals and metalloids in wastewater, there is no particular management needed if the soils are calcareous, i.e., contain appreciable levels of calcite that renders most metals immobile. However, metal ions may be a problem in acid soils, which need specific management measures such as liming, avoiding use of fertilizers with acidic reactions, and selection of crops that do

TABLE 5   Contribution of Irrigation with Recycled Wastewater in Terms of Nutrient Addition to the Soil

Nutrient	Concentration (mg/L)	Fertilizer Contribution (kg/ha)	
		Irrigation at 3,000 m³/ha	Irrigation at 5,000 m³/ha
Nitrogen	16–62	48–186	80–310
Phosphorus	4–24	12–72	20–120
Potassium	2–69	6–207	10–345
Calcium	18–208	54–624	90–1040
Magnesium	9–110	27–330	45–550

Source: Lazarova and Bahri.[20]

not accumulate the metals of concern.[22] In case of irrigation with wastewater containing elevated levels of sodium, care should be taken to avoid soil structure deterioration. Application of a source of calcium such as gypsum is desirable. Procedures to determine the rate of gypsum application to mitigate the effects of sodium resulting from sodic wastewater irrigation are available.

The quality and depth of groundwater prior to wastewater irrigation determine the detrimental effects of salts, nitrates, and metals reaching groundwater. The deeper the groundwater, the longer it will take to have such effects. In case of shallow groundwater or coarse-textured soils, i.e., sandy soils, which are highly permeable, care must be taken to prevent groundwater pollution.

Although the fertilizer value of undiluted wastewater, in particular, is of great importance as nutrients in wastewater contribute to crop requirements, periodic monitoring is required to estimate the nutrient loads in wastewater and adjust fertilizer applications accordingly.[20] Excessive nutrients not only can cause nutrient imbalances, undesirable vegetative growth, and delayed or uneven maturity, but can also reduce crop quality while polluting groundwater and surface water. The amount of nutrients applied via wastewater irrigation can vary considerably if it is raw, treated, or diluted with stream water. The contribution in terms of nutrient addition to the soil from irrigation with recycled wastewater is given in Table 5.

It becomes obvious that irrigation with wastewater has a variety of implications steering farmers' decision-making process in view of crop selection as well as soil and water management. Most of them are limiting the choice of options. However, while a wrong choice, for example, in view of salinity management or any other phytotoxic hazard usually results in a quick learning process for the farmer, hazards affecting farmers' and especially consumers' health might remain less obvious and hidden among various confounding factors, such as poor sanitation at home. With economic interests being of paramount importance in most cases, it is not surprising that farmers might opt for those options that yield highest returns while keeping investments as low as possible. The result is the common picture of high-value exotic vegetables irrigated with low-cost watering cans or via flood irrigation, both imposing high risks for human health. Increasing awareness about these risks and providing incentives and regulations to encourage alternative crop choices in spite of possible disadvantages will create the conditions that would favor the utilization of farm-based interventions for the safe and productive use of wastewater in irrigated agriculture.

Conclusions

In the arid and semi-arid areas of the world, wastewater is mainly used for agriculture because of the competition between agriculture and municipal sectors. While wastewater is managed safely and used in treated form mostly for landscape irrigation in developed countries, the situation is different in developing countries, where wastewater is rarely used in treated form but mostly in partly treated, diluted, or untreated forms to produce a range of crops, mostly vegetables as a market-ready product. Such wastewater use in developing countries continues to offset local water scarcity despite official restrictions and potential health implications. Thus, it is important to consider a range of risk management practices to eliminate or minimize the health and environmental impacts, particularly in situations where wastewater is used for irrigation in partly treated, diluted, or untreated forms. On-farm management practices can play a major role to eliminate or minimize the health and environmental impacts. The following aspects are the key features and have to be considered: 1) selection and diversification of wastewater-irrigated crops to reduce possible health risks, accounting for their market value and tolerance against ambient stress from salts, metals, and metalloids; 2) irrigation water management addressing water access, on-farm treatment, type of irrigation, application rates, and irrigation scheduling; and 3) soil- and health-based interventions considering soil characteristics, soil preparation practices, and application of fertilizers and amendments. Increasing awareness about health and environmental risks, and knowledge and implementation of best farm-based interventions by the wastewater-irrigating farmers will help create the conditions that would favor the safe and productive use of wastewater in agriculture.

References

1. Qadir, M.; Wichelns, D.; Raschid-Sally, L.; Minhas, P.S.; Drechsel, P.; Bahri, A.; McCornick, P. Agricultural use of marginal-quality water—Opportunities and challenges. In *Water for Food, Water for Life: A Comprehensive Assessment of Water Management in Agriculture*; Molden, D., Ed.; Earthscan: London, 2007; 425–457.

2. Jiménez, B.; Asano, T. Water reclamation and reuse around the world. In *Water Reuse: An International Survey of Current Practice, Issues and Needs*; Jimenez, B., Asano, T., Eds.; IWA Publishing: London, 2008; 3–26.

3. Drechsel, P.; Scott, C.A.; Raschid-Sally, L.; Redwood, M.; Bahri, A., Eds. *Wastewater Irrigation and Health: Assessing and Mitigating Risks in Low-Income Countries*; Earthscan–International Development Research Centre (IDRC)–International Water Management Institute (IWMI), Earth-scan: London, 2010.

4. Asano, T. *Wastewater Reclamation and Reuse*; Water Quality Management Library Series. Vol. 10; Technomic Publishing Co., Inc.: Lancaster, PA, 1998.

5. Keraita, B.N.; Jimenez, B.; Drechsel, P. Extent and implications of agricultural reuse of untreated, partly treated and diluted wastewater in developing countries. *CAB Rev.* 3 (058), 2008; 1–15.

6. Qadir, M.; Scott, C.A. Non-pathogenic tradeoffs of wastewater irrigation. In *Wastewater Irrigation and Health: Assessing and Mitigating Risks in Low-Income Countries*; Drechsel, P., Scott, C.A., Raschid-Sally, L., Redwood, M., Bahri, A., Eds.; Earthscan–International Development Research Centre (IDRC)–International Water Management Institute (IWMI), Earthscan: London, 2010; 101–126.

7. Jiménez, B.; Drechsel, P.; Koné, D.; Bahri, A.; Raschid-Sally, L.; Qadir, M. Wastewater, sludge and excreta use in developing countries: An overview. In *Wastewater Irrigation and Health: Assessing and Mitigating Risks in Low-Income Countries*; Drechsel, P., Scott, C.A., Raschid-Sally, L., Redwood, M., Bahri, A., Eds.; Earthscan–International Development Research Centre (IDRC)–International Water Management Institute (IWMI), Earthscan: London, 2010; 3–27.

8. WHO. *Guidelines for the Safe Use of Wastewater, Excreta and Grey Water: Wastewater Use in Agriculture*; WHO, FAO, UNEP: Geneva, 2006; Vol. 2.

9. Maas, E.V.; Hoffman, G.J. Crop salt tolerance—Current assessment. *J. Irrig. Drainage Div.* **1977**, *103*, 115–134.

10. Qadir, M.; Drechsel, P. Managing salts while irrigating with wastewater. *CAB Rev.* 5 (016), 2010; 1–14.

11. Qadir, M.; Oster, J.D. Crop and irrigation management strategies for saline-sodic soils and waters aimed at environmentally sustainable agriculture. *Sci. Tot. Environ.* **2004**, *323*, 1–19.

12. Hamilton, A.J.; Stagnitti, F.; Xiong, X.; Kreidl, S.L.; Benke, K.K.; Maher, P. Wastewater irrigation: The state of play. *Vadose Zone J.* **2007**, *6* (4), 823–840.
13. Shannon, M.C. Adaptation of plants to salinity. *Adv. Agron.* **1997**, *60*, 76–120.
14. Maas, E.V.; Grattan, S.R. Crop yields as affected by salinity. In *Agricultural Drainage*; Skaggs, R.W., van Schilfgaarde, J., Eds.; ASA-CSSA-SSSA: Madison, 1999; 55–108.
15. Qadir, M.; Tubeileh, A.; Akhtar, J.; Larbi, A.; Minhas, P.S.; Khan, M.A. Productivity enhancement of salt-affected environments through crop diversification. *Land Degrad. Dev.* **2008**, *19*, 429–453.
16. Hamilton, A.J.; Boland, A.-M.; Stevens, D.P.; Kelly, J.; Radcliffe, J.; Dillon, P.J.; Paulin, B. Position of the Australian horticultural industry with respect to the use of reclaimed water. *Agric. Water Manage.* **2005**, *71*, 181–209.
17. Simmons, R.W.; Qadir, M.; Drechsel, P. Farm-based measures for reducing human and environmental health risks from chemical constituents in wastewater. In *Wastewater Irrigation and Health: Assessing and Mitigating Risks in Low-Income Countries*; Drechsel, P., Scott, C.A., Raschid-Sally, L., Redwood, M., Bahri, A., Eds.; Earthscan–International Development Research Centre (IDRC)–International Water Management Institute (IWMI), Earthscan: London, 2010; 209–238.
18. Keraita, B.; Konradsen, F.; Drechsel, P. Farm-based measures for reducing microbiological health risks for consumers from informal wastewater-irrigated agriculture. In *Wastewater Irrigation and Health: Assessing and Mitigation Risks in Low-Income Countries*; Drechsel, P., Scott, C.A., Raschid-Sally, L., Redwood, M., Bahri, A., Eds.; Earthscan–IDRC–IWMI, Earthscan: London, 2010; 189–207.
19. Pescod, M.B., Ed. *Wastewater Treatment and Use in Agriculture*; Irrigation and Drainage Paper No. 47; Food and Agriculture Organization of the United Nations: Rome, Italy, 1992.
20. Lazarova, V.; Bahri, A. *Water Reuse for Irrigation: Agriculture, Landscapes, and Turf Grass*; CRC Press: Boca Raton, FL, 2005.
21. Qadir, M.; Raschid-Sally, L.; Drechsel, P. Wastewater use in agriculture: Agronomic considerations. In *Encyclopedia of Water Science*; Trimble, S.W., Ed.; Taylor & Francis: New York, 2008; 1296–1299.
22. Ayers, R.S.; Westcot, D.W. *Water Quality for Agriculture*; Irrigation and Drainage Paper 29 (Rev. 1); FAO: Rome, 1985.
23. Rhoades, J.D. Use of saline drainage water for irrigation. In *Agricultural Drainage*; Skaggs, R.W., van Schilfgaarde, J., Eds.; American Society of Agronomy (ASA)–Crop Science Society of America (CSSA) –Soil Science Society of America (SSSA): Madison, WI, 1999; 615–657.
24. Qadir, M.; Minhas, P.S. Wastewater use in agriculture: Saline and sodic waters. In *Encyclopedia of Water Science*; Trimble, S.W., Ed.; Taylor & Francis: New York, 2008; 1307–1310.
25. Minhas, P.S.; Samra, J.S. *Wastewater Use in Peri-urban Agriculture: Impacts and Opportunities*; Central Soil Salinity Research Institute: Karnal, 2004.
26. Kalavrouziotis, I.K.; Drakatos, P.A., 2002. Irrigation of certain Mediterranean plants with heavy metals. *Int. J. Environ. Pollut.* 2004, *18*, 294–300.
27. Qadir, M.; Ghafoor, A.; Murtaza, G. Cadmium concentration in vegetables grown on urban soils irrigated with untreated municipal sewage. *Environ. Dev. Sustain.* **2000**, *2*, 11–19.

70

Wetlands: Biodiversity

Role of Hydrology 709
A Landscape Perspective 709
Wetland Flora...709
Wetland Fauna 711
Conclusions 712
References... 712

Jean-Claude
Lefeuvre and
Virginie Bouchard

Role of Hydrology

Wetlands differ significantly in their water source and seasonal hydrologic regime. Hydrological patterns (i.e., flooding frequency, duration and hydroperiod) influence physical and chemical characteristics (e.g., salinity, oxygen and other gas diffusion rates, reduction–oxidation potential, nutrient solubility) of a wetland. In return, these internal parameters and processes control flora and fauna distribution as well as ecosystem functions. Plants, animals and microbes are often oriented in predictable ways along the hydrological gradient (Figure 1). Conversely, the biotic component affects the hydrology by eventually modifying flow or water level in a wetland.[1,2] Species also influence nutrient cycles and other ecosystem functions.[3]

A Landscape Perspective

Although the hydrology is part of the ecological signature of an individual wetland, wetlands are neither considered as aquatic nor terrestrial systems. They have characteristics from both systems and are defined as ecotones placed under this dual influence.[4] Because wetlands are located at the interface of multiple systems, they assure vital functions (e.g., wildlife habitats) beneficial at the landscape level. Reduction of wetland area often reduces biodiversity in the land-scape.[2,5] Increases in biodiversity occur when wetlands are created or restored in a disturbed landscape.[6]

Wetland Flora

Wetland plants are adapted to a variety of stressful abiotic conditions (e.g., immersion, wave abrasion, water level fluctuation, low oxygen conditions). Identical adaptations to common environmental features have led taxonomically distinct species to sometimes look similar in terms of morphology, life cycle and life forms.[7] Traditionally, wetland plants have been classified into groups of different life forms, primarily in relation with hydrological conditions. Helophytes are defined as plant species with over-wintering buds in water or in the submerged bottom.[8] They are differentiated from hydrophytes in that their vegetative organs are partially raised above water level.[8] Hejny's classification is based on relatively stable vegetative features that determine the ability of wetland plants to survive two unfavorable conditions, cold and drought.[7] This classification uses the types of photosynthetic organs present in both the growth and flowering phases. Other classifications include both life form and growth form. When a species has a range of growth form, it is classified under the form showing the greatest achievement of its potential.[7]

Plant communities in wetlands can be more or less homogeneous, mosaic-like, or distributed along a gradient resulting in a clear zonation of species. A gradient exists if one or several habitat parameters change gradually in space. This phenomenon is common in fresh water marshes that present a gradient of water depth and water saturated soils (Figure 1). Such a gradient is often accompanied by differences in peat accumulation that is influenced by waves or currents. General principles of the zonation of aquatic plants have been largely described.[7,9,10] Littoral vegetation can belong to several types of communities, which derived from the general principle that, from deeper water to the shore, we may expect successively submergent, floating, and emergent macrophytes. The most important habitat factor is water depth, depending on slope and peat accumulation.[10] Other factors may be poor irradiance caused by high turbidity or exposure to waves or flow.[7,10]

Riparian ecosystems are found along streams and rivers that occasionally flood beyond confined channels or where riparian sites are created by channel meandering in the stream network. Riparian or bottomland hardwood forests contain unique tree species that are flood tolerant. Species distribution is associated with floodplain topography, flooding frequency, and flooding duration.[11,12] In southeastern U.S. bottomland, seasonally flooded forests are colonized by *Platanus occidentalis* (sycamore), *Ulmus americana* (American elm), *Populus deltoides* (cottonwood) and are flooded between 2% and 25% of the growing season (Figure 1). Other species such as *Fraxinus pennsylvanica* (green ash), *Celtis laevigata* (sugarberry) and *Carya aquatica* (water hickory) colonized bottomlands that are flooded by less than 2% of the growing season.[11] Freshwater marshes are dominated with emergent macrophytes rooted in

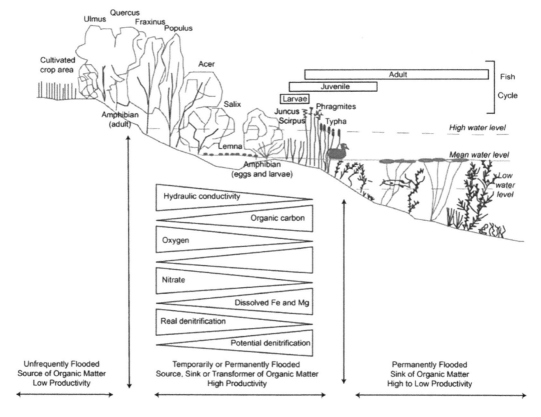

FIGURE 1 Species distribution along the hydrological gradient in a freshwater marsh.

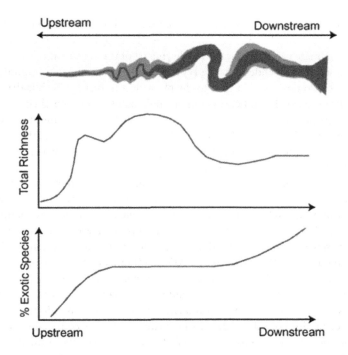

FIGURE 2 Total richness and invasive species distribution along a stream longitudinal gradient.
Source: Planty-Tabacchi.[14]

the bottom with aerial leaves (i.e., helophytes). Species such as *Typha* (cattail), *Phragmites* (reed grass) and *Scirpus* (bulrush) are often clonal. A plant community is usually organized in sequence of patches that are dominated by one species. The second plant groups are the rooted plants with floating leaves (*Nymphaeid*). Lotus (*Nelumbo*) and water lilies (*Nymphea*) have very identical morphology (i.e., similar leaves and flowers) but a genetic analysis showed that lotus is more closely related with plane-tree than with water lilies.[10] Submerged plants include elodeids (i.e., cauline species whose whole life cycle can be completed below the water surface or where only the flowers are emergent) and isoetids (i.e., species growing on the bottom whose whole life cycle can be completed without contact with the surface). Submerged species include species such as coontail (*Ceratophyllum demersum*) and water milfoil (*Myriophyllum* spp.). Plant species found in salt marshes are called halophytes (i.e., plants which complete their cycle in saline environments). A saltmarsh can be divided into low, middle, and upper marsh, according to flooding frequency and duration. Each zone is dominated by different plant species according to their tolerance to saline immersion.[10]

A dominant competitive species—often a clonal species—can modify the theoretical zonation. Change in water chemistry (i.e., eutrophication) or hydrology may favor a particular species over the natural plant community. Highly competitive species are often invasive and aggressive in displacing native species. The expansion of *Phragmites australis* into tidal wetlands of North America causes a reduction in biodiversity as many native species of plants are replaced by a more cosmopolitan species.[13] In riparian ecosystems, biodiversity is usually higher in the intermediate zones, whereas it is lower upstream and downstream (Figure 2). The percentage of exotic species is low in upstream areas, but can represent up to 40 in downstream zones (Figure 1).[14]

Wetland Fauna

Diversity of vertebrate and invertebrate species is the result of a diverse community composed of resident and transient species, which use the space differently and at various times of day and year. The density and variety of animal populations at a particular wetland site is also explained by climatic events that affect geographic areas on a large scale. For example, the population of waterfowl during winter is largely dependent on climate variations in the northern part of the continents.[15]

Resident species are often dependent on the type of vegetation. Animal communities are generally distributed along a zonation pattern, parallel to the plant communities, which is driven by the hydrological gradient. A few species depend entirely on a single plant species for their survival. One beetle (*Donacia* spp.) may depend on reeds, at least during its larval stage where another beetle (*Gale-rucella* spp.) uses only water lilies as their habitat and diet. For many other residents, their habitats extend to several plant communities during their life span. It is generally the case for many vertebrates such as amphibians, rodents, passerines, and waterfowl.

Many amphibian species depend on wetland or riparian zones for reproduction and larval stage. Likewise, wet meadows are necessary as a reproduction zone and nursery for a number of freshwater fishes. About 220 animals and 600 plant species are threatened by a serious reduction of wetlands in California, and the state's high rate of wetland loss (91% since the 1780s) is partly responsible.[15] Many waterfowl species are sensitive to areas of reduction, patch size and distribution, wetland density, and proximity to other wetlands.[16] When the Marais Poitevin (France), one of the principal wintering and passages sites for waterfowl in the Western Europe, underwent agricultural intensification in the 1980s, the population of ducks and waders declined tremendously. This decline was partly due to a 50% reduction of wet meadows between 1970 and 1995, primarily caused by the conversion to arable farmlands.[17] Thus, maintenance of biodiversity depends on the existence of inter connections between wetlands, and between aquatic and terrestrial ecosystems. In fact, some authors have pointed out that increase in biodiversity occurs when wetlands are created or restored in a disturbed landscape.[6,18]

Conclusions

Despite the importance of wetlands, conservation efforts have ignored them for a long time. It is urgent to conserve the existing wetlands, and also to restore and create wetland ecosystems. A wide range of local, state, federal, and private programs are available to support the national policies of wetland "No Net Loss" in the U.S., and around the world. From a biodiversity perspective, on-going wetland protection policies may not be working because restored or created wetlands are often very different from natural wetlands.[19] Created wetlands often result in the exchange of one type of wetland for another, and result in the loss of biodiversity and functions at the landscape level.[19] We know now that it takes more than water to restore a wet- land,[20] even if an important place should be given to the ability of self-design of wetland ecosystems.[21]

References

1. Naiman, R.J.; Johnston, C.A.; Kelley, J.C. Alteration of north American streams by beaver. Bioscience **1988**, *38*, 753–762.
2. National Research Council. *Wetlands: Characteristics and Boundaries;* National Academy Press: Washington, DC, 1995; 306 pp.
3. Naiman, R.J.; Pinay, G.; Johnston, C.A.; Pastor, J. Beaver influences on the long-term biogeochemical characteristics of boreal forest drainage networks. Ecology **1994**, *75*, 905–921.
4. Hansen, A.J; Di Castri, F; Eds.; Landscape boundaries. In *Consequences for Biotic Diversity and Ecological Flows*; Ecological Studies; Springer: Berlin, 1983; 452 pp.
5. Gibbs, J.P. Wetland loss and biodiversity conservation. Conser. Biol. **2000**, *14*, 314–317.
6. Weller, M.W. *Freshwater Marshes,* 3rd Ed.; University of Minnesota Press: Minneapolis, US, 1994; 192 pp.
7. Westlake, D.F.; Kvet, J.; Szczepanski, A.; Eds.; *The Primary Ecology of Wetlands;* Cambridge University Press, 1998; 568 pp.

8. Raunkiaer, C. *The Fife Forms of Plants and Statistical Plant Geography*; Oxford University Press: Oxford, England, 1934; 632 pp.

9. Grevilliot, F.; Krebs, L.; Muller, S. Comparative importance and interference of hydrological conditions and soil nutrient gradients in floristic biodiversity in flood meadows. Biodivers. Conserv. **1998**, *7*, 1495–1520.

10. Mitsch, W.J.; Gosselink, J.G. *Wetlands*, 3rd Ed.; Wiley: New York, 2000; 920 pp.

11. Clark, J.R.; Benforado, J.; Eds.; *Wetlands of Bottomland Hardwood Forest*; Elsevier: Amsterdam, the Netherlands, 1981; 401 pp.

12. Keogh, T.M.; Keddy, P.A.; Fraser, L.H. Patterns of tree species richness in forested wetlands. Wetlands **1999**, *19*, 639–647.

13. Chambers, R.M.; Meyerson, L.A.; Saltonstall, K. Expansion of Phragmites Australis into Tidal wetlands of North America. Aquat. Bot. **1999**, *64*, 261–273.

14. Planty-Tabacchi, A.M. Invasions Des Corridors Riverains Fluviaux Par Des Espèces Végétales D'origine Étrangère. These University Paul Sabatier Toulouse III, 1993; France, 177 pp.

15. Hudson, W.E., Ed.; *Landscape Linkages and Biodiversity*; Island Press: Washington, DC, 1991.

16. Leibowitz, S.C.; Abbruzzese, B.; Adams, P.R.; Hughes, L.E.; Frish, J. A Synoptic Approach to Cumulative Impact Assessment: A Proposed Methodology, 1992; EPA/600/R-92/167.

17. Duncan, P.; Hewison, A.J.M.; Houte, S.; Rosoux, R.; Tour-nebize, T.; Dubs, F.; Burel, F.; Bretagnolle, V. Long-term changes in agricultural practices and wildfowling in an internationally important wetland, and their effects on the guild of wintering ducks. J. Appl. Ecol. **1999**, *36*, 11–23.

18. Hickman, S. Improvement of habitat quality for resting and migrating birds at the des plaines river wetlands demonstration project. Ecol. Engng. **1994**, *3*, 485–494.

19. Whigham, D.F. Ecological issues related to wetland preservation, restoration, creation and assessment. Sci. Total Environ. **1999**, *240*, 31–40.

20. Zedler, J.B. Progress in wetland restoration ecology. Trends Ecol. Evol. **2000**, *15* (10), 402–407.

21. Mitsch, W.J.; Wu, X.; Nairn, R.W.; Weihe, P.E.; Wang, N.; Deal, R.; Boucher, C.E. Creating and restoring wetlands: a whole ecosystem experiment in self-design. BioScience **1998**, *48*, 1019–1030.

71

Wetlands: Carbon Sequestration

Introduction 715
Carbon Sequestration with Primary Production 715
Carbon Sequestration in Soils 717
Impact of Human Activities on Carbon Storage 717
Creation of Wetlands to Sequester Carbon 718
Conclusions 718
References 718

Virginie Bouchard
and Matthew
Cochran

Introduction

The increase in the concentration of "greenhouse gases" in the troposphere and its relation to human activities is now well-documented.[1] Carbon fixation via photosynthesis and carbon release during decomposition have always been two important processes regulating the concentrations of CO_2 and CH_4 in the atmosphere, even in prehistoric and pre-industrial times. In addition to these internal processes, wetland ecosystems receive and release organic carbon through hydrologically-driven mass fluxes. Thus, the carbon pool within any wetland ecosystem is in balance between primary production, microbial decomposition, and carbon fluxes within interconnected ecosystems (Figure 1). Wetlands play a particularly complex role in controlling greenhouses gases, as these ecosystems are intimately associated with all aspects of the production and consumption of both CO_2 and CH_4 (Figure 2).

Carbon Sequestration with Primary Production

The concept of primary productivity is directly related to the ideas of energy flow in ecosystems. A portion of the photosynthetically active radiation (PAR), which is radiation in the 400–700 nm wave band, received by an ecosystem is absorbed by autotrophic organisms (photosynthetic plants and microorganisms). The absorbed energy is reradiated, lost as latent heat, or stored by the activity of photosynthesis in organic substances. This last flow of energy corresponds to net primary production (NPP). Fundamental ecological questions relating to the global carbon budget, the location of the missing carbon sink, and predictions of global climate change rely on obtaining good estimates of NPP.

Wetland ecosystems can have very high standing biomass values and correspondingly high NPP.[2,3] The annual aboveground NPP of macrophytes is reported to be up to 5 kg dry matter in the most productive sites.[2] This production varies according to species, wetland type, and latitude[2,3] and is often well correlated with the maximum aboveground standing crop. Belowground production is much more difficult to estimate because roots and rhizomes grow and die at different rates and times and because materials are translocated to and from shoots. The ratio of belowground to aboveground production can vary from 0.2 to 2.5 according to various studies.[2] The connectivity of a wetland to hydrological fluxes

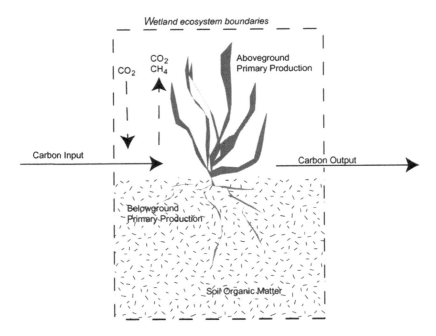

FIGURE 1   Conceptual model showing the fluxes of organic carbon in wetlands.

FIGURE 2   View of a freshwater wetland.

(i.e., saltmarshes flooded by tides, coastal freshwater marshes flooded by seiche, riparian bottomland forests flooded by floods) is one of the most important factors enhancing primary productivity.[4]

Net carbon capture from herbaceous vegetation is minimal compared to the long-term accumulation of carbon in bottomland hardwood forests. At the end of the growing season, most of the carbon trapped in the biomass of herbaceous macrophytes is found in dead plant litter and is released as CO_2 or CH_4 during decomposition or exported as dissolved or particulate organic matter to adjacent systems.

Bottomland hardwood forests store carbon in tree biomass for a much longer period of time. The productivity of bottomland forest ranges between 200 g and 2000 g dry organic matter per year.[5]

Carbon Sequestration in Soils

Carbon fixed in photosynthesis either remains in the sediment as carbon accretion or is decomposed to CO_2 and CH_4 by a suite of fermentative microbes involved in soil organic matter decomposition. A diversity of biological, chemical, and physical mechanisms is also known to selectively "protect" different pools of soil organic matter from decomposition by soil microorganism. Most of the recalcitrant carbon destined for sediment accretion is derived from heavily lignified biomass. Organic soil is a result of the anaerobic conditions created by standing water or poorly drained conditions.

Carbon is even better protected in acidic environments, in marine sediments and under low temperatures. Peat accumulation is a result of reduced oxidation of the biomass produced in wetlands. The northern peatlands have accumulated 25–38 Pmol C since the last glaciation, equivalent to 50%–70% of the total amount of carbon currently present in the atmosphere.[6] Peat accumulation is greater in bogs than in fens due to differences in nutrient availability. Decomposition rates are enhanced in minerotrophic fens receiving nutrients from adjacent mineral soils relative to ombrotrophic bogs, which are fed only by rainwater.

Canadian wetland soils store about 154 Gt C, representing an average of 60% more carbon than what is stored in Canadian forests (95 Gt C in biomass and soils), which in turn is two orders of magnitude larger than the agricultural soil carbon pool[7] (Table 1). Peatlands are a sink for between 20 and 30 g Cm^{-2} yr^{-1}, which means that Canadian peatlands sequester about 0.03 Gt Cyr^{-1} (Table 1). The sequestration rate of peatlands is currently smaller than the rate computed for boreal forests. However, the major difference between forests and peatlands is that a forest sink is transient and cannot be sustained continuously. If peatlands also have a theoretical limit to growth, it is reached only after many thousands of years.[8] However, the fluxes of CO_2 and CH_4 from peatlands vary highly according to seasons, years, species and sites. Inter-annual measurement schemes—such as those that have been developed for forest ecosystems with the Ameriflux and Euroflux programs—are needed are needed to establish more precise fluxes of CO_2.

Impact of Human Activities on Carbon Storage

Rates of plant production are limited by phosphorus supply in freshwater systems and by nitrogen supply in saltwater and terrestrial systems.[9,10] An excess of nutrients leads to an increase of both primary production and decomposition in terrestrial systems.[11] Decomposition rates increase, however, at a slower rate than primary production, leading to a potential increase of carbon storage in soil.[11] In wetlands, little is known about the effect of excess nutrients on the denitrification process, which is directly linked to the amount of soil organic matter.[12]

Wetlands have been drained on all continents due to human development. Rapid changes occur in organic soils after drainage as aerobic microbial decomposition is enhanced, releasing large amounts of CO_2 into the atmosphere. Carbon losses can lead to subsidence of the soil profile, changes in bulk density, and decreases in carbon content of the remaining soil.

TABLE 1 Summary of Carbon Stocks and Fluxes in Canadian Wetlands/Peatlands, Forests and Agricultural Fields

Land Use	Area (km²)	C Stock (Gt C)	Fluxes (Gt C yr⁻¹)	Potential Duration (Years)
Peatlands/wetlands soils	1.24×10^6	154	−0.03	1000 years (?)
Forests	5.15×10^6	95	−0.2	50
Agriculture fields	0.45×10^6	6	0.0034	50

Source: Adapted from Roulet.[7]

Creation of Wetlands to Sequester Carbon

Wetlands—particularly bottomland hardwood forests— are considered as potentially excellent "carbon sinks" because they take carbon dioxide out of the atmosphere and store it in living plant tissues and soil organic matter. Wetland creation might provide an opportunity to positively address wetland habitat losses while also addressing global warming [8,13] However, a question, still remains: are we capable of constructing wetlands that function in similar ways as natural wetlands? How well these created and constructed wetlands mimic natural wetlands is still being debated.[14,15]

Various studies have indicated that created wetlands might not function in the same capacity as adjacent reference wetlands.[16] In created *Spartina alterniflora* salt marshes, vegetation rapidly achieves 100% cover, although soil nitrogen and organic matter are slow to accumulate. Salt marshes constructed in North Carolina 25 years ago have lower soil organic carbon (SOC) and lower total N reservoirs than a 2000-year-old natural marsh.[16]

Their C accumulation rates are similar to those of reference sites, but N accumulation rates are higher, thus C/N ratios have declined over time. In Oregon, 95 restored freshwater marshes had lower soil organic matter than natural marshes and no evidence of accumulation.[17]

Because bottomland hardwood forests combine both a long-term carbon storage in tree biomass and a slow release of soil organic carbon under flooded conditions, they are considered as excellent ecosystems that could potentially be used to sequester carbon. Restoration of bottomland hardwood forests on marginal agricultural lands in the Mississippi River Valley offers significant net carbon sequestration in the south-central region of the U.S.[13] However hardwood plantings have problems with slow initial growth and excessive early mortality rates. The use of herbaceous weed control and bedding may offer the potential to overcome these difficulties.[13] Under the initiative of the U.S. Department of Energy, an 80 acre pilot study is currently underway in Louisiana. Abandoned marginally productive agricultural fields were planted in January 1997 with seven bottomland hardwood species. Following planting, the site will be monitored for planting success and survival. When the trees attain heights >4.5 ft, growth data and carbon sequestration per acre will be calculated.[13]

Conclusions

The primary production in wetland ecosystems is greatly enhanced by hydrological fluxes, which leads to high biomass, primary production and accretion rates. The carbon that is either fixed stored in standing biomass, released by soil decomposing microorganisms, or stored in soil sediments. With the continual destruction of wetlands worldwide, more of this sequestered carbon is released to the atmosphere. The protection of wetlands will preserve the amount of carbon already stored in these ecosystems.

The creation and restoration of new wetlands will certainly contribute to sequestration of carbon and should be considered as one way to mitigate greenhouse emissions.

References

1. Houghton, J.T., Meira Filho, L.G., Bruce, J., Hoesung, L., Callander, B.A., Haites, E., Harris, N., Maskell, K., Eds.; IPCC (Intergovernemental Panel on Climate Change), climate change 1994. In *Radiative Forcing of Climate Change and an Evaluation of the IPCC IS92 Emission Scenarios;* Cambridge University Press: Cambridge, U.K., 1995.
2. Westlake, D.F., Kvet, J., Szczepanski, A., Eds.; *The Primary Ecology of Wetlands;* Cambridge University Press: Cambridge, 1998; 568 pp.
3. Mitsch, W.J.; Gosselink, J.G. *Wetlands,* 3rd Ed.; Wiley: New York, 2000; 920 pp.
4. Odum, W.E.; Odum, E.P.; Odum, H.T. Natures pulsing paradigm. Estuaries **1995**, *18* (4), 547–555.
5. Conner, J. Effects of forest management practices on southern forested wetland productivity. Wetland **1994**, *14,* 27–40.

6. Gorham, E. Northern peatlands: role in the carbon cycle and probable responses to climatic warming. Ecol. Appl. **1991**, *1*, 182–195.

7. Roulet, N.T. Peatlands, carbon storage, greenhouse gases, and the kyoto protocol: prospects and significance for Canada. Wetlands **2000**, *20* (4), 605–615.

8. Clymo, R.S. Models of peat growth. SUO **1993**, *43*, 127–136.

9. Tamm, C.O. *Nitrogen in Terrestrial Ecosystems;* Springer: Berlin, 1991.

10. Vitousek, P.M.; Howarth, R.W. Nitrogen limitation on land and in the sea: how can it occur? Biogeochemistry **1991**, *75*, 87–115.

11. Vitousek, P.M.; Aber, J.D.; Howarth, R.W.; Likens, G.E.; Matson, P.A.; Schindler, D.W.; Schlesinger, W.H.; Tilman, D.G. Human alterations of the global nitrogen cycle: sources and consequences. Ecol. Appl. **1997**, *7* (3), 737–750.

12. Davidsson, T.E.; Stahl, M. The influence of organic carbon on nitrogen transformation in five wetland soils. Soil Sci. Soc. Am. J. **2000**, *64*, 1129–1136.

13. Williams, J.R. Addressing global warming and biodiversity through forest restoration and coastal wetlands creation. Sci. Total Environ. **1999**, *240*, 1–9.

14. Mitsch, W.J.; Wu, X.; Nairn, R.W.; Weihe, P.E.; Wang, N.; Deal, R.; Boucher, C.E. Creating and restoring wetlands: a whole ecosystem experiment in self-design. Bio Science **1998**, *48*, 1019–1030.

15. Zedler, J.B. Progress in wetland restoration ecology. Trends Ecol. Evol. **2000**, *15* (10), 402–407.

16. Craft, C.; Reader, J.; Sacco, J.N.; Broome, S.W. Twenty-five years of ecosystem development of constructed Spartina Alterniflora (Loisel) marshes. Ecol. Appl. **1999**, *9* (4), 1405–1419.

17. Shaffer, P.; Ernst, T. Distribution of soil organic matter in freshwater emergent/open water wetlands in the Portland, Oregon metropolitan area. Wetlands **1999**, *19*, 505–516.

Index

Page numbers followed by f and t indicate figures and tables, respectively.

A

Acartia tonsa, 638
Accounting framework, 424
Acetamide detection, 62
Acetogenesis, 570
Acidification, 69
    and acid rains, 114
Acidification of rivers and lakes, 87–98
    definition, 87
    environmental problems caused by, 96
    historical perspectives, 96
    measurement, 87–88
    solutions
        environmental legislation, 97
        technology, 97–98
    sources of freshwater
        anthropogenic, 90f, 94–96
        natural, 89f, 90–92
        redox reactions in water sediment systems, 92–94
Acidithiobacillus ferrooxidans, 83, 84, 90
Acidithiobacillus thiooxidans, 84
Acid mine drainage (AMD), 90, 95
    and groundwater pollution, 40
Acid-neutralizing capacity (ANC), 88
Acidogenesis, 570
Acid rain, 106
    program, 97
Acid sulfate soils (ASSs), 91–92
Action Plan, 500
Activated sludge process (ASP), 565, 566f
    operating parameters in, 566–568
Activated sludge system, 566–568
    attached growth processes, 568
Activated sludge wastewater treatment, 248, 248f
Adaptive water management (AWM), 230–231
Advanced Spaceborne Thermal Emission and
    Reflection Radiometer (ASTER), 493t
Advanced very high resolution radiometer
    (AVHRR), 493t
Advanced visible and near infrared radiometer
    (AVNIR), 493t
Advanced wastewater treatment (AWT), 74–75
Aeration, tanks, 565–566
Aerial photography, 495
Aerobacter aerogenes, 671
Aerobic biological waste treatment processes, 565–566
Aerobic processes, 564
Aerobic trickling filter, 569f
Agricultural and Environmental Information System
    for Windows (AEGIS/WIN), 494
Agricultural pollution, 106–107
Agricultural production systems simulator (APSIM),
    186, 187t
Agricultural water consumption estimation
    dynamic crop models, 186–187, 187t
    environmental extended multi-region input output
        model, 188
    hybrid models, 187
    hydrological models, 186, 187t
Agriculture, water quality, 169
Agrostis spp., 49
Airborne electromagnetic methods, for saltwater
    detection, 142
Air-stream bottom aerator (ASB), operation mode of,
    453, 453f
Alachlor, 60t
Aldicarb, 60t
Aldicarb sulfone, 60t
Aldicarb sulfoxide, 60t
Alexandria Lake Maryut, 301–313
    current state of
        hydraulic functioning, 304–305
        hydrological functioning, 304–305
        social considerations and governance, 307
        water quality and ecosystems, 305, 307
    integrated action plan for
        defined, 307–308
        model development, 308
        scenario identification, 308
    location map of, 302f
Alfalfa (Medicago sativa L.), 157
Algal blooms, 112, 325

Algal control measures, 75
American Carbon Registry, 419
Amide herbicides, 61
Aminosulfonyl acids, 61
Ammonia-rich wastewater, 573
Anaerobic ammonium oxidation, 573
Anaerobic processes, 564
Anaerobic wastewater treatment, 568–570
    advantages, 569t
    digestion process, 568–570
    disadvantages, 569t
    operating parameters in anaerobic reactors,
        570–572
        pH, 570–571
        UASB reactor, 570, 571f
Anammox, 573
Anilide herbicides, 61
Anoxic processes, 564
Anthropocentric approach, estuarine ecosystem,
        641–642, 641f
Anthropogenic factors for acidification, 90f, 94–96, 95t
    acid mine drainage, 95
    air pollution with acidifying compounds, 94–95
Anthropogenic pollutants, 105
Anti-erosion drop replacement, 524
Application efficiency, 166
AquaCrop, 186, 187t
Aquatic ecosystems, 65
Aquatic ecosystems and pesticides
    herbicides, 680
    impacts, 682
    measuring concentrations, 681–682
    occurrence, 679–681
    off-site movement, 679–680
    transport, 679–681
    usage, 679
Aquifers, 39, 271
    coastal, 133
    depletion, 271–272
Aral Sea, degradation of, 340–343
Aral Sea disaster, 317–321, 318t, 319t, 321f
    Aralsk and, 320
    average water supply and demand in, 317, 318t
    environmental problems, 320
    human tragedy, 321
    irrigation and cotton, 319–320
    Muynak and, 320
    overview, 317
Aralsk, Aral Sea disaster and, 320
Arctic soil, 552
Arsenic, 369
    content in soil, 369
    in drinking water, 369, 372
    groundwater contamination by, 41
    inorganic forms, 369
Arsenic-contaminated groundwater
    in Assam, 375

in Bangladesh, 376, 377t, 378, 378f, 379t
in Bihar, 374
in GMB Plain, 369, 370, 371f
in Haryana, 375
incidents of, 370, 370f
in India and Bangladesh, 377t
in Jharkhand, 375
in Kolkata, 373–374
in Manipur, 375
in Punjab, 376
in Uttar Pradesh, 374
in West Bengal, 370, 372–373, 372f, 373f
Artificial destratification, 451t
Artificial impoundments, 65
Arundo donax (Giant Reed), 327
Assam, groundwater arsenic contamination in, 375
Atmospheric pollution, 113
Atrazine, 60t
Attached growth process, 565
Avena sativa (Oat), 51

B

Bacillus thuringiensis (Bt), 9
Bacteria, in mine drainage systems, 84
Bahiagrass (Paspalum notatum Fluegge), 49
Baltic Sea, oil pollution in, 349–366
    observations of, 352–355
    oil spill drift, numerical modeling of, 359–364,
        360–364f
    operational satellite monitoring of, 355–359
    problems and solutions of, 364–365
    sources of, 350–352
    statistics of, 352–355
        tendencies of, 352–355
Bamboo-reinforced tanks, 127
Bamboo roofs, 126
Bangladesh, groundwater arsenic contamination, 376,
        377t, 378, 378f, 379t
Basin irrigation efficiency, 167f
Bell's vireo (Vireo bellii pusillus), 331
Bentgrass species (Agrostis spp.), 49
Bernoulli's equation, 245
Best management practices (BMPs), 482, 486
    evaluation of, 428–429
Beta vulgaris (Sugar beet), 51, 157
Bihar, groundwater arsenic contamination in, 374
Bioassays, evaluation of contaminants, 638
Biochemical oxygen demand (BOD), 535
Biodiversity, and wildlife, 331
Biofilters, 589
Biological aerated filters, 589
Biological control, 75
Biological degradation of organics, 564–565
Biological integrity sampling, 73
Biological methods, for contaminant remediation, 42
Biological oxygen demand (BOD), 307

reduction in municipal wastewater, 254–257, 254f–257f
Biological phosphorus removal, 573–574
Biological pollution, in coastal waters
    algal bloom, 24–25
    eutrophication, 24–25
    invasive species, 25
Biological removal of nitrogen, 572–574
    anaerobic ammonium oxidation, 573
    Canon process, 573
    combined nitrogen removal, 573
    Sharon process, 573
Biological treatment processes, 574
Biomanipulation, 75, 451t
Biotic community, 3
Bladder cancer, by NO_3-N exposure, 46
Blue baby syndrome, 45
"Blue Carbon" ecosystem, 217
Bluegrass (Poa pratensis L.), 49
Botswana, 474
Boundary element method (BEM), 402f, 403
Brassica oleracea (Broccoli), 51
Brine, 172t
Britain, work on field drainage in, 433
Broccoli (Brassica oleracea L.), 51
Budget airlines, 205

C

Cadmium (Cd) contamination, in coastal waters, 19
Ca/(HCO_3 + SO_4) ratios
    for saltwater detection, 143
Calibration, groundwater modeling, 394f, 395–396
California, wastewater reuse in, 176
Ca/Mg ratios, for saltwater detection, 143
Canadian Arctic wastewater treatment systems, 543
    background information, 545–550
    current knowledge and practice, 544–545
    methods, 550
    paulatuk treatment wetland, 545–546, 546t, 547f
    results, 550–552
Canon process, 573
Capping, 529t
Captan, 61
Carbamate pesticides, 59
Carbofuran (Furadan), 60t
Carbonaceous Biochemical Oxygen Demand (cBOD₅), 546, 550
Carbon sequestration in wetlands
    creation as "carbon sinks," 718
    human activities, impact of, 717
    net carbon capture, 716–717
    organic carbon fluxes, 715–716, 716f
    primary productivity, 715–717
    soils, 717
Carbon tetrachloride, 60, 60t, 61
Carboxylic herbicides, 61

Carcinogens, 176
Carry-over soil moisture (SMco), 165
Cartridge filter, 589
Carya aquatica, 710
Catchment surface, 126
cBOD₅, *see* Carbonaceous Biochemical Oxygen Demand (cBOD₅)
Celtis laevigata, 710
Center pivots, 155, 156
    irrigation, 159
Chemical methods, for contaminant remediation, 42
Chemical pollutants, 111
Chernobyl accident, 107
Chesapeake Bay, 323–326, 324f
    human population, growth in, 325
    meaning of name, 324
    physical conditions, 324–325
    sedimentation and, 325
Chimney effect, 296
Chlordane, 60t
Chlorinated hydrocarbons, 60
Chlorination of storage tanks, 127
Chlorophyll meter readings, 51
Chlorpyrifos, 60
Christiansen uniformity coefficient (UCC), 152
Citizen monitoring, 73–74
Cladium jamaicense, 215
Clay soils, water movement, 434
Cl/Br ratios, for saltwater detection, 143
CleanSeaNet, 352, 355
Clean Water Act (CWA), 37, 250, 429, 481, 482, 681
    overview, 482f
Clogging, 619
Clonal propagation, 328
Clouds earth's radiation energy system (CERES), 493t
Coastal aquifers, 133
Coastal waters, pollution in, 17–29
    biological pollution
        algal bloom, 24–25
        eutrophication, 24–25
        fertilizers, 26
        heavy metals, 18–20
        invasive species, 25
        light pollution, 28
        marine debris, 27
        metalloids, 18–20
        noise pollution, 27–28
        oils, 26–27
        organic compounds, 22–24
        pesticides, 26
        plastics, 27
        radionuclides, 20–22
        sewage effluents, 26
CODESA-3D, for saltwater intrusion, 146
Coefficient of variation (CV), 152
Coliform counts in wastewater, 176
Colonial times, 227

Combined nitrogen removal, 573
Competitive water markets, 279–283
    scarcity rents, 281–282
    shadow prices, 281–283
Compost wetlands, 85
Comprehensive Everglades Restoration Plan, 653
Computer models, for saltwater intrusion, 146
Conceptualization, groundwater modeling, 393, 394f
Constitutive equations, for saltwater intrusion, 144–145
Constructed wetlands (CWs), 553–555
    classification, 603
    definition, 603
    HF systems, 604–607, 604f, 607f
    hybrid systems, 609
    various types of wastewater, uses in, 609–610, 610t
    VF systems, 607–609, 608f
Continuous simulation models, water quality, 428
Controlled-release fertilizers (CRFs), 50
Convention on Wetlands of International
        Importance, 497
Conveyance systems, 126–127
Corn (Zea mays L.), 47
Courant number, 403
Critical loads, 88
Crop model (CM) techniques, 493, 494f
Cropping system adaptations, 189
Crops, water use, 165
Crown runoff, 203
Cryptosporidium oocysts, 242
Cryptosporidium parvum, 248
CWs, *see* Constructed wetlands (CWs)
Cyanide, groundwater contamination, 41
Cyanobacterial dominance, eutrophic freshwater
    consequences of blooms, 632
    factors, 631–632
    monitoring and management, 632–633
Cylindrospermopsis, 632
Cyprus, 474

D

DAI-ECD technique, 669–671, 670f
Dalapon, 60t
Darcy's law, 399
    for saltwater intrusion, 145
Dead Sea, degradation of, 338
Dead zones, 45
Decision scaling (DS), 513
Decision support system engine (DSSE), 494
Decision support system for agrotechnology transfer
        (DSSAT), 186, 187, 187t, 494
Decreasing block tariff (DBT) pricing strategies, 287
Deep sea mining, 114
Depletion, defined, 165
Designated uses, of navigable waters, 484
Deterministic models, 511
Deterministic water quality models, 428

Diazinon, 60
Digestion process, 568–570
Dinitroaniline herbicides, 61
Dinoseb, 60t
Diquat, 60t
Direct contamination of surface water, 66
Direct current (DC) resistivity, for saltwater
        detection, 142
Dirty water, 106
Disc filters, 599–600
Discretization, groundwater modeling, 394–395, 394f
Dissolved organic N (DON), 48
    leaching from agricultural soil, 48
Dissolved oxygen, concentration, 567
Distichlis spicata, 215
Distribution reservoir, 67
Distribution uniformity (DU), 168
Diuron, 61
Diversifying cropping systems, for reducing NO_3-N
        leaching, 51
Dominant paradigm today, 228–229, 230f
Double pumping, 146
Drain
    depth, 442
    spacing, 442
Drainage, 433
    artificial, 434
    climatic conditions effect on, 435
    definition of, 433
    downstream effects of, 433–436
    of drier soils, 434
    experimental studies on, 434
    factors influencing response to, 434
    and flood events, 433
    main drainage systems, 433
    of permeable soils, 434
    purpose of, 433–434
    role in crop production, 433
    shallow, 433
    simulation model, 434
    subsurface/groundwater, 433
        reduced peak flows from, 434
    surface, 433
        higher peak flows from, 434
    topsoil texture and, 435
Drainage, soil salinity management and
    conditions, 439–440
    requirement
        saline soils, 440, 441f
        sodic soils, 441
    system design, 441–443
        drainage wells, 442
        drain depth, 442
        drain spacing, 442
        relief drains, 442
        saline seeps, 442–443
Drainage lakes, 66

Drainage wells, 442
Dredging, 450t, 529t
Drinking water
    access to, 661, 662t
    analysis of trace organic pollutants, 669–670, 669f
    characteristics of bacteria and pathogens,
        671–672, 672t
    DAI-ECD technique, 669–671, 670f
    microbiological research methods, 672–673
    mineral components, 662–663, 663t
    organic compounds, health significance,
        668–669, 668t
    organic pollutants, 665–667, 667t
    suitability, 662
    toxic elements, 664–665, 664t
Drinking Water Directive in 1983, 270
Drip irrigation systems, 151–154, 705
    hydraulic design of, 152
    for optimal return, water conservation, and
        environmental protection, 153–154
    uniformity of water application and design
        considerations, 151–152
Dry reservoirs, 445
DS, *see* Decision scaling (DS)
Dynamic crop models, 186–187, 187t

E

Earth Observing System (EOS), 492
Ecocentric approach, estuarine ecosystem,
    641–642, 641f
Eco-engineering decision scaling (EEDS), 514t
Eco-filtration systems
    CWs, 597
    engineering applications, 597–598
    green bridge technology, 599
    soil biotechnology, 597
    soil scape filter, 598–599
Ecological network analysis (ENA), 424
Economically optimum N level (EON), 46
Ecosystem functioning, of Alexandria Lake Maryut,
    305, 307, 309, 311–312
EEDS, *see* Eco-engineering decision scaling (EEDS)
Egypt, 463, 474
Electrical conductivity, 171, 172t
Elodea nuttallii, 8
Elodeids, 711
Empirical water quality models, 427
ENA, *see* Ecological network analysis (ENA)
Endothall, 60t
Endrin, 60t
Energy-water nexus, 423–424, 425f
Enhanced Thematic Mappers (ETM), 492, 493t
Environmental degradation, 519
Environmental extended multi-region input-output
    model, 188
Environmentally sound technologies (ESTs), 555

Environmental policy integrated climate (EPIC), 186, 187t
Environmental problems, nitrogen leaching and,
    408–409
Environmental Protection Agency (EPA), 175, 176,
    482–485, 487, 488
Environmental services programs, payments, 418–419
Equity and fairness, 286
Error analysis, groundwater modeling, 395–396
Escherichia coli, 671, 672
ESTs, *see* Environmentally sound technologies (ESTs)
Estuaries
    biomarkers, 642–643
    community structure
        estuarine quality paradox, 638–639
        tolerance and monitoring, 639–640
    controlling pollution, 643
    environmental assessment
        chemical analyses, 637–638
        reference values, 641–642
        tolerance and monitoring, 639–640
    environmental problems, 636–637
    environmental quality standards and guidelines,
        643–644
    functional characteristics, 642
    risk assessment, 644
    worldwide distribution, 635
Estuarine quality paradox, 638–639
Estuarine turbidity maximum (ETM), 636
Ethiopia, 465
Ethylene dibromide (EDB), 60, 60t, 61
Eulerian-Lagrangian Localized Adjoint Method, 404
Eulerian methods, 403
Eurytemora affinis, 640
Eutrophication, 24–25, 68, 115
    definition, 693
    management strategy, 694
    model framework, 76
    phytoplankton, growth of, 694–698
    problems, 693–694
    solutions, 698–699, 699f
Evaporation, 168t
Evapotranspiration (ET), 169, 329
Everglades
    environmental impacts, 652
    landscape, 653
    origin, 651
    primitive attempts, 651–652
    restoration, 652–653
External contamination, sources of, 127
Extraction wells, 43

F

Facultative ponds, 536–537
    BOD removal, 536f
    kinetics, 537–538
Fecal bacteria, 538

Fecal Enterococcus, 673
Federal water policy, 270
Federal Water Pollution Control Act (FWPCA)
      (1948), 481
FEMWATER, saltwater intrusion, 146
Ferrihydrite, 84
Ferrocement tanks, 127
Fertilizers, coastal waters, 26
Field experiments, 188–189
Filter strips, 159
Finite differences (FD), 401–402, 402f
Finite element method (FEM), 402–403, 402f
Finite volumes, 402, 402f
First-flush (or foul-flush), 207
      device, 126
First-order plug-flow kinetics, 605–606, 606t
Fish catch, 451t
Fish depletion, 70
Flocculation, 619
Flood control reservoir, 67
Flooding, 329
Flood irrigation, 704
Flow-through reservoirs, 445
Flushing, 450t
Fonofos, 60
Food and agriculture organization, 269
Food Quality Protection Act (FQPA), 428
Food-to-microorganism ratio, 567–568
Fraxinus pennsylvanica, 710
Free-water surface (FWS) wetlands, 622–623, 623f
Frequency-domain electromagnetic methods (FDEMs)
      for saltwater detection, 141
Freshwater diversion projects, 189–190
Freshwater marsh, species distribution, 710–711, 710f
Freshwater pollution, 105
Freshwater/saline lakes, 67
Full cost prices, 284, 284f
Full economic cost, 284
Full supply cost, 284
Fulvic acids, 176
Fumigants, 61
Furrow-irrigation erosion, 156
FWS, *see* Free-water surface (FWS) wetlands

G

Ganga-Meghna-Brahmaputra (GMB) Plain, arsenic
      contamination, 369
Garden insecticide, 60
Generic groundwater model, 399–401, 400f, 401f
Geochemical investigation, saltwater intrusion,
      142–143
Geographic information system (GIS)
      inventorying and monitoring, 492
      research and development, 492–494, 493t, 494f, 494t
      types of, 494t
      for watershed management, 491–496, 493t

Geological investigation, saltwater intrusion, 133–147
Geophysical investigation
      of saltwater intrusion, 141–142
            advantages and disadvantages, 141
Ghana, 464
Ghyben-Herzberg relation, 135–137, 140f, 144
      saltwater detection using, 140–141, 140f
Giant reed *(Arundo donax)*
      biology and ecology, 327–329, 328f
            biodiversity and wildlife, 331
            control methods, restoration, revegetation, 331–333
            effects on streams/water resources, 329
            flooding, 329, 330f
            water use, 329–331
            wildfire, 331, 332f
      hand clearing methods, 332
      mechanical clearing methods, 332
Giardia lamblia, 248
Global Carbon System, 217
The Global Hydro-economic Model, 190
Globalization, 266
Global policy issues, 267
Global positioning systems (GPS), 494f
Global surface water
      concerns, 268–269
      issues, 268–269
Global water concerns, 269
Global water resource policy, 273–274
Glyphosate, 60t
Goethite, 84
Gold mining, and groundwater contamination, 41
Gompertz model, 537
Granule deterioration, 572
Grazing management, 655
Great Lakes, degradation, 339
Green Bridge technology, 599
Green revolution, 185
Ground catchment techniques, 124, 125f, 126
Groundwater, 46
      concerns, 268–269
      contaminants in, 45
      contamination, 269–271
            by arsenic, 41
            cyanide and, 41
            nitrate-nitrogen in, 45
            from nitrogen fertilizers, 45–52
            uranium and, 41
      depletion, 271–272
      drained lakes, 66
      impact analysis, 176
      issues, 268–269
      movement, 39
      non-point sources of, management of, 62
      point sources of, management of, 62
      pollution, 176
      resources, 38–39
      saltwater intrusion in, 133–147

Groundwater modeling, 389–398
   generic models, 392
   overview, 389–390
   phenomena, 390–391, 391f
   process, 393–397, 394f, 395f
      calibration, 394f, 395–396
      conceptualization, 393, 394f
      discretization, 394–395, 394f
      error analysis, 395–396
      model selection, 396
      predictions and uncertainty, 397
   site-specific models, 392–393
   usage of, 392–393, 392f
Groundwater problems, numerical methods
   boundary element method (BEM), 402f, 403
   finite differences (FD), 401–402, 402f
   finite element method (FEM), 402–403, 402f
   generic numerical method, solving groundwater
      flow, 399–401, 400f, 401f
   integrated finite differences (IFD), 402
   simulating solute transport, 403–404, 403f
Gutters for rainwater storage, 127

H

Halogenated compounds (volatile), 61
Halophytes, 711
Hand clearing methods for A. donax, 332
Haryana, groundwater arsenic contamination
   in, 375
Health, guidelines for irrigation, 176
Health advisory levels (HALs), 59
Heavy metals
   in coastal waters, 18–20
   road runoff, 196–200, 197f, 198t, 199t, 200f
Helophytes, 709
Helsinki Convention, 352
Heptachlor, 60t
Heptachlor epoxide, 60t
Herbicides, 61, 332–333
Heterotrophic bacteria, 537
Highest astronomical tide (HAT), 217
High resolution visible and middle infrared (HRVIR),
   492, 493t
H isotopes, for saltwater detection, 143
Hybrid models, 187
Hydraulic functioning, of Alexandria Lake Maryut,
   304–305, 308–309
Hydraulic mission, 227–228
Hydraulic retention time (HRT), 245–247
Hydrocarbons, chlorinated, 60
Hydrogeomorphic approach (HGM), wetland
   classification, 687–689
Hydrological functioning, of Alexandria Lake Maryut,
   304–305
Hydrological models, 186, 187t
Hydrological patterns, wetlands, 709

Hydrologic cycle, 269
   issues, 267
Hydrolysis, 568–569
Hydrophytes, 709
Hydroxy herbicides, 61
Hypolimnion water removal, 450t
Hypoxia, 47

I

Increasing block tariff (IBT) pricing strategies, 287–288
Increasingly urban world, 232–233, 233f
India, 465
Indo-German Watershed Project, 268
Indoor water storage system, 125f
Industrial pollution, 106
Industrial wastewaters, 175, 577–578
Infestations of A. donax, 327
Infiltration, 528
Injection wells, 43
Input-output analysis (IOA), 423–424
Insecticides, 60–61
Instantaneous field of view (IFOV), 492, 493t
Instituto Nacional de Investigaciones Forestales
   Agricolas y Pecuaris (INIFAP), 410
In-stream flow requirements, 169–170
Integrated environmental management, 301–313
Integrated finite differences (IFD), 402
Integrated membrane system (IMS), 596
Integrated pest management (IPM), strategies, 9
Integrated urban water management (IUWM), 232
Integrated water resources management (IWRM), 226,
   228–229, 230f
Intelligent Decision Support System (IDSS), 494
Inter-annual measurement schemes, 717
International Council for the Exploration of the Sea
   (ICES), 643
IOA, *see* Input-output analysis (IOA)
Iran, 463
Iron-oxidizing bacteria, 84
Irrigation
   importance of, 155
   management, 62–63
   method, 166–167
   overview, 179
   soil salinity and, 179–182
      causes, 180–181
      deleterious effects, 179–180
      management strategies, 181–182
   sprinkler, 155
   surface, 155, 156
Irrigation, wastewater
   challenges, 701
   crop selection, 702–703, 702t
   frequency, 705
   maximum allowable levels, metals and metalloids,
      703–704, 703t

Irrigation, wastewater (*cont.*)
 methods used, 704–705, 704t
 pathogen reductions, 705t
 soil-based nutrient interventions, 705–706, 706t
Irrigation and river flows
 basin irrigation efficiency, 167f, 168–169
 depletion, 165–166
 environmental concerns
  in-stream flow requirements, 169–170
  salt loading pick-up, 169
  water quality implications for agriculture, 169
 hydrograph modification
  irrigation efficiencies, 168
  irrigation methods, 166–167
  irrigation return flows, 166
  reservoir storage, 166
 hydrologic studies, 165
Irrigation efficiency, 165, 168
 basin, 167f, 168–169
Irrigation erosion, 155–161
 sprinkler-irrigation systems and, 159–161
 surface irrigation and, 157–159
Irrigation furrows, 156, 156f
Irrigation return flows (IRF), 166, 167f
Irrigation system, subsurface drip irrigation (SDI), 505–510
Isoetids, 711
Israel, 463, 465
 saltwater intrusion in, 134–135
 wastewater reuse in, 176
Italy, 474
 saltwater intrusion in, 135
Iva frutescens, 215
IWRM, *See* Integrated water resources management (IWRM)

J

Jarosite, 84
Jharkhand, groundwater arsenic contamination in, 375
Jordan, 463
Juncus roemerianus, 215

K

Kara-Bogaz-Gol Bay, degradation of, 343–345
Kolkata, groundwater arsenic contamination in, 373–374
Kuwait, 463

L

Lagrangian method, 404
Lake Chad, degradation of, 338–339
Lake Mead, degradation of, 339–340
Lake Ohlin, degradation of, 338
Lake recultivation, 445–457
 methods of, 448–449, 450t–451t, 452–457
 phases of, 448–449, 449f
Lake(s), 66
 aeration, 75
 degradation of, 338–340
 sampling, 73
Lakes and reservoirs, 65–78
 classification of, 66–67
  freshwater/saline lakes, 67
  trophic status, 67
 conventions for protection, 77
  The Protocol on Water and Health, 77
  Ramsar Convention, 77
 pollution, sources of, 72
 pollution issues of, 71t
 problems associated with, 68
  acidification, 69
  eutrophication, 68
  fish depletion, 70
  sedimentation, 69
  stratification, 70–71
  toxic materials, 69
 protective and restorative measures, 74–76
  eutrophication model framework, 76
 restoration measures, 76t
 water quality monitoring, 72–74
  biological integrity sampling, 73
  citizen monitoring, 73–74
  lake sampling, 73
  tributary mass load sampling, 73
  tributary water quality sampling, 73
Lake Victoria, degradation of, 339
Land, and river runoffs, 112
Land Information System (LANDIS), 494
Land surface catchments, 124, 125f
Land Use Analysis System (LUCAS), 494
Latium perenne (Ryegrass), 49
Leptospirillum ferrooxidans, 83
Lightning, 331
Light pollution, in coastal waters, 28
Lindane, 60t
Linear imaging self scanner system (LISS), 492, 493t
Linuron, 61
Littoral vegetation, 710
Loire estuary, 635, 636f
Long-range transboundary air pollution (LRTAP), 97

M

Macrophyte reintroduction, 450t
Macropores, 61
Maneb, 61
Manipur, groundwater arsenic contamination, 375
Manure application, and NO_3-N concentrations, 48
Marine debris, 27, 115–116
Marine pollution, 111–116
 acidification and acid rains, 114

atmospheric pollution, 113
   deep sea mining, 114
   eutrophication, 115
   land and river runoffs, 112
   noise pollution, 116
   oil and ship pollution, 113
   plastic debris, 115–116
   radioactive waste, 114–115
MARPOL Convention, 352
Mass balance, 695
Mathematical modeling, saltwater intrusion, 143–146
Mathematical water quality models, 427
Maturation ponds, 538–539
   for fecal coliform removal, 540–541
Maximum admissible concentrations (MACs),
       drinking water, 662
Maximum contaminant levels (MCL), 37, 59, 60t
   of contaminants from mining, 38t
   for nitrate, 46
Maximum contaminant levels (MCLs), 243
Maya civilization's downfall, 226
Meandering, 526
Mean high water (MHHW), 217
Mean high water (MHW), 216
Mean tide level (MTL), 216
Mechanical clearing methods for *A. donax*, 332
Medicago sativa L. (Alfalfa), 157
Mediterranean-type climates, 329
Membrane filtration, 590–593, 591f
Mercury (Hg), contamination in coastal waters, 19, 20
Metalloids, in coastal waters, 18–20
Methane (CH_4)
   consumption, 293–294
   emissions
      water table position, 294–295, 294f
   production, 293–294, 296
   soil temperature
      bubble ebullition, 295–296
      diffusion, 295–296
      transport through plants, 295–296
      vascular plants, 296
Methanogens, 570
Methemoglobinemia, 45, 46
Method of characteristics (MOC), 404
Methoxychlor, 60t
Mexico
   saltwater intrusion in, 135
   wastewater reuse in, 176
Mexico Nitrogen Index, 410
Michaelis-Menten equation, 695–696
Mine drainage, 81, 82f
   causes of, 82
   chemistry of, 83
   control of, 84–85
      prevention, 85
      treatment, 85
   environmental impacts of, 84

microbiology related to, 84
   mineralogy of, 83–84
   problems by, 81–82
   total acidity of, 85
Mining
   for coal, 37, 38
   groundwater pollution by, 37
      abandoned mine sites and, 40
      acid mine drainage and, 40
      biological methods for remediation, 39, 43
      chemical methods for remediation, 43
      containment of contaminants in, 42
      gold mining and, 41
      groundwater analysis, interpretation of, 41–42
      groundwater resources, 38–39
      hydrogeological characteristics of site and,
         39, 39t
      metal contaminants, 40
      organic contamination, 40
      physical methods for remediation, 43
      and pump-and-treat method, 42
      and remediation strategies, 42–43, 42f
      and in situ remediation, 42
      tests for constituents, 41, 41t
      transport of contaminants, 38
      water contaminants, 39–41, 40t
      water movement, 39
   impact on groundwater supplies, 37
   surface, 37
Mixed liquor, 565
Mixed-liquor suspended solids (MLSS), 566
Mixed-liquor volatile suspended solids (MLVSS), 566
MOCDENSE, for saltwater intrusion, 146
Modeling treatment wetlands, 555–556
Moderate resolution imaging spectrometer (MODIS),
     492, 493t
Moisture, carry-over, 166
Monitoring wells, for saltwater intrusion, 140
Monod equation, 537–538
Morocco, 463
Motile algae, 537
Mud for roof catchment, 126
Multipurpose reservoir, 67
Multiregional nexus network, 424
Municipal sewage, 175
Municipal solid waste (MSW), recycling of, 263–264
Municipal wastewater, 253–264
   composition of, 253–254, 253t
   recycling of, 263–264
   treatment methods for
      biological oxygen demand, reduction of,
         254–257, 254f–257f
      nitrogen concentration, reduction of, 260–263,
         261f–263f
      phosphorous concentration, reduction, 257–260,
         258f–260f
Muynak, Aral Sea disaster and, 295–296

N

Na/Cl ratios, for saltwater detection, 143
National Oceanic and Atmospheric Administration (NOAA), 643
National pesticide assessment, 59
National Pollutant Discharge Elimination System (NPDES), 37, 482
    permit program, 483, 483f
National Water Quality Assessment (NAWQA), 60
National Water-Quality Assessment (NAWQA) Program, 680
National Wetland Policy
    implementation strategies, 498–499, 499t
    Ramsar Convention, 499, 500t
National Wetland Strategy, 500
Natural resources degradation, 491
Natural riverbed protection, 526
Natural sources of freshwater acidification, 90–92, 90f
    geochemical factors, 91–92
    geological factors, 90–91
Nephalometric turbidity units (NTU), 242
Netherlands, saltwater intrusion, 134
New York Nitrate Leaching Index, 409
Nexus analysis, 423–424, 425f
Nitrapyrin, use of, 51
Nitrate Leaching Hazard Index, 409
Nitrate leaching index, 407–411
    assessment, 409
    environmental problems, 408–409
    quick tools and indicators, 409–410
Nitrate-nitrogen (NO_3-N), in groundwater, 45
    agricultural practices contributing to
        containerized horticultural crops, 50
        grasslands/turf, 48–50
        manure application, 48
        nitrogen fertilizer, 48
        row crops, 47–48
    elevated levels of, 45–46
    environmental impacts, 46–47
    health problems by, 45, 46
    by leaching of N fertilizer, 45–46
    nitrogen sources of, 46
    occurrence of, 46
    strategies, for reducing NO_3-N leaching, 51–52
        cover crops, 51–52
        diversified crop rotation, 51
        grassland/turf management, 52
        nitrification inhibitors, 50–51
        reduced tillage, 51–52
        soil testing and plant monitoring, 50
Nitrate pollution, 106
Nitrification inhibitors, 50–51
Nitrogen (N), 47, 407
    concentration, reduction in municipal wastewater, 260–263, 261f–263f

losses of measurements, need for, 416–417
    and NO_3-N contamination, 46
Nitrogen cycle, 695, 695f
Nitrogen fertilizer, NO_3-N contamination from, 46
Nitrogen oxides (N_xO_y), 106
Nitrogen trading tools (NTTs), 415–420, 416
NO_3 test, late-spring, 50
Noise pollution, 116
    in coastal waters, 27–28
Non-motile algae, 537
Non-point source, of water pollution, 71
Non-potable uses, of water, 124
Non-saline water, 172t
Nonspecific biomarkers, 642–643
Nuclear atmospheric weapons tests, 107
Nutrient reduction, 75–76
Nutrient removal, WSPS, 539
Nutrient tracking tool, 420

O

Oat (*Avena sativa* L.), 51
Ocean color and tem (OCTS), 492, 493t
Oil and ship pollution, 113
Oil pollution, in Baltic Sea, 349–366
    observations of, 352–355
    oil spill drift, numerical modeling of, 359–364, 360–364f
    operational satellite monitoring of, 355–359
    problems and solutions of, 364–365
    sources of, 350–352
    statistics of, 352–355
    tendencies of, 352–355
Oils, in coastal waters, 26–27
Oil spills drift, numerical modeling of, 359–364, 360–364f
O isotopes, for saltwater detection, 143
Oman, 463, 474
Operational satellite monitoring, of oil pollution, 355–359
Opportunity cost, 284
Optimizing model, 279–283
    scarcity rents, 281–282
    shadow prices, 281–283
Organic compounds, in coastal waters, 22–24
Organic loading rate, 568
Ovarian cancer, by NO_3-N exposure, 46
Owens Lake, degradation of, 339
Owner-operated rainwater harvesting, 123
Oxamyl (Vydate), 60t
Oxygenation, 446–447, 450t

P

Paleoenvironmental data and fly-ash particle analysis, 88
Palestine, saltwater intrusion, 134–135
Paspalum notatum (Bahiagrass), 49

Pathogen, in sewage effluent, 176
Paulatuk treatment wetland, 545–546, 546t, 547f
Pb contamination, in coastal waters, 19
Peat accumulation, 717
Peclet number, 404
Percolation, 166, 167, 175
Permethrin, 61
Pesticide
    in coastal waters, 26
    risk assessment, 428
Pesticide groundwater database (PGWDB), 60
Pesticide impacts, on aquatic communities balance, 3
    examples, 7–9
    measuring impacts, 4–5
    recent advances and outstanding issues, 9–11
    risk assessment and, 5–7
    risk reduction, 9
Pesticide-plant combinations, 62
Pesticides in groundwater
    groundwater contamination, 62
        management of, 62
    irrigation management, 62–63
    maximum contaminant levels (MCL), 59, 60t
    in soils and water
        fumigants, 61
        fungicides, 61
        herbicides, 61–62
        insecticides, 60–61
    use of, 59–60
Pest management, 62
Phenylurea herbicides, 61
Phosphorous concentration reduction, municipal
    wastewater, 257–260, 258f–260f
Phosphorus cycle, 696, 696f
Phosphorus-loading models, 76
Photosynthetically active radiation (PAR), 715
Phragmites, 711
Phragmites australis, 215, 605, 605f, 711
Physical methods, for contaminant remediation, 42
Physical water transfer
    freshwater diversion projects, 189–190
    via food trade, 190
Phytoplankton growth, 694–698, 697f
Picloram, 60t, 61
Pipe drainage, impact of, on downstream peak flows,
    435, 435f
Plastic(s)
    in coastal waters, 27
    debris, 115–116
Platanus occidentalis, 710
Pleated filtration panel filters, 600
PNECoral, 643
Poa pratensis (Bluegrass), 49
Policy *vs.* wise use, 497, 498
Pollution
    biological, 24–25
    in coastal waters, 17–29

    light, 28
    noise, 27–28
    wastewater treatment, 577
Polyacrylamide (PAM), 159, 161
Polychlorinated biphenyls (PCB), 111
Polycyclic aromatic hydrocarbons (PAHs), 195
Polyethylene tanks, 127
Polyphosphate-accumulating organisms (PAOs), 573
Pomatomus saltatrix, 6
Populus deltoides, 710
Potato (*Solanum tuberosum* L.), 48, 51
Potential acid sulfate soils, 92
Precipitation, 451t
Predicted no-effects concentrations (PNECs), 638
Pre-modern times, 226
Preplant N tests, 50–51
Pre-side-dress N tests, 50–51
Pressure sand filters, 587–588
Principal response curves (PRCs), 5
Probabilistic hazard assessment (PHA), 9
Probabilistic risk assessment (PRA), 3, 10
Protective and restorative measures, 74–76
The Protocol on Water and Health, 77
Publicly owned treatment works (POTWs), 247
Public water policy, 267
Pump-and-treat method, 42
Pumping, to manage saltwater intrusion, 146
Punjab, groundwater arsenic contamination in, 375
Pyramid Lake, degradation of, 339
Pyrethroid, insecticides, 61
Pyrite, oxidation of, 82–83

Q

Qinghai Hu Lake, degradation of, 338
Quaternary N herbicides, 61

R

Radioactive waste, 114–115
Radionuclides, 107, 114
    in coastal waters, 20–22
Rainwater harvesting
    advantages of, 123–124
    design/maintenance of
        catchment surface, 126
        conveyance systems, 126–127
        storage tanks, 127
    importance of, 123
    types of
        land surface catchments, 124, 124f, 125f, 126
        large systems, 124, 125f
        rooftop collection systems for high-rise
            buildings, 124
        simple rooftop collection systems, 124, 124f
        stormwater collection in urbanized
            catchment, 126

Rainwater pipes, 126
Ramsar Convention, 77
Random walk method, 404
Range and pasture lands
　biological characteristics, 657
　dissolved chemicals, 657
　livestock impacts
　　soil, 656
　　vegetation, 656
　protecting water quality
　　limit livestock use of waterways, 658
　　limit runoff and erosion, 658
　　wetland/riparian sites, 658
　suspended sediment, 655
Rapid gravity filters, 587
RDM, *see* Robust decision making (RDM)
Recommended Maximum Concentrations (RMCs), 703t
Recultivation, 521
Redox reactions
　in water-sediment systems, 92–94
　　biological factors, 92–93, 93t
　　climatic factors, 93–94, 94t
　　surface inland waters, 93t
Reduced tillage, 51–52
Registration, Evaluation, and Authorization of
　　Chemicals (REACH), 644
Rehabilitation, 520, 530t
Reinfestation of A. donax, 331
Relative sea-level rise (RSLR), 219
Relief drains, 442
Remediation, 530t
Remote sensing (RS)
　advances in, 493t
　inventorying and monitoring, 492
　research and development, 492–494, 493t, 494f,
　　494t
　satellites with sensors, 493t
　in watershed management, 491–496, 493t
Renaturization, 524
Reservoirs, 66–67
　dry, 445
　flow-through, 445
　retention, 445
　storage, 166
　tillage, 160
Restoration, 530t
　of native terrestrial plants, 530t
Retention, 528
　reservoirs, 445
Revegetation, 331–333
Revenue stability, 287
Reverse osmosis, 594–596, 596f
Revitalization, 530t
Rewashing, 450t
Rhizome or culm, 327
Riparian ecosystems/bottomland hardwood forests,
　　327, 328, 710

Riplox method, 454
River
　ecosystems, 528
　functionality analysis, 524
　recultivation, 521–528
　　multitasking for, 523, 523f
　　planning and development, 522f
　restoration, 519–521
River pollution, 105–108
　removal of pollutants, 107–108
　sources, 106–107
　　acid rain, 106
　　agricultural pollution, 106–107
　　industrial pollution, 106
　　radionuclides, 107
Road runoff samples, 196–200, 197f, 198t, 199t, 200f
　formaldehyde (HCHO) levels, 198, 200f
　maximum, minimum, and mean concentrations of
　　analytes, 199t
　mean concentrations of analytes in, 198t
　organochlorine pesticides, 198
　pesticides determined in, 199t
　ranges of concentrations in, 198t
Robust decision making (RDM), 514t
Robust reservoirs management, 512
Rock-soil-water interactions, 91–92
Roof catchment, materials used for, 126
Roof runoff, 200–203, 201t, 202t, 203t
　concentrations of analytes, 202, 202t
　detected and determined ions, 203, 203t
　material type, 201, 201t
　pH of roof runoff, 200, 201f
　types of roofing, 201, 202t
Rooftop collection systems
　for high-rise buildings in urban area, 124
　simple, 124, 124f
Row crops, 47–48
Runoff farming, 130
Runoff water, 195
　wet deposition, 195–196, 196f
Ryegrass (Latium perenne L.), 49

S

Safe Drinking Water Act (SDWA), 37, 242, 250
Safe Water Drinking Act of 1974, 270
Saline seeps, 442–443
Saline soils, drainage requirement, 440–441, 441f
Saline waters
　classification of, 172t
　effect on soil, 171
　irrigation with, 171–172
　low-salt and salty waters, mixing, 173
Salinity Control Act of 1974, 169
Salinity control projects, 169
Salt loading pick-up, 169
Salt marshes

blue carbon ecosystems, 217
geomorphological classification, 215, 216f
hot spot, 219
hydrologic regime, 218
interior platform, 218, 218f
marsh morphology, 218
sub-environments, 215, 216f
Salt(s)
concentration, 172t
in soil, 171
Saltwater intrusion
in groundwater, 133–147
combating, 146–147
computer models, 146
Israel and Palestinian territories, 134–135
Italy, 135
mathematical modeling, 143–146
mechanisms, 135–140
Mexico, 135
monitoring and exploration of, 140–143
Netherlands, The, 134
planning and management, 147
transition zones, 139–140, 139f, 144
into unconfined aquifer, 136f
United States, 134
vertical cross sections of, 136f
Saudi Arabia, 463, 474
Scarcity rents, 281–282
Scheduling and Network Analysis Program
(SNAP), 494
School of Environmental Studies (SOES),
survey of, 370
Schwertmannite, 84
Scirpus, 711
SEAWAT, for saltwater intrusion, 146
Secale cereale (Winter rye), 51
Sedimentation, 69
Sediment ponds, 159
Sediment Quality Triad, 637
Sediments
isolation, 451t
nutrient deposition/deactivation in, 451t
Sediment trap, 126
Seed and Fertilizer Approach, 491
Seepage lakes, 66
Serratia marcescens, 671
Service stability, 287
Seston removal/catch, 451t
Sewage effluents, in coastal waters, 26
Sewage effluents for irrigation
in California, 176
coliform counts in, 176
in Israel, 176
long-term effects of, 176
in Mexico, 176
in Middle East, 175
monitoring guidelines for, 176–177

objectives of, 175
reuse standards, 176
untreated, 176
Shadow prices, 281–283
Sharon process, 573
SHARP, for saltwater intrusion, 146
Shifting double cropping system, 186
Signal-to-noise ratio, 492, 493t
Simazine, 60t
Simulation of solute transport, 403–404, 403f
Single cropping system, 186
Site-specific models, groundwater modeling, 392–393
Slow sand bed filters, 587
Slow sand filter (SSF), 242
Sludge volume index (SVI), 567
Smith, John, 325
Smoothing natural riverbed, 526
Socioeconomic system, 423, 424f
Sodic soils, drainage requirement, 441
Sodium salts, 171
Soil
matrix, 175
primary function of, 175
retention, 527
Soil and water assessment tool (SWAT), 186,
187, 187t
Soil-aquifer recharge systems with dewatering, 176
Soil salinity
irrigation and, 179–182
causes, 180–181
deleterious effects, 179–180
management strategies, 181–182
Soil scape filter, 598–599
Soil temperature
bubble ebullition, 295–296
diffusion, 295–296
transport through plants, 295–296
vascular plants, 296
Solanum tuberosum (Potato), 48, 51
Solid retention time (SRT), 566–567
Solute transport, simulation of, 403–404, 403f
South Florida Water Management District, 134
South-north water transfer (SNWT) project, 189
Soybeans (*Glycine max* L. Merr.), 47
Spain, 463, 474
Sparging, 450t, 452
Spartina alterniflora, 215, 218, 219, 718
Spartina patens, 215
Species invasion, 25
Specific biomarkers, 642–643
Sprinkler-irrigation erosion, 159–161
Sprinkler irrigation system, 166, 167f
Sprinklers, 62, 172
Standard water treatment, 243, 243f
State-of-the-art k-C* model, 606
Stochastic (or random) models, water quality, 428
Stone backfill, 527

Storage, 528
   and conservation reservoirs, 67
   matrix, 400
   tanks for rainwater, 126
   variation (groundwater), 399
   of water, 165
Stormwater, collection in urbanized catchment, 126
Stratification, 70–71
Streamflow depletion, 165
Stream longitudinal gradient, invasive species
   distribution, 711, 711f
Submerged aquatic vegetation (SAV), 325
Subsurface drip irrigation (SDI), 505–510
   air entry and flushing, 509
   chemical injection, 508–509
   defined, 505
   development of, 505
   laterals and emitters, 508
   lateral type, spacing, and depth, 507
   maintenance, 509–510
   operation, 509
   pumps, filtration, and pressure regulation, 508
   site, water supply, and crop, 506
   special requirements, 507
   system components, 508–509
   system design, 506
SubWet 2.0, 556
Sugar beet (*Beta vulgaris* L.), 51, 157
Sulfur dioxide (SO$_2$), 106
Surface irrigation, 157–159, 166, 167f
Surface Mining Control and Reclamation Act
   (SMCRA), 37–38
Surface runoff, 126
   limitation, 528
Surface waters, 66
   contamination, 269–271
   direct contamination of, 66
   runoff, 112
Suspended growth processes, 565
Suspended sediment concentration (SSC), 219
Sustainable development goals (SDGs), 226
SUTRA, for saltwater intrusion, 146
Swedish Forest Agency, 98
Synthetic aperture radar (SAR), 350, 355–356, 493t
   advanced, 350
Syria, 468
Systeme Hydrologique European (SHE) model, 186, 187t

T

Tailwater, 166
Technical Guidance Document (TGD), 638
Terbufos, 60
Thematic Mapper (TM), 492
Thermally impaired waters, 487
Thiocarbamate herbicides, 60
Thiocarbamate pesticides, 59

Throughfall, 203–207
   from airports, 205–207, 205f–207f
   from farming areas, 204–205
Tile drainage, nitrate-nitrogen concentrations in, 47
Time-domain electromagnetic methods (TDEMs)
   for saltwater detection, 142
Total dissolved solids (TDS)
   for saltwater intrusion, 139, 145
Total maximum daily load (TMDL), 429,
   481–482, 655
   in achieving water quality standards, 486
   application of, 488–489
   nonpoint source state management
      water quality standards before TMDL process,
      485–486
   nonpoint source state management plans, 485–486
   point source regulation and
      NPDES permit program, 483, 483f
      pollutant discharge, 482–483
      section 402 dredge and fill permit program, 484
   section 297 impaired waters list, 486–487
   state water planning and, 487–488
   water quality standards in point source permitting
      before TMDL process, 484–485
Total organic carbon (TOC), 206
Toxaphene, 60t
Toxic chemicals, 176
Toxicity, 572
Toxic materials, 69
Toxins, 111
Transition zone, between saltwater and freshwater
   regions, 139–140, 139f, 144
Triazines, 61
Tributary mass load sampling, 73
Tributary water quality sampling, 73
Tributyltin (TBT), 19
Trickle irrigation, 166
Trickling filters, 589–590, 590f
Trihalomethanes (THMs), 665–666, 666f
Triticum aestivum (Wheat), 51, 157
Trophic status, lakes and reservoirs, 67
Trunk runoff, 203
Tunisia, 463, 474
Turf, NO$_3$-N contamination from, 48–50
Typha, 711

U

UASB reactor, 570, 571f
Ulmus americana, 710
Ultrafiltration, 593–594
United Arab Emirates, 463
United Nations Children's Fund (UNICEF), 372
United Nations Educational, Scientific and Cultural
   Organization (UNESCO), 272
United States, 463, 465
   saltwater intrusion in, 134

United States Environmental Protection Agency (USEPA), 270, 408
Unit filter run volume (UFRV), 244
Up-coning, 137–139, 138f
Upgrading existing wastewater treatment plants, 578
Uranium, groundwater contamination, 41
U.S. Department of Agriculture (USDA), 272
U.S. Environmental Protection Agency, 243
Utility-operated rainwater harvesting, 123
Uttar Pradesh, groundwater arsenic contamination, 374

V

Variable infiltration capacity model (VIC), 186, 187t
Vegetated submergedbed (VSB) wetlands, 623, 623f
Verdale-Simi Fire, 332f
Vireo bellii pusillus, 331
Virtual water flow
    freshwater diversion projects, 189–190
    via food trade, 190
Visible infrared scanner (VIRS), 493t
Volatile organochlorine compounds (VOCCs), 198
VSB, *see* Vegetated submergedbed (VSB) wetlands

W

Waste composition, 571
Waste load allocation, 487
Waste stabilization pond system (WSPS)
    design of, 539
        anaerobic ponds, 540
        design parameters, 539–540
        facultative ponds, 540
        Helminth egg removal, 541
        maturation ponds for fecal coliform removal, 540–541
        water flows and BOD concentrations, 540
    nutrient removal in, 539
    oxygen tension in, 538
    processes in, 536
        anaerobic ponds, 536
        facultative ponds, 536–537
        facultative ponds, kinetics, 537–538
        maturation ponds, 538–539
    types of, 535–536
    water quality and, 541
Wastewater, 461–462
    collection, 248
    exposure, motivating safe practices along, 472–473
    history, 247
    industrial, 461–462
    industrial waste, 249–250
    issues, 250
    liquid disposal, 249
    monitoring, 248
    policy interventions and risk reduction, 468
        agricultural communities, 469–470

        farmers and families, 468–469
        farm product consumers, 470, 472
    policy issues
        in developed countries, 466
        in developing countries, 466–467
    policy requirement, 465
    public policy examples, 474–475
    as resource in water-scarce settings, 462
        in developed countries, 463
        in developing countries, 464–465
    solids disposal, 249
    sources, 248
    standards, 248
    treatment, 248–249, 248f
        and non-treatment alternatives, 467–468
Wastewater irrigation
    challenges, 701
    crop selection, 702–703, 702t
    frequency, 705
    maximum allowable levels, metals and metalloids, 703–704, 703t
    methods used, 704–705, 704t
    pathogen reductions, 705t
    soil-based nutrient interventions, 705–706, 706t
Wastewater treatment, 248–249, 248f, 562–563
    activated sludge process
        operating parameters in, 566–568
        solid retention time, 566–567
    aerobic biological waste treatment processes, 565–566
        activated sludge process, 565, 566f
        aeration tanks, 565–566
        secondary clarifiers, 566
    anaerobic wastewater treatment
        granule deterioration, 572
        loading rate, 571
        retention time, 572
        temperature, 571
        toxicity, 572
        waste composition, 571
    biological phosphorus removal, 573–574
    biological removal of nitrogen, 572
        anaerobic ammonium oxidation, 573
        Canon process, 573
        combined nitrogen removal, 573
        Sharon process, 573
    biological treatment options, 563–565
        aerobic processes, 564
        anaerobic processes, 564
        anoxic processes, 564
        attached growth process, 565
        suspended growth processes, 565
    Canada, 543
    efficiency matrix, 579, 580t
    extreme cold climate, 543
    industrial wastewaters, 577–578
    pollution problems, 577

Wastewater treatment (*cont.*)
    pretreatment, 563
    primary treatment, 563
    secondary clarification, 563
    secondary treatment, 563
    selection of, 579–581
    survey of, 579t
    tertiary treatment, 563
Wastewater treatment, wetlands
    CWs *vs.* natural, 622
    design, 625–626
    FWS, 622–623, 625, 626
    influent concentrations, 624t
    operation and maintenance, 626
    treatment process, 623–625, 624f
    VSB, 622–623, 625, 626
    working, 621
Water, 241–247
    aquatic ecosystems and pesticides
        herbicides, 680
        impacts, 682
        measuring concentrations, 681–682
        occurrence, 679–681
        off-site movement, 679–680
        transport, 679–681
        usage, 679
    history, 242
    hydraulics, 245–247
    loss, 329
    monitoring, 242–243
    nutrient deposition/deactivation in, 451t
    safe drinking
        access to, 661, 662t
        analysis of trace organic pollutants,
            669–670, 669f
        characteristics of bacteria and pathogens,
            671–672, 672t
        DAI-ECD technique, 669–671, 670f
        microbiological research methods, 672–673
        mineral components, 662–663, 663t
        organic compounds, health significance,
            668–669, 668t
        organic pollutants, 665–667, 667t
        suitability, 662
        toxic elements, 664–665, 664t
    sources, 243
    standards, 242–243
    standard water treatment, 243–245, 243f
    transmission, 243
    use, 329–331
Water–energy–food nexus, 231–232
Water–energy nexus, 423–424, 425f
Water flows and BOD concentrations, 540
Water harvesting
    classification, 129–130

    definition, 129
    domestic uses, 130
    livestock drinking water, 130
    runoff farming, 130
    usage, 130–131
Water leaching index, 409
Water prices, 277
    competitive markets and, 278–279
    multiple-criteria framework, 286f
    optimizing model, 279–283
        scarcity rents, 281–282
        shadow prices, 281–283
    structuring, 283
        economic efficiency, 285–286
        equity and fairness, 286
        full cost, 284
        full economic cost, 284
        full supply cost, 284
        opportunity cost, 284
        simplicity, 286–287
        stability and quality, 287
        substitutability, 284–285
        sustainability, 287
        water tariffs and pricing strategies, 287–289
Water quality
    for agriculture, 169
    criteria, 484
    monitoring, 72–74
        biological integrity sampling, 73
        citizen monitoring, 73–74
        lake sampling, 73
        trading programs, 417–418
        tributary mass load sampling, 73
    tributary water quality sampling, 73
    and WSPS, 541
Water Quality Act (1965), 481
Water quality modeling, 427
    BMPS, 428–429
    classification, 427–428
    large-scale systems behavior, 429
    overview, 427
    risk assessment of pesticides, 428
    sources/impacts of pollutants, evaluation of, 429
    uses, 428
Water resources, 266
    ecological water budgets, 267
    global water management, 267–268
    hydrologic cycle issues, 267
    political control, 267–268
    robust and rigorous methodology, 268
Water resources management evolution, 225–234
    adaptive management, 230–231
    colonial times, 227
    five driving forces, 225
    hydraulic paradigm, 227–228

increasingly urban world, 232–233, 233f
IWRM – dominant paradigm today, 228–229, 230f
'lock-in' effects, 231
Nexus approach, 231–232
pre-modern times, 226
Watersheds
 management
  Geographic information system (GIS) for,
   491–496, 493t
  remote sensing (RS), 491–496, 493t
Water supply buffer, 123
Water-sustainable agricultural adaptations
 cropping system adaptations, 189
 field experiments, 188–189
Water systems modeling, 511
Water treatment; *see also individual filters*
 classification of solids, 584–585, 584f, 585f
 definition, 584
 filtration mechanisms, 585–587, 586f
 granular filters
  pressure sand, 587–588
  rapid gravity, 587
  slow sand bed, 587
 process, 584
 types, 588f
Water utilities, 241–250
Well-posed initial and boundary value problem, for
  saltwater intrusion, 145–146
West Bengal, groundwater arsenic contamination, 370,
  372–373, 372f, 373f
Wet deposition, roof runoff, 195–196
Wetlands
 carbon sequestration
  creation as "carbon sinks," 718
  human activities, impact of, 717
  net carbon capture, 716–717
  organic carbon fluxes, 715–716, 716f
  primary productivity, 715–717
  soils, 717
 current extent, 689, 689t
 CWs
  classification, 603
  definition, 603
  HF systems, 604–607, 604f, 607f
  hybrid systems, 609
  various types of wastewater, uses in, 609–610,
   610t
  VF systems, 607–609, 608f
 definitions, 685–686
 fauna, 711–712
 flora, 709–711
 functions, 685

greenhouse gases, 715
guidelines, 498–500
hydrological patterns, 709
landscape, 709
national policies in support of, 712
North America, 687f
policy *vs.* wise use, 497, 498
sedimentation problems
 accretion, 617
 ecological engineering solutions, 619–620
 phosphorus accumulation, 618
 reduced conductivity, bed, 617
 size distribution, particles, 618–619
types, 686–689
wastewater treatment
 CWs *vs.* natural, 622
 design, 625–626
 FWS, 622–623, 625, 626
 influent concentrations, 624t
 operation and maintenance, 626
 treatment process, 623–625, 624f
 VSB, 622–623, 625, 626
 working, 621
Wheal Jane metal mine, Cornwall, 40
Wheat (*T. aestivum* L.), 51, 157
Wildfire, 331, 332f
Wildlife, biodiversity and, 331
Windhoek, Namibia, 583–584
Winter cover crops, for reducing NO_3-N leaching,
  51–52
Winter rye (*Secale cereale* L.), 51
WISER project, 639–640
Wise use *vs.* policy, 497, 498
World Health Organization (WHO), 370, 374, 468, 469
World War II, 228, 265

Y

Yellow boy, 83
Yellow River, 383–385
 distributions of runoff, 384
 harnessing of, 385
 irrigation area, 384
 overview, 383
 terrain of, 383

Z

Zea mays, 47
Zinc (Zn) contamination, in coastal waters, 19
Zineb, 61

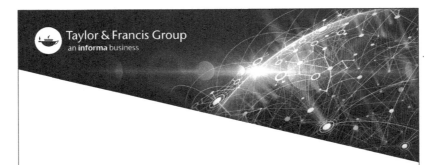